# THE THEORY OF DIFFERENCE SCHEMES

## MONOGRAPHS AND TEXTBOOKS IN PURE AND APPLIED MATHEMATICS

1. *K. Yano*, Integral Formulas in Riemannian Geometry (1970)
2. *S. Kobayashi*, Hyperbolic Manifolds and Holomorphic Mappings (1970)
3. *V. S. Vladimirov*, Equations of Mathematical Physics (A. Jeffrey, ed.; A. Littlewood, trans.) (1970)
4. *B. N. Pshenichnyi*, Necessary Conditions for an Extremum (L. Neustadt, translation ed.; K. Makowski, trans.) (1971)
5. *L. Narici et al.*, Functional Analysis and Valuation Theory (1971)
6. *S. S. Passman*, Infinite Group Rings (1971)
7. *L. Dornhoff*, Group Representation Theory. Part A: Ordinary Representation Theory. Part B: Modular Representation Theory (1971, 1972)
8. *W. Boothby and G. L. Weiss, eds.*, Symmetric Spaces (1972)
9. *Y. Matsushima*, Differentiable Manifolds (E. T. Kobayashi, trans.) (1972)
10. *L. E. Ward, Jr.*, Topology (1972)
11. *A. Babakhanian*, Cohomological Methods in Group Theory (1972)
12. *R. Gilmer*, Multiplicative Ideal Theory (1972)
13. *J. Yeh*, Stochastic Processes and the Wiener Integral (1973)
14. *J. Barros-Neto*, Introduction to the Theory of Distributions (1973)
15. *R. Larsen*, Functional Analysis (1973)
16. *K. Yano and S. Ishihara*, Tangent and Cotangent Bundles (1973)
17. *C. Procesi*, Rings with Polynomial Identities (1973)
18. *R. Hermann*, Geometry, Physics, and Systems (1973)
19. *N. R. Wallach*, Harmonic Analysis on Homogeneous Spaces (1973)
20. *J. Dieudonné*, Introduction to the Theory of Formal Groups (1973)
21. *I. Vaisman*, Cohomology and Differential Forms (1973)
22. *B.-Y. Chen*, Geometry of Submanifolds (1973)
23. *M. Marcus*, Finite Dimensional Multilinear Algebra (in two parts) (1973, 1975)
24. *R. Larsen*, Banach Algebras (1973)
25. *R. O. Kujala and A. L. Vitter, eds.*, Value Distribution Theory: Part A; Part B: Deficit and Bezout Estimates by Wilhelm Stoll (1973)
26. *K. B. Stolarsky*, Algebraic Numbers and Diophantine Approximation (1974)
27. *A. R. Magid*, The Separable Galois Theory of Commutative Rings (1974)
28. *B. R. McDonald*, Finite Rings with Identity (1974)
29. *J. Satake*, Linear Algebra (S. Koh et al., trans.) (1975)
30. *J. S. Golan*, Localization of Noncommutative Rings (1975)
31. *G. Klambauer*, Mathematical Analysis (1975)
32. *M. K. Agoston*, Algebraic Topology (1976)
33. *K. R. Goodearl*, Ring Theory (1976)
34. *L. E. Mansfield*, Linear Algebra with Geometric Applications (1976)
35. *N. J. Pullman*, Matrix Theory and Its Applications (1976)
36. *B. R. McDonald*, Geometric Algebra Over Local Rings (1976)
37. *C. W. Groetsch*, Generalized Inverses of Linear Operators (1977)
38. *J. E. Kuczkowski and J. L. Gersting*, Abstract Algebra (1977)
39. *C. O. Christenson and W. L. Voxman*, Aspects of Topology (1977)
40. *M. Nagata*, Field Theory (1977)
41. *R. L. Long*, Algebraic Number Theory (1977)
42. *W. F. Pfeffer*, Integrals and Measures (1977)
43. *R. L. Wheeden and A. Zygmund*, Measure and Integral (1977)
44. *J. H. Curtiss*, Introduction to Functions of a Complex Variable (1978)
45. *K. Hrbacek and T. Jech*, Introduction to Set Theory (1978)
46. *W. S. Massey*, Homology and Cohomology Theory (1978)
47. *M. Marcus*, Introduction to Modern Algebra (1978)
48. *E. C. Young*, Vector and Tensor Analysis (1978)
49. *S. B. Nadler, Jr.*, Hyperspaces of Sets (1978)
50. *S. K. Segal*, Topics in Group Kings (1978)
51. *A. C. M. van Rooij*, Non-Archimedean Functional Analysis (1978)
52. *L. Corwin and R. Szczarba*, Calculus in Vector Spaces (1979)
53. *C. Sadosky*, Interpolation of Operators and Singular Integrals (1979)
54. *J. Cronin*, Differential Equations (1980)
55. *C. W. Groetsch*, Elements of Applicable Functional Analysis (1980)

56. *I. Vaisman*, Foundations of Three-Dimensional Euclidean Geometry (1980)
57. *H. I. Freedan*, Deterministic Mathematical Models in Population Ecology (1980)
58. *S. B. Chae*, Lebesgue Integration (1980)
59. *C. S. Rees et al.*, Theory and Applications of Fourier Analysis (1981)
60. *L. Nachbin*, Introduction to Functional Analysis (R. M. Aron, trans.) (1981)
61. *G. Orzech and M. Orzech*, Plane Algebraic Curves (1981)
62. *R. Johnsonbaugh and W. E. Pfaffenberger*, Foundations of Mathematical Analysis (1981)
63. *W. L. Voxman and R. H. Goetschel*, Advanced Calculus (1981)
64. *L. J. Corwin and R. H. Szczarba*, Multivariable Calculus (1982)
65. *V. I. Istrăţescu*, Introduction to Linear Operator Theory (1981)
66. *R. D. Järvinen*, Finite and Infinite Dimensional Linear Spaces (1981)
67. *J. K. Beem and P. E. Ehrlich*, Global Lorentzian Geometry (1981)
68. *D. L. Armacost*, The Structure of Locally Compact Abelian Groups (1981)
69. *J. W. Brewer and M. K. Smith, eds.*, Emmy Noether: A Tribute (1981)
70. *K. H. Kim*, Boolean Matrix Theory and Applications (1982)
71. *T. W. Wieting*, The Mathematical Theory of Chromatic Plane Ornaments (1982)
72. *D. B.Gauld*, Differential Topology (1982)
73. *R. L. Faber*, Foundations of Euclidean and Non-Euclidean Geometry (1983)
74. *M. Carmeli*, Statistical Theory and Random Matrices (1983)
75. *J. H. Carruth et al.*, The Theory of Topological Semigroups (1983)
76. *R. L. Faber*, Differential Geometry and Relativity Theory (1983)
77. *S. Barnett*, Polynomials and Linear Control Systems (1983)
78. *G. Karpilovsky*, Commutative Group Algebras (1983)
79. *F. Van Oystaeyen and A. Verschoren*, Relative Invariants of Rings (1983)
80. *I. Vaisman*, A First Course in Differential Geometry (1984)
81. *G. W. Swan*, Applications of Optimal Control Theory in Biomedicine (1984)
82. *T. Petrie and J. D. Randall*, Transformation Groups on Manifolds (1984)
83. *K. Goebel and S. Reich*, Uniform Convexity, Hyperbolic Geometry, and Nonexpansive Mappings (1984)
84. *T. Albu and C. Năstăsescu*, Relative Finiteness in Module Theory (1984)
85. *K. Hrbacek and T. Jech*, Introduction to Set Theory: Second Edition (1984)
86. *F. Van Oystaeyen and A. Verschoren*, Relative Invariants of Rings (1984)
87. *B. R. McDonald*, Linear Algebra Over Commutative Rings (1984)
88. *M. Namba*, Geometry of Projective Algebraic Curves (1984)
89. *G. F. Webb*, Theory of Nonlinear Age-Dependent Population Dynamics (1985)
90. *M. R. Bremner et al.*, Tables of Dominant Weight Multiplicities for Representations of Simple Lie Algebras (1985)
91. *A. E. Fekete*, Real Linear Algebra (1985)
92. *S. B. Chae*, Holomorphy and Calculus in Normed Spaces (1985)
93. *A. J. Jerri*, Introduction to Integral Equations with Applications (1985)
94. *G. Karpilovsky*, Projective Representations of Finite Groups (1985)
95. *L. Narici and E. Beckenstein*, Topological Vector Spaces (1985)
96. *J. Weeks*, The Shape of Space (1985)
97. *P. R. Gribik and K. O. Kortanek*, Extremal Methods of Operations Research (1985)
98. *J.-A. Chao and W. A. Woyczynski, eds.*, Probability Theory and Harmonic Analysis (1986)
99. *G. D. Crown et al.*, Abstract Algebra (1986)
100. *J. H. Carruth et al.*, The Theory of Topological Semigroups, Volume 2 (1986)
101. *R. S. Doran and V. A. Belfi*, Characterizations of C*-Algebras (1986)
102. *M. W. Jeter*, Mathematical Programming (1986)
103. *M. Altman*, A Unified Theory of Nonlinear Operator and Evolution Equations with Applications (1986)
104. *A. Verschoren*, Relative Invariants of Sheaves (1987)
105. *R. A. Usmani*, Applied Linear Algebra (1987)
106. *P. Blass and J. Lang*, Zariski Surfaces and Differential Equations in Characteristic $p > 0$ (1987)
107. *J. A. Reneke et al.*, Structured Hereditary Systems (1987)
108. *H. Busemann and B. B. Phadke*, Spaces with Distinguished Geodesics (1987)
109. *R. Harte*, Invertibility and Singularity for Bounded Linear Operators (1988)
110. *G. S. Ladde et al.*, Oscillation Theory of Differential Equations with Deviating Arguments (1987)
111. *L. Dudkin et al.*, Iterative Aggregation Theory (1987)
112. *T. Okubo*, Differential Geometry (1987)

113. *D. L. Stancl and M. L. Stancl,* Real Analysis with Point-Set Topology (1987)
114. *T. C. Gard,* Introduction to Stochastic Differential Equations (1988)
115. *S. S. Abhyankar,* Enumerative Combinatorics of Young Tableaux (1988)
116. *H. Strade and R. Farnsteiner,* Modular Lie Algebras and Their Representations (1988)
117. *J. A. Huckaba,* Commutative Rings with Zero Divisors (1988)
118. *W. D. Wallis,* Combinatorial Designs (1988)
119. *W. Więsław,* Topological Fields (1988)
120. *G. Karpilovsky,* Field Theory (1988)
121. *S. Caenepeel and F. Van Oystaeyen,* Brauer Groups and the Cohomology of Graded Rings (1989)
122. *W. Kozlowski,* Modular Function Spaces (1988)
123. *E. Lowen-Colebunders,* Function Classes of Cauchy Continuous Maps (1989)
124. *M. Pavel,* Fundamentals of Pattern Recognition (1989)
125. *V. Lakshmikantham et al.,* Stability Analysis of Nonlinear Systems (1989)
126. *R. Sivaramakrishnan,* The Classical Theory of Arithmetic Functions (1989)
127. *N. A. Watson,* Parabolic Equations on an Infinite Strip (1989)
128. *K. J. Hastings,* Introduction to the Mathematics of Operations Research (1989)
129. *B. Fine,* Algebraic Theory of the Bianchi Groups (1989)
130. *D. N. Dikranjan et al.,* Topological Groups (1989)
131. *J. C. Morgan II,* Point Set Theory (1990)
132. *P. Biler and A. Witkowski,* Problems in Mathematical Analysis (1990)
133. *H. J. Sussmann,* Nonlinear Controllability and Optimal Control (1990)
134. *J.-P. Florens et al.,* Elements of Bayesian Statistics (1990)
135. *N. Shell,* Topological Fields and Near Valuations (1990)
136. *B. F. Doolin and C. F. Martin,* Introduction to Differential Geometry for Engineers (1990)
137. *S. S. Holland, Jr.,* Applied Analysis by the Hilbert Space Method (1990)
138. *J. Okniński,* Semigroup Algebras (1990)
139. *K. Zhu,* Operator Theory in Function Spaces (1990)
140. *G. B. Price,* An Introduction to Multicomplex Spaces and Functions (1991)
141. *R. B. Darst,* Introduction to Linear Programming (1991)
142. *P. L. Sachdev,* Nonlinear Ordinary Differential Equations and Their Applications (1991)
143. *T. Husain,* Orthogonal Schauder Bases (1991)
144. *J. Foran,* Fundamentals of Real Analysis (1991)
145. *W. C. Brown,* Matrices and Vector Spaces (1991)
146. *M. M. Rao and Z. D. Ren,* Theory of Orlicz Spaces (1991)
147. *J. S. Golan and T. Head,* Modules and the Structures of Rings (1991)
148. *C. Small,* Arithmetic of Finite Fields (1991)
149. *K. Yang,* Complex Algebraic Geometry (1991)
150. *D. G. Hoffman et al.,* Coding Theory (1991)
151. *M. O. González,* Classical Complex Analysis (1992)
152. *M. O. González,* Complex Analysis (1992)
153. *L. W. Baggett,* Functional Analysis (1992)
154. *M. Sniedovich,* Dynamic Programming (1992)
155. *R. P. Agarwal,* Difference Equations and Inequalities (1992)
156. *C. Brezinski,* Biorthogonality and Its Applications to Numerical Analysis (1992)
157. *C. Swartz,* An Introduction to Functional Analysis (1992)
158. *S. B. Nadler, Jr.,* Continuum Theory (1992)
159. *M. A. Al-Gwaiz,* Theory of Distributions (1992)
160. *E. Perry,* Geometry: Axiomatic Developments with Problem Solving (1992)
161. *E. Castillo and M. R. Ruiz-Cobo,* Functional Equations and Modelling in Science and Engineering (1992)
162. *A. J. Jerri,* Integral and Discrete Transforms with Applications and Error Analysis (1992)
163. *A. Charlier et al.,* Tensors and the Clifford Algebra (1992)
164. *P. Biler and T. Nadzieja,* Problems and Examples in Differential Equations (1992)
165. *E. Hansen,* Global Optimization Using Interval Analysis (1992)
166. *S. Guerre-Delabrière,* Classical Sequences in Banach Spaces (1992)
167. *Y. C. Wong,* Introductory Theory of Topological Vector Spaces (1992)
168. *S. H. Kulkarni and B. V. Limaye,* Real Function Algebras (1992)
169. *W. C. Brown,* Matrices Over Commutative Rings (1993)
170. *J. Loustau and M. Dillon,* Linear Geometry with Computer Graphics (1993)
171. *W. V. Petryshyn,* Approximation-Solvability of Nonlinear Functional and Differential Equations (1993)

172. *E. C. Young*, Vector and Tensor Analysis: Second Edition (1993)
173. *T. A. Bick*, Elementary Boundary Value Problems (1993)
174. *M. Pavel*, Fundamentals of Pattern Recognition: Second Edition (1993)
175. *S. A. Albeverio et al.*, Noncommutative Distributions (1993)
176. *W. Fulks*, Complex Variables (1993)
177. *M. M. Rao*, Conditional Measures and Applications (1993)
178. *A. Janicki and A. Weron*, Simulation and Chaotic Behavior of $\alpha$-Stable Stochastic Processes (1994)
179. *P. Neittaanmäki and D. Tiba*, Optimal Control of Nonlinear Parabolic Systems (1994)
180. *J. Cronin*, Differential Equations: Introduction and Qualitative Theory, Second Edition (1994)
181. *S. Heikkilä and V. Lakshmikantham*, Monotone Iterative Techniques for Discontinuous Nonlinear Differential Equations (1994)
182. *X. Mao*, Exponential Stability of Stochastic Differential Equations (1994)
183. *B. S. Thomson*, Symmetric Properties of Real Functions (1994)
184. *J. E. Rubio*, Optimization and Nonstandard Analysis (1994)
185. *J. L. Bueso et al.*, Compatibility, Stability, and Sheaves (1995)
186. *A. N. Michel and K. Wang*, Qualitative Theory of Dynamical Systems (1995)
187. *M. R. Damel*, Theory of Lattice-Ordered Groups (1995)
188. *Z. Naniewicz and P. D. Panagiotopoulos*, Mathematical Theory of Hemivariational Inequalities and Applications (1995)
189. *L. J. Corwin and R. H. Szczarba*, Calculus in Vector Spaces: Second Edition (1995)
190. *L. H. Erbe et al.*, Oscillation Theory for Functional Differential Equations (1995)
191. *S. Agaian et al.*, Binary Polynomial Transforms and Nonlinear Digital Filters (1995)
192. *M. I. Gil'*, Norm Estimations for Operation-Valued Functions and Applications (1995)
193. *P. A. Grillet*, Semigroups: An Introduction to the Structure Theory (1995)
194. *S. Kichenassamy*, Nonlinear Wave Equations (1996)
195. *V. F. Krotov*, Global Methods in Optimal Control Theory (1996)
196. *K. I. Beidar et al.*, Rings with Generalized Identities (1996)
197. *V. I. Arnautov et al.*, Introduction to the Theory of Topological Rings and Modules (1996)
198. *G. Sierksma*, Linear and Integer Programming (1996)
199. *R. Lasser*, Introduction to Fourier Series (1996)
200. *V. Sima*, Algorithms for Linear-Quadratic Optimization (1996)
201. *D. Redmond*, Number Theory (1996)
202. *J. K. Beem et al.*, Global Lorentzian Geometry: Second Edition (1996)
203. *M. Fontana et al.*, Prüfer Domains (1997)
204. *H. Tanabe*, Functional Analytic Methods for Partial Differential Equations (1997)
205. *C. Q. Zhang*, Integer Flows and Cycle Covers of Graphs (1997)
206. *E. Spiegel and C. J. O'Donnell*, Incidence Algebras (1997)
207. *B. Jakubczyk and W. Respondek*, Geometry of Feedback and Optimal Control (1998)
208. *T. W. Haynes et al.*, Fundamentals of Domination in Graphs (1998)
209. *T. W. Haynes et al.*, Domination in Graphs: Advanced Topics (1998)
210. *L. A. D'Alotto et al.*, A Unified Signal Algebra Approach to Two-Dimensional Parallel Digital Signal Processing (1998)
211. *F. Halter-Koch*, Ideal Systems (1998)
212. *N. K. Govil et al.*, Approximation Theory (1998)
213. *R. Cross*, Multivalued Linear Operators (1998)
214. *A. A. Martynyuk*, Stability by Liapunov's Matrix Function Method with Applications (1998)
215. *A. Favini and A. Yagi*, Degenerate Differential Equations in Banach Spaces (1999)
216. *A. Illanes and S. Nadler, Jr.*, Hyperspaces: Fundamentals and Recent Advances (1999)
217. *G. Kato and D. Struppa*, Fundamentals of Algebraic Microlocal Analysis (1999)
218. *G. X.-Z. Yuan*, KKM Theory and Applications in Nonlinear Analysis (1999)
219. *D. Motreanu and N. H. Pavel*, Tangency, Flow Invariance for Differential Equations, and Optimization Problems (1999)
220. *K. Hrbacek and T. Jech*, Introduction to Set Theory, Third Edition (1999)
221. *G. E. Kolosov*, Optimal Design of Control Systems (1999)
222. *N. L. Johnson*, Subplane Covered Nets (2000)
223. *B. Fine and G. Rosenberger*, Algebraic Generalizations of Discrete Groups (1999)
224. *M. Väth*, Volterra and Integral Equations of Vector Functions (2000)
225. *S. S. Miller and P. T. Mocanu*, Differential Subordinations (2000)

226. *R. Li et al.*, Generalized Difference Methods for Differential Equations: Numerical Analysis of Finite Volume Methods (2000)
227. *H. Li and F. Van Oystaeyen*, A Primer of Algebraic Geometry (2000)
228. *R. P. Agarwal*, Difference Equations and Inequalities: Theory, Methods, and Applications, Second Edition (2000)
229. *A. B. Kharazishvili*, Strange Functions in Real Analysis (2000)
230. *J. M. Appell et al.*, Partial Integral Operators and Integro-Differential Equations (2000)
231. *A. I. Prilepko et al.*, Methods for Solving Inverse Problems in Mathematical Physics (2000)
232. *F. Van Oystaeyen*, Algebraic Geometry for Associative Algebras (2000)
233. *D. L. Jagerman*, Difference Equations with Applications to Queues (2000)
234. *D. R. Hankerson et al.*, Coding Theory and Cryptography: The Essentials, Second Edition, Revised and Expanded (2000)
235. *S. Dăscălescu et al.*, Hopf Algebras: An Introduction (2001)
236. *R. Hagen et al.*, C*-Algebras and Numerical Analysis (2001)
237. *Y. Talpaert*, Differential Geometry: With Applications to Mechanics and Physics (2001)
238. *R. H. Villarreal*, Monomial Algebras (2001)
239. *A. N. Michel et al.*, Qualitative Theory of Dynamical Systems, Second Edition (2001)
240. *A. A. Samarskii*, The Theory of Difference Schemes (2001)

Additional Volumes in Preparation

# THE THEORY OF DIFFERENCE SCHEMES

Alexander A. Samarskii

*Moscow M. V. Lomonosov State University*
*Moscow, Russia*

MARCEL DEKKER, INC.    NEW YORK • BASEL

**ISBN: 0-8247-0468-1**

This book is printed on acid-free paper.

**Headquarters**
Marcel Dekker, Inc.
270 Madison Avenue, New York, NY 10016
tel: 212-696-9000; fax: 212-685-4540

**Eastern Hemisphere Distribution**
Marcel Dekker AG
Hutgasse 4, Postfach 812, CH-4001 Basel, Switzerland
tel: 41-61-261-8482; fax: 41-61-261-8896

**World Wide Web**
http://www.dekker.com

The publisher offers discounts on this book when ordered in bulk quantities. For more information, write to Special Sales/Professional Marketing at the headquarters address above.



Current printing (last digit):
10 9 8 7 6 5 4 3 2 1

**PRINTED IN THE UNITED STATES OF AMERICA**

# Preface

In the early 1960s, when computers first became freely available to academic scientists, they spawned a whole new area of mathematics. Suddenly, for example, the solution of a system of 30 linear equations in 30 unknowns reduced to a problem that could be solved in a straightforward fashion on a computer using Gaussian elimination. A whole host of other applications, whose computational efforts were previously prohibitive, became feasible with the new technology. In addition, the availability of the technology itself spurred mathematicians to invent new methods of solution for existing and emerging problems. As we begin the first decade of this century, the exploitation of computers in mathematics has become very diverse. Some researchers use "blackbox" library routines to solve problems arising from physical situations. Others are engaged in developing software for such routines. Yet others are carrying out fundamental research in the capacities of existing methods and the design of new ones. But perhaps the best evidence of the effect the computer has had on mathematics in recent years can be seen in the undergraduate curriculum. Most university degrees in mathematics would now be regarded as incomplete without courses in discrete mathematics and numerical analysis – both subjects that have come to prominence because of modern computer technology.

A rough description of the term **modeling** is the process of applying methods well-developed in computational mathematics to real-life situations. Quite often, the interest in modeling centers as much on getting an appropriate mathematical description of the problem as it does on the computational methods employed. Nevertheless there is a fruitful interplay between modeling and computational mathematics, as will be evident from a perusal of the first few chapters of this book.

One can now say that a new method arose for theoretical investigations of rather general and complex processes which allow for mathematical descriptions or mathematical modeling. This is the method of **numerical experiment**, which makes it possible better to understand real-world phenomena using exploratory devices of computational mathematics. When investigating some physical process by means of this method, an early step is the mathematical statement of a problem or the well-founded choice of a mathematical model capable of describing this process. It is preceded by a proper choice of physical approximations, that is, to decide which factors should be taken into account and which may be neglected or omitted in a general setting. This choice is the privilege of physicists.

What is a mathematical model? The group of unknown physical quantities which interest us and the group of available data are closely interconnected. This link may be embodied in algebraic or differential equations. A proper choice of the mathematical model facilitates solving these equations and providing the subsidiary information on the coefficients of equations as well as on the initial and boundary data.

Mathematical physics deals with a variety of mathematical models arising in physics. Equations of mathematical physics are mainly partial differential equations, integral, and integro-differential equations. Usually these equations reflect the conservation laws of the basic physical quantities (energy, angular momentum, mass, etc.) and, as a rule, turn out to be nonlinear.

After writing a system of equations capable of describing the physical process of interest, it is necessary for the investigator to consider the resulting mathematical model by methods of the general theory of differential and integral equations and to make sure that a problem has been completely posed and the available data are sufficient and consistent, to derive the conditions under which the problem has a unique solution, to find out whether this solution may be written in explicit form, and whether particular solutions are possible. Particular solutions are important in giving the preliminaries regarding the nature of the physical process. They also can serve as "goodness-of-fit" tests for the desired quality of numerical methods. At the second stage, one is to construct an approximate (numerical) method and a computer algorithm for solving the problem. The third stage is computer programming of the algorithm under such a choice. At the fourth stage the computational procedures need to be investigated and implemented. The fifth stage focuses on the review of their delivery and necessary corrections to the approved mathematical model governing what can happen.

It may turn out that the mathematical model is too rough, meaning numerical results of computations are not consistent with physical experiments, or the model is extremely cumbersome for everyday use and its solution can be obtained with a prescribed accuracy on the basis of simpler models. Then the same work should be started all over again and the remaining stages should be repeated once again.

The stages of numerical experiments for theoretical investigations of physical problems were explained above. Greatest attention is being paid to a new technology for research with rather complicated mathematical models.

Classical methods of mathematical physics are employed at the first stage. Numerous physical problems lead to mathematical models having no advanced methods for solving them. Quite often in practice, the user is forced to solve such nonlinear problems of mathematical physics for which even the theorems of existence and uniqueness have not yet been proven and some relevant issues are still open.

For the moment, we are interested only in the second stage of numerical experiments. A computational algorithm usually means a sequence of logic and arithmetic operations that enables us to solve a problem. A computational algorithm is developed to solve the problem with any degree of accuracy $\varepsilon > 0$ in a finite number of operations $Q(\varepsilon)$. This is the general requirement, thus raising many mathematical questions. What does the expression "with any degree of accuracy" mean? A possibility, at least in principle, of obtaining a solution with a prescribed accuracy should be substantiated. However, sometimes it may happen that $Q(\varepsilon)$ is finite for some $\varepsilon$, but it is so great that in practice it is unrealistic to produce a solution with such a degree of accuracy. For any problem an infinite number of various computational algorithms may be constructed possessing, for example, similar asymptotic properties, so that $Q(\varepsilon)$ will be of the same order in $\varepsilon$ as $\varepsilon \to 0$. Not all such methods should be involved, but only those suitable for computational restrictions. Naturally, the methods should be used that require the minimal execution time for solving the problem with a prescribed accuracy. The execution time must be reasonable, that is, measurable in minutes or hours if such calculations are encountered very rarely. Of course, the computing process must be as inexpensive as possible. The time for solving the problem depends on the algorithms, quality of the program and computer performance. It is difficult to evaluate the latter when only a priori estimates of the algorithm quality are available. The same applies equally well to the total number of logic operations regardless of how computers will be chosen. Therefore, in theoretical investigations it is fairly common to compare computer algorithms by means of the number of arithmetic operations $Q_0(\varepsilon)$. An economical algorithm having the minimal number of operations is sought in the class of possible algorithms (for instance, with the same asymptotics for $Q_0(\varepsilon)$ as $\varepsilon \to 0$). This is one of the main problems in the theory of numerical methods.

At the initial stage of such an analysis of algorithms it is usually supposed that a computing process is ideal, that is, computations are carried out with an infinite number of significant digits. But any computer makes calculations with a finite speed and a finite number of digits. Not all numbers are accessible to computers; there are computer null and computer infinity. For instance, abnormal termination occurs when computer infinity arises during the course of execution. A computing process may become unstable, thus causing difficulties. In such cases rounding errors may accumulate to a considerable extent so that the algorithm will be useless in practical applications (several examples of unstable algorithms are available in Chapter 2). A real, that is, a proper computer algorithm should be stable and should not allow excessively large intermediate values leading to abnormal termination. From such reasoning it seems clear that an ideal computer algorithm may be optimal with respect to the number $Q_0(\varepsilon)$ and quite unfit for computers. These requirements necessitate seeking an optimal real computer algorithm.

Any numerical experiment is not a one-time calculation by standard formulas. First and foremost, it is the computation of a number of possibilities for various mathematical models. For instance, it is required to find the optimal conditions for a chemical process, that is, the conditions under which the reaction is completed most rapidly. A solution of this problem depends on a number of parameters (for instance, temperature, pressure, composition of the reacting mixture, etc.). In order to find the optimal workable conditions, it is necessary to carry out computations for different values of those parameters, thereby exhausting all possibilities. Of course, some situations exist in which an algorithm is to be used only several times or even once.

The statement of the problem of finding an optimal algorithm depends on how it is to be applied (for individual variants or a great number of variants).

Here various issues surrounding programming, organization and realization of computing are of little interest and will be excluded from further consideration. It is worth noting, however, that programming should be closely linked with further development of numerical algorihms. Mathematical-physics problems are fairly complicated and algorithms for solving them are extremely cumbersome. In mastering the difficulties involved, one might reasonably try to split them into blocks or "modules". Fortunately, quite often processes having different physical backgrounds will be described by the same equations (for instance, the processes of diffusion, heat conduction, and magnetization). One and the same mathematical model could be suitable for various physical problems even if each one has its own physical particulars. On the other hand, a mathematical model may change substantially a number of times during the course of numerical experiments and this obstacle necessitates improving the algorithm constructed and the corresponding program. Therefore, it becomes extremely important to create programs (a program package) in conformity with the modulus principle, whose use permits us efficiently to conduct numerical experiments and solve various types of problems of miscellaneous physical genesis. This is one of the modern trends in programming and solving major mathematical-physics problems leading to the development of new numerical methods.

Usually the finite difference method or the grid method is aimed at numerical solution of various problems in mathematical physics. Under such an approach the solution of partial differential equations amounts to solving systems of algebraic equations.

This book is devoted to the theory of difference methods (schemes) applied to typical problems in mathematical physics. There seem to be at least two widespread approaches within the theory:

(1) Composition of discrete (difference) approximations to equations of mathematical physics and verifying a priori quality characteristics of these approximations, mainly the error of approximation, stability, convergence, and accuracy of the difference schemes obtained;

(2) Solution of difference equations by direct or iterative methods selected on the basis of the economy criteria for the corresponding computational algorithms.

Because of the enormous range of difference approximations to an equation having similar asymptotic properties with respect to a grid step (the same order of accuracy or the number of necessary operations), their numerical realizations resulted in the appearance of different schemes for solving the basic problems in mathematical physics.

Of course, one strives to develop the best possible method, whose use permits us to obtain the desired solution in minimal computing time. Indeed, the search for such numerical procedures among admissible methods is the main goal of such theory. In designing an optimal method (its choice depends on the type of problems being solved) the set of admissible methods is gradually reduced by successive introduction of necessary constraints and requirements regarding approximation, stability, economy, convergence, etc. The following general principle plays a crucial role in such matters: a difference scheme (a discrete scheme) must reproduce properties of the initial differential equation as well as possible.

In practical implementations, in order to construct difference schemes of a desired quality one should formulate the general approaches, heuristic tricks, turns, and rules for later use. The design of any method for solving a problem is stipulated by the homogeneity and conservatism of the appropriate difference scheme. The meaning of conservatism is that the difference scheme reproduces some conservation law (a balance equation) on a grid. Conservatism of homogeneous schemes is a necessary condition for convergence in the class of discontinuous coefficients for stationary and nonstationary problems in mathematical physics.

The property of the difference scheme conservatism for linear equations is generally equivalent to the self-adjointness of the relevant difference operator (see Chapters 3–4).

The basic notions of the theory of difference schemes are the error of approximation, stability, convergence, and accuracy of difference scheme. A more detailed exposition of these concepts will appear in Chapter 2. They are illustrated by considering a number of difference schemes for ordinary differential equations. In the same chapter we also outline the approach to the general formulations without regard to the particular form of the difference operator.

The question of the accuracy of the scheme, being of principal importance in the theory, amounts to studying the error of approximation and stability of the scheme. Stability analysis necessitates imposing a priori estimates for the difference problem solution in light of available input data. This is a problem in itself and needs investigation.

On the other hand, neither is the estimation of the approximation error a trivial issue. Even the simplest example of the scheme on a nonequidistant grid for a second-order equation shows that it is desirable to evaluate the approximation error not in the $C$-norm, but in a weaker norm of special type (in one of the negative norms), thus imposing the need for a priori estimates for a solution of a difference problem from the right-hand side in a weaker norm. Various estimates of this sort arising in Chapter 3 apply equally well to establishing the convergence of homogeneous difference schemes in the case of discontinuous coefficients.

This book includes many remarkable examples illustrating different approaches to the stability analysis of difference scheme. Thus, in Chapter

5 which is devoted to difference schemes for nonstationary equations with constant coefficients we discover asymptotic stability of difference schemes for the heat conduction equation, a property intrinsic to the differential equation.

Chapters 2–5 are concerned with concrete difference schemes for equations of elliptic, parabolic, and hyperbolic types. Chapter 3 focuses on homogeneous difference schemes for ordinary differential equations, by means of which we try to solve the canonical problem of the theory of difference schemes in which a primary family of difference schemes is specified (in such a case the availability of the family is provided by pattern functionals) and schemes of a desired quality should be selected within the primary family. This problem is solved in Chapter 3 using a particular form of the scheme and its solution leads us to conservative homogeneous schemes.

Chapter 4 provides the general theory of difference schemes in which it seems reasonable to eliminate constraints on the structure and implicit forms of difference operators. Such a theory treats difference schemes as operator equations (an analog of grid approximations for elliptic and integral equations) and operator-difference equations (difference equations in $t$ with operator coefficients), which are analogs of difference schemes for time-dependent equations of mathematical physics (for instance, equations of parabolic and hyperbolic types). The operators of schemes are viewed in such a setting as linear operators in an abstract normed vector space $H_h$ depending on the vector parameter $h$ (an analog of a grid step in $\mathbf{x} = (x_1, x_2, \dots, x_p)$) equipped with the norm $|h| > 0$. Thus, in the sequel two types of schemes will be given special investigation.

The operator scheme is associated with an operator equation of the first kind

$$A\mathbf{y} = \mathbf{f},$$

which can always be parametrized by a real variable $h$, where $A = A_h$, $A_h\colon H_h \mapsto H_h$ (the operator $A_h$ depending on $h$ really acts from $H_h$ into $H_h$), $\mathbf{f} = \mathbf{f}_h \in H_h$ is a known vector and $\mathbf{y} = \mathbf{y}_k \in H_h$ is the vector of unknowns.

The two-layer operator-difference scheme we are interested in acquires the canonical form

$$B\,\frac{y^{j+1} - y^j}{\tau} + A\,y^j = \varphi^j, \qquad j = 0, 1, \dots, \qquad y^0 \in H_h,$$

where $\tau$ is the step in $t$: $t_j = j\tau$, $j = 0, 1, \dots$; $A$, $B$: $H_h \mapsto H_h$ are operators depending on $h$ and $\tau$ and, generally, on $t_j$; $y^j = y_{h,\tau}(t_j) \in H_h$ and $\varphi^j = \varphi_{h,\tau}(t_j) \in H_h$ are, respectively, the sought and given functions of the discrete argument $t_j = j\tau$ with the values in the space $H_h$ (below we omit the subscripts $h$ and $\tau$).

Multilayer schemes, that is, the schemes containing the values $y(t)$ at some moments $t = t_j, t_{j+1}, t_{j+2}, \ldots$ may be reduced to two-layer schemes, where $A$ and $B$ are operator matrices.

Stability theory is the central part of the theory of difference schemes. Recent years have seen a great number of papers dedicated to investigating stability of such schemes. Many works are based on applications of spectral methods and include ineffective results given certain restrictions on the structure of difference operators. For schemes with non-self-adjoint operators the spectral theory guides only the choice of necessary stability conditions, but sufficient conditions and a priori estimates are of no less importance. An energy approach connected with the above definitions of the scheme permits one to carry out an exhaustive stability analysis for operators in a prescribed Hilbert space $H_h$.

Clearly, stability is an intrinsic property of schemes regardless of approximations and interrelations between the resulting schemes and relevant differential equations. Because of this, any stability condition should be imposed as the relationship between the operators $A$ and $B$. More specifically, let a family of schemes specified by the restrictions on the operators $A$ and $B$ be given: $A = A^* > 0$ or $(Ay, v) = (y, Av)$ and $(Ay, y) > 0$ for any $y, v \in H$, where $( \, , \, )$ is an inner product in $H$, $B > 0$ and $B \neq B^*$ ($B$ is non-self-adjoint). The problem statement consists of extracting from that family a set of schemes that are stable with respect to the initial data, having the form

$$B \, \frac{y^{j+1} - y^j}{\tau} + A\, y^j = 0 \,, \qquad j = 0, 1, \ldots \,, \qquad y^0 \in H_h \,,$$

and satisfying the inequality

$$\| y^j \|_A \le \| y^0 \|_A \,, \qquad \| y \|_A = \sqrt{(Ay, y)} \,.$$

The meaning of stability here is the validity of the preceding estimate.

A necessary and sufficient condition for a two-layer scheme to be stable can be written as the operator inequality

$$B \ge 0.5\, \tau A$$

or

$$(By, y) \ge 0.5\, \tau\, (Ay, y) \quad \text{for any} \quad y \in H \,.$$

This condition can easily be verified in the case of discrete schemes for equations of mathematical physics. It allows one to extract from the primary family of schemes the set of stable schemes, within which one should look for schemes with a prescribed accuracy, volume of computations, and other desirable properties and parameters.

One of the important corollaries to stability theory is the general method of regularization in the class of stable schemes (by changing the operators $A$ and $B$) for the design of schemes of a desired quality.

Chapter 6 includes a priori estimates expressing stability of two-layer and three-layer schemes in terms of the initial data and the right-hand side of the corresponding equations. It is worth noting here that relevant elements of functional analysis and linear algebra, such as the operator norm, self-adjoint operator, operator inequality, and others are much involved in the theory of difference schemes. For the reader's convenience the necessary prerequisities for reading the book are available in Chapters 1–2.

The book includes many good examples illustrating the practical use of general stability theory with regard to particular schemes to assist the users in subsequent implementations. Stability is probably the most pressing problem in any algorithm, since it is a necessary rather than a sufficient condition for accuracy.

Despite the great generality of the research presented in this book, it is of a constructive nature and gives the reader an understanding of relevant special cases as well as providing one with insight into the general theory. I hope it proves to be useful and stimulating.

One of the popular branches of modern mathematics is the theory of difference schemes for the numerical solution of the differential equations of mathematical physics. Difference schemes are also widely used in the general theory of differential equations as an apparatus available for proving existence theorems and investigating the differential properties of solutions.

The theory of difference schemes has a number of special problems.

In the final analysis, of greatest importance from the viewpoint of numerical analysis is the design of algorithms permitting one to obtain a solution of a differential equation on a computer with a prescribed accuracy in a finite number of operations. The user can encounter in this connection the question of the quality of an algorithm, that is, the manner in which the accuracy of the algorithm depends on

1) the available information on the original problem,
2) the amount of calculation (viz. the machine time spent in solving the problem with a prescribed accuracy).

Experience with computers has stimulated the formulation of a number of special (for the theory of difference methods) problems:

1) the determination of the attainable order of accuracy of difference schemes for various classes of problems,
2) the design of schemes for the solution of a wide class of problems with a certain guaranteed accuracy,
3) the construction of schemes with increased accuracy in narrower classes of problems,
4) the development of methods for investigating stability and convergence of difference schemes,
5) the formulation of the general principles for constructing stable difference schemes and economizing the amount of calculation (economical schemes)

and others.

The main purpose of the final chapters is to show how the results of the general theory of difference schemes are aimed at stating principles for constructing difference schemes of a prescribed quality. This approach requires forsaking a more detailed description of the structure of difference operators for concrete classes of differential equations and presenting the theory in the language of functional analysis. The difference schemes relating to analogs of nonstationary equations of mathematical physics are treated in this connection as difference (with respect to the variable $t$) equations with operator coefficients in an abstract space of any dimension. The difference schemes for elliptic equations are viewed as operator equations of the first kind. It should be emphasized, however, that the indicated notions of schemes have a much more general meaning.

Stability theory is quite applicable to formulate a general principle for regularizing difference schemes in order to obtain stable schemes of a prescribed quality.

The theory of iterative methods for solving the equation $Au = f$, where $A \in (H \mapsto H)$ is a linear operator in a Hilbert space $H$, is treated as a contemporary part of the general stability theory of operator-difference schemes. Our main concern is with obtaining effective estimates for the rate of convergence of the iterations and with choosing an optimal set of iteration parameters. Special attention is being paid to a class of implicit schemes with a factorized operator $B$ on the upper level of the form $B = (E + \omega R_1)(E + \omega R_2)$, where $E$ is the identity operator, $\omega > 0$ is a numerical parameter, and $R_1$ and $R_2 = R_1^*$ are adjoint or "triangular" (with a triangular matrix) linear operators. A formula for the parameter $\omega$ is obtained through such an analysis from the condition that the number of iterations be minimized.

An estimate of the convergence rate for the method of minimal corrections is derived in the case when $A$ is a non-self-adjoint operator and in others situations.

Also, we consider the total approximation method as a constructive method for creating economical difference schemes for the multidimensional equations of mathematical physics. The notion of additive scheme is introduced as a system of operator difference equations that approximates the original differential equation in the total sense. Two quite general heuristic methods (proposed earlier by the author) for obtaining additive economical schemes are discussed in full details. The additive schemes require a new technique for investigating convergence and a new type of a priori estimates that take into account the definition of the property of approximation.

We have not had the chance to discuss some works on difference methods of an applied character, although such works best illustrate the real possibilities of difference methods and provide constant sources of stimulation for the formulation of new problems.

When economical schemes for multidimensional problems in mathematical physics are developed in Chapter 9, we shall need a revised concept of approximation error, thereby changing the definition of scheme. The notion of summed (in $t$) approximation in Section 3 of Chapter 9 is of a constructive nature, making it possible to produce economical schemes for various problems.

Last, but not least, I wish to thank those who have assisted me in this enterprise.

Alexander Samarskiĭ

# Contents

# 1

# Preliminaries

Contemporary methods for solving problems of mathematical physics are gaining increasing popularity. They are being used more and more in solving applied problems not only by professional mathematicians, but also by investigators and users working in other branches of science, engineering and technology. In order to make this book accessible not only to specialists, but also to graduate and post-graduate students, we give a complete account of notions and definitions which will be used in the sequel. The concepts and theorems presented below are of an auxiliary nature and are included for references rather than for primary study. For this reason the majority of statements are quoted without proofs. We will also cite bibliographical sources for further, more detailed, information.

## 1.1 DIFFERENCE EQUATIONS

**1. Preliminary comments.** By applying approximate methods the problem of solving differential equations leads to the systems of linear algebraic equations:

$$\mathrm{A}\,\mathbf{u} = \mathbf{f}$$

where $\mathrm{A} = (a_{ij})$ is a square $(N \times N)$-matrix, $\mathbf{u}$ is the vector of unknowns and $\mathbf{f}$ is the right-hand side vector.

Most of the well-developed methods available for solving such a system falls within the categories of direct or "exact-fitted" methods and iterative or successive-approximate methods which are gaining increasing popularity.

The starting point in more a detailed exploration is the simplest systems of linear algebraic equations, namely, difference equations with special matrices in simplified form, for example, with tridiagonal matrices.

The history of difference equations contains plenty of good examples when such equations emerged during the course of direct descriptions of real processes in science and technology. Equations of mathematical physics, that is, partial differential equations are sources for a broad class of difference equations approximating integral and differential equations and give a substantial contribution to the continuing development of the theory of difference schemes. The main feature of difference equations is stipulated by the fact that the matrices of the high-order corresponding systems (about $10^4$–$10^5$) are sparse.

It is natural from the viewpoint of applications to treat various difference equations regardless of the initial differential equations which have induced them. All of the resulting schemes and properties are invariant to a concrete differential equation.

In the present section a direct method for solving the boundary-value problems associated with second-order difference equations will be the subject of special investigations.

**2. Examples of difference equations.** Undoubtedly, the reader has already encountered the simplest examples of first-order difference equations in connection with the formulae for the terms of an arithmetic or a geometric progressions: $a_{k+1} = a_k + d$ or $a_{k+1} - 2\,a_k + a_{k-1} = 0$ and $a_{k+1} = q\,a_k$, respectively, where the argument of the members $a_k = a(k)$ takes only positive integer values.

We briefly touch upon the basic concepts of the grid methods. The basic notions such as grids and grid functions will be studied in more detail in Chapter 2. A discrete set of points (**nodes**) is called a **grid** or **lattice.** Let a **grid function** $y_i = y(i)$ of the integer-valued argument $i = 0, \pm 1, \pm 2, \ldots,$ be given. The **right and left differences** are defined at a point $i$ to be:

$$\Delta\, y_i := y(i+1) - y(i) = y_{i+1} - y_i\,,$$
$$\nabla y_i := y(i) - y(i-1) = y_i - y_{i-1}\,.$$

By definition, $\Delta\, y_{i-1} = y_i - y_{i-1} = \nabla\, y_i$. It should be noted that the expressions above may be viewed as formal analogs of the first derivative

$L u = u'$ known from differential calculus. The same procedure works for the second, the third and other differences

$$\Delta^2 y_i = \Delta(\Delta y_i) = \Delta(y_{i+1} - y_i)$$
$$= (y_{i+2} - y_{i+1}) - (y_{i+1} - y_i) = y_{i+2} - 2y_{i+1} + y_i ,$$
$$\dots\dots\dots\dots\dots\dots\dots\dots\dots\dots$$
$$\Delta^m y_i = \Delta(\Delta^{m-1} y_i) .$$

We see that one more right point is captured every time when the operator $\Delta$ is applied. Consequently, applying $\Delta$ for $m$ times we justify that $\Delta^m y_i$ contains the values $y_i, y_{i+1}, \dots, y_{i+m}$ at the points $i, i+1, \dots, i+m$.

A very simple rule could be useful: the left difference operation at any point $i$ coincides with the right difference operation at the point $i-1$ so that

$$\Delta\nabla y_i = \Delta_i^2 y_{i-1} = y_{i+1} - 2y_i + y_{i-1} .$$

Let a linear equation with the entering differences and coefficients $\alpha_0(i), \alpha_1(i), \dots, \alpha_m(i)$ be composed:

$$\alpha_0(i)\,\Delta^m y_i + \alpha_1(i)\,\Delta^{m-1} y_i + \dots + \alpha_{m-1}(i)\,\Delta y_i + \alpha_m(i)\,y_i = f_i .$$

Substituting the expressions for differences $\Delta^k y_i$, $k = 1, 2, \dots, m$, one can modify it to an **$m$th order linear difference equation** related to an unknown $y_i$:

$$a_0(i)\,y_{i+m} + a_1(i)\,y_{i+m-1} + \dots + a_{m-1}(i)\,y_{i+1} + a_m(i)\,y_i = f_i ,$$
$$a_0(i) \neq 0 , \qquad a_m(i) \neq 0 .$$

This definition is a formal analog of an $m$th order ordinary differential equation

$$\alpha_0(x)\,\frac{d^m u}{d x^m} + \alpha_1(x)\,\frac{d^{m-1} u}{d x^{m-1}}, + \dots + \alpha_{m-1}(x)\,\frac{du}{dx} + \alpha_m(x)\,u = f(x) ,$$
$$\alpha_0(x) \neq 0 .$$

As one possible example we consider the simplest ordinary differential equation

$$\frac{du}{dx} = f(x)$$

and offer below some of its approximations with new notations $u_i = u(x_i)$, $u_{i+1} = u(x_{i+1})$ and $x_{i+1} = x_i + h$, where $h > 0$ is the distance from $x_i$ to $x_i + h$, namely

$$\left(\frac{du}{dx}\right)_{x=x_i} \sim \frac{u(x_i+h)-u(x_i)}{h} = \frac{u_{i+1}-u_i}{h} = \frac{\Delta\, u_i}{h},$$

$$\left(\frac{du}{dx}\right)_{x=x_i} \sim \frac{u_i-u_{i-1}}{h} = \frac{\nabla\, u_i}{h}$$

or

$$\left(\frac{du}{dx}\right)_{x=x_i} \sim \frac{u_{i+1}-u_{i-1}}{2\,h} = \frac{\Delta\, u_i+\nabla\, u_i}{2\,h}.$$

All of the preceding difference expressions approach $\frac{du}{dx}$ as $h \to 0$. The symbol $\sim$ means, as usual, approximation or correspondence. In what follows we say that the expression

$$\frac{\Delta\, u_i}{h} = \frac{u_{i+1}-u_i}{h}$$

approximates the first derivative $\frac{du}{dx} = u'$.

The very definition implies that the equation

$$\frac{\Delta\, y_i}{h} = f_i\,, \qquad f_i = f(x_i)\,,$$

is a first-order difference equation in one or another form

$$\Delta\, y_i = h\, f_i \qquad \text{or} \qquad y_{i+1} = y_i + h\, f_i\,.$$

There is no difficulty to solve this equation for a given initial value $y_0$. It should be noted here that a second-order difference equation also may appear in approximating a first-order differential equation, for example, in connection with these expressions

$$u(x_{i+1}) = u(x_i) + h\; u'(x_i) + \tfrac{1}{2}\, h^2\; u''(x_i) + \frac{1}{6}\, h^3\; u'''(x_i) + O(h^4)\,,$$

$$u(x_{i-1}) = u(x_i) - h\; u'(x_i) + \tfrac{1}{2}\, h^2\; u''(x_i) - \tfrac{1}{6}\, h^3\; u'''(x_i) + O(h^4)\,,$$

the sum of which gives

$$\frac{u_{i+1} - 2\,u_i + u_{i-1}}{h^2} = u_i'' + O(h^2)\,.$$

Neglecting $O(h^2)$ we derive rough expressions for $u''$ within the notations:

$$\left(\frac{d^2u}{dx^2}\right)_{x=x_i} \sim \frac{u_{i+1} - 2\,u_i + u_{i-1}}{h^2} = \frac{\Delta^2\,u_{i-1}}{h^2} = \frac{\Delta\nabla\,u_i}{h^2}\,.$$

Substituting the first expression into the expansion $u_{i+1} = u_i + h\,u_i' + \frac{1}{2}\,h^2\,u_i'' + O(h^3)$ instead of $u_i''$, we deduce that

$$(1) \qquad u_i' = \frac{u_{i+1} - u_i}{h} - \frac{h}{2}\,\frac{u_{i+1} - 2\,u_i + u_{i-1}}{h^2} + O(h^2)\,.$$

Replacing in (1) $u_i'$ by $f_i$ we cancel $O(h^2)$ and multiply the resulting equation by $2h$. As a final result we get the second-order difference equation

$$\Delta\nabla\,y_i - 2\,\Delta\,y_i = -2\,h\,f_i\,.$$

Its modification gives the approximation of the first-order differential equation

$$\frac{dy}{dx} = f.$$

Difference equations possess remarkable properties which will be given special investigation in the near future.

**3. The first-order difference equations and inequalities.** Of our concern is the first-order difference equation

$$(2) \qquad b\,\Delta\,y_i + a\,y_i = f_i\,,$$

which is a formal analog of the first-order differential equation

$$b\,\frac{du}{dt} + a\,u = f.$$

Alternative forms of equation (2) are

$$b\,(y_{i+1} - y_i) + a\,y_i = f_i \qquad \text{or} \qquad b\,y_{i+1} = c\,y_i + f_i\,, \qquad c = b - a\,.$$

In the general case $b = b_i$, $a = a_i$ and $c = c_i$, that is, we deal with known functions of the argument $i$. If $b_i \neq 0$, then

$$y_{i+1} = q_i\, y_i + \varphi_i\,.$$

It is evident that

$$q_i = \frac{c_i}{b_i}\,, \qquad \varphi_i = \frac{f_i}{b_i}\,, \qquad b_i \neq 0\,.$$

Thus, we see that this problem has a unique solution if the value $y$ is given for some $i$. For the sake of simplicity let $y_0$ be known in advance for $i = 0$. With this, one can determine all the values $y_1$, $y_2$, ... by the recurrence formula just established. In the case $q_i = q =$ const and $\varphi_i = 0$ this provides support for the view that the whole collection of $y_i$ constitutes a geometric progression. If $q_i = q$ and $\varphi_i \neq 0$, then

$$y_{i+1} = q\, y_i + \varphi_i = q\,(q\, y_{i-1} + \varphi_{i-1}) + \varphi_i = q^2\, y_{i-1} + \varphi_i + q\,\varphi_{i-1}\,.$$

The outcome of this is

$$(3) \qquad y_{i+1} = q^{i+1}\, y_0 + \varphi_i + q\,\varphi_{i-1} + \cdots + q^{i-1}\,\varphi_1 + q^i\,\varphi_0$$
$$= q^{i+1}\, y_0 + \sum_{k=0}^{i} q^{i-k}\,\varphi_k\,.$$

Adopting those ideas, it performs no difficulty to find a solution to the equation $y_{i+1} = q_i\, y_i + \varphi_i$, $i = 0, 1, 2, \ldots$, in the case $q_i \neq$ const.

In tackling the first-order difference inequalities

$$(4) \qquad y_{i+1} \leq q\, y_i + f_i\,, \qquad i = 0,\, 1,\, 2,\, \ldots\,,$$

with known members $q \geq 0$ and $f_i$ and a given value $y_0$ we prefer an alternative form of writing

$$(5) \qquad v_{i+1} = q\, v_i + f_i\,, \qquad v_0 = y_0\,.$$

Its solution can be most readily found with further reference to the relation $y_i \leq v_i$. Indeed, subtracting equality (5) from inequality (4) we get

$$y_{i+1} - v_{i+1} \leq q\,(y_i - v_i) \leq q^2\,(y_{i-1} - v_{i-1}) \leq \cdots \leq q^{i+1}\,(y_0 - v_0) = 0\,,$$

yielding $y_{i+1} \leq v_{i+1}$ for any $q$, where $v_i$ has been expressed in $q$, $v_0$, $f_i$ by analogy with (3).

**4. The second-order difference equations. The Cauchy problem. Boundary-value problems.** The second-order difference equation transforms into a more transparent form

$$(6) \qquad A_i\, y_{i-1} - C_i\, y_i + B_i\, y_{i+1} = -F_i\,, \qquad i = 1, 2, \dots,$$
$$A_i \neq 0\,, \qquad B_i \neq 0\,,$$

which in the notation $\Delta\, y_i = y_{i+1} - y_i$ becomes is modified as follows

$$(7) \qquad B_i\, \Delta\, y_i - A_i\, \Delta\, y_{i-1} - (C_i - B_i - A_i)\, y_i = -F_i\,.$$

By virtue of the relations

$$\Delta\, y_i - \nabla\, y_i = \Delta\, y_i - \Delta\, y_{i-1} = \Delta^2\, y_{i-1} = y_{i+1} - 2\, y_i + y_{i-1}\,,$$
$$\Delta\, y_{i-1} = -\Delta^2\, y_{i-1} + \Delta\, y_i\,,$$

equation (7), in turn, can be rewritten as

$$A_i\, \Delta^2\, y_{i-1} + (B_i - A_i)\, \Delta\, y_i - (C_i - A_i - B_i)\, y_i = -F_i\,, \quad A_i \neq 0\,,$$

or, what amounts to the same,

$$B_i\, \Delta^2\, y_{i-1} + (B_i - A_i)\, \Delta\, y_{i-1} - (C_i - A_i - B_i)\, y_i = -F_i\,.$$

The latter difference equation clarifies that (6) is an analog of a second-order differential equation.

It is necessary to specify two conditions for the complete posing of this or that problem. The assigned values of $y$ and $\Delta\, y$ suit us perfectly and lie in the background a widespread classification which will be used in the sequel. When equation (6) is put together with the values $y_i$ and $\Delta\, y_i$ given at one point, they are referred to as the **Cauchy problem**. Combination of two conditions at different nonneighboring points with equation (6) leads to a **boundary-value problem.**

For the sake of definiteness, we concentrate primarily on Cauchy problem with given $y_0$ and $\Delta\, y_0$. Knowing $y_0$ and $y_1 = y_0 + \Delta\, y_0$, $y_i$ for $i = 2, 3, \dots$ serves as a basis for the formulae

$$y_{i+1} = \frac{C_i\, y_i - A_i\, y_{i-1} - F_i}{B_i}\,, \qquad B_i \neq 0\,.$$

Therefore, it is concluded that problem (6) has a unique solution for given value $y_0$ and $y_1$.

For the second-order difference equations capable of describing the basic mathematical-physics problems, boundary-value problems with additional conditions given at different points are more typical. For example, if we know the value $y_0$ for $i = 0$ and the value $y_N$ for $i = N$, the corresponding boundary-value problem can be formulated as follows: it is necessary to find the solution $y_i$, $0 < i < N$, of problem (6) satisfying the boundary conditions

$$(8) \qquad y_0 = \mu_1, \qquad y_N = \mu_2$$

with known numbers $\mu_1$ and $\mu_2$.

Common boundary conditions may be specified by

$$(8') \qquad y_0 = \varkappa_1 y_1 + \mu_1, \qquad y_N = \varkappa_2 y_{N-1} + \mu_2,$$

that is, at the boundary nodes $i = 0$ and $i = N$ not only the function values, but also the first difference values or linear combinations of the function and difference values are yet to be known. Substituting $y_1 = y_0 + \Delta y_0$ into the first condition (8′) yields

$$(8'') \qquad \varkappa_1 \Delta y_0 - (1 - \varkappa_1) y_0 = -\mu_1$$

and needs certain clarification. The case $\varkappa_1 = 0$ corresponds to the **first kind boundary condition**: $y_0$ is given at the boundary node $i = 0$. When $\varkappa_1 = 1$ we deal with the **second kind boundary condition**: $\Delta y_0$ is given at the same node. All the cases with $\varkappa_1 \neq 0; 1$ reflect the **third kind boundary condition** as a linear combination of the function and the first difference at the node $i = 0$.

Due to serious achievements of the Russian and foreign mathematicians in applied mathematics the majority of mathematical-physics problems may be reduced to computational algorithms, at every step of which 3-point equations like (6) with conditions (8′) must be solved.

Moreover, a lot of rather complicated problems in numerical analysis gives rise to the canonical problem, where a square $(N+1) \times (N+1)$-matrix of the corresponding system acquires a tridiagonal form

$$\begin{pmatrix} 1 & -\varkappa_1 & 0 & \dots & 0 & 0 & 0 & \dots & 0 & 0 & 0 \\ A_1 & -C_1 & B_1 & \dots & 0 & 0 & 0 & \dots & 0 & 0 & 0 \\ \dots & \dots & \dots & \dots & \dots & \dots & \dots & \dots & \dots & \dots & \dots \\ 0 & 0 & 0 & \dots & A_i & -C_i & B_i & \dots & 0 & 0 & 0 \\ \dots & \dots & \dots & \dots & \dots & \dots & \dots & \dots & \dots & \dots & \dots \\ 0 & 0 & 0 & \dots & 0 & 0 & 0 & \dots & A_{N-1} & -C_{N-1} & B_{N-1} \\ 0 & 0 & 0 & \dots & 0 & 0 & 0 & \dots & 0 & -\varkappa_2 & 1 \end{pmatrix}.$$

In the case of boundary conditions of the second or third kinds its order is $N+1$, while for the system (6) with the supplementary conditions (8) the order is $N-1$. All the matrices of interest possess the main feature: they have nonzero elements only on the three diagonals (the main and two adjacent ones).

With the aid of effective Gauss method for solving linear equations with such matrices a direct method known as the elimination method has been designed and unveils its potential in solving difference equations.

**5. The elimination method.** The problem we must solve take now the form

$$(9)\qquad \begin{aligned} &A_i\, y_{i-1} - C_i\, y_i + B_i\, y_{i+1} = -F_i\,, \qquad i = 1,\, 2,\, \dots,\, N-1\,;\\ &y_0 = \varkappa_1\, y_1 + \mu_1\,, \qquad\qquad y_N = \varkappa_2\, y_{N-1} + \mu_2\,, \end{aligned}$$

where $A_i \neq 0$ and $B_i \neq 0$ for all $i = 1,\, 2,\, \dots,\, N-1$.

Other ideas are connected with reduction of the original second-order difference equation (9) to three first-order ones, which may be, generally speaking, nonlinear. First of all, the recurrence relation with indeterminate coefficients $\alpha_i$ and $\beta_i$ is supposed to be valid:

$$(10)\qquad y_i = \alpha_{i+1}\, y_{i+1} + \beta_{i+1}\,.$$

Substituting $y_{i-1} = \alpha_i\, y_i + \beta_i$ into (9) yields

$$(A_i\, \alpha_i - C_i)\, y_i + A_i\, \beta_i + B_i\, y_{i+1} = -F_i\,,$$

which leads, because of (10), to

$$\big[ (A_i\, \alpha_i - C_i)\, \alpha_{i+1} + B_i \big]\, y_{i+1} + A_i\, \beta_i + (A_i\, \alpha_i - C_i)\, \beta_{i+1} = -F_i\,.$$

If the conditions

$$(A_i\, \alpha_i - C_i)\, \alpha_{i+1} + B_i = 0\,, \qquad A_i\, \beta_i + (A_i\, \alpha_i - C_i)\, \beta_{i+1} + F_i = 0$$

are fulfilled simultaneously, then the equation in view holds true for any $y_i$. Thus, assuming $C_i - \alpha_i\, A_i \neq 0$, we establish the recurrence formulae for determination of both $\alpha_{i+1}$ and $\beta_{i+1}$:

$$(11)\qquad \alpha_{i+1} = \frac{B_i}{C_i - \alpha_i\, A_i}\,, \qquad i = 1, 2, \dots, N-1\,,$$

$$(12)\qquad \beta_{i+1} = \frac{A_i\, \beta_i + F_i}{C_i - \alpha_i\, A_i}\,, \qquad i = 1, 2, \dots, N-1\,,$$

under agreement (10).

By going through the matter chronologically the forward elimination path for finding $\alpha_i$, $\beta_i$ and the backward path for finding $y_i$ do arise during the course of the elimination method. Let us clarify the situation. With knowledge of $\alpha_i$ and $\beta_i$ one can determine, according to (10), all the values $y_i$ moving from $i+1$ to $i$. To find $\alpha_i$ and $\beta_i$, we succeed the reverse order (from $i$ to $i+1$) according to formulae (11)–(12).

Two relevant aspects are worth noting in this context:
1) equations (11)–(12) for finding $\alpha_i$ and $\beta_i$, being nonlinear, reproduce the relationships between the values at two adjacent points;
2) for each unknown $\alpha$, $\beta$ or $y$, it is necessary to solve the corresponding Cauchy problem.

These requirements necessitate imposing auxiliary values which can serve as the initial conditions. The assignment of boundary conditions is aimed at specifying these or those values. Having substituted $i = 0$ into (10), we get $y_0 = \alpha_1 y_1 + \beta_1$. On the other hand, $y_0 = \varkappa_1 y_1 + \mu_1$, giving

$$\alpha_1 = \varkappa_1 , \tag{13}$$

$$\beta_1 = \mu_1 . \tag{14}$$

Thus, we set up for $\alpha$ and $\beta$ the Cauchy problems described by (11), (13) and (12), (14), respectively, thereby completing the forward path of the elimination.

At the second stage, knowing $\alpha_i$ and $\beta_i$, the boundary value $y_N$ is recovered from the system of the equations

$$\begin{cases} y_N & = \varkappa_2 \, y_{N-1} + \mu_2 , \\ y_{N-1} & = \alpha_N \, y_N + \beta_N \end{cases}$$

under the constraint $1 - \alpha_N \varkappa_2 \neq 0$.

Hence, the initial condition in question becomes for (10)

$$y_N = \frac{\mu_2 + \varkappa_2 \beta_N}{1 - \alpha_N \varkappa_2} . \tag{15}$$

The computational formulae (10) and (15) constitute what is called the backward elimination path.

The algorithm presented below as the sequence of applied formulae is called the **right elimination method** and is showing the gateway for the general case:

$$\overset{(\rightarrow)}{\alpha}_{i+1} = \frac{B_i}{C_i - \alpha_i\, A_i}\,, \qquad i = 1, 2, \ldots, N-1\,, \qquad \alpha_1 = \varkappa_1\,,$$

$$\overset{(\rightarrow)}{\beta}_{i+1} = \frac{A_i\, \beta_i + F_i}{C_i - \alpha_i\, A_i}\,, \qquad i = 1, 2, \ldots, N-1\,, \qquad \beta_1 = \mu_1\,,$$

$$y_N = \frac{\mu_2 + \varkappa_2\, \beta_N}{1 - \alpha_N\, \varkappa_2}\,,$$

$$\overset{(\leftarrow)}{y}_i = \alpha_{i+1}\, y_{i+1} + \beta_{i+1}\,, \qquad i = N-1, N-2, \ldots, 1, 0\,.$$

Here the symbols $(\rightarrow)$ and $(\leftarrow)$ indicate the directions of index count: either from $i$ to $i+1$ or from $i+1$ to $i$.

In connection with the preceding algorithm, it is natural to raise the question of correctness and stability providing a possibility of applying the method and obtaining a solution with a prescribed accuracy. Special investigations give definite answers to to these questions.

**6. Stability of the elimination method.** Let us stress that the conditions $C_i - \alpha_i\, A_i \neq 0$ and $1 - \alpha_N\, \varkappa_2 \neq 0$ cannot be excluded or relaxed during the course of the right elimination method. Just for this reason the restrictions on coefficients for well-posed and stability conditions needs investigation.

Common practice involves sufficient conditions

$$(16) \qquad \begin{aligned} &|C_i| \geq |A_i| + |B_i|\,, \qquad i = 1, 2, \ldots, N-1\,, \\ &|\varkappa_\alpha| \leq 1\,, \quad \alpha = 1, 2\,, \qquad |\varkappa_1| + |\varkappa_2| < 2\,, \end{aligned}$$

yielding $|\alpha_i| < 1$ for all $i = 1, 2, \ldots, N$.

The proof is carried out by induction. Assuming $|\alpha_i| \leq 1$ we will show that $|\alpha_{i+1}| \leq 1$. Since $|\alpha_1| = |\varkappa_1| \leq 1$, the same propertry will cover all $i = 2, 3, \ldots, N$ on account the chain of the inequalities:

$$|C_i - \alpha_i\, A_i| - |B_i| \geq |C_i| - |\alpha_i| \cdot |A_i| - |B_i| \geq |A_i| \cdot (1 - |\alpha_i|) \geq 0\,,$$

from which it follows that $|C_i - \alpha_i\, A_i| > 0$, since $B_i \neq 0$.

Granted $|\alpha_i| \leq 1$, observe that

$$|\alpha_{i+1}| = \frac{|B_i|}{|C_i - A_i\, \alpha_i|} \leq 1$$

as well as $|\alpha_{i+1}| < 1$ are ensured by $|\alpha_i| < 1$. The assumption $|\alpha_1| = |\varkappa_1| < 1$ implies $|\alpha_i| < 1$ for all $i = 1, 2, \dots, n$. The lower estimate holds true for the denominator in formula (15):

$$|1 - \alpha_N \varkappa_2| \geq 1 - |\alpha_N| \cdot |\varkappa_2| \geq 1 - |\varkappa_2| > 0 ,$$

because either $|\varkappa_2| < 1$ or $|\alpha_N| < 1$.

Thus, we have proved that under conditions (16) the right elimination algorithm is correct, meaning nonzero denominators in formulae (11), (12) and (15). So, under conditions (16) problem (9) has a unique solution given by formulae (10)–(15).

It is necessary to point out that calculations by these formulae may induce accumulation of rounding errors arising in arithmetic operations. As a result we actually solve the same problem but with perturbed coefficients $\widetilde{A}_i$, $\widetilde{B}_i$, $\widetilde{C}_i$, $\widetilde{\varkappa}_1$, $\widetilde{\varkappa}_2$ and right parts $\widetilde{F}_i$, $\widetilde{\mu}_1$, $\widetilde{\mu}_2$. If $N$ is sufficiently large, the growth of rounding errors may cause large deviations of the computational solution $\tilde{y}_i$ from the proper solution $y_i$.

The trivial example shows how instability may arise in the process of calculations of $y_i$ by the formula $y_{i+1} = q\, y_i$, $q > 1$. One expects that, for any $y_0$, there exists a number $n_0$ such that overflow occur for $y_n = q^n\, y_0$, $n = n_0$, thus causing a abnormal termination. An important obstacle in dealing with this problem is that $\tilde{y}_i$ satisfies the equation $\tilde{y}_{i+1} = q\, \tilde{y}_i + \eta$ with a rounding error $\eta$. Indeed, for the error $\delta\, y_i = \tilde{y}_i - y_i$ the equation is valid:

$$\delta\, y_{i+1} = q\, \delta\, y_i + \eta , \qquad \delta\, y_0 = \eta ,$$

from which it follows that the value

$$\delta\, y_i = q^i\, \eta + \frac{q^i - 1}{q - 1}\, \eta , \qquad q > 1 ,$$

increases exponentially along with increasing $i$.

Returning to the right elimination method, we show that the conditions $|\alpha_i| \leq 1$ guarantee that the error $\delta\, y_{i+1} = \tilde{y}_{i+1} - y_{i+1}$ arising when computing $y_i$ does not increase. Indeed, the equations

$$y_i = \alpha_{i+1}\, y_{i+1} + \beta_{i+1} , \qquad \tilde{y}_i = \alpha_{i+1}\, \tilde{y}_{i+1} + \beta_{i+1}$$

imply that

$$\delta\, y_i = \alpha_{i+1}\, \delta\, y_{i+1} , \qquad |\delta\, y_i| \leq |\alpha_{i+1}| \cdot |\delta\, y_{i+1}| ,$$

meaning $|\delta\, y_i| \leq |\delta\, y_{i+1}|$ because $|\alpha_{i+1}| \leq 1$.

Taking into account perturbations of the coefficients $\alpha_{i+1}$ and $\beta_{i+1}$, another conclusion can be drawn that the accuracy $\varepsilon$ in determination of the solution $y_i$ of problem (9) is

$$\max_{1 \leq i \leq N} |\delta\, y_i| \approx \varepsilon_0\, N^2 ,$$

where $\varepsilon_0$ is a rounding error and $N$ is the total number of grid nodes.

**7. The left elimination method. The counter elimination method.** Still using the framework of the right elimination method (formulae (10)–(15)) in reverse order, we obtain through such an analysis the computational formulae of the left elimination method:

$$\overset{(\leftarrow)}{\xi}_i = \frac{A_i}{C_i - \xi_{i+1}\, B_i}\,, \qquad i = N-1, N-2, \dots, 2, 1\,, \quad \xi_N = \varkappa_2\,; \tag{17}$$

$$\overset{(\leftarrow)}{\eta}_i = \frac{B_i\, \eta_{i+1} + F_i}{C_i - \xi_{i+1}\, B_i}\,, \qquad i = N-1, N-2, \dots, 2, 1\,, \quad \eta_N = \mu_2\,; \tag{18}$$

$$\overset{(\rightarrow)}{y}_{i+1} = \xi_{i+1}\, y_i + \eta_{i+1}\,, \quad i = 0, 1, \dots, N-1\,, \tag{19}$$

and

$$y_0 = \frac{\mu_1 + \varkappa_1\, \eta_1}{1 - \xi_1\, \varkappa_1}\,. \tag{20}$$

Indeed, with the relation $y_{i+1} = \xi_{i+1}\, y_i + \eta_{i+1}$ granted we eliminate successively $y_{i+1}$, $y_i = \xi_i\, y_{i-1} + \eta_i$ from equation (9). The outcome of this is

$$\begin{aligned} -F_i &= A_i\, y_{i-1} + (B_i\, \xi_{i+1} - C_i)\, y_i + B_i\, \eta_{i+1} \\ &= \big[ A_i - (C_i - B_i\, \xi_{i+1})\, \xi_i \big]\, y_{i-1} + B_i\, \eta_{i+1} - (C_i - B_i\, \xi_{i+1})\, \eta_i\,. \end{aligned}$$

Equation (9) is valid if we agree to consider

$$A_i - (C_i - B_i\, \xi_{i+1})\, \xi_i = 0\,, \qquad -F_i = B_i\, \eta_{i+1} - (C_i - B_i\, \xi_{i+1})\, \eta_i\,,$$

from which formulae (17) and (18) immediately follow The value $y_0$ can be found from the condition $y_0 = \varkappa_1\, y_1 + \mu_1$ and the formula $y_1 = \xi_1\, y_1 + \eta_0$. The presented inequalities

$$|C_i - B_i\, \xi_{i+1}| \geq |C_i| - |B_i| \cdot |\xi_{i+1}|\,, \qquad |1 - \xi_1\, \varkappa_1| \geq 1 - |\xi_1| \cdot |\varkappa_1|$$

confirm that the correctness and stability of the left elimination method are ensured by conditions (16), since $|\xi_i| \leq 1$ for all $i = 1, 2, \dots, N$.

Joint use of the left and right elimination methods refers to the counter elimination method. The essence of this method is to consider a fixed inner node $i = i_0$, $0 < i_0 < N$, and to compute the coefficients $\alpha_i$ and $\beta_i$ for $0 \le i \le i_0 + 1$ according to formulae (10)–(15):

$$\alpha_{i+1} = \frac{B_i}{C_i - \alpha_i A_i}, \qquad i = 1, 2, \ldots, i_0, \qquad \alpha_1 = \varkappa_1,$$

$$\beta_{i+1} = \frac{A_i \beta_i + F_i}{C_i - \alpha_i A_i}, \qquad i = 1, 2, \ldots, i_0, \qquad \beta_1 = \mu_1,$$

and for $i_0 \le i \le N$ the coefficients $\xi_i$ and $\eta_i$ can be found by formulae (17)–(20):

$$\xi_i = \frac{A_i}{C_i - \xi_{i+1} B_i}, \qquad i = N-1, N-2, \ldots, i_0, \qquad \xi_N = \varkappa_2,$$

$$\eta_i = \frac{B_i \eta_{i+1} + F_i}{C_i - \xi_{i+1} B_i}, \qquad i = N-1, N-2, \ldots, i_0, \qquad \eta_N = \mu_2.$$

The next step is to put the first solution of the form (10) together with the second one of the form (9) the node $i = i_0$. The outcome of this is

$$y_{i_0} = \alpha_{i_0+1} y_{i_0+1} + \beta_{i_0+1}, \qquad y_{i_0+1} = \xi_{i_0+1} y_{i_0} + \eta_{i_0+1},$$

yielding

$$y_{i_0} = \frac{B_{i_0+1} + \alpha_{i_0+1} \eta_{i_0+1}}{1 - \alpha_{i_0+1} \eta_{i_0+1}}.$$

As far as $1 - \alpha_{i_0+1} \eta_{i_0+1} > 0$, this formula is meaningful because under conditions (16) at least one of the modules: either $|\alpha_{i_0+1}|$ or $|\eta_{i_0+1}|$ is less than 1. With knowledge of $y_{i_0}$ the remaining values $y_i$ arise in the process of parallel calculations: for $i < i_0$ by formula (10) and for $i > i_0$ by formula (19).

Of course, the counter elimination method could be especially effective in an attempt to determine $y_i$ merely at only one node $i = i_0$.

**8. Maximum principle.** To make our exposition more transparent, the case of interest is related to the first kind boundary-value problem with $\varkappa_1 = 0$ and $\varkappa_2 = 0$:

$$\begin{aligned} &\mathcal{L}[y_i] = A_i y_{i-1} - C_i y_i + B_i y_{i+1} = -F_i, \\ &i = 1, 2, \ldots, N-1, \qquad y_0 = \mu_1, \quad y_N = \mu_2. \end{aligned} \tag{21}$$

**Theorem 1** (The maximum principle). *Let the conditions*

$$(22) \qquad A_i > 0\,, \qquad B_i > 0\,, \qquad D_i = C_i - A_i - B_i \geq 0$$

*be fulfilled for all* $i = 1, 2, \dots, N-1$ *and the grid function* $y_i$, $y_i \not\equiv const$, *satisfy at all the inner nodes* $i = 1, 2, \dots, N-1$ *the condition* $\mathcal{L}[y_i] \geq 0$ (*or* $\mathcal{L}[y_i] \leq 0$). *Then* $y_i$ *cannot take the maximal positive value (or the minimal negative value) at the inner points, that is, for* $i = 1, 2, \dots, N-1$.

**Proof** Given a grid function $y_i$, let $\mathcal{L}[y_i] \geq 0$ for all $i = 1, 2, \dots, N-1$ and $y_i$ attain its maximum at one of the inner nodes $i = i_*$, $0 < i_* < N$, so that

$$y_{i_*} = \max_{0 \leq i \leq N} y_i = M_0 > 0\,.$$

As $y_i \not\equiv$ const, there exists an inner point $i_0$ (may be, coincident with $i_*$) such that $y_{i_0} = y_{i_*} = M_0 > 0$ and at one of the adjacent points, say for $i = i_0 - 1$, we have $y_{i_0 - 1} < M_0$. We may attempt $\mathcal{L}[y_i]$ in simplified form

$$\mathcal{L}[y_i] = B_i\,(y_{i+1} - y_i) - A_i\,(y_i - y_{i-1}) - (C_i - A_i - B_i)\,y_i\,,$$

leaving us under conditions (22) s at the point $i = i_0$ with

$$\begin{aligned}\mathcal{L}[\,y_{i_0}\,] &= B_{i_0}\,(y_{i_0+1} - y_{i_0}) - A_{i_0}\,(y_{i_0} - y_{i_0-1}) - (C_{i_0} - A_{i_0} - B_{i_0})\,y_{i_0} \\ &\leq -B_{i_0}\,(y_{i_0} - y_{i_0+1}) - A_{i_0}\,(y_{i_0} - y_{i_0-1}) < 0\end{aligned}$$

by virtue of the relations

$$y_{i_0} \geq y_{i_0+1}\,, \qquad y_{i_0} > y_{i_0-1}\,, \qquad A_{i_0} > 0\,, \qquad B_{i_0} > 0\,.$$

The result obtained is not consistent with the condition: $\mathcal{L}[y_i] \geq 0$, valid for all $i = 1, 2, \dots, N-1$ including $i = i_0$. Thus, we proved the first statement of the theorem. The second one can be established in a same way with further replacement of $y_i$ by $-y_i$.

**Corollary 1** *Under conditions* (22) *and*

$$\mathcal{L}[y_i] \leq 0\,, \qquad i = 1,\,2,\,\dots,\,N-1\,; \qquad y_0 \geq 0,\; y_N \geq 0\,,$$

*the function* $y_i$ *is nonnegative:* $y_i \geq 0$ *for all* $i = 0, 1, \dots, N$. *The case* $\mathcal{L}[y_i] \geq 0$, $y_0 \leq 0$ *and* $y_N \leq 0$ *leads to* $y_i \leq 0$ *for other subscripts* $i = 1, 2, \dots, N-1$.

Indeed, let $\mathcal{L}[y_i] \leq 0$ and $y_i < 0$ at least at one point $i = i_*$. Then $y_i$ should attain its minimal negative value at an inner point $i = i_0$, $0 < i_0 < N$. But this fact contradicts Theorem 1.

**Corollary 2** *Under conditions* (22) *the problem*

$$(23) \qquad \mathcal{L}\,[y_i] = 0\,, \qquad i = 1,\, 2,\, \dots\,,\, N-1\,; \qquad y_0 = 0\,,\; y_N = 0\,,$$

*has the unique solution* $y_i \equiv 0$, *thereby justifyingg that problem* (21) *is uniquely solvable for any ingredients* $F_i$, $\mu_1$ *and* $\mu_2$.

In fact, assuming that the solution $y_i$ of problem (23) becomes nonzero at least at one point $i = i_*$ we come to a contradiction with the maximum principle: if $y_{i_*} > 0$, then $y_i$ attains its maximal positive value at some point $i_0$, $0 < i_0 < N$, violating with Theorem 1; the case $y_{i_*} < 0$ may be viewed on the same footing.

**Theorem 2** (Comparison theorem). *Let conditions* (22) *hold*, $y_i$ *be the solution of problem* (21) *and* $\bar{y}_i$ *be the solution to the following problem:*

$$\mathcal{L}[\bar{y}_i] = -\bar{F}_i\,, \qquad i = 1, 2, \dots, N-1\,; \qquad \bar{y}_0 = \bar{\mu}_1\,, \qquad \bar{y}_N = \bar{\mu}_2\,,$$

*with*

$$|F_i| \le \bar{F}_i\,, \qquad i = 1, 2, \dots, N-1\,; \qquad |\mu_1| \le \bar{\mu}_1\,, \qquad |\mu_2| < \bar{\mu}_2\,.$$

*Then the relations occur:*

$$|y_i| \le \bar{y}_i\,, \qquad i = 0, 1, \dots, N\,.$$

**Proof** Due to Corollary 1 we have $\bar{y}_i \ge 0$ for $0 \le i \le N$, since $\mathcal{L}\,[\,\bar{y}_i\,] = -\bar{F}_i \le 0$ and $\bar{y}_0 \ge 0$, $\bar{y}_N \ge 0$. Observe that the functions $u_i = \bar{y}_i - y_i$ and $v_i = \bar{y}_i + y_i$ satisfy equation (21) with the right parts $\bar{F}_i - F_i \ge 0$ and $\bar{F}_i + F_i \ge 0$, and boundary values $u_0 = \bar{\mu}_1 - \mu_1 \ge 0$, $u_N = \bar{\mu}_2 - \mu_2 \ge 0$ and $v_0 = \bar{\mu}_1 + \mu_1 \ge 0$, $v_N = \bar{\mu}_2 + \mu_2 \ge 0$, respectively. By applying Corollary 1 to such a setting we get $u_i \ge 0$, $v_i \ge 0$ or $-\bar{y}_i \le y_i \le \bar{y}_i$, meaning $|y_i| \le \bar{y}_i$.

The function $\bar{y}_i$ is called a **majorant** for the solution of problem (21). A first step towards the solution of problem (21) is connected with a majorant $\bar{y}_i$: $\|y_C \le \|\bar{y}_C$.

**Corollary 3** *The solution to the problem*

$$\mathcal{L}\,[\,y_i\,] = 0\,, \qquad 0 < i < N\,; \qquad y_0 = \mu_1\,, \qquad y_N = \mu_2\,,$$

*can be majorized as*

$$\|y_C := \max_{0 \le i \le N} |y_i| \le \max\left(|\mu_1|,\, |\mu_2|\right).$$

The problem of auxiliary character such as

$$\mathcal{L}[\bar{y}_i] = 0, \qquad 0 < i < N; \qquad \bar{y}_0 = \bar{y}_N = \bar{\mu} = \max\left(|\mu_1|, |\mu_2|\right),$$

will complement our studiea, for which the comparison theorem gives $\|y_C \leq \|\bar{y}_C$. On the same grounds, Theorem 1 implies $\|\bar{y}_C \leq \bar{\mu}$, since the function $\bar{y}_i \geq 0$ takes its maximal positive value only at a boundary point: either for $i = 0$ or for $i = N$.

**Theorem 3** *Let the conditions*

$$(24) \qquad |A_i| > 0, \qquad |B_i| > 0, \qquad \overline{D}_i = |C_i| - |A_i| - |B_i| > 0$$

*hold for all* $i = 1, 2, \ldots, N-1$. *A solution of the problem*

$$(25) \quad \mathcal{L}[y_i] = -F_i, \qquad i = 1, 2, \ldots, N-1; \qquad y_0 = 0, \qquad y_N = 0,$$

*admits the estimate*

$$(26) \qquad \|y_C \leq \left\| \frac{F}{\overline{D}} \right\|_C .$$

To prove the desired estimate, the intention is to use an alternative form of equation (21)

$$(27) \qquad C_i\, y_i = A_i\, y_{i-1} + B_i\, y_{i+1} + F_i .$$

Let $|y_i|$ take its maximal value $|y_{i_0}| > 0$ at a point $i = i_0$, $0 < i_0 < N$, so that $|y_{i_0}| \geq |y_i|$ for all $i = 0, 1, \ldots, N$. We are led by equation (27) for $i = i_0$ to

$$\begin{aligned} |C_{i_0}\, y_{i_0}| = |C_{i_0}| \cdot |y_{i_0}| &= |A_{i_0}\, y_{i_0-1} + B_{i_0}\, y_{i_0+1} + F_{i_0}| \\ &\leq |A_{i_0}| \cdot |y_{i_0-1}| + |B_{i_0}| \cdot |y_{i_0+1}| + |F_{i_0}| \\ &\leq \left(|A_{i_0}| + |B_{i_0}|\right) \cdot |y_{i_0}| + |F_{i_0}| . \end{aligned}$$

Because of this

$$\left(|C_{i_0}| - |A_{i_0}| - |B_{i_0}|\right) \cdot |y_{i_0}| = \overline{D}_{i_0} \cdot |y_{i_0}| \leq |F_{i_0}|$$

and, consequently,

$$|y_{i_0}| = \max_{0 \le i \le N} |y_i| \le \frac{|F_{i_0}|}{D_{i_0}} \le \left\| \frac{F}{D} \right\|_C .$$

Before going further, problem (25) with coefficients subject to conditions (22) for $0 < i < N$ may be of help in achieving the final aims:

$$A_i > 0\,, \qquad B_i > 0\,, \qquad D_i = C_i - A_i - B_i \ge 0\,.$$

When the condition $D_i > 0$ fails to be true $y_i$ arranges itself as a sum $y_i = \overset{\circ}{y}_i + u_i$, where $\overset{\circ}{y}_i$ solves

(28) $$B_i\,(\overset{\circ}{y}_{i+1} - \overset{\circ}{y}_i) - A_i\,(\overset{\circ}{y}_i - \overset{\circ}{y}_{i-1}) = -F_i\,, \qquad 0 < i < N\,,$$

$$\overset{\circ}{y}_0 = \overset{\circ}{y}_N = 0\,.$$

The statement of the problem for $u_i$ is

(29) $$L\,[u_i] = B_i\,u_{i+1} - C_i\,u_i + A_i\,u_{i-1} = -D_i\,\overset{\circ}{y}_i\,, \qquad 0 < i < N\,,$$

$$u_0 = u_N = 0\,.$$

whose solution satisfies the inequality

(30) $$\|u\|_C = \max_{0 \le i \le N} |u_i| \le \|\overset{\circ}{y}\|_C\,.$$

This result is a corollary to the next lemma with $\varphi_i = \overset{\circ}{y}_i$ involved.

**Lemma.** *For the solution of problem (25) with coefficients subject to conditions (22) and the right part $F_i = D_i\,\varphi_i$, $i = 1, 2, \dots, N-1$, the estimate is valid:*

(31) $$\|y_C \le \|\varphi_C\,.$$

**Proof** In the case $D_i \equiv 0$ the estimate is obvious, since $y_i \equiv 0$ due to Corollary 2 to Theorem 1. By relating $D_i > 0$ at least at one point we have occasion to use the function $Y_i \ge 0$, being a solution to the problem

(32) $$\mathcal{L}\,[\,Y_i\,] = -D_i\,|\varphi_i|\,, \qquad i = 1, 2, \dots, N-1\,; \qquad Y_0 = Y_N = 0\,.$$

The solution of problem (32) can be most readily evaluated with the aid of the relation $\|y_C \leq \|Y_C$, valid an account of Theorem 2. Let the maximum of the function $Y_i$ be arrived at a point $i = i_0$. Then

$$B_{i_0}\left(Y_{i_0+1} - Y_{i_0}\right) \leq 0\,, \qquad A_i\left(Y_{i_0} - Y_{i_0-1}\right) \geq 0$$

and (32) implies the double inequality

$$D_{i_0}\,Y_{i_0} \leq D_{i_0}\,|\varphi_{i_0}| \leq D_{i_0}\,\|\varphi_C\,.$$

In the case $D_{i_0} > 0$, it follows from the foregoing that

$$\|Y_C \leq \|\varphi_C\,. \tag{33}$$

For $D_{i_0} = 0$ we deduce from (32) that

$$B_{i_0}\left(Y_{i_0+1} - Y_{i_0}\right) = A_{i_0}\left(Y_{i_0} - Y_{i_0-1}\right).$$

With the relations $Y_{i_0} \geq Y_{i_0-1}$ and $Y_{i_0} \geq Y_{i_0+1}$ in view, we find that

$$Y_{i_0+1} = Y_{i_0} = Y_{i_0-1}\,,$$

it being understood that the same maximal value is attained at the adjacent to $i_0$ points.

By merely setting $i = i_1 = i_0 + 1$ (or $i_1 = i_0 - 1$) we follow established practice and obtain, as a final result, the inequality

$$D_{i_1}\,Y_{i_1} \leq D_{i_1}\cdot\|\varphi_C\,,$$

giving either inequality (33) or the equality $Y_{i_1+1} = Y_{i_1-1} = Y_{i_1}$. As $D_i \not\equiv 0$, we get $D_{i_\alpha} > 0$ and inequality (33) for some $i = i_\alpha$, thereby completing the proof.

**Theorem 4** *The solution $y_i$ of problem (25) with coefficients (22) admits the estimate*

$$\|y_C \leq 2\,\|\overset{\circ}{y}_C \tag{34,}$$

*where $\overset{\circ}{y}$ is the solution of problem (28).*

**Proof** Recall that the difference $u_i = y_i - \overset{\circ}{y}_i$ is the solution of problem (29) with the right part $F_i = D_i \overset{\circ}{y}_i$. The inequality

$$\|y_C = \|\overset{\circ}{y} + u_C \le \|\overset{\circ}{y}_C + \|u_C$$

in combination with estimate (30) leads to estimate (34).

This provides reason enough to reduce proper evaluation of the solution of the general problem (25) to that of the simpler equation (28), whose solution $\overset{\circ}{y}_i$ can be found in explicit form.

**9. Maximum principle for the third kind boundary-value problem.** The maximum principle and its corollaries remain valid for the general boundary-value problem (6), (8′), whose statement is

$$(21^*)\qquad \begin{aligned} &\mathcal{L}[y_i] = -F_i\,, \qquad i = 0, 1, 2, \ldots, N\,;\\ &\mathcal{L}[y_0] = -y_0 + \varkappa_1 y_1\,, \quad F_0 = \mu_1\,,\\ &\mathcal{L}[y_N] = -y_N + \varkappa_2 y_{N-1}\,, \quad F_N = \mu_2\,, \end{aligned}$$

with the members $C_0 = 1$, $A_0 = 0$, $B_0 = \varkappa_1$, $C_N = 1$, $A_N = \varkappa_2$ and $B_N = 0$.

**Theorem 1** (The maximum principle). *Let under the set of the restrictions*

$$(22^*)\qquad \begin{aligned} &A_i > 0,\ B_i > 0\,, \qquad C_i \ge A_i + B_i\,, \qquad i = 1, 2, \ldots, N-1\,;\\ &0 \le \varkappa_1 \le 1\,, \qquad 0 \le \varkappa_2 \le 1\,, \qquad 0 < \varkappa_1 + \varkappa_2 < 2\,, \end{aligned}$$

*the function $y_i$, $y_i \not\equiv$ const, satisfy the conditions*

$$\mathcal{L}[y_i] \ge 0 \qquad (\mathcal{L}[y_i] \le 0)\,, \qquad i = 0, 1, 2, \ldots, N\,.$$

*Then $y_i$ cannot take the maximal positive (the minimal negative) value at any node $i = 0, 1, 2, \ldots, N$, that is, $y_i \le 0$ ($y_i \ge 0$).*

The proof is analogous to that of Theorem 1 from Section 8. It is necessary only to consider, in addition, the following two cases:

a) if $i_0 = 0$, that is, $\max_i y_i = y_0 = M_0 > 0$ and $y_1 < M_0$, then

$$\mathcal{L}[y_0] = -y_0 + \varkappa_1 y_1 = -(1-\varkappa_1)\, y_0 - \varkappa_1 (y_0 - y_1) < 0$$

for $0 \le \varkappa_1 \le 1$;

b) if $i_0 = N$, that is, $\max_i y_i = y_N = M_0 > 0$, but $y_{N-1} < M_0$, then

$$\mathcal{L}[y_N] = -y_N + \varkappa_2 y_{N-1} = -(1 - \varkappa_2) y_N - \varkappa_2 (y_N - y_{N-1}) < 0$$

for $0 \le \varkappa_2 \le 1$.

In both cases we came to a contradiction with the condition $\mathcal{L}[y_i] \ge 0$ for all $i = 0, 1, 2, \dots, N$. It is therefore concluded that $y_i \le 0$, since if $y_i > 0$ at least at one point $i = i_*$, its maximal positive value should be attained at some point $i = i_0$ (for example, at $i_0 = i_*$) that in principle is impossible.

**Corollary 2a** *Under conditions* (22*) *the problem*

$$\mathcal{L}[y_i] = 0, \qquad i = 0, 1, \dots, N$$

*has only the trivial solution.*

There is no difficulty to reformulate the remaining assertions from the previous sections, but we do not dwell on precise statements.

**10. Solution estimation for difference boundary-value problems by the elimination method.** In tackling the first boundary-value problem difference equation (21) has the tridiagonal matrix of order $N - 1$

$$\begin{pmatrix} -C_1 & B_1 & 0 & \dots & 0 & 0 & 0 \\ A_2 & -C_2 & B_2 & \dots & 0 & 0 & 0 \\ \vdots & \vdots & \vdots & \ddots & \vdots & \vdots & \vdots \\ 0 & 0 & 0 & \dots & A_{N-2} & -C_{N-2} & B_{N-2} \\ 0 & 0 & 0 & \dots & 0 & A_{N-1} & -C_{N-1} \end{pmatrix},$$

being symmetric for the case

$$B_i = A_{i+1}. \tag{35}$$

Difference equations with a symmetric matrix are typical in numerical solution of boundary-value problems associated with self-adjoint differential equations of second order. In what follows we will show that the condition $B_i = A_{i+1}$ is necessary and sufficient for the operator $\mathcal{L}[y_i]$ be self-adjoint.

As can readily be observed, any difference equation of the form

$$B_i y_{i+1} - C_i y_i + A_i y_{i-1} = -F_i, \qquad y_0 = \mu_1, \qquad y_N = \mu_2, \tag{36}$$

$$A_i \ne 0, \qquad B_i \ne 0, \qquad i = 1, 2, \dots, N - 1,$$

can be made self-adjoint:

$$\Lambda y_i = a_{i+1}(y_{i+1} - y_i) - a_i(y_i - y_{i-1}) - d_i y_i = -\varphi_i ,$$
$$i = 1, 2, \ldots, N-1 .$$

Indeed, we multiply equation (36) by the function $\eta_i \neq 0$ and involve the conditions $A_i \eta_i = a_i$ and $B_i \eta_i = a_{i+1}$, whose use permits us to establish $A_{i+1} \eta_{i+1} = B_i \eta_i = a_{i+1}$ and

$$\eta_{i+1} = \frac{B_i}{A_{i+1}} \eta_i = \eta_1 \prod_{k=1}^{i} \frac{B_k}{A_{k+1}}$$

where $\eta_1$ is an arbitrary constant. As a result we get the equation $\Lambda y_i = -\varphi_i$ with $\varphi_i = \eta_i F_i$ and $a_i = A_i \eta_i$, $d_i = (C_i - A_i - B_i)\eta_i = c_i - a_i - a_{i+1}$, $c_i = C_i \eta_i$.

By applying the elimination formulae to the problem

$$\Lambda y_i = a_i y_{i-1} - c_i y_i + a_{i+1} y_{i+1} = -\varphi_i , \tag{37}$$
$$i = 1, 2, \ldots, N-1 , \qquad y_0 = 0 , \qquad y_N = 0 ,$$

we deduce the estimate

$$\|y\|_C \leq \sum_{i=1}^{N-1} \frac{1}{|\alpha_{i+1}|} \sum_{k=1}^{i} |\varphi_k| \tag{38}$$

under the natural premises

$$|a_i| > 0 , \qquad |c_i| \geq |a_i| + |a_{i+1}| . \tag{39}$$

The computational formulae of the right elimination method help derive estimate (38):

$$\begin{aligned} y_i &= \alpha_{i+1} y_{i+1} + \beta_{i+1} , && i = 0, 1, \ldots, N-1 , && y_N = 0 , \\ \alpha_{i+1} &= \frac{a_{i+1}}{c_i - \alpha_i a_i} , && i = 1, 2, \ldots, N-1 , && \alpha_1 = 0 , \\ \beta_{i+1} &= \frac{a_i \beta_i + \varphi_i}{c_i - \alpha_i a_i} , && i = 1, 2, \ldots, N-1 , && \beta_1 = 0 . \end{aligned} \tag{40}$$

Since $|\alpha_{i+1}| \le 1$ under conditions (39), formulae (40) imply that

$$|y_i| \le |\alpha_{i+1}|\,|y_{i+1}| + |\beta_{i+1}| \le |y_{i+1}| + |\beta_{i+1}|,$$

yielding, in turn,

$$(41)\qquad |y_i| \le \sum_{k=i+1}^{N} |\beta_k|, \qquad 0 \le i \le N-1, \qquad y_N = 0.$$

Substitution $\gamma_i = a_i\,\beta_i$ leads to the useful relations

$$\gamma_{i+1} = \alpha_{i+1}\,(\gamma_i + \varphi_i),$$

$$|\gamma_{i+1}| \le |\gamma_i| + |\varphi_i| \le |\gamma_1| + \sum_{k=1}^{i} |\varphi_k| = \sum_{k=1}^{i} |\varphi_k|$$

and, consequently,

$$|\beta_i| \le \frac{1}{|a_i|} \sum_{k=1}^{i-1} |\varphi_k|, \qquad i = 2, 3, \dots, N, \qquad \beta_1 = 0.$$

Combination of (41) with the foregoing allows us to establish (38).

Before going further, observe that a solution to the problem

$$(42)\qquad \Lambda\, y_i = -\varphi_i, \qquad i = 1, 2, \dots, N-1; \qquad y_0 = \mu_1, \qquad y_N = \mu_2,$$

can be majorized by

$$\|y\|_C \le \max\left(|\mu_1|, |\mu_2|\right) + \sum_{i=1}^{N-1} \frac{1}{|a_{i+1}|} \sum_{k=1}^{i} |\varphi_k|.$$

As stated in Corollary 3, a solution to the problem

$$\Lambda\, \tilde{y}_i = 0, \qquad i = 1, 2, \dots, N-1; \qquad \tilde{y}_0 = \mu_1, \qquad \tilde{y}_N = \mu_2,$$

with $a_i > 0$ and $c_i \ge a_i + a_{i+1}$ can be most readily evaluated by the quantities

$$(43)\qquad \|\tilde{y}\|_C \le \max\left(|\mu_1|, |\mu_2|\right).$$

The solution of problem (42) arranges itself as a sum $y_i = \tilde{y}_i + v_i$, where the second summand is a solution of problem (37). Applying estimates (38) and (43), and the inequality $|y_i| \le |\tilde{y}_i| + |v_i|$ we come to the desired result.

**11. The second-order difference equations with constant coefficients.** If the coefficients of the difference equation

$$\mathcal{L}\,[y_i] = A_i\, y_{i-1} - C_i\, y_i + B_i\, y_{i+1} = 0 \tag{44}$$

do not depend on $i$, meaning $A_i = a$, $C_i = c$ and $B_i = b$ for all $i = 1, 2, \ldots,$ a solution of the following equation

$$b\, y_{i+1} - c\, y_i + a\, y_{i-1} = 0\,, \qquad b \neq 0\,, \quad a \neq 0\,, \tag{45}$$

may be found in explicit form. In so doing, let $y_i^{(1)}$ and $y_i^{(2)}$ be two solutions of difference equation (44) that are linearly independent if the equalities

$$C_1\, y_i^{(1)} + C_2\, y_i^{(2)} = 0\,, \qquad i = 0, 1, 2, \ldots\,,$$

hold only for $C_1 = C_2 = 0$. This condition can be replaced by the usual requirement for the determinant of the system of algebraic equations

$$\begin{cases} C_1\, y_i^{(1)} + C_2\, y_i^{(2)} = 0\,, & \\ C_1\, y_{i+m}^{(1)} + C_2\, y_{i+m}^{(2)} = 0\,, & m = 1, 2, \ldots\,, \end{cases}$$

saying that, for all $i$ and $m$

$$\Delta_{i,\,i+m} = \begin{vmatrix} y_i^{(1)} & y_i^{(2)} \\ y_{i+m}^{(1)} & y_{i+m}^{(2)} \end{vmatrix} \neq 0\,.$$

By pursuing the parallel we see, in particular, that the condition

$$\Delta_{i,\,i+1} = \begin{vmatrix} y_i^{(1)} & y_i^{(2)} \\ y_{i+1}^{(1)} & y_{i+1}^{(2)} \end{vmatrix} = \begin{vmatrix} y_i^{(1)} & y_i^{(2)} \\ \Delta\, y_i^{(1)} & \Delta\, y_i^{(2)} \end{vmatrix} \neq 0$$

is analogous to that of the linear independence for solutions $\bar{u}(x)$ and $\bar{\bar{u}}(x)$ of an ordinary second-order differential equation:

$$\begin{vmatrix} \bar{u}(x) & \bar{\bar{u}}(x) \\ (\bar{u}(x))' & (\bar{\bar{u}}(x))' \end{vmatrix} \neq 0\,.$$

Eliminating $y_{i+1}^{(1)}$ and $y_{i+1}^{(2)}$ with the aid of equation (44) we get

$$\Delta_{i,\,i+1} = \frac{1}{B_i}\left( B_i\, y_{i+1}^{(2)}\, y_i^{(1)} - B_i\, y_{i+1}^{(1)}\, y_i^{(2)} \right) = -\frac{A_i}{B_i}\left( y_i^{(1)}\, y_{i-1}^{(2)} - y_{i-1}^{(1)}\, y_i^{(2)} \right),$$

giving

$$\Delta_{i,\,i+1} = -\frac{A_i}{B_i}\,\Delta_{i,\,i-1}\,.$$

Due to this fact the condition $\Delta_{i_1,\,i_1+1} \neq 0$ for some $i = i_1$ yields $\Delta_{i,\,i+1} \neq 0$ for all possibilities of $i$.

Taking into account the equation $\mathcal{L}\,[y_i] = 0$ we see that, for any $m > 1$, $\Delta_{i,\,i+m}$ can be expressed through $\Delta_{i,\,i+1}$ and, hence, $\Delta_{i,\,i+m} \neq 0$ is ensured by the condition $\Delta_{i,\,i+1} \neq 0$.

If $y_i^{(1)}$ and $y_i^{(2)}$ are linearly independent solutions of the homogeneous equation (44), then its general solution can be designed as a linear combination of $y_i^{(1)}$ and $y_i^{(2)}$ with arbitrary constants $C_1$ and $C_2$:

$$y_i = C_1\,y_i^{(1)} + C_2\,y_i^{(2)}.$$

The constants are free to be chosen from the initial or boundary conditions, since $\Delta_{i,i+m} \neq 0$ for all admissible subscripts $i$ and $m$.

The general solution of any nonhomogeneous difference equation $\mathcal{L}\,[y_i] = -F_i$ is representable by

$$y_i = C_1\,y_i^{(1)} + C_2\,y_i^{(2)} + \tilde{y}_i\,,$$

where $\tilde{y}_i$ is a particular solution to the equation $\mathcal{L}\,[\tilde{y}_i] = -F_i$.

In the case of the Cauchy problem with assigned values $y_0$ and $y_1$, we have at our disposal the system of algebraic equations for constants $C_1$ and $C_2$:

$$\begin{cases} C_1\,y_0^{(1)} + C_2\,y_0^{(2)} = y_0\,, \\ C_1\,y_1^{(1)} + C_2\,y_1^{(2)} = y_1\,. \end{cases}$$

As far as $y_i^{(1)}$ and $y_i^{(2)}$ are linearly independent, the condition $\Delta_{0,1} \neq 0$ provides a possibility of finding $C_1$ and $C_2$. In the case of the boundary-value problem with $y_0 = \mu_1$ and $y_N = \mu_2$ incorporated, the constants $C_1$ and $C_2$ can uniquely be determined under the agreement $\Delta_{0,\,N} \neq 0$.

If the coefficients involved in equation (44) are constant, that is, $A_i = a$, $C_i = c$ and $B_i = b$, then particular solutions can be found in explicit form. This can be done by attempting particular solutions to equation (45) in the form $y_k = q^k$, while the number $q \neq 0$ remains as yet unknown. Substituting $y_k$ into (45) we obtain the quadratic equation for $q$:

$$b\,q^2 - c\,q + a = 0 \tag{46}$$

whose roots are

$$q_{1,2} = \frac{c \pm \sqrt{c^2 - 4ab}}{2b}.$$

Three possibilities of interest in accordance with the discriminant sign are:

1) If $D = c^2 - 4ab > 0$, then the preceding equation has two distinct real roots

$$q_1 = \frac{c + \sqrt{D}}{2b}, \qquad q_2 = \frac{c - \sqrt{D}}{2b}.$$

Two different particular solutions $y_k^{(1)} = q_1^k$ and $y_k^{(2)} = q_2^k$ corresponding to $q_1$ and $q_2$ are linearly independent, since

$$\Delta_{k,k+1} = \begin{vmatrix} q_1^k & q_2^k \\ q_1^{k+1} & q_2^{k+1} \end{vmatrix} = (q_2 - q_1)\, q_1^k\, q_2^k \neq 0,$$

and constitute what is called the general solution of (45):

$$y_k = C_1\, q_1^k + C_2\, q_2^k$$

where $C_1$ and $C_2$ are arbitrary constants.

2) If $D = c^2 - 4ab = 0$, then $q_1 = q_2 = c/(2b) = q_0$ and $y_k^{(1)} = q_0^k$ and $y_k^{(2)} = k\, q_0^k$ can be declared to be linearly independent particular solutions. Indeed, substituting $y_k^{(2)} = k\, q_0^k$ into (45) we obtain by minor changes

$$b\, y_{k+1}^{(2)} - c\, y_k^{(2)} + a\, y_{k-1}^{(2)} = \left[\, b\,(k+1)\, q_0^2 - c\, k\, q_0 + a\,(k-1) \,\right]\, q_0^{k-1}$$

$$= q_0^{k-1}\, k\left(b\, q_0^2 - c\, q_0 + a\right) + q_0^{k-1}\,(b\, q_0^2 - a) = q_0^{k-1}\,(b\, q_0^2 - a) = 0,$$

since $b\, q_0^2 - a = b\,[\,c/(2b)\,]^2 - a = D/(4b) = 0$. The discriminant

$$\Delta_{k,k+1} = \begin{vmatrix} q_0^k & k\, q_0^k \\ q_0^{k+1} & (k+1)\, q_0^{k+1} \end{vmatrix} = q_0^{2k+1} \neq 0$$

assures us that $q_0^k$ and $k\, q_0^k$ are linearly independent and can be chosen as a basis for constructing the general solution of (45):

$$y_k = (C_1 + k\, C_2)\, q_0^k, \qquad q_0 = \frac{c}{2b}.$$

3) If $D = c^2 - 4ab < 0$, equation (46) possesses two complex-conjugate roots

$$q_1 = \frac{c + \sqrt{|D|}\,i}{2b} = \rho\,(\cos\varphi + i\,\sin\,\varphi) = \rho\,e^{i\varphi},$$

$$q_2 = \frac{c - \sqrt{|D|}\,i}{2b} = \rho\,(\cos\varphi - i\,\sin\,\varphi) = \rho\,e^{-i\varphi},$$

where

$$\rho = \sqrt{\frac{a}{b}}\,, \qquad \varphi = \operatorname{arctg}\,\frac{\sqrt{|D|}}{c}\,, \qquad i = \sqrt{-1}\,.$$

The appropriate functions

$$q_1^k = \rho^k\,e^{i\,k\,\varphi} = \rho^k\,\left(\cos\,k\varphi + i\,\sin\,k\varphi\right),$$

$$q_2^k = \rho^k\,e^{-i\,k\,\varphi} = \rho^k\,\left(\cos\,k\varphi - i\,\sin\,k\varphi\right)$$

or

$$y_k^{(1)} = \rho^k\,\cos\,k\varphi\,, \qquad y_k^{(2)} = \rho^k\,\sin\,k\varphi$$

are just particular solutions to equation (45). In this case both pairs of solutions are linearly independent in connection with the linear independence of the functions sin $k\varphi$ and cos $k\varphi$: $\Delta_{k,\,k+1} \neq 0$. Then the general solution acquires the form

$$y_k = \rho^k\,\left(C_1\,\cos\,k\varphi + C_2\,\sin\,k\varphi\right).$$

In order to illustrate our approach a little better, it seems worthwhile giving several simple examples.

**Example 1** We are interested in giving the general solution to the equation

$$y_{k+1} - 2\,p\,y_k + y_{k-1} = 0\,, \qquad p > 0\,.$$

Still using the above framework with $D = 4\,(p^2 - 1)$, the following three cases need investigation.

a) Let $p < 1$. Setting $p = \cos\alpha$, $\alpha \neq 0$, and $y_k = q^k$ we obtain the quadratic equation for $q$:

$$q^2 - 2\cos\alpha\;q + 1 = 0$$

with discriminant $D = 4(\cos^2 \alpha - 1) = -4\sin^2 \alpha < 0$ and roots $q_{1,2} = e^{\pm i\alpha}$. Then $q_{1,2}^k = e^{\pm ik\alpha}$, particular solutions are given by

$$y_k^{(1)} = \cos k\alpha\,, \qquad y_k^{(2)} = \sin k\alpha$$

and the general solution can be written as

$$y_k = C_1 \cos k\alpha + C_2 \sin k\alpha$$

with arbitrary constants $C_1$ and $C_2$.
b) Let $p > 1$. Setting $p = \operatorname{ch}\alpha$ and $y_k = q^k$ we get

$$q^2 - 2\operatorname{ch}\alpha\, q + 1 = 0$$

with discriminant $D = 4(\operatorname{ch}^2\alpha - 1) > 0$ and roots $q_{1,2} = \operatorname{ch}\alpha \pm \operatorname{sh}\alpha = e^{\pm\alpha}$. Then $q_{1,2}^k = e^{\pm k\alpha}$, particular solutions become

$$y_k^{(1)} = \operatorname{ch} k\alpha\,, \qquad y_k^{(2)} = \operatorname{sh} k\alpha$$

and the general solution can be written as

$$y_k = C_1 \operatorname{ch} k\alpha + C_2 \operatorname{sh} k\alpha$$

with arbitrary constants $C_1$ and $C_2$.
c) Let $p = 1$. In this case, for $y_k = q^k$, we have $q^2 - 2q + 1 = 0$ and $q_{1,2} = 1$. The particular solutions $y_k^{(1)} = 1$ and $y_k^{(2)} = k$ form the general solution $y_k = C_1 + C_2 k$ as a linear function.

**Example 2** It is required to calculate the integral

$$I_k(\varphi) = \int_0^\pi \frac{\cos k\psi - \cos k\varphi}{\cos\psi - \cos\phi}\, d\psi\,, \qquad k = 0, 1, 2, \ldots.$$

First of all we note that

$$I_0(\varphi) = 0\,, \qquad I_1(\varphi) = \int_0^\pi d\psi = \pi\,.$$

We claim that $I_k$ is just the solution of the Cauchy problem for the second-order difference equation:

$$I_{k+1} - 2\cos\varphi\, I_k + I_{k-1} = 0\,, \qquad k = 1, 2, \ldots\,, \qquad I_0 = 0\,, \qquad I_1 = \pi\,,$$

assuming $\varphi$ to be arbitrary fixed. Indeed, the chain of the identities occur:

$$\begin{aligned}
&\big[\cos(k+1)\psi - \cos(k+1)\varphi\big] + \big[\cos(k-1)\psi - \cos(k-1)\varphi\big] \\
&\qquad = 2\cos k\psi \cos\psi - 2\cos k\varphi \cos\varphi \\
&\qquad = 2\left(\cos k\psi - \cos k\varphi\right)\cos\varphi + 2\left(\cos\psi - \cos\varphi\right)\cos k\psi\,,
\end{aligned}$$

from which the relations immediately follow

$$\begin{aligned}
I_{k+1} + I_{k-1} &= 2\cos\varphi\, I_k + 2\int_0^\pi \frac{\left(\cos\psi - \cos\varphi\right)\cos k\psi}{\cos\psi - \cos\varphi}\,d\psi \\
&= 2\cos\varphi\, I_k + 2\int_0^\pi \cos k\psi\, d\psi = 2\cos\varphi\, I_k\,, \qquad k > 1\,,
\end{aligned}$$

yielding $I_{k+1} - 2\cos\varphi\, I_k + I_{k-1} = 0$.

From Example 1, case a) we know that

$$I_k(\varphi) = C_1\cos k\varphi + C_2\sin k\varphi\,.$$

The initial conditions for $k = 0$ and $k = 1$ give $C_1 = 0$ and $C_2 = \pi/\sin\varphi$ in connection with the available information that $C_1\cos\varphi + C_2\sin\varphi = \pi$. The outcome of this is

$$I_k(\varphi) = \pi\,\frac{\sin k\varphi}{\sin\varphi}\,, \qquad k = 0, 1, 2, \ldots\,.$$

**12. Formulae of "difference differentiation" by parts of the product and sum.** The formula for differentiating the product of two real functions $u(x)$ and $v(x)$

$$\frac{d}{dx}\left(u(x)\,v(x)\right) = u(x)\,\frac{dv}{dx} + v(x)\,\frac{du}{dx}$$

is well-known from differential calculus. In an attempt to establish a grid analog of this correlation, we consider any two grid functions $y_i$ and $v_i$ of the discrete variable $i = 0, \pm 1, \pm 2, \ldots$. The following formulae of "difference differentiation" are valid:

$$\begin{aligned}
&\Delta\,(y_i\,v_i) = y_i\,\Delta\,v_i + v_{i+1}\,\Delta\,y_i = y_{i+1}\,\Delta\,v_i + v_i\,\Delta\,y_i\,, \\
&\nabla\,(y_i\,v_i) = y_{i-1}\nabla\,v_i + v_i\,\nabla y_i = y_i\,\nabla\,v_i + v_{i-1}\,\nabla\,y_i
\end{aligned} \tag{47}$$

with the well-established notations of the right and left differences: $\Delta y_i = y_{i+1} - y_i$ and $\nabla y_i = y_i - y_{i-1}$, so that $\nabla y_{i+1} = \Delta y_i$.

With this relation established, we find in a step-by-step fashion that

$$\begin{aligned} y_i \Delta v_i + v_{i+1} \Delta y_i &= y_i (v_{i+1} - v_i) + v_{i+1} (y_{i+1} - y_i) \\ &= y_{i+1} v_{i+1} - y_i v_i = \Delta (y_i v_i) . \end{aligned}$$

Likewise, one can check the second formula based on another relation

$$\nabla (y_i v_i) = \Delta (y_{i-1} v_{i-1}) .$$

An important role in the theory of difference schemes is played by the identities serving on this basis as grid analogs of integration by parts:

$$\int_a^b u' v \, dx = u v \Big|_a^b - \int_a^b u v' \, dx .$$

For any functions $y_i = y(i)$ and $v_i = v(i)$ defined on the grid $\omega = \{i = 0, 1, 2, \dots, N\}$, it will be sensible to introduce analogs of the integral $(u, v)_0 := \int_a^b uv \, dx$:

$$(y, v) := \sum_{i=1}^{N-1} y_i v_i , \qquad (y, v] := \sum_{i=1}^{N} y_i v_i , \qquad [y, v) := \sum_{i=0}^{N-1} y_i v_i .$$

With these, the summation by parts formula hholds:

$$(y, \Delta v) = -(v, \nabla y] + y_N v_N - y_0 v_1 . \tag{48}$$

With this aim, we replace $y_i \Delta v_i$ by its expression (47):

$$y_i \Delta v_i = \Delta (y_i v_i) - v_{i+1} \nabla y_{i+1}$$

in the sum

$$\begin{aligned} (y, \Delta v) = \sum_{i=1}^{N-1} y_i \Delta v_i &= \sum_{i=1}^{N-1} \Delta (y_i v_i) - \sum_{i=1}^{N-1} v_{i+1} \nabla y_{i+1} \\ &= y_N v_N - y_1 v_1 - \sum_{i'=2}^{N} v_{i'} \nabla y_{i'} , \end{aligned}$$

where $i' = i+1$. For $y_1 = y_0 + (y_1 - y_0) = y_0 + \nabla y_1$ we arrive at

$$(y,\, \Delta v) = y_N\, v_N - y_0\, v_1 - \sum_{i=1}^{N} v_i \nabla y_i = -(v,\, \Delta y] + y_N\, v_N - v_1\, y_0\,.$$

If $y_i$ vanishes at the boundary grid nodes $i = 0$ and $i = N$, that is, either $y_0 = 0$, $y_N = 0$ or $u_0 = 0$, $v_N = 0$, then (48) can be rewritten in the form

$$(y,\, \Delta v) \;=\; -(v,\, \nabla y]\,. \tag{49}$$

The identities obtained above are frequently encountered in difference transformations and calculating various finite sums and series. We give below some examples of such applications.

**Example 1** It is necessary to calculate the sum $S_N = \sum_{i=1}^{N} i\, 2^i$. By setting $y_i = i$ and $\Delta v_i = 2^i$ so that

$$v_{i+1} = v_i + 2^i = \sum_{k=0}^{i} 2^k + v_0 = 2^{i+1} - 1 + v_0$$

and choosing $v_0$ in such a way that $v_{N+1} = 0$ we are led by formula (48) to

$$\begin{aligned}\sum_{i=1}^{N} i\; 2^i &= \sum_{i=1}^{N} y_i\, \Delta v_i = -\sum_{i=1}^{N+1} v_i \nabla y_i + y_{N+1}\, v_{N+1} - y_0\, v_1 \\ &= -\sum_{i=1}^{N+1} \left(2^i - 2^{N+1}\right) = -2\left(2^{N+1} - 1\right) + 2^{N+1}\,(N+1) \\ &= (N-1)\; 2^{N+1} + 2\,.\end{aligned}$$

**Example 2** Of interest is the sum $S_N = \sum_{i=1}^{N-1} i\, a^i$. In this case $y_i = i$, $\Delta v_i = a^i$, $v_i = (a^i - a^N)\,/\,(a-1)$, $v_N = 0$ and

$$S_N \;=\; \frac{1}{(a-1)^2}\,\left[a^N\left(N\,(a-1) - a\right) + a\right].$$

**13. Green's difference formulae.** For the simplest operator $L\,u = u''$ the following identities are valid:

$$\int_a^b u\, v''\, dx = -\int_a^b u'\, v'\, dx + u\, v'\,\Big|_a^b\,,$$

$$\int_a^b \left(u\, v'' - u''\, v\right)\, dx \;=\; \left(u\, v' - u'\, v\right)\Big|_a^b\,,$$

which are called the **first** and the **second Green formulae**. Usually the first formula can be modified into a more general form, namely

$$(50) \qquad \int_a^b u\,Lv\,dx = -\int_a^b k\,u'\,v'\,dx - \int_a^b q\,u\,v\,dx + k\,u\,v'\,\Big|_a^b$$

with $L\,v := (k\,v')' - q\,v$.

Making here the mutual replacements of $u(x)$ and $v(x)$ and subtracting the resulting equation from (50), we establish the second Green formula in a more general form

$$(51) \qquad \int_a^b (u\,Lv - v\,Lu)\,dx = k\,(u\,v' - u'\,v)\,\Big|_a^b .$$

If, in addition, $u$ and $v$ vanish at the end points $x = a$ and $x = b$, then all preceding substitutions are equal to 0, thereby reducing formulae (50)–(51) to

$$(52) \qquad (u,\,L\,v)_0 = -(k\,u',\,v')_0 - (q\,u,\,v)_0\,, \qquad (u,\,L\,v)_0 = (v,\,L\,u)_0$$

where as usually $(u, v)_0 = \int_a^b u\,v\,dx$.

In particular, we might have

$$(u,\,L\,u)_0 = -(k,\,(u')^2)_0 - (q,\,u^2)_0\,.$$

Observe that the equality $(u,\,L\,v)_0 = (v,\,L\,u)_0$ means the self-adjointness of the operator $L$.

In the further development of the difference analogs of formulae (50), (51) and (52) $v_i = \Delta\,u_{i-1} = \nabla\,u_i$ is put together with (48):

$$(53) \qquad (y,\,\Delta\nabla\,u) = -\,(\nabla\,u, \nabla\,y] + y_N\,\nabla u_N - y_0\,\nabla u_1\,.$$

Likewise, setting $v_i = a_i \nabla u_i$ one obtains instead of (53) one more useful result

$$(54) \qquad (y,\,\Delta\,(a\,\nabla\,u)) = -(a\,\nabla u, \nabla y] + y_N\,a_N\,\nabla u_N - y_0\,a_1\,\nabla u_1\,.$$

Applying (54) to the difference operator

$$(55) \quad \Lambda\,u_i = \Delta\,(a_i\,\nabla\,u_i) - d_i\,u_i = a_{i+1}\,(u_{i+1} - u_i) - a_i\,(u_i - u_{i-1}) - d_i\,u_i$$

yields the first Green formula

$$(y,\, \Lambda\, u) = -\,(a \nabla u, \nabla y] - (du,\, y) + (a\, y \nabla u)_N - y_0\, (a \nabla u)_1\,, \tag{56}$$

which can be viewed as an analog of (50). After interchanging in (56) the positions of $u_i$ and $y_i$ we get the equation

$$(u,\, \Lambda\, y) = -(a \nabla y,\, \nabla u] - (dy,\, u) + (a\, y \nabla y)_N - u_0\, (a \nabla y)_1\,,$$

which will be subtracted at the next stage from (56). The final expression leads to the second Green formula

$$(y,\, \Lambda\, u) - (u,\, \Lambda\, y) = a\, (y \nabla u - u \nabla y)_N - a_1 (y_0 \nabla u_1 - u_0 \nabla y_1)\,. \tag{57}$$

In the particular case when $a_i \equiv 1$ and $d_i \equiv 0$, that is, for

$$\widetilde{\Lambda}\, y_i := \Delta \nabla\, y_i = \Delta^2\, y_{i-1}$$

Green's difference formula admits an alternative form

$$\begin{aligned}(y,\, \Delta \nabla\, u) &= (u,\, \Delta \nabla\, y) + (y \nabla u - u \nabla y)_N - (y\, \Delta\, u - u\, \Delta\, y)_0 \\ &= (u,\, \Delta \nabla\, y) + u_N\, y_{N-1} - y_N\, u_{N-1} + u_0\, y_1 - y_0\, u_1\,,\end{aligned}$$

which allows a more simpler writing of the ensuing formulae. When $y_0 = 0$ and $y_N = 0$, the first Green formula becomes

$$(y,\, \Lambda\, u) = -(a \nabla u,\, \nabla y) - (d\, u,\, y)\,, \tag{58}$$

giving for $u = y$

$$(y,\, \Lambda\, y) = -(a \nabla y, \nabla y) - (d,\, y^2)\,. \tag{59}$$

For any $y$ and $u$ subject to the homogeneous boundary conditions $y_0 = y_N = 0$ and $u_0 = u_N = 0$, the second Green formula takes a very elegant form

$$(\Lambda\, y,\, u) = (y,\, \Lambda\, u)\,. \tag{60}$$

## 1.2 SOME VARIANTS OF THE ELIMINATION METHOD

**1. The flow variant of the elimination method for difference problems with widely varying coefficients.** In numerical solution of thermal-flow hydrodynamics and magnito-hydrodynamics problems in which the coefficients of heat conductivity and electric conductivity depend on the thermodynamic parameters of the medium the elimination method for the corresponding difference equations leads to unsatisfactory results. In thermal problems adiabatic cells with infinitely large thermal-flow reveal themseves. The nonconductive cells or ideal conductivity may appear in magnetic problems, thus causing obstacles in connection with widely varying coefficients and considerable accuracy losses. In mastering the difficulties involved, the flow variant of the elimination method is aimed at solving a revised supplementary problem relating to the heat flow.

The statement of the boundary-value problem is

$$(61)\qquad a_i\, y_{i-1} - c_i\, y_i + a_{i+1}\, y_{i+1} = -f_i\,, \qquad i = 1, 2, \ldots, N-1\,,$$

$$(62)\qquad y_0 = \varkappa_1\, y_1 + \nu_1\,, \qquad y_N = \varkappa_2\, y_{N-1} + \nu_2\,,$$

with the members

$$(63)\qquad \begin{aligned} & c_i = a_{i+1} + a_i + d_i\,, \qquad d_i > 0,\ 0 \le a_i \le \infty\,, \\ & 1 \ge \varkappa_1,\ \varkappa_2 \ge 0\,, \qquad \varkappa_1 + \varkappa_2 < 2\,. \end{aligned}$$

Having stipulated condition (63), the computational formulae of the right elimination with regard to problem (61)–(62) can be written as

$$(64)\qquad \begin{aligned} & y_i = \dot\alpha_{i+1}\, y_{i+1} + \dot\beta_{i+1}\,, \qquad i = 0, 1, 2, \ldots, N-1\,, \\ & \dot\alpha_{i+1} = \frac{a_{i+1}}{a_{i+1} + a_i\,(1 - \dot\alpha_i) + d_i}\,, \\ & \dot\beta_{i+1} = \frac{\dot\alpha_{i+1}}{a_{i+1}}\,(a_i\, \dot\beta_i + f_i)\,, \qquad i = 1, 2, \ldots, N-1\,. \end{aligned}$$

A new unknown difference function is called a flow and is defined by

$$(65)\qquad w_i := a_i\,(y_{i-1} - y_i)\,,$$

making it possible to rearrange the problem statement and boundary conditions (61)–(62) as

$$(66)\quad w_i - w_{i+1} - d_i\, y_i = -f_i\,, \qquad i = 1, 2, \dots, N-1\,,$$

$$(67)\quad a_1\,(1 - \varkappa_1)\, y_1 + w_1 = a_1\,\nu_1\,, \quad a_N\,(1 - \varkappa_2)\, y_N - \varkappa_2 w_N = a_N\,\nu_2\,.$$

Substituting $y_i = y_{i+1} + w_{i+1}/a_{i+1}$ from (65) into the first formula (64) yields

$$(68)\qquad a_{i+1}\,(1 - \dot{\alpha}_{i+1})\, y_{i+1} + w_{i+1} = a_{i+1}\,\dot{\beta}_{i+1}\,.$$

To make our exposition more transparent, we introduce $\alpha_i = a_i\,(1 - \dot{\alpha}_i)$ and $\gamma_i = a_i\,\dot{\beta}_i$. With these, equation (8) admits the simplified form

$$(69)\qquad \alpha_i\, y_i + w_i = \gamma_i\,,$$

showing the new notations to be sensible ones. Having completed the elimination of $y_i$ from (66) and (69), we arrive at

$$(70)\qquad w_i = \frac{\alpha_i}{\alpha_i + d_i}\, w_{i+1} + \frac{d_i\,\gamma_i - \alpha_i\, f_i}{\alpha_i + d_i}\,.$$

In so doing $\alpha_i$ and $\gamma_i$ are recovered from the recurrence relations

$$\alpha_{i+1} = a_{i+1}\,(1 - \dot{\alpha}_{i+1}) = \frac{a_{i+1}\,[\,a_i\,(1 - \dot{\alpha}_i) + d_i\,]}{a_{i+1} + a_i\,(1 - \dot{\alpha}_i) + d_i}$$

or

$$\alpha_{i+1} = \frac{\alpha_i + d_i}{1 + (\alpha_i + d_i)\,/\,a_{i+1}}\,,$$

$$(71)\qquad \gamma_{i+1} - a_{i+1}\,\dot{\beta}_{i+1} = \dot{\alpha}_{i+1}\,(\,a_i\,\dot{\beta}_i + f_i\,) = \frac{a_{i+1}\,(\gamma_i + f_i)}{a_{i+1} + a_i\,(1 - \dot{\alpha}_i) + d_i}$$

or

$$(72)\qquad \gamma_{i+1} = \frac{\gamma_i + f_i}{1 + (\alpha_i + d_i)\,/\,a_{i+1}}\,.$$

By comparing the first boundary condition (67) with (69) for $i = 1$ we find that

$$(73)\qquad \alpha_1 = a_1\,(1 - \varkappa_1)\,, \qquad \gamma_1 = a_1\,\nu_1\,.$$

We note in passing that formulae (71)–(72) just established are very flexible and more convenient for computational procedures in the case $a_i >> 1$.

The requirement $a_i << 1$ necessitates making some modifications for stability of this or that difference scheme. As a final result of minor changes, the recurrence formulae have the representations

$$(71^*) \qquad \alpha_{i+1} = \frac{a_{i+1}\,(\alpha_i + d_i)}{a_{i+1} + (\alpha_i + d_i)}\,,$$

$$(72^*) \qquad \gamma_{i+1} = \frac{a_{i+1}\,(\gamma_i + f_i)}{a_{i+1} + (\alpha_i + d_i)}\,.$$

Under conditions (63) formulae (71) and (71*) imply that $\alpha_i \geq 0$. This provides support for the view that the coefficient $\alpha_i\,/\,(\alpha_i + d_i)$ in (70) is always less than 1, providing stability in the process of calculations of the flow $w_i$.

For determination of $y_i$ we rely on the appropriate formulae: for $a_i >> 1$

$$(74) \qquad y_i = \dot{\alpha}_{i+1}\, y_{i+1} + \dot{\beta}_{i+1} = \left(1 - \frac{\alpha_{i+1}}{a_{i+1}}\right) y_{i+1} + \frac{\gamma_{i+1}}{a_{i+1}}$$

and for $a_i << 1$

$$(75) \qquad y_i = \frac{a_{i+1}}{a_{i+1} + \alpha_i + d_i}\, y_{i+1} + \frac{\gamma_i + f_i}{a_{i+1} + \alpha_i + d_i}\,.$$

Formulae (74)–(75) show that the elimination method is stable. The values $w_N$ and $y_N$ should be known before proceeding to the applications of (70), (74) and (75). For this reason we involve here the second boundary condition (67) and relation (69) for $i = N$:

$$y_N = \frac{a_N\, \nu_2 + \gamma_N\, \varkappa_2}{a_N\,(1 - \varkappa_2) + \varkappa_2\, \alpha_N}\,, \qquad w_N = \frac{\gamma_N\, a_N\,(1 - \varkappa_2) - \alpha_N\, a_N\, \nu_2}{a_N\,(1 - \varkappa_2) + \varkappa_2\, \alpha_N}\,.$$

Observe that under conditions (63) the denominators in relevant expressions are always positive.

To distinguish two essentially different approaches, the final algorithm is of rather complicated structure and consists of the following steps:

- specification of $\alpha_1 = a_1\,(1 - \varkappa_1)$ and $\gamma_1 = a_1\, \nu_1$;

• successive refinement of the coefficients for $i = 1, 2, \dots, N-1$ with the aid of the relations

$$\alpha_{i+1} = \begin{cases} (\alpha_i + d_i)\,(1 + (\alpha_i + d_i)\,a_{i+1}^{-1})^{-1}, & a_{i+1} \ge 1\,, \\ a_{i+1}\,(\alpha_i + d_i)\,\big(a_{i+1} + (\alpha_i + d_i)\big)^{-1}, & a_{i+1} < 1\,, \end{cases}$$

$$\gamma_{i+1} = \begin{cases} (\gamma_i + f_i)\,\big(1 + (\alpha_i + d_i)\,a_{i+1}^{-1}\big)^{-1}, & a_{i+1} \ge 1\,, \\ a_{i+1}\,(\gamma_i + f_i)\,\big(a_{i+1} + (\alpha_i + d_i)\big)^{-1}, & a_{i+1} < 1\,; \end{cases}$$

• revision of proper boundary conditions

$$y_N = \begin{cases} \dfrac{\nu_2 + \gamma_N\,\varkappa_2\,a_N^{-1}}{1 - \varkappa_2 + \varkappa_2\,\alpha_N\,a_N^{-1}}\,, & a_N \ge 1\,, \\[2ex] \dfrac{a_N\,\nu_2 + \gamma_N\,\varkappa_2}{a_N\,(1 - \varkappa_2) + \varkappa_2\,\alpha_N}\,, & a_N < 1\,, \end{cases}$$

$$w_N = \begin{cases} \dfrac{\gamma_N\,(1 - \varkappa_2) - \alpha_N\,\nu_2}{1 - \varkappa_2 + \varkappa_2\,\alpha_N\,a_N^{-1}}\,, & a_N \ge 1\,, \\[2ex] \dfrac{\gamma_N\,(1 - \varkappa_2)\,a_N - \alpha_N\,\nu_2\,a_N}{a_N\,(1 - \varkappa_2) + \varkappa_2\,\alpha_N}\,, & a_N < 1\,. \end{cases}$$

• successive determination of solutions for $i = N-1,\ N-2,\ \dots,\ 0$:

$$w_i = \frac{\alpha_i}{\alpha_i + d_i}\,w_{i+1} + \frac{d_i\,\gamma_i - \alpha_i\,f_i}{\alpha_i + d_i}\,,$$

$$y_i = \begin{cases} \left(1 - \dfrac{\alpha_{i+1}}{a_{i+1}}\right) y_{i+1} + \dfrac{\gamma_{i+1}}{a_{i+1}}\,, & a_{i+1} \ge 1\,, \\[2ex] \dfrac{a_{i+1}}{a_{i+1} + \alpha_i + d_i}\,y_{i+1} + \dfrac{\gamma_i + f_i}{a_{i+1} + \alpha_i + d_i}\,, & a_{i+1} < 1\,. \end{cases}$$

Let us stress that the computational algorithm given above is stable.

**2. Cyclic elimination method.** We now focus the reader's attention on periodic solutions to difference schemes or systems of difference schemes being used in approximating partial and ordinary differential equations in spherical or cylindrical coordinates. A system of equations such as

$$\begin{aligned} a_1\,y_N - c_1\,y_1 + b_1\,y_2 &= -f_1\,, \\ a_i\,y_{i-1} - c_i\,y_i + b_i\,y_{i+1} &= -f_i\,, \qquad i = 2, 3, \dots, N-1\,, \\ a_N\,y_{N-1} - c_N\,y_N + b_N\,y_1 &= -f_N\,, \end{aligned} \tag{76}$$

was quite applicable in determination of the periodic solution $y_{i+N} = y_i$ of problem (61): $a_i\, y_{i-1} - c_i\, y_i + b_i\, y_{i+1} = -f_i$, provided that the conditions of periodicity

$$a_{i+N} = a_i\,, \qquad b_{i+N} = b_i\,, \qquad c_{i+N} = c_i\,, \qquad f_{i+N} = f_i$$

and the additional conditions

$$(77) \qquad a_i > 0\,, \qquad b_i > 0\,, \qquad c_i > a_i + b_i$$

hold.

We give below without proving the algorithm of the cyclic elimination method which will be used in the sequel:

$$(78) \qquad \alpha_{i+1} = \frac{b_i}{c_i - a_i\,\alpha_i}\,, \qquad \beta_{i+1} = \frac{f_i + a_i\,\beta_i}{c_i - a_i\,\alpha_i}\,, \qquad \gamma_{i+1} = \frac{a_i\,\gamma_i}{c_i - a_i\,\alpha_i}\,,$$

$$i = 2, 3, \ldots, N\,;$$

$$\alpha_2 = \frac{b_1}{c_1}\,, \qquad \beta_2 = \frac{f_1}{c_1}\,, \qquad \gamma_2 = \frac{a_1}{c_1}\,,$$

$$(79) \qquad p_i = \alpha_{i+1}\, p_{i+1} + \beta_{i+1}\,, \qquad q_i = \alpha_{i+1}\, q_{i+1} + \gamma_{i+1}\,,$$

$$i = N-2, N-3, \ldots, 1\,;$$

$$p_{N-1} = \beta_N\,, \qquad q_{N-1} = \alpha_N + \gamma_N\,;$$

$$y_N = \frac{\beta_{N+1} + \alpha_{N+1}\, p_1}{1 - \alpha_{N+1}\, q_1 - \gamma_{N+1}}\,,$$

$$(80) \qquad y_i = p_i + y_N\, q_i\,, \quad i = 1, 2, \ldots, N-1\,.$$

This algorithm is stable because the solutions of (79) are found by the right elimination method being stable under conditions (77) with the denominator $1 - \alpha_{N+1}\, q_1 - \gamma_{N+1} > 0$. Indeed, it follows from (77)–(78) that $\alpha_i < 1$, $\gamma_i > 0$ and $\alpha_2 + \gamma_2 < 1$. Assuming $\alpha_i + \gamma_i < 1$ we get

$$(81) \qquad \alpha_{i+1} + \gamma_{i+1} = \frac{b_i + a_i\,\gamma_i}{c_i - a_i\,\alpha_i} < \frac{b_i + a_i - a_i\,\alpha_i}{c_i - a_i\,\alpha_i} < 1\,.$$

Combination of (79) with (81) gives $q_{N-1} < 1$ and $q_i < 1$, thereby justifying the relation $1 - \alpha_{N+1}\, q_1 - \gamma_{N+1} > 0$.

**3. Factorization method.** At the final stage other ideas are connected with the boundary-value problem

$$(82)\qquad \begin{aligned} &L\,y_k = A_k\,y_{k-1} - C_k\,y_k + B_k\,y_{k+1} = -F_k\,, \quad k = 1, 2, \dots, N-1\,,\\ &y_0 = \varkappa_1\,y_1 + \nu_1\,, \quad y_N = \varkappa_2\,y_{N-1} + \nu_2\,. \end{aligned}$$

By introducing the transition operator $T$: $T\,y_k = y_{k+1}$ and the identity operator $E$: $E\,y_k = y_k$ and obvious rearranging of the left-hand side of (82) as a product

$$\begin{aligned} (b_k\,T - A_k)\,(\alpha_k\,T - E)\,y_{k-1} &= (b_k\,T - A_k)\,(\alpha_k\,y_k - y_{k-1})\\ &= b_k\,\alpha_{k+1}\,y_{k+1} - (A_k\,\alpha_k + b_k)\,y_k + A_k\,y_{k-1}\,, \end{aligned}$$

where

$$\alpha_{k+1}\,b_k = B_k\,, \qquad A_k\,\alpha_k + b_k = C_k\,,$$

we find during the course of the elimination of $b_k = C_k - A_k\,\alpha_k$ that

$$(83)\qquad \alpha_{k+1} = \frac{B_k}{C_k - A_k\,\alpha_k}\,, \qquad k = 1,\, 2,\, \dots,\, N-1\,.$$

The factorized equation

$$(\,b_k\,T - A_k\,)\,(\,\alpha_k\,y_k - y_{k-1}\,) = -F_k$$

can be solved using the following algorithm:

- recovery of the function $\beta_k$ from the equation

$$(\,b_k\,T - A_k\,)\,\beta_k = b_k\,\beta_{k+1} - A_k\,\beta_k = F_k$$

or

$$(84)\qquad \beta_{k+1} = \frac{A_k\,\beta_k + F_k}{C_k - A_k\,\alpha_k}\,, \qquad k = 1,\, 2,\, \dots,\, N-1\,;$$

- determination of $y_k$ by the right elimination method. The outcome of this procedure is

$$(85)\qquad y_k = \alpha_{k+1}\,y_{k+1} + \beta_{k+1}\,, \qquad k = 0,\, 1,\, \dots,\, N-1\,,$$

with the coefficients $\alpha_k$ and $\beta_k$ emerging from (83)–(84). The complete posing of the problem consists of (83)–(85), which are supplemented with the initial conditions

$$(86) \qquad \alpha_1 = \varkappa_1 , \qquad \beta_1 = \nu_1 , \qquad y_N = \frac{\varkappa_2 \beta_N + \nu_2}{1 - \varkappa_2 \alpha_N} .$$

With the right and left differences $\Delta v_k = v_{k+1} - v_k$ and $\nabla v_k = v_k - v_{k-1}$ in the usual sense, the operator

$$L y_k = A_k y_{k-1} - C_k y_k + B_k y_{k+1}$$

can be factorized by letting

$$L = L_1 L_2 , \qquad L_1 = b_k \Delta + \gamma_k , \qquad L_2 = \nabla + (\alpha_k - 1) .$$

The equality $L_1 L_2 y_k = L y_k$ holds true if

$$\gamma_k = b_k - A_k , \qquad \alpha_{k+1} b_k = B_k , \qquad A_k \alpha_k + B_k \alpha_{k+1}^{-1} = C_k .$$

The same procedure works for eliminating $\gamma_k$ and $b_k$ on the basis of formulae (83)–(86).

# 2

# Basic Concepts of the Theory of Difference Schemes

In this chapter elementary examples illustrate the basic concepts of the theory of difference schemes: approximation, stability and convergence. The reader is already familiar with several methods available for investigating stability and convergence such as the method of separation of variables and the method of energy inequalities. In Section 4 difference equations are treated as operator equations in an abstract space. These provide a means of studying a wide range of interesting methods in a unified manner. We begin our exposition with a discussion of examples that make it possible to draw fairly accurate outlines of the possible theory regarding these questions and with a listing of the basic results of the book together with a development desired for them.

## 2.1 RELEVANT ELEMENTS OF FUNCTIONAL ANALYSIS

Our account of the theory of difference schemes is mostly based on elementary notions from functional analysis. In what follows we list briefly widespread tools adopted in the theory of linear operators which will be used in the body of this book.

**1. Linear operators.** Let $X$ and $Y$ be normed vector spaces and $\mathcal{D}$ be a subspace of the space $X$. If to each vector $x \in \mathcal{D}$ there corresponds by an approved rule governing what can happen a vector $y = Ax \in Y$, we say that an **operator** $A$ is given on $\mathcal{D}$ (or in $X$) with values in $Y$. The set $\mathcal{D}$ is called the **domain** of the operator $A$ and is denoted by $\mathcal{D}(A)$. The set of all vectors of the form $y = Ax$, $x \in \mathcal{D}(A)$, is called the **range** of the operator $A$ and is denoted by $\mathcal{R}(A)$. It is customary to use also the well-established notation $A(x)$ instead of $Ax$.

Two operators $A$ and $B$ are said to be **equal** if their domains coincide and for all $x \in \mathcal{D}(A) = \mathcal{D}(B)$ the condition $Ax = Bx$ holds true.

An operator $A$ is called **linear** if it is:

1) **additive**, meaning that for all $x_1$, $x_2 \in \mathcal{D}(A)$

$$A(x_1 + x_2) = Ax_1 + Ax_2\,;$$

2) **homogeneous**, meaning that for all $x \in \mathcal{D}(A)$ and any number $\lambda$

$$A(\lambda x) = \lambda\ A\, x\,.$$

A linear operator $A$ is said to be **bounded** if there is a constant $M > 0$ such that for any $x \in \mathcal{D}(A)$

$$\| Ax \|_2 \leq M \, \| x \|_1 \qquad (1)$$

(here $\| \cdot \|_1$ and $\| \cdot \|_2$ are admissible forms of the functional norms on the spaces $X$ and $Y$, respectively).

The minimal constant $M$ satisfying condition (1) is called the **norm of the operator** $A$ and is denoted by $\|A\|_{X\to Y}$ or simply $\| A \|$.

It follows from the definition of norm that

$$\| A \| = \sup_{\| x \|_1 = 1} \| Ax \|_2 \qquad \text{or} \qquad \| A \| = \sup_{x \neq 0} \frac{\| Ax \|_2}{\| x \|_1}\,. \qquad (2)$$

It is worth noting here that in a finite-dimensional space any linear operator is bounded. All of the linear bounded operators from $X$ into $Y$ constitute what is called a normed vector space, since the norm $\| A \|$ of an operator $A$ satisfies all of the axioms of the norm:

1) $\| A \| \geq 0$; if $\| A \| = 0$, then $\| Ax \|_2 = 0$ for all $x$ and $A = 0$;
2) $\| \lambda A \| = | \lambda | \cdot \| A \|$;
3) $\| A + B \| \leq \| A \| + \| B \|$.

We will denote by $X \mapsto X$ the set of linear bounded operators with the domain coinciding with $X$ and the range belonging to $X$. On the set $X \to X$

it is possible to introduce the operation of multiplication $AB$ of operators $A$ and $B$: $(AB)x = A(Bx)$. Clearly, $AB$ is a linear bounded operator in light of the obvious relation $\| AB \| \leq \| A \| \cdot \| B \|$. If $(AB)x = (BA)x$ for all $x \in X$, then operators $A$ and $B$ are said to be **commuting**. In that case we will write $AB = BA$.

While solving equations of the form $Ax = y$ we shall need yet the notion of the inverse $A^{-1}$. Let $A$ be an operator from the space $X$ into the space $Y$. By definition, this means that $\mathcal{D}(A) = X$ and $\mathcal{R}(A) = Y$. If to each $y \in Y$ there corresponds only one element $x \in X$, for which $Ax = y$, then this correspondence specifies an operator $A^{-1}$, known as the **inverse** for $A$, with the domain $Y$ and range $X$. By the definition of inverse operator, we have for any $x \in X$ and any $y \in Y$

$$A^{-1}(Ax) = x\,, \qquad A(A^{-1}y) = y\,.$$

It is easy to show that if an operator $A$ is linear, then so is the inverse $A^{-1}$ (if it exists).

**Lemma 1** *In order that an additive operator $A$ with $\mathcal{D}(A) = X$ and $\mathcal{R}(A) = Y$ possess an inverse, it is necessary and sufficient that $Ax = 0$ only if $x = 0$.*

**Theorem 1** *Let $A$ be a linear operator from $X$ into $Y$. In order that the inverse operator $A^{-1}$ exist and be bounded, as an operator from $Y$ into $X$, it is necessary and sufficient that there is a constant $\delta > 0$ such that for all $x \in X$*

$$\| Ax \|_2 \geq \delta \, \| x \|_1 \tag{3}$$

*($\| \cdot \|_1$ is the norm on the space $X$ and $\| \cdot \|_2$ is the norm on the space $Y$). Moreover, the estimate $\| A^{-1} \| \leq 1/\delta$ is valid.*

**2. Linear bounded operators in a real Hilbert space.** Let $H$ be a real Hilbert space equipped with an inner product $(x, y)$ and associated norm $\| x \| = \sqrt{(x,x)}$. We consider bounded linear operators defined on the space $H$ $(\mathcal{D}(A) = H)$. Before giving further motivations, it will be convenient to introduce several definitions. We call an operator $A$

(a) nonnegative if

$$(Ax, x) \geq 0 \quad \text{for all} \quad x \in H\,; \tag{4}$$

(b) positive if

$$(Ax, x) > 0 \quad \text{for all} \quad x \in H \quad \text{except for} \quad x = 0\,; \tag{5}$$

(c) lower semibounded if

$$(Ax, x) \geq -c_* \, \| x \|^2 \quad \text{for any} \quad x \in H \,, \tag{6}$$

where $c_*$ is a positive number;

(d) positive definite if

$$(Ax, x) \geq \delta \, \| x \|^2 \quad \text{for any} \quad x \in H \,, \tag{7}$$

where $\delta$ is a positive number.

For an arbitrary nonnegative operator $A$ and $x \in H$, the number $(Ax, x)$ is called the **energy of the operator** $A$. Further comparison of operators $A$ and $B$ will be carried out by means of the energy. If $((A - B)x, x) \geq 0$ for all $x$, we write $A \geq B$. In particular, inequalities (4)–(7) can be replaced by the following operator inequalities:

$$\begin{aligned} & A \geq 0 \,, && \text{meaning} && (Ax, x) \geq 0 \,, \\ & A > 0 \,, && \text{meaning} && (Ax, x) > 0 \,, \\ & A \geq -c_* \, E \,, && \text{meaning} && (Ax, x) \geq -c_* \, \| x \|^2, \\ & A \geq \delta E \,, && \text{meaning} && (Ax, x) \geq \delta \, \| x \|^2 \,, \end{aligned} \tag{8}$$

where $E$ stands for the identity operator leaving a vector $x$ unchanged: $Ex = x$.

It is straightforward to verify that the relation established on the set of linear operators $(H \mapsto H)$ possesses the following properties:

(a) $A \geq B$ and $C \geq D$ imply $A + C \geq B + D$;
(b) $A \geq 0$ and $\lambda \geq 0$ imply $\lambda A \geq 0$;
(c) $A \geq B$ and $B \geq C$ imply $A \geq C$;
(d) If $A > 0$ and $A^{-1}$ exists, then $A^{-1} > 0$.

If $A$ is a linear operator defined on $H$, then the operator $A^*$ on $H$ subject to the condition $(Ax, y) = (x, A^* y)$, $x, y \in H$, is called the **adjoint** operator to $A$. If $A$ is a linear bounded operator, then its own adjoint is uniquely defined and falls within the category of linear bounded operators with the norm $\| A^* \| = \| A \|$. A linear bounded operator $A$ is called **self-adjoint** if $A^* = A$, that is, $(Ax, y) = (x, Ay)$ for all $x, y \in H$. If $A$ is an arbitrary linear operator, then $A^*A$ and $AA^*$ are self-adjoint nonnegative operators:

$$(A^*Ax, y) = (Ax, Ay) = (x, A^*Ay), \qquad (A^*Ax, x) = \| Ax \|^2 \geq 0,$$

$$(AA^*x, x) = \| A^*x \|^2 \geq 0.$$

It is worth noting the obvious relations $(A^*)^* = A$ and $(A^*)^{-1} = (A^{-1})^*$.

Any nonnegative operator $A$ in a complex Hilbert space $\widetilde{H}$ is self-adjoint:

$$\text{if} \quad (Ax, x) \geq 0 \quad \text{for all} \quad x \in \widetilde{H}, \quad \text{then} \quad A = A^*.$$

For real Hilbert spaces this statement fails to be true. As far as only real Hilbert spaces are considered, we will use the operator inequalities for non-self-adjoint operators as well.

**Theorem 2** *The product $AB$ of two commuting nonnegative self-adjoint operators $A$ and $B$ is also a nonnegative self-adjoint operator.*

An operator $B$ is called a **square root of an operator** $A$ if $B^2 = A$.

**Theorem 3** *There exists a unique nonnegative self-adjoint square root $B$ of any nonnegative self-adjoint operator $A$ commuting with any operator which commutes with $A$.*

We denote by $A^{1/2}$ the square root of an operator $A$.

Let $A$ be a positive self-adjoint linear operator. By introducing on the space $H$ the inner product $(x, y)_A = (Ax, y)$ and the associated norm $\| x \|_A = \sqrt{(x, x)_A}$ we obtain a Hilbert space $H_A$, which is usually called the **energetic space** $H_A$. It is easy to show that the inner product

$$(x, y)_A = (Ax, y)$$

satisfies all of the axioms of the inner product:

(1) $(x, y)_A = (y, x)_A$;
(2) $(x + y, z)_A = (x, z)_A + (y, z)_A$;
(3) $(\lambda x, y)_A = \lambda (x, y)_A$;
(4) $(x, x)_A > 0$ for $x \neq 0$ and $(x, x) = 0$ only for $x = 0$.

Axioms (2) and (3) are met by virtue of the linearity property. The validity of (4) is stipulated by the fact that the operator $A$ is positive. The meaning of the self-adjointness of the operator $A$ is that we should have

$(x, y)_A = (y, x)_A$ or $(Ax, y) = (x, Ay) = (Ay, x)$. The axioms of the inner product imply the Cauchy–Bunyakovskiĭ inequality

$$|(x, y)_A| \leq \|x\|_A \cdot \|y\|_A$$

and the triangle inequality

$$\|x + y\|_A \leq \|x\|_A + \|y\|_A .$$

This profound result is covered by the following assertion.

**Lemma 2** *For any positive self-adjoint operator $A$ in a real Hilbert space the generalized Cauchy–Bunyakovskiĭ inequality holds:*

$$(Ax, y)^2 \leq (Ax, x)\,(Ay, y) . \tag{9}$$

**Remark** The preceding inequality remains valid in the case when $A$ is a nonnegative operator.

If $A$ is a self-adjoint operator for which $A^{-1}$ exists, its "negative" norm can be defined by

$$\|\varphi\|_{A^{-1}} = \sqrt{(A^{-1}\varphi, \varphi)} . \tag{10}$$

In this line, we claim that

$$\|\varphi\|_{A^{-1}} = \sup_{x \neq 0} \frac{|(\varphi, x)|}{\|x\|_A} . \tag{10'}$$

Indeed, we deduce from inequality (9) that

$$|(\varphi, x)| = |(A^{-1}\varphi, Ax)| \leq \|\varphi\|_{A^{-1}} \cdot \|x\|_A ,$$

yielding

$$\sup_{x \neq 0} \frac{|(\varphi, x)|}{\|x\|_A} \leq \sup_{x \neq 0} \frac{\|\varphi\|_{A^{-1}} \cdot \|x\|_A}{\|x\|_A} = \|\varphi\|_{A^{-1}} .$$

On the other hand, for $x = A^{-1}\varphi$ it is plain to show that

$$\frac{|(\varphi, x)|}{\|x\|_A} = \frac{(\varphi, A^{-1}\varphi)}{\sqrt{(AA^{-1}\varphi, A^{-1}\varphi)}} = \|\varphi\|_{A^{-1}} ,$$

which confirms the equivalence of (10) and (10′).

In the sequel sufficient conditions for the existence of a bounded inverse operator $A^{-1}$ defined in the entire space $H$, $\mathcal{D}(A^{-1}) = H$, will be of great importance for us.

We note in passing that Lemma 1 and Theorem 1 guarantee the existence of an inverse operator defined only on $\mathcal{R}(A)$, the range of $A$, which is not obliged to coincide with $H$. If the range of an operator $A$ happens to be the entire space $H$, $\mathcal{R}(A) = H$, then the conditions of Lemma 1 or Theorem 1 ensure the existence of an operator $A^{-1}$ with $\mathcal{D}(A^{-1}) = H$. In particular, a positive operator $A$ with the range $\mathcal{R}(A) = H$ possesses an inverse $A^{-1}$ with $\mathcal{D}(A^{-1}) = H$, since the condition $(Ax, x) > 0$ for all $x \neq 0$ implies that $Ax \neq 0$ for $x \neq 0$ and Lemma 1 applies equally well to such a setting.

**Theorem 4** *Let $A$ be a linear bounded operator in Hilbert space $H$, $\mathcal{D}(A) = H$. In order that the operator $A$ possess an inverse operator $A^{-1}$ with the domain $\mathcal{D}(A^{-1}) = H$, it is necessary and sufficient the existence of a constant $\delta > 0$ such that for all $x \in H$ the following inequalities hold:*

$$\| Ax \| \geq \delta \, \| x \| , \qquad \| A^* x \| \geq \delta \, \| x \| .$$

*Moreover, the estimate $\| A^{-1} \| \leq 1/\delta$ is true.*

**Corollary** *Let $A$ be a positive definite linear bounded operator with the domain $\mathcal{D}(A) = H$. Then there exists a bounded inverse operator $A^{-1}$ with the domain $\mathcal{D}(A^{-1}) = H$.*

Indeed, with the relation $A \geq \delta E$, $\delta > 0$ in view, we arrive at the chains of the relations

$$\| Ax \| \cdot \| x \| \geq (Ax, x) \geq \delta \| x \|^2 ,$$

$$\| A^* x \| \cdot \| x \| \geq | (A^* x, x) | = | (x, Ax) | = (Ax, x) \geq \delta \| x \|^2 ,$$

thereby providing the relations $\| Ax \| \geq \delta \| x \|$ and $\| A^* x \| \geq \delta \| x \|$ as well as the validity of the conditions of Theorem 4. The norm of the inverse admits the estimate $A^{-1} \leq 1/\delta$.

**Remark** For the existence of an inverse $A^{-1}$ in a finite-dimensional Hilbert space it suffices to require the positiveness of the operator $A$, since the condition $A > 0$ implies the existence of a constant $\delta > 0$ such that

$$(Ax, x) \geq \delta \| x \|^2$$

for all $x$. As a matter of fact,

$$(Ax, x) = (A_0 x, x),$$

where $A_0 = (A + A^*)/2$ is a self-adjoint operator. Hence, $(Ax, x) \geq \delta \| x \|^2$, where $\delta$ is the smallest eigenvalue of $A_0$. The number $\delta$ cannot vanish due to the positiveness of the operator $A$.

For the sake of simplicity we will assume in the body of this book that $H$ is a finite-dimensional space.

Recall that the norm of an operator $A$ is defined by

$$\| A \| = \sup_{\| x \| = 1} \| Ax \| .$$

With this, for any self-adjoint operator $A$ the following relations occur:

$$(11) \qquad \| A \| = \sup_{\| x \| = 1} |(Ax, x)| = \| A \| = \sup_{\| x \| \neq 0} \frac{|(Ax, x)|}{\| x \|} .$$

**Lemma 3** *If $S = S^*$ is a linear bounded operator and $n$ is a positive integer, then*

$$\| S^n \| = \| S \|^n .$$

**Proof** We proceed to prove this assertion by induction on $n$. Let $n = 2$. Then

$$\| S^2 \| = \sup_{\| x \| = 1} |(S^2 x, x)| = \sup_{\| x \| = 1} \| Sx \|^2 = \| S \|^2 ,$$

meaning $\| S^2 \| = \| S \|^2$. Assuming that relation (12) is valid with $n = k - 1$ and $n = k$, we are going to show that it continues to hold for $n = k + 1$, $k > 1$. Indeed,

$$\| S^{2k} \| = \sup_{\| x \| = 1} |(S^{2k} x, x)| = \sup_{\| x \| = 1} |(S^{k+1} x, S^{k-1} x)|$$

$$\leq \sup_{\| x \| = 1} \| S^{k+1} x \| \cdot \| S^{k-1} x \| \leq \| S^{k+1} \| \cdot \| S^{k-1} \|$$

or, what amounts to the same,

$$\| S^{k+1} \| \cdot \| S^{k-1} \| \geq \| S^{2k} \| = \| S^k \|^2 = \| S \|^{2k} .$$

This provides support for the view that $\| S^{k+1} \| \geq \| S \|^{k+1}$, since

$$\| S^{k-1} \| = \| S \|^{k-1} .$$

On the other hand,

$$\| S^{k+1} \| \leq \| S \|^{k+1} ,$$

so that

$$\| S^{k+1} \| = \| S \|^{k+1} .$$

As far as (12) holds for $n = 1$ and $n = 2$, it will be true for any $n$.

**Lemma 4** *If $A$ is a self-adjoint positive bounded operator, then the estimate is valid:*

$$\| Ay \|^2 \le \| A \| \, (Ay, y) \, . \tag{13}$$

Since $A^* = A > 0$, there always exists the operator $A^{1/2}$. By merely setting $v = A^{1/2y}$ we obtain

$$(Ay, Ay) = (Av, v) \le \| A \| \cdot \| v \|^2 = \| A \| \, (Ay, y) \, .$$

**3. Linear operators in finite-dimensional spaces.** It is supposed that an $n$-dimensional vector space $R_n$ is equipped with an inner product $( \cdot \, , \; )$ and associated norm $\| x \| = \sqrt{(x, x)}$. By the definition of finite-dimensional space, any vector $\mathbf{x} \in R_n$ can uniquely be represented as a linear combination $\mathbf{x} = c_1 \, \boldsymbol{\xi}_1 + \cdots c_n \, \boldsymbol{\xi}_n$ of linearly independent vectors $\boldsymbol{\xi}_1 , \ldots , \boldsymbol{\xi}_n$, which constitute a basis for the space $R_n$. The numbers $c_k$ are called the coordinates of the vector $\mathbf{x}$. One can always choose as a basis an orthogonal and normed system of vectors $\boldsymbol{\xi}_1, \ldots , \boldsymbol{\xi}_n$:

$$(\boldsymbol{\xi}_i \, , \boldsymbol{\xi}_k) = \delta_{ik} = \begin{cases} 0 \, , & i \neq k \, , \\ 1 \, , & i = k \, , \end{cases}$$

by means of which it is possible to write $c_k = (\mathbf{x}, \boldsymbol{\xi}_k)$.

Let $A$ be a linear operator in the space $R_n$. Any operator $A$ in the basis $\boldsymbol{\xi}_1, \ldots , \boldsymbol{\xi}_n$ can be put in correspondence with an $n \times n$ matrix $\mathfrak{A} = (a_{ik})$, whose element $a_{ik}$ is the $i$th component of the vector $A\xi_k$. Conversely, any matrix $\mathfrak{A} = (a_{ik})$, $i$, $k = 1, \ldots , n$, specifies a linear operator.

The matrix of a self-adjoint operator in any orthonormal basis is a symmetric matrix.

Let us dwell on the properties of eigenvalues and eigenvectors of a linear self-adjoint operator $A$. A number $\lambda$ such that there exists a vector $\boldsymbol{\xi} \neq 0$ with $A\boldsymbol{\xi} = \lambda \boldsymbol{\xi}$ is called an **eigenvalue** of the operator $A$. This vector $\boldsymbol{\xi}$ is called an **eigenvector** corresponding to the given eigenvalue $\lambda$.

1. A self-adjoint operator $A$ in the space $R_n$ possesses $n$ mutually orthogonal eigenvectors $\boldsymbol{\xi}_1, \ldots , \boldsymbol{\xi}_n$. We assume that all the $\boldsymbol{\xi}_k$'s are normalized, that is, $\| \boldsymbol{\xi}_k \| = 1$ for $k = 1, \ldots , n$. Then $(\boldsymbol{\xi}_i, \, \boldsymbol{\xi}_k) = \delta_{ik}$. The corresponding eigenvalues are ordered with respect to absolute values:

$$| \lambda_1 | \le | \lambda_2 | \le \cdots \le | \lambda_n | \, .$$

2. If a linear operator $A$ given on $R_n$ possesses $n$ mutually orthogonal eigenvalues, then $A$ is a self-adjoint operator: $A = A^*$.

3. If $A^* = A \geq 0$, then all the eigenvalues of $A$ are nonnegative.

4. Any vector $\mathbf{x} \in R_n$ is representable by the eigenvectors of any operator $A = A^*$:

$$\mathbf{x} = \sum_{k=1}^{n} c_k \,\boldsymbol{\xi}_k \,, \qquad c_k = (\mathbf{x}, \boldsymbol{\xi}_k) \qquad \text{and} \qquad \|\,\mathbf{x}\,\|^2 = \sum_{k=1}^{n} c_k^2 \,.$$

5. Let $A^* = A > 0$ ($A$ is self-adjoint and nonnegative). Then

$$\lambda_1 \,\|\,\mathbf{x}\,\|^2 \leq (A\mathbf{x}, \mathbf{x}) \leq \lambda_n \|\,\mathbf{x}\,\|^2$$

and

$$\lambda_1 \,\|\,\mathbf{x}\,\| \leq \|\,A\mathbf{x}\,\| \leq \lambda_n \|\,\mathbf{x}\,\| \quad \text{for all} \quad x \in H \,,$$

where $\lambda_1 \geq 0$ and $\lambda_n \geq 0$ are, respectively, the smallest and greatest eigenvalues of $A$. The norm of a self-adjoint nonnegative operator in the space $R_n$ is equal to its greatest eigenvalue: $\|\,A\,\| = \lambda_n$.

6. If self-adjoint operators $A$ and $B$ are commuting ($AB = BA$), then they possess a common system of eigenvectors.

7. Let self-adjoint operators $A$ and $B$ be commuting ($AB = BA$). Then the operator $AB$ possesses the same system of eigenvectors as the operators $A$ and $B$ and $\lambda_{AB}^{(k)} = \lambda_A^{(k)} \lambda_B^{(k)}$, $k = 1, 2, \ldots, n$, where $\lambda_A^{(k)}$, $\lambda_B^{(k)}$, and $\lambda_{AB}^{(k)}$ are the $k$th eigenvalues of $A$, $B$ and $AB = BA$, respectively. By the same token,

$$\lambda_{A+B}^{(k)} = \lambda_A^{(k)} + \lambda_B^{(k)} \,.$$

## 2.2 DIFFERENCE APPROXIMATION OF ELEMENTARY DIFFERENTIAL OPERATORS

**1. Grids and grid functions.** The composition of a difference scheme approximating a differential equation of interest amouts to performing the following operations:

- to substitute the domain of discrete variation of an argument for the domain of continuous variation;
- to replace a differential operator by some difference operator and impose difference analogs of boundary conditions and initial data.

Following these procedures, we are led to a system of algebraic equations, thereby reducing numerical solution of an initial (linear) differential equation to solving an algebraic system.

We would like to discuss the questions raised above in more detail. Obviously, in numerical solution of mathematical problems it is unrealistic to reproduce a difference solution for all the values of the argument varying in a certain domain of a prescribed Euclidean space. The traditional way of covering this is to select some finite set of points in this domain and look for an approximate solution only at those points. Any such set of points is called a **grid** and the isolated points are termed the **grid nodes**.

Any such function defined at nodal points is called a **grid function**. Thus, the first step in this direction is to replace the domain of continuous variation of an argument by a grid, that is, by the domain of discrete variation of the same argument. In other words, we have approximated the space of solutions to a differential equation by the space of grid functions. For this reason the properties of a difference solution and, in particular, its proximity to an exact solution depend essentially on a proper choice of the grid. It seems worthwhile giving several simple grids to help motivate what is done.

**Example 1.** ***An equidistant grid on a segment.*** The segment $[0,1]$ of unit length is splitted into $N$ equal intervals. The spacing between the adjacent nodes $x_i - x_{i-1} = h = \frac{1}{N}$ is termed a **grid step** or simply **step**. The splitting points $x_i = ih$ constitute what is called the set of grid nodes $\omega_h = \{x_i = ih,\ i = 1, \dots, N-1\}$ and generate one possible grid on this segment (see Fig. 1):

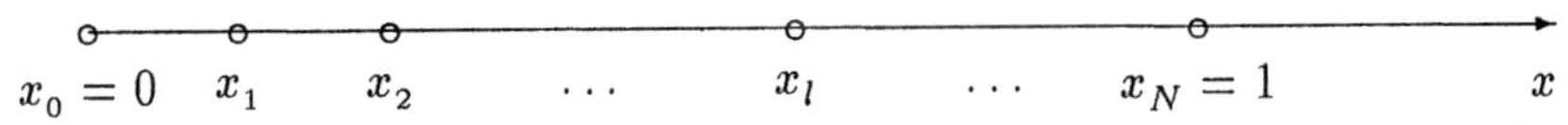


**Figure 1.**

When the boundary points $x_0 = 0$ and $x_N = 1$ are put together with the grid $\omega_h$, it will be denoted by $\bar{\omega}_h = \{x_i = ih,\ i = 0, 1, \dots, N-1, N\}$.

On the segment $[0,1]$ we are working with a new function $y_h(x_i)$ of the discrete argument instead of a given function $y(x)$ of the continuous argument. The values of such a function are calculated at the grid nodes $x_i$ and the function itself depends on the step $h$ as on the parameter.

**Example 2.** ***An equidistant grid in a plane.*** We now consider a set of functions $u(x,t)$ of two arguments in the rectangle $\bar{D} = \{0 \le x \le 1,\ 0 \le t \le T\}$ and split up the segment $[0,1]$ on the $x$-axis and the segment $[0,T]$ on the $t$-axis into $N_1$ and $N_2$ parts with steps $h = 1/N_1$ and $\tau =$

$T/N_2$, respectively. After that, we draw the straight lines parallel to the appropriate axes through the splitting points. The points of the intersection of those lines are adopted, as usual, as the nodes $(x_i, t_j)$, which constitute a widespread grid

$$\bar{\omega}_{h\tau} = \{(x_i, t_j) \in \bar{\mathcal{D}}\}$$

with steps $h$ and $\tau$ along the directions $Ox$ and $Ot$, respectively (see Fig. 2).

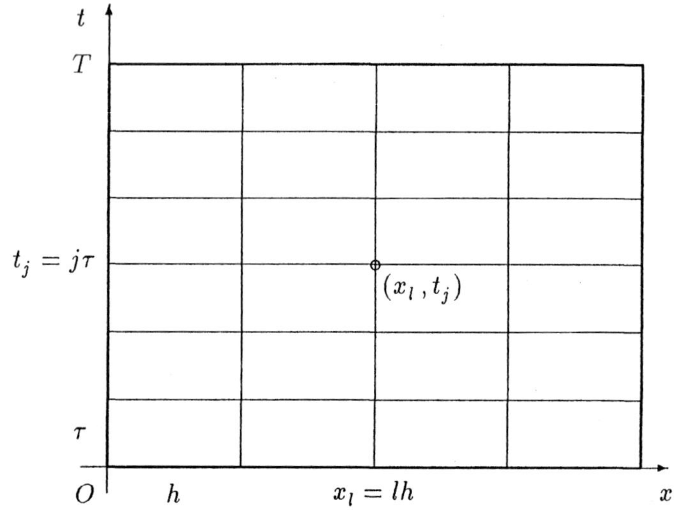


**Figure 2.**

The nodes lying on one and the same straight line (horizontal or vertical), the distance between which is equal to the grid step ($h$ or $\tau$), are called **adjacent grid nodes**.

**Example 3.** ***A non-equidistant grid on a segment.*** The next example is devoted to the segment $0 \le x \le 1$ with $N$ subintervals without concern for how the points $0 < x_1 < x_2 < \ldots < x_{N-1} < 1$ will be chosen. The nodes $\{x_i,\ i = 0, \ldots, N,\ x_0 = 0,\ x_N = 1\}$ constitute what is called a non-equidistant grid $\hat{\omega}_h[0,1]$. The distance between the adjacent nodes, being a grid step, equals $h_i = x_i - x_{i-1}$ and depends on the subscript $i$. Any distance of this type falls within the category of grid functions. The steps should satisfy the normalization condition $\sum_{i=1}^{N} h_i = 1$.

**Example 4.** ***A grid in a two-dimensional domain.*** Our final example is connected with a complex domain $G$ in the plane $x = (x_1, x_2)$ of rather complicated configuration with the boundary $\Gamma$. One way of proceeding is to subdivide into sets of rectangles by equally spaced grid lines parallel to $Ox_2$ defined by $x_1^{(i_1)} = i_1 h_1, i_1 = 0, \pm 1, \pm 2, \dots, h_1 > 0$, and equally spaced grid lines parallel to $Ox_1$ defined by $x_2^{(i_2)} = i_2 h_2$, $i_2 = 0, \pm 1, \pm 2, \dots, h_2 > 0$, as shown in Fig. 3. As a final result we obtain the grid with the nodes $(i_1 h_1, i_2 h_2)$, $i_1$, $i_2 = 0, \pm 1, \pm 2, \dots$ in the plane $(x_1, x_2)$. It seems clear that the lattice so constructed is equidistant in each of the directions $Ox_1$ and $Ox_2$. We are only interested in the representative grid nodes from the domain $\bar{G} = G + \Gamma$, including the boundary $\Gamma$. The nodes $(i_1 h_1, i_2 h_2)$ inside $G$ are called **inner nodes** and the notation $\omega_h$ is used for the whole set of such nodal points. The points of the intersection of the straight lines $x_1^{(i_1)} = i_1 h_1$ and $x_2^{(i_2)} = i_2 h_2$, $i_1, i_2 = 0, \pm 1, \pm 2, \dots 0$, with the boundary $\Gamma$ are known as **boundary nodes**. The set of all boundary nodes is denoted by $\gamma_h$. In Fig. 3 the boundary and inner nodes are quoted with the marks $*$ and $\circ$, respectively.

As can readily be observed, there are boundary nodes with the distance to the nearest inner nodes smaller than $h_1$ or $h_2$. In spite of the obvious fact that the grid in the plane is equidistant in $x_1$ and $x_2$ both, the grid $\bar{\omega}_h = \omega_h + \gamma_h$ in the domain $\bar{G}$ is non-equidistant near the boundary. This case will be the subject of special investigations in Chapter 4.

The approach we have described above depends for its success on replacing the domain $\bar{G}$ of the argument $x$ by the grid $\bar{\omega}_h$, that is, by a finite-dimensional set of points $x_i$ belonging to the domain $\bar{G}$. The grid functions $y(x_i)$ will be quite applicable in place of the functions $u(x)$ of the continuous argument $x \in \bar{G}$ with $x_i$ being a node of the grid $\bar{\omega}_h = \{x_i\}$. Also, we may attempt the grid function $y(x_i)$ in vector form. This is due to the fact that by enumerating the nodes in some order $x_1, x_2, \dots x_N$ the values of the grid function at those nodes arrange themselves as the components of the vector

$$Y = (y_1, \dots, y_i, \dots y_N).$$

If the domain $G$ of forming the grid is finite, then so is the dimension $N$ of the vector $Y$. In the case of an infinite domain $G$ the grid consists of an infinite number of nodes and thereby $Y$ becomes an infinite-dimensional vector. Within the framework of this book the sets of grids $\omega_h$ depending on the step $h$ as on the parameter will be given special investigation, so that the relevant grid functions $y_h(x)$ will depend on the parameter $h$ or on

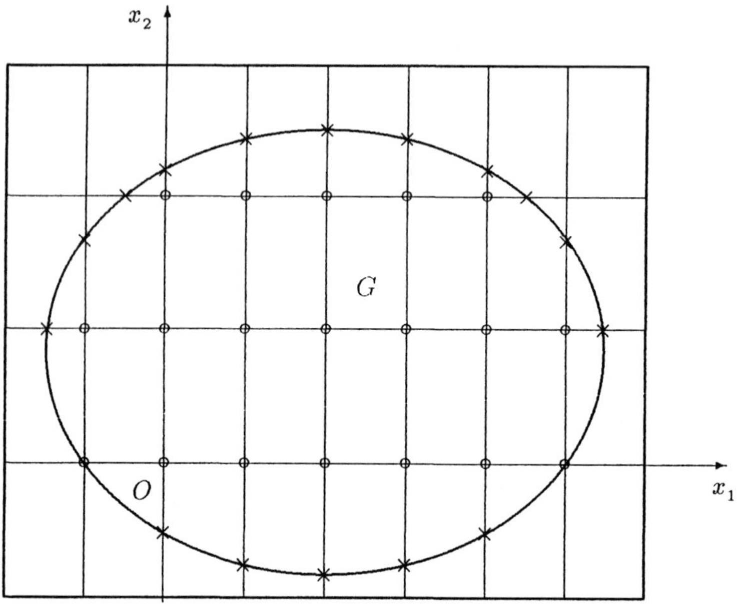


**Figure 3.**

the number $N$ in the case of an equidistant grid. In dealing with various non-equidistant grids $\omega_h$ we mean by step $h$ the vector $h = (h_1, h_2, \dots, h_N)$ with components $h_1, h_2, \dots, h_N$.

The same observation remains valid with multidimensional domains $G$. Here $x = (x_1, \dots, x_p)$ and, on the same grounds, $h = (h_1, h_2, \dots, h_p)$ if the grid $\omega_h$ is equidistant in each of the arguments $x_1, \dots, x_p$.

Throughout the entire chapter, the functions $u(x)$ of the continuous argument $x \in G$ are the elements of some functional space $H_0$. The space $H_h$ comprises all of the grid functions $y_h(x)$, providing a possibility to replace within the framework of the finite difference method the space $H_0$ by the space $H_h$ of grid functions $y_h(x)$. Recall that although the fixed notation $\| \cdot \|$ is usually adopted, there is a wide variety of possible choices of the functional form of $\| \cdot \|$.

In a common setting the set $\{H_h\}$ of spaces of grid functions depending on the parameter $h$ corresponds to the set of grids $\{\omega_h\}$, making it possible to introduce in the vector space $H_h$ the norm $\| \cdot \|_h$, which is a grid analog of the norm $\| \cdot \|_0$ of the initial space $H_0$. We give below two norms in the

space $H_h$ for the grids $\bar{\omega}_h = \{x_i = ih\}$ on the segment $0 \le x \le 1$. These norms are widely used in many applications to numerical analysis (here the subscript $h$ of $y_h$ is omitted):
(1) the grid analog of the norm on the space $C$:

$$\| y \|_C = \max_{x \in \bar{\omega}_h} | y(x) | \quad \text{or} \quad \| y \|_C = \max_{0 \le i \le N} | y_i | ;$$

(2) the grid analog of the norm on the space $L_2$:

$$\| y \| = \left( \sum_{i=1}^{N-1} y_i^2 \, h \right)^{1/2} \quad \text{or} \quad \| y \| = \left( \sum_{i=1}^{N} y_i^2 \, h \right)^{1/2} .$$

In what follows we will use, as a rule, several norms associated with inner products in the space $H_h$ (the grid analogs of the $L_2$ and $W_2^1$-norms are available in Chapter 1).

Given a solution $u(x)$ of the original continuous problem, $u \in H_0$, and a solution $y_h$ of the appropriate approximate (difference) problem, $y_h \in H_h$, the main goal of the possible theory of approximate methods is the accurate account of the proximity between $y_h$ and $u$. It is worth noting here that in a common setting the vectors $y_h$ and $u$ belong to different spaces. In this context, two interesting possibilities reduce to the following ones:

- the grid function $y_h$ defined at the nodes $\omega_h(G)$ should be extended (for instance, via the linear interpolation) to all of the remaining points $x$ of the domain $G$. As a final result we get a function $\tilde{y}(x, h)$ of the continuous argument $x \in G$ for which the difference $\tilde{y}(x, h) - u(x)$ belongs to the space $H_0$. The proximity of $y_h$ to $u$ is well-characterized by the number $\| \tilde{y}(x, h) - u(x) \|_0$, where $\| \cdot \|_0$ is the norm of the space $H_0$;
- the space $H_0$ is mapped onto the space $H_h$, thereby putting every function $u(x) \in H_0$ in correspondence with a grid function $u_h(x)$, $x \in \omega_h$, so that $u_h = \mathcal{P}_h \, u \in H_h$, where $\mathcal{P}_h$ is a linear operator from $H_0$ into $H_h$. It is possible to establish this correspondence in a number of different ways by approval of different operators $\mathcal{P}_h$. If $u(x)$ is a continuous function, we might accept $u_h(x) = u(x)$ for $x \in \omega_h$. Sometimes $u_h(x_i)$ is determined at a node $x_i \in \omega_h$ as the integral mean value of $u(x)$ over some neighborhood of this node (for instance, of diameter $O(h)$). In the sequel we will always assume that $u(x)$ is a continuous function and keep $u_h(x_i) = u(x_i)$ for all $x_i \in \omega_h$ unless otherwise is explicitly stated.

For a difference function $u_h$, we turn to the difference $y_h - u_h$, which gives a vector of the space $H_h$. The nearness of $y_h$ to $u$ is well-characterized by the number $\| y_h - u_h \|_h$, where $\| \cdot \|_h$ is the norm of the space $H_h$. It seems natural to require that the norm $\| \cdot \|_h$ should approximate the norm $\| \cdot \|_0$ in the following sense: for any vector $u \in H_0$

$$\lim_{h \to 0} \| u_h \|_h = \| u \|_0 .$$

This condition is known as the **condition of concordance** of the norms on the spaces $H_h$ and $H_0$.

Under the second approach mentioned above we proceed to the accurate account of the errors of difference methods in the space of grid functions. In the most cases the spaces involved appear to be finite-dimensional.

As we will see later, it is possible to present the principal aspects of the theory of difference schemes with further treatment of $H_h$ as an abstract vector space of arbitrary dimension.

After preliminary discussions of the simplest examples illustrating some ways of producing grids and, thereby, of forming the spaces $H_h$ of grid functions we concentrate primarily on the problem of the difference approximation of differential operators.

**2. The difference approximation of elementary differential operators.** Let a linear operator $L$ assign the values to a function $v = v(x)$. By replacing the derivatives built into $Lv$ by their difference counterparts we derive the difference expression $L_h v_h$, which is a linear combination of the values of the grid function $v_h$ on some set of grid nodes known as a **pattern**:

$$L_h v_h(x) = \sum_{\xi \in Patt(x)} A_h(x, \xi) \, v_h(\xi)$$

or

$$(L_h v_h)_i = \sum_{x_j \in Patt(x_i)} A_h(x_i, x_j) \, v_h(x_j) ,$$

where $A_h(x, \xi)$ are the coefficients, $h$ is a grid spacing and $Patt(x)$ is a pattern at a point $x$. Any replacement of $Lv$ by $L_h v_h$ is called the **approximation of a differential operator by a difference operator** or the **difference approximation of an operator** $L$.

The way this approach is used in practice is connected with preliminary studies of difference approximations of the operator $L$ locally, that is, at an arbitrary fixed point of the space. If $v(x)$ is a continuous function, then $v_h(x) = v(x)$. Before giving further constructions of a difference approximation of the operator $L$, it is necessary to choose a pattern of proper form, that is, to compose the set of nodes adjacent to $x$ at which the values of the grid function $v(x)$ are aimed at approximating the operator $L$.

In this section we consider several examples of difference approximations for elementary differential operators.

**Example 1** $Lv = \dfrac{dv}{dx}$. Let us fix some point $x$ on the $Ox$-axis by capturing the neighboring points $x-h$ and $x+h$, where $h > 0$, and try to approximate $Lv$. Also, it will be sensible to introduce the following expressions:

$$L_h^+ v \equiv \frac{v(x+h)-v(x)}{h} \equiv v_x \,, \tag{1}$$

$$L_h^- v \equiv \frac{v(x)-v(x-h)}{h} \equiv v_{\bar{x}} \,. \tag{2}$$

Expressions (1) and (2) are called the **right difference derivative** and the **left difference derivative** and are denoted by $v_x$ and $v_{\bar{x}}$, respectively. The difference expressions $L_h^+$ and $L_h^-$ are defined at two points. In other words, these are based on the two-point patterns $x$, $x+h$ and $x-h$, $x$. Moreover, any linear combination of expressions (1)–(2) such as

$$L_h^{(\sigma)} v \equiv \sigma\, v_x + (1-\sigma)\, v_{\bar{x}} \,, \tag{3}$$

where $\sigma$ is a real number, can be adopted as the difference approximation of the derivative $dv/dx$. In particular, for $\sigma = 0.5$ we get the **central two-sided difference derivative**

$$v_{\overset{\circ}{x}} = \frac{1}{2}\,(v_x + v_{\bar{x}}) = \frac{v(x+h)-v(x-h)}{2\,h} \,. \tag{4}$$

It turns out that there is an uncountable set of difference expressions approximating $Lv = v'$ and this is something one might expect. The following question is of significant importance: what is the error of one or another difference approximation and how does the difference $\psi(x) = L_h\, v(x) - Lv(x)$ behave at a point $x$ as $h \to 0$? The quantity $\psi(x) = L_h\, v(x) - Lv(x)$ refers to the **error of the difference approximation to** $Lv$ **at a point** $x$. We next develop $v(x)$ in the series by Taylor's formula

$$v(x \pm h) = v(x) \pm h\, v'(x) + \frac{h^2}{2}\, v''(x) + O(h^3) \,,$$

assuming $v(x)$ to be a sufficiently smooth function in some neighborhood $(x-h_0, x+h_0)$ of the point $x$ and $h < h_0$, where the number $h_0$ is kept fixed. Substituting the preceding series into (1), (2), and (4) yields

$$
\begin{aligned}
v_x &= \frac{v(x+h)-v(x)}{h} = v'(x) + \frac{h}{2}\, v''(x) + O(h^2)\,,\\
v_{\bar{x}} &= \frac{v(x)-v(x-h)}{h} = v'(x) - \frac{h}{2}\, v''(x) + O(h^2)\,,\\
v_{\overset{\circ}{x}} &= \frac{v(x+h)-v(x-h)}{2\,h} = v'(x) + O(h^2)\,,
\end{aligned}
\tag{5}
$$

thereby justifying that

$$
\begin{aligned}
\psi &= v_x - v'(x) = O(h)\,,\\
\psi &= v_{\bar{x}} - v'(x) = O(h)\,,\\
\psi &= v_{\overset{\circ}{x}} - v'(x) = O(h^2)\,.
\end{aligned}
$$

Let $V$ be a class of sufficiently smooth functions $v \in V$ defined in a neighborhood $Patt(x, h_0)$ of a point $x$ containing for $h < h_0$ the pattern $Patt(x, h)$ of a difference operator $L_h$. We say that $L_h$ **approximates the differential operator $L$ with order $m > 0$ at a point $x$** if

$$
\psi(x) = L_h\, v(x) - L\, v(x) = O(h^m)\,.
$$

Thus, the right and left difference derivatives generate approximations of order 1 to $Lv = v'$, while the central difference derivative approximates to the second order the same.

**Example 2** $Lv = v'' = \dfrac{d^2 v}{dx^2}$. In order to construct a difference approximation of the second derivative, it is necessary to rely on the three-point pattern $(x-h, x, x+h)$. In that case we have

$$
L_h\, v = \frac{v(x+h) - 2\,v(x) + v(x-h)}{h^2}\,.
\tag{6}
$$

Observe that the right difference derivative at a point $x$ is identical with the left difference derivative at the point $x+h$, that is, the relation $v_x(x) = v_{\bar{x}}(x+h)$ occurs, permitting us to rewrite (6) as

$$
L_h\, v = \frac{v_x(x) - v_{\bar{x}}(x)}{h} = \frac{1}{h}\left[v_{\bar{x}}(x+h) - v_{\bar{x}}(x)\right] = v_{\bar{x}x}(x)\,.
\tag{7}
$$

With the aid of the expansion of the function $v(x)$ in Taylor's series it is not difficult to show that the order of approximation is equal to 2, meaning $v_{\bar{x}x} - v''(x) = O(h^2)$ by virtue of the expansion

$$v_{\bar{x}x} = v'' + \frac{h^2}{12}\, v^{(4)} + O(h^4)\,. \tag{8}$$

**Example 3** $Lv = v^{(4)}$. We now deal with the five-point pattern consisting of the points $(x-2h, x-h, x, x+h, x+2h)$ and accept $L_h\, v = v_{\bar{x}x\bar{x}x}$. By applying formula (6) to $v_{\bar{x}x}$ we derive the expression for $v_{\bar{x}x\bar{x}x}$, which will be needed in the sequel:

$$\begin{aligned} L_h\, v = v_{\bar{x}x\bar{x}x} &= \frac{1}{h^2}\left[v_{\bar{x}x}(x+h) - 2\, v_{\bar{x}x}(x) + v_{\bar{x}x}(x-h)\right] \\ &= \frac{1}{h^4}\left[v(x+2h) - 4\, v(x+h) + 6\, v(x) - 4\, v(x-h) + v(x-2h)\right]. \end{aligned}$$

It is straightforward to verify that $L_h$ provides an approximation of order 2 to $L$, so that $v_{\bar{x}x\bar{x}x} - v^{(4)} = \frac{h^2}{6} v^{(6)} + O(h^4)$. Indeed, this can be done using the expansion in Taylor's series

$$v(x+\sigma k h) = v(x) + \sum_{s=1}^{7} \frac{\sigma^s\, k^s\, h^s}{s\,!}\, \frac{d^s v(x)}{dx^s} + O(h^8)\,, \qquad \sigma = \pm 1\,,$$

for $k = 1, 2$ and exploiting the obvious fact that the sum $v(x+kh) + v(x-kh)$ contains only even powers.

The expansion of the approximation error $\psi = L_h\, v - Lv$ in powers of $h$ is aimed at achieving the order of approximation as high as possible. Indeed, we might have

$$v_{\bar{x}x} - v'' - \frac{h^2}{12}\, v^{(4)} + O(h^4) = \frac{h^2}{12}\, v_{\bar{x}x\bar{x}x} + O(h^4)\,,$$

whence it follows that $Lv = v''$ is approximated to fourth order by the operator

$$L'_h\, v = v_{\bar{x}x} - \frac{h^2}{12}\, v_{\bar{x}x\bar{x}x}$$

on the pattern $(x-2h, x-h, x, x+h, x+2h)$.

In principle one can continue further the process of raising the order of approximation further and achieve any order in the class of sufficiently smooth functions $v \in V$. During this process the pattern, that is, the total number of the nodes will increase. However, the way of achieving a higher-order difference approximation we have described above is of little practical significance, since the quality of resulting operators becomes worse in the sense of computational volume, the conditions of existence for the inverse operator, stability, etc. Later in this chapter we will survey some devices that can be used to obtain higher orders.

In the sequel an auxiliary lemma is needed.

**Lemma** *The formulae are valid:*

$$(9)\quad v_{\bar{x}x} = \frac{v(x+h) - 2\,v(x) + v(x-h)}{h^2} = v''(\xi)\,, \quad \xi = x + \theta h\,, \quad |\,\theta\,| \le 1\,,$$

*if* $v \in C^{(2)}[x-h, x+h]$,

$$(10)\qquad v_{\bar{x}x} = v''(x) + \frac{h^2}{12}\, v^{(4)}(\xi)\,, \quad \xi = x + \theta_1 h\,, \quad |\,\theta_1\,| \le 1\,,$$

*if* $v \in C^{(4)}[x-h, x+h]$. *Here* $C^{(k)}[a,b]$ *stands for the class of functions with the kth continuous derivative on the segment* $a \le x \le b$.

**Proof** We have occasion to use Taylor's formula with the remainder term in integral form

$$(11)\qquad v(x) = v(a) + (x-a)\,v'(a) + \cdots + \frac{(x-a)^r}{r\,!}\, v^{(r)} + R_{r+1}(x)\,,$$

where

$$(12)\qquad R_{r+1}(x) = \frac{1}{r\,!} \int_a^x (x-\xi)^r\, v^{(r+1)}(\xi)\; d\xi$$

$$= \frac{(x-a)^{r+1}}{r\,!} \int_0^1 (1-s)^r\, v^{(r+1)}\big(a + s\,(x-a)\big)\; ds\,.$$

Applying Lagrange's formula to the last integral reveals the remainder term $R_{r+1}(x)$ to be

$$R_{r+1}(x) = \frac{(x-a)^{r+1}}{(r+1)\,!}\; v^{(r+1)}(\xi)\,,$$

where $\xi$ is the mean value of $x$ on the segment $[a, x]$:

$$\xi = a + \theta\,(x - a), \quad 0 \le \theta \le 1, \quad \int_0^1 (1-s)^r \, ds = \frac{1}{r+1} \,.$$

Upon substituting $x + h$ for $x$ and $x$ for $a$ into formula (11) we find that

$$v(x+h) = v(x) + h\,v'(x) + h^2 \int_0^1 (1-s)\,v''(x+sh)\,ds\,, \tag{13}$$

$$v(x+h) = v(x) + h\,v'(x) + \frac{h^2}{2}\,v''(x) + \frac{h^3}{6}\,v'''(x) + \frac{h^4}{6} \int_0^1 (1-s)^3\,v^{(4)}(x+sh)\,ds \tag{14}$$

for $r = 1$ and $r = 3$, respectively. Replacing here $h$ by $-h$ and then $s$ by $-s$ gives for later use the new formulae

$$v(x-h) = v(x) - h\,v'(x) + h^2 \int_{-1}^0 (1+s)\,v''(x+sh)\,ds\,, \tag{15}$$

$$v(x-h) = v(x) - h\,v'(x) + \frac{h^2}{2}\,v''(x) - \frac{h^3}{6}\,v'''(x) + \frac{h^4}{6} \int_{-1}^0 (1+s)^3\,v^{(4)}(x+sh)\,ds\,. \tag{16}$$

Adding (13) and (15), placing the term $2\,v(x)$ on the left and then dividing the resulting expression by $h^2$, we finally get

$$v_{\bar{x}x} = \frac{v(x+h) + v(x-h) - 2\,v(x)}{h^2} = \int_{-1}^1 g_2(s)\,v''(x+sh)\,ds\,,$$

where

$$g_2(s) = \begin{cases} 1+s & \text{for} \quad -1 \le s < 0\,, \\ 1-s & \text{for} \quad 0 \le s \le 1\,. \end{cases}$$

Since $g_2(s) \geq 0$, Lagrange's formula provides support for the representation

$$v_{\bar{x}x} = v''(x+\theta h) \int_{-1}^{1} g_2(s)\, ds = v''(x+\theta h) = v''(\xi), \quad -1 \leq \theta \leq 1,$$

where $\xi$ is the middle point of the segment $[x-h, x+h]$. The second formula (10) can be derived in the same manner as before.

Below we follow these procedures in a step-by-step fashion: collect (14) and (16), carry $2\,v(x)$ over to the left-hand side and divide the resulting expression by $h^2$. Simple algebra gives

$$v_{\bar{x}x} = v''(x) + \frac{h^2}{6} \int_{-1}^{1} g_4(s)\, v^{(4)}(x+sh)\, ds,$$

where

$$g_4(s) = \begin{cases} (1+s)^3 & \text{for} \quad -1 \leq s \leq 0, \\ (1-s)^3 & \text{for} \quad 0 \leq s \leq 1, \end{cases} \qquad \int_{-1}^{1} g_4(s)\, ds = \frac{1}{2}.$$

Because $g_4(s) \geq 0$ and $v^{(4)}(x)$ is continuous, applying Lagrange's formula yields

$$v_{\bar{x}x} = v''(x) + \frac{h^2}{12}\, v^{(4)}(x+\theta h), \qquad |\,\theta\,| \leq 1.$$

**Remark 1** By exactly the same reasoning as before, we can establish

$$(17) \quad v_{\bar{x}x} = v''(x) + \frac{h^2}{12}\, v^{(4)}(x) + \frac{h^4}{360}\, v^{(6)}(\xi), \quad \xi = x+\theta h, \quad |\,\theta\,| \leq 1,$$

if $v(x)$ belongs to the class $C^{(6)}[x-h, x+h]$.

**Remark 2** A similar result is still valid for the derivative

$$(18) \qquad v_{\bar{x}x\bar{x}x} = \frac{1}{h^4}\, \big[v(x+2h) - 4\,v(x+h) + 6\,v(x) - 4\,v(x-h) + v(x-2h)\big] = v^{(4)}(\xi),$$

where $\xi = x+\theta h$, $|\theta| \leq 2$, is the middle point of the segment $[x-2h, x+2h]$ and $v \in C^{(4)}[x-2h, x+2h]$. This expression will be proved if we succeed in showing that

$$v_{\bar{x}x\bar{x}x} = \frac{1}{6} \int_{-2}^{2} g(s)\, v^{(4)}(x+sh)\, ds,$$

where

$$g(s) = \begin{cases} 8\,(1+0.5\,s)^3 & \text{for} \quad -2 \le s \le -1\,, \\ 8\,(1+0.5\,s)^3 - 4\,(1+s)^3 & \text{for} \quad -1 \le s \le 0\,, \\ 8\,(1-0.5\,s)^3 - 4\,(1-s)^3 & \text{for} \quad 0 \le s \le 1\,, \\ 8\,(1-0.5\,s)^3 & \text{for} \quad 1 \le s \le 2\,. \end{cases}$$

The reader is invited to do it on his/her own.

**Example 4** $L\,v = \dfrac{\partial v}{\partial t} - \dfrac{\partial^2 v}{\partial x^2}$, $v = v(x,t)$. Let $(x,t)$ be a fixed point in the plane $(x,t)$ and let $h > 0$ and $\tau > 0$ be two arbitrary numbers taken as steps in the sequel. A difference approximation $L_{h\tau}$ of the operator $L$ is connected with a proper choice of the pattern.

We begin by placing approximations of the simplest type for which the pattern consists of the four points (Fig. 4.a), so that $L_{h\tau}$ is certainly expressed by

$$L_{h\tau}^{(0)}\,v = \frac{v(x,t+\tau) - v(x,t)}{\tau} - \frac{v(x+h,t) - 2\,v(x,t) + v(x-h,t)}{h^2}\,. \tag{19}$$

A suitably chosen symbolism may symplify the form of writing various difference expressions and makes our exposition more transparent. This is acceptable if we agree to consider

$$v = v(x,t)\,, \qquad \hat{v} = v(x,t+\tau)\,, \qquad \check{v} = v(x,t-\tau)\,.$$

Within these notations, the difference derivative with respect to $t$ becomes

$$v_t = \frac{v(x,t+\tau) - v(x,t)}{\tau} = \frac{\hat{v} - v}{\tau}\,. \tag{20}$$

By virtue of relations (7) and (20) we write down (19) in the form

$$L_{h\tau}^{(0)}\,v = v_t - v_{\bar{x}x}\,. \tag{19'}$$

In the preceding constructions of $L_{h\tau}^{(0)}$ we have taken the value of $v_{\bar{x}x}$ at the moment $t$, that is, on the lower layer. On the pattern depicted in Fig. 4.b it

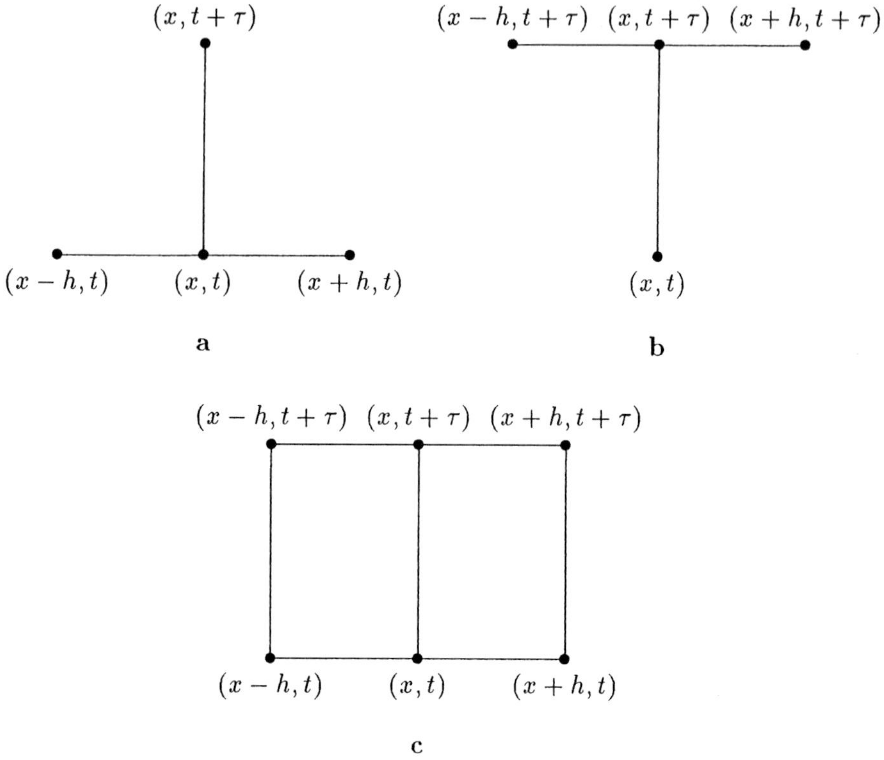


**Figure 4.**

is reasonable to place $v_{\bar{x}x}$ at the moment $t + \tau$, that is, on the upper layer, so that

$$L_{h\tau}^{(1)} v = v_t - \hat{v}_{\bar{x}x} \,. \tag{21}$$

By a linear combination of (19′) and (21) we get a one-parameter family of difference operators on the six-point pattern shown in Fig. 4.c for $\sigma \neq 0$ and $\sigma \neq 1$:

$$L_{h\tau}^{(\sigma)} v = v_t - \left(\sigma\, \hat{v}_{\bar{x}x} + (1 - \sigma)\, v_{\bar{x}x}\right) . \tag{22}$$

In the estimation of the order of approximation we substitute into the formulae for $L_{h\tau}^{(0)}\, v$, $L_{h\tau}^{(1)}\, v$ and $L_{h\tau}^{(\sigma)}\, v$ the following expressions:

$$\begin{aligned} v_t &= \frac{\partial v(x,t)}{\partial t} + \frac{\tau}{2}\, \frac{\partial^2 v(x,t)}{\partial t^2} + O(\tau^2) \\ &= \frac{\partial v(x, t+\tau/2)}{\partial t} + O(\tau^2)\,, \\ v_{\bar{x}x} &= \frac{\partial^2 v(x,t)}{\partial x^2} + \frac{h^2}{12}\, \frac{\partial^4 v(x,t)}{\partial x^4} + O(h^4) \\ &= \frac{\partial^2 v(x, t+\tau/2)}{\partial x^2} - \frac{\tau}{2}\, \frac{\partial^3 v(x, t+\tau/2)}{\partial x^2\, \partial t} + O(h^2+\tau^2)\,, \\ \check{v}_{\bar{x}x} &= \frac{\partial^2 v(x, t+\tau/2)}{\partial x^2} + \frac{\tau}{2}\, \frac{\partial^3 v(x, t+\tau/2)}{\partial x^2\, \partial t} + O(h^2+\tau^2)\,. \end{aligned}$$

The outcomes of such manipulations are:

$$\begin{aligned} 1) \quad L_{h\tau}^{(0)}\, v &= \frac{\partial v(x,t)}{\partial t} - \frac{\partial^2 v(x,t)}{\partial x^2} + O(h^2+\tau) \\ &= Lv(x,t) + O(h^2+\tau)\,, \end{aligned}$$

$$\text{meaning,} \quad \psi^{(0)} = L_{h\tau}^{(0)}\, v - Lv(x,t) = O(h^2+\tau)\,;$$

$$\begin{aligned} 2) \quad L_{h\tau}^{(1)}\, v &= \frac{\partial v(x, t+\tau)}{\partial t} - \frac{\partial^2 v(x, t+\tau)}{\partial x^2} + O(h^2+\tau) \\ &= Lv(x, t+\tau) = O(h^2+\tau)\,, \end{aligned}$$

$$\text{meaning,} \quad \psi^{(1)} = L_{h\tau}^{(1)}\, v - Lv(x, t+\tau) = O(h^2+\tau)\,;$$

$$\begin{aligned} 3) \quad L_{h\tau}^{(0.5)}\, v &= \frac{\partial v(x, t+\tau/2)}{\partial t} - \frac{\partial^2 v(x, t+\tau/2)}{\partial x^2} + O(h^2+\tau^2) \\ &= Lv(x, t+\tau/2) = O(h^2+\tau^2)\,, \end{aligned}$$

$$\text{meaning,} \quad \psi^{(0.5)} = L_{h\tau}^{(0.5)}\, v - Lv(x, t+\tau/2) = O(h^2+\tau^2)\,.$$

All this enables us to conclude that the operator $L_{h\tau}^{(\sigma)}$ provides the approximation of order 2 with respect to $h$ for any $\sigma$, the approximation of order 1 with respect to $\tau$ for $\sigma = 0$ and $\sigma = 1$, and the approximation of order 2 with respect to $\tau$ for $\sigma = 0.5$.

**Example 5** $L\,v = \dfrac{\partial^2 v}{\partial t^2} - \dfrac{\partial^2 v}{\partial x^2}$ . In that case the values of a grid function at three instants of time $t-\tau$, $t$ and $t+\tau$ are aimed at selecting the difference operator $L_{h\tau}$. The five-point pattern plotted in Fig. 5.a,b,c is the best possible for our purposes.

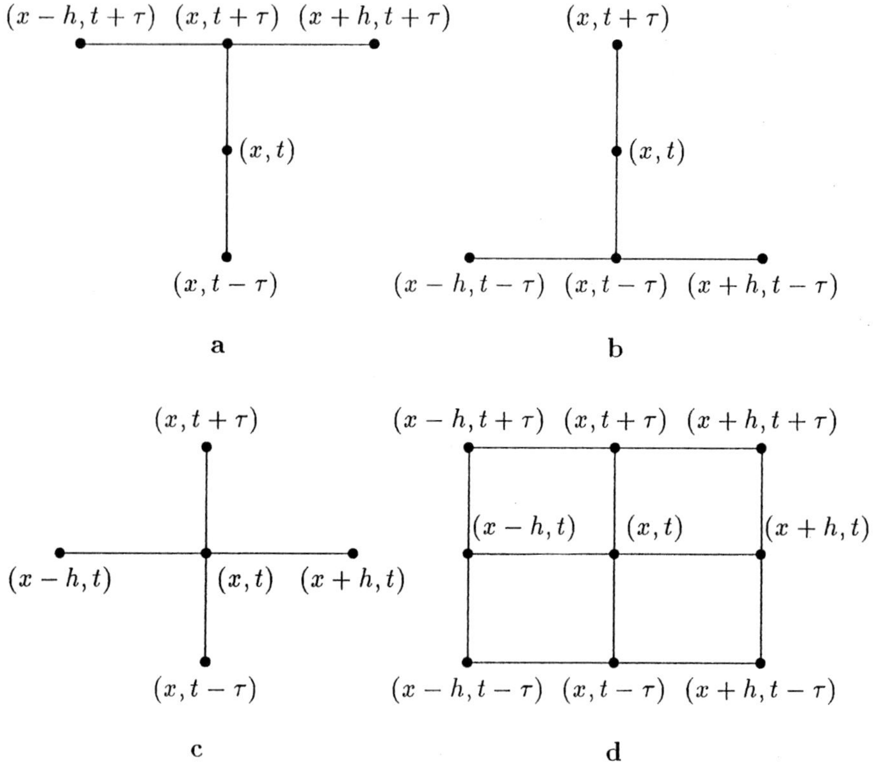


**Figure 5.**

One of the possible approximations on the pattern 5.c with using the value $v_{\bar{x}x}$ on the middle layer $t$ is of the form

(23) $$L_{h\tau}\, v = v_{\bar{t}t} - v_{\bar{x}x}\,,$$

where

$$v_{\bar{t}t}(x,t) = \frac{v(x,t+\tau) - 2\,v(x,t) + v(x,t-\tau)}{\tau^2}\,.$$

In a similar manner one can write down on the pattern 5.a the operator

$$L_{h\tau}\, v = v_{\bar{t}t} - \hat{v}_{\bar{x}x}\,. \tag{24}$$

On the other hand, a two-parameter family of difference operators

$$L_{h\tau}^{(\sigma_1,\,\sigma_2)}\, v = v_{\bar{t}t} - \left(\sigma_1\, \hat{v}_{\bar{x}x} + (1-\sigma_1-\sigma_2)\, v_{\bar{x}x} + \sigma_2\, \check{v}_{\bar{x}x}\right) \tag{25}$$

is based on the nine-point pattern 5.d, whence (23) follows for $\sigma_1 = \sigma_2 = 0$ and (24) follows for $\sigma_2 = 0$ and $\sigma_1 = 1$. We have occasion to use the asymptotic formulae

$$v_{\bar{t}t} = \frac{\partial^2 v(x,t)}{\partial t^2} + O(\tau^2), \qquad v_{\bar{x}x} = \frac{\partial^2 v(x,t)}{\partial x^2} + O(h^2)\,,$$

thereby justifying that operator (23) provides an approximation of $O(h^2 + \tau^2)$. Operator (25) has the same order of approximation for $\sigma_1 = \sigma_2 = \sigma$, where $\sigma$ is an arbitrary number.

We note in passing that a key role of the parameters $\sigma_1$ and $\sigma_2$ just as the parameter $\sigma$ in the previous example is connected not only with the approximation order, but also with stability of the appropriate difference scheme. This important property will appear in subsequent discussions in Chapter 5, Section 1.

**Example 6.** $L\, v = v''$. ***An irregular pattern (a non-equidistant grid).*** Granted two numbers $h_- > 0$ and $h_+ > 0$, other ideas are connected with the three-point pattern $(x - h_-, x, x + h_+)$. When $h_- \neq h_+$, any such pattern is said to be **irregular**, since the grid including this pattern turns out to be **non-equidistant**. Being concerned with new members

$$v_{\bar{x}} = \frac{v(x) - v(x-h_-)}{h_-}\,, \qquad v_x = \frac{v(x+h_+) - v(x)}{h_+}\,, \qquad \hbar = \frac{h_- + h_+}{2}\,,$$

we refer to the operator $L_h$ with the values

$$L_h\, v = \frac{1}{\hbar}\left[\frac{v(x+h_+) - v(x)}{h_+} - \frac{v(x) - v(x-h_-)}{h_-}\right] = \frac{v_x - v_{\bar{x}}}{\hbar}\,. \tag{26}$$

For $h_- = h_+ = \hbar$ the preceding is identical with expression (7) (see Example 2). Plain calculations of the local approximation error at a point $x$ show that

$$\psi(x) = L_h\, v(x) - L\, v(x)\,.$$

Having stipulated the condition $h_\pm < 2\hbar$, the well-established expansions of a sufficiently smooth function $v(x)$ in a neighborhood of the node $x$ such as

$$v(x+h_+) = v(x) + h_+ v'(x) + \frac{h_+^2}{2} v''(x) + \frac{h_+^3}{6} v'''(x) + O(h_+^4),$$

$$v(x-h_-) = v(x) - h_- v'(x) + \frac{h_-^2}{2} v''(x) - \frac{h_-^3}{6} v'''(x) + O(h_-^4)$$

lead to the following ones:

$$v_x = v'(x) + \frac{h_+}{2} v''(x) + \frac{h_+^2}{6} v'''(x) + O(h_+^3),$$

$$v_{\bar{x}} = v'(x) - \frac{h_-}{2} v''(x) + \frac{h_-^2}{6} v'''(x) + O(h_-^3),$$

$$L_h v = \frac{v_x - v_{\bar{x}}}{\hbar} = v''(x) + \frac{h_+^2 - h_-^2}{6\hbar} v'''(x) + O(\hbar^2).$$

With these relations in view, we derive the usefull expressions for $\psi(x)$

$$\psi = L_h v - L v = \frac{h_+ - h_-}{3} v''' + O(\hbar^2) = O(\hbar). \tag{27}$$

Thus, operator (26) provides the local approximation of order 1 on any irregular pattern with $h_- \neq h_+$.

**3. The error of approximation on a grid.** So far we have considered the local difference approximation; meaning the approximation at a point. Just in this sense we spoke about the order of approximation in the preceding section. Usually some estimates of the difference approximation order on the whole grid are needed in various constructions.

In preparation for this, let $\omega_h$ be a grid in a domain $G$ of the Euclidean space $\{x = (x_1, \dots, x_p)\}$, $H_h$ be a vector space of grid functions defined on the grid $\omega_h$ and let the space $H_0$ comprise all of the smooth functions $v(x)$, whose norms are defined by $\|\cdot\|_0$ and $\|\cdot\|_h$, respectively. In the sequel we take for granted the following assumptions:

(1) there exists an operator $\mathcal{P}_h$ such that $\mathcal{P}_h u = u_h \in H_h$ for any $u \in H_0$;

(2) the norms $\|\cdot\|_0$ and $\|\cdot\|_h$ are concordant, that is,

$$\lim_{|h|\to 0} \|\mathcal{P}_h u\|_h = \|u\|_0,$$

where $|h|$ stands for the norm of a vector $h$.

The next step is to introduce an operator $L$ in the space $H_0$ and an operator $L_h$ carrying a grid function $v_h$ into a grid function $L_h v_h$ on the grid $\omega_h$ ($L_h$: $H_h \mapsto H_h$).

A grid function

$$\psi_h = L_h v_h - (Lv)_h ,$$

where $v_h = \mathcal{P}_h v$, $(Lv)_h = \mathcal{P}_h(Lv)$ and $v$ is an arbitrary function (vector, element) of the space $H_0$, is called the **error of approximation** of the operator $L$ by the difference operator $L_h$.

If $\| \psi_h \|_h \to 0$ as $h \to 0$, we say that the difference operator $L_h$ **approximates** the differential operator $L$.

Likewise, the difference operator $L_h$ **approximates the differential operator $L$ with order** $m > 0$ if

$$\| \psi_h \|_h = \| L_h v_h - (Lv)_h \|_h = O(| h |^m) \tag{28}$$

or $\| L_h v_h - (Lv)_h \|_h \le M \, | h |^m$, where $M$ is a positive constant independent of $| h |$.

**Remark 1** We give below several examples of projectors $\mathcal{P}_h$ onto the set of grid functions:

(1) for a continuous function $v(x)$, $v_h = \mathcal{P}_h \, v(x) = v(x)$, $x \in \omega_h$;

(2)

$$v_h = \mathcal{P}_h \, v = \frac{1}{2h} \int\limits_{x-h}^{x+h} v(t) \, dt = \frac{1}{2} \int\limits_{-1}^{1} v(x + sh) \, ds$$

if $v(x)$ is a summable function, etc.

**Remark 2** The length $| h | = (h_1^2 + \cdots + h_p^2)^{1/2}$ of a vector $h = (h_1, \ldots, h_p)$ with components $h_1, \ldots, h_p$ may be taken as the quantity involved in the above definitions. Observe that approximations with respect to $h_\alpha$, $\alpha = 1, 2, \ldots, p$, may be different in order. If so, instead of (28) we might have

$$\| L_h v_h - (Lv)_h \|_h \le M \sum_{\alpha=1}^{p} h_\alpha^{m_\alpha} , \qquad m_\alpha > 0 .$$

Choosing among the numbers $m_1, \ldots, m_p$ the minimal one and denoting it by $m$, we get estimate (28).

**Remark 3** If the grid $\omega_h$ is non-equidistant, that is, $h = (h_1, \dots, h_N)$, where $N$ is the total number of nodes, it is sensible to deal with $h = \max_{1 \le i \le N} h_i$ or the mean square value $|\, h \,|$.

Several examples add interest and help in understanding the outlined theory.

**Example 1.** ***The difference approximation on a non-equidistant grid.*** In working in the space $H_0 = C^{(4)}[0,1]$ comprising all the functions defined on the segment $0 \le x \le 1$ we refer to the operator $L$ with the values

$$L\, v = \frac{d^2 v}{dx^2}$$

and take on this segment an arbitrary non-equidistant grid

$$\hat{\bar{\omega}}_h = \{x_i,\ i = 0, 1, \dots, N,\ x_0 = 0,\ x_N = 1\}\,.$$

On account of Example 6 in the preceding section the difference operator

$$(L_h\, v)_i = \frac{1}{\hbar_i} \left( \frac{v_{i+1} - v_i}{h_{i+1}} - \frac{v_i - v_{i-1}}{h_i} \right), \qquad v_i = v(x_i)\,, \quad \hbar_i = \frac{h_i + h_{i+1}}{2}\,,$$

defined at the node $x_i$ on the irregular three-point pattern $(x_{i-1}, x_i, x_{i+1})$, is associated with the operator $Lv$. With this, we rewrite $L_h\, v$ as

$$(L_h\, v)_i = v_{\bar{x}\,\hat{x},\, i} = v_{\bar{x}\,\hat{x}}$$

with more compact notations

$$v_{\bar{x},\, i} = \frac{v_i - v_{i-1}}{h_i}\,, \qquad v_{x,\, i} = v_{\bar{x},\, i+1} = \frac{v_{i+1} - v_i}{h_{i+1}}\,, \qquad v_{\hat{x},i} = \frac{v_{i+1} - v_i}{\hbar_i}\,.$$

In Section 1.2 we have found the local approximation error taking now the form

$$\psi_i = (L_h\, v)_i - (Lv)_i = \frac{h_{i+1} - h_i}{3}\; v_i''' + O(\hbar_i^2)\,, \qquad i = 1, 2, \dots, N-1\,.$$

From here it is easily seen that the operator $L_h$ is of order 1 with respect to the grid norm of the space $C$:

$$\|\, \psi \,\|_C = \max_{1 \le i \le N-1} |\, \psi_i \,| = O(h)\,, \qquad h = \max_{1 \le i \le N} h_i\,.$$

The same is still valid in the grid norm of the space $L_2$:

$$\|\psi\|_{L_2} = \left( \sum_{i=1}^{N-1} \hbar_i \, \psi_i^2 \right)^{1/2} = O(h) .$$

We claim that in an alternative norm known as the negative norm

$$\| \psi \|_{(-1)} = \left[ \sum_{i=1}^{N-1} h_i \left( \sum_{k=1}^{i} \hbar_k \, \psi_k \right)^2 \right]^{1/2}$$

the local approximation error $\psi$ is of order 2 and behaves like

$$\| \psi \|_{(-1)} = O(h^2) , \qquad h = \max_{1 \leq i \leq N} h_i . \tag{29}$$

Indeed, rewriting $\psi$ as

$$\psi_i = \frac{h_{i+1}^2 - h_i^2}{6 \, \hbar_i} \, v_i''' + O(\hbar_i^2)$$

and taking into account that $v_i''' = v_{i+1}''' + O(h_{i+1})$, we establish the decomposition

$$\psi_i = \frac{h_{i+1}^2 \, v_{i+1}''' - h_i^2 \, v_i'''}{6 \, \hbar_i} + \psi_i^* = \overset{\circ}{\psi}_i + \psi_i^*$$

with $\psi_i^* = O(h^2)$ in any suitable norm. The main term $\overset{\circ}{\psi}_i$ in the established decomposition $\psi_i = \overset{\circ}{\psi}_i + \psi_i^*$ is of "divergent form", thereby justifying that

$$S_i = \sum_{k=1}^{i} \hbar_k \, \overset{\circ}{\psi}_k = \sum_{k=1}^{i} \frac{h_{k+1}^2 \, v_{k+1}''' - h_k^2 \, v_k'''}{6} = \frac{h_{i+1}^2 \, v_{i+1}''' - h_1^2 \, v_1'''}{6} .$$

Since $|S_i| \leq M h^2$, we find that

$$\| \overset{\circ}{\psi} \|_{(-1)} = \left( \sum_{i=1}^{N-1} h_i \, S_i^2 \right)^{1/2} = O(h^2) ,$$

which in combination with the relations

$$\| \psi \|_{(-1)} \leq \| \overset{\circ}{\psi} \|_{(-1)} + \| \psi^* \|_{(-1)} , \qquad \| \psi^* \|_{(-1)} = O(h^2)$$

implies that $\| \psi \|_{(-1)} < Mh^2$. This means that the approximation error is of order 2 in the negative norm $\| \cdot \|_{(-1)}$. Observe that the norm $\| \cdot \|_{(-1)}$ is concordant with the norm $\| u \|_0 = \left[\int_0^1 dx \left(\int_0^x u(\xi)\, d\xi\right)^2\right]^{1/2}$, so that $\| u_h \|_{(-1)} \to \| u \|_0$ as $h \to 0$.

Another conclusion can be drawn from this example. Of course, the study of the local approximation is unsufficient for determination of the order of the difference approximation and proper evaluation of the quality of a difference operator.

In the estimation of the approximation error the well-founded choice of the norm depends on the structure of an operator and needs investigation in every particular case. A precise relationship between an operator and a norm in the process of searching the error of approximation will be established in the general case in Section 4. Its concretization for the example of interest leads naturally to the appearance of the negative norm $\| \cdot \|_{(-1)}$. A similar obstacle occurs in trying to construct difference approximations of the operator $L\,u = (k\,u')'$, where $k(x)$ is a piecewise continuous function (see Chapter 3).

Where searching a solution $u(x,t)$ to a nonstationary equation (for instance, to the heat conduction equation), it is not unreasonable to separate the variable $t$ (the time). Some consensus of opinion is that the sought function $u(x,t)$ as a function of the argument $x$ is an element of the space $H_0$. Let $\omega_h$ be a grid in a domain $G$ of the space $\{x = (x_1, \dots, x_p)\}$ and let $\omega_\tau$ be a grid on the segment $0 \le t \le t_0$. The grid function $y(x,t) = y_{h\tau}(x,t)$ is defined on the grid $\omega_{h\tau} = \omega_h \times \omega_\tau = \{(x,t),\ x \in \omega_h,\ y \in \omega_\tau\}$ and gives as a function of the argument $x \in \omega_h$ a vector of the space $H_h$ with the norm $\| \cdot \|_h$. In dealing with $y(x,t)$ on the grid $\omega_{h\tau}$ many scientists generally prefer either

$$(30) \qquad \| y \|_{h\tau} = \max_{t \in \omega_\tau} \| y(t) \|_h$$

or one of the alternative norms

$$(31) \qquad \| y \|_{h\tau} = \sum_{t \in \omega_\tau} \tau \, \| y(t) \|_h , \qquad \| y \|_{h\tau} = \Big( \sum_{t \in \omega_\tau} \tau \, \| y(t) \|_h^{\,2} \Big)^{1/2} .$$

Let $L_{h\tau}\, v_{h\tau}$ be a difference approximation of the operator $Lu$, where $u = u(x,t)$. The operator $L_{h\tau}$ assigns really the values to the grid functions $v_{h\tau}(x,t)$ defined on the grid $\omega_{h\tau}$. If $v(x,t)$ as a function of the argument $x$ belongs to the space $H_0$, then $v_h(x,t) = \mathcal{P}_h\, v(x,t) \in H_h$ for any $t \in [0, t_0]$. For any continuous in $t$ function $v(x,t)$ and all $t \in \omega_\tau$, we put $v_{h\tau}(x,t) =$

$v_h(x,t)$, thereby specifying $v_{h\tau}$ on the grid $\omega_{h\tau}$ and reducing the error of approximation to

$$\psi_{h\tau}(x,t) = L_{h\tau}\, v_{h\tau}(x,t) - (Lv)_{h\tau}(x,t)\,, \qquad (x,t) \in \omega_{h\tau}\,.$$

We say that $L_{h\tau}$ **approximates** $L$ **with order** $m > 0$ **in** $x$ **and with order** $n > 0$ **in** $t$ if in the class of sufficiently smooth functions $v(x,t)$ one of the estimates

$$\|\,\psi_{h\tau}(x,t)\,\|_{h\tau} = O(|\,h\,|^m + \tau^n)\,, \qquad \|\,\psi_{h\tau}\,\|_{h\tau} \le M\,(|\,h\,|^m + \tau^n)$$

is valid with a positive constant $M$ independent of $|\,h\,|$ and $\tau$ both.

**Example 2**

$$L\,v = \frac{\partial v}{\partial t} - \frac{\partial^2 v}{\partial x^2}\,, \qquad 0 < x < 1\,, \quad 0 < t \le t_0\,, \qquad L_{h\tau}\, v = v_t - v_{\bar{x}x}\,.$$

The difference operator $L_{h\tau}$ defined at all inner nodes of the grid

$$\omega_{h\tau} = \{\,(x_i\,,t_j),\ x_i = ih,\ t_j = j\tau,\ 0 < i < N,\ 0 < j < j_0\,,\ j_0 = t_0/\tau\,\}$$

complements subsequent studies and proves to be useful in many aspects. If $v(x,t)$ possesses two derivatives with respect to $t$ and four derivatives with respect to $x$ ($v \in C_2^4$) that are continuous in the rectangle $\{0 \le x \le 1,\ 0 \le t \le t_0\}$, then

$$\psi_{h\tau}(x,t) = L_{h\tau}\, v_{h\tau} - (L\,v)_{h\tau} = O(h^2 + \tau)$$

at each of the inner nodes of the grid $\omega_{h\tau}$. This is also consistent with the results of Section 1.2. Hence, $L_{h\tau}$ approximates $L$ to second order in $x$ and first order in $t$ in either of the norms (30) and (31), where $\|\,\psi\,\|_h = \max_{x\in\omega_h} |\,\psi\,|$ or $\|\,\psi\,\|_h = \left(\sum_{i=1}^{N-1} \psi_i^2\, h\right)^{1/2}$, etc. Thus, in the case of interest the approximation on the grid follows from the local approximation.

So far we have studied the error of the difference approximation on the functions $v$ from certain classes $V$. In particular, in the preceding examples the class of sufficiently smooth functions stands for $V$.

In what follows it is supposed that $v = u$ is a solution of a differential equation, say

$$L\,u = u'' = -f\,.$$

The meaning of $L_h\, v = v_{\bar{x}x}$ as the difference approximation of the operator $L\, v = v''$ reveals that

$$L_h\, v = v'' + \frac{h^2}{12}\, v^{(4)} + \frac{h^4}{360}\, v^{(6)}(x+\theta h)\,, \qquad -1 \le \theta \le 1\,.$$

We substitute here $v = u$, $u'' = -f$, $u^{(4)} = -f''$ and accept $f'' = 0$, implying that $f^{(k)} \equiv 0$ for all $k > 2$. Then $L_h\, u = u'' = Lu$, that is, the approximation error happens to be from the class of solutions to the equation $Lu = -f$, where $f$ is a linear function identically equal to zero: $\psi \equiv 0$. In that case the approximmation is said to be **exact**. But if $f'' \neq 0$, the accuracy will be improvable once we attempt the operator in the form

$$\widetilde{L}_h\, v = L_h\, v + \frac{h^2}{12}\, f''\,,$$

thereby providing the approximation $\widetilde{\psi} = \widetilde{L}_h\, u - L\, u = O(h^4)$.

Thus, the accurate account of the error of the difference approximation on a solution of the differential equation helps raise the order of approximation.

**4. The statement of a difference problem.** In the preceding sections we were interested in approximate substitutions of difference operators for differential ones. However, many problems of mathematical physics involve not only differential equations, but also the supplementary conditions (boundary and initial) which guide a proper choice of a unique solution from the collection of possible solutions.

The complete posing of a difference problem necessitates specifying the difference analogs of those conditions in addition to the approximation of the governing differential equation. The set of difference equations approximating the differential equation in hand and the supplementary boundary and initial conditions constitute what is called a **difference scheme**. In order to clarify the essence of the matter, we give below several examples.

**Example 1.** ***The Cauchy problem for an ordinary differential equation:***

$$u' = f(x)\,, \qquad x > 0\,, \qquad u(0) = u_0\,. \tag{32}$$

We proceed, as usual, on the simplest uniform grid $\omega_h = \{x_i = ih,\ i = 1, 2, \dots\}$ and put the difference problem

$$\begin{cases} y_x = \varphi\,, \\ y_0 = u_0\,, \end{cases} \qquad \text{or} \qquad \frac{y_{i+1} - y_i}{h} = \varphi_i\,, \qquad i = 0, 1, \dots\,; \qquad y_0 = u_0\,,$$

in correspondence with problem (32). Here the right-hand side $\varphi_i$ can be specified in a number of different ways, for instance, by the formulae

$$\varphi_i = f(x_i), \qquad \varphi_i = \frac{f(x_i) + f(x_{i+1})}{2},$$

provided the condition $\varphi_i - f_i = O(h)$ holds.

Thus, given $y_0 = u_0$, the solutions can be found from the recurrence relation $y_{i+1} = y_i + h\,\varphi_i$, $i = 0, 1, 2 \ldots$.

**Example 2.** ***The Cauchy problem for a system of first-order differential equations:***

$$\frac{du}{dt} + A\,u = 0\,, \qquad t > 0\,, \qquad u(0) = u_0\,,$$

where $A = (a_{ik})$ is a square $n \times n$-matrix and $u = (u_1, \ldots, u_n)$ is an $n$-dimensional vector. As we will see later, it seems reasonable to introduce the grid $\omega_\tau = \{t_j = j\tau,\ j = 0, 1, 2, \ldots\}$ with step $\tau$. One of the possibilities is Euler's difference scheme

$$\frac{y^{j+1} - y^j}{\tau} + A\,y^j = 0\,, \qquad j = 0, 1, 2, \ldots, \qquad y^j = (y_1^j, y_2^j, \ldots, y_n^j)\,.$$

The difference problem under consideration will be completely posed if the subsidiary information is available on the vector $y^j$ for $j = 0$ with the initial condition $y^0 = u_0$. The value $y^{j+1}$ is successively calculated by the explicit formula

$$y^{j+1} = y^j - \tau A\,y^j\,.$$

This is one way of gaining experience with Euler's scheme that can be accepted and used in theory and practice.

**Example 3.** ***The boundary-value problem:***

$$u''(x) = -f(x)\,, \quad 0 < x < 1\,, \quad u(0) = \mu_1\,, \quad u(1) = \mu_2\,. \tag{33}$$

We take once again the equidistant grid

$$\bar{\omega}_h = \{x_i = ih,\ i = 0, 1, \ldots, N,\ hN = 1\}$$

and set up on it the difference problem

$$y_{\bar{x}x} = -\varphi \qquad \text{or} \qquad \frac{y_{j+1} - 2\,y_i - y_{j-1}}{h^2} = -\varphi_i\,, \tag{$33_h$}$$

$$i = 1, 2, \ldots, N-1\,, \qquad y_0 = \mu_1\,, \quad y_N = \mu_2\,.$$

Thus emerged the system of algebraic equations with a tridiagonal matrix. Because of this form, the elimination method may be useful (see Chapter 1, Section 1).

**Example 4.** ***The first boundary-value problem for the heat conduction equation:***

$$(34) \qquad L\,u = \frac{\partial u}{\partial t} - \frac{\partial^2 u}{\partial x^2} = f(x,t)\,, \qquad 0 < x < 1\,, \quad 0 < t \le t_0\,,$$

$$u(0,t) = \mu_1(t)\,, \qquad u(1,t) = \mu_2(t)\,, \qquad u(x,0) = u_0(x)\,.$$

Choosing the equidistant grid $\omega_{h\tau} = \{(x_i = ih,\ t_j = j\tau),\ i = 0, 1, \dots, N_1,\ j = 0, 1, \dots, N_2\}$ and the simplest four-point pattern from Example 4 in Section 1.2, we are now in a position to set up the difference problem

$$y_t = y_{\bar{x}x} + \varphi\,,$$

which admits the index form

$$(34_h) \qquad \frac{y_i^{j+1} - y_i^j}{\tau} = \frac{y_{i-1}^j - 2\,y_i^j + y_{i+1}^j}{h^2} + \varphi_i^j\,,$$

$$1 \le i \le N_1 - 1\,, \qquad 0 \le j \le N_2 - 1\,,$$

$$y_0^j = \mu_1(t_j)\,, \quad y_{N_1}^j = \mu_2(t_j)\,, \quad y_i^0 = u_0(x_i)\,.$$

Here the right-hand side $\varphi$ can be defined in a number of different ways:

$$\varphi_i^j = f(x_i, t_j)\,, \qquad \varphi_i^j = f(x_i, t_{j+1/2})\,, \qquad \text{etc.}$$

The difference problem $(34_h)$ illustrates the implementation of the so-called **explicit scheme** in which the values of the solution on the upper layer $y^{j+1}$ are expressed through the values on the current layer by the explicit formulae

$$y^{j+1} = y^j + \tau\,(y_{\bar{x}x}^j + \varphi^j)\,.$$

Common practice involves also the **implicit scheme**

$$y_t = \hat{y}_{\bar{x}x} + \varphi\,, \qquad y(x,0) = u_0(x)\,, \qquad x \in \omega_h\,,$$

$$y(0,t) = \mu_1(t)\,, \quad y(1,t) = \mu_2(t)\,, \qquad t \in \omega_\tau\,,$$

which finds a wide range of applications. In order to have for later use the values $\hat{y} = y^{j+1}$ on the $(j+1)$th layer, we must solve by the elimination method the system of algebraic equations with a tridiagonal matrix

$$\frac{y^{j+1}}{\tau} - y_{\bar{x}x}^{j+1} = F^j\,, \qquad F^j = \frac{y^j}{\tau} + \varphi^j\,,$$

or

$$y_{i-1}^{j+1} - (2 + \frac{h^2}{\tau})\, y_i^{j+1} + y_{i+1}^{j+1} = -h^2\, F_i^j\,, \qquad 0 < i < N\,,$$

$$y_0^{j+1} = \mu_1(t_{j+1})\,, \qquad y_N^{j+1} = \mu_2(t_{j+1})\,.$$

So far we have considered the first kind boundary conditions approximated exactly on grids. In the case of the third kind boundary conditions the question of their approximation needs investigation. In the next section we will say a little more about this.

**5. Convergence and accuracy of schemes.** While solving a problem by an approximate method the accuracy which is provided by this method should be properly evaluated and predicted before proceeding to further constructions. In this regard, the question of convergence and accuracy of difference schemes arises naturally.

For convenience in analysis, we look for in a domain $G$ with the boundary $\Gamma$ a solution to the linear differential equation

$$(35) \qquad L\,u = f(x)\,, \qquad x \in G\,,$$

subject to additional (boundary or initial) conditions

$$(36) \qquad l\,u = \mu(x)\,, \qquad x \in \Gamma\,,$$

where $f(x)$ and $\mu(x)$ are given functions (the input data of the problem) and $l$ is a linear differential operator, under the agreement that a solution of problem (35)–(36) exists and is unique.

Within the framework of the present chapter the domain $G + \Gamma$ of continuous variation of an argument (point) is replaced by some discrete set of points (nodes) $x_i$ known as a grid.

Let $h$ be a vector parameter related to the distribution density of the nodes of the grid and let $\omega_h$ and $\gamma_h$ be the sets of its inner and boundary nodes. With these ingredients, the difference problem

$$(37) \qquad L_h\, y_h = \varphi_h\,, \quad x \in \omega_h\,, \qquad l_h\, y_h = \chi_h \quad \text{for} \quad x \in \gamma_h\,,$$

where $\varphi_h(x)$ and $\chi_h(x)$ are given grid functions, is put in correspondence with problem (35)–(36). Here the operators $L_h$ and $l_h$ assign the values to grid functions defined for $x \in \bar{\omega}_h = \omega_h + \gamma_h$. A solution $y_h$ of problem (37) is a grid function of the nodes of the grid $\bar{\omega}_h$. Varying $h$ and thereby composing different grids $\bar{\omega}_h$, we constitute the set of solutions $\{y_h\}$ depending on the parameter $h$. In this context, the family of difference problems (37) corresponding to different values of the parameter $h$ is of great interest and refers to a **difference scheme**.

The main goal of any approximate method is to solve an original (continuous) problem with a prescribed accuracy $\varepsilon > 0$ in a finite number of operations. In order to clarify whether it is possible in principle to approximate a solution $u$ of problem (35)–(36) by a solution $y_h$ of problem (37) with any prescribed accuracy $\varepsilon > 0$ depending on the step $h(\varepsilon)$, we follow established practice. This is concerned with further comparison of $y_h$ with $u(x)$ in the space of grid functions $H_h$. Let $u_h$ be a value of the function $u(x)$ on the grid $\omega_h$, so that $u_h \in H_h$. The error $z_h = y_h - u_h$ of the difference scheme (37) needs more a detailed exploration for a complete and rigorous treatment.

The condition for $z_h$ can be derived upon substituting $y_h = z_h + u_h$ into (37). Through such an analysis the problem of the same type arises as problem (37):

$$L_h z_h = \psi_h, \quad x \in \omega_h, \qquad l_h z_h = \nu_h, \quad x \in \gamma_h, \tag{38}$$

where $\psi_h$ and $\nu_h$ are residuals equal to $\psi_h = \varphi_h - L_h u_h$ and $\nu_h = \chi_h - l_h u_h$.

The right-hand sides $\psi_h$ and $\nu_h$ of problem (38) are called the error of approximation of equation (35) by the difference equation (37) and the error of approximation of condition (36) by the difference condition $l_h y_h = \chi_h$ on a solution of problem (35)–(36) or, briefly, $\psi_h$ is the error of approximation for the equation $L_h y_h = \varphi_h$ on a solution $u(x)$ to equation (35) and $\nu_h$ is the error of approximation for the condition $l_h y_h = \chi_h$ on a solution of problem (35)–(36).

The accurate account of the error $z_h$ of a scheme as well as of the errors of approximations $\psi_h$ and $\nu_h$ is carried out in suitable norms $\|\cdot\|_{(1_h)}$, $\|\cdot\|_{(2_h)}$ and $\|\cdot\|_{(3_h)}$ on the space of grid functions.

We say that a solution of the difference problem (37) converges to a solution of problem (35)–(36) (scheme (37) is said to be convergent) if

$$\| z_h \|_{(1_h)} = \| y_h - u_h \|_{(1_h)} \longrightarrow 0 \quad \text{as} \quad |h| \to 0$$

or

$$\| z_h \|_{(1_h)} = \| \rho(|h|) \|, \quad \text{where} \quad \rho(|h|) \to 0 \quad \text{as} \quad |h| \to 0 .$$

The difference scheme (37) is said to be convergent with the rate $O(|h|^n)$ or is of the $n$th accuracy order (of accuracy $O(|h|^n)$) if for all sufficiently small $|h| \le h_0$ the inequality

$$\| z_h \|_{(1_h)} = \| y_h - u_h \|_{(1_h)} \le M \, |h|^n$$

holds with constant $M > 0$ independent of $|h|$ and $n > 0$ both.

We say the difference scheme (37) is of the $n$th approximation order if

$$\| \psi_h \|_{(2_h)} = O(|h|^n), \qquad \| \nu_h \|_{(3_h)} = O(|h|^n).$$

Denoting by $f_h$ and $(Lu)_h$ the values of $f(x)$ and $Lu(x)$ on the grid $\omega_h$ and taking into account that $(f - Lu)_h = 0$, we can rewrite $\psi_h$ as

$$\begin{aligned} \psi_h &= (\varphi_h - L_h u_h) - (f_h - (Lu)_h) \\ &= (\varphi_h - f_h) + ((Lu)_h - L_h u_h) \\ &= \psi_h^{(1)} + \psi_h^{(2)}, \end{aligned}$$

thus rearranging the approximation error $\psi_h$ of a scheme as a sum of the error of approximation $\psi_h^{(1)} = \varphi_h - f_h$ of the right-hand side and the error of approximation $\psi_h^{(2)} = (Lu)_h - L_h u_h$ of the differential operator.

Since $\psi_h$ is the error of approximation in the class of solutions to a differential equation, the condition $\| \psi_h \|_{(2_h)} = O(|h|^n)$ holds true if neither $\psi_h^{(1)}$ nor $\psi_h^{(2)}$ is of order $n$. An example in Section 1.3 confirms this statement.

We now should raise the question: how does the accuracy order of a scheme depend on the approximation order on a solution? Because the error $z_h = y_h - u_h$ solves problem (38) with the right-hand side $\psi_h$ (and $\nu_h$), the link between the order of accuracy and the order of approximation is stipulated by the character of dependence of the difference problem solution upon the right-hand side. Let $z_h$ depend on $\psi_h$ and $\nu_h$ continuously and uniformly in $h$. In other words, if a scheme is stable, its order of accuracy coincides with the order of approximation.

A rigorous definition of stability of a difference scheme will be formulated in the next section. The improvement of the approximation order for a difference scheme on a solution of a differential equation will be of great importance since the scientists wish the order to be as high as possible.

**6. The attainable order of approximation of a difference scheme.** As we have stated in Section 2.3, the order of approximation to a differential operator on a solution of a differential equation can be made higher without enlarging a pattern. For convenience in analysis, we take into consideration only two different difference schemes which will be associated with problem (32) for $u' = f(x)$ and $u(0) = u_0$. Our starting point is the difference scheme of the form

$$y_{\bar{x}} = \varphi, \qquad y_0 = u_0.$$

At the next stage we look for the residual $\psi = u_x - \varphi$ on a solution $u = u(x)$ to the equation $u' = f(x)$. The traditional way of covering this is to develop Taylor's series

$$u(x+h) = u(x) + h\,u' + \frac{h^2}{2}\,u'' + \frac{h^3}{6}\,u''' + O(h^4)$$

with the forthcoming substitutions $u' = f$, $u'' = f'$, ... and further reference to expansions

$$\begin{aligned} \psi &= u' + \frac{h}{2}\,u'' + \frac{h^2}{6}\,u''' - \varphi + O(h^3) \\ &= f(x) - \varphi(x) + \frac{h}{2}\,u'' + \frac{h^2}{6}\,u''' + O(h^3) \\ &= f(x) - \varphi(x) + \frac{h}{2}\,f'(x) + \frac{h^2}{6}\,u''' + O(h^3)\,. \end{aligned}$$

Upon substituting $\varphi = f(x) + \frac{1}{2}\,h\,f'$ or $\varphi_i = \left(f + \frac{1}{2}\,h\,f_x\right)_i$ we obtain the scheme of order 2 on the solution $u = u(x)$, thus demonstrating that the residual $\psi$ behaves like $O(h^2)$.

By virtue of the relations

$$u''' = f'' = f_{\bar{x}x} + O(h^2)$$

and

$$f' = f_x - \frac{1}{2}\,h\,f_{\bar{x}x} + O(h^2)$$

we find that

$$\psi = f + \frac{h}{2}\,f_x - \frac{h^2}{12}\,f_{\bar{x}x} + O(h^3) - \varphi\,,$$

thereby clarifying that the scheme $y_x = \varphi$ with the right-hand side

$$\varphi = f + \frac{h}{2}\,f_x - \frac{h^2}{12}\,f_{\bar{x}x}$$

is of order 3 on the solution $u = u(x)$ ($\psi = O(h^3)$).

For the boundary-value problem

$$u'' - q\,u = -f(x)\,, \qquad 0 < x < 1\,,$$

$$u(0) = 0\,, \quad u(1) = 0\,, \quad q = \text{const}\,,$$

we focus our attention on the three-point difference scheme

$$y_{\bar{x}x} - dy = -\varphi(x)\,, \qquad x = ih\,, \quad i = 1, 2, \dots N-1\,, \qquad y_0 = y_N = 0\,.$$

We will show that its order of approximation on a solution $u = u(x)$ can be made higher without enlarging a pattern under a proper choice of $d$ and $\varphi$.

For a solution $u = u(x)$ of the original problem, the residual $\psi = u_{\bar{x}x} - du + \varphi$ will be involved in the expansion

$$u_{\bar{x}x} = u'' + \frac{h^2}{12}\, u^{(4)} + O(h^4)$$

with regard to the equation $u'' = q\,u - f(x)$, permitting us to deduce that

$$\psi = (q-d)\,u + (\varphi - f) + \frac{h^2}{12}\, u^{(4)} + O(h^4)\,,$$

thus causing the behaviour like $O(h^2)$ for $d = q$ and $\varphi = f$. Substitution of $u^{(4)} = q''u - f'' = q\,(q\,u - f) - f_{\bar{x}x} + O(h^2)$ into the formula for $\psi$ yields

$$\psi = \left[\varphi - f - \frac{h^2}{12}\,(f_{\bar{x}x} + qf)\right] - \left[d - \left(q + \frac{q^2\,h^2}{12}\right)\right] u + O(h^4)\,.$$

A higher-order approximation $\psi = O(h^4)$ can be achieved on the solution $u(x)$ of the initial equation by merely setting

$$d = q + \frac{q^2\,h^2}{12}\,, \qquad \varphi = f + \frac{h^2}{12}\,(f_{\bar{x}x} + q\,f)\,.$$

We now turn to the question of approximations of boundary and initial conditions on a solution of the original problem. This question is intimately connected with the statement of a difference problem.

**7. Approximations of boundary and initial conditions.** From Section 2.5 it seems clear that the accuracy of a scheme depends on the order of approximation on a solution of the original problem. By an approximation we mean not only the equation itself, but also the supplementary (boundary and initial) conditions.

In this section we give several examples of raising the order of approximation for boundary and initial conditions without upgrading a pattern.

**Example 1.** *The third boundary-value problem for an ordinary second-order differential equation:*

$$(39)\qquad \begin{aligned} &\frac{d^2 u}{dx^2} - q\,u = -f(x)\,, \qquad q = \text{const}\,, \quad 0 < x < 1\,, \\ &\frac{du(0)}{dx} = \sigma\,u(0) - \mu_1, \qquad u(1) = \mu_2\,. \end{aligned}$$

On the equidistant grid $\bar{\omega}_h = \{x_i = ih,\ 0 \leq i \leq N\}$ we may attempt the difference equation in the form

$$(40)\qquad y_{\bar{x}x} - q\,y = -\varphi$$

with $\varphi_i = f(x_i)$, where $f(x)$ is a continuous function.

At the point $x = 1$ the boundary condition is satisfied exactly:

$$(41)\qquad y(1) = y_N = \mu_2\,.$$

Furthermore, we replace the first derivative $u'(0)$ by the first difference derivative $y_{x,0} = (y_1 - y_0)/h$ and impose the boundary condition at the point $x = 0$ such as

$$(42)\qquad y_{x,\,0} = \sigma\,y_0 - \mu_1 \qquad \text{or} \qquad l_h\,y = \mu_1\,,$$

where an operator $l_h$ is defined on the two-point pattern $(0, h)$. Substituting here $y = z + u$, where $u$ is a solution of problem (29), we establish for the error $z$ the condition

$$z_{x,\,0} = \sigma\,z_0 - \nu_1\,,$$

where the error of approximation $\nu_1$ for the boundary condition on a solution is equal to $\nu_1 = \mu_1 + u_{x,0} - \sigma u_0$.

Developing $u(x)$ about the node $x = 0$ in Taylor's series

$$u_1 = u_0 + h\,u_0' + \frac{h^2}{2}\,u_0'' + O(h^3)\,,$$

we find that

$$(43)\qquad \begin{aligned} u_{x,0} &= u_0' + \tfrac{1}{2}\,u_0'' + O(h^2)\,, \\ \nu_1 &= \big[\mu_1 + u'(0) - \sigma\,u(0)\big] + \tfrac{1}{2}\,h\,u''(0) + O(h^2) \\ &= \tfrac{1}{2}\,h\,u''(0) + O(h^2)\,, \end{aligned}$$

since $\mu_1 + u'(0) - \sigma\, u(0) = 0$. This provides support for the view that $\nu_1 = O(h)$. It is necessary to make some changes in condition (42) to achieve the order of approximation $O(h^2)$. This is due to the fact that $u(x)$ is just the solution of the original problem (39). From the governing differential equation the value $u''(0)$ can be expressed by

$$u''(0) = q\, u(0) - f(0)\,. \tag{44}$$

Substitution of (44) into (43) yields

$$u_{x,0} - \tfrac{1}{2}\, h\, \big(q\, u(0) - f(0)\big) = u'(0) + O(h^2)\,, \tag{45}$$

thereby justifying that the left-hand side of (45) approximates to second order the derivative $u'(x)$ at the point $x = 0$ on the solution to the equation $u'' - q\, u = -f$.

From here and (42) it follows that the approximation of the boundary condition

$$y_{x,0} = \bar{\sigma}\, y_0 - \bar{\mu}_1\,, \qquad \bar{\sigma} = \sigma + \tfrac{1}{2}\, h\, q, \qquad \bar{\mu}_1 = \mu_1 + \tfrac{1}{2}\, h\, f(0)\,, \tag{46}$$

is of order 2 on a solution of problem (39).

It is worth emphasizing here that we have succeeded in raising the order of approximation without enlarging the total number of grid nodes which will be needed in this connection for approximating the boundary condition.

**Example 2.** ***The boundary-value problem for the heat conduction equation:***

$$\begin{gathered}
\frac{\partial u}{\partial t} = \frac{\partial^2 u}{\partial x^2} + f(x,t)\,, \qquad 0 < x < 1\,, \quad 0 < t \le t_0\,, \\
u(x,0) = u_0(x)\,, \\
\frac{\partial u(0,t)}{\partial x} = \sigma\, u(0,t) - \mu_1(t)\,, \qquad u(1,t) = \mu_2(t)\,.
\end{gathered} \tag{47}$$

On the grid $\omega_{h\tau}$ arising from Section 2.1 it is simple to follow the explicit scheme of accuracy $O(h^2 + \tau)$:

$$y_t = y_{\bar{x}x} + \varphi\,, \qquad y(x,0) = u_0(x)\,, \quad y(1,t) = \mu_2(t)\,, \tag{48}$$

where $\varphi = \varphi_i^j = f(x_i, t_j)$. In giving the difference approximation of the same order for the boundary condition at the point $x = 0$ we have occasion to use

$$u_{x,0} = \frac{\partial u(0,t)}{\partial x} + \frac{h}{2}\,\frac{\partial^2 u(0,t)}{\partial x^2} + O(h^2)\,.$$

By having recourse to the heat conduction equation for $x = 0$ we establish a precise relationship

$$\frac{\partial^2 u(0,t)}{\partial x^2} = \frac{\partial u(0,t)}{\partial t} - f(0,t)\,,$$

which implies that

$$u_x(0,t) - \frac{h}{2}\left(\frac{\partial u(0,t)}{\partial t} - f(0,t)\right) = \frac{\partial u(0,t)}{\partial x} + O(h^2)\,.$$

The expression on the left-hand side of this equality approximates the derivative $\partial u/\partial x$ for $x = 0$ to $O(h^2)$. Replacing $\frac{\partial u}{\partial t}\big|_{x=0}$ by the appropriate difference derivative

$$u_{t,0} = \frac{u(0,t+\tau) - u(0,t)}{\tau}$$

we impose the difference boundary condition at the point $x = 0$:

$$\text{(49)} \qquad y_{x,0} = \tfrac{1}{2}\,h\,y_{t,0} + \sigma\,y_0 - \bar{\mu}_1\,, \qquad \bar{\mu}_1 = \mu_1 + \tfrac{1}{2}\,h\,f(0,t)\,,$$

whose approximation on a solution of problem (47) is of accuracy $O(h + \tau^2)$. In the case of the implicit scheme $y_t = \hat{y}_{\bar{x}x} + \varphi$ the following condition is acceptable to be an alternative:

$$\text{(50)} \qquad \hat{y}_{x,0} = \tfrac{1}{2}\,h\,y_{t,0} + \sigma\,\hat{y}_0 - \hat{\bar{\mu}}_1\,, \qquad \bar{\mu}_1 = \mu_1 + \tfrac{1}{2}\,h\,f(0,t)\,.$$

**Example 3.** *The second-order hyperbolic equation:*

$$\text{(51)} \qquad \frac{\partial^2 u}{\partial t^2} = \frac{\partial^2 u}{\partial x^2} + f(x,t)\,, \qquad 0 < x < 1\,, \quad 0 < t \le t_0\,,$$

$$u(0,t) = u_1(t)\,, \quad u(1,t) = u_2(t)\,,$$

$$u(x,0) = u_0(x)\,, \quad \frac{\partial u(0,t)}{\partial x} = \bar{u}_0(x)\,.$$

Evidently, a special attention in approximating problem (51) is being paid to the difference form of writing the initial condition for the derivative $\partial u/\partial t$. On any equidistant in $x$ and $t$ grid $\omega_{h\tau}$ with steps $h$ and $\tau$ (see Section 2.1) the simplest approximation $u_t(x,0) = \bar{u}_0(x)$ gives the error of approximation $O(\tau)$. Plain calculations show that

$$u_t(x,0) = \frac{u(x,\tau) - u(x,0)}{\tau}$$

$$= \frac{\partial u(x,0)}{\partial t} + \frac{\tau}{2}\,\frac{\partial^2 x(x,0)}{\partial t^2} + O(\tau^2).$$

From the governing differential equation it follows that

$$\frac{\partial^2 u(x,0)}{\partial t^2} = \frac{\partial^2 u(x,0)}{\partial x^2} + f(x,0) = L\,u_0(x) + f(x,0),$$

$$L\,u_0 = \frac{d^2 u_0}{dx^2}.$$

This is due to the fact that

$$\frac{\partial^2 u(x,0)}{\partial x^2} = \frac{d^2 u_0(x)}{dx^2}.$$

In this line,

$$u_t(x,0) - \tfrac{1}{2}\tau\,(L\,u_0 + f(x,0)) = \frac{\partial u(x,0)}{\partial t} + O(\tau^2).$$

Therefore, the difference initial condition $y_t(x,0) = \tilde{u}_0(x)$, where

$$\tilde{u}_0(x) = \bar{u}(x) + \frac{1}{2}\,\tau(L\,u_0 + f(x,0)),$$

approximates to second order in $\tau$ the condition $\partial u(x,0)/\partial t = \bar{u}_0(x)$ on the solution of problem (41).

In this case the condition $u(x,0) = u_0(x)$ and the boundary conditions are approximated exactly. For instance, one of the schemes arising in Section 1.2 is good enough for the difference approximation of the initial equation. No doubt, we preassumed not only the existence and continuity of the derivatives involved in the equation on the boundary of the domain in view (at $x = 0$ or $t = 0$), but also the existence and boundedness of the third derivatives of a solution for raising the order of approximation of boundary and initial conditions.

**Example 4.** ***The three-layer difference scheme for the heat conduction equation.*** A special attention is being paid to the first boundary-value problem

$$(52)\qquad \frac{\partial u}{\partial t} = \frac{\partial^2 u}{\partial x^2} + f(x,t), \qquad 0 < x < 1, \quad 0 < t \le t_0,$$

$$u(0,t) = u_1(t), \quad u(1,t) = u_2(t), \qquad u(x,0) = u_0(x).$$

The usual practice in numerical analysis of the heat conduction equation (52) is connected with three-layer schemes. The values $y^{j-1}(x)$, $y^j(x)$ and $y^{j+1}(x)$ of a grid function on the three time layers $t_{j-1}$, $t_j$ and $t_{j+1}$ are aimed at constructing such schemes.

The three-layer symmetric scheme on the equidistant grid $\omega_{h\tau}$ with steps $h$ and $\tau$, being the most familiar one, comes first:

$$(53)\qquad \frac{y^{j+1} - y^{j-1}}{2\tau} = \Lambda\left(\sigma y^{j+1} + (1-2\sigma)\, y^j + \sigma y^{j-1}\right) + \varphi^j,$$

$$y_i^0 = u_0(x_i), \quad y_0^j = u_1^j, \quad y_N^j = u_2^j,$$

where $\Lambda y = y_{\bar{x}x}$, $\sigma$ is a real parameter and $\varphi^j = f(x_i, t_j)$.

Since the central difference derivative in $t$ approximates to the second order in $\tau$ $\frac{\partial u}{\partial t}\big|_{t=t_j}$ and $\Delta u = \frac{\partial^2 u}{\partial x^2} + O(h^2)$, scheme (53) approximates equation (52) with accuracy $O(h^2 + \tau^2)$. From what has been said above it is clear that problem (53) is overdetermined as it were. Applications of the three-layer scheme concerned necessitate imposing one more initial condition, for instance, by doing this on the first layer. But under such a condition the approximation $O(\tau^2 + h^2)$ should remain unchanged. There seem to be at least two ways of determining $y(x,\tau)$. One way of proceeding is to approve at the first step the two-layer scheme of accuracy $O(\tau^2 + h^2)$ in specifying $y(x,\tau)$:

$$\frac{y^1 - y^0}{\tau} = \frac{1}{2}\,\Lambda\,(y^1 + y^0) + \varphi^0.$$

Under the second approach the value of $y(x,\tau)$ arranges itself as a sum $y(x,\tau) = u_0(x) + \tau\mu(x)$ and $\mu$ is so chosen as to obtain the error $y(x,\tau) - u(x,\tau)$, not exceeding $O(\tau^2 + h^2)$. Substitution of the value $\frac{\partial u}{\partial t}\big|_{t=0}$, arising from the differential equation

$$\frac{\partial u}{\partial t}\Big|_{t=0} = L\,u_0 + f(x,0), \qquad L\,u_0 = \frac{d^2 u_0}{dx^2},$$

into the formula

$$u(x,\tau) - u_0(x) = \tau\, \frac{\partial u}{\partial t}\bigg|_{t=0} + \frac{\tau^2}{2}\, \frac{\partial^2 u}{\partial t^2}\bigg|_{t=0} + O(\tau^3)$$

yields $\mu = L\,u_0 + f(x,0)$, so that

$$y(x,\tau) = u_0(x) + \tau\,\left(u_0''(x) + f(x,0)\right).$$

## 2.3 STABILITY OF A DIFFERENCE SCHEME

**1. Examples of stable and unstable difference schemes.** As we have shown in the preceding section, the introduction of a difference scheme permits one to reduce the solution of a problem associated with a differential equation to a system of linear algebraic equations. In this situation the right-hand sides of equations, boundary and initial conditions, all of which we will call in the sequel the **input data**, are specified with a certain error. This procedure causes rounding errors to be inevitable in the process of numerical solution of a system. Because of this, it seems natural to require that small errors in specifying the input data should not increase during the process of the execution and result in wrong reasoning. The schemes in which initial errors grow during the course of calculations turn out to be unstable and, from the viewpoint of possible applications, cannot find response.

We will not attempt to encompass a wide variety of situations, but instead look in more detail at several exhaustive examples before formulating the definition of stability of a difference scheme with respect to the input data, the concept of which we have intuitively developed earlier.

**Example 1.** ***A stable scheme.*** Let

$$(1) \qquad u' = -\alpha\, u\,, \quad x > 0\,, \quad u(0) = u_0\,, \quad \alpha > 0\,.$$

It is straightforward to verify that the function $u(x) = u_0 \exp\{-\alpha x\}$ gives the exact solution of problem (1). This solution does not increase with increasing $x$: $|\,u(x)\,| < |\,u_0\,|$ for $\alpha > 0$, so that $u(x)$ continuously depends on $u_0$. An excellent start in this direction is to approximate problem (1) on the equidistant grid $\omega_h = \{x_i = ih,\ i = 0, 1, \dots\}$ by the difference problem

$$(2) \qquad \frac{y_i - y_{i-1}}{h} + \alpha\, y_i = 0\,, \qquad y_0 = u_0\,, \qquad i = 1, 2, \dots\,,$$

which can be rewritten as

$$y_i = s\, y_{i-1}\,, \qquad s = \frac{1}{1+\alpha h}\,, \qquad i = 1, 2, \dots, \qquad y_0 = u_0\,,$$

yielding $y_i = s^i y_0$.

We regard a point $\bar{x}$ to be fixed and take a sequence of steps $h$ so that $\bar{x}$ would always belong the set of grid nodes: $\bar{x} = i_0 h$. The attached number $i_0$ may be made arbitrarily large once we will refine the grid in any convenient way, that is, letting $h \to 0$. The value of $y$ at this point becomes

$$y(\bar{x}) = y_{i_0} = s^{i_0}\, y_0\,.$$

Since $|\,s\,| < 1$ for $\alpha > 0$ and any $h$, we thus have

$$|\,y(\bar{x})\,| \le |\,s\,|^{i_0}\, |\,y_0\,| < |\,y(0)\,|$$

for any $h$.

The last inequality implies that the solution of the difference problem (2) continuously depends on the input data. In such cases we say that a difference scheme is stable with respect to the input data.

**Example 2.** ***An unstable scheme.*** For problem (1) we rely on the scheme

$$\text{(3)} \qquad \begin{aligned} &\sigma\, \frac{y_i - y_{i-1}}{h} + (1-\sigma)\, \frac{y_{i+1} - y_i}{h} + \alpha\, y_i = 0\,, \\ &y_0 = u_0\,, \qquad y_1 = \bar{u}_0\,, \qquad i = 1, 2, \dots, \end{aligned}$$

where $\sigma > 1$ is a numerical parameter. Observe that this scheme is a three-point one, since the difference equation is of order 2, so there is some reason to assign the value $y_1$ in addition to the value $y_0$. The approximation order of scheme (3) is no less than 1 regardless of the choice of the parameter $\sigma$. With $\bar{u}_0 = (1-\alpha h)u_0$, a proper evaluation of the deviation gives $\bar{u} - u(h) = O(h^2)$. We look for particular solutions to the difference equation (3) in the form $y_i = s^i$. Substituting $y_i = s^i$ into (3) gives the quadratic equation related to an unknown $s$:

$$\text{(4)} \qquad (\sigma - 1)\, s^2 - (2\sigma - 1 + \alpha h)\, s + \sigma = 0\,,$$

which possesses two distinct roots

$$s_{1,2} = \frac{2\,\sigma - 1 + \alpha h \pm \sqrt{1 + 2\,(2\sigma - 1)\,\alpha h + \alpha^2 h^2}}{2\,(\sigma - 1)}\,.$$

It is well-known that the general solution to equation (3) is of the form

$$y_i = A\,s_1^i + B\,s_2^i\,. \tag{5}$$

Putting $i = 0$ and $i = 1$ and using the assigned values $y_0 = u_0$ and $y_1 = \bar{u}_0$, it is easy to calculate the constants $A$ and $B$:

$$A = \frac{\bar{u}_0 - s_2\,u_0}{s_1 - s_2}\,, \qquad B = \frac{s_1\,u_0 - \bar{u}_0}{s_1 - s_2}\,.$$

It is clear that $s_1\,s_2 > 1$, since $\sigma > \sigma - 1 > 0$. We claim that $s_2 < 1$ for any value of $\alpha h$. Indeed, for $\sigma > 1$

$$2\,(\sigma - 1) - (2\sigma - 1 + \alpha h - \sqrt{1 + 2\,(2\sigma - 1)\,\alpha h + \alpha^2 h^2})$$

$$= \sqrt{(1 + \alpha h)^2 + 4\,(\sigma - 1)\,\alpha h} - (1 + \alpha h) > 0\,.$$

Let us stress also that $s_1 > 1$ for any value of $\alpha h$ on account of the bound $s_1\,s_2 > 1$ in view.

From the well-established representation $y_i = A\,s_1^i + B\,s_2^i$ it is easily seen that $y_i \to \infty$ as $i \to \infty$ if $A \neq 0$. But we can always select $y_1 = \bar{u}_0$ so that the condition $A = 0$ holds true for the choice $\bar{u}_0 = u_0 s_2$. The rounding errors arise inevitably in developing the solution $s_1^i$, thus causing the instability of the indicated type. In this scheme the solution increases along with increasing $x_i = ih$ if $h$ is kept fixed. Successive grid refinement, that is, successive refinement of $h$, leads to the growth of errors at a fixed point $\bar{x} = i_0\,h$, since $i_0 = \bar{x}/h$ increases along with decreasing $h$. A small change of input data results in an enormous change in the solution of the problem at any fixed point $\bar{x}$ as $h \to 0$.

We quote below the results of computations for problem (3) with $y_0 = 1$ and $y_1 = s_2$, where $s_2$ is the smallest root to the quadratic equation (4). Once supplemented with those initial conditions, the exact solution of problem (3) takes the form $y_i = Bs_2^i$ $(A = 0)$. Because of rounding errors, the first summand emerged in formula (5). This member increases along with increasing $i$, thus causing abnormal termination in computational procedures.

Modern computers allow the implementation of model problems. We have carried out the calculations for several variants:

1) $\sigma = 1.1, \quad \alpha h = 0.01, \quad s_1 = 11.11, \quad s_2 = 0.99;$

2) $\sigma = 1.1, \quad \alpha h = 0.1, \quad s_1 = 12.09, \quad s_2 = 0.91;$

3) $\sigma = 2, \quad \alpha h = 0.01, \quad s_1 = 2.02, \quad s_2 = 0.99;$

4) $\sigma = 2, \quad \alpha h = 0.1, \quad s_1 = 2.17, \quad s_2 = 0.92.$

Table 1 contains the values $y_i$ at several nodes of the grid $\omega_h$ for the four variants. In the final line of this table the numbers of grid nodes connected with abnormal termination are indicated by "infinitiy".

**Table 1**

| Variant 1 | | Variant 2 | | Variant 3 | | Variant 4 | |
|---|---|---|---|---|---|---|---|
| $i$ | $y_i$ | $i$ | $y_i$ | $i$ | $y_i$ | $i$ | $y_i$ |
| 6 | 0.952 | 7 | 0.567 | 32 | 0.724 | 28 | $8.97 \cdot 10^{-2}$ |
| 7 | 0.942 | 8 | 0.516 | 33 | 0.703 | 29 | $7.87 \cdot 10^{-2}$ |
| 11 | 0.900 | 10 | 0.420 | 37 | 0.260 | 32 | $3.78 \cdot 10^{-3}$ |
| 12 | 0.832 | 11 | 0.306 | 38 | $-0.196$ | 33 | $-7.41 \cdot 10^{-2}$ |
| 13 | 0.174 | 12 | $-0.641$ | 39 | $-1.11$ | 34 | $-0.237$ |
| 14 | $-7.05$ | 13 | $-1.17 \cdot 10^{1}$ | 40 | $-2.95$ | 40 | $-3.17 \cdot 10^{1}$ |
| 15 | $-8.72 \cdot 10^{1}$ | 14 | $-1.45 \cdot 10^{2}$ | 50 | $-4.10 \cdot 10^{3}$ | 50 | $-7.84 \cdot 10^{4}$ |
| 16 | $-9.77 \cdot 10^{2}$ | 15 | $-1.76 \cdot 10^{3}$ | 60 | $-4.64 \cdot 10^{6}$ | 60 | $-1.94 \cdot 10^{8}$ |
| 20 | $-1.49 \cdot 10^{7}$ | 20 | $-4.54 \cdot 10^{8}$ | 80 | $-5.92 \cdot 10^{12}$ | 80 | $-1.19 \cdot 10^{15}$ |
| 25 | $-2.52 \cdot 10^{12}$ | 25 | $-1.17 \cdot 10^{14}$ | 90 | $-6.69 \cdot 10^{15}$ | 90 | $-2.94 \cdot 10^{18}$ |
| 32 | $\infty$ | 30 | $\infty$ | 100 | $\infty$ | 92 | $\infty$ |

**2. The Cauchy problem for a system of differential equations of first order. Stability condition for Euler's scheme.** We illustrate those ideas with concern of the Cauchy problem for the system of differential equations of first order

$$\frac{d\mathbf{u}}{dt} + A\,\mathbf{u} = 0\,, \qquad t > 0\,, \qquad \mathbf{u}(0) = \mathbf{u}_0\,, \tag{6}$$

where $\mathbf{u} = \left(u^{(1)}, u^{(2)}, \ldots, u^{(n)}\right)$ is the vector of unknowns, $\mathbf{u}_0 = \left(u_0^{(1)}, u_0^{(2)}, \ldots, u_0^{(n)}\right)$ is an $n$-dimensional given vector and $A = (a_{ij})$ is a symmetric positive definite $n \times n$-matrix. In what follows a vector space $H_n$ is equipped with the inner product $(\mathbf{u}, \mathbf{v}) = \sum_{i=1}^{n} u^{(i)} v^{(i)}$ and associated norm $\|\mathbf{u}\| = \sqrt{(\mathbf{u}, \mathbf{u})}$. Under these structures, a linear self-adjoint operator $A\colon H_n \mapsto H_n$, $A = A^* > 0$, is associated with the matrix $A$ of interest.

We denote by $\{\lambda_k\,, \boldsymbol{\xi}_k\}$ the system of eigenvalues and orthonormal eigenvectors of the operator $A$:

$$A\,\boldsymbol{\xi}_k = \lambda_k\,\boldsymbol{\xi}_k\,, \qquad k = 1, 2, \ldots, n\,,$$

so that

$$0 < \lambda_1 \le \lambda_2 \le \cdots \le \lambda_n , \qquad (\boldsymbol{\xi}_k , \boldsymbol{\xi}_s) = \delta_{ks} = \begin{cases} 1, & k = s , \\ 0, & k \ne s . \end{cases}$$

With these entries, we look for a solution of problem (6) in the form

$$\mathbf{u}(t) = \sum_{s=1}^{n} \alpha_s(t)\, \boldsymbol{\xi}_s .$$

Substitution of this expression into equation (6) yields

$$\sum_{s=1}^{n} \left( \frac{d\alpha_s}{dt} + \lambda_s \alpha_s \right) \boldsymbol{\xi}_s = 0 ,$$

which implies that

$$\frac{d\alpha_k}{dt} + \lambda_k \alpha_k = 0 , \quad \alpha_k(t) = \alpha_k(0) \exp\{-\lambda_k t\} , \quad k = 1, 2, \ldots , n ,$$

providing the same grounds for the series

$$\mathbf{u}(t) = \sum_{k=1}^{n} \alpha_k(0) \exp\{-\lambda_k t\}\, \boldsymbol{\xi}_k . \tag{7}$$

Under the initial condition

$$\mathbf{u}(0) = \mathbf{u}_0 = \sum_{k=1}^{n} \alpha_k(0)\, \boldsymbol{\xi}_k ,$$

we arrive at the relations

$$\alpha_k(0) = (\mathbf{u}_0 , \boldsymbol{\xi}_k) , \qquad \| \mathbf{u}_0 \|^2 = \sum_{k=1}^{n} \alpha_k^2(0) .$$

From (7) it follows that

$$\begin{aligned} \| \mathbf{u}(t) \|^2 &= \sum_{k=1}^{n} \sum_{s=1}^{n} \alpha_k(0)\, \alpha_s(0) \exp\{-\lambda_k t\} \exp\{-\lambda_s t\} (\boldsymbol{\xi}_k , \boldsymbol{\xi}_s) \\ &= \sum_{k=1}^{n} \alpha_k^2(0) \exp\{-2\lambda_k t\} \le \exp\{-2\lambda_1 t\} \sum_{k=1}^{n} \alpha_k^2(0) \\ &= \exp\{-2\lambda_1 t\} \| \mathbf{u}_0 \|^2 , \end{aligned}$$

which implies that

$$(8)\qquad \| \mathbf{u}(t) \| \le \exp\{-\lambda_1 t\} \| \mathbf{u}_0 \|^2 .$$

From such reasoning it seems clear that the solution of the Cauchy problem decreases along with increasing $t$:

$$(9)\qquad \| \mathbf{u}(t) \| \le \| \mathbf{u}_0 \| \quad \text{for} \quad t > 0 .$$

Of special interest is Euler's scheme for the Cauchy problem

$$(10)\qquad \frac{\mathbf{y}_{j+1} - \mathbf{y}_j}{\tau} + A\,\mathbf{y}_j = 0 , \qquad j = 0, 1, \dots , \qquad \mathbf{y}_0 = \mathbf{u}_0 ,$$

where $\mathbf{y}_j = \mathbf{y}(t_j)$ and $t_j = j\tau$. We seek its solution as a sum

$$(11)\qquad \mathbf{y}_j = \sum_{k=1}^{n} a_k\, q_k^j\, \xi_k ,$$

so that $\mathbf{y}_0 = \mathbf{u}_0 = \sum_{k=1}^{n} a_k\, \xi_k$ and $a_k = (\mathbf{u}_0 , \xi_k)$. We note in passing that expression (11) satisfies equation (10) only if

$$\sum_{k=1}^{n} a_k \left( \frac{q_k - 1}{\tau} + \lambda_k \right) q_k^j\, \xi_k = 0$$

or, what amounts to the same,

$$\frac{q_k - 1}{\tau} + \lambda_k = 0 .$$

It follows from the foregoing that $q_k = 1 - \tau \lambda_k$.

By rearranging the norm as

$$\| y_j \|^2 = \sum_{k=1}^{n} | q_k^j |^2\, a_k^2$$

it is easy to derive the chain of the relations

$$\| \mathbf{y}_j \|^2 \le \max_k | q_k^j |^2 \sum_{k=1}^{n} a_k^2 = \max_k | q_k^j |^2 \, \| \mathbf{y}_0 \|^2 < \| \mathbf{y}_0 \|^2 ,$$

which are valid under the conditions $\max_k | q_k | < 1$ and $\| \mathbf{y}_j \|^2 > \| \mathbf{y}_0 \|^2$ if $\min_k | q_k | > 1$. Being an alternative of (9), the inequality

$$(12)\qquad \| \mathbf{y}_j \| < \| \mathbf{y}_0 \|$$

expresses the stability of the problem (10) solution.

The condition $\max_k |q_k| < 1$ is fulfilled if $-1 < q_k = 1 - \tau\lambda_k < 1$, that is, $\tau\lambda_k < 2$ for all $k = 1, 2, \dots, n$. For the latter to hold, it is sufficient that

$$\tau < \frac{2}{\Delta}, \tag{13}$$

where $\Delta = \max_k \lambda_k = \lambda_n$.

By a rather similar argument we take $\mathbf{y}_0 = c\,\boldsymbol{\xi}_n$, leaving us with $a_k = 0$ for $k = 1, 2, \dots, n-1$, $a_n = c$ and the relations

$$\mathbf{y}_j = c\,q_n^j\,\boldsymbol{\xi}_n = q_n^j\,\mathbf{y}_0\,, \qquad \|\mathbf{y}_j\| = |q_n|^j\,\|\mathbf{y}_0\| > \|\mathbf{y}_0\|,$$

since $|q_n| = \tau\lambda_n - 1 = \tau\Delta - 1 > 1$ if $\tau\Delta > 2$. For instance, when $\tau\Delta = 2 + p$, $p > 0$, we draw the conclusions that $|q_n| = 1 + p$ and $|q_n^j| = (1+p)^j \to \infty$ as $j \to \infty$. In that case scheme (10) becomes unstable and makes the method so inefficient as to be unusable.

In mastering the difficulties involved, we impose the initial condition $\mathbf{y}_0 = c\,\boldsymbol{\xi}_1$, thereby providing

$$\mathbf{y}_j = q_1^j\,\mathbf{y}_0\,, \qquad \|\mathbf{y}_j\| = |q_1|^j\,\|\mathbf{y}_0\|,$$

where $q_1 = 1 - \tau\lambda_1$.

It may happen that $\tau > 2/\Delta$, but $\tau\lambda_1 < 2$ and $|q_1| < 1$, which assures us of the validity of the estimate

$$\|\mathbf{y}_j\| \le |q_1|^j\,\mathbf{y}_0 < \|\mathbf{y}_0\|. \tag{14}$$

Because of rounding errors, $\mathbf{y}_0$ is determined with some error $\boldsymbol{\varepsilon}$ still subject to the approved decomposition

$$\boldsymbol{\varepsilon} = \sum_{k=1}^{n} \varepsilon_k\,\boldsymbol{\xi}_k\,, \qquad \varepsilon_k = (\boldsymbol{\varepsilon}, \boldsymbol{\xi}_k),$$

which is not surprising. No matter how the value $\varepsilon_n$ is chosen, there always exists $j_0$ such that $|\varepsilon_n|\,|q_n|^{j_0} > M_\infty$, where $M_\infty$ is computer infinity, meaning that for $j = j_0$ abnormal termination occurs because $|q_n| > 1$. Thus, scheme (10) turns out to be unstable with respect to rounding errors under any initial condition in the case where $\tau\Delta > 2$.

For the solution $\mathbf{y}_j$ of problem (10), the requirement of having an estimate similar to inequality (8) necessitates imposing one more condition $1 - \tau\lambda_1 > \tau\lambda_n - 1$ or, what amounts to the same,

$$\tau < \frac{2}{\lambda_1 + \lambda_n}.$$

Assuming this to be the case, we deduce in light of the obvious relations $1 - \tau\lambda_1 < \exp\{-\lambda_1 \tau\}$ and $|q_1| > |q_k|$ for $k > 1$ that

$$\|\mathbf{y}_j\| \le \max_k |q_k|^j \|\mathbf{y}_0\| \le (1-\tau\lambda_1)^j \|\mathbf{y}_0\| < \exp\{-\lambda_1 t_j\} \|\mathbf{y}_0\|.$$

We have here one of many examples reinforcing the view that the general ideas of stability are sensible. Consider a system of two equations

$$\frac{du}{dt} + au + bv = 0, \qquad \frac{dv}{dt} + bu + av = 0$$

with the matrix $A = \begin{pmatrix} a & b \\ b & a \end{pmatrix}$, where $a = \frac{1}{2}(\Delta + \delta)$ and $b = \frac{1}{2}(\Delta - \delta)$. It is straightforward to verify that its eigenvalues and eigenvectors are equal to

$$\lambda_1 = \delta, \quad \lambda_2 = \Delta, \qquad \xi_1 = \left(\tfrac{1}{\sqrt{2}}, -\tfrac{1}{\sqrt{2}}\right), \quad \xi_2 = \left(\tfrac{1}{\sqrt{2}}, \tfrac{1}{\sqrt{2}}\right).$$

With these entries, scheme (10) reduces to

$$(15) \qquad \begin{aligned} &\frac{y_{j+1} - y_j}{\tau} + a\, y_j + b\, z_j = 0, \\ &\frac{z_{j+1} - z_j}{\tau} + b\, y_j + a\, z_j = 0. \end{aligned} \qquad j = 0, 1, 2, \dots,$$

Observe that the vector $\{y_0, z_0\}$ coincides with the first eigenvector $\xi_1$ if we agree to consider

$$(16) \qquad y_0 = \tfrac{1}{\sqrt{2}}, \qquad z_0 = -\tfrac{1}{\sqrt{2}}.$$

After scrutinising the available information we initiate the review of final results of calculations for problem (15)–(16) with the following values of parameters:

$$1)\quad \delta = 1, \quad \Delta = 2, \quad \tau = \frac{3}{\Delta};$$

$$2)\quad \delta = 10, \quad \Delta = 400, \quad \tau = \frac{6}{\Delta}.$$

In both cases $|1 - \tau\delta| < 1$ and, therefore, for a solution of problem (15)–(16) estimate (14) is valid, but $\tau > 2/\Delta$. Because of rounding errors, the computational process is unstable: for large $j$ the growth of its solution causes abnormal termination in the computer realization of the algorithm.

The final results of calculations are tabulated in Table 2 with the values of the functions $y_j$ and $z_j$ for several $j$. For variants 1 and 2 abnormal terminations have occurred at the 105th and 46th time steps, respectively.

**Table 2**

| Variant 1 | | | Variant 2 | | |
|---|---|---|---|---|---|
| $j$ | $y_j$ | $z_j$ | $j$ | $y_j$ | $z_j$ |
| 8 | $-5.52 \cdot 10^{-3}$ | $5.52 \cdot 10^{-3}$ | 8 | 0.227 | 0.227 |
| 9 | $2.76 \cdot 10^{-3}$ | $-2.76 \cdot 10^{-3}$ | 8 | 0.193 | −0.193 |
| 10 | $-1.38 \cdot 10^{-3}$ | $1.38 \cdot 10^{-3}$ | 10 | 0.164 | −0.164 |
| 17 | $1.08 \cdot 10^{-5}$ | $-1.08 \cdot 10^{-5}$ | 13 | 0.101 | −0.101 |
| 18 | $-5.46 \cdot 10^{-6}$ | $5.33 \cdot 10^{-6}$ | 14 | $8.49 \cdot 10^{-2}$ | $-8.61 \cdot 10^{-2}$ |
| 19 | $2.82 \cdot 10^{-6}$ | $-2.57 \cdot 10^{-6}$ | 15 | $7.56 \cdot 10^{-2}$ | $-6.96 \cdot 10^{-2}$ |
| 20 | $-1.60 \cdot 10^{-6}$ | $1.09 \cdot 10^{-6}$ | 16 | $4.69 \cdot 10^{-2}$ | $-7.65 \cdot 10^{-2}$ |
| 25 | $8.18 \cdot 10^{-6}$ | $8.10 \cdot 10^{-6}$ | 17 | 0.127 | $2.16 \cdot 10^{-2}$ |
| 26 | $-1.63 \cdot 10^{-5}$ | $-1.62 \cdot 10^{-5}$ | 18 | −0.326 | −0.415 |
| 27*) | $3.26 \cdot 10^{-6}$ | | 19 | 1.89 | 1.81 |
| 28 | $-6.51 \cdot 10^{-5}$ | | 20 | −9.23 | −9.30 |
| 30 | $-2.60 \cdot 10^{-4}$ | | 21*) | $4.63 \cdot 10^{1}$ | |
| 50 | $-2.73 \cdot 10^{2}$ | | 22 | $-2.32 \cdot 10^{2}$ | |
| 68 | $-7.16 \cdot 10^{7}$ | | 38 | $-3.53 \cdot 10^{13}$ | |
| 69 | $1.43 \cdot 10^{8}$ | | 39 | $1.77 \cdot 10^{14}$ | |
| 70 | $-2.86 \cdot 10^{8}$ | | 40 | $-8.83 \cdot 10^{14}$ | |
| 105 | $\infty$ | | 46 | $\infty$ | |

*) Once started with this number $j$ we have $z_j = y_j$ .

Observe that in variant 1 the accuracy in specifying the first eigenvector $\boldsymbol{\xi}_1$ is of significant importance. For the values

$$y_0 = 1\,, \qquad z_0 = -1\,, \tag{17}$$

abnormal termination occurred at the 197th time step. At the beginning the variables $y_j$ and $z_j$ continue down to some quantities of order $10^{-19}$ for $j \approx 65$–70 and then begin to grow. Unlike this tendency, the initial data (17) did not change essentially the results of computations for variant 2: abnormal termination occurred for $j = 47$.

If for $\delta = 1$ and $\Delta = 2$ we could take the value $\tau = 2.1/\Delta$, for which the condition $\tau \leq \frac{2}{\Delta}$ is slightly violated ($\tau$ is exceeded by 5%), the same still holds at the 760–790th time steps.

The preceding examples provide enough reason to conclude that the concept of stability with respect to the input data is identical with the concept of continuous dependence of the solution of a difference problem upon the input data. The two major aims, improved accuracy and better stability, have to be balanced against the need to avoid expending unnecessary computational effort. In the next few sections the advanced theory in such matters will be developed and presented in full details.

**3. On the concept of well-posedness for a difference problem.** There is another matter which is one of some interest. In conformity with statements of problems of mathematical physics, it is fairly common to call a problem **well-posed** if the following conditions are satisfied:

(1) the problem is uniquely solvable for any input data from some class;
(2) a solution of the problem continuously depends on the input data.

Being concerned with a solution $y_h$ and input data $\varphi_h$ of a difference problem depending on the grid step $h$ as the parameter, we introduce the concept of well-posedness for a difference problem in a similar manner. Varying $h$ we are in possession of two sequences of solutions $\{y_h\}$ and input data $\{\varphi_h\}$, thus causing not only a single difference problem, but also a family of problems depending on the parameter $h$. The concept of well-posedness is introduced for a family of difference problems (schemes) as $|h| \to 0$.

We say that a difference problem (scheme) is **well-posed** if for all sufficiently small $|h| \le h_0$

(1) a solution $y_h$ of the difference problem exists and is unique for all input data $\varphi_h$ from some family;
(2) the solution $y_h$ continuously depends on $\varphi_h$ and this dependence is uniform in $h$.

More precisely, the second condition implies that there exists a constant $M > 0$ independent of $h$ such as for sufficiently small $|h| \le h_0$ the inequality

$$\| \tilde{y}_h - y_h \|_{(1_h)} \le M \| \tilde{\varphi}_h - \varphi_h \|_{(2_h)} \tag{18}$$

holds, where $\tilde{y}_h$ is the solution of the problem with the input data $\tilde{\varphi}_h$, and $\| \cdot \|_{(1_h)}$ and $\| \cdot \|_{(2_h)}$ are suitable norms on the set of grid functions given on the grid $\omega_h$.

The property of the continuous dependence of the solution for a difference problem on the input data is expressed by inequality (18) and is treated as the stability of the scheme with respect to the input data or simply stability (see Section 4.2)

**4. Stability, approximation and convergence.** We now examine a continuous problem of the type

$$L u = f(x) \quad \text{for} \quad x \in G, \qquad l u = \mu(x) \quad \text{for} \quad x \in \Gamma, \tag{19}$$

which is approximated on a grid $\bar{\omega}_h = \omega_h + \gamma_h$ by a difference problem

$$(20) \qquad L_h\, y_h = \varphi_h \quad \text{for} \quad x \in \omega_h\,, \qquad l_h\, y_h = \tilde{\mu}_h \quad \text{for} \quad x \in \gamma_h\,.$$

The statement of the problem for the error $z_h = y_h - u_h$, where $u_h$ is the projection of the solution $u$ of problem (19) onto the grid $\omega_h$, is

$$(21) \qquad L_h\, z_h = \psi_h \quad \text{for} \quad x \in \omega_h\,, \qquad l_h\, z_h = \nu_h \quad \text{for} \quad x \in \gamma_h\,,$$

where $\psi_h$ and $\nu_h$ are the errors of approximation to the governing equation and the additional condition. Instead of (21) we may write down formally

$$\widetilde{L}_h\, z_h = \tilde{\psi}_h\,.$$

If the operator $\widetilde{L}_h$ is linear and the difference scheme is well-posed, we are led by (18) to

$$(22) \qquad \begin{aligned} &\| z_h \|_{(1_h)} \le M\, \| \tilde{\psi}_h \|_{(2_h)} \quad \text{or} \\ &\| z_h \|_{(1_h)} \le M\, \big( \| \psi_h \|_{(2_h)} + \| \nu_h \|_{(3_h)} \big)\,. \end{aligned}$$

With these relations established, we conclude that if the scheme is stable and approximates the original problem, then it is convergent. In other words, "convergence follows from approximation and stability" and the order of accuracy and the rate of convergence are connected with the order of approximation.

Everything just said means that in establishing convergence and in determining the order of accuracy of a scheme it is necessary to evaluate the error of approximation, discover stability and then derive estimates of the form (22) known as **a priori estimates**.

Being the elements of different spaces, the solution $z_h$ and the right-hand side $\psi_h$ of the difference problem should be evaluated in different norms.

The examples of suitable norms are available in the preceding sections in which the errors of solutions and approximations on the grid $\omega_h$ are investigated in the usual way. Unfortunately, it is impossible to get estimates of the form (22) directly for stability of concrete difference problems. Auxiliary mathematical tools and techniques such as the summation by parts formulae, the difference Green formulae and elementary analogs of embedding theorems are aimed at deriving various estimates for solutions of difference analogs of boundary-value problems associated with an ordinary second-order differential equation. In this particular case we will encounter typical situations related to stability, approximation and accuracy of difference schemes. Later we will elaborate on this for rather complicated problems. At the same time, subsequent discussions of them will involve issues which are different from those occurred before.

## 2.4 MATHEMATICAL APPARATUS IN THE THEORY OF DIFFERENCE SCHEMES

**1. Some difference formulae.** In the sequel, when dealing with various difference expressions, we shall need the formulae for difference differentiating of a product, for summation by parts and difference Green's formulae. In this section we derive these formulae within the framework similar to the appropriate apparatus of the differential calculus. Similar expressions were obtained in Section 2 of Chapter 1 in studying second-order difference operators, but there other notations have been used. It performs no difficulty to establish a relationship between formulae from Section 2 of Chapter 1 and those of the present section.

1) ***Formulae for difference differentiating of a product.*** As known from the differential calculus, the formula

$$(uv)' = u'v + uv'$$

holds for differentiating the product of two functions $u(x)$ and $v(x)$.

In Section 1 we have already introduced two types of difference derivatives for grid functions: the left and the right ones, which correspond to different formulae for difference differentiating of a product

$$(1) \qquad (uv)_x = u_x\, v + u^{(+1)} v_x = u_x\, v^{(+1)} + u\, v_x \,,$$

$$(2) \qquad (uv)_{\bar{x}} = u_{\bar{x}}\, v + u^{(-1)} v_{\bar{x}} = u_{\bar{x}}\, v^{(-1)} + u\, v_{\bar{x}} \,.$$

Here we retain the notations

$$f^{(\pm 1)} = f(x \pm h), \qquad f_x = \frac{f^{(+1)} - f}{h}, \qquad f_{\bar{x}} = \frac{f - f^{(-1)}}{h}.$$

In this context, we draw the reader's attention to the fact that these formulae involve the index shift. Let us prove, for instance, the first equality. With this aim, we write it in the index form

$$\frac{u_{i+1}\, v_{i+1} - u_i\, v_i}{h} = \frac{u_{i+1}\, v_{i+1} - u_i\, v_{i+1}}{h} + \frac{u_i\, v_{i+1} - u_i\, v_i}{h},$$

which assures us of the validity of the equality.

2) ***The summation by parts formulae.*** We recall the integration by parts formula

$$\int_0^1 uv'\, dx = uv\big|_0^1 - \int_0^1 u'v\, dx\,.$$

As in the previous case similar formulae of two types

$$(3) \qquad (u, v_x) = u_N v_N - u_0 v_1 - (u_{\bar{x}}, v],$$

$$(4) \qquad (u, v_{\bar{x}}) = u_N v_{N-1} - u_0 v_0 - [u_x, v)$$

are valid for grid functions keeping the notations

$$(5) \quad (u, v) = \sum_{i=1}^{N-1} u_i v_i h, \qquad (u, v] = \sum_{i=1}^{N} u_i v_i h, \qquad [u, v) = \sum_{i=0}^{N-1} u_i v_i h.$$

Let us prove, for instance, (3). First, applying formula (1) yields

$$(6) \qquad \begin{aligned} \sum_{i=1}^{N-1} (uv_x)_i h &= \sum_{i=1}^{N-1} (uv)_{x,i} h - \sum_{i=1}^{N-1} (u_x v^{(+1)})_i h \\ &= (uv)_N - (uv)_1 - \sum_{i=2}^{N} (u_{\bar{x}} v)_i h \\ &= (uv)_N - u_1 v_1 - \sum_{i=1}^{N} u_{\bar{x},i} v_i h + (u_{\bar{x}} v)_1 h. \end{aligned}$$

Second, it is obvious that $h(u_{\bar{x}} v)_1 = u_1 v_1 - u_0 v_1$. Substitution of the preceding into (6) along with (5) yields (3).

In what follows we will frequently employ a non-equidistant grid denoting it by $\hat{\omega}_h$ in contrast to an equidistant introduced in Section 1. On any such grid the formulae of the inner product and the difference summation by parts formulae are somewhat different:

$$(7) \qquad \begin{aligned} &(u, v)_* = \sum_{i=1}^{N-1} u_i v_i \hbar_i, \qquad (u, v) = \sum_{i=1}^{N-1} u_i v_i h_{i+1}, \\ &(u, v] = \sum_{i=1}^{N} u_i v_i h_i, \qquad (u, v_{\hat{x}})_* = u_N v_N - u_0 v_1 - (v, u_{\bar{x}}], \end{aligned}$$

where $\hbar_i = \frac{1}{2}(h_i + h_{i+1})$. The notations for the difference derivative on a non-equidistant grid

$$v_{\hat{x},i} = \frac{v_{i+1} - v_i}{\hbar_i}$$

and the inner product on a non-equidistant grid $(\ ,\ )_*$ allow a simpler writing of the ensuing formulae.

To prove (7), a simple observation that

$$v_{\hat{x}} = \frac{h_{i+1}}{\hbar_i}\, v_x\,, \qquad v_x = \frac{v_{i+1} - v_i}{h_{i+1}}\, v_x$$

may be useful. Indeed, substituting this expression into the inner product

$$(u, v_{\hat{x}})_* = (u, v_x), \qquad \text{where} \quad (u, w) = \sum_{i=1}^{N-1} u_i\, w_i\, h_{i+1}\,,$$

and following the proof of identity (3), we arrive at (7) as desired.

3) ***The first Green formula.*** The equality

$$\int_0^1 u\,(kv')'\, dx = -\int_0^1 ku'v'\, dx + kuv'|_0^1$$

is usually called the **first Green formula**.

For grid functions an analog of the Green formula can be obtained by the summation by parts formulae. This can be done by substituting

$$u = z, \qquad v = a y_{\bar{x}}$$

into (3). The outcome of this is the first difference Green formula

$$(8) \qquad (z, (a y_{\bar{x}})_x) = -(a y_{\bar{x}}\,, z_{\bar{x}}] + a z y_{\bar{x}}|_N - a_1\, y_{x,0}\, z_0\,.$$

If $z_0 = z_N = 0$, the last two terms in (8) vanish and the first Green formula appears in simplified form:

$$(8') \qquad (z, \Lambda y) = -(a y_{\bar{x}}\,, z_{\bar{x}}], \qquad \Lambda y = (a y_{\bar{x}})_x\,.$$

In particular, for $z = y$ the preceding becomes

$$(8'') \qquad (\Lambda y, y) = -(a, (y_{\bar{x}})^2], \qquad y_0 = y_N = 0\,.$$

In the case of a non-equidistant grid we obtain a similar result:

$$(z, (a y_{\bar{x}})_{\hat{x}})_* = -(a y_{\bar{x}}\,, z_{\bar{x}}] + a z y_{\bar{x}}|_N - a_1\, y_{x,0}\, z_0\,,$$

$$(z, (a y_{\bar{x}})_{\hat{x}})_* = -(a y_{\bar{x}}\,, z_x] \quad \text{for} \quad z_0 = z_N = 0\,.$$

4) ***The second Green formula.*** In the integral calculus the following formula

$$\int_0^1 u\,(kv')'\,dx - \int_0^1 v\,(ku')'\,dx = k\,(uv' - vu')|_0^1$$

is known as the second Green formula. Inserting $u = y$ and $v = az_{\bar{x}}$ in (3) yields

$$(9) \qquad \big(y, (az_{\bar{x}})_x\big) = -(ay_{\bar{x}}\,, z_{\bar{x}}] + ayz_{\bar{x}}|_N - a_1\, y_0\, z_{x,0}\,.$$

Substracting (9) from (8) we derive the **difference analog of the second Green formula**

$$(10) \qquad \big(z, (ay_{\bar{x}})_x\big) - \big(y, (az_{\bar{x}})_x\big) = a_N\,(zy_{\bar{x}} - z_{\bar{x}}\,y)_N - a_1\,(y_x\,z - z_x\,y)_0\,.$$

In just the same way as was done on a non-equidistant grid, we find that

$$(11) \qquad \big(z, (ay_{\bar{x}})_{\hat{x}}\big)_* - \big(y, (az_{\bar{x}})_{\hat{x}}\big)_* = a_N\,(zy_{\bar{x}} - z_{\bar{x}}\,y)_N - a_1\,(y_x\,z - z_x\,y)_0\,.$$

If $y$ and $z$ vanish at the points $x = 0$ and $x = 1$, the appropriate terms in the expressions being substituted into (3) are equal to zero, thereby reducing to

$$(10') \qquad (\Lambda y, z) = (y, \Lambda z), \qquad \Lambda y = (ay_{\bar{x}})_x\,,$$

$$(11') \qquad (\Lambda y, z)_* = (y, \Lambda z)_*\,, \qquad \Lambda y = (ay_{\bar{x}})_{\hat{x}}\,.$$

These formulae confirm that the operator $\Lambda$ is self-adjoint.

5) ***The Cauchy–Bunyakovskiĭ inequality and the ε-inequality.*** In the sequel we shall need the well-known Cauchy–Bunyakovskiĭ inequality

$$|(u, v)| \le \|\,u\,\| \cdot \|\,v\,\|\,,$$

where $(\;,\;)$ is the inner product in a vector space and $\|\,u\,\| = \sqrt{(u,u)}$ is the associated norm. To be more specific, $(\;,\;)$ designates one of the inner products we have introduced above.

In what follows we will frequently employ the inequality

$$|\,ab\,| \le \varepsilon\, a^2 + \frac{b^2}{4\,\varepsilon}\,,$$

where $\varepsilon > 0$ is an arbitrary number, which is sometimes referred to as the **$\varepsilon$-inequality**, permitting us to establish

$$(12) \qquad |(u,v)| \le \| u \| \cdot \| v \| \le \varepsilon \| u \|^2 + \frac{1}{4\,\varepsilon} \| v \|^2 .$$

**2. The search for eigenfunctions and eigenvalues in the example of the simplest difference problem.** The method of separation of variables being involved in the apparatus of mathematical physics applies equelly well to difference problems. Employing this method enables one to split up an original problem with several independent variables into a series of more simpler problems with a smaller number of variables. As a rule, in this situation eigenvalue problems with respect to separate coordinates do arise. Difference problems can be solved in a quite similar manner.

In this section we consider the problem of searching for eigenvalues of the simplest difference operator.

Results we are going to obtain here will be needed in the sequel because applying the method of separation of variables leads to problems just of the same type. In the next chapters we give various examples of employing this method for the discovery of stability and convergence of concrete difference schemes.

Before giving further motivations, we would like to recall the basic aspects concerned with the elementary problem of determining eigenfunctions and eigenvalues for the differential equation

$$(13) \qquad u''(x) + \lambda u(x) = 0, \qquad 0 < x < l, \quad u(0) = u(l) = 0 .$$

The nontrivial solutions of this problem, that is, the eigenfunctions $u_k$ and the appropriate eigenvalues $\lambda_k$ are expressed by

1.

$$u_k(x) = \sqrt{\frac{2}{l}}\, \sin\, \frac{k \pi x}{l}, \qquad \lambda_k = \frac{k^2 \pi^2}{l^2}, \qquad k = 1, 2, \ldots$$

2. The eigenfunctions $u_k$ constitute what is called an orthonormal system:

$$\int\limits_0^1 u_k(x)\, u_m(x)\, dx = \delta_{km} , \qquad \text{where} \quad \delta_{km} = \begin{cases} 0, & k \ne m, \\ 1, & k = m . \end{cases}$$

3. If $f(x)$ is twice differentiable and satisfies the nonhomogeneous boundary conditions $f(0) = f(l) = 0$, it arranges itself into a uniformly convergent series

$$f(x) = \sum_{k=1}^{\infty} f_k\, u_k(x)$$

where $f_k = \int_0^1 f(x)\, u_k(x)\, dx$; in so doing

$$\| f \|^2 = \int_0^1 f^2(x)\ dx = \sum_{k=1}^{\infty} f_k^2 .$$

We put the difference problem

$$(14) \qquad \begin{gathered} y_{\bar{x}x} + \lambda y = 0 \qquad x = ih, \quad 0 < i < N, \quad h = \frac{l}{N}, \\ y(0) = y(l) = 0, \quad y(x) \not\equiv 0, \end{gathered}$$

of searching for nontrivial solutions being the eigenfunctions and eigenvalues of problem (14) in correspondence with the differential problem (13). For this, the index form of (14) is

$$(15) \qquad y_{i+1} - 2\,(1 - \tfrac{1}{2} h^2 \lambda)\, y_i + y_{i-1} = 0, \qquad i = 1, 2, \ldots, N-1 .$$

We look for a solution of problem (14) in the form

$$y(x) = \sin\ \alpha x$$

with $\alpha$ to be determined. Along these lines,

$$y_{i+1} + y_{i-1} = \sin\ \alpha(x+h) + \sin\ \alpha(x-h) = 2\ \sin\ \alpha x\ \ \cos \alpha h .$$

Substituting the expression just established into (15) yields

$$2\ \sin\ \alpha x\ \ \cos \alpha h = 2\,(1 - \tfrac{1}{2}\, h^2 \lambda)\ \sin\ \alpha x .$$

Since only nontrivial solutions are those to be found, that is, $\sin\ \alpha x \not\equiv 0$, it follows from the foregoing that

$$1 - \tfrac{1}{2}\, h^2 \lambda = \cos\ \alpha h$$

and

$$\lambda = \frac{2}{h^2}\,(1 - \cos\ \alpha h) = \frac{4}{h^2}\ \sin^2\ \frac{\alpha h}{2} .$$

The value of the parameter $\alpha$ is chosen in such a way that the function $y(x) = \sin\ \alpha x$ satisfies the boundary conditions of problem (14):

$$y(0) = y(l) = 0 .$$

Observe, that the boundary condition at the point $x = 0$ is automatically fulfilled for any $\alpha$. At the point $x = l$ we have

$$\sin \alpha l = 0 ,$$

giving $\alpha = \alpha_k = k\pi/l$, $k = 1, 2, \dots, N-1$.

Thus, we have obtained the eigenfunctions and eigenvalues of problem (14). A brief survey of their properties is presented below.

1.

$$(16) \quad y^{(k)}(x) = \sin \frac{k\pi x}{l}, \qquad \lambda_k = \frac{4}{h^2} \sin^2 \frac{k\pi h}{2l}, \qquad k = 1, 2, \dots, N-1 .$$

2. The eigenvalues $\lambda_k$ are enumerated in increasing order and for the whole collection $\{\lambda_k\}$ the estimates holds:

$$(17) \qquad 0 < \lambda_1 = \frac{4}{h^2} \sin^2 \frac{\pi h}{2l} < \lambda_2 < \cdots < \lambda_{N-1} = \frac{4}{h^2} \sin^2 \frac{\pi h (N-1)}{2l}$$

$$= \frac{4}{h^2} \cos^2 \frac{\pi h}{2l} < \frac{4}{h^2} .$$

In particular, it follows from (17) that all the eigenvalues of problem (14) are positive.

3. Eigenfunctions of problem (14) $y^{(k)}$, $y^{(m)}$, corresponding to distinct eigenvalues, are orthogonal in the sense of the inner product (5):

$$(18) \qquad (y^{(k)}, y^{(m)}) = 0, \qquad k \neq m .$$

To prove this fact, we make use of the second Green formula for the homogeneous boundary conditions (10′):

$$0 = (y^{(k)}_{\bar{x}x}, y^{(m)}) - (y^{(k)}, y^{(m)}_{\bar{x}x}) = (\lambda_m - \lambda_k)\,(y^{(k)}, y^{(m)}) .$$

In light of our supposition, $y^{(k)}$ and $y^{(m)}$ are eigenfunctions corresponding to distinct eigenvalues, that is, $\lambda_k \neq \lambda_m$, the orthogonality of $y^{(k)}$ and $y^{(m)}$ follows from the preceding equality:

$$(y^{(k)}, y^{(m)}) = 0 .$$

4. The norm of an eigenfunction $y^{(k)}(x)$ is $\| y^{(k)} \| = \sqrt{l/2}$. Recall that the norm is understood in the sense of the inner product (5):

$$\| y \|^2 = (y, y) = \sum_{i=1}^{N-1} y_i^2 \, h .$$

We perform simple algebraic calculations and write only the final results

$$\| y^{(k)} \|^2 = \sum_{s=1}^{N-1} \left( y^{(k)}(x_s) \right)^2 h = \sum_{s=1}^{N-1} h \sin^2 \frac{\pi k x_s}{l} \tag{19}$$

$$= \sum_{s=1}^{N-1} \frac{h}{2} \left( 1 - \cos \frac{2\pi k x_s}{l} \right).$$

Using $q_k = \exp \left\{ i \, \frac{2\pi k h}{l} \right\}$ and taking into account that $q_k^s = \exp \left\{ i \, \frac{2\pi k}{l} \, x_s \right\}$ and $q_k^N = 1$, we find that

$$\sum_{s=1}^{N-1} h \cos \frac{2\pi k}{l} \, x_s = \operatorname{Re} \sum_{s=1}^{N-1} h \, q_k^s = \operatorname{Re} h \, \frac{q_k^N - q_k}{q_k - 1} = \operatorname{Re} h \, \frac{1 - q_k}{q_k - 1} = -h \, .$$

Substitution of this value into (19) yields

$$\| y^{(k)} \|^2 = \frac{h \, (N-1)}{2} + \frac{h}{2} = \frac{hN}{2} = \frac{l}{2} \, .$$

Thus, the set of grid functions

$$\mu^{(k)}(x) = \sqrt{\frac{2}{l}} \; y^{(k)}(x), \qquad k = 1, 2, \ldots, N-1 \, , \tag{20}$$

constitutes an orthogonal and normed system in the sense of the inner product $(\mu^{(k)}, \mu^{(m)}) = \delta_{km}$.

5. Let a function $f(x)$ with the values $f_0 = f_N = 0$ be given on a grid $\bar{\omega}_h$. Then, obviously, it can be represented as a sum of the eigenfunctions of problem (14):

$$f(x) = \sum_{k=1}^{N-1} f_k \, \mu^{(k)}(x) \tag{21}$$

where the coefficients are defined by the relations $f_k = \left( f(x), \mu^{(k)}(x) \right)$. What is more, in that case

$$\| f \|^2 = \sum_{k=1}^{N-1} f_k^2 \tag{22}$$

Indeed,

$$\| f \|^2 = \sum_{i=1}^{N-1} h \, f^2(x_i) = (f^2, 1) = \left( \sum_{k=1}^{N-1} f_k \, \mu^{(k)}, \sum_{k=1}^{N-1} f_k \, \mu^{(k)} \right)$$

$$= \left( \sum_{k,m=1}^{N-1} f_k \, f_m \, \mu^{(k)} \mu^{(m)}, 1 \right) = \sum_{k,m=1}^{N-1} f_k \, f_m \left( \mu^{(k)}, \mu^{(m)} \right) = \sum_{k=1}^{N-1} f_k^2 \, ,$$

since $\left( \mu^{(k)}, \mu^{(m)} \right) = \delta_{km}$ as established before.

**3. Eigenvalue problems with boundary conditions of the second and third kinds.** We now turn to the **second eigenvalue problem**

$$(23)\qquad u'' + \lambda u = 0, \qquad 0 < x < l, \quad u'(0) = u'(l) = 0, \quad u(x) \not\equiv 0 .$$

With the relations $u_{x,0} = u'(0) + \frac{1}{2}hu''(0) + O(h^2) = -\frac{1}{2}h\lambda u(0) + O(h^2)$ and $u_{x,0} + \frac{1}{2}h\lambda u(0) = O(h^2)$ in view, we approximate the boundary conditions with accuracy $O(h^2)$, leading to the second boundary-value problem on eigenvalues:

$$(24)\qquad \begin{aligned} &y_{\bar{x}x} + \lambda y(x) = 0\,, \qquad x = ih, \quad i = 1, 2, \ldots, N-1, \quad h = \tfrac{l}{N}\,, \\ &y_{x,0} + \tfrac{1}{2}\, h\lambda y_0 = 0, \qquad -y_{\bar{x},N} + \tfrac{1}{2}\, h\lambda y_N = 0\,. \end{aligned}$$

In such a setting it is required to find the values of the parameter $\lambda$ such that these homogeneous equations have nontrivial solutions $y(x) \not\equiv 0$. In contrast to the first boundary-value problem, here the parameter $\lambda$ enters not only the governing equation, but also the boundary conditions. The introduction of new sensible notations

$$(25)\qquad Av = \begin{cases} -\dfrac{2}{h}\, v_x & \text{for} \quad x = 0\,, \\ -v_{\bar{x}x} & \text{for} \quad x = ih, \quad 0 < i < N\,,\,, \\ \dfrac{2}{h}\, v_{\bar{x}} & \text{for} \quad x = l\,, \end{cases}$$

is connected with setting problem (24) in the operator form

$$(26)\qquad Ay = \lambda y\,,$$

where the operator $A$ acts in the space $H = \bar{\Omega}$ comprising all the functions $y(x)$ defined on the grid $\bar{\omega}_h = \{x_i = ih,\ i = 0, 1, \ldots, N\}$.

We introduce in that space the inner product

$$[y, v] = \sum_{i=1}^{N-1} y_i\, v_i\, h + \tfrac{h}{2}\,(y_0\, v_0 + y_N\, v_N)$$

and show that the operator $A$ is self-adjoint and nonnegative. By definition, this means that

$$[Ay, v] = [y, Av]\,,$$

$$[Ay, y] \geq 0 \qquad \text{for any} \quad y \in H\,.$$

Indeed,

$$[Ay, v] = (-y_{\bar{x}x}\,, v) + \frac{h}{2}\Big(-\frac{2}{h}\,v_0\,y_{x,0} + \frac{2}{h}\,v_N\,y_{\bar{x},N}\Big)\,.$$

Making use of the second Green formula (10), we get

$$\begin{aligned}[Ay, v] &= (y, -v_{\bar{x}x}) + (v y_x - y v_x)_0 - (v y_{\bar{x}} - y v_{\bar{x}})_N + (-v_0\,y_{x,0} + v_N\,y_{\bar{x},N})\\ &= (y, -v_{\bar{x}x}) + (-y_0\,v_{x,0} + y_N\,v_{\bar{x},N}) = [y, Av]\,,\end{aligned}$$

that is, $A = A^*$. The first Green formula (8) gives for $z = y$

$$\begin{aligned}[Ay, y] &= (y, -y_{\bar{x}x}) + (-y_0\,y_{x,0} + y_N\,y_{\bar{x},N})\\ &= (y_{\bar{x}}, -y_{\bar{x}}] - (y y_{\bar{x}})_N + (y y_x)_0 + (-y_0\,y_{x,0} + y_N\,y_{\bar{x},N})\\ &= (y_{\bar{x}}, y_{\bar{x}}] \ge 0\,.\end{aligned}$$

Problem (26) can be solved by appeal to the general theory (see Section 1). Let us determine the eigenvalues $\lambda_k$ and eigenfunctions $\mu_k(x)$ of problem (24) accepting a solution of problem (24) in the form

$$y = \mu(x) = A\cos\,\alpha x\,, \qquad A \ne 0\,.$$

Upon substituting $\mu(x)$ into equation (24) we get

$$\lambda = \frac{4}{h^2}\,\sin^2\frac{\alpha h}{2}\,.$$

We now require the boundary conditions to be satisfied at the points $x = 0$ and $x = l$. The condition $\cos\,\alpha h - 1 + \frac{1}{2}\,h^2\lambda = 0$ is automatically fulfilled at the point $x = 0$. The case where $x = l$ is not simple to follow and needs investigation:

$$\big(1 - \tfrac{1}{2}\,h^2\lambda\big)\,\cos\,\alpha l - \cos\,\alpha(l - h) = 0$$

or

$$\cos\,\alpha h\,\cos\,\alpha l - \cos\,\alpha l\,\cos\,\alpha h + \sin\,\alpha l\,\sin\,\alpha h = \sin\,\alpha l\,\sin\,\alpha h = 0\,,$$

implying that $\alpha l = \pi k$, $k = 0, 1, 2, \ldots, N$. Thus, we have determined the eigenvalues

$$\lambda_0 = 0\,, \qquad \lambda_k = \frac{4}{h^2}\,\sin^2\frac{\pi k h}{2l}\,, \qquad k = 1, 2, \ldots, N\,, \tag{27}$$

and the orthonormal eigenfunctions $\{\mu_k(x)\}$, for which

$$[\mu_k\,,\mu_m] = \delta_{km} = \begin{cases} 1, & k = m, \\ 0, & k \neq m, \end{cases}$$

where

$$\tag{28} \begin{gathered} \mu_0 = \sqrt{\frac{1}{l}}\,, \qquad \mu_N = \sqrt{\frac{1}{l}}\,\cos\frac{\pi N x}{l}\,, \\ \mu_k = \sqrt{\frac{1}{l}}\,\cos\frac{\pi k x}{l}\,, \qquad k = 1, 2, \dots, N-1\,. \end{gathered}$$

The normalizing multiplier $A_k$ can be recovered from the condition

$$\|[\mu_k]\|^2 = [\mu_k\,,\mu_k] = A_k^2\left[\cos\frac{\pi k x}{l},\,\cos\frac{\pi k x}{l}\right],$$

where the sum

$$\left(\cos\frac{\pi k x}{l},\,\cos\frac{\pi k x}{l}\right) = \sum_{s=1}^{N-1} h\,\cos^2\frac{\pi k x_s}{l}$$

is calculated by analogy with the preceding section. The orthonormality of the eigenfunctions $\{\mu_k\}$ follows from the general theory, since $A = A^* \geq 0$ and all the eigenvalues are simple.

Any grid function $f(x)$ defined on the grid $\bar\omega_h$ arranges itself into a series of $\{\mu_k(x)\}$:

$$f(x) = \sum_{k=0}^{N} f_k\,\mu_k(x)\,, \qquad f_k = [f, \mu_k]\,, \qquad k = 0, 1, \dots, N\,,$$

under the inner product structure $[f, f] = \sum_{k=0}^{N} f_k^2$.

The statement of the **third boundary-value problem on eigenvalues** is

$$\tag{29} \begin{aligned} & u'' + \lambda u(x) = 0\,, && 0 < x < l\,, \\ & u'(0) = \sigma_1\,u(0)\,, && \sigma_1 > 0\,, \\ & -u'(l) = \sigma_2\,u(l)\,, && \sigma_2 > 0\,, \quad u(x) \not\equiv 0\,. \end{aligned}$$

As can readily be observed, the difference scheme of second-order approximation acquires the form

$$(30)\qquad y_{\bar{x}x} + \lambda\, y = 0\,, \qquad x = ih\,, \quad i = 1, 2, \dots, N-1\,,$$

$$(y_{x,0} - \sigma_1\, y_0) + \tfrac{1}{2}\, h\lambda\, y_0 = 0\,,$$

$$-(y_{\bar{x},N} + \sigma_2\, y_N) + \tfrac{1}{2}\, h\lambda\, y_N = 0\,, \quad y(x) \not\equiv 0\,,$$

yielding

$$u_{x,0} - \sigma_1\, u_0 + \tfrac{1}{2}\, h\lambda\, u_0 = u_0' + \tfrac{1}{2}\, h u_0'' + O(h^2) - \sigma_1\, u_0 + \tfrac{1}{2}\, h\lambda\, u_0 = O(h^2)\,.$$

It will be sensible to introduce the operator $A$ with the values

$$(31)\qquad Ay = \begin{cases} -\dfrac{1}{0.5h}\,(y_x - \sigma_1\, y) & \text{for} \quad x = 0\,, \\ -y_{\bar{x}x} & \text{for} \quad x = ih\,, \quad 0 < i < N\,, \\ \dfrac{1}{0.5h}\,(y_{\bar{x}} + \sigma_2\, y) & \text{for} \quad x = l\,, \end{cases}$$

by means of which problem (30) admits an alternative form of writing

$$(32)\qquad A\mu = \lambda\, \mu\,.$$

The domain of operator (31) coincides with the entire space $H$, the domain of operator (25), $A + A^* > 0$ and, what is more,

$$[Ay, y] = (\, y_{\bar{x}}\,, y_{\bar{x}}\,] + \sigma_1\, y_0^2 + \sigma_2\, y_N^2\,,$$

it being understood that $A > 0$ if either at least one of the coefficients $\sigma_1$ and $\sigma_2$ becomes nonzero or $\sigma_1 + \sigma_2 > 0$, but $\sigma_1 \geq 0$ and $\sigma_2 \geq 0$. In that case $[Ay, y] = 0$ only for $y(x) \equiv 0$.

Unlike the preceding two problems, we come to nothing in trying to derive the explicit formula for $\alpha$. The parameter $\alpha$ should be recovered from the equation

$$(33)\qquad \left(\sin^2 \alpha h - h^2 \sigma_1\, \sigma_2\right) \operatorname{tg} \alpha l = h\,(\sigma_1 + \sigma_2) \sin \alpha h\,,$$

but the eigenvalues are expressed, as usual, by the formulae

$$(34)\qquad \lambda_k = \frac{4}{h^2}\, \sin^2 \frac{\alpha_k\, h}{2}\,, \qquad k = 0, 1, 2, \dots, N\,.$$

The eigenfunctions are determined within a constant by

$$(35)\qquad \mu_k(x) = \cos\, \alpha_k\,(l - x) + \frac{h\, \sigma_1}{\sin\, \alpha_k\, h}\, \sin\, \alpha_k\,(l - x)\,.$$

**4. Difference analogs of the embedding theorems.** In the estimation of various properties of difference schemes such as stability, convergence, etc. we shall need yet inequalities corresponding to the simplest Sobolev embedding theorems. In this respect the appropriate results have been obtained with the following lemmas.

**Lemma 1** *For any grid function $y(x)$ defined on the grid*

$$\bar{\omega}_h = \{ x_i = ih,\ 0 \le i \le N,\ x_0 = 0,\ x_N = 1 \}$$

*and vanishing at the points $x = 0$ and $x = 1$ the inequality holds:*

$$\| y \|_C \le \tfrac{1}{2} \, \| y_{\bar{x}} ]| \,, \tag{36}$$

*where* $\| y \|_C = \max_{x \in \bar{\omega}_h} | y(x) |$ *and* $\| y_{\bar{x}} ]| = (y_{\bar{x}},\, y_{\bar{x}}]^{1/2}$.

**Proof** The function $y(x)$ given on the grid $\bar{\omega}_h$ can be expressed in the form of the identity

$$y^2(x) = (1 - x)\, y^2(x) + x\, y^2(x)\,. \tag{37}$$

With the assigned values $y(0) = y(1) = 0$, one can write down

$$y^2(x) = \Big( \sum_{x'=h}^{x} y_{\bar{x}}(x')\, h \Big)^2 \quad \text{or} \quad y^2(x) = \Big( \sum_{x'=x+h}^{1} y_{\bar{x}}(x')\, h \Big)^2 .$$

Upon substituting these equalities into (37) we find that

$$y^2(x) = (1 - x) \Big( \sum_{x'=h}^{x} h\, y_{\bar{x}}(x') \Big)^2 + x \Big( \sum_{x'=x+h}^{1} h\, y_{\bar{x}}(x') \Big)^2 .$$

Let us estimate the sums on the right-hand side of the preceding relation with the aid of the Cauchy–Bunyakovskiĭ inequality:

$$y^2(x) \le (1 - x) \sum_{x'=h}^{x} h \sum_{x'=h}^{x} y_{\bar{x}}^2(x')\, h + x \sum_{x'=x+h}^{1} h \sum_{x'=x+h}^{1} y_{\bar{x}}^2(x')\, h$$

$$= x\,(1 - x) \sum_{x'=h}^{1} y_{\bar{x}}^2(x')\, h = x\,(1 - x)\, \| y_{\bar{x}} ]|^2,$$

where $y_{\bar{x}}^2 = (y_{\bar{x}})^2$. The maximum of the expression $x\,(1 - x)$ on the segment $[0,\, 1]$ is attained at the point $x = \frac{1}{2}$ and equals $\frac{1}{4}$. Therefore,

$$y^2(x) \le \tfrac{1}{4} \, \| y_{\bar{x}} ]|^2$$

and, consequently,

$$\| y \|_C \le \tfrac{1}{2} \, \| y_{\bar{x}} ]| \,.$$

**Remark 1** Lemma 1 remains valid on an arbitrary non-equidistant grid $\hat{\omega}_h$.

**Remark 2** In the sequel we will also use an inequality of the type (36) for segments of arbitrary length $l$. Such an inequality can be derived from (36) by merely substituting $x' = l\,x$. Then $x'$ varies on the interval $(0,\, l)$ and

$$y_{\bar{x}'} = y_{\bar{x}} \,/\, l\,, \qquad h' = h\,l\,.$$

By inserting $y_{\bar{x}} = y_{\bar{x}'}\,l$ and $h = h'\,/\,l$ in (24) we arrive at the chain of the relations

$$\|\, y_{\bar{x}}\,]|^2 = \sum_{i=1}^{N} (y_{\bar{x}'})_i^2\, l^2\, l^{-1}\, h' = l \sum_{i=1}^{N} (y_{\bar{x}'})_i^2\, h' = l\, \|\, y_{\bar{x}'}\,]|^2 .$$

Consequently, on any interval of length $l$

$$|\, y(x)\,| \leq \|\, y\,\|_C \leq \tfrac{1}{2}\, \sqrt{l}\, \|\, y_{\bar{x}}\,]| \,. \tag{38}$$

**Remark 3** Inequalities (36) and (38) are valid for all the functions vanishing at both ends of the interval in view. Being concerned with the function $y(x)$ vanishing only on the boundary, one can derive another inequality

$$\|\, y\,\|_C \leq \sqrt{l}\, \|\, y_{\bar{x}}\,]| \,. \tag{39}$$

Inequalities (36), (38) and (39) fail to be true, in general, for arbitrary functions. However, it is plain to show that in this case the inequalities of alternative forms occur:

$$\|\, y\,\|_C^2 \leq 2\, (\, l\, \|\, y_{\bar{x}}\,]|^2 + y_0^2)\,, \qquad \|\, y\,\|_C^2 \leq 2\, (\, l\, \|\, y_{\bar{x}}\,]|^2 + y_N^2)\,.$$

**Lemma 2** *For any function $y(x)$ defined on an arbitrary grid*

$$\hat{\bar{\omega}}_h = \{x_i,\, x_0 = 0,\, x_N = l\}$$

*and vanishing at the points $x = 0$ and $x = l$ the inequality holds:*

$$\|\, y\,\|^2 \leq \frac{l^2}{4}\, \|\, y_{\bar{x}}\,]|^2\,, \qquad y_0 = y_N = 0\,. \tag{40}$$

Indeed, it is easy to check that $\|\, y\,\|^2 \leq l\, \|\, y\,\|_C^2$. Substituting this inequality into (36) yields (40). In the case of an equidistant grid estimate (40) can be improved and the reader is invited to do it on his/her own.

**Lemma 3** *Any function $y(x)$ defined on the equidistant grid*

$$\bar{\omega}_h = \{ x_i = ih,\ i = 0,\ 1,\ \ldots,\ N,\ x_0 = 0,\ x_N = l \}$$

*and vanishing at the points $x = 0$ and $x = l$ admits the estimates*

$$\frac{h^2}{4} \, \| y_{\bar{x}} ]|^2 \le \| y \|^2 \le \frac{l^2}{8} \, \| y_{\bar{x}} ]|^2 . \tag{41}$$

**Proof** We have occasion to use the expansion of $y(x)$ with respect to the eigenfunctions of problem (14):

$$y(x) = \sum_{k=1}^{N-1} c_k \, \mu^{(k)}(x) , \qquad c_k = (y(x),\ \mu^{(k)}(x)) , \qquad \| y \|^2 = \sum_{k=1}^{N-1} c_k^2 .$$

By the first Green formula (8) we thus have

$$(-\Lambda \, y,\ y) = \| y_{\bar{x}} ]|^2, \tag{42}$$

$$\text{where} \quad \Lambda \, y = y_{\bar{x}x} , \qquad \| y_{\bar{x}} ]|^2 = ( 1,\ (y_{\bar{x}})^2 ] .$$

By definition, $\Lambda \, \mu^{(k)} = -\lambda_k \, \mu^{(k)}$ and, therefore,

$$-\Lambda \, y = \sum_{k=1}^{N-1} c_k \, \lambda_k \, \mu^{(k)}(x) .$$

Let us substitute this expression into (42) and take into account that $\{ \mu^{(k)} \}$ is an orthonormal system. As a final result we obtain

$$\| y_{\bar{x}} ]|^2 = -(\Lambda \, y,\ y) = \sum_{k=1}^{N-1} \lambda_k \, c_k^2 .$$

We deduce from here that $\lambda_1 \, \| y \|^2 \le \| y_{\bar{x}} ]|^2 \le \lambda_{N-1} \, \| y \|^2$, where

$$\lambda_1 = \frac{4}{h^2} \sin^2 \frac{\pi h}{2 \, l} , \qquad \lambda_{N-1} = \frac{4}{h^2} \cos^2 \frac{\pi h}{2 \, l} .$$

The next goal of our studies is to construct a lower bound for $\lambda_1$ with respect to $\alpha = \pi h \, / \, (2l)$:

$$\lambda_1 = \frac{\pi^2}{l^2} \left( \frac{\sin \alpha}{\alpha} \right)^2 .$$

Since $h \le 0.5 \, l$, the quantity $\alpha$ is varying on the half-interval $( 0,\ \pi \, / \, 4 ]$. It is easy to check that the minimum of the function for $\alpha \in ( 0,\ \pi \, / \, 4 ]$ is attained at the point $\alpha = \pi \, / \, 4$, that is, $\lambda_1(h)$ attains its minimum in the case where $h = 0.5 \, l$. This serves to motivate that $\lambda_1 \ge 8 \, / \, l^2$. Taking into account also that $\lambda_{N-1} < 4 \, / \, h^2$, we come to (41) and finish the proof of the lemma.

**5. The method of energy inequalities.** One of the general and very effective ways of constructing a priori estimates is the method of energy inequalities. We bring several examples illustrating how to use this method in deriving a priori estimates in difference problems and find, for instance, the rate of convergence of a difference scheme on the basis of these estimates.

In this section we consider the simplest model problem

$$(43)\qquad u''(x) + f(x) = 0\,, \qquad 0 < x < 1\,, \qquad u(0) = u(1) = 0\,.$$

**Example 1** Assuming that an equidistant grid $\bar{\omega}_h$ is given on the segment $[0, 1]$, we now consider the difference approximation of problem (43)

$$(44)\qquad y_{\bar{x}x} + f(x) = 0\,, \qquad x \in \omega_h\,, \qquad y_0 = y_N = 0\,.$$

Multipling equation (44) by $hy$ and summing up the resulting equality over the grid nodes of $\omega_h$, we eventually get

$$(45)\qquad \sum_{i=1}^{N-1} (y_{\bar{x}x})_i\, y_i\, h + \sum_{i=1}^{N-1} f_i\, y_i\, h = 0\,.$$

We now rewrite (45) in terms of inner products, whose use permits us to reduce it to the following one:

$$(46)\qquad (y_{\bar{x}x}, y) + (f, y) = 0\,.$$

Via transform of the first summand in (46) by the Green difference formula (8″) we find that

$$(47)\qquad -(\,y_{\bar{x}}, y_{\bar{x}}\,] + (f, y) = 0 \qquad \text{or} \qquad \|\, y_{\bar{x}}\,]|^2 = (f, y)\,.$$

The estimation of the inner product $(f, y)$ can be done using the Cauchy–Bunyakowskiĭ inequality (12): $|\,(f, y)\,| \le \|\, f\,\| \cdot \|\, y\,\|$. By Lemma 3,

$$\|\, y\,\| \le \|\, y_{\bar{x}}\,]| / \sqrt{8}\,.$$

Putting these together with (47) we deduce in agreement with Lemma 1 that

$$\|\, y_{\bar{x}}\,]| \le \|\, f\,\| / \sqrt{8}\,.$$

We finally get one possible a priori estimate for the solution of problem (44):

$$(48)\qquad \|\, y\,\|_C \le \|\, f\,\| / (4\sqrt{2})\,.$$

This inequality will be aimed at estimating the rate of convergence of scheme (44). We write beforehand the equation for the accuracy of scheme (44): $z = y - u$, where $u$ is a solution of problem (43) and $y$ is a solution of the difference problem (44). Upon substituting $y = z + u$ into (44) it is plain to set up the problem for $z$:

$$z_{\bar{x}x} + \psi(x) = 0\,, \qquad x \in \omega_h\,, \qquad z_0 = z_N = 0\,. \tag{49}$$

Here $\psi(x) = u_{\bar{x}x} + f(x)$ is the approximation error of scheme (44). For all sufficiently smooth functions $u(x)$ it is well-known that $\psi(x)$ is a quantity of order $O(h^2)$, thus causing the same type of the problem for the function $z(x)$ as occured for the function $y(x)$. Because of this fact, estimate (48) is still valid for $z(x)$:

$$\| z \|_C \leq \| \psi \| \,/\, (4\sqrt{2})\,. \tag{50}$$

However, $\psi = O(h^2)$ and, consequently, $\| z \|_C = \| y - u \|_C \leq M\, h^2$, where $M$ is a positive constant independent of step $h$. In agreement with the above definitions (see Section 1) estimate (50) provides the uniform convergence of a solution of the difference problem (44) to a solution of the differential problem (43) with the rate $O(h^2)$.

So far we have established an estimate for the rate of convergence in a very simple problem. It is possible to obtain a similar result for this problem by means of several other methods that might be even much more simpler. However, the indisputable merit of the well-developed method of energy inequalities is its universal applicability: it can be translated without essential changes to the multidimensional case, the case of variable coefficients, difference schemes for parabolic and hyperbolic equations and other situations.

Let us show, for example, that this method leads without any difficulties to the desired result for the case of a non-equidistant grid.

**Example 2** Let a non-equidistant grid $\hat{\omega}_h$ be given on the segment $[\,0,\, 1\,]$. On this grid problem (43) can be approximated as follows:

$$y_{\bar{x}\hat{x}} + f(x) = 0\,, \qquad x \in \hat{\omega}_h\,, \qquad y_0 = y_N = 0\,, \tag{51}$$

(for the notations see Section 1.3, Example 1).

For problem (51) one can derive an a priori estimate of the same type as estimate (48) for problem (44). However, in this case such an estimate fails to provide with a quite reliable idea on the speed of convergence of scheme (51). We have shown in Section 1.3 that the local approximation error $\psi = u_{\bar{x}\hat{x}} + f$ of scheme (51) is a quantity of order $O(\bar{h}_i)$ and

$$\|\psi\|_C \le M\, h_{\max}\,. \tag{52}$$

Estimate (52) indicates an effective reduction of the speed of convergence of scheme (51) on the non-equidistant grid $\hat{\omega}_h$ in comparison with scheme (44) on the equidistant grid. However, as we have stated above, if the error of approximation is evaluated not in the grid norm of the spaces $C$ or $L_2$, but in a specially constructed negative norm $\|\cdot\|_{(-1)}$, then the error of approximation on any non-equidistant grid will be of the same order $O(h^2)$. Namely, the negative norm

$$\|\psi\|_{(-1)} := \left[\sum_{i=1}^{N-1} h_i \left(\sum_{k=i}^{N-1} \bar{h}_k\, \psi_k\right)^2\right]^{1/2} = O(h^2)$$

is good enough for our purposes. What has been said above implies that in the further derivation of a priori estimates for problem (51) the right-hand side should be evaluated in the negative norm $\|\cdot\|_{(-1)}$.

Let us obtain this a priori estimate by multiplying equation (51) by $y_i \bar{h}_i$ and summing over all grid nodes of $\hat{\omega}_h$. In terms of the inner products the resulting expression can be written as

$$(y_{\bar{x}\hat{x}},\, y)_* + (f,\, y)_* = 0\,. \tag{53}$$

Via transform of the first summand in (53) by the Green difference formula (8) it is not difficult to establish the relation

$$(\, y_{\bar{x}},\, y_{\bar{x}}] = (f,\, y)_*\,. \tag{54}$$

As we will see later, it will be sensible to deal with the function $\eta(x)$ specified by the relations

$$\eta_{\hat{x},i} = f_i\,, \qquad i = 1, 2, \ldots, N-1\,, \qquad \eta_N = 0\,. \tag{55}$$

While solving problem (55) it reduces to $-\eta(x_i) = \sum_{k=1}^{N-1} f_k \bar{h}_k$.

The inner product on the right-hand side of (54) is modified by the summation by parts formula (7) into

$$(f,\, y)_* = (\eta_{\hat{x}},\, y)_* = -(\,\eta,\, y_{\bar{x}}]\,.$$

On account of the Cauchy–Bunyakovskiĭ inequality,

$$|(f, y)_*| = |(\eta, y_{\bar{x}})| \leq \|\eta\| \cdot \|y_{\bar{x}}]| \,.$$

Substituting this estimate into (54) and eliminating $\|y_{\bar{x}}]|$ from both sides of the resulting inequality, we arrive at the relations

$$\|y_{\bar{x}}]| \leq \|\eta\| = \left[ \sum_{i=1}^{N-1} h_i \left( \sum_{k=i}^{N-1} \bar{h}_k f_k \right)^2 \right]^{1/2} = \|f\|_{(-1)} \,.$$

By Lemma 1,

$$\|y\|_C \leq \|y_{\bar{x}}]|/2$$

and, consequently, we derive the relation

$$\|y\|_C \leq \tfrac{1}{2} \|f\|_{(-1)} \,, \tag{56}$$

thereby justifying the desired estimate. Further development is connected with the accurate account of the error $z = y - u$, where $y$ is a solution of problem (51) and $u$ is a solution of the original differential problem (43). Upon substituting $y = z + u$ into (51) the problem arises for the error $z$:

$$z_{\bar{x}\hat{x}} + \psi = 0 \,, \qquad x \in \hat{\omega}_h \,, \qquad z_0 = z_N = 0 \,. \tag{57}$$

Applying estimate (56) to problem (57) yields

$$\|z\|_C \leq \tfrac{1}{2} \|\psi\|_{(-1)} \,.$$

However, we have stated in Section 1.3 that $\|\psi\|_{(-1)} \leq M h^2$, where $h = \max_{1\leq i \leq N} h_i$. Therefore, scheme (51) on an arbitrary non-equidistant grid $\hat{\omega}_h$ converges in the space $C$ with the rate $O(h^2)$.

## 2.5 DIFFERENCE SCHEMES AS OPERATOR EQUATIONS. GENERAL FORMULATIONS

Difference schemes for the simplest differential equations have been considered in preceding sections, the basic topics in the theory of difference schemes have been introduced for them as well as all the tricks and turns available for investigating stability and convergence of such schemes have been demonstrated with a great success.

In this section a unified interpretation of difference equations as operator equations in an abstract space is carried out and, after this, the corresponding definitions of approximation, stability and convergence are presented. This approach is quite applicable in mathematical physics for stationary problems.

**1. Difference schemes as operator equations.** After replacing differential equations by difference equations on a certain grid $\omega_h$ we obtain a system of linear algebraic equations that can be written in matrix form. The outcome of this is

$$\mathfrak{A}\, Y = \Phi\,, \tag{1}$$

where $\mathfrak{A}$ is a square $N \times N$-matrix, $Y = (y_1, y_2, \ldots, y_N)$ is the vector of unknowns and $\Phi = (\varphi_1, \varphi_2, \ldots, \varphi_N)$ is a known right-hand side including the right-hand sides of boundary conditions. With every matrix $\mathfrak{A}$ one can associate a linear operator $A$ acting from $R_N$ into $R_N$. With this correspondence in view, equation (1) takes the form

$$A\, y = \varphi\,, \tag{2}$$

where the unknown vector $y$ is sought, while the right-hand side $\varphi \in R_N$ is a given vector. The operator $A$ maps onto itself the space of grid functions defined on $\omega_h$ and satisfying the homogeneous boundary conditions. Several examples can add interest and aid in understanding.

**Example 1.** ***The first boundary-value problem.*** Given on the segment $[0, 1]$ an equidistant grid $\bar\omega_h = \{ x_i = i\,h\,,\ i = 0, 1, \ldots, N\,,\ h = 1/N \}$, let us look for a solution of the first boundary-value problem

$$\Lambda\, y = y_{\bar x x} = -f(x)\,, \qquad 0 < x = i\,h < 1\,, \qquad y_0 = u_1\,, \qquad y_N = u_2\,, \tag{3}$$

or

$$\frac{1}{h^2}\,(y_{i-1} - 2\,y_i + y_{i+1}) = -f_i\,, \qquad i = 1, 2, \ldots, N-1\,, \tag{3'}$$
$$y_0 = u_1, \qquad y_N = u_2$$

As the first step towards the solution of this problem, we form the vector $Y = (y_1, y_2, \ldots, y_{N-1})$, making it possible to rewrite equation (3) in the form (1) with the $(N-1) \times (N-1)$-matrix

$$\mathfrak{A} = \begin{pmatrix} 2 & -1 & 0 & 0 & \ldots & 0 & 0 \\ -1 & 2 & -1 & 0 & \ldots & 0 & 0 \\ 0 & -1 & 2 & -1 & \ldots & 0 & 0 \\ \vdots & \vdots & \vdots & \vdots & \ddots & \vdots & \vdots \\ 0 & 0 & 0 & 0 & \ldots & -1 & 2 \end{pmatrix}.$$

The right-hand side vector $\Phi = (\varphi_1, \varphi_2, \ldots, \varphi_{N-1})$ includes the right-hand sides of the boundary conditions (3)

$$\varphi_i = f_i\,, \quad i = 2, \ldots, N-2\,, \quad \varphi_1 = f_1 + \frac{u_1}{h^2}\,, \quad \varphi_{N-1} = f_{N-1} + \frac{u_2}{h^2}\,,$$

so that $\varphi_i$ may differ from $f_i$ only at the grid nodes adjacent to the boundary points, that is, for $i = 1$ and $i = N-1$.

The matrix $\mathfrak{A}$ specifies an operator $A = -\widetilde{\Lambda}$, carrying a grid function $y(x_i)$, that is, a vector of the $(N-1)$-dimensional space into a vector of the same space (into the grid function $(-\widetilde{\Lambda}\, y)_i$). The operator $\widetilde{\Lambda}$ coincides with the operator $\Lambda$ on all grid functions vanishing at the boundary nodes (for $i = 0$ and $i = N$), so that $(\widetilde{\Lambda}\, y)_i = (\Lambda\, y)_i$ for $i = 2, 3, \ldots, N-2$ and

$$(4) \qquad (\widetilde{\Lambda}\, y)_1 = \frac{-2\, y_1 + y_2}{h^2}\,, \qquad (\widetilde{\Lambda}\, y)_{N-1} = \frac{y_{N-2} - 2\, y_{N-1}}{h^2}\,.$$

Let $\Omega_h$ be the set of grid functions defined at the inner nodes of the grid $\omega_h$. The set so constructed is certainly linear. Once equipped with the inner product $(y, v) = \sum_{i=1}^{N-1} y_i\, v_i\, h$ and associated norm $\|\, y\,\| = \sqrt{(y, y)}$, the space $\Omega_h$ becomes a normed vector space. The above operator $A$ is linear and maps $\Omega_h$ into itself, meaning that the domain and range of the operator $A$ coincide with the entire space $\Omega_h$.

Let $\overset{\circ}{\Omega}_h$ be the space of all grid functions defined for all nodes of the grid $\bar{\omega}_h$ and vanishing at the boundary nodes, that is, for $x \in \gamma_h$. Then the operator $A$ may be treated as an operator from $\overset{\circ}{\Omega}$ into $\Omega_h$. Obviously, $A = A_h$, $y = y_h$ and $\varphi = \varphi_h$ depend solely on the grid step $h$. Just for this reason the subject of subsequent discussions is a family of equations depending on the parameter $h$ rather than the single equation (2). A family of such equations constitutes what is called an operator-difference scheme (see Section 1.2).

While solving the operator equations (2) we establish the basic properties of the operator $A$ such as self-adjointness, positive definiteness, the lower bound of the operator and its norm and more. The operator $A$ constructed in Example 1 will be frequently encountered in the sequel. Before stating the main results, will be sensible to list its basic properties.

The operator $A$ is self-adjoint, that is, $(A\, y, v) = (y, A\, v)$ for any $y, v \in \Omega_h$. As a matter of fact, $(A\, y, v) = (-\widetilde{\Lambda}\, y, v)$. Making use of the second Green formula (Section 3) and taking into account that $\widetilde{\Lambda}$ coincides with $\Lambda$ on the set of grid functions vanishing at the boundary nodes, we establish the relation $(\widetilde{\Lambda}\, y, v) = (y, \widetilde{\Lambda}\, v)$, implying that $A = A^*$.

The operator $A$ is positive definite, that is, $(A\,y,\,y) \geq 8\,\|\,y\,\|^2$. This is a consequence of Lemma 3 in Section 3.4.

The norm of the operator $A$ is equal to

$$\|\,A\,\| = \frac{4}{h^2}\,\cos^2\frac{\pi h}{2} < \frac{4}{h^2}\,. \tag{5}$$

Indeed, the norm of a self-adjoint positive operator in a finite-dimensional space $\Omega_h$ is equal to its greatest eigenvalue: $\|\,A\,\| = \lambda_{N-1}$. In that case, in complete agreement with the results of Section 4.2, we might have

$$\lambda_{N-1} = \tfrac{4}{h^2}\,\cos^2\frac{\pi h}{2}$$

and, therefore, formula (5) holds true. In addition,

$$(A\,y,\,y) \leq \|\,A\,\| \cdot \|\,y\,\|^2\,.$$

The next quantity we will introduce is an operator with the values

$$A\,y = -(a y_{\bar{x}})_x + dy\,, \qquad 0 < c_1 \leq a \leq c_2\,, \qquad 0 \leq d \leq c_3\,. \tag{6}$$

Due to the second Green formula it is self-adjoint. In turn, the first Green formula assures us of the validity of the relation

$$(A\,y,\,y) = (a,\,y_{\bar{x}}^2] + (dy,\,y)\,, \tag{7}$$

where $y_{\bar{x}}^2 = (y_{\bar{x}})^2$. This implies that

$$(A\,y,\,y) \geq c_1\,(1,\,y_{\bar{x}}^2] = c_1\|\,y_{\bar{x}}\,]|^2 \geq 8\,c_1\|\,y\,\|^2\,, \tag{8}$$

meaning that $A$ is positive definite. It seems clear from formulae (7) and (8) that its norm satisfies the estimate $\|\,A\,\| < 4\,c_2\,/h^2 + c_3$.

**Remark** Here the operator $A = A_h$ depends only on the grid step $h$ playing for the moment the role of parameter. When $\bar{\omega}_h = \{x_i \in [0,\,1],\, i = 0,\,1,\,\ldots,\,N,\, x_0 = 0,\, x_N = 1\}$ is a non-equidistant grid, its step $h_i = x_i - x_{i-1}$ itself becomes a grid function or an $N$-dimensional vector $h = (h_1,\,h_2,\,\ldots,\,h_N)$. The operator $A$ admits the form $A = -\widetilde{\Lambda}$, where

$$\Lambda\,y_i = \frac{1}{\bar{h}_i}\left(\frac{y_{i+1}-y_i}{h_{i+1}} - \frac{y_i - y_{i-1}}{h_i}\right) = y_{\bar{x}\hat{x},i}\,, \qquad \bar{h}_i = \frac{1}{2}\,(h_i + h_{i+1})\,.$$

In this case we write, as before, $A = A_h$ with further reference that $h$ is a vector of the dimension $N$, that is, an element of the space $\Omega_h^+$ consisting of all functions defined on the grid

$$\omega_h^+ = \{\, x_i \in (\,0,\, 1\,],\; i = 1, 2, \ldots, N,\; x_N = 1\,\}\,.$$

**Example 2.** ***The third boundary-value problem.*** Given the same grid $\bar{\omega}_h$ as in Example 1, we now consider the difference boundary-value problem of the third kind

$$(9) \qquad \begin{aligned} &\Lambda\, y = y_{\bar{x}x} = -f(x)\,, && 0 < x = i\,h < 1\,,\\ &y_{x,\,0} = \sigma_1\, y_0 - \mu_1\,, && -y_{\bar{x},N} = \sigma_2\, y_N - \mu_2\,. \end{aligned}$$

Let $\bar{\Omega}_h$ be a set of all functions defined on the grid $\bar{\omega}_h = \{\, x_i = i\,h,\; i = 0,\, 1,\, \ldots,\, N\,\}$. We begin by specifying the operator $\widetilde{\Lambda}$ by the relations

$$(\widetilde{\Lambda}\, y)_i = \begin{cases} \dfrac{1}{0.5\,h}\,(y_{x,\,0} - \sigma_1\, y_0) = \Lambda^-\, y\,, & i = 0\,,\\ \Lambda\, y = y_{\bar{x}x}\,, & i = 1, 2, \ldots, N-1\,,\\ -\dfrac{1}{0.5\,h}\,(y_{\bar{x},N} + \sigma_2 y_N) = \Lambda^+\, y\,, & i = N\,. \end{cases}$$

By merely setting $A = -\widetilde{\Lambda}$ problem (9) is recast as

$$(10) \qquad A\, y = \varphi\,,$$

where

$$\varphi_i = \begin{cases} \dfrac{1}{0.5\,h}\,\mu_1\,, & i = 0\,,\\ f_i\,, & i = 1, 2, \ldots, N-1\,,\\ \dfrac{1}{0.5\,h}\,\mu_2\,, & i = N\,. \end{cases}$$

The linear operator $A$ maps $\bar{\Omega}_h$ onto $\bar{\Omega}_h$. To make our exposition more transparent, it will be convenient to introduce the inner product

$$[y,\, v] = \sum_{i=1}^{N-1} y_i\, v_i\, h + \tfrac{1}{2}\, h\, (y_0\, v_0 + y_N\, v_N)$$

and the associated norm $\|\,[y]\,\| = \sqrt{[y,\, y]}$. The operator $A$ is self-adjoint, that is, $[y,\, A\, v] = [v,\, A\, y]$, where

$$[y,\, A\, v] = -(y,\, \Lambda\, v) - \frac{1}{2}\, h\, (\, y_0\, \Lambda^-\, v + y_N\, \Lambda^+\, v\,)$$

and $(y, \omega) = \sum_{i=1}^{N-1} y_i \, \omega_i \, h$. Making use of the Green formula from Section 3 and substituting the relevant expressions for $\Lambda^- v$ and $\Lambda^+ v$, we find that

$$-(y, \Lambda v) = (-\Lambda y, v) - (y_{x,0} \, v_0 - y_{\bar{x},N} \, v_N) + (y_0 \, v_{x,0} - y_N \, v_{\bar{x},N}),$$

$$-0.5\, h \, (y_0 \, \Lambda^- v + y_N \, \Lambda^+ v) = -y_0 \, (v_{x,0} - \sigma_1 \, v_0) + y_N \, (v_{\bar{x},N} + \sigma_2 \, v_N).$$

Since $y_{x,0} = \sigma_1 y_0 + 0.5\, h \, \Lambda^- y$ and $y_{\bar{x},N} = -\sigma_2 \, y_N - 0.5\, h \, \Lambda^+ y$, we arrive at the chain of the relations

$$[\, y, A v\,] = -(\Lambda y, v) - 0.5\, h \, (v_0 \, \Lambda^- y + v_N \, \Lambda^+ y) = [\, A y, v\,]$$

as required.

We are going to show that if $\sigma_1 \geq c_1 > 0$ and $\sigma_2 \geq c_1 > 0$, then the operator $A$ is positive definite:

(11)
$$[\, A y, y\,] \geq \frac{2\, c_1}{1 + c_1} \, | [y] |^2 .$$

With this aim, we follow the same procedures as we did in the proof of Lemma 1 in Section 3. Namely, those ideas are connected with attempting the function $y^2(x)$ in the form

$$y^2(x) = \Big( y_0 + \sum_{x'=h}^{x} y_{\bar{x}}(x') \, h \Big)^2 = y_0^2 + 2\, y_0 \sum_{x'=h}^{x} y_{\bar{x}}(x') \, h + \Big( \sum_{x'=h}^{x} y_{\bar{x}}(x') \, h \Big)^2$$

and applying then the $\varepsilon$-inequality. The outcome of this is

$$y^2(x) \leq (1 + \varepsilon) \, y_0^2 + \Big( 1 + \frac{1}{\varepsilon} \Big) \Big( \sum_{x'=h}^{x} y_{\bar{x}}(x') \, h \Big)^2 .$$

By the Cauchy–Bunyakovskiĭ inequality,

$$\Big( \sum_{x'=h}^{x} y_{\bar{x}}(x') \, h \Big)^2 \leq x \Big( \sum_{x'=h}^{x} y_{\bar{x}}^2(x') \, h \Big) \leq x \, (1, y_{\bar{x}}^2 \,],$$

implying that

$$y^2(x) \leq (1 + \varepsilon) \, y_0^2 + \Big( 1 + \frac{1}{\varepsilon} \Big) \, x \, (1, y_{\bar{x}}^2 \,].$$

Likewise, one can prove another inequality

$$y^2(x) \le (1+\varepsilon)\, y_N^2 + \left(1+\frac{1}{\varepsilon}\right)(1-x)\,(1,\, y_{\bar{x}}^2\,] .$$

By the same token,

$$y^2(x_i) \le \tfrac{1}{2}(1+\varepsilon)\,(y_0^2 + y_N^2) + \tfrac{1}{2}\left(1+\frac{1}{\varepsilon}\right)(1,\, y_{\bar{x}}^2\,] ,$$

$$|\,[y]\,|^2 \le \tfrac{1}{2}(1+\varepsilon)\,(y_0^2 + y_N^2) + \tfrac{1}{2}\left(1+\frac{1}{\varepsilon}\right)(1,\, y_{\bar{x}}^2\,] .$$

Putting $\varepsilon = c_1$ and using the identity

$$[A\,y,\, y] = (1,\, y_{\bar{x}}^2\,] + \sigma_1\, y_0^2 + \sigma_2\, y_N^2 , \tag{12}$$

we establish relation (11).

It is plain to deduce for the norm of the operator $A$ the estimate

$$\|\,A\,\| \le \frac{4}{h^2}\,(1+\tfrac{1}{2}\,c_2 h), \qquad \text{where} \qquad c_2 = \max(\sigma_1,\, \sigma_2) . \tag{13}$$

Indeed,

$$(\,y_{\bar{x},i}\,)^2 \le \frac{2}{h^2}(\,y_i^2 + y_{i-1}^2\,)$$

and

$$(\,1,\, y_{\bar{x}x}^2\,] \le \frac{4}{h^2}\,|\,[y]\,|^2,$$

where $|\,[y]\,|^2 = \sum_{i=1}^{N-1} y_i^2\, h + \frac{1}{2}\, h\,(\,y_0^2 + y_N^2\,)$. By virtue of the relations

$$[A\,y,\, y] \le \|\,A\,\| \cdot |\,[y]\,|^2 ,$$

$$\sigma_1\, y_0^2 + \sigma_2\, y_N^2 = \frac{2}{h}\,(\,\frac{1}{2}\, h\,\sigma_1\, y_0^2 + \frac{1}{2}\, h\,\sigma_2\, y_N^2\,) \le \frac{2\,c_2}{h}\,|\,[y]\,|^2$$

we get from (12) the desired estimate (13).

**Example 3.** ***Non-self-adjoint operators.*** Let $\bar{\omega}_h = \{ x_i = i h, i = 0, 1, \ldots, N, h = 1 / N \}$ be a grid on the segment $0 \le x \le 1$ and let the difference operators

$$\Lambda^- y = y_{\bar{x}}, \qquad \Lambda^+ y = -y_{\bar{x}}, \tag{14}$$

mapping the set $\overset{\circ}{\Omega}_h$ of grid functions defined on $\bar{\omega}_h$ and vanishing for $i = 0$ and $i = N$ into $\Omega_h$, be such that

$$(\Lambda^- y)_i = \begin{cases} y_1 / h, & i = 1, \\ (y_i - y_{i-1}) / h, & i = 2, 3, \ldots, N-1, \end{cases}$$

$$(\Lambda^+ y)_i = \begin{cases} -(y_{i+1} - y_i) / h, & i = 1, 2, \ldots, N-2, \\ y_{N-1} / h, & i = N-1. \end{cases}$$

These formulae show that the operators $\Lambda^-$ and $\Lambda^+$ can be treated under such an approach as operators from $\Omega_h$ into $\Omega_h$.

Let $(y, v) = \sum_{i=1}^{N-1} y_i v_i h$ be the inner product in $\overset{\circ}{\Omega}_h$ and $\Omega_h$. It seems clear that the operators $\Lambda^-$ and $\Lambda^+$ are mutually adjoint:

$$(\Lambda^- y, v) = (y, \Lambda^+ v) \qquad \text{for any} \qquad y, v \in \Omega_h. \tag{15}$$

Indeed, by the summation by parts formula we have $-(y, v_x) = (y_{\bar{x}}, v)$ if $y = v = 0$ for $i = 0$ and $i = N$. This just implies the fact that $\Lambda^-$ and $\Lambda^+$ are operators adjoint to each other. We note in passing that

$$\Lambda^- + \Lambda^+ = -(y_x - y_{\bar{x}}) = -h \Lambda y,$$

where $\Lambda y = y_{\bar{x}x}$, that is, $\Lambda^- + \Lambda^+ = -h \Lambda$. Therefore, the operators $\Lambda^-$ and $\Lambda^+$ are positive definite:

$$(\Lambda^- y, y) = (\Lambda^+ y, y) = \frac{h}{2} (-\Lambda y, y) = \frac{h}{2} \| y_{\bar{x}} ]|^2 \ge 4 h \| y \|^2.$$

The last inequality holds true on account of Lemma 3 in Section 3.4.

Later we will deal with operators from $\Omega_h$ into $\Omega_h$ of the structure

$$\begin{aligned} R_1 y &= \frac{1}{h} y_{\bar{x}}, & R_2 y &= \frac{1}{h} (-y_x), \\ R_1 &= \frac{1}{h} \Lambda^-, & R_2 &= \frac{1}{h} \Lambda^+. \end{aligned} \tag{16}$$

They are mutually adjoint, that is, $(R_1 y, v) = (y, R_2 v)$ and

$$(R_1 y, y) = (R_2 y, y) = \tfrac{1}{2} \| y_{\bar{x}} ]|^2 \geq 4 \| y \|^2.$$

The useful relation

$$\| R_1 y \|^2 = h^{-2} \sum_{i=1}^{N-1} y_{\bar{x}}^2 h \leq 4 h^{-4} \| y \|^2$$

provides the validity of the following estimates:

$$(17) \qquad \| R \|_1 \leq \frac{2}{h^2}, \qquad \| R \|_2 \leq \frac{2}{h^2}.$$

Since $R_1 = R_2^*$,

$$(R_1 R_2 y, y) = \| R_2 y \|^2 = \frac{1}{h^2} \sum_{i=1}^{N-1} y_{x,i}^2 h \leq \frac{1}{h^2} \sum_{i=1}^{N} y_{\bar{x},i}^2 h = \frac{1}{h^2} \| y_{\bar{x}} ]|^2,$$

yielding

$$(18) \qquad \| R_2 y \|^2 \leq \frac{1}{h^2} (R y, y), \qquad R = R_1 + R_2.$$

The difference operator $R_3 y = \frac{1}{h^2} y_{\overset{\circ}{h}} = \frac{1}{2} (R_1 - R_2) y$, $y \in \overset{\circ}{\Omega}_h$, is obviously skew-symmetric due to the relationships

$$R_3^* = \tfrac{1}{2} (R_1^* - R_2^*) = \tfrac{1}{2} (R_2 - R_1) = -R_3$$

and, consequently, $(R_3 y, y) = 0$. Its norm satisfies the estimate

$$\| R_3 \| \leq \frac{1}{2} (\| R_1 \| + \| R_2 \|) \leq \frac{2}{h^2}.$$

It is not difficult to refine the estimate for $\| R_3 \|$ with the aid of relations

$$\| y_{\mathring{x}} \|^2 = \frac{1}{4 h^2} \sum_{i=1}^{N-1} (y_{i+1} - y_{i-1})^2 h \leq \frac{1}{2 h^2} \sum_{i=1}^{N-1} (y_{i+1}^2 + y_{i-1}^2) h \leq \frac{\| y \|^2}{h^2},$$

giving $\| R_3 \| \leq 1 / h^2$.

Non-self-adjoint difference operators appear, for example, in the approximation of second-order elliptic operators with the first derivatives. The operator $L u = u''(x) + b u'(x)$, $x \in [0, 1]$, $b = \text{const}$, is approximated by the difference operators $\Lambda_1 y = y_{\bar{x}x} + b\, y_x$ for $b > 0$ or $\Lambda_2 y = y_{\bar{x}x} + b\, y_{\bar{x}}$ for $b < 0$ with $y \in \hat{\Omega}_h$.

Let $\widetilde{\Lambda}\, y$ be an operator from $\Omega_h$ into $\Omega_h$ coinciding with $\Lambda\, y$ for $y \in \overset{\circ}{\Omega}_h$. The operators $A_1 = -\widetilde{\Lambda}_1$ and $A_2 = -\widetilde{\Lambda}_2$, acting from $\Omega_h$ into $\Omega_h$, are positive definite for any $h$. Indeed,

$$(19)\qquad \begin{aligned} (A_1 y, y) &= (-y_{\bar{x}x}, y) - b(y_x, y) = (1 + \tfrac{1}{2} h b)\, \| y_{\bar{x}} ]|^2, \\ (A_2 y, y) &= (1 - \tfrac{1}{2} h b)\, \| y_{\bar{x}} ]|^2 = (1 + \tfrac{1}{2} h\, |b|)\, \| y_{\bar{x}} ]|^2. \end{aligned}$$

We conclude from here that

$$(A_\alpha y, y) \ge 8(1 + \tfrac{1}{2} h\, |b|)\, \| y \|^2$$

for $\alpha = 1, 2$.

Be re-ordering $A_1 = A + b\,\Lambda^+$ and $A_2 = A + |b|\,\Lambda^-$, where $A = -\Lambda$, $\Lambda\, y = y_{\bar{x}x}$ and $\| \Lambda^{\pm} \| \le 2/h$, $\| A \| \le 4/h^2$, we are led due to the triangle inequality for the norms to

$$(20)\qquad \begin{aligned} \| A_1 \| &\le \| A \| + b\, \| \Lambda^+ \| < \frac{4}{h^2} (1 + \tfrac{1}{2} h b), \qquad b > 0, \\ \| A_2 \| &\le \| A \| + b\, \| \Lambda^- \| < \frac{4}{h^2} (1 + \tfrac{1}{2} h\, |b|), \qquad b < 0. \end{aligned}$$

We note in passing that if we approximate the operator $L u = u'' + b\, u'$ by the expression $\Lambda\, y = y_{\bar{x}x} + b\, y_{\bar{x}}$ for $b > 0$, then the member $1 - \frac{1}{2} h b$ arises in place of $1 + \frac{1}{2} h b$ in (19) and thereby the operator $-\Lambda$ will be positive definite only for $h < 2/b$.

If $u'(x)$ is approximated by the central difference derivative $u_{\overset{\circ}{x}}$ for any sign $b$, we have the operator $A_3 y = -\widetilde{\Lambda}_3 y$, where $\Lambda_3 y = y_{\bar{x}x} + b\, y_{\overset{\circ}{x}}$. This operator $A_3 y = -y_{\bar{x}x} - h b R_3 y$ is of second-approximation order and satisfies the relations

$$(A_3 y, y) = (-y_{\bar{x}x}, y) - h b (R_3 y, y) = \| y_{\bar{x}} ]|^2$$

and

$$\| A_3 \| \le \| A \| + h\, |b|\, \| R_3 \| < \frac{4}{h^2} (1 + h\, |b|).$$

Although the complete theory could be recast in the general case, we confine ourselves to simplest examples. In subsequent chapters the difference operators approximating elliptic operators (in particular, the Laplace operator) in rectangular domains will be studied by means of similar methods. If the initial differential operator is self-adjoint and positive definite, one should construct the difference operator also to possess these properties in the grid space. This can be achieved, for example, by employing the balance method (the integro-interpolation method from Chapter 3) or the variational method in designing difference schemes.

We learn from the examples under consideration that the difference equations can be treated as operator equations with operators in a finite-dimensional normed vector space. A feature of such operators is that they map the entire space into itself as further developments occur.

We proceed to the description of the theory of difference schemes treated as operator equations, the meaning of which we have discussed above.

**2. Stability of a difference scheme.** Let two normed vector spaces $\mathcal{B}_h^{(1)}$ and $\mathcal{B}_h^{(2)}$ be given with parameter $h$ being a vector of some normed space with the norm $|h| > 0$. In dealing with a linear operator $A_h$ with the domain $\mathcal{D}(A_h) = \mathcal{B}_h^{(1)}$ and range $\mathcal{R}(A_h) \subseteq \mathcal{B}_h^{(2)}$ we consider the equation

$$(21) \qquad A_h\, y_h = \varphi_h\,, \qquad y_h \in \mathcal{B}_h^{(1)}, \qquad \varphi_h \in \mathcal{B}_h^{(2)},$$

where $\varphi_h$ is a given vector. Varying the parameter $h$, we obtain the set of solutions $\{ y_h \}$ to equation (21). We call the operator equation (21) depending on the parameter $h$ a **difference scheme**.

Let $\| \cdot \|_{(1_h)}$ and $\| \cdot \|_{(2_h)}$ be the norms on the spaces $\mathcal{B}_h^{(1)}$ and $\mathcal{B}_h^{(2)}$, respectively. Scheme (21) is said to be correct or problem (21) is said to be well-posed if for all sufficiently small $|h| \le h_0$

(1) a solution $y_h$ of (21) exists and is unique for all $\varphi_h \in \mathcal{B}_h^{(2)}$ (scheme (21) is uniquely solvable),

(2) this solution continuously depends on $\varphi_h$ and this dependence is uniform in $h$ (scheme (21) is stable). In other words, there exists a positive constant $M$ independent of $h$ and $\varphi_h$ such that a solution to equation (21) admits for any $\varphi_h \in \mathcal{B}_h^{(2)}$ the estimate

$$(22) \qquad \| y_h \|_{(1_h)} \le M \, \| \varphi_h \|_{(2_h)} \,.$$

The meaning of the solvability of scheme (21) is that there exists an inverse operator $A_h^{-1}$ such that

$$(23) \qquad y_h = A_h^{-1}\, \varphi_h \,.$$

Stability of the scheme is to be understood as the property that the inverse $A_h^{-1}$ acting from $\mathcal{B}_h^{(2)}$ into $\mathcal{B}_h^{(1)}$ is uniformly bounded in $h$:

(24) $\quad \| A_h^{-1} \| \leq M \quad$ where $\quad M > 0 \quad$ does not depend on $\quad h$.

Combination of (23) and (24) gives estimate (22):

$$\| y_h \|_{(1_h)} \leq \| A_h^{-1} \| \cdot \| \varphi_h \|_{(2_h)} \leq M \| \varphi_h \|_{(2_h)} .$$

That is to say, the meaning of stability of scheme (21) is that a solution (21) depends continuously on the right-hand side and this dependence is uniform in the parameter $h$. This implies that a small change of the right-hand side results in a small change of the solution. If the scheme is solvable and stable, it is correct. Note that the uniqueness of the scheme (21) solution is a consequence of its solvability and stability and, hence, we might get rid of the uniqueness requirement in condition (1). Indeed, assume to the contrary that there were two solutions to equation (21), say $\bar{y}_h$ and $\bar{\bar{y}}_h \neq \bar{y}_h$. By the linearity property of the operator $A_h$, their difference $z_h = \bar{\bar{y}}_h - \bar{y}_h$ should satisfy the homogeneous equation

$$A_h z_h = A_h ( \bar{y}_h - \bar{\bar{y}}_h ) = A_h \bar{y}_h - A_h \bar{\bar{y}}_h = \varphi_h - \varphi_h = 0 .$$

Because scheme (21) is stable, inequality (22) holds true and, therefore,

$$\| z_h \|_{(1_h)} = \| \bar{\bar{y}}_h - \bar{y}_h \|_{(1_h)} \leq M \, \| 0 \|_{(2_h)} = 0 ,$$

implying that $\bar{y}_h = \bar{\bar{y}}_h$.

To prove the stability of (21), we need an a priori estimate of the form (22). A derivation of some a priori estimates for the operator equation (21) will be carried out in Section 4. A difference scheme $A_h y_h = \varphi_h$ is said to be **ill-posed** if at least one of the conditions (1)–(2) we have imposed above is violated.

Suppose that a solution $y_h$ of problem (21) exists for any $\varphi_h \in \mathcal{B}_h^{(2)}$, so that $y_h^{-1} = A_h^{-1} \varphi_h$. Since $\mathcal{B}_h^{(1)}$ and $\mathcal{B}_h^{(2)}$ are finite-dimensional spaces, the inverse operator $A_h^{-1}$ from $\mathcal{B}_h^{(2)}$ into $\mathcal{B}_h^{(1)}$ is bounded and its norm equals $\| A_h^{-1} \| = M_h$, where $M_h$ is a positive constant depending on the parameter $h$. If the scheme is stable, there exists a constant $M > 0$ independent of $h$ such that $M_h \leq M$ for all $| h | \leq h_0$. The meaning of instability of scheme (21) and, hence, of its ill-posedness is that $M_h \to \infty$ as $| h | \to 0$, it being understood that the above constant $M$ does not exist. For an ill-posed problem only an estimate of the form

(25) $$\| y \|_{(1_h)} \leq M_h \, \| \varphi_h \|_{(2_h)}$$

may be true.

Here the meaning of weak stability is that condition (25) should be valid. In other words, this asserts that there exists a domain for $|h|$, for example, $0 < h_* \le |h| < h_0$, where (22) is satisfied with constant $M$, depending on $h_*$.

The definition of well-posedness and ill-posedness of a scheme is closely connected with the selection rules for norms $\|\cdot\|_{(1_h)}$ and $\|\cdot\|_{(2_h)}$. It may happen that for some choices of these norms estimate (25) is fulfilled, while for the others estimate (22) is true. It is worth noting here that for the scheme $Ay = -y_{\bar{x}x} = \varphi$ for $y \in \overset{\circ}{\Omega}_h$, as stated in Section 3, estimate (22) is still valid in the norms

$$\|y_h\|_{(1_h)} = \|y_h\|_C = \max_{1\le i\le N-1} |y_h(x_i)|, \qquad M = 1,$$

$$\|\varphi_h\|_{(2_h)} = \left[\sum_{i=1}^{N-1} h\left(\sum_{k=i}^{N-1} h\varphi_k\right)^2\right]^{1/2}.$$

**3. Convergence and approximation.** Let $\mathcal{B}^{(1)}$ and $\mathcal{B}^{(2)}$ be normed vector spaces with norms $\|\cdot\|_{(1)}$ and $\|\cdot\|_{(2)}$, respectively. One assumes, in addition, that

(1) there exist linear operators $\mathcal{P}_h^{(1)}$ from $\mathcal{B}^{(1)}$ into $\mathcal{B}_h^{(1)}$ and $\mathcal{P}_h^{(2)}$ from $\mathcal{B}^{(2)}$ into $\mathcal{B}_h^{(2)}$ known as projectors such that

$$\mathcal{P}_h^{(1)} u = u_h \in \mathcal{B}_h^{(1)} \quad \text{if } u \in \mathcal{B}^{(1)} \quad \text{and} \quad \mathcal{P}_h^{(2)} f = f_h \in \mathcal{B}_h^{(2)} \quad \text{if } f \in \mathcal{B}^{(2)};$$

(2) the conditions of concordance of norms are satisfied:

$$(26) \qquad \lim_{|h|\to 0} \|\mathcal{P}_h^{(1)} u\|_{(1_h)} = \|u\|_{(1)}, \qquad \lim_{|h|\to 0} \|\mathcal{P}_h^{(2)} f\|_{(2_h)} = \|f\|_{(2)}.$$

Given a vector $y_h$ of the space $\mathcal{B}_h^{(1)}$, we study the convergence of $\{y_h\}$ as $|h| \to 0$ to a fixed element $u$ from $\mathcal{B}^{(1)}$.

1) A sequence $\{y_h\}$, where $y_h \in \mathcal{B}_h^{(1)}$, is said to be convergent to an element $u \in \mathcal{B}^{(1)}$ if

$$(27) \qquad \lim_{|h|\to 0} \|y_h - \mathcal{P}_h^{(1)} u\|_{(1_h)} = 0.$$

2) A sequence $\{y_h\}$ is said to be convergent to $u_h \in \mathcal{B}^{(1)}$ with the rate $O(|h|^n)$, $n > 0$, or $\{y_h\}$ approximates $u$ with accuracy $O(|h|^n)$ if for all sufficiently small $|h| \le h_0$ the estimate is valid:

$$(28) \qquad \|y_h - u_h\|_{(1_h)} \le M\,|h|^n, \qquad u_h = \mathcal{P}_h^{(1)} u,$$

where $M > 0$ is a constant independent of $h$.

**Remark 1** Convergence conditions (27) imply that

$$\lim_{|h|\to 0} \| y_h \|_{(1_h)} = \| u \|_{(1)} . \tag{29}$$

Indeed, with the aid of the triangle inequality and the relation $u_h = \mathcal{P}_h^{(1)} u$ one can write down

$$\begin{aligned} \| y_h \|_{(1_h)} &= \| ( y_h - u_h ) + u_h \|_{(1_h)} \\ &\le \| y_h - u_h \|_{(1_h)} + \| u_h \|_{(1_h)} \\ &\le \| y_h - u_h \|_{(1_h)} + \| \mathcal{P}_h^{(1)} u \|_{(1_h)} . \end{aligned}$$

Passing to the limit as $| h | \to 0$ and taking into account (27) and (26), we obtain

$$\lim_{|h|\to 0} \| y_h \|_{(1_h)} \le \| u \|_{(1)} .$$

A similar reasoning yields

$$\| u \|_{(1)} \le \lim_{|h|\to 0} \| y_h \|_{(1_h)} ,$$

giving (29).

**Remark 2** The sequence $\{ y_h \}$ can converge only to a single element $u \in \mathcal{B}^{(1)}$. On the contrary, let there exist two limit elements $\bar{u}, \bar{\bar{u}} \in \mathcal{B}^{(1)}$, $\bar{u} \ne \bar{\bar{u}}$ ($\{ y_h \}$ converges to each of them), that is,

$$\lim_{|h|\to 0} \| y_h - \bar{u}_h \|_{(1_h)} = \lim_{|h|\to 0} \| y_h - \bar{\bar{u}}_h \|_{(1_h)} = 0 .$$

We are going to show that $\mathbf{u} \ne \bar{u}$ by considering the difference

$$\bar{u}_h - \mathbf{u}_h = ( \bar{u}_h - y_h ) + ( y_h - \mathbf{u}_h )$$

and appealing to the triangle inequality

$$\| \bar{u}_h - \mathbf{u}_h \|_{(1_h)} \le \| y_h - \bar{u}_h \|_{(1_h)} + \| y_h - \mathbf{u}_h \|_{(1_h)} .$$

Passing to the limit as $| h | \to 0$ and taking into account the convergence of $y_h$ to both $\bar{u}$ and $\mathbf{u}$ and the norm concordance condition (26), we deduce that $\| \bar{u} - \mathbf{u} \|_{(1)} = 0$, implying that $\bar{u} = \mathbf{u}$.

Let $y_h$ be a solution of problem (21). We say that
1) scheme (21) converges if there exists an element $u \in \mathcal{B}^{(1)}$ such that the limit relation (27) occurs;
2) the scheme is of accuracy $O(\,|h|^2)$ if there exists an element $u \in \mathcal{B}^{(1)}$ such that for $|h| \leq h_0$ relation (28) occurs.

Let us define the notion of the approximation error on an element $u \in \mathcal{B}^{(1)}$. To this end, we must write down the equation for the difference $z_h = y_h - u_h$. Substitution of $y_h = z_h + u_h$ into (21) gives

$$(30) \qquad A_h z_h = \psi_h , \qquad \psi_h = \varphi_h - A_h u_h , \qquad \psi_h \in \mathcal{B}_h^{(2)} .$$

In this context, we call the right-hand side $\psi_h = \psi_h(u)$ depending on the choice of an element $u$ from $\mathcal{B}^{(1)}$ the error of approximation on the element $u \in \mathcal{B}^{(1)}$ for scheme (21). Obviously, $\psi_h(u)$ is the residual emerging as a result of replacing $y_h$ by the element $u_h \in \mathcal{P}_h^{(1)} u$ in (21).

We say that
1) scheme (21) generates an approximation on an element $u \in \mathcal{B}^{(1)}$ if

$$(31) \qquad \lim_{|h| \to 0} \| \psi_h(u) \|_{(2_h)} = \lim_{|h| \to 0} \| \varphi_h - A_h u_h \|_{(2_h)} = 0 ;$$

2) scheme (21) is of the $n$th approximation order on an element $u \in \mathcal{B}^{(1)}$ if for sufficiently small $|h| \leq h_0$

$$(32) \qquad \| \psi_h(u) \|_{(2_h)} \leq M \,|h|^n \qquad \text{or} \qquad \| \psi_h(u) \|_{(2_h)} = O(\,|h|^n),$$

where $M$ is a positive constant independent of $h$, $h > 0$.

Our next goal is to establish direct links between stability, approximation on an element $u \in \mathcal{B}^{(1)}$ and convergence to $u$ for scheme (21). If scheme (21) is correct, then problem (30) for $z_h$ is well-posed. Because of this fact, its solution obeys the estimate

$$(33) \qquad \| z_h \|_{(1_h)} \leq M \, \| \psi_h \|_{(2_h)} .$$

In this respect, we obtain the profound result with the following theorem.

**Theorem 1** *If scheme* (21) *is correct and generates an approximation on an element* $u \in \mathcal{B}^{(1)}$, *then it converges. More precisely, a solution* $y_h$ *of problem* (21) *converges to this element* $u \in \mathcal{B}^{(1)}$ *as* $|h| \to 0$ *and, in addition, the order of accuracy of scheme* (21) *coincides with the order* *of approximation, that is, approximation and correctness (stability) of a scheme imply its convergence.*

Until now we spoke about convergence of a scheme and approximation on a fixed element $u$ of the space $\mathcal{B}^{(1)}$. However, if $u$ belongs to the domain of a linear operator $\mathcal{A}$ from $\mathcal{B}^{(1)}$ into $\mathcal{B}^{(2)}$, then $\mathcal{A}u = f$, $f \in \mathcal{B}^{(2)}$. Hence, $u$ can be adopted as a solution to the equation

$$(34) \qquad \mathcal{A}u = f\,, \qquad u \in \mathcal{B}^{(1)}, \qquad f \in \mathcal{B}^{(2)},$$

and, therefore, one can speak about the approximation of this equation by a difference scheme. The only reason we did not introduce equation (34) for it is that no restriction on the operator $\mathcal{A}$ was imposed in the definitions. Everywhere we have dealt only with an element $u \in \mathcal{B}^{(1)}$.

However, if $u$ is a solution to some equation like (34), then one can speak, as usual, about the approximation of equation (34) by scheme (21) on a solution of equation (34), about the convergence to a solution of equation (34), etc.

Once started with the notion of approximation to an element $f$ from $\mathcal{B}^{(2)}$ by a set $\{\varphi_h\}$ from $\{\mathcal{B}_h^{(2)}\}$, we can speak about the approximation of $f$ by the elements $\varphi_h$ as well as about the approximation of an operator $\mathcal{A}$ by $A_h$:

1) $\varphi_h$ approximates $f$ with order $h$ if

$$(35) \qquad \| \varphi_h - \mathcal{P}_h^{(2)} f \|_{(2_h)} = O(\,|h|^n\,);$$

2) an operator $A_h$ approximates an operator $\mathcal{A}$ with order $n$ if for any $u \in \mathcal{B}^{(1)}$

$$(36) \qquad \| A_h u_h - \mathcal{P}_h^{(2)}(\mathcal{A}u) \|_{(2_h)} = \| A_h(\mathcal{P}_h^{(1)}u) - \mathcal{P}_h^{(2)}(\mathcal{A}u) \|_{(2_h)} = O(\,|h|^n\,).$$

Obviously, if conditions (35) and (36) are satisfied, then scheme (21) is of the $n$th order of approximation on the solution $u$ to equation (34). With the relation $\mathcal{P}_h^{(2)}(f - \mathcal{A}u) = 0$ in view,

$$\psi_h(u) = \varphi_h - A_h u_h$$

$$= (\varphi_h - \mathcal{P}_h^{(2)} f) - (A_h(\mathcal{P}_h^{(1)}u) - \mathcal{P}_h^{(2)}(\mathcal{A}u)),$$

$$\| \psi_h(u) \|_{(2_h)} \le \| \varphi_h - \mathcal{P}_h^{(2)} f \|_{(2_h)} + \| A_h(\mathcal{P}_h^{(1)}u) - \mathcal{P}_h^{(2)}(\mathcal{A}u) \|_{(2_h)}$$

$$\le M\,|h|^n,$$

provided conditions (35)–(36) hold. From such reasoning it seems clear that the converse assertion, in general, fails to be true, that is, relations (32) do not imply conditions (35) and (36).

Let us stress once again that in order to estimate the order of accuracy of a scheme, it is necessary to estimate its order of accuracy only on a solution of the original problem.

So far we have always preassumed that the operator $A_h$ is linear (scheme (21) is linear). If $A_h$ is nonlinear (scheme (21) is nonlinear), the preceding arguments need minor changes only in the concept of stability. In dealing with a nonlinear scheme

$$(21^*) \qquad A_h\, y_h = \varphi_h\,, \qquad \varphi_h \in \mathcal{B}_h^{(2)};$$

where $y_h$ is a solution and $\bar{y}_h$ is a solution with the right-hand side $\bar{\varphi}_h \in \mathcal{B}_h^{(2)}$, scheme (21*) is said to be **stable** if there are positive constants $h_0 > 0$ and $M > 0$ independent of the parameter $h$ and disregarding to the choice of $\varphi_h$ and $\bar{\varphi}_h$ such that for $|\,h\,| < h_0$ the inequality

$$(22^*) \qquad \|\,\bar{y}_h - y_h\,\|_{(1_h)} \le M \|\,\bar{\varphi}_h - \varphi_h\,\|_{(2_h)}$$

is satisfied for any $\varphi_h, \bar{\varphi}_h \in \mathcal{B}_h^{(2)}$. We note in passing that for a linear scheme with $\bar{\varphi}_h = 0$ and $\bar{y}_h = 0$ this implies (22). All of the above definitions of approximation and convergence remain valid. Theorem 1 is also true. However, its proof follows another reasoning: instead of (30) it is more cinvenient to write down

$$A_h\, u_h = \bar{\varphi}_h\,, \qquad \bar{\varphi}_h = \varphi_h - \psi_h\,, \qquad \psi_h = \varphi_h - A_h u_h\,,$$

where $\psi_h$ is the approximation error (residual) on the element $u \in \mathcal{B}^{(1)}$. Denote by $y_h$ a solution of equation (21*), $\bar{y}_h = u_h$ and make use of the stability condition (22*). We obtain in this direction estimate (33)

$$\|\,z\,\|_{(1_h)} = \|\,y_h - u_h\,\|_{(1_h)} \le M\ \|\,\psi_h\,\|_{(2_h)} = M\ \|\,\varphi_h - A_h\, u_h\,\|_{(2_h)}\,,$$

thereby completing the proof of Theorem 1 for a nonlinear scheme as well.

**4. Some a priori estimates.** We now consider several simplest a priori estimates for a solution to equation (21), the form of which depends on the subsidiary information on the operator of a scheme. These estimates are typical for difference elliptic problems.

For simplicity of writing we will omit the subscript $h$ if this does not cause an ambiguity. The equation we must solve is of the form

$$(37) \qquad A\, y = \varphi\,,$$

where $A$ is a linear bounded operator defined in a real Hilbert space $H$, $\varphi$ is given and $y$ is sought in the space $H$.

We will assume that problem (37) is solvable for any right-hand sides $\varphi \in H$: there exists an operator $A^{-1}$ with the domain $\mathcal{D}(A^{-1}) = H$. All the constants below are supposed to be independent of $h$. In what follows the space $H$ is equipped with an inner product $(\ ,\ )$ and associated norm $\| x \| = \sqrt{(x, x)}$. The writing $A = A^* > 0$ means that $A$ is a self-adjoint positive operator. Set

$$\| \varphi ]|_{A^{-1}} = \sqrt{(A^{-1}\varphi, \varphi)}, \qquad \| y \|_A = \sqrt{(A y, y)}, \qquad A = A^* > 0 .$$

An a priori estimate depends on the nature of the subsidiary information on the operator $A$.

Let $H$ be a finite-dimensional space.

1) Consider first the simplest case when $A$ is a non-self-adjoint positive definite operator:

$$(38)\quad A \geq \delta E, \quad \delta > 0 \qquad \text{or} \qquad (A y, y) \geq \delta \| y \|^2 \qquad \text{for any} \quad y \in H,$$

where $E$ is the identity operator. Then the inverse $A^{-1}$ is bounded in norm by the constant $1 / \delta$:

$$(39) \qquad \| A^{-1} \| \leq \frac{1}{\delta} .$$

Indeed,

$$\begin{aligned} 0 \leq (Ay, y) - \delta \| y \|^2 &= (A^{-1}x, x) - \delta \| A^{-1}x \|^2 \\ &\leq \| A^{-1}x \| \cdot \| x \| - \delta \| A^{-1}x \|^2 \\ &= \delta \| A^{-1}x \| \left( \frac{1}{\delta} \| x \| - \| A^{-1}x \| \right), \qquad x = Ay, \end{aligned}$$

implying that $\| A^{-1}x \| \leq \delta^{-1} \| x \|$ or $\| A^{-1} \| \leq \delta^{-1}$. Since $y = A^{-1}\varphi$ and $\| y \| \leq \| A^{-1} \| \cdot \| \varphi \|$, the solution to equation (37) admits the estimate

$$(40) \qquad \| y \| \leq \frac{1}{\delta} \| \varphi \| \qquad \text{for} \qquad A \geq \delta E, \quad \delta > 0 .$$

2) The precise estimate appears to be useful:

$$\| y \|_A = \| \varphi \|_{A^{-1}} \qquad \text{for} \qquad A = A^* > 0 .$$

To make sure of it, we take the inner product of (37) and $y = A^{-1}\varphi$:

$$(A y, y) = (\varphi, y) = (\varphi, A^{-1}\varphi)$$

or $\| y \|_A^2 = \| \varphi \|_{A^{-1}}^2$.
3) If $A > 0$, then

$$\| y \|_{A_0} \le \| \varphi \|_{A_0^{-1}}, \qquad A_0 = \tfrac{1}{2}(A + A^*). \tag{41}$$

Furthermore, taking the inner product of (37) and $y$ we obtain

$$(A y, y) = (\varphi, y), \qquad A = A_0 + A_1, \tag{42}$$

where $A_0 = \frac{1}{2}(A + A^*)$ is the symmetric part and $A_0 = \frac{1}{2}(A - A^*)$ is the skew-symmetric part of the operator $A$. This provides the sufficient background for the relation $(A y, y) = (A_0 y, y)$ due to the fact that $(A_1 y, y) = 0$. As far as $A_0 > 0$, the inverse operator $A_0^{-1}$ exists, because the space $H$ is finite-dimensional*) and, therefore, by the generalized Cauchy–Bunyakovskiĭ inequality one can write down

$$(\varphi, y) = (A_0^{-1}\varphi, A_0 y) \le \| \varphi \|_{A_0^{-1}} \cdot \| y \|_{A_0}.$$

Substitution of this inequality into (42) leads to the relations

$$(A_0 y, y) = \| y \|_{A_0}^2 \le \| y \|_{A_0} \cdot \| \varphi \|_{A_0^{-1}},$$

from which the desired estimate (41) immediately follows.
4) If $A \ge \delta E$ with $\delta > 0$, then

$$\| y \|_{A_0} \le \frac{1}{\sqrt{\delta}} \| \varphi \|. \tag{43}$$

To prove this fact, we make use of estimate (41) and the inequality

$$\| \varphi \|_{A_0^{-1}} \le \delta^{-1/2} \| \varphi \|,$$

which is a consequence of

$$A_0 \ge \delta E \quad \text{and} \quad \| \varphi \|_{A_0^{-1}}^2 = (A_0^{-1}\varphi, \varphi) \le \| A_0^{-1} \| \cdot \| \varphi \|^2 \le \frac{1}{\delta} \| \varphi \|^2.$$

---

*)If $H$ is an infinite-dimensional space, one should require instead of $A > 0$ that $A \ge \delta E$ with $\delta > 0$.

5) Assuming $A$ to be a non-self-adjoint operator subject to the inequality $A \geq \gamma \mathring{A}$ with $\gamma > 0$ and $\mathring{A}^* = \mathring{A} > 0$, we derive for a solution to equation (37) the estimate

$$\| y \|_{\mathring{A}} \leq \frac{1}{\gamma} \| \varphi \|_{\mathring{A}^{-1}} . \tag{44}$$

The identity $( A y, y ) = ( \varphi, y )$ yields the chain of the relations

$$\gamma ( \mathring{A} y, y ) \leq ( A y, y ) = ( \varphi, y ) \leq \| \varphi \|_{\mathring{A}^{-1}} \cdot \| y \|_{\mathring{A}} ,$$

thereby providing the validity of (44).

6) For a solution to equation (37) we thus have

$$\| \mathring{A} y \| \leq \frac{1}{\gamma} \| \varphi \| \tag{45}$$

under the following conditions:

$$A^* = A \geq \gamma \mathring{A} , \qquad \mathring{A} = \mathring{A}^* > 0 , \qquad A \mathring{A} = \mathring{A} A , \qquad \gamma > 0 .$$

It suffices to show that $\| A y \| \geq \gamma \| \mathring{A} y \|$ and to involve in further reasoning equation (37), giving $\| A y \| = \| \varphi \|$. The conditions $A \geq \gamma \mathring{A}$ and $A \mathring{A} = \mathring{A} A$ together imply the chain of the relations

$$\begin{aligned}
\| A y \|^2 &= ( A y, A y ) = ( A ( A^{1/2} y ), ( A^{1/2} y )) \\
&\geq \gamma ( \mathring{A} ( A^{1/2} y ), ( A^{1/2} y )) = \gamma ( A \mathring{A} y, y ) \\
&= \gamma ( A ( \mathring{A}^{1/2} y ), ( \mathring{A}^{1/2} y )) \geq \gamma^2 ( \mathring{A} ( \mathring{A}^{1/2} y ), ( \mathring{A}^{1/2} y )) \\
&= \gamma^2 \| \mathring{A} y \|^2 ,
\end{aligned}$$

meaning $\| A y \| \geq \gamma \| \mathring{A} y \|$. Here we exploit the fact that the operators $A$ and $A^{1/2}$ as well as the operators $\mathring{A}$ and $\mathring{A}^{1/2}$ commute with each other.

**5. Negative norms.** In a priori estimates (41) and (42) we were dealing with the negative norm $\| \varphi \|_{A^{-1}}$. Just for this reason the question of computing negative norms for the simplest operators will be of great importance. First of all we quote a result of simple observations:

$$\| \varphi \|_{A^{-1}} \le \frac{1}{\sqrt{\gamma}} \| \varphi \|_{\overset{\circ}{A}^{-1}} \quad \text{if} \quad A \ge \gamma \overset{\circ}{A}, \tag{46}$$

where $A = A^* > 0$ and $\overset{\circ}{A} = \overset{\circ}{A}{}^* > 0$. This follows from the equivalence of the operator inequalities

$$A \ge \gamma \overset{\circ}{A} \quad \text{and} \quad \overset{\circ}{A}{}^{-1} \ge \gamma A^{-1}. \tag{47}$$

Our first goal is to prove the equivalence between the inequalities $Q \ge 0$ and $L^* Q L \ge 0$, where $Q$, $L$: $H \mapsto H$ and $L^{-1}$ exists. Indeed,

$$(L^* Q L y, y) = (Q L y, L y) = (Q v, v),$$

where $v = L y$ and $y = L^{-1} v$. Accepting $A \ge \gamma \overset{\circ}{A}$ or $Q = A - \gamma \overset{\circ}{A} \ge 0$ and setting $L = L^* = \overset{\circ}{A}{}^{-1/2}$, we obtain $C - \gamma E \ge 0$, where

$$C = \overset{\circ}{A}{}^{-1/2} A \overset{\circ}{A}{}^{-1/2} = C^* > 0 .$$

Since the operator $C^{-1/2} = (C^{-1/2})^* > 0$ does exist, the inequality $C - \gamma E \ge 0$ is equivalent to

$$C^{-1/2} (C - \gamma E) C^{-1/2} = E - \gamma C^{-1} = E - \gamma \overset{\circ}{A}{}^{1/2} A^{-1} \overset{\circ}{A}{}^{1/2} \ge 0 .$$

By merely setting $L = L^* = \overset{\circ}{A}{}^{-1/2}$ we get $\overset{\circ}{A}{}^{-1} - \gamma A^{-1} \ge 0$, what means that inequalities (47) are equivalent as required.

**Example 4** Consider the first boundary-value problem

$$\begin{aligned} \Lambda y &= (a y_{\bar{x}})_x = -\varphi(x_i), \quad x_i = i h, \quad i = 1, 2, \ldots, N-1, \\ h N &= 1, y_0 = y_N = 0, \quad a_i \ge c_1 > 0, \quad i = 1, 2, \ldots, N, \end{aligned} \tag{48}$$

on the uniform grid $\bar{\omega}_h = \{ x_i = i h, \ i = 0, 1, 2, \ldots, N, \ h = 1 / N \}$. One assumes, as usual, that $H_h = \overset{\circ}{\Omega}_h$ is the space of all grid functions defined on $\bar{\omega}_h$ and vanishing for $i = 0$ and $i = N$ under the inner product structure $(y, v) = \sum_{i=1}^{N-1} y_i v_i h$ in $H$.

As a first step towards the solution of this problem, the operator $A$ is defined by the relation $Ay = -\widetilde{\Lambda} y$, where $\widetilde{\Lambda} y = \Lambda y$ for $y \in \Omega$, so that

$$(49)\quad (Ay)_i = -(a y_{\bar{x}})_{x,i}, \quad i = 1, 2, \ldots, N-1, \quad (Ay)_0 = (Ay)_N = 0$$

and the vector $\varphi$ is taken to be $\varphi = (0, \varphi_1, \varphi_2, \ldots, \varphi_{N-1}, 0)$. The next step is to recast the problem in view as an operator equation

$$Ay = \varphi.$$

It is easy to see from Green's formulae

$$(y, (a y_{\bar{x}})_x) = -(a, (y_{\bar{x}})^2]$$

and

$$(v, (a y_{\bar{x}})_x) = (y, (a v_{\bar{x}})_x),$$

where $y, v \in \overset{\circ}{\Omega}$, that the operator $A$ is self-adjoint and positive definite, that is, $A \geq \delta E$, $\delta = 8 c_1$. Recall that we have established in Section 3 from Chapter 2 that

$$(Ay, y) = (a, (y_{\bar{x}})^2] \geq c_1 (1, (y_{\bar{x}})^2] \geq 8 c_1 \| y \|^2 .$$

This implies that the inverse $A^{-1}$ exists and $(A^{-1})^* = A^{-1} > 0$. More exactly,

$$(50)\qquad \frac{1}{\| A \|} E \leq A^{-1} \leq \frac{1}{\delta} E .$$

Let us show that the negative norm $\| \varphi \|_{A^{-1}}$ of operator (49) is representable by

$$(51)\qquad \| \varphi \|^2_{A^{-1}} = \sum_{i=1}^{N} \frac{h}{a_i} S_i^2 - \left( \sum_{i=1}^{N} \frac{h}{a_i} S_i \right)^2 \left( \sum_{i=1}^{N} \frac{h}{a_i} \right)^{-1},$$

$$(52)\qquad S_i = \sum_{k=1}^{i-1} h \varphi_k , \qquad i = 2, 3, \ldots, N, \qquad S_1 = 0 .$$

All this enables us to derive for this norm the estimate

$$(53) \qquad \| \varphi \|_{A^{-1}} \le \frac{1}{\sqrt{c_1}} \left\| \sum_{k=1}^{i-1} h \varphi_k \right\| .$$

For its proof the right-hand side of equation (48) is expressed in the form $\varphi_i = S_{x,i}$, $i = 1, 2, \ldots, N-1$, where $S_i$ is specified by conditions (52). The equation $(a y_{\bar{x}})_x + \varphi = (a y_{\bar{x}} + S)_x = 0$ implies that $a_i y_{\bar{x},i} + S_i = c =$ const for all $i = 1, 2, \ldots, N$. Whence $y_i - y_{i-1} = h(c - S_i)/a_i$. Summing up over $i = 1, 2, \ldots, N$ and taking into account that $y_0 = y_N$, we deduce that

$$c = \left( \sum_{i=1}^{N} \frac{h}{a_i} S_i \right)^2 \left( \sum_{i=1}^{N} \frac{h}{a_i} \right)^{-1} .$$

On the other hand,

$$\| \varphi \|_{A^{-1}}^2 = (A^{-1} \varphi, \varphi) = (y, \varphi) = (y, S_x) = -(S, y_{\bar{x}}]$$
$$= -\sum_{i=1}^{N} h\, y_{\bar{x},i} S_i .$$

Substituting $y_{\bar{x},i} = (c - S_i)/a_i$ yields

$$\| \varphi \|_{A^{-1}}^2 = -c \sum_{i=1}^{N} \frac{h S_i}{a_i} + \sum_{i=1}^{N} \frac{h}{a_i} S_i^2$$
$$= \sum_{i=1}^{N} \frac{h}{a_i} S_i^2 - \left( \sum_{i=1}^{N} \frac{h S_i}{a_i} \right)^2 \left( \sum_{i=1}^{N} \frac{h}{a_i} \right)^{-1}$$

as required. Estimate (53) is simple to follow:

$$\| \varphi \|_{A^{-1}}^2 \le \sum_{i=1}^{N} \frac{h}{a_i} S_i^2 \le \frac{1}{c_1} \sum_{i=1}^{N} h S_i^2$$
$$= \frac{1}{c_1} \sum_{i=1}^{N-1} h S_{i+1}^2 = \frac{1}{c_1} \left\| \sum_{k=1}^{i} h \varphi_k \right\|^2 .$$

**Example 5** Consider now the third boundary-value problem (9). As in Example 2 of Section 1 it will be convenient to introduce the space $H_h = \bar{\Omega}_h$ of the dimension $N+1$ consisting of all grid functions defined on the uniform grid $\bar{\omega}_h = \{ x_i = ih,\ i = 0, 1, 2, \ldots, N,\ h = 1/N \}$ under the inner product structure $[y, v] = \sum_{i=1}^{N-1} y_i v_i h + \frac{1}{2} h (y_0 v_0 + y_N v_N)$. Then problem (9) reduces to an operator equation of the form

$$A y = \varphi ,$$

where $y$ and $\varphi$ are the vectors of the dimension $N+1$:

$$y = \{ y_0, y_1, \dots, y_{N-1}, y_N \}, \quad \varphi = \left\{ \frac{1}{0.5h} \mu_1, f_1, \dots, f_{N-1}, \frac{1}{0.5h} \mu_2 \right\}$$

and the operator $A$ acts in accordance with the rule

$$(54) \qquad (A y)_i = \begin{cases} -\dfrac{1}{0.5h} (y_{x,0} - \sigma_1 y_0), & i = 0, \\ -y_{\bar{x}x,i}, & 1 \le i \le N-1, \\ -\dfrac{1}{0.5h} (y_{\bar{x},N} - \sigma_2 y_N), & i = N. \end{cases}$$

The negative norm of operator (54) is expressed by the collection of formulae

$$(55) \quad \| \varphi \|_{A^{-1}}^2 = [A^{-1}\varphi, \varphi]$$

$$= \sum_{i=1}^{N} h S_i^2 + \frac{1}{\sigma_2} S_{N+1}^2 - \left( \sum_{i=1}^{N} h S_i + \frac{1}{\sigma_2} S_{N+1} \right)^2 \Big/ \left( 1 + \frac{1}{\sigma_1} + \frac{1}{\sigma_2} \right)^2,$$

$$(56) \quad S_1 = \tfrac{1}{2} h \varphi_0, \qquad S_i = \tfrac{1}{2} h \varphi_0 + \sum_{k=1}^{i-1} h \varphi_k, \qquad i = 2, 3, \dots, N,$$

$$S_{N+1} = \tfrac{1}{2} h (\varphi_0 + \varphi_N) + \sum_{k=1}^{N-1} h \varphi_k$$

and admits the cotimate

$$(57) \qquad \| \varphi \|_{A^{-1}}^2 \le \| \frac{1}{2} h \varphi_0 + \sum_{k=1}^{i} \varphi_k ] \|^2 + \frac{1}{\sigma_2} [1, \varphi]^2.$$

In establishing these relations the grid should be enlarged by recording two "artifical" points $x_{-1} = -h$ and $x_{N+1} = 1 + h$ and assigning the values $y(x_{-1}) = y_{-1} = 0$ and $y(x_{N+1}) = y_{N+1} = 0$. All this enables us to impose the boundary condition in (9) for $i = 0$ as follows:

$$-\frac{(y_1 - y_0) - a_0(y_0 - y_{-1})}{h^2} = \tilde{\varphi}_0, \qquad a_0 = h \sigma_1, \qquad \tilde{\varphi}_0 = \frac{1}{2} \varphi_0 .$$

In just the same way as before, we write down the boundary condition in (9) for $i = N$:

$$-\frac{a_{N+1}(y_{N+1} - y_N) - (y_N - y_{N-1})}{h^2} = \tilde{\varphi}_N ,$$

$$a_{N+1} = h\,\sigma_2 , \qquad \tilde{\varphi}_N = \frac{1}{2}\,\varphi_N .$$

In such a setting problem (9) turns out to be equivalent to the first boundary-value problem

$$(58) \qquad -(a\,y_{\bar{x}})_{x,\,i} = \tilde{\varphi}_i , \qquad i = 0,\, 1,\, \ldots,\, N , \qquad y_{-1} = y_{N+1} = 0 ,$$

where $\tilde{\varphi}_i = \varphi_i$, $i = 1, 2, \ldots, N-1$, $\tilde{\varphi}_0 = \frac{1}{2}\varphi_0$, $\tilde{\varphi}_N = \frac{1}{2}\varphi_N$, $a_i = 1$, $i = 1, 2, \ldots, N$ and $a_0 = h\,\sigma_1$, $a_{N+1} = h\,\sigma_2$. If $y$ is a solution of problem (9), then

$$\begin{aligned}
[A^{-1}\varphi,\, \varphi] &= [y,\, \varphi] = \tfrac{1}{2}\,h\,(y_0\,\varphi_0 + y_N\,\varphi_N) + \sum_{i=1}^{N-1} y_i\,\varphi_i\,h \\
&= \sum_{i=1}^{N-1} y_i\,\varphi_i\,h + h\,(y_0\,\tilde{\varphi}_0 + y_N\,\tilde{\varphi}_N) = \sum_{i=0}^{N} y_i\,\tilde{\varphi}_i\,h \\
&= \sum_{i=0}^{N} (A^{-1}\tilde{\varphi})_i\,\tilde{\varphi}_i\,h .
\end{aligned}$$

By applying successively formulae (51)–(52) to problem (58) and estimate (53) we arrive at (55) and (57).

**Corollary** *If $\sigma_2 \geq c_1 > 0$, then operator (54) admits the estimate*

$$(59) \qquad \|\,y\,\|^2_{A^{-1}} \leq \left\| \left[ \sum_{k=1}^{i-1} h\,\varphi_k + \frac{1}{2}\,h\,\varphi_0 \right] \right|^2 + \frac{1}{c_1}\,[1,\, \varphi]^2 .$$

**6. Operator equations of divergent type.** We now deal with operators of the special structure known as divergent or conservative operators:

$$(60) \qquad A = T^* S\, T ,$$

where $T$, $S$ and $T^*$ are linear bounded operators. Operators of this type are frequently encountered in this book in the approximation of differential operators having the form $L\,u = \operatorname{div}(\,k \operatorname{grad} u\,)$. Let $H$ be a vector space with the inner product $(y, v)$ and associated norm $\| y \| = \sqrt{(y, y)}$ and let $H_1$ be a vector space with the inner product $(y, v]$ and associated norm $\| y \| = \sqrt{(y, y]}$. Let, in addition, $T$ be an operator from $H$ into $H_1$, $S$ be an operator from $H_1$ into $H_1$ and $T^*$ be an operator from $H_1$ into $H$. Then the operator $A$ really acts from $H$ into $H$: $A: H \mapsto H$. The operators $T$ and $T^*$ are mutually adjoint in the following sense:

$$(T y, v] = (y, T^* v) \quad \text{for all} \quad y \in H, \ v \in H_1 .$$

Before going further, we give below several examples illustrating how to construct factorized operators (60) for the simplest difference schemes.

**Example 6** Of our initial concern is the first boundary-value problem

$$(61) \qquad (a y_{\bar{x}})_x = -\varphi , \qquad 0 < x = i h < 1 ,$$
$$a \geq c_1 > 0 , \qquad y_0 = 0 , \qquad y_N = 0 .$$

In this case $H \in \Omega_h$ is the space of all grid functions defined at the inner nodes of the uniform grid $\bar{\omega}_h = \{ x_i = i h , i = 0, 1, 2, \dots , N , h N = 1 \}$, that is, for $0 < i < N$ and $H_1 = \Omega_h^+$ is the space of all grid functions defined at the nodes of the uniform grid $\bar{\omega}_h^+ = \{ x_i = i h , i = 1, 2, \dots , N , h N = 1 \}$. The operator $A: H \mapsto H$ equals $A y = -\tilde{\Lambda} y$, where $\tilde{\Lambda} y$ takes the form

$$(\tilde{\Lambda} y)_i = (a y_{\bar{x}})_{x, i} , \qquad i = 2, 3, \dots, N - 2 ,$$

$$(\tilde{\Lambda} y)_1 = \frac{a_2 (y_2 - y_1) - a_1 y_1}{h^2} ,$$

$$(\tilde{\Lambda} y)_{N-1} = \frac{-a_N y_{N-1} - a_{N-1} (y_{N-1} - y_{N-2})}{h^2} .$$

The inner products in the spaces $H$ and $H_1$ are defined by

$$(y, v) = \sum_{i=1}^{N-1} y_i v_i h \quad \text{for} \quad y_i , v_i \in H ,$$

$$(y, v] = \sum_{i=1}^{N} y_i v_i h \quad \text{for } y_i, v_i \in H_1 .$$

Under these structures, we specify the operators $Ty$, $T^* v$ and $Sv$ by the relations

$$(T y)_i = \frac{y_i - y_{i-1}}{h} = y_{\bar{x}, i} , \qquad i = 2, 3, \dots, N - 1 ,$$

$$(T y)_1 = \frac{y_1}{h} , \qquad (T y)_N = - \frac{y_{N-1}}{h} ,$$

so that $v = T\,y \in H_1$ if $y \in H$;

$$(T^* v)_i = -\frac{v_{i+1} - v_i}{h}\,, \qquad i = 1, 2, \ldots, N-1\,,$$

so that $T^* v \in H$ if $v \in H_1$. Finally, $(S\,v)_i = a_i\,v_i$ for $i = 1, 2, \ldots, N$, that is, $S\,v \in H_1$ if $v \in H_1$ and $(S\,v,\, v] \ge c_1 \| v \|^2$. From such reasoning it seems clear that

$$(S\,T\,y)_i = a_i\,y_{\bar{x},\,i}\,, \qquad i = 2,\,3,\,\ldots,\,N-1\,, \qquad (S\,T\,y)_1 = a_1\,y_1\,/\,h\,,$$

$$(S\,T\,y)_N = -a_N\,y_{N-1}\,/\,h\,,$$

$$(T^* S\,T\,y)_i = -(a\,y_{\bar{x}})_{\bar{x},\,i}\,, \qquad i = 1, 2, \ldots, N\,, \qquad \text{for} \qquad y_0 = y_N = 0\,,$$

which serve to motivate that the operator of problem (61) can be factorized in the form (60) so as to have instead of (37)

$$T^* S\,T\,y = \varphi\,. \tag{62}$$

The fact that the operators $T$ and $T^*$ are mutually adjoint is a corollary to the summation by parts formula: $(y,\, v_x) = -(v,\, y_{\bar{x}}]$, so that

$$(y,\, T^* v) = -(v,\, T\,y]\,.$$

**Example 7** The third boundary-value problem comes next:

$$(a\,y_{\bar{x}})_x = -\varphi\,, \qquad 0 < x = i\,h < 1\,, \qquad a \ge c_1 > 0\,, \tag{63}$$

$$a_1\,y_{\bar{x},\,1} = \sigma_1\,y_0 - \mu_1\,, \qquad -a_N\,y_{\bar{x},\,N} = \sigma\,2y_N - \mu_2\,,$$

$$\sigma_1 \ge c_1 > 0\,, \qquad \sigma_2 \ge c_1 > 0\,.$$

In this case the operator $A$ is of the form (see Example 2 in Section 1):

$$(A\,y)_i = \begin{cases} -\dfrac{1}{0.5\,h}\,(a_1\,y_{\bar{x},\,1} - \sigma_1\,y_0)\,, & i = 0\,, \\ -(a y_{\bar{x}})_x\,, & i = 1, 2, \ldots, N-1\,, \\ \dfrac{1}{0.5\,h}\,(a_N\,y_{\bar{x},N} + \sigma_2\,y_N)\,, & i = N\,. \end{cases} \tag{64}$$

The operator (64) can be made of the same type as the preceding operator (60) if we impose another grid

$$\bar{\omega}_h^1 = \{x_0,\, x_{1/2},\, \ldots,\, x_{i-1/2},\, \ldots,\, x_{N-1/2},\, x_N\}\,, \qquad x_{i-1/2} = (i - \tfrac{1}{2})\,h\,,$$

and work in the space $H_1$ of all grid functions defined on the grid $\bar{\omega}_h^1$ under the inner product structure $(y, v] = \sum_{i=1}^{N} h y_{i-1/2} v_{i-1/2} + y_N v_N + y_0 v_0$ and the norm $\| y]| = \sqrt{(y, y]}$. As before, $H$ is the space of all grid functions defined on the uniform grid $\bar{\omega}_h = \{ x_i = ih, i = 0, 1, \dots, N, h = 1/N \}$ with the inner product $(y, v) = \sum_{i=1}^{N-1} y_i v_i h + \frac{1}{2} h (y_0 v_0 + y_N v_N)$ and the associated norm $\| y \| = \sqrt{(y, y)}$. We specify the operators $T$: $H \mapsto H_1$ and $T^*$: $H_1 \mapsto H$ as follows:

$$(Ty)_0 = y_0, \quad (Ty)_{i-1/2} = \frac{y_i - y_{i-1}}{h}, \quad i = 1, 2, \dots, N, \quad (Ty)_N = -y_N,$$

$$(T^* y)_0 = -\frac{v_{1/2} - v_0}{0.5h}, \qquad (T^* v)_N = -\frac{v_N - v_{N-1/2}}{0.5h},$$

$$(T^* v)_i = -\frac{v_{i+1/2} - v_{i-1/2}}{h}, \quad i = 1, 2, \dots, N-1, \quad y \in H, \quad v \in H_1.$$

This provides support for the view that $A = T^* S T$, where the operator $S$: $H_1 \mapsto H_1$ is defined by the relations

$$(Sy)_0 = \sigma_1 y_0, \qquad (Sy)_{i-1/2} = a_i y_{i-1/2}, \qquad 1 \le i \le N,$$
$$(Sy)_N = \sigma_2 y_N.$$

Obviously, $S$ is a self-adjoint operator in $H_1$ and $(Sy, y] \ge c_1 \| y]|^2$ with constant $c_1 = \min(a_i, \sigma_1, \sigma_2)$. Let us show that the operators $T$ and $T^*$ are mutually adjoint in the following sense: $(T^* v, y) = (v, Ty]$ for $y \in H$, $v \in H_1$. Indeed,

$$
\begin{aligned}
(T^* v, y) &= -\sum_{i=1}^{N-1} (v_{i+1/2} - v_{i-1/2}) y_i - y_0 (v_{1/2} - v_0) \\
&\quad - y_N (v_N - v_{N-1/2}) \\
&= \sum_{i=1}^{N} v_{i-1/2} (y_i - y_{i-1}) + y_0 v_0 - y_N v_N \\
&= (v, Ty].
\end{aligned}
$$

We are now in a position to derive some a priory estimates for the equation $Ay = T^* S T y = \varphi$. Under the natural premise $S \ge c_1 E$, $c_1 > 0$, we obtain

$$(Ay, y) = (T^*STy, y) = (STy, Ty] \geq c_1 ||Ty]|^2,$$

meaning $A \geq \gamma A_0$, where $A_0 = T^*T$ and $\gamma = c_1$. The operator $A_0$ is self-adjoint:

$$(A_0 y, z) = (T^*Ty, z) = (y, T^*Tz) = (y, A_0 z).$$

Therefore, estimate (44) holds true and takes the form

$$||Ty]| \leq \frac{1}{c_1} ||(T^*)^{-1}\varphi-]| . \tag{65}$$

It is worth bearing in mind here that the inverses $T^{-1}$ and $(T^*)^{-1}$ do exist. Indeed,

$$||y||^2_{A_0} = (A_0 y, y) = ||Ty]|^2,$$

$$||\varphi||^2_{A_0^{-1}} = (T^{-1}(T^*)^{-1}\varphi, \varphi) = ||(T^*)^{-1}\varphi]|^2.$$

Estimate (65) can be simplified in the case when the right-hand side $\varphi$ of equation (37) is of the special form $\varphi = T^*\eta$ and $A = T^*ST$. The inner product of (37) and $y$ discovers the relationships

$$(T^*STy, y) = (T^*\eta, y) = (Ty, \eta].$$

Putting these together with the inequalities

$$(T^*STy, y) \geq c_1 ||Ty]|^2, \qquad (Ty, \eta] \leq ||Ty]| \cdot ||\eta]|,$$

we derive the estimate $||Ty]| \leq c_1{}^{-1} ||\eta]|$.

We have restricted ourselves to the simplest examples demonstrating how the a priori estimates that can be obtained through such an analysis for the operator equation $Ay = \varphi$ apply equally well to important problems arising in theory and practice.

# 3

# Homogeneous Difference Schemes

This chapter presents the theory of homogeneous difference schemes for the solution of equations with variable coefficients

$$L\,u + f(x) = 0\,, \qquad L\,u \;=\; \frac{d}{dx}\left( k(x)\,\frac{du}{dx} \right) - g(x)\,u\,.$$

A special attention is being paid to various forms of homogeneous difference schemes and their approximation and convergence for discontinuous coefficients $k$, $q$ and $f$ as well as for non-equidistant grids. Not much is known in such cases. Later in this chapter we will survey some devices that can be used to obtain simpler forms and higher orders.

## 3.1 HOMOGENEOUS SCHEMES FOR SECOND-ORDER EQUATIONS WITH VARIABLE COEFFICIENTS

**1. Introduction.** Modern computers permit implementation of highly accurate difference schemes. Just for this reason, it is unreasonable to develop difference methods and create high quality software for solving particular problems. An actual problem consists of constructing difference schemes capable of describing classes of problems that are determined by a given type of the governing differential equation, a class of boundary and initial conditions as well as of forming a functional space comprising the coefficients of the differential equation. Naturally, such universal difference schemes should possess the convergence and stability properties on any sequence of grids and for any original problem from the given class. The requirement of universality of computational algorithms for solving a class of problems necessitates imposing the notion of homogeneous difference scheme. By a **homogeneous difference scheme** we mean one whose form is independent of a concrete problem from this class and the choice of a grid. At all grid nodes the difference equations take the same form for any problem from this class. The coefficients of a homogeneous difference scheme are treated as functionals of the coefficients of the differential equation.

For instance, of great interest are "through" or "continuous" execution schemes available for solving the diffusion equation with discontinuous diffusion coefficients by means of the same formulae (software). No selection of points or lines of discontinuities of the coefficients applies here. This means that the scheme remains unchanged in a neighborhood of discontinuities and the computations at all grid nodes can be carried out by the same formulae without concern of discontinuity or continuity of the diffusion coefficient.

Homogeneous "through" execution schemes are quite applicable in the cases where the diffusion coefficient is found as an approximate solution of other equations. For instance, such schemes are aimed at solving the equations of gas dynamics in a heat conducting gas when the diffusion coefficient depends on the density and has discontinuities on the shock waves.

In the theory of difference schemes with a **primary family** of schemes the coefficients of a homogeneous difference scheme are expressed through the coefficients of the initial differential equation by means of the so-called **pattern functionals**; the arbitrariness in the choice of these functionals is limited by the requirements of approximation, solvability, etc. There are various ways of taking care of these restrictions. The availability of a primary family of homogeneous difference schemes is ensured by a family of admissible pattern functionals known in advance.

Let us clarify the subject of investigation in a more simpler situation through the use of difference operators acting on functions of only one variable $x_i = i\,h$, $i = 0, \pm 1, \ldots$. One way of proceeding is connected with the following two steps. A difference operator is defined beforehand on an **integral pattern**, that is, on a set of the type $\mathfrak{M}_0 = \{-m_1, -m_1 + 1, \ldots, -1, 0, 1, \ldots, m_2\}$, where $m_1$, $m_2$ are positive integers. The next step is the transition to the real grid $\omega_h = \{x_i = i\,h,\ i = 0, \pm 1, \ldots\}$

with step $h$. Let $\bar{k}(s)$ be a vector function defined for $-m_1 \le s \le m_2$ and called the coefficient pattern. In the sequel we are dealing with **pattern functionals**

$$A_j^h\,[\bar{k}(s)]\,, \qquad -m_1 \le j \le m_2\,, \qquad B^h\,[\bar{k}(s)]\,,$$

which usually depend on the parameter $h$ and are defined for the vector functions $\bar{k}(s)$, $s \in [-m_1, m_2]$. By a linear with respect to a grid function $y^h$ **homogeneous difference scheme** is meant $\left(L_h^{(k)} y^h\right)_i = 0$, where

$$\left(L_h^{(k)} y^h\right)_i = \sum_{m=-m_1}^{m_2} A_m^h [\,k\,(x_i + s\,h)]\; y^h(x_i + m\,h) + B^h[\,k\,(x_i + s\,h)]\,.$$

Omitting the subscript $i$ one can rewrite the preceding as

$$L_h^{(k)} y^h = \sum_{m=-m_1}^{m_2} A_m^h [\,k\,(x + s\,h)]\; y^h(x + m\,h) + B^h[\,k\,(x + s\,h)]\,.$$

The principal question in the theory of homogeneous difference schemes is connected with further design of admissible schemes within a primary family for solving a class of typical problems as wide as possible and choosing the most efficient ones (in accuracy, volume of computations, etc.).

## 3.2 CONSERVATIVE SCHEMES

**1. An example of the scheme which is divergent in the case of discontinuous coefficients.** We now consider problem (1) of Section 2.1 with $q \equiv 0$ and $f \equiv 0$ incorporated:

(1) $$(ku')' = 0\,, \qquad 0 < x < 1\,, \qquad u(0) = 1\,, \qquad u(1) = 0\,.$$

As one might expect, the derivative $(ku')'$ should be replaced by $ku'' + k'u'$. As a first step towards the construction of a second-order approximation, it will be sensible to carry out the forthcoming substitutions

$$u'' \sim u_{\bar{x}x}\,, \quad k' \sim k_{\overset{\circ}{x}} = \frac{k_{i+1} - k_{i-1}}{2\,h}\,, \quad u' \sim u_{\overset{\circ}{x}} = \frac{u_{i+1} - u_{i-1}}{2\,h}\,.$$

Within these notations, a reasonable form of the difference scheme is

(2) $$k_i\,\frac{y_{i+1} - 2\,y_i + y_{i-1}}{h^2} + \frac{k_{i+1} - k_{i-1}}{2\,h}\;\frac{y_{i+1} - y_{i-1}}{2\,h} = 0\,,$$

$$0 < i < N, \qquad y_0 = 1, \qquad y_N = 0\,.$$

Reducing scheme (2) to the form (4) from Section 2.1 we find that

$$(3)\quad a_i = k_i - \frac{k_{i+1} - k_{i-1}}{4}\,, \qquad b_i = k_i + \frac{k_{i+1} - k_{i-1}}{4}\,, \qquad d_i = \varphi_i = 0\,,$$

whence it follows that scheme (2) belongs to family (4) from the preceding section.

Conditions (5) and (6) from Section 2.1 hold true, since on segments, where the function $k(x)$ is smooth enough, the relations occur:

$$a_i = k_i - \tfrac{1}{2}\,h\,k_i' + O(h^2)\,, \qquad b_i = k_i + \tfrac{1}{2}\,h\,k_i' + O(h^2)\,,$$

$$\tfrac{1}{2}(a_i + b_i) = k_i\,, \qquad b_i - a_i = \tfrac{1}{2}(k_{i+1} - k_{i-1}) = h\,k_i' + O(h^3)\,,$$

so that $a_i > 0$ and $b_i > 0$ for sufficiently small $h$.

We are going to show that scheme (2) is divergent even in the class of step coefficients

$$(4)\qquad k(x) = \begin{cases} k_1\,, & 0 < x < \xi\,, \\ k_2\,, & \xi < x < 1\,, \end{cases}$$

with $\xi$ being an irrational number such that $\xi = x_n + \theta h$, $x_n = nh$, $0 < \theta < 1$.

The exact solution of problem (1), (4) subject to the continuity conditions is of the form

$$(5)\quad u(x) = \begin{cases} 1 - \alpha_0\,x\,, & 0 \le x \le \xi, \quad \alpha_0 = (\varkappa + (1-\varkappa)\,\xi)^{-1}\,, \\ \beta_0\,(1-x)\,, & \xi \le x \le 1\,, \quad \beta_0 = \varkappa\,\alpha_0\,, \quad \varkappa = k_1/k_2\,. \end{cases}$$

We proceed to solve the difference problem (2), (4) in the usual way. Since $a_i = b_i = k_1$ for $0 < i < n$ and $a_i = b_i = k_2$ for $n+1 < i < N$, equation (2) reduces to $y_{i-1} - 2\,y_i + y_{i+1} = 0$ for $i \ne n$ and $i \ne n+1$. Whence it follows that

$$(6)\qquad y_i = y(x_i) = \begin{cases} 1 - \alpha\,x_i\,, & 0 \le x \le x_n\,, \\ \beta\,(1 - x_i)\,, & x_{n+1} \le x \le 1\,. \end{cases}$$

The coefficients $\alpha$ and $\beta$ can be most readily recovered from the appropriate equations with $i = n$ and $i = n+1$ incorporated:

$$(7)\qquad \begin{aligned} &b_n\,\big[\beta\,(1 - x_{n+1}) - (1 - \alpha\,x_n)\big] + a_n\,\alpha\,h = 0\,, \\ &b_{n+1}\,\beta\,h + a_{n+1}\,\big[\beta\,(1 - x_{n+1}) - (1 - \alpha\,x_n)\big] = 0\,. \end{aligned}$$

From (3) and (4) we find that $a_n = \frac{1}{4}(5k_1 - k_2)$, $a_{n+1} = \frac{1}{4}(k_1 + 3k_2)$, $b_n = \frac{1}{4}(3k_1 + k_2)$ and $b_{n+1} = \frac{1}{4}(5k_2 - k_1)$. Solving equations (7) with respect to $\alpha$ and $\beta$ and recalling the expressions $x_n = \xi - \theta h$ and $x_{n+1} = \xi + (1-\theta)h$, we determine the values ascribed to the parameters

$$(8) \qquad \begin{cases} \beta = \mu\,\alpha\,, & \alpha = \dfrac{1}{\mu + (1-\mu)\,\xi + h\,(\lambda - \theta - (1-\theta)\,\mu)}\,, \\[2ex] & \mu = \dfrac{3+\varkappa}{5-\varkappa}\,\lambda\,, \qquad \lambda = \dfrac{5\varkappa - 1}{3\varkappa + 1}\,. \end{cases}$$

The passage to the limit as $h \to 0$ yields

$$\lim_{h\to 0} \alpha = \bar{\alpha}_0\,, \qquad \lim_{h\to 0} \beta = \bar{\beta}_0\,,$$

where

$$(9) \qquad \bar{\alpha}_0 = \big(\mu + (1-\mu)\,\xi\big)^{-1}, \qquad \bar{\beta}_0 = \mu\,\bar{\alpha}_0\,.$$

Via the linear interpolation we extend the functions (6) on the whole segment $0 \le x \le 1$. Under such an approach we have at our disposal a new function $\tilde{y}(x,h)$, $x \in [0,1]$, which coincides with $y_i$ at the grid nodes $x_i = ih$ and possesses the limiting function as $h \to 0$:

$$(10) \qquad \bar{u}(x) = \lim_{h\to 0} \tilde{y}(x,h) = \begin{cases} 1 - \bar{\alpha}_0\,x, & 0 \le x \le \xi\,, \\ \bar{\beta}_0\,(1-x), & \xi \le x \le 1\,. \end{cases}$$

Comparison of $\bar{u}(x)$ with the exact solution $u(x)$ specified by (5), where the coefficients $\bar{\alpha}_0$, $\bar{\beta}_0$ are determined by formula (9), shows that $\bar{u}(x) = u(x)$ for $\bar{\alpha}_0 = \alpha_0$ and $\bar{\beta}_0 = \beta_0$. But it is possible only if $\varkappa = 1$ or $k_1 = k_2$. This provides support for the view that, as $h \to 0$, solution (6) of the difference problem (2), (5) approaches the function $\bar{u}(x)$ other than the exact solution $u(x)$ of problem (1) in the case $k_1 \ne k_2$. Due to this fact scheme (2) is divergent.

Before stating the main results, it will be sensible to clarify a physical sense of the function $\bar{u}(x)$, which solves problem (1) subject to the conditions $[\bar{u}] = 0$ and $[k\,\bar{u}'] = -\bar{\alpha}_0\,(\mu - \varkappa)\,k_2 = q$ at the point $x = \xi$. Here $q$ stands for the capacity of a point heat source (sink) at the point $x = \xi$. Being dependent on $\varkappa$, the quantity $q$ varies very widely. Specifically, $q \to \pm\infty$ as $\varkappa \to 5 \pm 0$. Thus, the physical reason for the convergence of scheme (2) is that the heat balance (the conservation law of heat) is violated, thus causing the appearance of the extra heat source (for $q < 0$) or sink (for $q > 0$) at the point $x = \xi$.

We call various schemes, in which conservation laws fail to be true, **nonconservative** or **disbalanced**.

The above example shows that in designing difference schemes it is very desirable to reproduce the appropriate conservative law on a grid. The schemes with this property are said to be **conservative**. In subsequent sections the general method for constructing conservative schemes, which are convergent in the class of discontinuity coefficients, will be appreciated. Before we undertake the complete description of this method, it is worth noting two things.

Quite often, in practical implementations without concern of theoretical estimations of the desired quality of a scheme, the convergence of the scheme is verified by experiments in which the grids are successively refined. Sometimes this approach may cause erroneous conclusions on convergence of the scheme on account of some nearness between a solution of the difference problem and some limiting function $\bar{u}(x)$ during the course of successive grid refinement. In the above example the function $\bar{u}(x)$ may deviate, generally speaking, from the solution $u(x)$ of the original problem as large as we like. That is why the method of successive grid refinement is employed with caution. Anyhow it cannot exclude theoretical investigations at least in model problems.

The method of **test functions** is quite applicable in verifying convergence and determining the order of accuracy and is stipulated by a proper choice of the function $U(x)$. Such a function is free to be chosen in any convenient way so as to provide the validity of the continuity conditions at every discontinuity point of coefficients. By inserting it in equation (1) of Section 1 we are led to the right-hand side $f = (kU')' - qU$ and the boundary values $\mu_1 = U(0)$ and $\mu_2 = U(1)$. The solution of such a problem relies on scheme (4) of Section 1 and then the difference solution will be compared with a known function $U(x)$ on various grids.

One more thing is worth noting here. It would be erroneous to think that every scheme, which is convergent in the case of smooth coefficients, is obliged to converge in the case of discontinuity coefficients. Further explorations are connected with a family of schemes converging in the class of discontinuity coefficients, thus expanding possibilities. Let us stress that in the sequel we will deal with such schemes only.

**2. The integro-interpolational method for constructing homogeneous difference schemes.** Various physical processes (heat conduction or diffusion, vibrations, gas dynamics, etc.) are well-characterized by some integral conservation laws (of heat, mass, momentum, energy, etc.). Usually the derivation of a differential equation of mathematical physics is on an integral relation (a balance equation) which expresses such a conservation law for a small volume. Letting the volume to zero and assuming that the derivatives involved in the balance equation are continuous, one can write down the appropriate differential equation.

From a physical point of view, the finite difference method is mostly based based on the further replacement of a continuous medium by its discrete model. Adopting those ideas, it is natural to require that the principal characteristics of a physical process should be in full force. Such characteristics are certainly conservation laws. Difference schemes, which express various conservation laws on grids, are said to be **conservative** or **divergent**. For conservative schemes the relevant conservative laws in the entire grid domain (integral conservative laws) do follow as an algebraic corollary to difference equations.

In this view, it seems reasonable to construct conservative difference schemes with the aid of balance equations for an elementary volume (cell) of a grid domain. Integrals and derivatives involved in these equations should be replaced by approximate difference expressions, thereby completing the design of a homogeneous difference scheme. The way of obtaining homogeneous difference schemes is called the **integro-interpolational method** or **balance method**.

In what follows the illustration of this method is concerned with equation (1) from (Section 1) capable of describing the stationary distribution of temperature over a homogeneous bar $0 \leq x \leq 1$. The equation of the heat balance can be written on the segment $x_{i-1/2} \leq x \leq x_{i+1/2}$ as

$$(11) \qquad w_{i-1/2} - w_{i+1/2} + \int\limits_{x_{i-1/2}}^{x_{i+1/2}} f(x)\, dx = \int\limits_{x_{i-1/2}}^{x_{i+1/2}} q(x)\, u(x)\, dx,$$

$$w = -k\, u',$$

where $w(x)$ is the heat flow, $q(x)\, u(x)$ is the capacity of heat sinks (sources for $q < 0$) proportional to the temperature and $f(x)$ is the distribution density of external heat sources (sinks).

The existence of a heat sink owes a debt to the heat exchange with the external environment on the lateral surface of the bar. The quantity $w_{i-1/2}$ is equal to the amount of heat being supplied to the segment $x_{i-1/2} \leq x \leq x_{i+1/2}$ through the cross-section $x = x_{i-1/2}$, while $w_{i+1/2}$ refers in a similar fashion to the heat transfer from this segment through the cut $x = x_{i+1/2}$. The third member on the left-hand side of (11) reflects the amount of heat being emitted on the segment $[x_{i-1/2}, x_{i+1/2}]$ by heat sources with the distribution density $f(x)$. The integral on the right-hand side of (11) is the amount of heat being transferred to the external environment by the heat exchange on the lateral surface.

In order to develop a difference equation from (11), we substitute linear combinations of the values of $u$ at the grid nodes in place of $w$ and the integral with $u$ that can be obtained through the interpolations in some neighborhood of the node $x_i$. The simplest interpolation gives

$$u = \text{const} = u_i \quad \text{for} \quad x_{i-1/2} \leq x \leq x_{i+1/2} ,$$

$$(12) \qquad \int\limits_{x_{i-1/2}}^{x_{i+1/2}} q(x)\, u(x)\, dx \approx h\, d_i\, u_i , \qquad d_i = \frac{1}{h} \int\limits_{x_{i-1/2}}^{x_{i+1/2}} q(x)\, dx ,$$

where $d_i$ is the mean value of the function $q(x)$ on the segment $x_{i-1/2} \leq x \leq x_{i+1/2}$ of length $h$. Upon integrating the equality $u' = w/k$ over the segment $x_{i-1} \leq x \leq x_i$ we arrive at

$$u_{i-1} - u_i = \int\limits_{x_{i-1}}^{x_i} \frac{w(x)}{k(x)}\, dx ,$$

for which the substitution $w(x) = \tilde{w}_{i-1/2} = \text{const}$ on the segment $x_{i-1} \leq x \leq x_i$ yields

$$u_{i-1} - u_i \approx \tilde{w}_{i-1/2} \int\limits_{x_{i-1}}^{x_i} \frac{dx}{k(x)} .$$

From what has been said above it is clear that the approximate value $\tilde{w}_{i-1/2}$ of the flow is

$$(13) \quad \tilde{w}_{i-1/2} = -a_i\, \frac{u_i - u_{i-1}}{h} = -a_i\, u_{\bar{x},i} , \qquad a_i = \left( \frac{1}{h} \int\limits_{x_{i-1}}^{x_i} \frac{dx}{k(x)} \right)^{-1} .$$

Here $\int_{x_{i-1}}^{x_i} dx/k(x)$ is the heat resistance of the segment $[x_{i-1}, x_i]$.

Substituting (12) and (13) into (11) and denoting by $y_i$ the unknown function, we obtain the conservative difference scheme

$$(14) \qquad \frac{1}{h} \left( a_{i+1}\, \frac{y_{i+1} - y_i}{h} - a_i\, \frac{y_i - y_{i-1}}{h} \right) - d_i\, y_i = -\varphi_i ,$$

where

$$a_i = \overset{\circ}{a}_i = \left( \frac{1}{h} \int\limits_{x_{i-1}}^{x_i} \frac{dx}{k(x)} \right)^{-1} = \left( \int\limits_{-1}^{0} \frac{ds}{k\,(x_i + sh)} \right)^{-1},$$

(15) $$d_i = \overset{\circ}{d}_i = \int\limits_{-1/2}^{1/2} q(x_i + sh)\, ds\,,$$

$$\varphi_i = \overset{\circ}{\varphi}_i = \int\limits_{-1/2}^{1/2} f(x_i + sh)\, ds\,.$$

We have written the difference equation (14) at a fixed node $x = x_i$. With an arbitrarily chosen node $x_i$ it is plain to derive equation (14) at all inner nodes of the grid. Since at all the nodes $x_i$, $i = 1, 2, \dots, N-1$, the coefficients $a_i$ and $b_i$ are specified by the same formulae (15), scheme (14)–(15) is treated as a homogeneous conservative scheme. Because of this, we may omit the subscript $i$ in formulae (14)–(15) and write down an alternative form of scheme (14):

$$\left(a\, y_{\bar{x}}\right)_x - dy = -\varphi\,.$$

In the general case the coefficient $a_i$ built into the formula for the heat flow is some functional of the values of $k(x)$ on the segment $[x_{i-1}, x_i]$.

Observe that the conservation law in the entire grid domain $\omega_h$ known as the "integral conservation law" is an algebraic corollary to equation (14) for any conservative scheme of the form (14) with arbitrary ingredients $a$, $d$ and $\varphi$. Indeed, with the notation $\bar{\omega}_{i-1/2} = -a_i\,(y_i - y_{i-1})/h$ for the difference expression of the heat flow at the point $x = x_{i-1/2}$, we can rearrange (14) as $\bar{\omega}_{i-1/2} - \bar{\omega}_{i+1/2} + h\,\varphi_i = h\, d_i\, y_i$, which becomes after summation over $i = 1, 2, \dots, N-1$ the difference conservation heat law within the entire grid domain:

$$\bar{\omega}_{1/2} - \bar{\omega}_{N-1/2} + \sum_{i=1}^{N-1} h\,\varphi_i = \sum_{i=1}^{N-1} h\, d_i\, y_i\,,$$

meaning the difference approximation of the integral conservation law for equation (1) from Section 1.

**3. Homogeneous conservative schemes.** In the preceding section we have designed the conservative scheme (14) by means of the integro-interpolational method. In the general case the coefficients $a$, $d$ and $\varphi$ of scheme (14) are some functionals of the coefficients $k(x)$, $q(x)$ and $f(x)$ involved in the differential equation

$$
\begin{aligned}
a(x) &= A\big[k(x+sh)\big]\,, \\
d(x) &= F\big[q(x+sh)\big]\,, \qquad (16)\\
\varphi(x) &= F\big[f(x+sh)\big]\,.
\end{aligned}
$$

The domain of the pattern functional $A[\bar{k}(s)]$ is $Q^{(0)}[-1,1]$ (function $\bar{k}(s) \in Q^{(0)}[-1,1]$), while the domain of $F[\bar{f}(s)]$ is $Q^{(0)}[-1/2,1/2]$. In other words, the functional $A[\bar{k}(s)]$ (or $F[\bar{f}(s)]$) is defined for all piecewise continuous functions $\bar{k}(s)$ (or $\bar{f}(s)$) given on the segment $-1 \le s \le 1$ (or $-1/2 \le s \le 1/2$). In trying to recover the coefficient $a(x)$ the intention is to use formula (16) with $\bar{k}(s) = k(x+sh)$. This is consistent with the passage from the pattern $-1 \le s \le 1$, on which the function $\bar{k}(s)$ is defined, to the pattern $x - h \le x' \le x + h$, so that the function $k(x')$ should provide a possibility of subsequent calculations of $a(x)$.

At the next stage we consider the homogeneous conservative scheme

$$
\begin{gathered}
\left(a y_{\bar{x}}\right)_x - d y = -\varphi(x)\,, \qquad x \in \omega_h\,, \\
y(0) = u_1\,, \qquad y(1) = u_2\,, \qquad a \ge c_1 > 0\,, \qquad d \ge 0\,,
\end{gathered} \qquad (17)
$$

for which the coefficients are expressed by formulae (16). Comparison of conservative schemes (17) or (14) with the three-point scheme of general form (4) from Section 1 shows that scheme (17) is governed and constructed in accordance with the rule $b_i = a_{i+1}$. We will not pursue analysis of this: the ideas needed to do so have been covered.

The requirement that scheme (4) of Section 1 should be conservative ("divergent") is equivalent to being self-adjoint of the appropriate difference operator. To make sure of it, we refer to a second-order operator

$$
(\Lambda y)_i = a_i\, y_{i-1} - c_i\, y_i + b_i\, y_{i+1}
$$

in the space $\overset{\circ}{\Omega}_h$ of all grid functions $y = \{y_i\}$ defined on the grid $\bar{\omega}_h$ and vanishing on the boundary: $y_0 = y_N = 0$. The space $\overset{\circ}{\Omega}_h$ is equipped with the inner product $(y,v) = \sum_{i=1}^{N-1} y_i\, v_i\, h$. Being elements of the space $\overset{\circ}{\Omega}_h$, any functions $y, v \in \overset{\circ}{\Omega}_h$ are subject to the identity:

$$\sum_{i=1}^{N-1} \big(a_i\, y_{i-1} - c_i\, y_i + b_i\, y_{i+1}\big)\, v_i = \sum_{i=1}^{N-1} \big(b_{i-1}\, v_{i-1} - c_i\, v_i + a_{i+1}\, v_{i+1}\big)\, y_i\,.$$

This provides enough reason to conclude that the condition

$$(\Lambda y, v) = (y, \Lambda v)$$

holds for arbitrary $y, v \in \overset{\circ}{\Omega}_h$ if and only if $b_i = a_{i+1}$, $i = 1, \dots, N-1$ (see Section 2, Chapter 1). The condition $b_i = a_{i+1}$ for scheme (4) of Section 1 means that we should have $B\big[k(x+sh)\big] = A\big[k\big(x+(s+1)h\big)\big]$ or $B\big[\bar{k}(s)\big] = A\big[\bar{k}(s+1)\big]$ for any $\bar{k}(s) \in Q^{(0)}[-1,1]$. Evidently, it is possible only in the case when the functional $A[\bar{k}(s)]$ is independent of the values of $\bar{k}(s)$ on the segment $0 \le s \le 1$. The same applies equally well to the functional $B[\bar{k}(s)]$ and the function $\bar{k}(s)$ on the segment $-1 \le s \le 0$, so that

$$a(x) = A\big[k(x+sh)\big] \quad \text{for} \quad -1 \le s \le 0\,.$$

Conditions (5) of Section 1 relating to the second-order local approximation for the conservative scheme (17) acquire the form

$$\frac{a(x+h) - a(x)}{h} = k'(x) + O(h^2),$$

$$\text{(18)} \qquad \frac{a(x+h) + a(x)}{2} = k(x) + O(h^2)\,,$$

$$d(x) = q(x) + O(h^2), \qquad \varphi(x) = f(x) + O(h^2)\,,$$

implying that $a(x) = k(x) - \frac{1}{2}\, h\, k'(x) + O(h^2)$ or $a(x) = k(x - \frac{1}{2}\, h) + O(h^2)$.

In Section 3.2 the integro-interpolational method was aimed at constructing the homogeneous conservative scheme (16) with the coefficients $a$, $d$ and $\varphi$ of the special form (15), namely with pattern functionals such that

$$\text{(15')} \qquad A\big[\bar{k}(s)\big] = \left(\int_{-1}^{0} \frac{ds}{\bar{k}(s)}\right)^{-1}, \qquad F\big[\bar{f}(s)\big] = \int_{-1/2}^{1/2} \bar{f}(s)\, ds\,.$$

In this case the coefficients $a$, $d$ and $\varphi$ are calculated by integrating the functions $k(x)$, $q(x)$ and $f(x)$ (see (15)).

In practice it is convenient to employ more simpler formulae for determination of $a$, $d$ and $\varphi$ with using the values of $k$, $q$ and $f$ at the isolated points. Usually a pattern consisting of one or two points permits one to make considerable simplifications:

$$
(19)\qquad
\begin{aligned}
& a_i = k_{i-1/2} = k(x_i - 0.5\,h) && \left(A\big[\bar{k}(s)\big] = \bar{k}(-0.5)\right), \\
& d_i = q_i\,, \qquad \varphi_i = f_i && \left(F\big[\bar{f}(s)\big] = \bar{f}(0)\right)
\end{aligned}
$$

or

$$
a_i = \tfrac{1}{2}\,(k_i + k_{i-1}) \qquad \left(A\big[\bar{k}(s)\big] = \tfrac{1}{2}\,\big(\bar{k}(-1) + \bar{k}(0)\big)\right),
$$

$$
a_i = \frac{2\,k_i\,k_{i-1}}{k_i + k_{i-1}} \qquad \left(\frac{1}{\left(A\big[\bar{k}(s)\big]\right.} = \frac{1}{2}\left(\frac{1}{\bar{k}(0)} + \frac{1}{\bar{k}(-1)}\right)\right).
$$

Conditions (18) are certainly true for all the schemes we have mentioned above.

If the coefficient $k(x)$ is discontinuous at the middle nodal points $x = x_{i-1/2}$ and the coefficients $q(x)$ and $f(x)$ are discontinuous at the points $x = x_i$, the half-sums of the left and right limiting values have to be substituted into formulae (19). As a final result we get

$$
\begin{aligned}
a_i &= \tfrac{1}{2}\,\big(k(x_{i-1/2} - 0) + k(x_{i-1/2} + 0)\big)\,, \\
d_i &= \tfrac{1}{2}\,\big(q(x_i - 0) + q(x_i + 0)\big)\,, \\
\varphi_i &= \tfrac{1}{2}\,\big(f(x_i - 0) + f(x_i + 0)\big)\,.
\end{aligned}
$$

We note in passing that formulae (19) and some others for determination of the coefficients $a$, $d$ and $\varphi$ can be derived through the approximations to integrals (15) and other members

$$
\frac{1}{h}\int\limits_{x_{i-1}}^{x_i} \frac{dx}{k(x)} \sim \frac{1}{k_{i-1/2}}\,, \qquad \frac{1}{h}\int\limits_{x_{i-1}}^{x_i} \frac{dx}{k(x)} \sim \frac{1}{2}\left(\frac{1}{k_i} + \frac{1}{k_{i-1}}\right).
$$

**4. A primary family of conservative schemes.** We spoke above about the family of the homogeneous conservative schemes (17), whose description is connected with some class of pattern functionals $A\big[\bar{k}(s)\big]$ and $F\big[\bar{f}(s)\big]$. For convenience in analysis, it is supposed that $F[\bar{f}(s)]$ is a linear nonnegative functional such that

(1) $F[c_1 \bar{f}_1(s) + c_2 \bar{f}_2(s)] = c_1 F[\bar{f}_1(s)] + c_2 F[\bar{f}_2(s)]$, $-1/2 \le s \le 1/2$, where $c_1$ and $c_2$ are arbitrary constants;

(2) $F[\bar{f}(s)] \ge 0$ for $\bar{f}(s) \ge 0$.

In spite of the fact that $A[\bar{k}(s)]$ is usually a nonlinear functional (see scheme (15′)), we may assume for the sake of simplicity that $A[\bar{k}(s)]$ is a linear nonnegative functional and consider, in addition to schemes (16)–(17), those with the coefficient $a(x)$ still subject to the relationship (cf. (15′))

$$\frac{1}{a(x)} = A\left[\frac{1}{k(x+sh)}\right]. \tag{16'}$$

Conditions for the second-order approximation (18) imply some restrictions on the pattern functionals $A[\bar{k}(s)]$ and $F[\bar{f}(s)]$. In preparation for this, plain calculations give

$$\begin{aligned}\varphi(x) = F[f(x+sh)] &= F[f(x) + s\,h\,f'(x) + O(h^2)] \\ &= f(x)\,F[1] + h\,f'(x)\,F[s] + O(h^2)\,.\end{aligned}$$

Since $f(x)$ is taken arbitrarily, it follows from the foregoing and (18) that

$$F[1] = 1\,, \qquad F[s] = 0\,. \tag{20}$$

By the same token,

$$\begin{aligned}a(x) &= A\left[k(x) + s\,h\,k'(x) + \tfrac{1}{2}\,s^2\,h^2\,k''(x) + O(h^3)\right] \\ &= k(x)\,A[1] + h\,k'(x)\,A[s] + \tfrac{1}{2}\,h^2\,k''(x)\,A[s^2] + O(h^3),\end{aligned}$$

$$\begin{aligned}a(x+h) &= k(x)\,A[1] + h\,k'(x)\,A[1+s] \\ &\quad + \tfrac{1}{2}\,h^2\,k''(x)\,A[(1+s)^2] + O(h^3),\end{aligned}$$

$$\begin{aligned}\frac{a(x+h) - a(x)}{h} &= \big(A[1+s] - A[s]\big)\,k'(x) \\ &\quad + \tfrac{1}{2}\,h\Big(A[(1+s)^2] - A[s^2]\Big)\,k''(x) + O(h^2),\end{aligned}$$

$$\frac{a(x+h) + a(x)}{2} = A[1]\,k(x) + \tfrac{1}{2}\,h\,\big(A[1+s] + A[s]\big)\,k'(x) + O(h^2).$$

Comparision of the resulting expansions with (18) permits us to assign the values

$$(21) \qquad A[1] = 1\,, \qquad A[s] = -\tfrac{1}{2}$$

by virtue of the relations

$$A[1+s] + A[s] = A[1] + 2\,A[s] = 0\,,$$

$$A\big[(1+s)^2\big] - A[s^2] = A\big[(1+s)^2 - s^2\big] = A[1+2s]$$

$$= A[1] + 2A[s] = 0\,.$$

To decide for yourself whether conditions (18) related to schemes (17), (15′) are met or not, we should take into account that

$$a(x+h) \pm a(x) = a(x)\,a(x+h)\left(\frac{1}{a(x)} \pm \frac{1}{a(x+h)}\right),$$

$$\frac{1}{2}\left(\frac{1}{a(x+h)} + \frac{1}{a(x)}\right) = \frac{1}{k(x)} + O(h^2)\,,$$

$$\frac{1}{h}\left(\frac{1}{a(x+h)} - \frac{1}{a(x)}\right) = \left(\frac{1}{k(x)}\right)' + O(h^2)\,,$$

$$a(x)\,a(x+h) = k^2(x) + O(h^2)\,.$$

We contrived to do it and give an answer to this question. When this is the case, the functional $A\big[\bar{k}(s)\big]$ also satisfies conditions (21). As far as the functionals

$$F\big[\bar{f}(s)\big] = \int_{-1/2}^{1/2} \bar{f}(s)\;ds\,, \qquad A\big[\bar{f}(s)\big] = \int_{-1}^{0} \bar{f}(s)\;ds$$

are concerned, we thus have

$$F[1] = \int_{-1/2}^{1/2} ds = 1\,, \qquad F[s] = \int_{-1/2}^{1/2} s\;ds = 0\,,$$

$$A[1] = \int_{-1}^{0} ds = 1\,, \qquad A[s] = \int_{-1}^{0} s\;ds = -\tfrac{1}{2}\,,$$

provided conditions (20)–(21) hold.

In what follows we deal everywhere with the primary family of homogeneous conservative schemes (16), (17) and (16′), (17) as well as with linear nonnegative pattern functionals $A[\bar{k}(s)]$ and $F[\bar{f}(s)]$ still subject to conditions (20) and (21) of second-order approximation.

In the sequel scheme (17), (15) will be called the best possible.

## 3.3 CONVERGENCE AND ACCURACY OF HOMOGENEOUS CONSERVATIVE SCHEMES

**1. The error of approximation in the class of smooth coefficients.** The main point of the theory is the accurate account of the accuracy of the uniform scheme (16)–(17) in the class of continuous and discontinuous functions $k(x)$, $q(x)$ and $f(x)$. In preparation for this, let $u(x)$ be an exact solution of the original problem

$$(1)\qquad \begin{gathered} \big(k(x)\,u'(x)\big)' - q(x)\,u(x) = -f(x)\,, \qquad 0 < x < 1\,, \\ u(0) = u_1\,, \quad u(1) = u_2\,, \quad k(x) \geq c_1 > 0\,, \quad q(x) \geq 0\,, \end{gathered}$$

and let $y = y(x)$ be a solution of the difference problem

$$(2)\qquad \begin{gathered} \big(a y_{\bar{x}}\big)_x - d(x)\,y = -\varphi(x)\,, \qquad x = ih\,, \quad i = 1, 2, \ldots, N-1\,, \\ y_0 = u_1\,, \qquad y_N = u_2\,, \qquad a(x) \geq c_1 > 0\,, \qquad d(x) \geq 0\,, \end{gathered}$$

which belongs to the primary family of conservative schemes designed in Section 2.4.

Common practice involves the error $z(x) = y(x) - u(x)$ being a grid function for $x \in \bar{\omega}_h$. Inserting the representation $y(x) = z(x) + u(x)$ in (2) and assuming $u(x)$ to be a known function, we may set up the difference problem for the error

$$(3)\qquad \begin{gathered} \Lambda\, z = \big(a z_{\bar{x}}\big)_x - d\, z = -\psi(x)\,, \qquad x = ih\,, \quad i = 1, 2, \ldots, N-1\,, \\ z(0) = z(1) = 0\,, \qquad a \geq c_1 > 0\,, \qquad d \geq 0\,, \end{gathered}$$

where the residual

$$(4)\qquad \psi(x) = \Lambda\, u + \varphi(x) = \big(a u_{\bar{x}}\big)_x - du + \varphi$$

is the error of approximation of equation (1) by the difference scheme (2) on the solution $u = u(x)$ of problem (1).

In Section 1.4 we have established conditions of the second-order local approximation for the conservative scheme (2) with linear nonnegative pattern functionals $A\big[\bar{k}(s)\big]$ and $F\big[\bar{f}(s)\big]$ such as:

$$(5) \qquad A[1] = 1\,, \qquad A[s] = -\tfrac{1}{2}\,, \qquad F[1] = 1\,, \qquad F[s] = 0\,.$$

As can readily be observed, these conditions remain valid for any scheme from the primary family.

In order to evaluate the order of accuracy for scheme (2), it is necessary to make the accurate account of the error $z = y - u$ being viewed as a solution of problem (3). Moreover, the desirable estimate should be expressed in terms of the right-hand side $\psi$. In this direction the error of approximation to $\psi(x)$ is considered first. If $k(x) \in C^{(3)}$ and $q(x), f(x) \in C^{(2)}$, then

$$\begin{aligned} \psi(x) &= \Lambda\, u + \varphi - (Lu + f) \\ &= \big[(a u_{\bar{x}})_x - (k u')'\big] - (d - q)\, u + (\varphi - f) = O(h^2)\,. \end{aligned}$$

This means that scheme (2) provides a local approximation of order 2, so that $\|\,\psi\,\|_C \leq M h^2$, where $M = \text{const} > 0$ is independent of $h$.

In the sequel we succeed in showing that

$$\|\,\psi\,\|_* = \sum_{i=1}^{N-1} h \left| \sum_{k=1}^{i} h\, \psi_k \right| \leq M\, h^2$$

if $k(x)$, $q(x)$, $f(x) \in C^{(2)}$, meaning that $k(x)$ possesses two (but not three) continuous derivatives.

**2. The error of approximation in the class of discontinuous coefficients.** Our aim here is to justify that the error of approximation (4) can always be represented in the form

$$(6) \qquad \psi = \eta_x + \psi^*\,,$$

$$(7) \qquad \eta_i = (a u_{\bar{x}})_i - (k u')_{i-1/2}\,,$$

$$(8) \qquad \psi_i^* = \Big(\varphi_i - \int_{-1/2}^{1/2} f(x_i + sh)\, ds\Big) - \Big(d_i\, u_i - \int_{-1/2}^{1/2} q(x_i + sh)\, u(x_i + sh)\, ds\Big)\,.$$

This can be done with the aid of the balance equation upon integrating equation (1) over $x$ from $x_{i-1/2}$ to $x_{i+1/2}$:

$$\frac{1}{h}\left[(ku')_{i+1/2}-(ku')_{i-1/2}\right]-\frac{1}{h}\int\limits_{x_{i-1/2}}^{x_{i+1/2}} q(x)\,u(x)\,dx+\frac{1}{h}\int\limits_{x_{i-1/2}}^{x_{i+1/2}} f(x)\,dx=0\,.$$

With the new variable $s=(x-x_i)/h$, the balance equation becomes

$$\left((ku')_{i-1/2}\right)_x-\int\limits_{-1/2}^{1/2} q(x_i+sh)\,u(x_i+sh)\,ds+\int\limits_{-1/2}^{1/2} f(x_i+sh)\,ds=0\,.$$

Upon subtracting the last equation from (4) we establish formulae (6)–(8) and deduce that if $k(x)$, $q(x)$, $f(x)\in C^{(2)}$, then

$$\eta_i=O(h^2)\,,\qquad \psi_i^*=O(h^2)\,. \tag{9}$$

For this, we proceed as usual. What is available are the useful expansions

$$u_i=u_{i-1/2}+\frac{1}{2}\,h\,u'_{i-1/2}+\frac{h^2}{8}\,u''_{i-1/2}+O(h^3)\,,$$

$$u_{i-1}=u_{i-1/2}-\frac{1}{2}\,h\,u'_{i-1/2}+\frac{h^2}{8}\,u''_{i-1/2}+O(h^3)\,,$$

$$u_{\bar{x},i}=u'_{i-1/2}+O(h^2)\,,$$

$$\eta_i=(a_i-k_{i-1/2})\,u'_{i-1/2}+O(h^2)=O(h^2)\,,$$

which emerge from the chain of the relations

$$\begin{aligned}a_i&=A\big[k(x_i+sh)\big]=A\big[k\big(x_{i-1/2}+(s+\tfrac{1}{2})\,h\big)\big]\\&=A\big[k_{i-1/2}+h\,k'_{i-1/2}(s+\tfrac{1}{2})+O(h^2)\big]\\&=k_{i-1/2}+h\,k'_{i-1/2}\,A[s+\tfrac{1}{2}]+O(h^2)=k_{i-1/2}+O(h^2)\,.\end{aligned}$$

A simple observation that for any $v(x)\in C^{(2)}$

$$\int\limits_{-1/2}^{1/2} v(x_i+sh)\,ds=v_i+O(h^2)$$

helps motivate what is done with $\psi_i^* = (q_i\, u_i - d_i\, u_i) + (\varphi_i - f_i) + O(h^2) = O(h^2)$.

We now turn to the case of discontinuity coefficients $k(x)$, $q(x)$ and $f(x)$ and assume without loss of generality that $k$, $q$ and $f$ have discontinuities of the first kind only at a single point $x = \xi \in (0,1)$, so that

$$\xi = x_n + \theta h\,, \qquad 0 \le \theta = \theta(h) \le 1\,, \qquad 0 < n < N\,.$$

The solution $u = u(x)$ to equation (1) for $x = \xi$ satisfies the continuity conditions for the function $u(x)$ and the flow $k(x)u'(x)$:

$$[u] = u(\xi + 0) - u(\xi - 0) = 0\,, \quad [ku'] = 0 \quad \text{for} \quad x = \xi\,.$$

Under the assumptions $k(x)$, $q(x)$, $f(x) \in Q^{(2)}$ we thus have $u(x) \in Q^{(3)}$.

Before going further, it is straightforward to verify for $\eta_i$ that

$$\eta_i = \left(a u_{\bar{x}}\right)_i - \left(k u'\right)_{i-1/2} = O(h^2) \quad \text{for all} \quad i \ne n+1\,.$$

In dealing with $\eta_{n+1} = a_{n+1}\, u_{\bar{x},n+1} - (ku')_{n+1/2}$ we make use of the expansions in a neighborhood of the point $x = \xi$:

$$u_{n+1} = u(\xi) + (1-\theta)\, h\, u'(\xi + 0) + \frac{(1-\theta)^2}{2}\, h^2 u''(\xi + 0) + O(h^3)\,,$$

$$u_n = u(\xi) - \theta\, h\, u'(\xi - 0) + \frac{\theta^2}{2}\, h^2 u''(\xi - 0) + O(h^3)\,,$$

$$(ku')_{n+1/2} = \begin{cases} (ku')_{x=\xi+0} + (0.5-\theta)\, h\, (ku')'_{x=\xi+0} + O(h^2) & \text{for } \theta \le 0.5, \\[2ex] (ku')_{x=\xi-0} + (0.5-\theta)\, h\, (ku')'_{x=\xi-0} + O(h^2) & \text{for } \theta > 0.5, \end{cases}$$

by means of which an alternative form of writing $\eta_{n+1}$ can be re-ordered for later use:

$$\eta_{n+1} = a_{n+1}\left((1-\theta)\, u'_{\text{right}} + \theta\, u'_{\text{left}}\right) - w_0 + O(h)\,,$$

where $w_0 = (ku')_{\text{right}} = (ku')_{\text{left}}$, $v_{\text{left}} = v(\xi - 0)$ and $v_{\text{right}} = v(\xi + 0)$.

At the same time, the continuity condition $[ku'] = 0$ assures us of the validity of the relations

$$(1-\theta)\, u'_{\text{right}} + \theta\, u'_{\text{left}} = \left(\frac{\theta}{k_{\text{left}}} + \frac{1-\theta}{k_{\text{right}}}\right) w_0\,,$$

$$\eta_{n+1} = \left[a_{n+1}\left(\frac{\theta}{k_{\text{left}}} + \frac{1-\theta}{k_{\text{right}}}\right) - 1\right] w_0 + O(h)\,.$$

It seems worthwhile to focus the reader's attention on the best scheme (15) from Section 2 for which the chain of the relations occurs:

$$\frac{1}{\overset{\circ}{a}_{n+1}} = \int_{-1}^{0} \frac{ds}{k(x_{n+1}+sh)} = \int_{0}^{1} \frac{ds}{k(x_n+sh)}$$

$$= \int_{0}^{\theta} \frac{ds}{k(x_n+sh)} + \int_{\theta}^{1} \frac{ds}{k(x_n+sh)}$$

$$= \int_{0}^{\theta} \left( \frac{1}{k_{\text{left}}} + (s-\theta)\, h' \left(\frac{1}{k}\right)_{\text{left}} + O(h^2) \right) ds$$

$$+ \int_{\theta}^{1} \left( \frac{1}{k_{\text{right}}} + (s-\theta)\, h' \left(\frac{1}{k}\right)_{\text{right}} + O(h^2) \right) ds$$

$$= \frac{\theta}{k_{\text{left}}} + \frac{1-\theta}{k_{\text{right}}} + O(h)\,,$$

so that $\eta_{n+1} = \overset{\circ}{\eta}_{n+1} = O(h)$. For all other schemes $\eta_{n+1} = O(1)$.

In the estimation of $\psi^*$ it is necessary to distinguish two possibilities of interest: $\theta < 0.5$ and $\theta > 0.5$.

(1) Let $\theta < 0.5$. In every such case $\psi_i^* = O(h^2)$ for $i \neq n$, while $\psi_n^* = O(1)$ and only for the best scheme with $\varphi_i = \overset{\circ}{\varphi}_i$ and $d_i = \overset{\circ}{d}_i$ incorporated $\overset{\circ}{\psi}{}_i^* = \int_{-1/2}^{1/2} q(x_i+sh)\,\big(u(x_i+sh) - u_i\big)\, ds = O(h)$ is attained for $i = n$, since $u(x_n+sh) = u(\xi) + O(h)$ for any $s \in [-0.5, 0.5]$.

(2) If $\theta > 0.5$, then $\psi_i^* = O(h^2)$ for $i \neq n+1$, while $\psi_{i+1}^* = O(1)$ and $\overset{\circ}{\psi}{}_{n+1}^* = O(h)$.

It follows from the foregoing that

$$
(10)\qquad
\begin{aligned}
&\eta_i = O(h^2) && \text{for} \quad i \neq n+1, \quad \eta_{n+1} = O(1), \quad \overset{\circ}{\eta}_{n+1} = O(h)\,,\\
&\psi_i^* = O(h^2) && \text{for} \quad i \neq n, \quad \psi_n^* = O(1)\,,\\
&\overset{\circ}{\psi}{}_n^* = O(h) && \text{if} \quad \theta < 0.5\,,\\
&\psi_i^* = O(h^2) && \text{for} \quad i \neq n+1, \quad \psi_{n+1}^* = O(1)\,,\\
&\overset{\circ}{\psi}{}_{n+1}^* = O(h) && \text{if} \quad \theta > 0.5\,.
\end{aligned}
$$

Using these estimates behind we draw the conclusion that at the nodes $x = x_n$ and $x = x_{n+1}$ the function $\psi(x)$ can be expressed by

$$
(11)\quad \psi_n = \frac{\eta_{n+1}}{h} + O(1)\,, \qquad \psi_{n+1} = -\frac{\eta_{n+1}}{h} + O(1)\,, \qquad \eta_{n+1} = O(1)\,,
$$

thereby clarifying that at the nodes adjacent to the discontinuity point $x = \xi$ scheme (2) does not approximate equation (1) in light of the limit relations

$$
\psi_n = O\Big(\frac{1}{h}\Big) \to \infty\,, \qquad \psi_{n+1} = O\Big(\frac{1}{h}\Big) \to \infty
$$

as $h \to 0$. From asymptotic formulae (11) it is readily seen that the main summands in the expressions for $\psi_n$ and $\psi_{n+1}$ have equal modules and opposite signs, so that

$$
\psi_n + \psi_{n+1} = O(1)\,, \qquad h\,(\psi_n + \psi_{n+1}) = O(h)\,,
$$

meaning that the error of approximation is of dipole character in neighborhoods of discontinuity points of the coefficient $k(x)$. This provides enough reason to conclude that the conservative scheme (2) is of order 1 in the norm

$$
\|\,\psi\,\|_* = (1, |\,\eta\,|) + (1, |\,\mu\,|) = O(h)
$$

with $\mu_i = \sum_{k=1}^{i-1} h\psi_k^*$ for $i = 2, 3, \ldots, N$ and $\mu_1 = 0$.

**3. A priori estimates of the error.** We now estimate the error $z = y - u$, which is a solution of problem (3):

$$
\Lambda\, z = \big(a z_{\bar{x}}\big)_x - dz = -\psi(x), \qquad 0 < x = ih < 1, \quad z(0) = z(1) = 0\,,
$$

where $\psi(x)$ is the error of approximation having the form (6): $\psi = \eta_x + \psi^*$. We will show that the solution of problem (3) with the right-hand side (6) admits the estimate

$$\| z \|_C = \| y - u \|_C \leq \frac{2}{c_1} \left( (1, |\eta|] + (1, |\mu|] \right), \tag{12}$$

where $\mu_i = \sum_{k=1}^{i-1} h\, \psi_k^*$ for $i = 2, 3, \ldots, N$ and $\mu_1 = 0$. To that end, it suffices to carry out the proper evaluation of a solution of the auxiliary problem

$$\Lambda v = \left( a v_{\bar{x}} \right)_x - d v = -\eta_x \,, \qquad v(0) = v(1) = 0 \,. \tag{13}$$

Other ideas are connected with solving the difference problem

$$\left( a w_{\bar{x}} \right)_x = -\eta_x \,, \qquad w(0) = w(1) = 0 \,,$$

from which follows immediately that $a w_{\bar{x}} + \eta = \text{const} = c_0$. As its corollary, we establish the recurrence relation for $\omega_k$:

$$\omega_k = \omega_{k-1} - \frac{h\, \eta_k}{a_k} + \frac{h\, c_0}{a_k} \,, \qquad k = 1, 2, \ldots, N \,.$$

Summation over $k$ from 1 to $i$ yields

$$w_i = w_0 - \sum_{k=1}^{i} \frac{h\, \eta_k}{a_k} + c_0 A_i \,, \qquad A_i = \sum_{k=1}^{i} \frac{h}{a_k} \,.$$

Putting $i = N$ and keeping in mind that $w_0 = w_N = 0$, we find that

$$c_0 = \frac{1}{A_N} \sum_{k=1}^{N} \frac{h\, \eta_k}{a_k} \,,$$

yielding

$$w_i = - \sum_{k=1}^{i} \frac{h\, \eta_k}{a_k} + \frac{A_i}{A_N} \sum_{k=1}^{N} \frac{h\, \eta_k}{a_k}$$

or

$$w_i = -\left( 1 - \frac{A_i}{A_N} \right) \sum_{k=1}^{i} \frac{h\, \eta_k}{a_k} + \frac{A_i}{A_N} \sum_{k=i+1}^{N} \frac{h\, \eta_k}{a_k} \,.$$

As far as $A_i \le A_N$, the preceding decomposition implies that

$$| w_i | \le \left(1 - \frac{A_i}{A_N}\right) \sum_{k=1}^{i} \frac{h \, | \eta_k |}{a_k} + \frac{A_i}{A_N} \sum_{k=i+1}^{N} \frac{h \, | \eta_k |}{a_k}$$

$$\le \sum_{k=1}^{N} \frac{h \, | \eta_k |}{a_k} \le \frac{1}{c_1} \, (1, | \eta |] \, .$$

The difference $\xi_i = v_i - w_i$, solving the problem

$$\left(a \, \xi_{\bar{x}}\right)_x - d \, \xi = d \, w \, , \qquad \xi(0) = \xi(1) = 0 \, ,$$

with $a \ge c_1 > 0$ and $d \ge 0$, is aimed at the further estimation of $v_i$ in complete agreement with the lemma from Chapter 1, Section 2.8. The results of this section provide support for the view that $\| \xi \|_C \le \| w \|_C$ and, therefore,

$$\| v \|_C \le \| w \|_C + \| \xi \|_C \le 2 \, \| w \|_C \le \frac{2}{c_1} \, (1, | \eta |] \, .$$

Summarizing, we arrive at

$$\| v \|_C \le \frac{2}{c_1} \, (1, | \eta |] \, . \tag{14}$$

The function $\psi^*$ is representable by

$$\psi_i^* = \mu_x \, , \tag{15}$$

where $\mu_i = \sum_{k=1}^{i-1} h \psi_k^*$ for $i = 2, 3, \ldots, N$ and $\mu_1 = 0$. Because of this form, it is possible to write down $\psi = (\eta + \mu)_x$ and involve estimate (14) in subsequent reasonings:

$$\| z \|_C \le \frac{2}{c_1} \, (1, | \eta + \mu |] \le \frac{2}{c_1} \, \left( (1, | \eta |] + (1, | \mu |] \right) .$$

**4. On convergence and accuracy.** The results obtained in the preceding two sections may be of help in establishing the rate of convergence for scheme (2).

**Theorem 1** *The accuracy of any scheme (2) with coefficients (16) is of order 2 in the class of smooth coefficients $k(x)$, $q(x)$, $f(x) \in C^{(2)}[0,1]$:*

$$\| y - u \|_C \leq M\, h^2$$

*with constant $M > 0$ independent of $h$, whereas it is of order 1 in the class of discontinuity coefficients $k(x)$, $q(x)$, $f(x) \in Q^{(2)}[0,1]$:*

$$\| y - u \|_C \leq M\, h \,.$$

*The best scheme also has the second-order accuracy in the class of discontinuity coefficients.*

These assertions follow from the representation of the approximation error in the form (6)–(8) and a priori estimate (12). On the basis of the estimates for $\eta_i$ and $\psi_i^*$ obtained in Section 3.2 we find that

$$(1, |\,\eta\,|] = h\, |\, \eta_{n+1} \,| + O(h^2)\,, \tag{16}$$

$$\mu_i = O(h^2) \quad \text{for} \quad i \leq n\,,$$

$$\mu_i = h\big(\psi_n^* + \psi_{n+1}^*\big) + O(h^2) \quad \text{for} \quad i > n+1\,,$$

yielding

$$(1, |\,\mu\,|] = h\, |\, \psi_n^* + \psi_{n+1}^* \,|\, (1 - x_n) + O(h^2)\,. \tag{17}$$

Combination of relations (16)–(17) just established and estimate (12) provides the sufficient background for the validity of the assertions of the theorem in light of the asymptotic representations

$$\psi_n^* + \psi_{n+1}^* = O(1)\,,$$

$$\overset{\circ}{\psi}{}_n^* + \overset{\circ}{\psi}{}_{n+1}^* = O(h) \quad \text{for any} \quad \theta \in [0,1]\,,$$

$$\eta_{n+1} = O(1), \qquad \overset{\circ}{\eta}_{n+1} = O(h)\,.$$

In the case of smooth coefficients we achieve for any admissible scheme $\eta_i = O(h^2)$ and $\psi_i^* = O(h^2)$ for all $i = 1, 2, \ldots, N-1$ and, therefore, $(1, |\,\eta\,|] + (1, |\,\mu\,|] = O(h^2)$.

**Remark** It can be shown that the approximation in the class of smooth coefficients is necessary and sufficient for the convergence of the homogeneous scheme (2).

## 3.4 HOMOGENEOUS DIFFERENCE SCHEMES ON NON-EQUIDISTANT GRIDS

**1. Schemes on non-equidistant grids.** Quite often, in practical implementations difference schemes on non-equidistant grids are in common usage for solving differential equations. In Chapter 2, Section 1 we have produced for the simplest equation $u'' = -f$ a difference scheme on the non-equidistant grid

$$\hat{\omega}_h = \{x_i\,,\ i = 0, 1, \ldots, N,\ x_0 = 0,\ x_N = 1,\ h_i = x_i - x_{i-1}\}$$

and have determined for this scheme the error of approximation.

In order to develop a homogeneous conservative scheme on the non-equidistant grid $\hat{\omega}_h$, we write down the balance equation

$$(1)\quad w_{i-1/2} - w_{i+1/2} - \int\limits_{x_{i-1/2}}^{x_{i+1/2}} q(x)\,u(x)\,dx = - \int\limits_{x_{i-1/2}}^{x_{i+1/2}} f(x)\,dx, \qquad w = -k\,u',$$

on the interval $(x_{i-1/2}\,, x_{i+1/2})$ with the ends $x_{i-1/2} = x_i - 0.5\,h_i$ and $x_{i+1/2} = x_i + 0.5\,h_{i+1}$. By analogy with Section 2.2 the forthcoming substitutions

$$\int\limits_{x_{i-1/2}}^{x_{i+1/2}} q\,u\,dx \sim \hbar_i\,\overset{\circ}{d}_i\,u_i\,, \qquad \overset{\circ}{d}_i = \frac{1}{\hbar_i} \int\limits_{x_{i-1/2}}^{x_{i+1/2}} q(x)\,dx\,,$$

$$\hbar_i = \tfrac{1}{2}\,(h_i + h_{i+1})\,,$$

$$w_{i-1/2} \sim \tilde{w}_{i-1/2} = -\,\overset{\circ}{a}_i\,\frac{u_i - u_{i-1}}{h_i} = -\,\overset{\circ}{a}_i\,u_{\bar{x},i}\,,$$

$$\overset{\circ}{a}_i = \left(\frac{1}{h_i}\int\limits_{x_{i-1}}^{x_i} \frac{dx}{k(x)}\right)^{-1}$$

lead to the difference scheme

$$(2)\qquad \frac{1}{\hbar_i}\left[\overset{\circ}{a}_{i+1}\,\frac{y_{i+1}-y_i}{h_{i+1}}-\overset{\circ}{a}_i\,\frac{y_i-y_{i-1}}{h_i}\right]-\overset{\circ}{d}_i\,y_i=-\overset{\circ}{\varphi}_i\,,$$

$$\overset{\circ}{a}_i=\left(\frac{1}{h_i}\int\limits_{x_{i-1}}^{x_i}\frac{dx}{k(x)}\right)^{-1},\qquad \overset{\circ}{d}_i=\frac{1}{\hbar_i}\int\limits_{x_{i-1/2}}^{x_{i+1/2}}q(x)\,dx\,,$$

$$\overset{\circ}{\varphi}_i=\frac{1}{\hbar_i}\int\limits_{x_{i-1/2}}^{x_{i+1/2}}f(x)\,dx\,,$$

which is the **best** possible in the same sense as we spoke above about this on an equidistant grid. On the same grounds as before, the coefficients $\overset{\circ}{a}_i$, $\overset{\circ}{d}_i$ and $\overset{\circ}{\varphi}_i$ are representable by

$$\overset{\circ}{a}_i=\left(\int\limits_{-1}^{0}\frac{ds}{k(x_i+sh_i)}\right)^{-1},$$

$$(3)\qquad \overset{\circ}{d}_i=\frac{h_i}{\hbar_i}\int\limits_{-1/2}^{0}q(x_i+sh_i)\,ds+\frac{h_{i+1}}{\hbar_i}\int\limits_{0}^{1/2}q(x_i+sh_{i+1})\,ds\,,$$

$$\overset{\circ}{\varphi}_i=\frac{h_i}{\hbar_i}\int\limits_{-1/2}^{0}f(x_i+sh_i)\,ds+\frac{h_{i+1}}{\hbar_i}\int\limits_{0}^{1/2}f(x_i+sh_{i+1})\,ds\,.$$

Retaining the notations of Chapter 2, Section 1

$$y_{\bar{x},i}=\frac{y_i-y_{i-1}}{h_i},\qquad y_{x,i}=\frac{y_{i+1}-y_i}{h_{i+1}},\qquad y_{\hat{x},i}=\frac{y_{i+1}-y_i}{\hbar_i},$$

which allow a simpler writing of the ensuing formulae, we refer to the three-point scheme

$$(4)\qquad \begin{aligned}&(ay_{\bar{x}})_{\hat{x}}-dy=-\varphi\,, && x=x_i\in\hat{\omega}_h\,,\\ &y(0)=u_1\,,\quad y(1)=u_2\,, && a\ge c_1>0\,,\quad d\ge 0\,.\end{aligned}$$

When the available functions $k(x)$, $q(x)$ and $f(x)$ happen to be of the class $Q^{(0)}[0,1]$ and their discontinuity points are known, a non-equidistant grid can be made so that all discontinuity points of the coefficients $k$, $q$ and $f$ would be nodal points of such a grid. We denote by $\hat{\omega}_h(K)$ any such grid depending on concrete functions $k$, $q$ and $f$ and arising in subsequent discussions. It is easily seen from (3) that the simplest expressions for $a_i$, $d_i$ and $\varphi_i$ on $\hat{\omega}_h(K)$ are given by

$$(5)\quad a_i = k_{i-1/2}\,, \qquad d_i = \frac{h_i\, q_i^- + h_{i+1}\, q_i^+}{2\,\hbar_i}\,, \qquad \varphi_i = \frac{h_i\, f_i^- + h_{i+1}\, f_i^+}{2\,\hbar_i}\,,$$

where $f_i^\pm = f(x_i \pm 0)$, etc.

$$a_i = \tfrac{1}{2}\left(k_{i-1}^+ + k_i^-\right), \qquad a_i = \frac{2\,k_{i-1}^+\, k_i^-}{k_{i-1}^+ + k_i^-}\,,$$

$$d_i = \frac{h_i\, q_{i-1/2} + h_{i+1}\, q_{i+1/2}}{2\,\hbar_i}\,, \qquad \varphi_i = \frac{h_i\, f_{i-1/2} + h_{i+1}\, f_{i+1/2}}{2\,\hbar_i}\,.$$

In the case of continuous coefficients expressions (5) imply that $a_i = k_{i-1/2}$, $d_i = q_i$ and $\varphi_i = f_i$. If discontinuity points coincide with nodal points of the grid $\hat{\omega}_h$, that is, $x = x_{i-1/2}$, then the members $a_i$, $d_i$ and $\varphi_i$ can be found in simplified form:

$$(5')\qquad a_i = \frac{2\,k_i\, k_{i-1}}{k_i + k_{i-1}}\,, \qquad d_i = q_i\,, \qquad \varphi_i = f_i$$

or

$$a_i = \frac{2\,k_{i-1/2}^+\, k_{i-1/2}^-}{k_{i-1/2}^+ + k_{i-1/2}^-}\,,$$

$$d_i = \frac{h_{i+1}\, q_{i+1/2}^- + h_i\, q_{i-1/2}^+}{2\,\hbar_i}\,,$$

$$\varphi_i = \frac{h_{i+1}\, f_{i+1/2}^- + h_i\, f_{i-1/2}^+}{2\,\hbar_i}\,.$$

**2. The error of approximation.** We now investigate the error of approximation for scheme (4) on the non-equidistant grid $\hat{\omega}_h$ by considering the equation for the error $z = y - u$:

$$(6)\qquad \begin{gathered} (a z_{\bar{x}})_{\hat{x}} - d z = -\psi(x)\,, \qquad x \in \hat{\omega}_h\,, \\ z_0 = z_N = 0\,, \quad a \ge c_1 > 0\,, \quad d \ge 0\,, \end{gathered}$$

where $\psi(x) = (a u_{\bar{x}})_{\hat{x}} - du + \varphi(x)$ is the error of approximation.

On account of the balance equation (1) the error of approximation $\psi_i$ reduces to

$$\psi_i = \eta_{\hat{x},\,i} + \psi_i^*, \qquad \eta_i = (a u_{\bar{x}})_i - (k\,u')_{i-1/2}\,,$$

where

$$\text{(7)} \qquad \psi_i^* = (\varphi_i - \overset{\circ}{\varphi}_i) - d_i\,u_i + \frac{1}{\hbar_i} \int\limits_{x_{i-1/2}}^{x_{i+1/2}} q(x)\,u(x)\,dx\,.$$

In what follows the functions $k$, $q$ and $f$ will be smooth for $x_{i-1} < x < x_i$ and $x_i < x < x_{i+1}$ and possess discontinuities of the first kind at a single node $x_i$. Combination of the expansions

$$f(x_i + sh_i) = f_{i-0} + s\,h_i\,f'_{i-0} + O(h_i^2)$$

for $s < 0$ and $f(x_i + s\,h_i) = f_{i+0} + s\,h_{i+1}\,f'_{i+0} + O(h_{i+1}^2)$ for $s > 0$ with formula (3) for $\overset{\circ}{\varphi}_i$ gives

$$\overset{\circ}{\varphi}_i = \frac{h_i\,f_{i-0} + h_{i+1}\,f_{i+0}}{2\,\hbar_i} + \left( \frac{h_i^2\,f'_{i-1/2}}{8} \right)_{\hat{x},\,i} + O(\hbar_i^2)\,.$$

Likewise,

$$\frac{1}{\hbar_i} \int\limits_{x_{i-1/2}}^{x_{i+1/2}} q(x)\,u(x)\,dx = \frac{h_i\,q_{i-0} + h_{i+1}\,q_{i+0}}{2\,\hbar_i} + \left( \frac{h_i^2\,(qu)'_{i-1/2}}{8} \right)_{\hat{x},\,i} + O(\hbar_i^2)\,.$$

With the relations $a_i = k_{i-1/2}$ and $u_{\bar{x},\,i} = u'_{i-1/2} + O(h_i^2)$ in view, we deduce from the foregoing that for the scheme with coefficients (5)

$$\text{(8)} \qquad \psi_i = (\eta + \bar{\eta})_{\hat{x},\,i} + \psi_i^{**},$$

whose ingredients behave as follows:

$$\psi_i^{**} = O(\hbar_i^2)\,,$$

$$\text{(9)} \qquad \bar{\eta}_i = \frac{h_i^2\,(qu - f)'_{i-1/2}}{8} = O(h_i^2)\,,$$

$$\eta_i = a_i\,u_{\bar{x},i} - (ku')_{i-1/2} = O(h_i^2)\,.$$

**3. The order of accuracy on non-equidistant grids.** As before, the inner products are defined by

$$(y,v)_* = \sum_{i=1}^{N-1} y_i\, v_i\, \hbar_i\,, \qquad (y,v]_* = \sum_{i=1}^{N} y_i\, v_i\, h_i\,.$$

*For problem* (6) *with the right-hand side* (8) *the estimate*

$$\| z \|_C \le \frac{2}{c_1} \Big\{ (1, |\eta| + |\bar{\eta}|] + (1, |\mu|] \Big\} \tag{10}$$

*is valid with* $\mu_i = \sum_{k=1}^{i-1} \hbar_k\, \psi_k^*$ *for* $i = 2, 3, \dots, N$ *and* $\mu_1 = 0$.

The derivation of this estimate coincides with that of inequality (12) carried out in Section 3 by inserting $a_i/h_i$ in place of $a_i/h$.

In light of the asymptotic relations (9) estimate (10) implies that if $k$, $q$, $f \in Q^{(2)}[0,1]$, then scheme (4)–(5) is of second-order accuracy on the sequence of non-equidistant grids $\hat{\omega}_h(K)$:

$$\| z \|_C = \| y - u \|_C \le M\, \bar{h}^2,$$

where $\bar{h} = \sqrt{(1, h^2]}$ is the mean square step.

When only one coefficient $k(x) \in Q^{(2)}$ is discontinuous, while other coefficients $q$, $f \in C^{(2)}$ are continuous, any conservative scheme (4) generating a second-order approximation is of second-order accuracy on the sequence of non-equidistant grids $\hat{\omega}_h(K)$. This fact is an immediate implication of the expansions $\eta_i = a_i\, u_{\bar{x},i} - (k\,u')_{i-1/2} = O(h_i^2)$, valid for the aforementioned schemes, and $\psi_i^* = O(\hbar_i^2)$.

We are now interested in the question concerning the accuracy of scheme (4) with second-order approximation on an arbitrary non-equidistant grid. A discontinuity point $x = \xi$ is free to be chosen for the relevant coefficients:

$$x_n \le \xi \le x_{n+1}\,, \qquad \xi = x_n + \theta\, h\,, \qquad 0 \le \theta \le 1\,.$$

Being concerned with $\psi_i^*$ specified by (7) and involved in the representations

$$\psi_i = \eta_{\hat{x},i} + \psi_i^*\,, \qquad \eta_i = \big(a u_{\bar{x}}\big)_i - (k u')_{i-1/2}\,,$$

which are always valid, we follow the same procedures in the accuracy analysis as we did on equidistant grids in Section 3. This amounts to regarding $q$, $f \in C^{(2)}$ to be continuous and establishing the relations:

$$\eta_i = O(h_i^2) \quad \text{for} \quad i \neq n+1\,, \qquad \eta_{n+1} = O(1)\,.$$

Only for the best scheme (3) we might achieve $\eta_{n+1} = O(h_{n+1})$ and

$$\psi_i^* = \bar{\eta}_{\hat{x},i} + \psi_i^{**}, \qquad \bar{\eta}_i = \frac{h_i^2\,(qu-f)'_{i-1/2}}{8}\,, \qquad \psi_i^{**} = O(\hbar_i^2)$$

for all $i = 1, 2, \dots, N-1$.

Using estimate (10) behind we deduce that the best scheme (3)–(4) retains the second-order accuracy in the class of discontinuous coefficients on an arbitrary sequence of non-equidistant grids, while any scheme (4) appears to be of order 1.

*In the class of continuous coefficients $k$, $q$, $f \in C^{(2)}[0,1]$ any scheme (4) retains the second-order accuracy on an arbitrary sequence of non-equidistant grids.*

Before we undertake the proof of the last assertion, it is worth mentioning the estimate obtained in Chapter 2, Section 4 for the operator equation $A\,z = \psi$:

$$\|\, z \,\|_A = \|\, \psi \,\|_{A^{-1}} \leq \frac{1}{c_1}\, \|\, \psi \,\|_{\mathring{A}^{-1}}\,,$$

where $A \geq c_1 \mathring{A}$, $A = A^* > 0$ and $\mathring{A} = \mathring{A}^* > 0$. In the case of interest

$$A\,y = -(a y_{\bar{x}})_{\hat{x}} + d y\,, \qquad \mathring{A}\,y = -y_{\bar{x}\hat{x}}\,, \qquad y \in \mathring{\Omega}\,,$$

where $\mathring{\Omega} = H$ is the space of grid functions defined on $\omega_h$ and vanishing for $i = 0, N$. The negative norm $\|\,\psi\,\|_{\mathring{A}^{-1}}$ was estimated in Chapter 2, Section 4 as follows:

$$\|\,\psi\,\|_{\mathring{A}^{-1}} \leq \|\,\eta]| + \|\,\bar{\eta}]| + \|\,\mu]| = O(\bar{h}^2)\,.$$

Here $\bar{h}^2$ denotes, as usual, the mean square of $h_i^2$ and so the desired assertion follows immediately from the preceding.

**4. A greater gain in accuracy on a sequence of grids. The Runge method.** In a common setting we are dealing with a linear mathematical-physics equation

$$(11) \qquad L\,u = f(x)\,, \qquad x \in G\,,$$

regardless of additional conditions on the boundary $\Gamma$. Let $\omega_h$ be a grid in the domain G with step $h$, by means of which a difference scheme

$$(12) \qquad L_h\,u_h = f(x)\,, \qquad x \in \omega_h\,,$$

can be put in correspondence with problem (11).

When this scheme happens to be stable, so that $\|y_h\|_{(1_h)} \leq M \|\varphi_h\|_{(2_h)}$ with a positive constant $M$ independent of $h$ (see Chapter 2, Section 2), the approximation implies the convergence

$$(13) \qquad \|\, y_h - u_h \,\|_{(1_h)} \leq M \;\|\, \psi_h \,\|_{(2_h)}\,,$$

where $\psi_h = \varphi_h - L_h\,u_h = (\varphi_h - f_h) - (L_h\,u_h - (Lu)_h)$ is the error of approximation (residual) on the problem (11) solution $u = u(x)$. It is readily seen from (13) that the order of accuracy is not lower than the order of approximation and the behaviour of the residual $\|\,\psi_h\,\|_{(2_h)} = O(h^n)$ is a corollary to the estimate for the relevant deviation $\|\, y_h - u_h \,\|_{(1_h)} = O(h^n)$.

In order to improve the accuracy of the approximate solution, it is necessary to diminish the grid step $h$ or make the approximation order $n$ of the scheme higher. An elementary example illustrating how to raise the order of approximation on a solution is available in Chapter 1, Section 2.2. However, in trying to develop higher-accuracy schemes for many problems, especially for equations with variable coefficients, considerable obstacles of technical character do arise. Moreover, the transition to such schemes except the schemes for the heat conduction equation with constant coefficients and the Laplace equation which will be given special investigation in Chapters 4–5, should cause the essential growth of the volume of computations.

The gain in accuracy provided by refining the step $h$ is limited by requirements of economy. Such an approach is equivalent to minimizing the execution time necessary in this connection in obtaining the solution. But if the solution of the original problem $u$ and $f$ both are smooth functions of $x$, the accuracy of numerical solution can be increased by performing calculations for the same problem (12) on a sequence of grids $\omega_{h_1}, \ldots, \omega_{h_n}$.

In the sequel we assume that $u = u(x)$ possesses all necessary derivatives which do arise in the further development. In order to understand this approach a little better, we consider the simplest case, provided that the asymptotic expansion

$$y_h(x) = u_h(x) + \alpha_1(x)\, h^{k_1} + O(h^{k_2}), \qquad k_2 > k_1 > 0, \tag{14}$$

holds, where $\alpha_1(x)$ does not depend on $h$. In such a setting it is required to find a grid function $\tilde{y}_h(x)$ satisfying the relation

$$\tilde{y}_h(x) = u_h(x) + O(h^{k_2}). \tag{15}$$

The traditional way of covering this is to introduce two grids $\omega_{h_1}$ and $\omega_{h_2}$ with steps $h_1$ and $h_2$ and common nodes, the set of which is denoted by $\tilde{\omega}_h$, and to form the grid function

$$\tilde{y}_h(x) = c_1\, y_{h_1}(x) + c_2\, y_{h_2}(x), \qquad x \in \tilde{\omega}_h, \tag{16}$$

while $c_1$ and $c_2$ remain as yet unknown. Substitution of expansions (14) for $y_{h_1}$ and $y_{h_2}$ into (16) gives

$$\tilde{y}_h(x) = (c_1 + c_2)\, u_h(x) + \left(c_1\, h_1^{k_1} + c_2\, h_2^{k_1}\right) \alpha_1(x) + O(h^{k_2}),$$

whence it follows that $\tilde{y}_h - u_h = O(h^{k_2})$ if

$$c_1 + c_2 = 1, \qquad c_1\, h_1^{k_1} + c_2\, h_2^{k_1} = 0.$$

This is certainly so with

$$c_2 = 1 - c_1, \qquad c_1 = -\frac{h_2^{k_1}}{h_1^{k_1} - h_2^{k_1}}. \tag{17}$$

In particular, keeping $h_1 = h$ and $h_2 = \frac{1}{2}h$, we find $\tilde{\omega}_h = \omega_h$ and $c_1 = -1/(2^{k_1} - 1)$.

Thus, the improved accuracy of a grid solution on some set of nodes $\tilde{\omega}_h$ is connected with solving problem (12) twice (first on the grid $\omega_{h_1}$ and then on the grid $\omega_{h_2}$) and drawing up the linear combination (16) with coefficients (17). The grids $\omega_{h_1}$ and $\omega_{h_2}$ are chosen so that their intersection coincides with $\tilde{\omega}_h$. For instance, by applying a scheme of second-order accuracy satisfying the relation

$$y_h(x) = u_h(x) + \alpha_1\, h^2 + O(h^4)$$

and having $k_1 = 2$ and $k_2 = 4$ to the difference problem with steps $h_1 = h$ and $h_2 = \frac{1}{2}h$ we calculate the solutions $y_h$ and $y_{h/2}$ and the coefficients $c_1 = -\frac{1}{3}$ and $c_2 = \frac{4}{3}$, leading to the function in question:

$$\tilde{y}(x) = \tfrac{4}{3}\, y_{h/2} - \tfrac{1}{3}\, y_h \,,$$

which is defined on the grid $\omega_h$ and approximates the exact solution $u(x)$ with accuracy $O(h^4)$:

$$\tilde{y}_h = u_h + O(h^4) \quad \text{for} \quad x \in \omega_h \,.$$

One way of proceeding is to take for granted the expansion

$$y_h = u_h + \alpha_1(x)\, h^{k_1} + \alpha_2(x)\, h^{k_2} + O(h^{k_3})\,, \qquad k_3 > k_2 > k_1 > 0\,,$$

where the functions $\alpha_1(x)$ and $\alpha_2(x)$ are independent of $h$. Because of this, we must solve the difference problem concerned three times with steps $h_1$, $h_2$ and $h_3$, respectively, in an attempt to obtain a solution with the prescribed accuracy $O(h^{k_3})$. In what follows $y_{h_1}(x)$, $y_{h_2}(x)$ and $y_{h_3}(x)$ will stand for the appropriate solutions. Also, it will be sensible to introduce their linear combination

$$\tilde{y}_h(x) = c_1\, y_{h_1}(x) + c_2\, y_{h_2}(x) + c_3\, y_{h_3}(x)\,, \qquad x \in \tilde{\omega}_h \,.$$

As before, the grid $\tilde{\omega}_h$ is the intersection of the three grids $\omega_{h_1}$, $\omega_{h_2}$ and $\omega_{h_3}$. In particular, $\tilde{\omega}_h = \omega_h$ under the agreements $h_1 = h$, $h_2 = \frac{1}{2}h$ and $h_3 = \frac{1}{4}h$. When the approximation $\tilde{y}_h = u_h + O(h^{k_3})$ is accepted for later use, the following equations

$$c_1 + c_2 + c_3 = 1, \qquad c_1\, h_1^{k_1} + c_2\, h_2^{k_1} + c_3\, h_3^{k_1} = 0\,, \qquad c_1\, h_1^{k_2} + c_2\, h_2^{k_2} + c_3\, h_3^{k_2} = 0$$

constitute the system for finding the coefficients $c_1$, $c_2$, $c_3$. Evidently, the determinant of this system is nonzero.

In the general case the expansion of the error $y_h - u_h$ in powers of $h$ is of the form

$$y_h = u_h + \sum_{s=1}^{n-1} \alpha_s(x)\, h^s + \alpha_n(x, h)\, h^n, \tag{18}$$

where $\alpha_s(x)$, $s = 1, 2, \ldots, n-1$, are independent of $h$ and the absolute value of $\alpha_n(x, h)$ is bounded by a constant $M = \text{const} > 0$, which is independent of $h$. Our next step is to establish an a priori expansion of the form (18).

When $u(x)$, $f(x)$ and the coefficients of equation (11) are smooth enough, the expression for the error of approximation $\psi_h(x)$ is simple to follow:

$$\psi_h(x) = \sum_{s=1}^{n-1} \beta_s(x)\, h^s + \beta_n(x,h)\, h^n, \tag{19}$$

where $\beta_s(x)$, $1 \leq s \leq n-1$, are independent of $h$ and $|\beta_n(x,h)| \leq M$, $M = \text{const} > 0$ is also independent of $h$. From such reasoning it seems clear that any sufficiently smooth function $\alpha_s(x)$ admits the representation

$$L_h\, \alpha_s(x) = L\, \alpha_s(x) + u_h + \sum_{s=1}^{n-1} \gamma_s(x)\, h^s + \gamma_n(x,h)\, h^n. \tag{20}$$

In this view, it is not unreasonable to attempt the difference $y_h - u_h$ in the form (18). Applying the operator $L_h$ to identity (18) yields

$$L_h\,(y_h - u_h) = \sum_{s=1}^{n-1} L_h\, \alpha_s(x)\, h^s + L\alpha_n(x,h)\, h^n\,.$$

Using the relation $L_h(y_h - u_h) = \psi_h$ and formula (20) for $L_h \alpha_s$, we arrive at

$$L_h(y_h - u_h) = \sum_{s=1}^{n-1} \Big( L\, \alpha_s + \sum_{m=1}^{s-1} \gamma_m\, h^m \Big)\, h^s + O(h^n)\,,$$

giving

$$\psi_h = \sum_{s=1}^{n-1} \Big( L\, \alpha_s + \sum_{m=1}^{s-1} \gamma_m\, h^m \Big)\, h^s + O(h^n)\,. \tag{21}$$

Comparison of formulae (19) and (20) justifies the validity of expansion (18) if functions $\alpha_s(x)$ are solutions to the equations

$$L\alpha_s = \beta_s(x) - \sum_{m=1}^{s-1} \gamma_m(x) \qquad \text{for} \quad s = 1, 2, \ldots, n-1\,.$$

Observe that expansion (19) is not obliged to contain all of the powers of $h$, some of them may be omitted. In this case the appropriate coefficient equals $\beta_s \equiv 0$.

Adaptive grids may be of assistance in raising the order of accuracy without increasing the total number of nodes. If the subsidiary information on the behaviour of the solution of an original problem is available, a grid can be made so as to achieve a prescribed accuracy at minimum nodal points and, therefore, in a minimal number of the necessary operations. For instance, successive grid refinement will be appreciated in the region of widely varying coefficients and the right-hand side. Specifically, near the point (the line) of discontinuity of coefficients, representing the boundary between two media, the grid is refined in such a way to attain the minimum of the step near the boundary. After that, the step is being enlarged for instance, as a geometric progression in moving from the boundary. When no information about the behaviour of the solution is available, a preliminary computation proves to be useful on a sparse grid, after which we are working on a new grid with a more smaller step in the regions with large deviations of the solution. In practice non-equidistant grids are widely used. As shown in Section 3, it is possible to construct special grids $\hat{\omega}_h(K)$ so that all discontinuity points of the coefficient $k(x)$ involved in the equation

$$\big(k(x)\,u'\big)' - q(x)\;u = -f(x)$$

fall into nodal points of the grid $\hat{\omega}_h(K)$. Under such a choice, any homogeneous difference scheme $\big(a y_{\bar{x}}\big)_{\hat{x}} - dy = -\varphi$ generating approximation of order 2 (in the class of smooth coefficients) is of second-order accuracy in the class of discontinuity coefficients $k(x)$. The accuracy of difference schemes, which depends on the existing grids suitable to computer calculations for equations with variable coefficients, needs investigation for each concrete problem.

## 3.5 OTHER PROBLEMS

**1. The third boundary-value problem.** The main goal of our studies is a homogeneous difference scheme for the boundary-value problem of the third kind:

$$\begin{gathered} L\,u = \big(k(x)u'\big)' - q(x)\,u = -f(x), \\ 0 < x < 1, \quad k(x) \geq c_1 > 0, \quad q \geq 0\,, \\ k(0)\,u'(0) = \beta_1\,u(0) - \mu_1\,, \qquad -k(1)\,u'(1) = \beta_2\,u(1) - \mu_2\,, \\ \beta_1 \geq 0, \qquad \beta_2 \geq 0, \qquad \beta_1 + \beta_2 > 0\,. \end{gathered} \tag{1}$$

We approximate equation (1) in the usual way:

(2) $\quad \Lambda\,y = -\varphi(x), \qquad \Lambda\,y = \big(a y_{\bar{x}}\big)_x - dy, \qquad a \geq c_1 > 0, \quad d \geq 0\,,$

where $a$, $d$ and $\varphi$ satisfy the approximation conditions (18) of Section 2.

First, we consider the simplest approximation of the boundary condition at the point $x = 0$: $a_1\, y_{\bar{x},1} = \beta_1\, y_0 - \mu_1$ and calculate the error of approximation by inserting $y = z + u$:

$$a_1\, z_{\bar{x},1} = \beta_1\, z_0 - \nu_1, \qquad \nu_1 = a_1\, u_{\bar{x},1} - \beta_1\, u_0 + \mu_1 .$$

We obtain

$$\nu_1 = (k_0\, u_0' - \beta_1\, u_0 + \mu_1) + \tfrac{1}{2}\, h\, (k\, u')_0' + O(h^2) = \tfrac{1}{2}\, h\, (k\, u')_0' + O(h^2)$$

through the approximations

$$a_1 = k_0 + \tfrac{1}{2}\, h\, k_0' + O(h^2), \qquad u_{\bar{x},1} = u_0' + \tfrac{1}{2}\, h\, u_0'' + O(h^2) .$$

Substituting here $(ku')_0' = q_0\, u_0 - f_0$ emerged from equation (1) we find that

$$\nu_1 = \tfrac{1}{2}\, h\, (q_0\, u_0 - f_0) + O(h^2) ,$$

thereby justifying that the boundary condition

$$(3) \qquad a_1\, y_{\bar{x},1} = \bar{\beta}_1\, y_0 - \bar{\mu}_1, \qquad \bar{\beta}_1 = \beta_1 + \tfrac{1}{2}\, h\, q_0 , \qquad \bar{\mu}_1 = \mu_1 + \tfrac{1}{2}\, h f_0$$

provides an approximation of order 2 on the solution $u(x)$ of problem (1). In a similar manner we derive the difference boundary condition of second-order approximation at the point $x = 1$:

$$(4) \quad -a_N\, y_{\bar{x},N} = \bar{\beta}_2\, y_N - \bar{\mu}_2, \qquad \bar{\beta}_2 = \beta_2 + \tfrac{1}{2}\, h\, q_N , \qquad \bar{\mu}_2 = \mu_2 + \tfrac{1}{2}\, h\, f_N .$$

In this way, the third kind difference boundary-value problem (2)–(4) of second-order approximation on the solution of the original problem is put in correspondence with the original problem (1).

**2. A problem with periodicity conditions.** First, we study the elementary problem in which it is required to find on the segment $0 \le x \le 1$ a solution to the equation

$$(5) \qquad u''(x) - q_0\, u = -f(x), \qquad q_0 = \text{const} > 0, \quad 0 < x < 1 ,$$

satisfying the condition of periodicity

$$(6) \qquad u(x+1) = u(x) \qquad \text{for all} \quad x \in (0,1) .$$

Here $f(x)$ is a periodic function of period 1: $f(x+1) = f(x)$.

At any point $x \in (0,1)$ condition (6) is equivalent to the pair of continuity conditions at the single point $x = 0$:

(7) $$u(0+0) = u(1-0), \qquad u'(0+0) = u'(1-0).$$

Problem (5)–(6) has a unique solution having the estimate

$$\| u \|_C \le \| f \|_C / q_0$$

in complete agreement with the maximum principle.

For $q_0 = 0$ the statement of the problem is

$$u'' = -f(x), \qquad u(0+0) = u(1-0), \qquad u'(0+0) = u'(1-0),$$

which is solvable under the condition $\int_0^1 f(x)dx = 0$ and possesses a unique solution $u = u(x)$ only if

(8) $$\int_0^1 u(x)\, dx = 0.$$

Indeed, the general solution to the equation $u'' = -f(x)$ admits the form

$$u(x) = C_1\, x + C_2 - \int_0^x \left( \int_0^t f(\alpha)\, d\alpha \right) dt = C_1 x + C_2 - \int_0^x (x-t) f(t)\, dt$$

with arbitrary constants $C_1$ and $C_2$.

With this, conditions (7) provide

$$\int_0^1 f(t)\, dt = 0, \qquad C_1 = -\int_0^1 t\, f(t)\, dt = 0,$$

it being understood that the function $u(x)$ can be recovered from conditions (7) within a constant $C_2$. Under condition (8) we find that $C_2 = 0$, extracting a unique solution of the problem.

As a first step towards the solution of the original problem, we initiate the design of a difference scheme on an equidistant grid $\bar{\omega}_h = \{x_i = ih,\ i = 0, 1, \dots, N\}$ on the segment $0 \le x \le 1$ with step $h = 1/N$ and approximate equation (5) and the continuity conditions (7). The first of these conditions is satisfied if $y_0 = y_N$.

At the nodes $x_i = ih$, $i = 1, 2, \dots, N-1$, we write down the three-point equation

$$y_{\bar{x}x} - q_0\, y = -\varphi(x), \qquad x = ih, \quad i = 0, 1, \dots, N-1\,.$$

In this context, the difference derivatives

$$u_{\bar{x},N} = u'(1-0) - \tfrac{1}{2}\, h\, u''(1-0) + O(h^2)\,,$$

$$u_{x,0} = u'(0+0) + \tfrac{1}{2}\, h\, u''(0+0) + O(h^2)$$

will be given special investigation. Substituting here $u'' = q_0\, u_0 - f$ recovered from (5) we obtain

$$u_{\bar{x},N} + \tfrac{1}{2}\, h\big(q_0\, u(1) - f(1-0)\big) = u'(1-0) + O(h^2)\,,$$

$$u_{x,0} - \tfrac{1}{2}\, h\big(q_0\, u(0) - f(0+0)\big) = u'(0+0) + O(h^2)\,.$$

With these relations established, the second continuity condition $u'(0+0) = u'(1-0)$ is approximated to $O(h^2)$ by the difference equation

$$(9) \qquad u_{x,0} - \tfrac{1}{2}\, h\, q_0\, y_0 + \tfrac{1}{2}\, h\, f(0+0) = u_{\bar{x},N} + \tfrac{1}{2}\, h\, q_0\, y_N - \tfrac{1}{2}\, h\, f(1-0)\,.$$

By merely setting $y_{N+1} = y_1$ condition (9) becomes

$$u_{\bar{x}x,\,N} - q_0\, y_N = -\varphi_N\,, \qquad \varphi_N = \tfrac{1}{2}\left(f(1-0) + f(0+0)\right).$$

This is a way of establishing the correspondence between problem (5), (7) and the difference scheme

$$(10) \qquad y_{\bar{x}x} - q_0\, y = -\varphi(x), \qquad x = ih, \qquad i = 1, 2, \dots, N\,,$$

with the conditions of periodicity

$$(11) \qquad y_0 = y_N\,, \qquad y_1 = y_{N+1}\,.$$

We will elaborate on this for rather complicated cases and turn to the equation with variable coefficients

$$(12) \qquad (ku')' - q\, u = -f(x), \qquad 0 < x < 1\,,$$

where $k(x)$, $q(x)$ and $f(x)$ refer to periodic functions of period 1:

$$(13) \qquad k(x+1) = k(x), \qquad q(x+1) = q(x), \qquad f(x+1) = f(x)\,,$$

which are continuous at the point $x = 0$ ($x = 1$), so that $k(1-0) = k(0+0) = k(0)$, etc. Let

$$(14) \qquad k(x) \geq c_1 > 0\,, \qquad q(x) \geq c_1 > 0\,.$$

Assuming this to be the case, it is required to find a solution to equation (12) subject to the periodicity condition $u(x+1) = u(x)$, which is equivalent to the requirements

$$(15) \qquad u(0+0) = u(1-0)\,, \qquad k\,u'|_{x=0+0} = k\,u'|_{x=1-0}\,.$$

The maximum principle implies that problem (12)–(15) is uniquely solvable. With this in mind, we start from the scheme for $0 < x = ih < 1$:

$$\left(a\,y_{\bar{x}}\right)_x - dy = -\varphi(x)\,, \qquad x = ih\,, \qquad i = 1, 2, \ldots, N-1\,,$$

keeping $y_0 = y_N$. The coefficients $a$, $d$, $\varphi$ can be recovered from the conditions of second-order approximation in Section 2.4.

With the aid of the equalities

$$\left(a y_{\bar{x}}\right)_i = (k\,u')_{i-0} + \tfrac{1}{2}\,h\,(f - qu)_{i-1} + O(h^2)\,,$$

$$a_{i+1}\,u_{x,i} = (k\,u')_{i+0} - \tfrac{1}{2}\,h\,(f - qu)_{i+0} + O(h^2)$$

the condition $ku'|_{x=0+0} = ku'|_{x=1-0}$ is approximated to second order by the difference relation

$$a_1\,y_{x,0} - \tfrac{1}{2}\,h\big(q(0)\,y_0 - f(0+0)\big) = a_N\,y_{\bar{x},N} + \tfrac{1}{2}\,h\big(q(1-0)\,y_N - f(1-0)\big)\,.$$

Under the agreements $y_{N+1} = y_1$ and $a_{N+1} = a_1$ the preceding reduces to

$$\left(a y_{\bar{x}}\right)_x - dy = -\varphi(x)\,, \qquad x = x_N = 1\,,$$

with

$$d = d_N = \tfrac{1}{2}\,\big(q(0+0) + q(1-0)\big), \qquad \varphi = \varphi_N = \tfrac{1}{2}\,\big(f(0+0) + f(1-0)\big)\,,$$

leaving us with the periodic difference scheme

$$(16) \qquad \begin{gathered} \left(a y_{\bar{x}}\right)_x - dy = -\varphi(x)\,, \qquad x = i\,h, \qquad i = 1, 2, \ldots, N\,, \\ a \geq c_1 > 0\,, \qquad d \geq c_1 > 0\,, \end{gathered}$$

subject to the conditions

$$(17) \qquad y_0 = y_N\,, \qquad y_1 = y_{N+1}\,, \qquad a_1 = a_{N+1}\,.$$

For determination of $y_i$, $i = 1, 2, \dots N$, we obtain the system of equations

$$a_i\, y_{i-1} - (a_i + a_{i+1} + d_i\, h^2)\, y_i + a_{i+1}\, y_{i+1} = -\varphi_i\, h^2\,, \qquad i = 1, 2, \dots, N\,,$$

supplied by the conditions of periodicity: $y_0 = y_N$ and $y_{N+1} = y_1$. This system can be solved by the cyclic elimination method (for more detail see Chapter 1, Section 1.2).

Since $a \ge c_1 > 0$ and $d \ge c_1 > 0$, the maximum principle is still valid for problem (16)–(17), due to which

$$\| y \|_C \le \frac{1}{c_1} \| \varphi \|_C\,.$$

The inequality obtained permits us to derive for the error $z = y - u$ the estimate

$$\| z \|_C = O(h^2)\,,$$

since $\psi_i = O(h^2)$ for $i = 1, 2, \dots, N$. Thus, scheme (16)–(17) is of second-order accuracy in the space $C$ when $k(x) \in C^{(3)}$ and $q(x)$, $f(x) \in C^{(2)}$.

**3. Monotone schemes for an equation of general form.** The object of investigation is the boundary-value problem

$$(18) \qquad \begin{aligned} &L\, u = (k\, u')' + r(x)\, u' - q(x)\, u = -f(x)\,, \qquad 0 < x < 1\,, \\ &u(0) = u_1\,, \quad u(1) = u_2\,, \quad k(x) \ge c_1 > 0\,, \quad |\, r(x)\,| \le c_2\,, \quad q \ge 0\,. \end{aligned}$$

The main idea here is connected with the design of a new difference scheme of second-order approximation for which the maximum principle would be in full force for any step $h$. The meaning of this property is that we should have (see Chapter 1, Section 1)

$$(19) \qquad A_i\, y_{i-1} - C_i\, y_i + B_i\, y_{i+1} = -F_i\,, \qquad i = 1, 2, \dots, N-1\,,$$

where $A_i > 0$, $B_i > 0$ and $C_i - A_i - B_i = D_i \ge 0$.

Any such scheme is said to be **monotone**. As before, the operator $Lu = (ku')' - qu$ is approximated to second order by the homogeneous three-point scheme $\Lambda y = \left(a y_{\bar{x}}\right)_x - dy$.

The natural replacement of the central difference derivative $u'(x)$ by the first derivative $u_{\overset{\circ}{x}}$ leads to a scheme of second-order approximation. Such a scheme is monotone only for sufficiently small grid steps. Moreover, the elimination method can be applied only for sufficiently small $h$ under the restriction $h\,|\,r(x)\,| < 2\,k(x)$. If $u'$ is approximated by one-sided difference derivatives (the right one $u_x$ for $r > 0$ and the left one $u_{\bar{x}}$ for $r < 0$), we obtain a monotone scheme for which the maximum principle is certainly true for any step $h$, but it is of first-order approximation. This is unacceptable for us.

It is worth mentioning here that the sign of $r(x)$ has had a significant impact on construction of monotone schemes. One way of providing a second-order approximation and taking care of this sign is connected with a monotone scheme with one-sided first difference derivatives for the equation with perturbed coefficients

$$(20)\qquad \widetilde{L}\,u = -f, \qquad \widetilde{L}\,u = \varkappa\,(k\,u')' + r\,u' - q\,u\,,$$

where $\varkappa = 1/(1+R)$ and $R = 0.5\,h\,|\,r\,|/k$ is the Reynolds difference number. By obvious rearranging $r(x)$ as a sum

$$r = r^+ + r^-\,, \qquad r^+ = \tfrac{1}{2}\,(r + |\,r\,|) \geq 0, \qquad r^- = \tfrac{1}{2}\,(r - |\,r\,|) \leq 0\,,$$

the expression $ru'$ is approximated by the formula

$$(ru')_i = \Big(\frac{r}{k}\,(ku')\Big)_i \sim b_i^+\,a_{i+1}\,u_{x,i} + b_i^-\,a_i\,u_{\bar{x},i}\,,$$

where $b_i^\pm = F\big[\tilde{r}^\pm(x_i{+}sh)\big]\,,\ \tilde{r}^\pm = r^\pm/k$, $F$ is a pattern functional being used for calculations of the coefficients $d$ and $\varphi$. We may accept, for instance, $b^+ = r^+/k$ and $b^- = r^-/k$. As a result we get the homogeneous scheme

$$\widetilde{\Lambda} y = \varkappa\big(a y_{\bar{x}}\big)_x + b^+ a^{(+1)} y_x + b^- a\, y_{\bar{x}} - dy = -\varphi\,,$$

$$(21)\qquad y_0 = u_1\,,\quad y_N = u_2\,,\quad a^{(+1)} = a(x+h), \quad a \geq c_1 > 0,$$

$$\varkappa = \tfrac{1}{1+R}, \qquad R = \tfrac{|\,r\,|\,h}{2\,k}\,.$$

We are going to show that scheme (21) is monotone by observing that

$$(22)\qquad A_i\,y_{i-1} - C_i\,y_i + B_i\,y_{i+1} = \varphi_i\,, \qquad y_0 = u_1\,, \quad y_N = u_2\,,$$

where

$$A_i = \frac{a_i}{h^2}\,\big(\varkappa_i - h\,b_i^-\big), \qquad B_i = \frac{a_{i+1}}{h^2}\,\big(\varkappa_i + h\,b_i^+\big), \qquad C_i = A_i + B_i + d_i\,.$$

As can readily be observed from the foregoing, $A_i > 0$, $B_i > 0$ and $D_i \geq 0$, since $b_i^- \leq 0$, $b_i^+ \geq 0$ and $d_i \geq 0$.

Equations (22) can be solved by the elimination method for any $h$ and $r$. The error of approximation for scheme (21)

$$\psi = \varkappa \left(a u_{\bar{x}}\right)_x + b^+ a^{(+1)} u_x + b^- a u_{\bar{x}} - d u + \varphi - (L u + f)$$

is representable by $\psi = \psi^{(1)} + \psi^{(2)}$ with the members

$$\psi^{(1)} = \left[\left(a u_{\bar{x}}\right)_x - d u + \varphi\right] - \left[(k u')' - q u + f\right],$$

$$\psi^{(2)} = \left[(\varkappa - 1)\left(a u_{\bar{x}}\right)_x + b^+ a^{(+1)} u_x + b^- a u_{\bar{x}}\right] - r\, u'.$$

The first summand satisfies the estimate

$$\psi^{(1)} = O(h^2) \qquad \text{for} \quad k \in C^{(3)}, q, f \in C^{(2)}.$$

By virtue of the relations

$$b^+ = \tilde{r}^+ + O(h^2), \qquad b^- = \tilde{r}^- + O(h^2),$$

$$k\, \tilde{r}^{\pm} = r^{\pm}, \qquad r^+ + r^- = r, \quad r^+ - r^- = |\, r\,|,$$

$$a u_{\bar{x}} = k u' - \tfrac{1}{2}\, h (k u')' + O(h^2),$$

$$a^{(+1)} u_x = k u' + \tfrac{1}{2}\, h (k u')' + O(h^2),$$

$$\left(a u_{\bar{x}}\right)_x = (k u')' + O(h^2)$$

we obtain

$$b^+ a^{(+1)} u_x + b^- a\, u_{\bar{x}} = r u' + \frac{1}{2} h (k u')' \,\frac{|\, r\,|}{k} + O(h^2),$$

$$\psi^{(2)} = -\frac{R}{R+1}\,(k u')' + R\,(k u')' + O(h^2) = \frac{R^2}{R+1}\,(k u')' + O(h^2) = O(h^2)$$

exploiting the fact that $R = 0.5\, h\, |\, r\,| / k = O(h)$.

Because of this, the monotone scheme (21) generates an approximation of order 2:

$$\psi = O(h^2). \tag{23}$$

If $q \geq c_1 > 0$, then for a solution of the difference problem (21) the maximum principle provides the estimate $\| y \|_C \leq \frac{1}{c_1} \| \varphi \|_C$ with $y_0 = y_N = 0$, which implies on the basis of (23) the uniform convergence of scheme (21) with the rate $O(h^2)$.

The monotone function (21) is quite applicable when $r(x)$ is a fastly varying function of the variable $x$ and the condition $R < 1$ fails to hold at some points with no influence on the accuracy to a considerable extent.

The same procedure with a monotone scheme works on a non-equidistant grid.

One more method available for designing a monotone scheme is related to equation (18). In preparation for this, let us multiply equation (18) by the function $\mu(x)$ and require that

$$\mu\,(k\,u')' + r\,\mu\,u' = (k\,\mu\,u')'\,.$$

This is true only if $r\mu = k\mu'$, so that

$$\mu(x) = \mu_0\,\exp\left\{\int_0^x \tilde{r}(t)\,dt\right\}, \qquad \tilde{r} = \frac{r}{k}\,.$$

The equation

$$(\mu\,k\,u')' - \mu\,q\,u = -\mu\,f$$

can be approximated by the conservative monotone scheme

$$(\bar{\mu}\,a\,y_{\bar{x}})_x - \mu\,d\,y = -\mu\,\varphi, \qquad \bar{\mu} = \mu\,(x - \tfrac{1}{2}h)\,, \tag{24}$$

where, for instance, the members are taken to be $a_i = k_{i-1/2}$, $d_i = q_i$, $\varphi_i = f_i$.

The number $\mu(x)$ may be very large when the ratio $\tilde{r} = r/k$ grows. Dividing both sides of the difference equation

$$\frac{1}{h^2}\left[a_{i+1}\mu_{i+1/2}(y_{i+1} - y_i) - a_i\mu_{i-1/2}(y_i - y_{i-1})\right] - \mu_i\,q_i\,y_i = -\mu_i\,f_i \tag{24'}$$

by $\mu_i = \mu_0 \exp\left\{\int_0^{x_i} \tilde{r}(t)\,dt\right\}$, we obtain the nonconservative, but monotone scheme

$$\begin{gathered} \frac{1}{h}\left(\bar{b}_i\,y_{x,\,i} - \bar{a}_i\,y_{\bar{x},\,i}\right) - q_i\,y_i = -f_i\,, \qquad i = 1, 2, \ldots, N-1\,, \\ y_0 = u_1\,, \qquad y_N = u_2\,, \end{gathered} \tag{25}$$

with the coefficients

$$(25') \qquad \begin{aligned} \bar{b}_i &= a_{i+1} \exp\left\{ \int\limits_{x_i}^{x_{i+1/2}} \tilde{r}(t)\, dt \right\}, \\ \bar{a}_i &= a_i \exp\left\{ - \int\limits_{x_{i-1/2}}^{x_i} \tilde{r}(t)\, dt \right\}. \end{aligned}$$

It is straightforward to verify that the scheme concerned provides an approximation of order 2 due to the expansions

$$\frac{\bar{b}_i - \bar{a}_i}{h} = k_i' + r_i + O(h^2), \qquad \frac{\bar{b}_i + \bar{a}_i}{2} = k_i + O(h^2).$$

Replacing the integrals in (25′) to $O(h^2)$ by the expressions $\frac{h}{8}(3\tilde{r}_i + \tilde{r}_{i+1})$ and $\frac{h}{8}(3\tilde{r}_i + \tilde{r}_{i-1})$, respectively, we arrive at the monotone scheme with the coefficients

$$\bar{b}_i = a_{i+1} \exp\left\{ \tfrac{h}{8}(3\tilde{r}_i + \tilde{r}_{i+1}) \right\} \qquad \text{and} \qquad \bar{a}_i = a_i \exp\left\{ -\tfrac{h}{8}(3\tilde{r}_i + \tilde{r}_{i-1}) \right\},$$

generating an approximation of order 2.

**4. Difference schemes for a stationary equation in cylindrical coordinates.** The stationary diffusion or heat conduction equation

$$\operatorname{div}(k \operatorname{grad} u) - qu = -f(r, \varphi, z)$$

takes in the cylindrical coordinate system $(r, \varphi, z)$ the form

$$(26) \qquad \begin{gathered} \frac{1}{r} \frac{d}{dr}\left( r\, k(r) \frac{du}{dr} \right) - q(r)\, u = -f(r), \qquad 0 < r < R, \\ q(r) \geq 0, \qquad 0 < c_1 \leq k(r) \leq c_2, \end{gathered}$$

in the case when the solution $u = u(r)$ depends neither on $z$ nor on $\varphi$, that is, in the case of the axial symmetry.

When $r = 0$ we impose the boundedness condition $|\, u(0)\, | < \infty$ being equivalent to the requirements

$$(27) \qquad \lim_{r \to 0} r\, k(r) \frac{du}{dr} = 0.$$

When $r = R$ one of the usual boundary conditions applies here, for instance,

$$u(R) = \mu_2 . \tag{28}$$

Let $u_1(r)$ and $u_2(r)$ be linearly independent solutions to equation (26), $u_1(r)$ being bounded for $r \in [0, R]$. A brief survey of their properties is presented below.

1) If $q(0)$ and $f(0)$ are finite, then $u_1(0) \neq 0$ and $u_1'(0) = 0$.
2) If $q(r)$, $f(r) \in C^{(2)}[0, R]$ and $k(r) \in C^{(3)}[0, R]$, then the derivatives $u_1'$, $u_1''$, $u_1^{(3)}$ and $u_1^{(4)}$ are bounded for $0 \leq r \leq R$.
3) The second solution $u_2(r)$ of equation (26), which is linearly independent of $u_1(r)$, has a logarithmic singularity at the point $r = 0$.

Conditions (27) and (28) together provide the existence of a unique solution to equation (26). By virtue of property 1) condition (27) can be replaced by

$$u'(0) = 0 . \tag{29}$$

We proceed as usual and introduce on the segment $0 \leq r \leq R$ the equidistant grid $\omega_h = \{r_i = ih,\ i = 0, 1, \ldots, N,\ hN = R\}$.

Still using the framework of Section 2, a difference scheme for equation (26) can be obtained by the balance method:

$$\frac{w_{i+1/2} - w_{i-1/2}}{r_i\, h} - \frac{1}{r_i\, h} \int\limits_{r_{i-1/2}}^{r_{i+1/2}} q\, u\, r\, dr = -\frac{1}{r_i\, h} \int\limits_{r_{i-1/2}}^{r_{i+1/2}} f(r)\, r\, dr , \tag{30}$$

where

$$r_i = ih, \qquad i = 1, 2, \ldots, N-1 ,$$

$$h = \frac{R}{N}, \qquad r_{i\pm 1/2} = r_i \pm \frac{1}{2}\, h, \qquad w = r\, k(r)\, \frac{du}{dr} .$$

Approximating the flow $w$ by the expression

$$w_{i-1/2} \sim r_{i-1/2}\, a_i\, (u_i - u_{i-1})/h$$

and substituting $d_i\, u_i\, r_i\, h$ and $\varphi_i\, r_i\, h$, respectively, for the integrals in the balance equation (30), we arrive at the difference equation

$$\Lambda\, y_i = \frac{1}{r_i}\left(r_{i-1/2}\, a_i\, y_{\bar r, i}\right)_{r, i} - d_i\, y_i = -\varphi_i , \qquad i = 1, 2, \ldots, N-1, \tag{31}$$

where

$$y_{\bar{r},\,i} = \frac{y_i - y_{i-1}}{h}\,, \qquad y_{r,\,i} = \frac{y_{i+1} - y_i}{h}\,,$$

the coefficients $a_i$, $d_i$ and $\varphi_i$ are chosen so that

$$(32) \qquad a_i = k_{i-1/2} + O(h^2), \qquad d_i = q_i + O(h^2), \qquad \varphi_i = f_i + O(h^2)\,.$$

In the simplest case we accept

$$(33) \qquad a_i = k_{i-1/2}, \qquad d_i = q_i, \qquad \varphi_i = f_i\,.$$

Let us approximate the boundary condition at $r = 0$ that can be declared to be the condition of the zero flow at $r = 0$: $w(0) = 0$. We are going to show that the difference boundary condition

$$(34) \qquad a_1\, y_r(0) = \frac{h}{4}\left(q(0)y(0) - f(0)\right)$$

has the approximation error $O(h^2)$ on a solution to equation (26) satisfying the boundary condition (27).

Indeed, the residual for (34) is equal to

$$(35) \qquad \tilde{\nu} = a_1\, u_r(0) - \frac{h}{4}\left(q(0)\, y(0) - f(0)\right).$$

The forthcoming substitutions

$$a_1 = k_0 + \tfrac{1}{2}\, h\, k_0' + O(h^2), \qquad u_r(0) = u'(0) + \tfrac{1}{2}\, h\,'' u(0) + O(h^2)$$

yield

$$(36) \qquad \tilde{\nu} = (ku')_0 + \frac{1}{2}\, h\, (ku')_0' - \frac{h}{4}\, (q_0\, u_0 - f_0) + O(h^2)\,.$$

From equation (26) we deduce that

$$(k\, u')' = q\, u - f - \frac{k\, u'}{r}\,.$$

Since $u' \to 0$ as $r \to 0$, we have

$$\frac{k\, u'}{r} \to (k\, u')_0' \quad \text{as} \quad r \to 0$$

and

$$(37) \qquad (k\,u')'_0 = (q\,u - f)_0 - (k\,u')'_0 = \frac{1}{2}\,(qu - f)_0\,.$$

Substituting (37) into formula (36) and taking into account (29), we get $\tilde{\nu} = O(h^2)$.

The difference boundary condition (34) can be expressed by

$$\frac{a_1\,y_{r,0}}{h_*} - q_0\,y_0 = -f_0\,, \qquad h_* = \tfrac{h}{4}\,.$$

With these, the difference scheme

$$(38) \qquad \begin{aligned} &\Lambda\,y = \frac{1}{r}\left((r - \tfrac{1}{2}\,h)\,a\,y_{\bar r}\right)_r - dy = -\varphi\,, \qquad 0 < r = ih < 1\,, \\ &\frac{a_1\,y_{r,0}}{h_*} - q_0\,y_0 = -f_0\,, \qquad h_* = \frac{h}{4}\,, \quad y_N = \mu_2\,, \end{aligned}$$

is put in correspondence with problem (26)–(28).

The statement of the difference boundary-value problem for determination of $y_i$ is

$$(39) \qquad \begin{aligned} &A_i\,y_{i-1} - C_i\,y_i + A_{i+1}\,y_{i+1} = -F_i\,, \qquad i = 1, 2, \dots N-1\,, \\ &A_i = \frac{r_{i-1/2}\,a_i}{h^2}\,, \qquad C_i = A_i + A_{i+1} + r_i\,d_i\,, \qquad F_i = \varphi_i\,r_i\,, \end{aligned}$$

which are supplemented with the boundary conditions

$$(40) \qquad y_0 = \varkappa_1\,y_1 + \mu_1\,, \qquad y_N = \mu_2\,,$$

where

$$\varkappa_1 = a_1 \Big/ \left(a_1 + \frac{h^2}{4}\,q_0\right) \quad \text{and} \quad \mu_1 = \frac{h^2}{4}\,f_0 \Big/ \left(a_1 + \frac{h^2}{4}\,q_0\right).$$

This problem can be solved by the elimination method (see Chapter 1, Section 1), for which the stability conditions are satisfied, because $A_i > 0$, $C_i \ge A_i + A_{i+1}$ and $0 < \varkappa_1 \le 1$, $\varkappa_2 = 0$.

We now estimate the accuracy of scheme (38). Substituting into (38) $y = z + u$, where $u$ is a solution of problem (26)–(27) and $y$ is a solution of problem (38), we may set up the problem for the error $z = y - u$:

$$\Lambda z = \frac{1}{r}\left(\left(r - \tfrac{1}{2}h\right) a\, z_{\bar r}\right)_r - dz = -\psi\,, \qquad 0 < r = ih < 1\,,$$

(41)

$$\frac{a_1\, z_{r,0}}{h_*} - q_0\, z_0 = -\nu\,, \qquad z_N = 0\,,$$

where $\psi$ and $\nu$ are the errors of approximation to the equation

(42) $$\psi_i = \frac{1}{r_i}\left(r_{i-1/2}\, a_i\, u_{\bar r,i}\right)_{r,i} - d_i\, u_i + \varphi_i$$

and the boundary condition

(43) $$\nu = \frac{a_1\, u_{r,0}}{h_*} - q_0\, u_0 + f_0\,,$$

respectively.

The balance equation (30) with regard to $\psi$ gives

$$\psi_i = \frac{1}{r_i}\,\eta_{r,i} + \psi_i^*\,, \qquad \eta_i = r_{i-1/2}\left(a_i\, u_{\bar r,i} - (k\,u')_{i-1/2}\right),$$

$$\psi_i^* = \varphi_i - \frac{1}{h\,r_i}\int\limits_{r_{i-1}}^{r_{i+1/2}} f(r)\,r\;dr - \left(d_i\,u_i - \frac{1}{h\,r_i}\int\limits_{r_{i-1/2}}^{r_{i+1/2}} q\,u\,r\;dr\right).$$

Next, we set $r = r_i + sh$ and develop the expansion of integrals involved in the formula for $\psi_i^*$ in powers of $h$:

$$\frac{1}{h\,r_i}\int\limits_{r_{i-1/2}}^{r_{i+1/2}} f(r)\,r\;dr = \frac{1}{r_i}\int\limits_{-1/2}^{1/2} f(r_i + sh)\,(r_i + sh)\;ds$$

$$= \int\limits_{-1/2}^{1/2} f(r_i + sh)\;ds + \frac{h}{r_i}\int\limits_{-1/2}^{1/2}\left(f_i + shf_i' + O(h^2)\right) s\;ds$$

$$= f_i + \frac{h^2}{12\,r_i}\,f_i' + O(h^2)\,.$$

By the same token,

$$\frac{1}{h\,r_i}\int\limits_{r_{i-1/2}}^{r_{i+1/2}} q\,u\,r\,dr = q_i\,u_i + \frac{h^2}{12\,r_i}\,(q\,u)_i' + O(h^2)\,,$$

so that

$$\psi_i^* = \frac{h^2}{12\,r_i}\,(qu-f)_i' + O(h^2)\,, \tag{44}$$

giving $\|\,r\,\psi^*\,\|_C = O(h^2)$.

It is clear that the function $\overset{\circ}{\eta}_i = a_i\,u_{\bar{r},i} - (ku')_{i-1/2} = O(h^2)$, meaning

$$\eta_i = r_{i-1/2}\,\overset{\circ}{\eta}_i\,, \qquad \overset{\circ}{\eta}_i = O(h^2)\,. \tag{45}$$

Comparison of formulae (35) and (43) yields $\tilde{\nu} = h_*\nu$. This provides support for the view that $\nu = O(h)$, because $\tilde{\nu} = O(h^2)$ as stated above.

In order to estimate the error $z = y - u$, we shall need an auxiliary lemma.

**Lemma** *Let $z$ be a solution of problem* (41) *and let $v$ be a solution of the same problem for $d_i \equiv 0$, $i = 1, 2, \dots, N-1$, $q_0 = 0$. Then the inequality is valid:*

$$\|\,z\,\|_C = \max_{0\le i<N}\,|\,z_i\,| \le 2\,\|\,v\,\|_C\,. \tag{46}$$

To prove this assertion, it suffices to use only the lemma of Chapter 1, Section 1, Subsection 8 and to write down the equations for $z$ and $z - v$ in the form (41).

The function $v_i$ can be recovered in explicit form from the conditions

$$\frac{w_{i+1} - w_i}{h\,r_i} = -\psi_i\,, \quad i = 1, 2, \dots, N-1, \quad w_i = b_i\,v_{\bar{r},i}\,, \quad b_i = a_i\,r_{i-1/2}\,,$$

$$v_N = 0, \qquad \frac{w_1}{h_*\,r_{1/2}} = -\nu, \quad \text{that is,} \quad w_1 = -\frac{h^2}{8}\nu,$$

where $\psi_i$ is specified by formula (42) for $d_i \equiv 0$. Summing the equations $w_{k+1} = w_k - h\,r_k\,\psi_k$ over $k = 1, 2, \dots, j$ yields

$$w_{j+1} = w_1 - \sum_{k=1}^{j} h\,r_k\,\psi_k\,, \qquad w_1 = -\frac{h^2}{8}\,\nu\,. \tag{47}$$

Substitution of $w_{j+1} = b_{j+1}(v_{j+1} - v_j)/h$ into (47) leads to the relation

$$v_j = v_{j+1} - \frac{w_1 h}{b_{j+1}} + \frac{h}{b_{j+1}} \sum_{k=1}^{j} h\, r_k\, \psi_k\,. \tag{48}$$

Recalling that $v_N = 0$ and summing equalities (48) over $j = i, i+1, \dots, N-1$, we obtain

$$v_i = -w_1 \sum_{j=i}^{N-1} \frac{h}{b_{j+1}} + \sum_{j=i}^{N-1} \frac{h}{b_{j+1}} \sum_{k=1}^{j} h\, r_k\, \psi_k\,. \tag{49}$$

Substituting $\psi_k = \dfrac{1}{r_k}\eta_{r,k} + \psi_k^*$ into (49) yields

$$\frac{1}{b_{j+1}} \sum_{k=1}^{j} h\, r_k\, \psi_k = \frac{1}{b_{j+1}}(\eta_{j+1} - \eta_1) + \frac{1}{b_{j+1}} \sum_{k=1}^{j} h\, r_k\, \psi_k^*\,,$$

$$\frac{1}{b_{j+1}} \left| \sum_{k=1}^{j} h\, r_k\, \psi_k \right| \le \frac{1}{a_{j+1}}(|\,\mathring{\eta}_{j+1}\,| + |\,\mathring{\eta}_1\,|) + \frac{\|\, r\psi^*\,\|_C}{a_{j+1}\, r_{j+1/2}}\, r_j$$

$$\le \frac{1}{c_1}(|\,\mathring{\eta}_{j+1}\,| + |\,\mathring{\eta}_1\,|) + \frac{1}{c_1}\|\, r\psi^*\,\|_C$$

in light of the relations

$$\eta_{j+1} = r_{j+1/2}\, \mathring{\eta}_{j+1/2}\,, \qquad \sum_{k=1}^{j} h = r_j \le r_{j+1/2}\,.$$

Furthermore, we might have

$$\left|\frac{w_1}{b_{j+1}}\right| = \left|\frac{h^2}{8}\, \frac{\nu}{r_{j+1/2}\, a_{j+1}}\right| < \frac{h^2}{8}\, \frac{|\,\nu\,|}{r_{1/2}\, c_1} = \frac{h}{4\, c_1}|\,\nu\,|\,.$$

As a final result we get the inequality

$$\|\, v\,\|_C \le \frac{R\, h\, |\,\nu\,|}{4\, c_1} + \frac{1}{c_1} \sum_{i=0}^{N-1} h\, (|\,\mathring{\eta}_{j+1}\,| + |\,\mathring{h}_1\,|) + \frac{R}{c_1}\|\, r\, \psi^*\,\|_C\,. \tag{50}$$

By inserting in (50) the estimates

$$\nu = O(h), \qquad | \overset{\circ}{\eta}_i | = O(h^2), \qquad \| r \psi^* \|_C = O(h^2)$$

we conclude that $\| v \|_C = O(h^2)$ and, hence,

$$\| z \|_C \le 2 \| v \|_C \le M h^2 .$$

This means that scheme (38) is of second-order accuracy in the space $C$.

Let us turn to a scheme of the second type known as a "scheme on the current grid". We split up the segment $[0, R]$ into $N$ parts by the nodes (current points)

$$r_0 = 0, \quad \bar{r}_1 = \tfrac{1}{2} h, \quad \bar{r}_2 = \tfrac{3}{2} h, \quad \dots, \quad \bar{r}_i = \left(i - \tfrac{1}{2}\right) h, \quad \dots,$$
$$\bar{r}_{N-1} = \left(N - \tfrac{3}{2}\right) h, \quad \bar{r}_N = \left(N - \tfrac{1}{2}\right) h = R .$$

Denote by $\bar{y}_i = y(\bar{r}_i)$ the values of a grid function at those nodes.

The balance equation for (26), which is an analog of (30), is aimed at designing the difference scheme, making it possible to write on the interval $r_{i-1} = \bar{r}_{i-1/2} < r < \bar{r}_{i+1/2} = r_i$ the difference scheme

$$\text{(51)} \qquad \frac{1}{\bar{r}_i} \left( r_{i-1}\, a_{i-1}\, \bar{y}_{\bar{r},i} \right)_{r,i} - d_i\, \bar{y}_i + \varphi(\bar{r}_i) = 0, \qquad i = 2, 3, \dots, N-1,$$

where $r_i = ih$, $\bar{r}_i = \left(i - \frac{1}{2}\right)/h$ and $a$, $d$, $\varphi$ are chosen by analogy with (32)–(33), so that in the simplest case

$$\text{(52)} \qquad a_i = k(r_i), \qquad \varphi_i = f(\bar{r}_i), \qquad d_i = q(\bar{r}_i) .$$

The balance equation written on the interval $0 < r < r_1 = h$

$$\frac{w_1 - w_0}{\bar{r}_1\, h} + \frac{1}{\bar{r}_1\, h} \int\limits_0^h \left( f(r) - q(r)\, u(r) \right) r \, dr = 0, \qquad w(r) = r\, k(r)\, u'(r),$$

and the condition $w_0 = 0$ imply the difference equation

$$\text{(53)} \qquad \frac{r_1\, a_1\, \bar{y}_{r,1}}{\bar{r}_1\, h} - d_1\, \bar{y}_1 + \varphi_1 = 0$$

for $r = \bar{r}_1 = \frac{1}{2} h$, where $a_1$, $d_1$, $\varphi_1$ are specified by formulae (52).

The standard condition

$$y_N = \mu_2 \tag{54}$$

is imposed for $r = R$. As a final result we get the difference equation (51) subject to the boundary conditions (53) and (54).

Let $\bar{y} = y(\bar{r})$ be a solution of this problem. We begin by placing the following problem for the error $z = \bar{y} - \bar{u}$:

$$\begin{gathered}
\frac{1}{\bar{r}_i}\left(r_{i-1}\, a_{i-1}\, z_{\bar{r},i}\right)_{r,i} - d_i\, z_i + \psi(\bar{r}_i) = 0\,, \\
\bar{r}_i = \left(i - \tfrac{1}{2}\right) h, \qquad i = 2, 3, \ldots, N-1, \qquad z_N = 0\,, \\
\frac{r_1\, a_1\, z_{r,1}}{\bar{r}_1\, h} - d_1\, z_1 + \psi_1 = 0\,,
\end{gathered} \tag{55}$$

where $\psi$ is the approximation error equal to

$$\psi_i = \psi(\bar{r}_i) = \frac{1}{\bar{r}_i}\left(r_{i-1}\, a_{i-1} \bar{u}_{\bar{r},i}\right)_{r,i} - d_i\, \bar{u}_i + \varphi_i\,, \quad i = 2, 3, \ldots, N-1\,,$$

$$\psi_1 = \frac{r_1}{\bar{r}_1\, h}\, a_1\, \bar{u}_{r,1} - d_1\, \bar{u}_1 + \varphi_1$$

with

$$\bar{u}_i = u(\bar{r}_i) = u_{i-1/2}\,.$$

Combination of the resulting expressions and the balance equation on the interval $r_{i-1} \le r < r_i$ gives

$$\psi_i = \frac{1}{\bar{r}_i}\left(r_{i-1}\, \overset{\circ}{\eta}_i\right)_r + \psi_i^*\,,$$

$$\overset{\circ}{\eta}_i = a_{i-1}\, \bar{u}_{\bar{r},i} - k_{i-1}\, u'_{i-1}\,, \qquad i = 2, 3, \ldots, N, \qquad \overset{\circ}{\eta}_1 = 0\,,$$

$$\psi_i^* = \varphi_i - \overset{\circ}{\varphi}_i - \left(d_i - \overset{\circ}{d}_i\right) \bar{u}_i + \frac{1}{\bar{r}_i\, h} \int\limits_{r_{i-1}}^{r_i} r\, q(r) \left(u(r) - u(\bar{r}_i)\right) dr = 0,$$

where

$$\overset{\circ}{\varphi}_i = \frac{1}{\bar{r}_i\, h} \int\limits_{r_{i-1}}^{r_i} r\, f(r)\, dr, \qquad \overset{\circ}{d}_i = \frac{1}{\bar{r}_i\, h} \int\limits_{r_{i-1}}^{r_i} r\, q(r)\, dr\,.$$

It follows from the foregoing that

$$\psi_i^* = \frac{\mu_i}{r_i}, \qquad \mu_i = O(h^2).$$

An estimate similar to that established for problem (41) is valid for a solution of problem (55) with the right-hand side

$$\psi_i = \frac{1}{\bar{r}_i} \left(r_{i-1} \overset{\circ}{\eta}_i\right)_r + \psi_i = \frac{\mu_i}{r_i},$$

yielding

$$\| z \|_C = \| y - u \|_C = O(h^2).$$

This means that scheme (51), (53), (54) is of second-order accuracy if we agree to consider $k(x)$, $q(x)$, $f(x) \in C^{(2)}[0,1]$.

**5. Difference schemes for an equation in spherical coordinates.** If a solution to the equation

$$\operatorname{div}(k \operatorname{grad} u) - qu = -f(r, \theta, \varphi)$$

in the spherical coordinate system is centrally symmetric, that is, is independent of $\theta$ and $\varphi$, then the function $u = u(r)$ satisfies the equation

$$\text{(56)} \qquad \begin{gathered} \frac{1}{r^2} \frac{d}{dr} \left( r^2 k(r) \frac{du}{dr} \right) - q(r)\, u = -f(r), \qquad 0 < r < R, \\ 0 < c_1 \leq k(r) \leq c_2, \qquad q(r) \geq 0. \end{gathered}$$

In the general setting the function $u(r)$ is supposed to be bounded at the point $r = 0$. This property is equivalent to the condition

$$\text{(57)} \qquad r^2 k(r) \frac{du}{dr}\bigg|_{r=0} = 0.$$

At the point $r = R$ we may impose, for instance, the standard condition

$$\text{(58)} \qquad u(R) = \mu_2.$$

A bounded solution of problem (56)–(58) possesses the same properties as in the case of the axial symmetry (for more detail see problem (26)–(28)).

We introduce on the segment $0 \le r \le R$ the equidistant grid $\bar{\omega}_h = \{r_i = ih,\ i = 0, 1, \dots, N,\ hN = R\}$ and by analogy with (31) may attempt the scheme in the form

$$\text{(59)}\quad \Lambda\, y_i = \frac{1}{r_i^2} \left(r_{i-1/2}^2\, a_i\, y_{\bar{r},\, i}\right)_{r,\, i} - d_i\, y_i = -\varphi_i\,, \qquad i = 1, 2, \dots, N-1\,.$$

At the point $r = 0$ we impose the difference boundary condition

$$\text{(60)}\qquad \frac{a_1\, y_{r,\,0}}{h_*} - q_0\, y_0 = -f_0\,, \qquad h_* = \frac{h}{6}\,,$$

and for $i = N$ set

$$\text{(61)}\qquad y_N = \mu_2\,.$$

These equations admit the form (39) and can be solved by the elimination method. The main difference from the cylindrical case lies in the selection rules for $\varphi$ and $d$. To obtain the formulae for $\varphi$ and $d$, we consider the residual

$$\text{(62)}\qquad \psi_i = \Lambda\, u_i + \varphi_i = \frac{1}{r_i^2} \left(r_{i-1/2}^2\, a_i\, u_{\bar{r},i}\right)_{r,i} - d_i\, u_i + \varphi_i$$

and subtract from equality (62) the balance equation written on the segment $r_{i-1/2} \le r \le r_{i+1/2}$

$$0 = \frac{w_{i+1/2} - w_{i-1/2}}{h\, r_i^2} - \frac{1}{h\, r_i^2} \int\limits_{r_{i-1/2}}^{r_{i+1/2}} q\, u\, r^2\, dr + \frac{1}{h\, r_i^2} \int\limits_{r_{i-1/2}}^{r_{i+1/2}} f\, r^2\, dr,$$

$$w = r^2 k u'\,.$$

The outcome of this is

$$\psi_i = \frac{1}{r_i^2}\, \eta_{r,i} + \psi_i^*\,, \qquad \eta_i = r_{i-1/2}^2\, \mathring{\eta}_i\,, \qquad \mathring{\eta}_i = a_i\, u_{\bar{r},i} - (k\, u')_{i-1/2}\,,$$

where

$$\text{(63)}\quad \psi_i^* = \varphi_i - \frac{1}{h\, r_i^2} \int\limits_{r_{i-1/2}}^{r_{i+1/2}} f\, r^2\, dr - \left( d_i\, u_i - \frac{1}{h\, r_i^2} \int\limits_{r_{i-1/2}}^{r_{i+1/2}} q\, u\, r^2\, dr \right).$$

After replacing the variable by $r = r_i + sh$ we find that

$$\frac{1}{h\,r_i^2} \int\limits_{r_{i-1/2}}^{r_{i+1/2}} f r^2 \, dr = \frac{1}{r_i^2} \int\limits_{-1/2}^{1/2} f(r_i + sh)\,(r_i + sh)^2 \, ds$$

$$= \int\limits_{-1/2}^{1/2} f(r_i + sh) \, ds + \frac{2h}{r_i} \int\limits_{-1/2}^{1/2} s\, f(r_i + sh) \, ds + \frac{h^2}{r_i^2} \int\limits_{-1/2}^{1/2} f(r_i + sh)\, s^2 \, ds$$

$$= f_i + O(h^2) + \frac{h^2}{6\,r_i}\, f_i' + \frac{h^2}{12\,r_i^2}\, f_i + O(h^2)$$

$$= \left(1 + \frac{h^2}{12\,r_i^2}\right) f_i + \frac{h^2}{6\,r_i}\, f_i' + O(h^2)\,.$$

Also, a similar expression can be derived for the second integral by merely setting $qu$ in place of $f$. From here and formula (63) it is plain to show that

$$\psi_i^* = \frac{h^2}{6\,r_i}\,(qu - f)_i' + O(h^2)\,, \qquad \|\, r\,\psi^* \,\|_C = O(h^2)$$

if $\varphi_i$ and $d_i$ are specified by the formulae

$$(64) \quad \varphi_i = \left(1 + \frac{h^2}{12\,r_i^2}\right) f_i\,, \quad d_i = \left(1 + \frac{h^2}{12\,r_i^2}\right) q_i\,, \quad i = 1, 2, \ldots, N-1\,,$$

or by other formulae differing from (64) only by the quantity $O(h^2)$.

Arguing as in the preceding section, we verify that the residual in the boundary condition reduces to

$$(65) \qquad \nu = \frac{a_1\, u_{r,0}}{h^*} - q_0\, u_0 + f_0 = O(h)\,.$$

To develop those ideas, we contrived to do it with further reference to the problem for the error $z = y - u$

$$(66) \qquad \Lambda\, z = -\psi, \quad r \in \omega_h\,, \quad \frac{a_1\, z_{r,0}}{h^*} - q_0\, u_0 = -\nu, \quad z_N = 0\,.$$

A solution of this problem can be estimated in a similar way as was done in the preceding section, but with $b_i = r_{i-1/2}^2\, a_i$. In concluding this section we establish through such an analysis that scheme (59)–(61) converges uniformly at the rate $O(h^2)$:

$$\|\, z \,\|_C = \|\, y - u \,\|_C = O(h^2)\,.$$

## 3.6 DIFFERENCE GREEN'S FUNCTION

**1. Difference Green's function.** Further estimation of a solution of the boundary-value problem for a second-order difference equation will involve its representation in terms of Green's function. The boundary-value problem for the differential equation

$$
\begin{gathered}
L u = \frac{d}{dx}\left(k(x)\,\frac{du}{dx}\right) - q(x)\,u = -f(x), \qquad 0 < x < 1\,, \\
u(0) = 0, \qquad u(1) = 0, \qquad k(x) \ge c_1 > 0, \qquad q(x) \ge 0\,,
\end{gathered} \tag{1}
$$

can add interest and aid in understanding. As known, the solution of this problem arranges itself as an integral

$$
u(x) = \int_0^1 G(x,\xi)\,f(\xi)\,d\xi\,, \tag{2}
$$

where $G(x,\xi)$ is the source function or Green's function. Function (2) is a solution to equation (1) subject to the boundary conditions $u(0) = 0$ and $u(1) = 0$ if Green's function $G(x,\xi)$ as a function of $x$ for fixed $\xi$ satisfies the conditions

$$
\begin{gathered}
L_x\, G(x,\xi) = \frac{d}{dx}\left(k(x)\,\frac{dG(x,\xi)}{dx}\right) - q(x)\,G(x,\xi) = 0\,, \\
x \ne \xi, \qquad 0 < x < 1, \qquad G(0,\xi) = G(1,\xi) = 0\,, \\
[G] = G(\xi+0,\xi) - G(\xi-0,\xi) = 0, \qquad \left[k\,\frac{dG}{dx}\right] = -1 \quad \text{for} \quad x = \xi\,.
\end{gathered} \tag{3}
$$

The very definition implies that Green's function is nonnegative and symmetric:

$$
G(x,\xi) \ge 0, \qquad G(x,\xi) = G(\xi,x)\,.
$$

The function $G(x,\xi)$ so defined can be written in the explicit form

$$
G(x,\xi) = \begin{cases} \dfrac{\alpha(x)\,\beta(\xi)}{\alpha(1)} & \text{for} \quad x \le \xi\,, \\[2ex] \dfrac{\alpha(\xi)\,\beta(x)}{\alpha(1)} & \text{for} \quad x \ge \xi\,, \end{cases} \tag{4}
$$

where $\alpha(x)$ and $\beta(x)$ are solutions of the relevant Cauchy problems:

$$(5)\qquad \begin{aligned} &L\,\alpha = 0, \qquad 0 < x < 1, \qquad \alpha(0) = 0, \qquad k(0)\,\alpha'(0) = 1\,,\\ &L\,\beta = 0, \qquad 0 < x < 1, \qquad \beta(1) = 0, \qquad k(1)\,\beta'(1) = -1\,. \end{aligned}$$

The functions $\alpha(x)$ and $\beta(x)$ are linearly independent. This is due to the fact that the Wronskian is nonzero $\Delta(x) \neq 0$. Moreover, $\alpha(x) > 0$ for $x > 0$ and $\beta(x) > 0$ for $0 \leq x < 1$.

We now turn to a difference equation of second order. We learn from Chapter 1, Section 1 that any difference equation of second order $A_i\, y_{i-1} - C_i\, y_i + B_i\, y_{i+1} = -F_i$ can be treated as an equation of divergent type, meaning

$$a_{i+1}\,(y_{i+1} - y_i) - a_i\,(y_i - y_{i-1}) - d_i\, y_i = -\varphi_i\,.$$

By replacing here $\varphi_i$ by $h^2\varphi_i$ and $d_i$ by $h^2 d_i$ we obtain

$$\Lambda\, y_i = \frac{1}{h^2}\,\big[a_{i+1}\,(y_{i+1} - y_i) - a_i\,(y_i - y_{i-1})\big] - d_i\, y_i = -\varphi_i\,,$$

$$i = 1, 2, \ldots, N-1\,,$$

which is more convenient for the further comparision with the differential equation. For the moment, we write down this equation without subscripts

$$(6)\qquad \begin{aligned} &\Lambda\, y = \big(a\, y_{\bar{x}}\big)_x - d\, y = -\varphi(x)\,, \qquad x = ih\,,\\ &a(x) \geq c_1 > 0\,, \qquad d(x) \geq 0\,, \qquad i = 1, 2, \ldots, N-1\,. \end{aligned}$$

For $i = 0$ ($x = 0$) and $i = N$ ($x = 1$) the boundary conditions of the first kind are imposed as usual:

$$(7)\qquad y_0 = 0, \qquad y_N = 0\,.$$

Common practice in numerical analysis involves the inner products

$$(y, v) = \sum_{i=1}^{N-1} y_i v_i h, \qquad (y, v] = \sum_{i=1}^{N} y_i v_i h\,.$$

The traditional way of covering this is to seek a solution of problem (6) in the form

$$(8)\qquad y_i = \sum_{k=1}^{N-1} G_{ik}\,\varphi_k\, h = (G_{ik}\,, \varphi_k)$$

under the agreement that this expression solves the equation $\Lambda y_i = -\varphi_i$. From the equality $\Lambda y_i = \sum_{k=1}^{N-1} \Lambda_{(i)} G_{ik} \varphi_k h$ it is easily seen that equation (6) holds true only if $\Lambda_{(i)} G_{ik} = -\delta_{ik}/h$, where $\delta_{ik}$ is Kronecker's delta:

$$\delta_{ik} = \begin{cases} 1, & i = k, \\ 0, & i \neq k. \end{cases}$$

The equalities $y_0 = y_N = 0$ are certainly true for the choice $G_{0k} = G_{Nk} = 0$, thus formula (8) gives the solution of problem (6)–(7) if $G_{ik} = G(x_i, x_k)$ as a function of $i$ for fixed $k = 1, 2, \ldots, N-1$ satisfies the conditions

$$\text{(9)} \qquad \begin{aligned} &\Lambda_{(i)} G_{ik} = \left(a_i (G_{ik})_{\bar{x},i}\right)_{x,i} - d_i G_{ik} = -\frac{\delta_{ik}}{h}, \\ &i, k = 1, 2, \ldots, N-1, \qquad G_{0k} = G_{Nk} = 0. \end{aligned}$$

We must show that Green's function specified in such a way exists and find its explicit representation similar to expression (4). With this aim, the functions $\alpha_i$ and $\beta_i$ will be declared to be solutions of the corresponding Cauchy problems

$$\text{(10)} \qquad \begin{aligned} &\Lambda \alpha_i = 0, \qquad i = 1, 2, \ldots, N-1, \\ &\alpha_0 = 0, \qquad a_1 \alpha_{x,0} = a_1 \frac{\alpha_1 - \alpha_0}{h} = 1, \\ &\Lambda \beta_i = 0, \qquad i = 1, 2, \ldots, N-1, \\ &\beta_N = 0, \qquad a_N \beta_{\bar{x},N} = a_N \frac{\beta_N - \beta_{N-1}}{h} = -1. \end{aligned}$$

It is straightforward to verify that the functions $\alpha_i$ and $\beta_i$ possess the following properties:

1) $\alpha_i$ and $\beta_i$ are positive functions, $\alpha_i$ being monotonically increasing and $\beta_i$ being monotonically decreasing:

$$\alpha_i \leq \alpha_{i+1} \leq \alpha_N, \qquad \beta_i \leq \beta_{i-1} \leq \beta_0,$$

$\alpha_i > 0$ for $i = 1, 2, \ldots, N$ and $\beta_i > 0$ for $i = 0, 1, \ldots, N-1$. Indeed, conditions (10) imply that

$$a_i \alpha_{\bar{x},i} = 1 + \sum_{k=1}^{i-1} h d_k \alpha_k, \qquad \alpha_1 = \frac{h}{a_1} > 0.$$

If $\alpha_k > 0$ for $k = 1, 2 \ldots, i-1$, then $a_i \, \alpha_{\bar{x},i} > 0$ and $\alpha_i > \alpha_{i-1} > 0$. In a similar way we can be pretty sure that $\beta_{\bar{x},i} < 0$ and $0 < \beta_i < \beta_{i-1}$.

2) $\alpha_N = \beta_0$ or $\alpha(1) = \beta(0)$. From the second Green formula

$$(\alpha, \Lambda\, \beta) = (\beta, \Lambda\, \alpha) + a_N \left(\alpha\, \beta_{\bar{x}} - \beta\, \alpha_{\bar{x}}\right)_N - a_1 \left(\alpha\, \beta_x - \beta\, \alpha_x\right)_0$$

it follows immediately that $\alpha_N = \beta_0$ on account of conditions (10).

3) The determinant $\Delta_i = a_i \left(\alpha_{\bar{x},i}\, \beta_i - \alpha_i\, \beta_{\bar{x},i}\right) = \text{const } = \alpha_N$ is positive for $0 < i \le N$. Applying the second Green's formula in the domain $\{0 \le x_i = ih \le x_{i_0} = x\}$ we obtain

$$\begin{aligned} 0 &= \sum_{i=1}^{i_0-1} (\alpha\Lambda\beta - \beta\Lambda\alpha)_i\, h = a_{i_0} \left(\alpha\, \beta_{\bar{x}} - \beta\, \alpha_{\bar{x}}\right)_{i_0} - a_1 \left(\alpha\, \beta_x - \beta\, \alpha_x\right)_0 \\ &= -\Delta(x_{i_0}) + \beta(0)\, a_1\, \alpha_{\bar{x},i} = -\Delta(x_{i_0}) + \beta_0\,. \end{aligned}$$

Since $x_{i_0} = x$ is an arbitrary node of the grid $\omega_h$, we might have

$$\Delta(x) = \text{const } = \beta(0) = \alpha(1)\,.$$

Other ideas are connected with a possibility of arranging the Green function such as

$$G_{ik} = \begin{cases} \dfrac{\alpha_i\, \beta_k}{\alpha_N} & \text{for} \quad i \le k\,, \\[2ex] \dfrac{\alpha_k\, \beta_i}{\alpha_N} & \text{for} \quad i \ge k\,. \end{cases} \tag{11}$$

Whence it follows that $G_{0k} = G_{Nk} = 0$.

The preceding representation will be proved if we succeed in showing that the function specified by formulae (11) solves the equation $\Lambda_{(i)}\, G_{ik} = -\delta_{ik}/h$. Indeed, $\Lambda_{(i)}\, G_{ik} = 0$ for $i \ne k$ due to the facts that $\Lambda\alpha_i = 0$ and $\Lambda\beta_i = 0$. The expression $\Lambda_{(i)}\, G_{ik}$ for $i = k$ needs investigation:

$$\begin{aligned} (\Lambda_{(i)}\, G_{ik})_{i=k} = \frac{1}{\alpha_N\, h^2} \big[ & a_{k+1} \left(\alpha_k\, \beta_{k+1} - \alpha_k\, \beta_k\right) \\ & - a_k \left(\alpha_k\, \beta_k - \alpha_{k-1}\, \beta_k\right) \big] - d_k\, G_{kk}\,. \end{aligned} \tag{12}$$

From the condition $\Lambda_{k+1} = a_{k+1} \left(\alpha_{k+1}\, \beta_k - \alpha_k\, \beta_{k+1}\right)/h = \alpha_N$ we derive the expression $a_{k+1}\, \alpha_k\, \beta_{k+1} = a_{k+1}\, \alpha_{k+1}\, \beta_k - h\alpha_N$ and substitute it into the right-hand side of formula (12). As a final result we get

$$(\Lambda_{(i)}\, G_{ik})_{i=k} = \frac{\beta_k}{\alpha_N} \left(a\alpha_{\bar{x}}\right)_{x,\,k} - \frac{1}{h} - \frac{d_k\, \beta_k}{\alpha_N}\, \alpha_k = \frac{\beta_k}{\alpha_N}\, \Lambda\, \alpha_k - \frac{1}{h} = -\frac{1}{h}$$

as required.

Formula (11) means that $G_{ik} > 0$ for $i, k \neq 0, N$ and $G_{ik} = G_{ki}$. Moreover, $G_{ik}$ as a function of $k$ for fixed $i$ satisfies the conditions

$$\Lambda_{(k)} G_{ik} = -\frac{\delta_{ik}}{h}, \qquad G_{i0} = G_{iN} = 0.$$

In the case of the boundary conditions $y_0 = \varkappa_1 y_1$ and $y_N = \varkappa_2 y_{N-1}$ the Green function can be constructed in a similar way. Of interest is the special case where $d(x) \equiv 0$. Here the functions $\alpha = \overset{\circ}{\alpha}(x)$ and $\beta = \overset{\circ}{\beta}(x)$ can be determined from equations (10) in the explicit form:

$$\overset{\circ}{\alpha}_i = \sum_{s=1}^{i} \frac{h}{a_s}, \qquad \overset{\circ}{\beta}_i = \sum_{s=i+1}^{N} \frac{h}{a_s} \tag{13}$$

and the Green function related to the problem

$$\left(a y_{\bar{x}}\right)_x = -\varphi, \qquad x \in \omega_h, \qquad y_0 = 0, \qquad y_N = 0, \tag{14}$$

reduces to

$$G_{ik} = \begin{cases} \sum\limits_{s=1}^{i} \dfrac{h}{a_s} \sum\limits_{s=k+1}^{N} \dfrac{h}{a_s} \Big/ \sum\limits_{s=1}^{N} \dfrac{h}{a_s}, & i \leq k, \\[2ex] \sum\limits_{s=1}^{k} \dfrac{h}{a_s} \sum\limits_{s=i+1}^{N} \dfrac{h}{a_s} \Big/ \sum\limits_{s=1}^{N} \dfrac{h}{a_s}, & i \geq k. \end{cases} \tag{15}$$

For the best scheme (15), (17) arising in Section 2 $G_{ik}$ coincides on the grid $\omega_h$ with Green's function for a differential equation.

**2. A priori estimates.** The explicit representation (8) of a solution of problem (6)–(7) in terms of Green's function lies in the background of various a priori estimates of a solution in terms of the right-hand side. It is easily seen from (8) that

$$|y_i| \leq (G_{ik}, |\varphi_k|) = \sum_{k=1}^{N-1} G_{ik} |\varphi_k| h \tag{16}$$

and, therefore, a uniform estimate of $y_i$ can be obtained by estimating $\max_{i,k} G_{ik}$. In this line, $\varphi_i$ arranges itself as

$$\varphi = \eta_x, \qquad \eta_1 = 0, \qquad \eta_i = \sum_{s=1}^{i-1} h \varphi_s, \qquad i = 2, 3, \ldots, N. \tag{17}$$

Upon substituting these expressions into formula (8) we deduce by the summation by parts formula that

$$y(x) = \big(G(x,\xi), \eta_\xi\big) = -\big(G_{\bar{\xi}}(x,\xi), \eta(\xi)\big], \qquad x, \xi \in \omega_h. \tag{18}$$

The next step is connected with $\left| G_{\bar{\xi}}(x,\xi) \right|$.

**Lemma** *For the Green function $G(x,\xi)$ related to problem (6)–(7) the uniform estimates*

$$G(x,\xi) \le \frac{1}{c_1}, \tag{19}$$

$$\left| G_{\bar{x}}(x,\xi) \right| \le \frac{2}{c_1}, \qquad \left| G_{\bar{\xi}}(x,\xi) \right| \le \frac{2}{c_1} \tag{20}$$

*are valid for all $x$, $\xi \in \omega_h$.*

**Proof** 1) In Chapter 1, Section 1 we have obtained for a solution of problem (6)–(7) the estimate

$$\| y \|_C \le \sum_{i=1}^{N-1} \frac{1}{a_{i+1}} \sum_{s=1}^{i} | h^2 \varphi_s | \le \frac{1}{c_1} \sum_{i=1}^{N-1} \sum_{s=1}^{i} | h^2 \varphi_s | . \tag{21}$$

Replacing in (21) $y_i$ by $G_{ik}$ and $\varphi_s$ by $\delta_{sk}/h$, we establish the relations

$$\max_{i,\,k} G_{ik} \le \frac{1}{c_1} \sum_{i=1}^{N-1} h\, S_{ik} \le \frac{1 - x_k}{c_1} \le \frac{1}{c_1},$$

where

$$S_{ik} = \sum_{s=1}^{i} \delta_{sk}, \quad S_{ik} = 0 \quad \text{for} \quad i < k, \qquad S_{ik} = 1 \quad \text{for} \quad i \ge k .$$

2) Let $d(x) \equiv 0$ and the Green function $G = G_0(x,\xi)$ be specified by formulae (15). Assuming this to be the case, we find that

$$G_{0\bar{x}}(x,\xi) = \begin{cases} \dfrac{\overset{\circ}{\alpha}_{\bar{x}}(x)\, \overset{\circ}{\beta}(\xi)}{\overset{\circ}{\alpha}(1)} & \text{for} \quad x \le \xi, \\[2ex] \dfrac{\overset{\circ}{\alpha}(\xi)\, \overset{\circ}{\beta}_{\bar{x}}(x)}{\overset{\circ}{\alpha}(1)} & \text{for} \quad x > \xi. \end{cases}$$

By virtue of the relations

$$\overset{\circ}{\alpha}_{\bar{x},i} = \frac{1}{a_i} \le \frac{1}{c_1}, \qquad \overset{\circ}{\beta}_{\bar{x},i} = \frac{1}{a_i} \le \frac{1}{c_1}, \qquad \overset{\circ}{\alpha}(\xi) \le \overset{\circ}{\alpha}(1), \quad \overset{\circ}{\beta}(\xi) \le \overset{\circ}{\alpha}(1)$$

we arrive at

$$\left| G_{0\bar{x}}(x,\xi) \right| \le \frac{1}{c_1} \,. \tag{22}$$

By the same token,

$$\left| G_{0\bar{\xi}}(x,\xi) \right| \le \frac{1}{c_1} \,.$$

Let $v(x,\xi) = G_0(x,\xi) - G(x,\xi)$. The equations for $G_0$ and $G$ imply that

$$\Lambda_\alpha\, v(x,\xi) = -d(x)\, G_0(x,\xi), \qquad v(0,\xi) = v(1,\xi) = 0 \,.$$

We can calculate the left difference derivative of both sides of this equation with respect to $\xi$, whose use permits us to establish for $w(x,\xi) = v_{\bar{\xi}}$

$$\Lambda_x\, w(x,\xi) = \big(a(x)\, w_{\bar{x}}(x,\xi)\big)_x - d(x)\, w(x,\xi) = -d(x)\, G_{0\bar{\xi}}(x,\xi)\,,$$

$$w(0,\xi) = 0, \qquad w(1,\xi) = 0 \,.$$

By the lemma from Chapter 1, Section 1,

$$\max_x |\, w(x,\xi)\,| \le \max_x |\, G_{0\bar{\xi}}(x,\xi)\,| \le \frac{1}{c_1} \,. \tag{23}$$

Using estimates (22) and (23) behind, we derive from the obvious inequality

$$|\, G_{\bar{\xi}}(x,\xi)\,| \le |\, G_{0\bar{\xi}}(x,\xi)\,| + |\, w(x\,,\xi)\,|$$

the desired estimate (20).

**Theorem** *For a solution of problem (6)–(7) the estimates are valid:*

$$\|\, y\,\|_C \le \frac{2}{c_1} \left( 1,\, \left| \sum_{s=i}^{N-1} h\, \varphi_s \right| \right) = \frac{2}{c_1} \sum_{i=1}^{N-1} h \left| \sum_{s=i}^{N-1} h\, \varphi_s \right|, \tag{24}$$

$$\|\, y\,\|_C \le \frac{2}{c_1} \left( 1,\, \left| \sum_{s=1}^{i} h\, \varphi_s \right| \right) = \frac{2}{c_1} \sum_{i=1}^{N-1} h \left| \sum_{s=1}^{i} h\, \varphi_s \right|. \tag{25}$$

*When $\varphi(x)$ happens to be of the form $\varphi = \eta_x + \varphi^*$, a solution of problem (6)–(7) satisfies the estimate*

$$\|\, y\,\|_C \le \frac{2}{c_1} \left\{ (1, |\,\eta\,|] + (1, |\,\mu\,|] \right\}, \tag{26}$$

*where $\mu_i = \sum_{k=1}^{i-1} h\, \varphi_k^*$ for $i = 2, 3 \ldots, N$ and $\mu_1 = 0$.*

**Proof** Putting $\mu = \eta_x$ and applying then formula (18), we deduce by the preceding lemma that

$$|y_i(x)| \le (|G_{\bar{\xi}}(x,\xi)|, |\eta(\xi)|] \le \frac{2}{c_1}(1, |\eta(\xi)|]. \tag{27}$$

The function $\eta(x)$ can be recovered from the condition $\eta_x = \varphi$ within an arbitrary constant. A solution to the difference equation

$$\eta_{i+1} - \eta_i = h\,\varphi_i$$

can be expressed either by

$$\eta_i = \eta_N - \sum_{s=i}^{N-1} h\,\varphi_s = -\sum_{s=i}^{N-1} h\,\varphi_s \quad \text{if} \quad \eta_N = 0$$

or by

$$\eta_{i+1} = \eta_1 + \sum_{s=1}^{i} h\,\varphi_s = \sum_{s=1}^{i} h\,\varphi_s \qquad \text{if} \quad \eta_1 = 0\,.$$

Substitution of these formulae for $\eta_i$ into the right-hand side of inequality (27) leads to estimates (24) and (25).

To prove inequality (26), it suffices to set $\varphi^* = \mu_x$, so that $\varphi = (\eta+\mu)_x$ and to repeat the preceding arguments.

**Remark** Formula (16) entails the estimate

$$\|y\|_C \le \frac{1}{c_1}(1, |\varphi|) \le \frac{1}{c_1}\|\varphi\| \le \frac{1}{c_1}\|\varphi\|_C$$

in the norm $\|\varphi\| = \sqrt{(\varphi,\varphi)}$. It is plain to show the obvious inequalities

$$\Big\| \sum_{k=1}^{i} h\varphi_k \Big\| \le (1, |\varphi|) \le \|\varphi\| \le \|\varphi\|_C\,.$$

With the aid of the estimate

$$\|G_{\bar{x}}(x,\xi)\|_C \le \frac{2}{c_1}$$

an a priori estimate in the space $C$ for the difference derivative of a solution of the boundary-value problem (6)–(7) can be derived without difficulty. Indeed, the relations

$$|y_{\bar{x}}| = |(G_{\bar{x}}(x,\xi), \varphi(\xi))| \le \frac{2}{c_1}(1, |\varphi|)$$

lead to

$$\|y_{\bar{x}}\|_C \le \frac{2}{c_1}(1, |\varphi|)\,.$$

## 3.7 HIGHER–ACCURACY SCHEMES

**1. An exact scheme.** For equation (1) in Section 1 it is possible to make the design of a homogeneous conservative exact three-point scheme so that a solution $y_i$ of the difference problem is identical with the exact solution $u = u(x)$ of problem (1) from Section 1 at all the nodes of any grid $\omega_h$:

$$y_i = u(x_i) \qquad \text{for} \quad k, q, f \in Q^{(0)}[0,1].$$

Before giving further motivations, it will be convenient to set up problem (1) arising in Section 1 in the form

$$(1) \qquad L^{(p,q)} u = \frac{d}{dx} \left( \frac{1}{p(x)} \frac{du}{dx} \right) - q(x)\, u = -f(x), \qquad 0 < x < 1,$$

$$u(0) = u_1, \quad u(1) = u_2, \quad p(x) = k^{-1}(x), \quad 0 < p(x) < \frac{1}{c_1}, \quad q(x) \geq 0.$$

It is worth noting here that the best scheme (14)–(15) of Section 2 is exact for $q = f \equiv 0$. Indeed, the function

$$(2) \qquad u(x) = u_1 + c \int\limits_0^x p(t)\, dt, \qquad c = (u_2 - u_1) \Big( \int\limits_0^1 p(t)\, dt \Big)^{-1}$$

is just the solution of problem (1) for $q = f \equiv 0$. From such reasoning it seems clear that

$$u_{\bar{x},i} = \frac{c}{h} \int\limits_{x_{i-1}}^{x_i} p(t)\, dt = \frac{c}{\overset{\circ}{a}_i}, \qquad \overset{\circ}{a}_i\, u_{\bar{x},i} = c,$$

where $\overset{\circ}{a}_i = \Big( h^{-1} \int_{x_{i-1}}^{x_i} p(t)\, dt \Big)^{-1}$ and, therefore, function (2) solves the equation $(\overset{\circ}{a} u_{\bar{x}})_x = 0$.

We now turn to equation (1) on an equidistant grid $\bar{\omega}_h$. A key idea in the further development of an exact scheme is to express at any inner point (and, particular, at $x = x_i$) of the interval $(x_{i-1}, x_{i+1})$ a solution $u = u(x)$ to the second-order equation (1) in terms of the values $u_{i-1}$, $u_{i+1}$ and the right-hand side $f(x)$. This can be done using $u(x)$ in the form

$$(3) \qquad u(x) = A_i\, v_1^i(x) + B_i\, v_2^i(x) + v_3^i(x), \qquad x_{i-1} \leq x \leq x_{i+1},$$

where $A_i$ and $B_i$ are some numbers which will be specified in the sequel, $v_1^i(x)$ and $v_2^i(x)$ are linearly independent solutions to the homogeneous equation $L^{(p,q)}u = 0$ (pattern functions) and $v_3^i(x)$ is a particular solution to the nonhomogeneous equation (1) subject to the homogeneous conditions:

$$(4)\quad L^{(p,q)}v_3^i = f(x), \qquad x_{i-1} < x < x_{i+1}, \qquad v_3^i(x_{i+1}) = v_3^i(x_{i-1}) = 0\,.$$

The accepted view is that the **pattern functions** $v_1^i(x)$ and $v_2^i(x)$ will be declared to be solutions of the appropriate Cauchy problems

$$(5)\qquad L^{(p,q)}v_1^i = 0, \quad x_{i-1} < x < x_{i+1}\,,$$

$$v_1^i(x_{i-1}) = 0\,, \quad \frac{1}{p(x_{i-1})}\,(v_1^i)'(x_{i-1}) = 1\,,$$

$$(6)\qquad L^{(p,q)}v_2^i = 0\,, \quad x_{i-1} < x < x_{i+1},$$

$$v_2^i(x_{i+1}) = 0\,, \quad \frac{1}{p(x_{i+1})}\,(v_2^i)'(x_{i+1}) = -1\,.$$

By merely setting in (3) $x = x_{i-1}$ and $x = x_{i+1}$ we find that

$$(7)\qquad A_i = \frac{u(x_{i+1})}{v_1^i(x_{i+1})}, \qquad B_i = \frac{u(x_{i-1})}{v_2^i(x_{i-1})}\,.$$

Pattern functions so defined possess a number of nice properties (compare with Section 6), so there is some reason to be concerned about this:

1) $v_1^i(x)$ is positive and monotonically increasing for $x_{i-1} < x < x_{i+1}$;
   $v_2^i(x)$ is positive and monotonically decreasing for $x_{i-1} < x < x_{i+1}$;
2) the equality is certainly true:

$$(8)\qquad v_1^i(x_{i+1}) = v_2^i(x_{i-1})\,;$$

3) the relation occurs:

$$(9)\qquad v_1^i(x_{i+1}) = v_1^i(x_i) + v_2^i(x_i) + v_2^i(x_i)\int_{x_{i-1}}^{x_i} v_1^i\, q(x)\, dx + v_1^i(x_i)\int_{x_i}^{x_{i+1}} v_2^i\, q(x)\, dx\,;$$

4) the recurrence relation is valid:

$$(10)\qquad v_2^i(x_i) = v_1^{i+1}(x_{i+1})\,.$$

We will prove the indicated properties.

1) Property 1) follows directly from problem statement (5)–(6).

2) Taking into account (5) and (6), we have for $L = L^{(p,q)}$

$$0 = \int\limits_{x_{i-1}}^{x_{i+1}} \left(v_1^i \; Lv_2^i - v_2^i \; Lv_1^i\right) \, dx = \left(v_1^i \, \frac{1}{p} \, (v_2^i)' - v_2^i \, \frac{1}{p} \, (v_1^i)'\right)\Big|_{x_{i-1}}^{x_{i+1}}$$

$$= -v_1^i(x_{i+1}) + v_2^i(x_{i-1}) \, .$$

3) Applying Green's formula on the segment $[x_{i-1}\,, x_i]$ yields

$$0 = \int\limits_{x_{i-1}}^{x_i} \left(v_1^i Lv_2^i - v_2^i Lv_1^i\right) \, dx = \left(v_1^i \, \frac{1}{p} \, (v_2^i)' - v_2^i \, \frac{1}{p} \, (v_1^i)'\right)\Big|_{x_{i-1}}^{x_i}$$

$$= \frac{1}{p_i} \, (v_2^i)'(x_i) \, v_1^i(x_i) - \frac{1}{p_i} \, (v_1^i)'(x_i) \, v_2^i(x_{i-1}) + v_2^i(x_{i-1}) \, .$$

Upon substituting here the relations

$$\frac{1}{p_i} \, (v_1^i)'(x_i) = 1 + \int\limits_{x_{i-1}}^{x_i} q(x) \, v_1^i(x) \; dx \, ,$$

$$\frac{1}{p_i} \, (v_2^i)'(x_i) = -1 - \int\limits_{x_i}^{x_{i+1}} q(x) \, v_2^i(x) \; dx$$

we establish property 3).

4) Having stipulated the conditions

$$L^{(p,q)} v_1^{i+1} = 0 \, , \quad x_i < x < x_{i+2} \, , \quad v_1^{i+1}(x_i) = 0 \, , \quad \frac{1}{p} \, \frac{dv_1^{i+1}}{dx}\Big|_{x=x_i} = 1 \, ,$$

the function $v_1^{i+1}(x)$ does follow

$$0 = \int\limits_{x_i}^{x_{i+1}} \left(v_1^{i+1} L \, v_2^i - v_2^i \, L \, v_1^{i+1}\right) \, dx$$

$$= \left(v_1^{i+1} \, \frac{1}{p} \, (v_2^i)' - v_2^i \, \frac{1}{p} \, (v_1^{i+1})'\right)\Bigg|_{x_i}^{x_{i+1}}$$

$$= -v_1^{i+1}(x_{i+1}) + v_2^i(x_i) \, .$$

The function $v_3^i(x)$ admits an alternative form

$$v_3^i(x) = \int_{x_{i-1}}^{x_{i+1}} G(x,\xi)\, f(\xi)\, d\xi\,, \qquad x \in [x_{i-1}\,, x_{i+1}]\,, \tag{11}$$

where $G(x,\xi)$ is the Green function related to problem (4) (see (11) in Section 6):

$$G(x,\xi) = \begin{cases} \dfrac{v_1^i(\xi)\, v_2^i(x)}{v_1^i(x_{i+1})}\,, & x_{i-1} < \xi \le x\,, \\[2ex] \dfrac{v_1^i(x)\, v_2^i(\xi)}{v_1^i(x_{i+1})}\,, & x \le \xi < x_{i+1}\,. \end{cases} \tag{12}$$

By placing successively expression (12) and $x = x_i$ in (11) we obtain

$$\begin{aligned} v_3^i(x_i) = \frac{1}{v_1^i(x_{i+1})} \Bigg( & v_2^i(x_i) \int_{x_{i-1}}^{x_i} v_1^i(\xi)\, f(\xi)\, d\xi \\ & + v_1^i(x_i) \int_{x_i}^{x_{i+1}} v_2^i(\xi)\, f(\xi)\, d\xi \Bigg)\,. \end{aligned} \tag{13}$$

By virtue of relations (7), (9) and (13) we deduce from (3) that

$$\frac{1}{h}\left( \frac{u_{i+1} - u_i}{v_2^i(x_i)} - \frac{u_i - u_{i-1}}{v_1^i(x_i)} \right) - d_i\, u_i = -\varphi_i\,, \tag{14}$$

where

$$\begin{cases} d_i = \dfrac{1}{h\, v_1^i(x_i)} \displaystyle\int_{x_{i-1}}^{x_i} v_1^i(\xi)\, q(\xi)\, d\xi + \dfrac{1}{h\, v_2^i(x_i)} \displaystyle\int_{x_i}^{x_{i+1}} v_2^i(\xi)\, q(\xi)\, d\xi\,, \\[3ex] \varphi_i = \dfrac{1}{h\, v_1^i(x_i)} \displaystyle\int_{x_{i-1}}^{x_i} v_1^i(\xi)\, f(\xi)\, d\xi + \dfrac{1}{h\, v_2^i(x_i)} \displaystyle\int_{x_i}^{x_{i+1}} v_2^i(\xi)\, f(\xi)\, d\xi\,. \end{cases} \tag{15}$$

We now introduce a local coordinate system at a point $x = x_i$ by merely setting $x = x_i + sh$, $s = (x - x_i)/h$, in which the segment $[x_{i-1}, x_{i+1}]$ carries into the segment (pattern) $-1 \leq s \leq 1$ with the node $x = x_i$ corresponding to the point $s = 0$. Then we concentrate primarily on $v_1^i(x) = v_1^i(x_i + sh) = h\,\alpha^i(s,h)$ and $v_2^i(x) = v_2^i(x_i + sh) = h\,\beta^i(s,h)$, $-1 \leq s \leq 1$, $v_1^i(x_i) = ha_i$ and, because of (10), $v_2^i(x_i) = ha_{i+1}$. Being the pattern functions, $\alpha^i(s,h)$ and $\beta^i(s,h)$ are subject to the conditions

$$(16) \qquad L\,\alpha = \frac{d}{ds}\left(\frac{1}{\bar{p}(s)}\,\frac{d\alpha}{ds}\right) - h^2\,\bar{q}(s)\,\alpha - 0, \qquad -1 < s < 1,$$

$$\alpha(-1,h) = 0, \qquad \alpha'(-1,h) = \bar{p}(-1),$$

$$L\,\beta = 0, \quad -1 < s < 1, \quad \beta(1,h) = 0, \quad \beta'(1,h) = -\bar{p}(1),$$

where $\bar{p}(s) = p(x_i + sh)$ and $\bar{q}(s) = q(x_i + sh)$ depend only on the values $\bar{q}(s)$ and $\bar{p}(s)$ $\big(p(x)$ and $q(x)\big)$ on the segment $-1 \leq s \leq 1$ (on the segment $x_{i-1} \leq x \leq x_{i+1}$). Omitting subscript $i$ in formula (14) we obtain the homogeneous conservative scheme for $y(x) = u(x)$, $x \in \omega_h$:

$$(17) \qquad \left(a^{-1}\,y_{\bar{x}}\right)_x - dy = -\varphi, \qquad x \in \omega_h, \quad y(0) = u_1, \quad y(1) = u_2,$$

where

$$(18) \qquad \begin{cases} a(x) = \alpha(0,h) = A\big[p(x+sh), q(x+sh)\big], \\[2mm] d(x) = \dfrac{1}{a(x)} \displaystyle\int_{-1}^{0} \alpha(s,h)\,q(x+sh)\,ds \\[2mm] \qquad\qquad + \dfrac{1}{a(x+h)} \displaystyle\int_{0}^{1} \beta(s,h)\,q(x+sh)\,ds, \\[2mm] \varphi(x) = \dfrac{1}{a(x)} \displaystyle\int_{-1}^{0} \alpha(s,h)\,f(x+sh)\,ds \\[2mm] \qquad\qquad + \dfrac{1}{a(x+h)} \displaystyle\int_{0}^{1} \beta(s,h)\,f(x+sh)\,ds. \end{cases}$$

The coefficients $d(x)$, $\varphi(x)$ can be calculated by one and the same formula

$$\varphi(x) = F\left[p(x+sh), q(x+sh); f(x+sh)\right],$$

$$d(x) = F\left[p(x+sh), q(x+sh); q(x+sh)\right].$$

Pattern functionals are defined in the class of piecewise continuous functions:

$$A\left[\bar{p}(s), \bar{q}(s)\right] \qquad \text{for} \quad \bar{p}(s), \bar{q}(s) \in Q^{(0)}[-1,0] \qquad \text{and}$$

$$F\left[\bar{p}(s), \bar{q}(s);\ \bar{f}(s)\right] \qquad \text{for} \quad \bar{p}(s), \bar{q}(s), \bar{f}(s) \in Q^{(0)}[-1,1].$$

It is easily seen from (17)–(18) that the exact scheme does not belong to the family of schemes (16)–(17) in Section 2, whose pattern functionals $A[\bar{p}(s)]$ and $F[\bar{f}(s)]$ depend solely on a single function. In the case of equation (1) with constant coefficients $p(x) = p_0 = \text{const}$ and $q(x) = q_0 = \text{const}$ the pattern functions $\alpha(s,h)$ and $\beta(s,h)$ can be determined explicitly:

$$\alpha(s,h) = p_0 \, \frac{\text{sh}\,\big(\varkappa(1+s)h\big)}{\varkappa\, h}, \qquad \beta(s,h) = p_0 \, \frac{\text{sh}\,\big(\varkappa(1-s)h\big)}{\varkappa\, h},$$

$$\varkappa = \sqrt{p_0\, q_0}\,.$$

At the same time the coefficients $a(x)$ and $d(x)$ become constant values

$$a(x) = p_0 \, \frac{\text{sh}\,(\varkappa h)}{\varkappa\, h}, \qquad d(x) = \frac{2\, q_0}{\varkappa\, h} \ \text{th} \, \frac{\varkappa\, h}{2}.$$

**2. Schemes of arbitrary order accuracy.** There is no difficulty to construct a scheme of arbitrary order accuracy by means of appropriate expressions for coefficients of an exact scheme.

We now know from (16) that $\alpha(s,h)$ and $\beta(s,h)$ are analytic functions of the parameter $h^2$ and, therefore, can be expanded in the series

$$(19) \qquad \alpha(s,h) = \sum_{k=0}^{\infty} \alpha_k(s)\, h^{2k}, \qquad \beta(s,h) = \sum_{k=0}^{\infty} \beta_k(s)\, h^{2k},$$

where $\alpha_k(s)$ and $\beta_k(s)$ are calculated by the recurrence formulae

$$\alpha_k(s) = \int_{-1}^{s} \bar{p}(t) \left( \int_{-1}^{t} \alpha_{k-1}(\lambda)\, \bar{q}(\lambda)\, d\lambda \right) dt, \quad k > 0, \quad \alpha_0(s) = \int_{-1}^{s} \bar{p}(t)\, dt,$$

$$\beta_k(s) = \int_{s}^{1} \bar{p}(t) \left( \int_{t}^{1} \beta_{k-1}(\lambda)\, \bar{q}(\lambda)\, d\lambda \right) dt, \quad k > 0, \quad \beta_0(s) = \int_{s}^{1} \bar{p}(t)\, dt.$$

One trick we have encountered is to take only a finite number of the members in series (19):

$$\alpha^{(m)}(s,h) = \sum_{k=0}^{m} \alpha_k(s)\, h^{2k}, \qquad \beta^{(m)}(s,h) = \sum_{k=0}^{m} \beta_k(s)\, h^{2k},$$

making it possible to determine the coefficients $a^{(m)}$, $d^{(m)}$ and $\varphi^{(m)}$ by formulae (18), where the polynomials $\alpha^{(m)}$ and $\beta^{(m)}$ stand for $\alpha$ and $\beta$. As a final result we get a scheme of accuracy $O(h^{2m+2})$ in the class of piecewise continuous functions $k(x)$, $q(x)$, $f(x) \in Q^{(0)}[0,1]$. Any such scheme is called a **truncated scheme of rank** $m$.

For $m=0$ it refers to schemes of zero rank and accuracy $O(h^2)$ with regard to $k$, $q$, $f \in Q^{(0)}$. When providing current manipulations, the expressions for $d$ and $\varphi$ involved in this scheme

$$a^{(0)} = \frac{1}{\overset{\circ}{a}} = \int_{-1}^{0} p(x+sh)\, ds\,, \qquad p = \frac{1}{k}\,,$$

$$d^{(0)} = \overset{\circ}{d} + (h\, d_*)_x\,, \qquad \varphi^{(0)} = \overset{\circ}{\varphi} + (h\,\varphi_*)_x\,,$$

$$d_* = \overset{\circ}{a} \int_{-1}^{0} \frac{ds}{k\,(x+sh)} \int_{-1/2}^{s} q(x+th)\, dt\,,$$

$$\varphi_* = \overset{\circ}{a} \int_{-1}^{0} \frac{ds}{k\,(x+sh)} \int_{-1/2}^{s} f(x+th)\, dt$$

are different from those being used in the best scheme (14)–(15) of Section 2. Truncated schemes provide a possibility to attain any order of accuracy for arbitrary piecewise continuous functions $k(x)$, $q(x)$ and $f(x)$ and appear to be useful in many aspects.

An exact scheme and truncated schemes can be designed on an arbitrary non-equidistant grid $\hat{\omega}_h$ by the same methods as we employed before.

In practice the use of truncated schemes in the case of equation (1) with variable coefficients necessitates carrying out calculations of multiple integrals on each interval of the grid. Replacing those integrals by finite sums we are able to create more simpler schemes of accuracy $O(h^4)$ and $O(h^6)$, whose coefficients can be expressed through the values of $k$, $q$ and $f$

at some isolated points on each segment $[x_{i-1}, x_{i+1}]$. Such schemes permit one to retain the order of accuracy in the case of discontinuous coefficients $k$, $q$, $f$ on the grids $\hat{\omega}_h(K)$ when discontinuity points fall into the nodes of the grid $\hat{\omega}_h(K)$.

Exact and truncated schemes are quite applicable in the estimation of the accuracy of schemes (16)–(17) in Section 2. This approach allows to weaken or get rid of the smoothness of the functions $k$, $q$, $f$ involved in the estimation of the accuracy order of schemes (16)–(17) in Section 2.

## 3.8 METHODS FOR DESIGNING DIFFERENCE SCHEMES

**1. General remarks.** From such reasoning it seams clear that in the space of grid functions difference schemes should retain the basic properties of differential equations such as self-adjointness, the validity of certain a priori estimates (for instance, the maximum principle), etc. Moreover, a scheme which interests us must satisfy, first at all, the requirements of solvability, stability, approximation and, hence, accuracy of a certain order; at last, an algorithm must be expedient in computer resources. The expediency depends not only on a scheme, but also on making a substantiated choice of the method for solving difference equations and a grid which, generally speaking, may be non-equidistant and depends on the behaviour of a solution.

The construction of schemes with the indicated properties and a desired quality is one of the outlines of the possible theory.

As a result, a considerable amount of effort has been expended in designing various methods for providing difference approximations of differential equations. The simplest and, in a certain sense, natural method is connected with selecting a suitable pattern and imposing on this pattern a difference equation with undetermined coefficients which may depend on nodal points and step. Requirements of solvability and approximation of a certain order cause some limitations on a proper choice of coefficients. However, those constraints are rather mild and we get an infinite set (for instance, a multi-parameter family) of schemes. There is some consensus of opinion that this is acceptable if we wish to get more and more properties of schemes such as homogeneity, conservatism, etc., leaving us with narrower classes of admissible schemes.

Actually this way has been demonstrated in Sections 1–3. Several methods find a wide range of applications in designing difference schemes of a desirable quality, among them

1) the integro-interpolational method (see Section 2);

2) the variational difference methods (the Ritz method and Bubnov–Galerkin methods) and the finite element method;
3) the method of the summator identity;
4) the method of approximating a variational functional.

**2. The integro-interpolational method (IIM).** In Section 2 we have already studied the IIM, but its possibilities and potential have not been illustrated in full measure. Here we consider other ways of its applications by appeal to the problem

$$(1)\qquad (ku')' - q(x)u = -f(x)\,, \qquad 0 < x < 1\,,$$

$$(2)\qquad k\,u' - \sigma_1\,u = -\mu_1\,, \quad x = 0\,, \qquad -k\,u' - \sigma_2\,u = -\mu_2\,, \quad x = 1\,,$$

$$0 < c_1 \le k(x) \le c_2\,, \quad \sigma_1 \ge 0, \quad \sigma_2 \ge 0, \quad q(x) \ge 0\,.$$

We introduce an equidistant grid on the segment $0 \le x \le 1$

$$\bar{\omega}_h = \{x_i = ih,\ i = 0, 1, \ldots, N,\ hN = 1\}\,.$$

Integration of equation (1) is accomplished over the segment $x_i \le x \le x_{i+1}$, leading to

$$(3)\qquad w_{i+1} - w_i = \int\limits_{x_i}^{x_{i+1}} (qu - f)\,dx = \Phi_{i+1}\,, \qquad w = ku'\,.$$

Here, in contrast to Section 2, the flow $-w$ is taken at the same node $x_i$ as the unknown function $u$, which is sought. Therefore, the intention is to use instead of $w_{i+1/2}$ the approximation $(w_{i+1} + w_i)/2$ by accepting

$$(4)\qquad \tfrac{1}{2}\,(w_{i+1} + w_i) \approx a_{i+1}\,u_{\bar{x},\,i+1}\,,$$

where $a_{i+1}$ is some functional of $k(x)$ on the segment $x_i \le x \le x_{i+1}$ satisfying the relation $a_i = k(x_{i-1/2}) + O(h^2)$. In the course of the elimination of $w_i$ from (3) and (4) we find that

$$w_{i+1} \approx a_{i+1}\,u_{\bar{x},\,i+1} + \tfrac{1}{2}\,\Phi_{i+1}\,.$$

Subtracting from here the relevant expression

$$w_i \approx a_i\,u_{\bar{x},i} + \tfrac{1}{2}\,\Phi_i$$

and taking into account (3), we establish

$$(5) \qquad \left(a u_{\bar{x}}\right)_{x,i} \approx \frac{\Phi_i + \Phi_{i+1}}{2h} = \frac{1}{2h} \int\limits_{x_{i-1}}^{x_{i+1}} (qu - f)\, dx\,.$$

The approximation of the integral on the right-hand side of (5) can be done using various quadrature formulae, for instance, by the formula of trapezoids

$$\frac{1}{2h} \int\limits_{x_{i-1}}^{x_{i+1}} (qu - f)\, dx \approx (qu - f)_i$$

or

$$\frac{1}{2h} \int\limits_{x_{i-1}}^{x_{i+1}} (qu - f)\, dx \approx \frac{1}{4} \left((qu - f)_{i-1} + 2\,(qu - f)_i + (qu - f)_{i+1}\right)$$

$$= (qu - f)_i + \frac{h^2}{4} (qu - f)_{\bar{x}x,i}\,.$$

With these, we arrive at the schemes of accuracy $O(h^2)$:

$$(6) \qquad \left(a y_{\bar{x}}\right)_x - q\,y = -f\,,$$

$$(7) \qquad \left(a y_{\bar{x}} - \frac{h^2}{4} (q\,y)_{\bar{x}}\right)_x - qy = -\left(f + \frac{h^2}{4} f_{\bar{x}x}\right).$$

To approximate the boundary condition, for instance, at the point $x = 0$, we apply (3) for $i = 0$

$$w_1 - w_0 = \int\limits_0^h (qu - f)\, dx$$

and then adopt here

$$w_1 \approx a_1\, u_{x,0} + \tfrac{1}{2} \int\limits_0^h (qu - f)\, dx\,, \qquad w_0 = (ku')_0 = \sigma_1\, u_0 - \mu_1\,,$$

so that

$$a_1\, u_{x,0} - (\sigma_1\, u_0 - \mu_1) \approx \tfrac{1}{2} \int\limits_0^h (qu - f)\, dx\,.$$

Making use of the approximation

$$\int_0^h (qu - f)\, dx \approx (qu - f)_0\, h\,,$$

we impose the boundary condition for $y_i$ with the approximation error $O(h^2)$ at the point $x = 0$:

$$(8) \qquad a_1\, y_{x,0} = \bar{\sigma}_1\, y_0 - \mu_1\,, \qquad \bar{\sigma}_1 = \sigma_1 + \tfrac{1}{2}\, h\, q_0, \qquad \bar{\mu}_1 = \mu_1 - \tfrac{1}{2}\, h\, f_0\,.$$

The boundary condition (2) of the third kind can be established at the point $x = 1$ in a similar way.

So far we have studied some versions of the IIM on the basis of the balance equation (the balance method). We now consider the second method for the design of homogeneous difference schemes by means of the IIM, in the framework of which equation (1) has to be integrated twice: first, we integrate equation (1) from $x_i$ to $x$:

$$(9) \qquad w(x) - w(x_i) = \int_{x_i}^{x} (qu - f)\, dx\,.$$

Second, we integrate the preceding over $x$ from $x_i$ to $x_{i+1}$ and from $x_{i-1}$ to $x_i$:

$$(10) \qquad \int_{x_i}^{x_{i+1}} w(x)\, dx - w_i\, h = \int_{x_i}^{x_{i+1}} dx \left( \int_{x_i}^{x} (qu - f)\, dt \right),$$

$$(11) \qquad \int_{x_{i-1}}^{x_i} w(x)\, dx - w_i\, h = \int_{x_{i-1}}^{x_i} dx \left( \int_{x_i}^{x} (qu - f)\, dt \right).$$

The interchange of the integration order leads to

$$\int_{x_i}^{x_{i+1}} dx \left( \int_{x_i}^{x} (qu - f)\, dt \right) = \int_{x_i}^{x_{i+1}} (x_{i+1} - t)\, (qu - f)\, dt\,,$$

$$\int_{x_{i-1}}^{x_i} dx \left( \int_{x_i}^{x} (qu - f)\, dt \right) = - \int_{x_{i-1}}^{x_i} (t - x_{i-1})\, (qu - f)\, dt\,.$$

The function $u(x)$ is linearly interpolated on each of the segments $[x_{i-1}, x_i]$ and $[x_i, x_{i+1}]$, so that

$$u(x) \approx \begin{cases} u_i + (x - x_i)\, u_{x,i} & \text{for} \quad x_i \le x \le x_{i+1}, \\ u_i + (x - x_i)\, u_{\bar{x},i} & \text{for} \quad x_{i-1} \le x \le x_i, \end{cases}$$

which simplifies the huge job done with the integrals

$$\int\limits_{x_i}^{x_{i+1}} w\, dx = \int\limits_{x_i}^{x_{i+1}} k u'\, dx \approx u_{x,i} \int\limits_{x_i}^{x_{i+1}} k(x)\, dx,$$

$$\int\limits_{x_{i-1}}^{x_i} w\, dx \approx u_{\bar{x},i} \int\limits_{x_{i-1}}^{x_i} k(x)\, dx,$$

$$- \int\limits_{x_{i-1}}^{x_i} q u (t - x_{i-1})\, dt \approx -u_i \int\limits_{x_{i-1}}^{x_i} q(t)\,(t - x_{i-1})\, dt$$

$$+ u_{\bar{x},i} \int\limits_{x_{i-1}}^{x_i} q(t)\,(x_i - t)\,(t - x_{i-1})\, dt,$$

$$\int\limits_{x_i}^{x_{i+1}} (x_{i+1} - t)\, q u\, dt \approx u_i \int\limits_{x_i}^{x_{i+1}} q(t)\,(x_{i+1} - t)\, dt$$

$$+ u_{x,i} \int\limits_{x_i}^{x_{i+1}} q(t)\,(x_{i+1} - t)\,(t - x_i)\, dt.$$

Substituting the resulting expressions into identities (10)–(11) and subtracting the second identity from the first one, we arrive at the difference scheme

$$(ay_{\bar{x}})_x - dy = -\varphi\,, \tag{12}$$

$$a_i = \frac{1}{h}\int\limits_{x_{i-1}}^{x_i} k(t)\,dt - \frac{1}{h}\int\limits_{x_{i-1}}^{x_i} q(t)\,(x_i - t)\,(t - x_{i-1})\,dt\,,$$

$$d_i = \frac{1}{h^2}\left(\int\limits_{x_{i-1}}^{x_i} (t - x_{i-1})\,q(t)\,dt + \int\limits_{x_i}^{x_{i+1}} (x_{i+1} - t)\,q(t)\,dt\right).$$

The right-hand side $\varphi_i$ can be determined by the same formula as we used for $d_i$ with $f(t)$ standing in place of $q(t)$. The formulae for $a_i$ and $d_i$ can be rewritten as

$$(13)\qquad \begin{aligned} a_i &= \int\limits_{-1}^{0} k(x_i + sh)\,ds + h^2\int\limits_{-1}^{0} s(1+s)\,q(x_i + sh)\,ds\,,\\ d_i &= \int\limits_{-1}^{0} (1+s)\,q(x_i + sh)\,ds + \int\limits_{0}^{1} (1-s)\,q(x_i + sh)\,ds\,. \end{aligned}$$

If $k$ and $q$ are constants, then

$$a_i = k - \frac{h^2}{6}\,q\,, \qquad d_i = q\,. \tag{13'}$$

In the sequel we will show that scheme (12) is identical with the scheme emerged in variational difference methods (the finite element method).

In what follows we share our practical experience of the design of difference schemes for problems with lumped parameters by means of the IIM. Suppose, for instance, that a single heat source of capacity $Q$ is located at a point $x = \xi$ so that a solution of problem (1)–(2) satisfies the conditions

$$[u] = 0\,, \qquad \left[k\,\frac{du}{dx}\right] = -Q \qquad \text{for} \quad x = \xi\,. \tag{14}$$

Let $\xi = x_n + \theta h$, $0 \le \theta \le 1$, that is, $x_n \le \xi \le x_{n+1}$, $n > 0$. The balance equation (3) on the segment $x_n \le x \le x_{n+1}$ is of the form

$$w_{n+1} - w_n + [w] = \int_{x_n}^{x_{n+1}} (qu - f)\, dx$$

or

$$w_{n+1} - w_n = \Phi_{n+1} + Q\,. \tag{15}$$

On the remaining segments $[x_i\,, x_{i+1}]$, $i \ne n$, identity (3) holds true. As a final result, instead of (12), we get the scheme

$$(a y_{\bar{x}})_{x,\,i} - q_i\, y_i = -\varphi_i\,,$$

where

$$\varphi_i = f_i\,, \qquad i \ne n, \quad i \ne n+1\,,$$

$$\varphi_n = f_n + \frac{Q}{2\,h}\,, \qquad \varphi_{n+1} = f_{n+1} + \frac{Q}{2\,h}\,. \tag{16}$$

In the physical language, this is a way of saying that the source is spread over two intervals.

Rewriting the same identity on the segment $[x_{i-1/2}\,, x_{i+1/2}]$ reveals another scheme

$$(a y_{\bar{x}})_x - dy = -\varphi\,, \qquad \varphi_{n+1} = f_{n+1}\,, \quad \varphi_n = f_n + \frac{Q}{n} \tag{17}$$

for $0 \le \theta < \frac{1}{2}$,

$$\varphi_{n+1} = f_{n+1} + \frac{Q}{n}\,, \qquad \varphi_n = f_n$$

for $\frac{1}{2} \le \theta < 1$.

Let us stress that the integro-interpolational method is a rather flexible and general tool in designing difference schemes relating to stationary and nonstationary problems with one or several spatial variables.

**3. Variational difference methods (the Ritz method and the Bubnov–Galerkin method).** The Ritz and the Bubnov–Galerkin variational methods have had considerable impact on complex numerical modeling problems and designs of difference schemes.

Let $A$ be a self-adjoint positive definite linear operator in Hilbert space $H$ equipped with an inner product $(\,,\,)$ and let $f$ be a given element of the space $H$. The problem of minimizing the functional

$$I[u] = (Au, u) - 2\,(u, f) \tag{18}$$

is equivalent to the problem of solving the equation

$$Au = f\,. \tag{19}$$

The element $u_0 \in H$ satisfying the equation $Au_0 = f$ and realizing

$$\min I[u] = I[u_0]$$

is unique.

The main idea behind the Ritz method is to take into consideration a sequence of finite-dimensional spaces $V_n$ with basis functions $\varphi_i^{(n)}$, $i = 1, 2, \dots, n$, and look for an element $u_n \in V_n$, minimizing the functional $I[u]$ in the space $V_n$.

Still using the framework of this method, we may attempt an approximate solution $u_n$ in the form

$$u_n = \sum_{j=1}^{n} y_j\,\varphi_j \tag{20}$$

with unknown coefficients $y_1, y_2, \dots, y_n$. By inserting this expression in the formula for $I[u]$ we find that

$$I[u_n] = \sum_{i,j=1}^{n} \alpha_{ij}\,y_i\,y_j - 2\sum_{j=1}^{n} \beta_j\,y_j\,, \tag{21}$$

where

$$\alpha_{ij} = (A\varphi_i\,, \varphi_j), \qquad \beta_i = (f, \varphi_i)\,. \tag{22}$$

Since $A = A^*$ is a self-adjoint operator, we have $\alpha_{ij} = \alpha_{ji}$. The functional $I[u_n]$ is a function of $n$ coefficients $y_1, y_2, \dots, y_n$. By equating the derivatives $\partial I[u_n]/\partial y_i$ to zero and using the symmetry of coefficients $\alpha_{ij} = \alpha_{ji}$, we obtain $n$ equations for determination of $y_i$:

$$\sum_{j=1}^{n} \alpha_{ij}\,y_j - \beta_i = 0, \qquad i = 1, 2, \dots, n\,. \tag{23}$$

The Ritz method proves to be useful in studying the problem

$$(24) \qquad (ku')' - qu = -f(x), \qquad 0 < x < 1, \quad u(0) = 0, \quad u(1) = 0 .$$

Other ideas are connected with the function

$$(25) \qquad \eta(s) = \begin{cases} 0, & s < -1, \quad s > 1, \\ 1 + s, & -1 < s < 0, \\ 1 - s, & 0 < s < 1, \end{cases}$$

which guides the choice of the functions

$$(26) \qquad \varphi_i(x) = \eta\Big(\frac{x - x_i}{h}\Big) = \eta_i(x),$$

where $x_i = ih$, $i = 1, 2, \ldots, N-1$, is a node of the grid $\bar{\omega}_h = \{x_i = ih, \ i = 0, 1, \ldots, N, \ hN = 1\}$. It follows from the foregoing that

$$(25') \qquad \eta_i(x) = \begin{cases} 0 & \text{for} \quad x < x_{i-1} \quad \text{and} \quad x > x_{i+1}, \\[2mm] \dfrac{x - x_{i-1}}{h} & \text{for} \quad x_{i-1} < x < x_i, \\[2mm] \dfrac{x_{i+1} - x}{h} & \text{for} \quad x_i < x < x_{i+1}, \end{cases}$$

and, hence,

$$(27) \qquad \frac{d\eta_i}{dx} = \begin{cases} 0 & \text{for} \quad x < x_{i-1} \quad \text{and} \quad x > x_{i+1}, \\[2mm] \dfrac{1}{h} & \text{for} \quad x_{i-1} < x < x_i, \\[2mm] -\dfrac{1}{h} & \text{for} \quad x_i < x < x_{i+1}. \end{cases}$$

Upon substituting

$$A\,u = -\frac{d}{dx}\left(k\,\frac{du}{dx}\right) + qu$$

into (22) we get

$$(28) \qquad \alpha_{ij} = \int_0^1 \left(k\,\frac{d\eta_i}{dx}\,\frac{d\eta_j}{dx} + q\,\eta_i\,\eta_j\right) dx, \qquad \beta_i = \int_0^1 f(x)\,\eta_i(x)\,dx .$$

In light of the properties of the function $\eta_i(x)$ and its derivatives the matrix $\{\alpha_{ij}\}$ is tridiagonal, because only the elements with $j = i-1$, $j = i$ and $j = i+1$ are nonzero.

In the new notations

$$(29) \qquad a_i = -h\,\alpha_{i,i-1}\,, \qquad h^2 d_i = h\,\alpha_{i,i} + h\,(\alpha_{i,i-1} + \alpha_{i,i+1})$$

we obtain

$$(30) \qquad \begin{aligned} a_i &= \frac{1}{h} \int_{x_{i-1}}^{x_i} k(x)\,dx - \frac{1}{h} \int_{x_{i-1}}^{x_i} q(x)\,(x - x_{i-1})\,(x_i - x)\,dx\,, \\ d_i &= \frac{1}{h^2} \left( \int_{x_{i-1}}^{x_i} q(x)\,(x - x_{i-1})\,dx + \int_{x_i}^{x_{i+1}} q(x)\,(x_{i+1} - x)\,dx \right). \end{aligned}$$

Then the system of equations

$$\alpha_{i,\,i-1}\,y_{i-1} + \alpha_{i,\,i}\,y_i + \alpha_{i,\,i+1}\,y_{i+1} - \beta_i = 0$$

can be rewritten as

$$a_i\,y_{i-1} - (a_i + a_{i+1} + h^2 d_i)\,y_i + a_{i+1}\,y_{i+1} + h^2 \varphi_i = 0$$

or

$$(31) \qquad (a y_{\bar{x}})_x - dy = -\varphi\,,$$

where

$$(32) \qquad \varphi_i = \frac{1}{h^2} \left( \int_{x_{i-1}}^{x_i} f(x)\,(x - x_{i-1})\,dx + \int_{x_i}^{x_{i+1}} f(x)\,(x_{i+1} - x)\,dx \right),$$

thereby clarifying that $\varphi_i$ are calculated by means of the same formula as $d_i$.

So, the three-point scheme (30)–(32) constructed by the Ritz method is identical with scheme (12) obtained by means of the IIM. In contrast to the Ritz method the Bubnov–Galerkin method applies equally well to problems, whose ingredients have no fixed sign and turn out to be non-self-adjoint. In this case the coefficients $y_i$ of the approximate solution (20) are to be determined from the orthogonality conditions for the residual $Au_n - f$ with respect to all of the basis functions $\eta_i(x)$:

$$(A\,u_n - f, \eta_i) = 0\,, \qquad i = 1, 2, \dots, n\,. \tag{33}$$

A non-self-adjoint boundary-value problem acquires the form:

$$\begin{gathered}(ku')' + r(x)\,u' - q(x)\,u = -f(x)\,,\\ 0 < x < 1, \qquad u(0) = u(1) = 0,\\ k(x) > 0, \qquad q(x) \ge 0\,.\end{gathered} \tag{34}$$

We introduce the grid $\bar{\omega}_h = \{x_i = ih,\ i = 0, 1, \dots, N,\ hN = 1\}$. Then the dimension $n$ of the space $V_n$ equals $N - 1$. The functions

$$\eta_i(x) = \eta\Big(\frac{x - x_i}{h}\Big), \qquad i = 1, 2, \dots, N-1\,,$$

where the function $\eta(s)$ was specified by (25), provide the background for subsequent constructions.

In this context, condition (33) becomes

$$\sum_{j=1}^{N-1} \alpha_{ij}\, y_j - \beta_i = 0\,, \qquad i = 1, 2, \dots, N-1\,, \tag{35}$$

where

$$\begin{aligned}\alpha_{ij} &= \int_0^1 \Big(k(x)\,\frac{d\eta_i}{dx}\,\frac{d\eta_j}{dx} - r(x)\,\frac{d\eta_i}{dx}\,\eta_j(x) + q(x)\,\eta_i(x)\,\eta_j(x)\Big)\,dx\,,\\ \beta_i &= \int_0^1 f(x)\,\eta_i(x)\,dx\,, \qquad i, j = 1, 2, \dots, N-1\,.\end{aligned} \tag{36}$$

By definitions (25), (25′) of the function $\eta_i(x)$, the coefficients $\alpha_{i,j}$ are nonzero only for $j = i-1, i, i+1$. Retaining notations (30) for $a_i$ and $d_i$ and (32) for $\varphi_i$ and accepting

$$\begin{aligned}b_i^- &= \frac{1}{h^2}\int_{x_{i-1}}^{x_i} r(x)\,(x - x_{i-1})\,dx = \int_{-1}^{0} r(x_i + sh)\,(1+s)\,ds\,,\\ b_i^+ &= \frac{1}{h^2}\int_{x_i}^{x_{i+1}} r(x)\,(x_{i+1} - x)\,dx = \int_0^1 r(x_i + sh)\,(1-s)\,ds\,,\end{aligned} \tag{37}$$

we reduce the system of equations (35) with the members $b_i^+$ and $b_i^-$ to

(38)
$$\frac{1}{h^2}\left[a_{i+1}(y_{i+1}-y_i)-a_i(y_i-y_{i-1})\right]$$
$$+\frac{b_i^-(y_i-y_{i-1})}{h}+\frac{b_i^+(y_{i+1}-y_i)}{h}-d_i\,y_i=-\varphi_i\,,$$
$$i=1,2,\dots,N-1\,.$$

Thus, we arrive at the difference scheme

(39)
$$\begin{aligned}&\left(a y_{\bar{x}}\right)_x+b^-y_{\bar{x}}+b^+y_x-dy=-\varphi(x), \qquad 0<x=ih<1\,,\\ &y_0=0, \qquad y_N=0\,,\end{aligned}$$

whose coefficients can be recovered from (30), (31) and (37).

For $r(x)\equiv 0$ this scheme is identical with scheme (31)–(32) obtained by means of the Ritz method. In the case of constant coefficients $k(x)$, $r(x)$ and $q(x)$

$$a_i=k-\frac{h^2}{6}\,q, \quad d_i=d=q, \quad b_i^-=b_i^+=\frac{r}{2}, \quad b_i^-y_{\bar{x}}+b_i^+y_x=r\,y_{\overset{\circ}{x}}\,.$$

When the coordinate functions $\varphi_i(x)=\eta\big((x-x_i)/h\big)$ are chosen by an approved rule as suggested before, the Ritz and the Bubnov–Galerkin methods coincide with the finite element method.

**4. The method of approximating a quadratic functional.** The boundary-value problem

$$Lu=(ku')'-q(x)u=-f(x), \qquad 0<x<1, \quad u(0)=0, \quad u(1)=0\,,$$

is equivalent to the problem of searching a minimizing element for the functional (see Section 3)

(40)
$$I[u]=\int_0^1\left[k(u')^2+q\,u^2\right]\,dx-2\int_0^1 fu\;dx\,.$$

Recall that the equation $Lu=-f(x)$ is Euler's equation related to such a functional $I[u]$.

We introduce the grid $\bar{\omega}_h = \{x_i = ih,\ i = 0, 1, \dots, N,\ hN = 1\}$ and approximate on it the functional $I[u]$ by appeal to one of the auxiliary modifications

$$I[u] = \sum_{i=1}^{N} \int_{x_{i-1}}^{x_i} k(u')^2 \, dx + \sum_{i=1}^{N} \int_{x_{i-1}}^{x_i} (qu^2 - 2fu) \, dx \, .$$

With this in mind, we approximate the integrals

$$\int_{x_{i-1}}^{x_i} k(u')^2 \, dx \approx a_i \, (u_{\bar{x},i})^2 h \, ,$$

$$\int_{x_{i-1}}^{x_i} (qu^2 - 2\,fu) \, dx \approx \frac{h}{2} \Big( (qu^2 - 2\,fu)_i + (qu^2 - 2\,fu)_{i-1} \Big) \, ,$$

where $a_i$ is a functional depending on $k(x)$ on the segment $x_{i-1} \le x \le x_i$. There are many ways of taking care of these restrictions. For instance, we might agree to consider

$$a_i = \frac{1}{h} \int_{x_{i-1}}^{x_i} k(x) \, dx \, , \qquad \text{etc} \, .$$

Thus, instead of $I[u]$ we deal with the functional

$$I_h[u] = \sum_{i=1}^{N} a_i \, (u_{\bar{x},i})^2 \, h + \sum_{i=1}^{N-1} (q_i \, y_i^2 - 2 \, f_i \, y_i) \, h \, , \tag{41}$$

where $y$ is an arbitrary grid function vanishing for $i = 0$ and $i = N$: $y_0 = y_N = 0$. The functional $I_h[u]$ is a function of $N - 1$ variables $y_1$, $y_2$, $\dots$, $y_{N-1}$. Equating the first derivatives

$$\frac{\partial I_h}{\partial y_i} = 2 \, a_{i+1} \, y_{x,i} (-1) + 2 \, a_i \, y_{\bar{x},i} + 2 \, q_i \, y_i \, h - 2 \, f_i \, h$$

to zero leads to the difference equations

$$\big( a y_{\bar{x}} \big)_x - q \, y = -f(x) \, , \qquad x = ih \, . \tag{42}$$

Adopting those ideas to problem (1), (2), (14) concerning a point heat source, an excellent start in this direction is to replace the function $f(x)$ involved in formula (40) by $f(x)+\delta(x-\xi)\,Q$, where $\delta(x-\xi)$ is Dirac's delta-function. Recall that $\delta(x-\xi)=0$ if $x\neq\xi$, $\delta(x-\xi)=\infty$ if $x=\xi$ and $\int_{\xi-\varepsilon}^{\xi+\varepsilon}\delta(x-\xi)\,dx=1$ for any $\varepsilon>0$. As a final result we get

$$I[u]=\int\limits_0^1\left[k(u')^2+q\,u^2-2\,fu\right]\,dx-2\,Q\,u(\xi)\,. \tag{43}$$

If $\xi=x_n+\theta h$, $0\leq\theta\leq 1$, then $u(\xi)$ must be replaced by $u_n$ for $\theta<\frac{1}{2}$ and by $u_{n+1}$ for $\theta>\frac{1}{2}$, whose use permits us to establish

$$I_h[u]=\sum_{i=1}^{N}a_i\,(y_{\bar{x},i})^2h+\sum_{i=1}^{N-1}(d_i\,y_i^2-2\,\varphi_i\,y_i)\,h\,, \tag{44}$$

where

$$\varphi_i=\begin{cases} f_i+\dfrac{Q}{h}\,\delta_{i,n} & \text{for}\quad \theta<\frac{1}{2}\,,\\[2ex] f_i+\dfrac{Q}{h}\,\delta_{i,n+1} & \text{for}\quad \theta>\frac{1}{2}\,,\end{cases} \tag{45}$$

and $\delta_{ik}$ is, as usual, Kronecker's delta. Equating $\partial I_h/\partial y_i$ to zero we arrive at the scheme $(ay_{\bar{x}})_x-dy=-\varphi$ with the right-hand side specified by formulae (45).

In the case of a non-equidistant grid $\hat{\bar{\omega}}_h=\{x_i\,,\ i=0,1,\dots,N,\ x_0=0,\ x_N=1\}$ we obtain instead of (44)

$$I_h[y]=\sum_{i=1}^{N}a_i\,(y_{\bar{x},i})^2h_i+\sum_{i=1}^{N-1}(d_i\,y_i^2-2\,\varphi_i\,y_i)\,\hbar_i\,.$$

In particular, it is always possible to choose a grid so that $\xi=x_n$ would be one of the nodal point and

$$\varphi_n=f_n+\frac{Q}{\hbar_n}\,,\qquad \varphi_i=f_i\,,\quad i\neq n\,.$$

We are led by equating the derivatives $\partial I_h[u]/\partial y_i$ to zero to the scheme

$$(ay_{\bar{x}})_{\hat{x},i}-d_i\,y_i=-\varphi_i\,,\qquad i=1,2,\dots,N-1\,. \tag{46}$$

**5. The method of the summator identities (the method of approximating an integral identity).** A solution of the problem

$$(47)\qquad \begin{gathered} Lu = (ku')' - qu = -f(x)\,, \qquad 0 < x < 1\,, \\ ku' = \sigma_1 u - \mu_1\,, \quad x = 0, \qquad -ku' = \sigma_2 u - \mu_2\,, \quad x = 1\,, \end{gathered}$$

satisfies the integral identity

$$(48)\qquad \begin{aligned} I[u,v] = \int\limits_0^1 (ku'v' + q\,u\,v - fv)\,dx + \sigma_1\,u(0)\,v(0) \\ + \sigma_2\,u(1)\,v(1) - \mu_1\,v(0) - \mu_2\,v(1) = 0\,, \end{aligned}$$

where $v = v(x)$ is an arbitrary function being continuous on the segment $0 \le x \le 1$ and having the summable derivative in the space $L_2[0,1]$. This identity will be used for the determination of the generalized solution of problem (47).

The design of a difference scheme on an equidistant grid $\bar{\omega}_h = \{x_i = ih,\ i = 0, 1, \dots, N,\ hN = 1\}$ is based on the approximation of the integral identity (48) by the summator identity for grid functions, for instance,

$$(49)\qquad \begin{aligned} I_h[u,v] = \sum_{i=1}^{N} a_i\,y_{\bar{x},i}\,v_{\bar{x},i}\,h + \sum_{i=1}^{N-1} (q_i\,y_i - f_i)\,v_i\,h \\ + \bar{\sigma}_1\,y_0\,v_0 + \bar{\sigma}_2\,y_N\,v_N - \bar{\mu}_1\,v_0 - \bar{\mu}_2\,v_N = 0\,, \end{aligned}$$

where $v_i$ is an arbitrary grid function. Here $a_i$ is any of the coefficients having the form $a_i = A[k(x_i + sh)]$, $-1 \le s \le 0$, and providing an approximation of order 2: $a_i = k_{i-1/2} + O(h^2)$. There is no difficulty to verify that

$$\bar{\sigma}_1 = \sigma_1 + \tfrac{1}{2}\,h\,q_0\,, \qquad \bar{\sigma}_2 = \sigma_2 + \tfrac{1}{2}\,h\,q_N\,,$$

$$\bar{\mu}_1 = \mu_1 + \tfrac{1}{2}\,h\,f_0\,, \qquad \bar{\mu}_2 = \mu_2 + \tfrac{1}{2}\,h\,f_N$$

if the trapezoids formula is in hand in computing the integrals

$$\int\limits_{x_{i-1}}^{x_i} (qu - f)\,dx$$

by means of the relation

$$\int_0^h (qu - f)\, dx = \frac{1}{2}\, h\, (qu - f)_0 + \frac{1}{2}\, h\, (qu - f)_1\,,$$

etc. Accepting, for instance, $v_i = \delta_{i,\, i_0}$, $0 < i_0 < N$, and recalling that $v_{\bar{x},\, i} = 0$ for $i < i_0$ and $i > i_0 + 1$, $v_{\bar{x},\, i_0} = 1/h$ and $v_{\bar{x},\, i_0+1} = -1/h$, we find for $i = i_0$ that

$$\left[ -\frac{a_{i+1}\, y_{x,i}}{h} + \frac{a_i\, y_{\bar{x},i}}{h} \right] h + \left( d_i\, y_i - f_i \right) h = 0$$

or $(a y_{\bar{x}})_x - dy = -f$.

If $v_i = \delta_{i,\, 0}$, then $v_{\bar{x},\, i} = (-1/h)\, \delta_{i,1}$ and identity (49) yields

$$(-1/h)\, a_1\, y_{\bar{x},\, 1} + \bar{\sigma}_1\, y_0 - \bar{\mu}_1 = 0$$

or $a_1\, y_{\bar{x},\, 1} = \bar{\sigma}_1\, y_0 - \bar{\mu}_1$. Likewise, for $v_i = \delta_{i,N}$

$$-\left( a\, y_{\bar{x}} \right)_N = \bar{\sigma}_2\, y_N - \bar{\mu}_2\,.$$

With these, we arrive at the difference boundary-value problem

$$\left( a y_{\bar{x}} \right)_x - dy = -f, \qquad 0 < x = ih < 1\,,$$
$$a_1\, y_{x,\, 0} = \bar{\sigma}_1\, y_0 - \bar{\mu}_1, \qquad -a_N\, y_{\bar{x},\, N} = \bar{\sigma}_2\, y_N - \bar{\mu}_2\,.$$

## 3.9 STABILITY WITH RESPECT TO COEFFICIENTS

**1. Stability of difference schemes with respect to coefficients.** In solving some or other problems for a differential equation it may happen that coefficients of the equation are specified not exactly, but with some error because they may be determined by means of some computational algorithms or physical measurements, etc. Coefficients of a homogeneous difference scheme are functionals of coefficients of the relevant differential equation. An error in determining coefficients of a scheme may be caused by various factors: an error in calculations of pattern functionals, an error in specifying coefficients of a differential equation, rounding errors, etc.

We say that a scheme is **stable with respect to coefficients (co-stable)** if a solution of the boundary-value problem has slight variations under small perturbations of the scheme coefficients. In order to avoid misunderstanding, we focus our attention on the scheme with coefficients $a$, $d$, $\varphi$

$$(1)\quad \Lambda y = \left(a y_{\bar{x}}\right)_x - d\, y = -\varphi, \qquad 0 < x = ih < 1, \quad y(0) = 0, \quad y(1) = 0\,,$$

and the same scheme with perturbed coefficients $\tilde{a}$, $\tilde{d}$, $\tilde{\varphi}$ (for the sake of simplicity the values $y(0)$ and $y(1)$ remain unperturbed)

$$(2)\quad \tilde{\Lambda}\tilde{y} = \left(\tilde{a}\tilde{y}_{\bar{x}}\right)_x - \tilde{d}\,\tilde{y} = -\tilde{\varphi}\,, \qquad 0 < x = ih < 1\,, \quad \tilde{y}(0) = 0, \quad \tilde{y}(1) = 0\,,$$

under the conditions

$$(3)\quad \begin{gathered} a(x) \ge c_1 > 0, \qquad \tilde{a}(x) \ge c_1 > 0, \\ d(x) \ge 0, \qquad \tilde{d}(x) \ge 0, \qquad c_1 = \text{const} > 0\,, \end{gathered}$$

with constant $c_1$ independent of a grid.

We estimate the difference $z = \tilde{y} - y$ in terms of perturbed coefficients. Substituting $\tilde{y} = z + y$ into (2) and taking into account (1), we get

$$(4)\qquad \tilde{\Lambda} z = \left(\tilde{a} z_{\bar{x}}\right)_x - \tilde{d} z = -\Psi, \qquad z_0 = z_N = 0\,,$$

where

$$(5)\qquad \Psi = \tilde{\varphi} - \varphi + (\tilde{\Lambda} - \Lambda)\, y = \tilde{\varphi} - \varphi + \left((\tilde{a} - a)\, y_{\bar{x}}\right)_x - (\tilde{d} - d)\, y\,.$$

From (4) it is easily seen that $\Psi$ is representable by

$$(6)\qquad \Psi = \mu_x + \eta_x\,,$$

where

$$(7)\qquad \mu = (\tilde{a} - a)\, y_{\bar{x}}$$

and $\eta$ is determined from the conditions $\eta_x = \tilde{\varphi} - \varphi - (\tilde{d} - d)y$ and $\eta_1 = 0$, so that

$$(8)\quad \eta_i = \sum_{k=1}^{i-1} h\left[(\tilde{\varphi}_k - \varphi_k) - (\tilde{d}_k - d_k)\, y_k\right], \qquad i = 2, 3, \dots, N, \quad \eta_1 = 0\,.$$

To evaluate a solution of problem (4)–(7) on the basis of (6)–(8), we rely on the estimate obtained in Section 3 for (4)–(7), making it possible to establish

$$\| z \|_C \le \frac{2}{c_1} \{ (1, | (\tilde{a} - a) \, y_{\bar{x}} \, | ] + (1, | \eta \, | ] \} .$$

We shall need the estimates for $y$ and $y_{\bar{x}}$ from Section 6 such as

$$\| y \|_C \le \frac{1}{c_1} (1, | \varphi \, |) , \qquad \| y_{\bar{x}} \|_C \le \frac{2}{c_1} (1, | \varphi \, |) .$$

Combination of the inequalities

$$(1, | \eta \, | ] \le (1, | \overset{\circ}{\eta} \, | ] + (1, | \tilde{d} - d \, |) \, \| y \|_C \le (1, | \overset{\circ}{\eta} \, | ] + \frac{1}{c_1} (1, | \tilde{d} - d \, |) \, (1, | \varphi \, |) ,$$

where $\overset{\circ}{\eta}_i = \sum_{k=1}^{i-1} h(\tilde{\varphi}_k - \varphi_k)$, $\overset{\circ}{\eta}_1 = 0$, $i = 2, 3, \ldots, N$, and

$$(1, | (\tilde{a} - a) \, y_{\bar{x}} \, |] \le \frac{2}{c_1} (1, | \varphi \, |) \, (1, | \tilde{a} - a \, | ]$$

gives the estimate

$$(9) \quad \| \tilde{y} - y \|_C \le \frac{2}{c_1} \Big\{ (1, | \overset{\circ}{\eta} \, | ] + \frac{2}{c_1} (1, | \varphi \, |) \, \big( (1, | \tilde{d} - d \, |) + (1, | \tilde{a} - a \, | ] \big) \Big\} ,$$

here $y_i$ is a solution of problem (1) and $\tilde{y}_i$ is a solution of problem (2), provided conditions (3) hold.

Relation (9) can be replaced by a more rough estimate

$$(10) \quad \| \tilde{y} - y \|_C \le \frac{2}{c_1} \Big\{ (1, \| \tilde{\varphi} - \varphi \| + \frac{2}{c_1} (1, | \varphi \, |) \, \big( (1, | \tilde{d} - d \, |) + (1, | \tilde{a} - a \, | ] \big) \Big\} .$$

If

$$(1, | \overset{\circ}{\eta} \, |) = \rho(h) , \qquad (1, | \tilde{a} - a \, | ] = \rho(h) , \qquad (1, | \tilde{d} - d \, |) = \rho(h) ,$$

where $\rho(h) \to 0$ as $h \to 0$, schemes (1) and (2) are said to be **co-equivalent** and for $\rho(h) = O(h^m)$ they are of the $m$th **order of the co-equivalence**. If schemes (1) and (2) are co-equivalent and scheme (1) is convergent, then so is scheme (2). This fact follows immediately from the inequality

$$\| \tilde{y} - u \|_C \le \| \tilde{y} - y \|_C + \| y - u \|_C \longrightarrow 0 \qquad \text{as} \quad h \to 0 .$$

The co-equivalence property of homogeneous schemes lies in the main idea behind a new approach to the further estimation of the order of accuracy of a scheme: on account of (9) or (10) its coefficients $a$, $d$, $\varphi$ should be compared with coefficients $\tilde{a}$, $\tilde{d}$, $\tilde{\varphi}$ of a simple specimen scheme, the accuracy order of which is well-known (see Section 7).

**2. Stability of the first kind operator equations with respect to coefficients.** Here we give the general formulation of the concept of stability with respect to coefficients for the difference scheme in Section 1 by having recourse to an operator equation of the first kind

$$Au = f\,, \qquad f \in H\,, \tag{11}$$

where $A$ is a linear operator acting from Hilbert space $H$ into $H$, $A$: $H \to H$, $f \in H$ is a given vector, $u \in H$ is the unknown vector.

Problem (11) is said to be **well-posed** if there exists a unique solution of equation (11) for any $f \in H$ and this solution continuously depends on the right-hand side $f$, so that

$$\| \tilde{u} - u \|_{(1)} \leq M_0 \| \tilde{f} - f \|_{(2)}\,, \tag{12}$$

where $\tilde{u}$ is a solution of equation (11) with perturbed right-hand side $\tilde{f}$

$$A\tilde{u} = \tilde{f}\,, \tag{13}$$

here $\| \cdot \|_{(1)}$ and $\| \cdot \|_{(2)}$ are some suitable norms on the set $H$.

A case in point that in the statement of problem (11) we must specify not only the right-hand side, but also the operator $A$. If, for instance, $A$ is a differential or difference operator, the coefficients of the equation should be known in advance.

It is natural to require a solution of problem (11) to depend continuously not only on perturbations of the right-hand side, but also on perturbations of the operator $A$ (for instance, on the coefficients of a difference operator). As in the case of difference schemes arising in Section 1, this property of operator equations is to be understood as **stability with respect to coefficients** or **co-stability of an operator equation**.

The stability of a solution to equation (11) with respect to perturbations of the right-hand side $f$ and perturbations of the operator $A$ is called **strong stability**. The problem statement here is as follows: with regard to the equations

$$Au = f\,, \qquad \tilde{A}\tilde{u} = \tilde{f}\,, \tag{14}$$

where $A$ and $\tilde{A}$ are linear operators, whose domains coincide with the entire space $H$, $f$ and $\tilde{f}$ are arbitrary vectors of the space $H$, it is required to estimate the perturbations of a solution

$$z = \tilde{u} - u \tag{15}$$

via perturbations of $f$ and $A$.

Suppose that the inverse operators $A^{-1}$ and $\widetilde{A}^{-1}$ exist. Moreover, we assume that $A$ and $\widetilde{A}$ are self-adjoint positive operators. Substitutions of $u = A^{-1} f$ and $\tilde{u} = \widetilde{A}^{-1} f$ into (15) yield

$$z = \widetilde{A}^{-1} f - A^{-1} f = \widetilde{A}^{-1} (\tilde{f} - f) + (\widetilde{A}^{-1} - A^{-1})\, f\,. \tag{16}$$

Applying $\widetilde{A}^{1/2}$ to both sides of equality (16) we find that

$$\widetilde{A}^{1/2} z = \widetilde{A}^{-1/2} (\tilde{f} - f) + \widetilde{A}^{1/2} (\widetilde{A}^{-1} - A^{-1})\, f\,.$$

We will estimate the vector $z$ in the norm $\| z \|_{\widetilde{A}} = \sqrt{(\widetilde{A} z, z)}$ of the space $H_{\widetilde{A}}$ together with the same things for $f$ and $\tilde{f}$ in the negative norm $\| f \|_{\widetilde{A}^{-1}} = \sqrt{(\widetilde{A}^{-1} f, f)}$ of the energy space $H_{\widetilde{A}^{-1}}$. Via transform

$$\widetilde{A}^{1/2} (\widetilde{A}^{-1} - A^{-1})\, f = (E - \widetilde{A}^{1/2} A^{-1} \widetilde{A}^{1/2}) (\widetilde{A}^{-1/2} f)$$

we obtain

$$\| \widetilde{A}^{1/2} (\widetilde{A}^{-1} - A^{-1})\, f \| \leq \| (E - \widetilde{A}^{1/2} A^{-1} \widetilde{A}^{1/2} \| \cdot \| \widetilde{A}^{-1/2} f \|\,. \tag{17}$$

As a measure of perturbations of the operator $A$ we adopt a relative variation of the energy $(Ax, x)$ of the operator $A$. The meaning of this is that we should have for all $x \in H$

$$\big| ((\widetilde{A} - A)\, x, x) \big| \leq \alpha\, (Ax, x), \qquad \alpha > 0\,, \tag{18}$$

whence it follows that

$$(1 - \alpha)\, A \leq \widetilde{A} \leq (1 + \alpha)\, A\,, \tag{19}$$

$$(1 - \alpha)\, \widetilde{A}^{-1} \leq A^{-1} \leq (1 + \alpha)\, \widetilde{A}^{-1}\,. \tag{20}$$

Observe that (20) is an immediate implication of (19). To make sure of it, we compose the difference $J = (1 + \alpha)(Ax, x) - (\widetilde{A} x, x)$ and insert $A^{1/2} x = y$:

$$J = (1 + \alpha)\, \| y \|^2 - (A^{-1/2} \widetilde{A} A^{-1/2} y, y) = (1 + \alpha)\, \| y \|^2 - (Dy, y)\,,$$

where $D = A^{-1/2}\widetilde{A}A^{-1/2}$, and, after this, pass to $D^{1/2}y = z$:

$$J = (1+\alpha)(D^{-1}z, z) - \|z\|^2$$
$$= (1+\alpha)(A^{1/2}\widetilde{A}^{-1}A^{1/2}z, z) - \|z\|^2$$
$$= (1+\alpha)(\widetilde{A}^{-1}v, v) - (A^{-1}v, v) = A^{1/2}z\,.$$

As far as $J \geq 0$, we have $(1+\alpha)\widetilde{A}^{-1} \geq A^{-1}$, thereby justifying that for any operators $A = A^* > 0$ and $\widetilde{A} = \widetilde{A}^* > 0$ the inequality $A^{-1} \leq (1+\alpha)\widetilde{A}^{-1}$ follows from the inequality $\widetilde{A} \leq (1+\alpha)A$. Moreover, we claim that inequalities (20) are equivalent to

$$(1-\alpha)E \leq C \leq (1+\alpha)E\,, \qquad C = \widetilde{A}^{1/2}A^{-1}\widetilde{A}^{1/2}\,.$$

Indeed,

$$(1+\alpha)(\widetilde{A}^{-1}x, x) - (A^{-1}x, x) = (1+\alpha)\|y\|^2 - (\widetilde{A}^{1/2}A^{-1}\widetilde{A}^{1/2}y, y)$$
$$= (1+\alpha)\|y\|^2 - (Cy, y) \geq 0\,.$$

Thus, (18) implies that $-\alpha E \leq E - C \leq \alpha E$ and $C = \widetilde{A}^{1/2}A^{-1}\widetilde{A}^{1/2}$.

By the definition of the norm of self-adjoint operator,

$$\|E - C\| = \|E - \widetilde{A}^{1/2}A^{-1}\widetilde{A}^{1/2}\| \leq \alpha\,.$$

Substituting this estimate into (17) we deduce from (16) that

$$\|z\|_{\widetilde{A}} \leq \|\tilde{f} - f\|_{\widetilde{A}^{-1}} + \alpha\,\|f\|_{\widetilde{A}^{-1}}$$

or, what amounts to the same,

$$\|\tilde{u} - u\|_{\widetilde{A}} \leq \|\tilde{f} - f\|_{\widetilde{A}^{-1}} + \alpha\,\|f\|_{\widetilde{A}^{-1}}\,.$$

If we are in possession of an operator $A_0 = A_0^*$ of rather simplified structure than the operator $\widetilde{A}$ satisfying the condition $\widetilde{A} \geq c_1 A_0$, $c_1 > 0$, then

$$\widetilde{A}^{-1} \leq \frac{1}{c_1}A_0^{-1}, \qquad \|f\|_{\widetilde{A}^{-1}} \leq \frac{1}{\sqrt{c_1}}\|f\|_{A_0^{-1}}\,,$$

provided the inverse $A_0^{-1}$ exists. Thus, we have proved the comparison theorem.

**Theorem** *Let $u$ be a solution to equation (11) and $\tilde{u}$ be a solution to equation (14), where $A$, $\widetilde{A}$ and $A_0$ are self-adjoint positive operators for which the inverse operators exist. If condition (18) and the inequality $\widetilde{A} \geq c_1 A_0$, $c_1 > 0$ hold, then the estimates are valid:*

$$\| \tilde{u} - u \|_{\widetilde{A}} \leq \| \tilde{f} - f \|_{\widetilde{A}^{-1}} + \alpha \| f \|_{\widetilde{A}^{-1}} , \tag{21}$$

$$\|\tilde{u} - u\|_{A_0} \leq \frac{1}{c_1} \|\tilde{f} - f\|_{A_0^{-1}} + \frac{\alpha}{c_1} \|f\|_{A_0^{-1}} . \tag{22}$$

The first summand on the right-hand side of (21) is the value of perturbations of the right-hand side $f$ and the second one involves the coefficient $\alpha$, which is the value of a relative perturbation of the operator.

**Example** Let $H$ be the set of all grid functions defined on $\bar{\omega}_h = \{x_i = ih,\ 0 \leq i \leq N\}$ and vanishing for $i = 0$ and $i = N$. We refer to the difference operators

$$Ay = -\left(a y_{\bar{x}}\right)_x + dy, \qquad a \geq c_1 > 0, \quad d \geq 0 ,$$

$$\widetilde{A}y = -\left(\tilde{a} y_{\bar{x}}\right)_x + \tilde{d}y, \qquad \tilde{a} \geq c_1 > 0, \quad \tilde{d} \geq 0 ,$$

$$A_0 y = -y_{\bar{x}x} .$$

By introducing the inner product in the usual way and applying Green's formulae we get the inequalities $\widetilde{A} \geq c_1 A_0$ and $A \geq c_1 A_0$. In conformity with Chapter 2, Section 5,

$$\|f\|_{A_0^{-1}} \leq \| f \|_{(-1)} = \left( \sum_{i=1}^{N-1} h \left( \sum_{k=i}^{N-1} h f_k \right)^2 \right)^{1/2} , \qquad f \in H .$$

With these relations established, estimate (22) reduces to

$$\| z_{\bar{x}} \| \leq \frac{1}{c_1} \| \tilde{f} - f \|_{(-1)} + \frac{\alpha}{c_1} \| f \|_{(-1)}$$

or, on account of the inequality $\| z \|_C \leq \frac{1}{2} \| z_{\bar{x}} \|$,

$$\| z \|_C = \| \tilde{y} - y \|_C \leq \frac{1}{2 c_1} \| \tilde{f} - f \|_{(-1)} + \frac{\alpha}{c_1} \| f \|_{(-1)} .$$

In concluding this chapter we clarify the meaning of condition (18) by observing that

$$(1 - \alpha) \left( (a, y_{\bar{x}}^2 ] + (d, y^2) \right) \leq (\tilde{a}, y_{\bar{x}}^2 ] + (\tilde{d}, y^2) \leq (1 + \alpha) \left( (a, y_{\bar{x}}^2 ] + (d, y^2) \right) .$$

From such reasoning it seems clear that the fulfilment of (18) is ensured by the inequalities

$$| \tilde{a} - a | \leq \alpha a , \qquad | d - \tilde{d} | \leq \alpha d .$$

# 4

# Difference Schemes for Elliptic Equations

This chapter is devoted to various difference approximations of second-order elliptic equations. In Sections 1–3 we present results of more a detailed exploration of the Dirichlet difference problem for Poisson's equation. The approximation technique for the Laplace operator and formulations of difference boundary conditions are described for regions of arbitrary shape. The maximum principle (Section 2) and all of its corollaries are established for grid equations of common structure. These tools are aimed at establishing the uniform convergence with the rate $O(|h|^2)$ for the difference scheme constructed in Section 1 for the case of an arbitrary domain. In Section 4 we study the properties of the difference Laplace operator and develop the difference operators corresponding to elliptic operators of general form with variable coefficients. In Section 5 higher-accuracy schemes are designed for Poisson's equation in a rectangle.

## 4.1 THE DIRICHLET DIFFERENCE PROBLEM FOR POISSON'S EQUATION

We now turn to the design of difference schemes for solving the Dirichlet problem in which it is required to find a continuous in $G + \Gamma$ function $u(x)$

solving the Poisson equation

$$(1) \qquad \Delta u = \sum_{\alpha=1}^{p} \frac{\partial^2 u}{\partial x_\alpha^2} = -f(x) , \qquad x \in G ,$$

subject to the boundary condition

$$u|_{\Gamma} = \mu(x) ,$$

where $x = (x_1 , x_2 \dots , x_p)$, $G$ is a $p$-dimensional finite domain with the boundary $\Gamma$.

**1. The difference approximation of the Laplace operator.** We begin by defining a difference analog of the Laplace operator in the plane $x = (x_1 , x_2)$:

$$(2) \qquad \Delta u = L_1 u + L_2 u , \qquad L_\alpha u = \frac{\partial^2 u}{\partial x_\alpha^2} , \qquad \alpha = 1, 2 .$$

In this direction the operator $L_1 u = \frac{\partial^2 u}{\partial x_1^2}$ or $L_2 u = \frac{\partial^2 u}{\partial x_2^2}$ is approximated at a point $x = (x_1 , x_2)$ by the three-point operator $\Lambda_1$ or $\Lambda_2$, respectively:

$$(3) \qquad L_1 v \sim \Lambda_1 v = v_{\bar{x}_1 x_1} = \frac{1}{h_1^2} \Big( v(x_1 + h_1 , x_2) - 2v(x_1 , x_2) + v(x_1 - h_1 , x_2) \Big) ,$$

$$(4) \qquad L_2 v \sim \Lambda_2 v = v_{\bar{x}_2 x_2} = \frac{1}{h_2^2} \Big( v(x_1 , x_2 + h_2) - 2v(x_1 , x_2) + v(x_1 , x_2 - h_2) \Big) ,$$

where approximation is denoted by the symbol $\sim$ and $h_1 > 0$, $h_2 > 0$ are the steps along the axes $x_1$ and $x_2$, respectively.

The operators $\Lambda_1$ and $\Lambda_2$ are specified on the regular three-point patterns

$$(x_1 - h_1 , x_2), \quad (x_1 , x_2) \quad (x_1 + h_1 , x_2)$$

and

$$(x_1 , x_2 - h_2), \quad (x_1 , x_2) \quad (x_1 , x_2 + h_2) ,$$

respectively. Taking into account (3) and (4), we replace the Laplace operator (2) by the difference operator

$$(5) \qquad \Lambda v = \Lambda_1 v + \Lambda_2 v = v_{\bar{x}_1 x_1} + v_{\bar{x}_2 x_2} ,$$

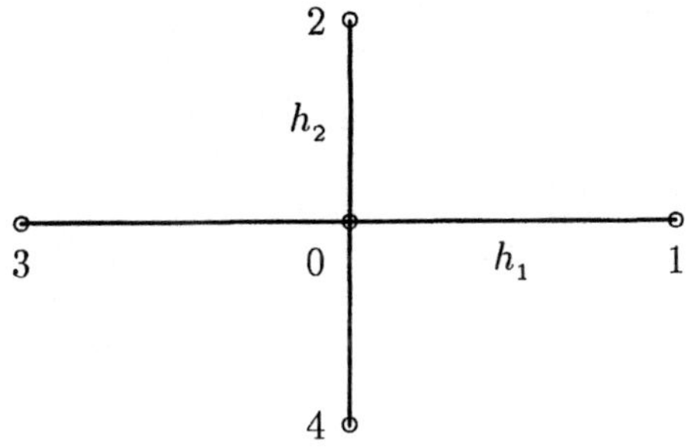


**Figure 6.** The regular "cross"-pattern

which is defined on the five-point pattern consisting of the nodes $(x_1 \pm h_1, x_2)$, $(x_1, x_2)$, $(x_1, x_2 \pm h_2)$. Any such regular pattern is called a "cross" pattern and is depicted in Fig. 6.

Here the symbol O corresponds to the point $(x_1, x_2)$, while the symbol 1 corresponds to the point $(x_1 + h_1, x_2)$, etc.

From formulae (3)–(5) and Fig. 6 it follows that

$$\Lambda v_0 = \frac{1}{h_1^2}(v_1 - 2v_0 + v_3) + \frac{1}{h_2^2}(v_2 - 2v_0 + v_4). \tag{6}$$

In particular, for $h_1 = h_2 = h$ (on any square pattern) we thus have

$$\Lambda v_0 = \frac{1}{h^2}(v_1 + v_2 + v_3 + v_4 - 4v_0). \tag{7}$$

The next step is to calculate the error of approximation of the Laplace operator (2) by the difference operator (5). Since for $\alpha = 1, 2$

$$\Lambda_\alpha v = \frac{\partial^2 v}{\partial x_\alpha^2} + \frac{h_\alpha^2}{12}\frac{\partial^4 v}{\partial x_\alpha^4} + O(h_\alpha^4) = L_\alpha v + \frac{h_\alpha^2}{12} L_\alpha^2 v + O(h_\alpha^4) \tag{8}$$

(see Chapter 2, Section 1), it is plain to derive the expression

$$\Lambda v - \Delta v = \frac{h_1^2}{12} L_1^2 v + \frac{h_2^2}{12} L_2^2 v + O(h_1^4 + h_2^4).$$

This provides support for the view that

$$\Lambda v - \Delta v = O(|h|^2), \qquad |h|^2 = h_1^2 + h_2^2,$$

if $v(x)$ is an arbitrary function having no less than four derivatives with respect to $x_\alpha$, $\alpha = 1, 2$, that are bounded at least in the rectangle $\{x_\alpha - \bar{h}_\alpha \le x'_\alpha \le x_\alpha + \bar{h}_\alpha;\ \alpha = 1, 2\}$ for $h_\alpha \le \bar{h}_\alpha$. So, the Laplace operator (2) is approximated to second order by the difference operator (5) on a regular "cross" pattern. A difference approximation of the $p$-dimensional ($p > 2$) Laplace operator

$$(9) \qquad L\,u = \sum_{\alpha=1}^{p} L_\alpha\, u\,, \qquad L_\alpha\, u = \frac{\partial^2 u}{\partial x_\alpha^2}\,,$$

can be arranged in just the same way. This can be done by replacing $L_\alpha$ by the three-point difference operator $\Lambda_\alpha$ and accepting the decomposition

$$(10) \qquad \Lambda\, v = \sum_{\alpha=1}^{p} \Lambda_\alpha\, v\,, \qquad \Lambda_\alpha\, v = v_{\bar{x}_\alpha x_\alpha}\,,$$

so that

$$(11) \qquad \Lambda_\alpha\, v = v_{\bar{x}_\alpha x_\alpha} = \frac{1}{h_\alpha^2}\left(v^{(+1_\alpha)} - 2\,v + v^{(-1_\alpha)}\right),$$

where $v^{(\pm 1_\alpha)} = v\big(x^{(\pm 1_\alpha)}\big)$. Here $x^{(+1_\alpha)}$ (or $x^{(-1_\alpha)}$) is a point into which the point $x = (x_1, \ldots, x_p)$ moves after the shift by one interval $h_\alpha$ along the direction $x_\alpha$ to the right (or to the left) (see Fig. 7).

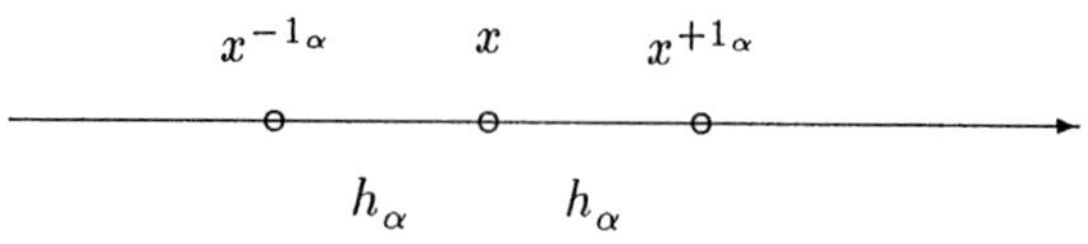


**Figure 7.**

Evidently, the pattern for operator (10) consists of $2p+1$ points: $x$, $x^{(\pm 1_\alpha)}$, $\alpha = 1, \ldots, p$ (7 points in the case $p = 3$) and the approximation here is of order 2.

**2. Approximation of the Laplace operator on an irregular "cross" pattern.** We now consider a difference approximation of the Laplace operator on an irregular "cross" pattern. In the two-dimensional case ($p = 2$) such a pattern consists of the five points

$$(x_1 - h_{1-}\,, x_2), \quad (x_1 + h_{1+}\,, x_2), \quad (x_1\,, x_2), \quad (x_1\,, x_2 - h_{2-}), \quad (x_1\,, x_2 + h_{2+}),$$

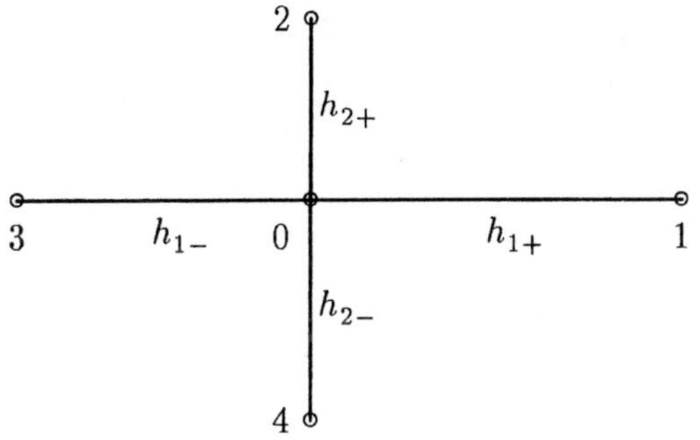


**Figure 8.** The irregular "cross" pattern

where $h_{1\pm} > 0$, $h_{2\pm} > 0$ and $h_{\alpha+} \neq h_{\alpha-}$ at least for one value of $\alpha$ (Fig. 8).

We approximate either of the operators $L_1$ and $L_2$ at three points

$$(x_1 - h_{1-}\,, x_2),\quad (x_1 + h_{1+}\,, x_2),\quad (x_1\,, x_2) \qquad \text{(the points } 3, 1, 0),$$

$$(x_1\,, x_2 - h_{2-}),\quad (x_1\,, x_2 + h_{2+}),\quad (x_1\,, x_2) \qquad \text{(the points } 4, 2, 0),$$

respectively. These approximations can be arranged via transformations (Chapter 2, Section 1) such as

$$(12)\qquad \begin{cases} L_1\, v \sim \Lambda_1^*\, v = \dfrac{1}{\hbar_1}\left[\dfrac{v\,(x_1 + h_{1+}\,, x_2) - v\,(x_1\,, x_2)}{h_{1+}} - \dfrac{v\,(x_1\,, x_2) - v\,(x_1 - h_{1-}\,, x_2)}{h_{1-}}\right], \\[2ex] L_2\, v \sim \Lambda_2^*\, v = \dfrac{1}{\hbar_2}\left[\dfrac{v\,(x_1\,, x_2 + h_{2+}) - v\,(x_1\,, x_2)}{h_{2+}} - \dfrac{v\,(x_1\,, x_2) - v\,(x_1\,, x_2 - h_{2-})}{h_{2-}}\right], \end{cases}$$

where $\hbar_\alpha = \frac{1}{2}(h_{\alpha-} + h_{\alpha+})$ for $\alpha = 1, 2$. On any irregular pattern the difference Laplace operator takes the form

$$(13)\qquad \Lambda^* v = \Lambda_1^* v + \Lambda_2^* v = v_{\bar{x}_1 \hat{x}_1} + v_{\bar{x}_2 \hat{x}_2}\,.$$

If, for instance, $h_{1-} = h_{1+} = h_1$, then $\Lambda_1^* v = \Lambda_1 v = v_{\bar{x}_1 \hat{x}_1}$, etc. The procedure of writing out the arguments is somewhat cumbersome. This is especially true for the case $p > 2$. To make our exposition more transparent, it will be sensible to introduce the notations

$$x^{(+1_1)} = (x_1 + h_{1+}, x_2), \qquad x^{(-1_1)} = (x_1 - h_{1-}, x_2),$$

$$x^{(\pm 1_2)} = (x_1, x_2 \pm h_{2\pm}),$$

$$v^{(+1_\alpha)} = v\left(x^{(+1_\alpha)}\right), \quad v = v(x), \quad v^{(-1_\alpha)} = v\left(x^{(-1_\alpha)}\right), \quad \alpha = 1, 2.$$

The disposition of the points $x$ and $x^{(\pm 1_\alpha)}$ is shown in Fig. 9.

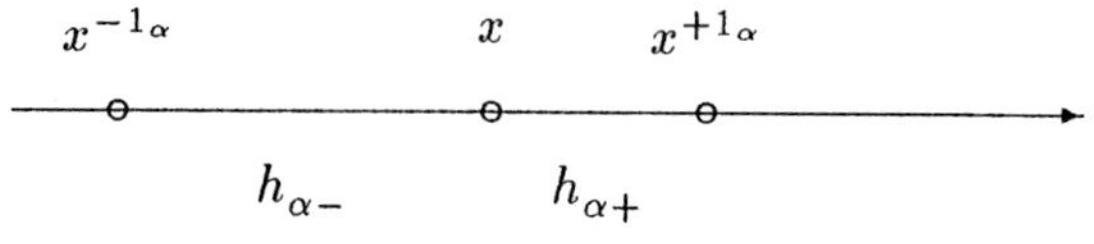


**Figure 9.**

The expression for $\Lambda_\alpha^*$ can be rewritten as

$$\text{(14)} \qquad \Lambda_\alpha^* v = v_{\bar{x}_\alpha \hat{x}_\alpha} = \frac{1}{\hbar_\alpha} \left[ \frac{v^{(+1_\alpha)} - v}{h_{\alpha+}} - \frac{v - v^{(-1_\alpha)}}{h_{\alpha-}} \right],$$

$$\hbar_\alpha = \tfrac{1}{2} (h_{\alpha-} + h_{\alpha+}), \qquad \alpha = 1, 2.$$

In Chapter 2, Section 1 (see formula (27)) we have approved the expression for $v_{\bar{x}\hat{x}} - v''$, whose use permits us to establish

$$\text{(15)} \qquad \Lambda_\alpha^* v - L_\alpha v = \frac{1}{3} \left( h_{\alpha+} - h_{\alpha-} \right) \frac{\partial^3 v}{\partial x_\alpha^3} + O(\hbar_\alpha^2).$$

Thus, on any irregular pattern the Laplace operator is approximated to first order by the difference operator $\Lambda^*$ specified by formula (13).

The approximation of the type (12) applies equally well to a non-equidistant grid and at the near-boundary nodes in the case of an arbitrary domain. We follow the second way of approximating the Laplace operator on an irregular pattern (see Fig. 8). The expression

$$(16)\quad \Lambda_\alpha^* v = \frac{1}{h_\alpha}\left[\frac{v^{(+1_\alpha)} - v}{h_{\alpha+}} - \frac{v - v^{(-1_\alpha)}}{h_{\alpha-}}\right], \qquad h_\alpha = \max(h_{\alpha-}, h_{\alpha+}),$$

reveals $\Lambda_\alpha^*$ instead of formula (14) as

$$\Lambda_\alpha^* v = \frac{\hbar_\alpha}{h_\alpha}\, v_{\bar{x}_\alpha \hat{x}_\alpha} .$$

We claim that in this case the operator $\Lambda_\alpha^*$ generates the local approximation of zero order

$$\psi_\alpha = \Lambda_\alpha^* u - \Lambda_\alpha u = O(1) .$$

Indeed, taking into account (15) we arrive at the chain of the relations

$$\psi_\alpha = \Lambda_\alpha^* u - \Lambda_\alpha u = -\left(1 - \frac{\hbar_\alpha}{h_\alpha}\right) L_\alpha u$$

$$+ \frac{\hbar_\alpha}{3\, h_\alpha}\,(h_{\alpha+} - h_{\alpha-})\,\frac{\partial^3 u}{\partial x_\alpha^3} + O(\hbar_\alpha^2)$$

$$= \pm \frac{h_{\alpha+} - h_{\alpha-}}{2\, h_\alpha}\; L_\alpha u + O(\hbar_\alpha) = O(1) .$$

We clarify the situation, in which approximation (16) is quite applicable, on the basis of one possible example.

**Example** For the boundary-value problem

$$u'' = -f(x), \qquad 0 < x < 1, \quad u(0) = 0, \quad u(1) = 0,$$

we form the grid

$$\hat{\omega}_h = \{x_i,\ x_1 = h_1,\ x_{i+1} = x_i + h,\ i = 1, 2, \dots, N-1,\ x_{N+1} = x_N + h_2\},$$

which is everywhere equidistant, but near the boundary $h_1 < h$, $h_2 < h$, $h_1 + h_2 + (N-1)h = 1$. At the regular nodes $x_i$, $1 < i < N$, we find that

$$u_i'' \sim u_{\bar{x}x,i} = \frac{u_{i+1} - 2\, u_i + u_{i-1}}{h^2} .$$

By the same token,

$$u_1'' \sim \Lambda^* u_1 = \frac{1}{h}\left[\frac{u_2-u_1}{h} - \frac{u_1-u_0}{h_1}\right],$$

$$u_N'' \sim \Lambda^* u_N = \frac{1}{h}\left[\frac{u_{N+1}-u_N}{h_2} - \frac{u_N-u_{N-1}}{h}\right].$$

With these relations established, we arrive at the difference scheme

$$(17)\qquad \begin{aligned} &y_{\bar{x}x} = -f(x), && x_i = h_1 + (i-1)\,h, && 1 < i < N, \\ &\Lambda^* y_1 = -f(x_1), && \Lambda^* y_N = -f(x_N), && y_0 = y_{N+1} = 0, \end{aligned}$$

which will be convenient to be expressed with respect to the error $z = y-u$:

$$(18)\qquad \Lambda\, z = -\psi(x), \qquad 0 < x < 1, \qquad z_0 = z_{N+1} = 0,$$

where $\Lambda\, z = z_{\bar{x}x}$ for $x_1 < x_i < x_N$, $\Lambda\, z_1 = \Lambda^* z_1$ and $\Lambda z_N = \Lambda^* z_N$, $\psi_i = O(h^2)$ for $i = 2, 3, \ldots, N-1$, $\psi_i = O(1)$ for $i = 1, N$.

In spite of the fact that this scheme generates no approximation at the near-boundary nodes $i = 1$ and $i = N$, scheme (17) is of second-order accuracy in the space $C$: $\| z \|_C = O(h^2)$. In order to obtain this estimate at the points $x = x_1, x_N$, we rewrite equation (18) as

$$\frac{1}{h}\left(\frac{z_2-z_1}{h} - \frac{z_1-\bar{z}_0}{h_1}\right) = 0, \qquad \frac{1}{h}\left(\frac{\bar{z}_{N+1}-z_N}{h_2} - \frac{z_N-z_{N-1}}{h}\right) = 0,$$

where $\bar{z}_0 = h\,h_1\psi_1$ and $\bar{z}_{N+1} = h\,h_2\psi_N$. Thus, problem (18) is equivalent to the following one:

$$z_{\bar{x}x} = -\psi(x), \qquad x_1 < x < x_N, \quad \Lambda^* z_1 = 0, \quad \Lambda^* z_N = 0,$$

$$z_0 = h\,h_1\psi_1, \qquad z_{N+1} = h\,h_2\psi_N.$$

On account of the a priori estimate

$$\| z \|_C \le \max\left(| z_0 |, | z_{N+1} |\right) + \sum_{i=1}^{N} h \sum_{k=1}^{i} h\, | \psi_k |,$$

arising in Section 2 of Chapter 1, it is easily verified that

$$\| z \|_C = \| y-u \|_C \le h\,h_1\, | \psi_1 | + h\,h_2\, | \psi_N | + \max_{1<i<N} | \psi_i | \le M\,h^2.$$

This means that scheme (17) provides an approximation of order 2.

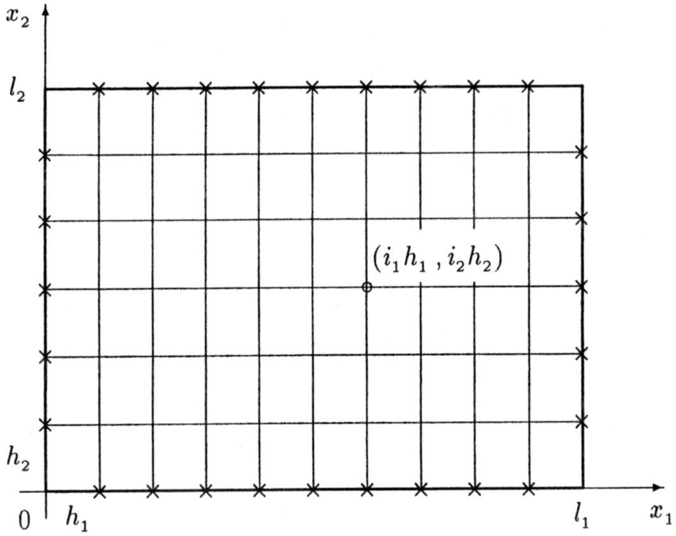


**Figure 10.**

**3. The Dirichlet difference problem in a rectangle.** Let now

$$\bar{G}_0 = (0 \le x_1 \le l_1 \,,\ 0 \le x_2 \le l_2)$$

be a rectangle of sides $l_1$, $l_2$ (see Fig.10) with boundary $\Gamma$.

Of our initial concern is the Dirichlet problem in the rectangle $\bar{G}_0 = \bar{G}_0 + \Gamma$ for the Poisson equation

$$(1') \qquad \Delta u = -f(x), \qquad x = (x_1, x_2) \in G_0 \,, \qquad u|_\Gamma = \mu(x) \,.$$

In order to form in $\bar{G}_0$ the grid $\bar{\omega}_h$ with steps $h_1 = l_1/N_1$ and $h_2 = l_2/N_2$, where $N_1$ and $N_2$ are positive integers, we draw up two families of straight lines such as

$$x_1^{(i_1)} = i_1 h_1, \quad i_1 = 0, 1, \dots, N_1, \qquad x_2^{(i_2)} = i_2 h_2, \quad i_2 = 0, 1, \dots, N_2 \,.$$

We call the points of intersection $x = (i_1h_1, i_2h_2)$ of those straight lines with the coordinates $i_1h_1$ and $i_2h_2$ **nodes**. If a nodal point $x = (i_1h_1, i_2h_2)$ is inside the rectangle, that is, $0 < i_1 < N_1$, $0 < i_2 < N_2$, it falls within the collection of **inner nodes**. Let $\omega_h$ be the set of all inner nodes. The total number of inner nodes is equal to $(N_1 - 1)(N_2 - 1)$.

The nodes on its boundary (for $i_1 = 0, N_1$ or $i_2 = 0, N_2$) except for four points $(0,0)$, $(0,l_2)$, $(l_1, 0)$, $(l_1, l_2)$ are called **boundary nodes** and are labeled crosses in Fig.10. They constitute the set $\gamma_h = \{(i_1 h_1, i_2 h_2)\}$. The set of all inner and boundary nodes is known as the **grid** $\bar{\omega}_h = \omega_h + \gamma_h$ in a rectangle $\bar{G}_0$.

Following established practice, at each of the inner nodes $x \in \omega_h$ we compose a five-point regular "cross" pattern, whose nodes $x^{\pm 1_\alpha}$, $\alpha = 1, 2$, belong to $\bar{\omega}_h$, that is, either to $\omega_h$ or to $\gamma_h$. For this reason at all inner nodes the Laplace operator $\Delta u$ can be replaced by the difference operator

$$\Lambda u = u_{\bar{x}_1 x_1} + u_{\bar{x}_2 x_2} .$$

In this view, it seems reasonable to approximate the right-hand side $-f(x)$ of equation (1′) by the grid function $-\varphi(x)$ so as to achieve the error $\varphi(x) - f(x) = O(|\, h^2 \,|)$, $f(x) \in C^{(2)}$. Assuming the function $f(x)$ to be continuous, in what follows we accept $\varphi(x) = f(x)$.

As a final result problem (1′) is associated with the Dirichlet difference problem relating to the determination of a grid function $y(x)$ defined on the grid $\bar{\omega}_h$, satisfying at the inner nodes, that is, on $\omega_h$ the equation

$$(19) \qquad \Lambda y = -f(x)\,, \qquad \Lambda y = y_{\bar{x}_1 x_1} + y_{\bar{x}_2 x_2}\,, \qquad x \in \omega_h\,,$$

and taking the assigned values on the boundary $\gamma_h$

$$(20) \qquad y(x) = \mu(x), \qquad x \in \gamma_h\,.$$

It is worth noting here that the grid $\bar{\omega}_h(\bar{G}_0)$ becomes **rectangular** for $h_1 \neq h_2$. In the case where $h_1 = h_2 = h$ it refers to a **square grid**. More a detailed expression for $\Lambda y$ on any square grid is of the form

$$\Lambda y = \frac{1}{h^2} \left( y^{(+1_1)} + y^{(-1_1)} + y^{(+1_2)} + y^{(-1_2)} - 4\, y \right).$$

Let $\varphi = 0$. The equation $\Lambda y = 0$ can be solved with respect to $y$:

$$y = \frac{1}{4} \left( y^{(+1_1)} + y^{(-1_1)} + y^{(+1_2)} + y^{(-1_2)} \right).$$

The value of $y$ at the center of the pattern is the arithmetic mean of the values of $y$ at the remaining four nodes of the pattern. This formula gives a difference analog of the formula for the mean value of a harmonic function.

It is easily seen from (19) and (20) that no values of $\mu(x)$ at the vertices of the rectangle appear in this matter. This feature has some influence on a proper choice of $\gamma_h$. For the third boundary-value problem and the scheme of accuracy $O(|h|^4)$ (see Section 5) the boundary $\gamma_h$ consists of all the nodes on the boundary of the rectangle including its vertices.

To evaluate the accuracy of the difference scheme (19)–(20), we pass to the difference $z = y - u$, where $y$ is a solution of problem (19)–(20) and $u$ is a solution of problem (1′). Substituting $y = z + u$ into (1′) we set up the problem for the error $z$

$$\text{(21)} \qquad \Lambda z = -\psi, \quad x \in \omega_h, \qquad z = 0, \quad x \in \gamma_h,$$

where $\psi = \Lambda u + f$ is the error of approximation of equation (1′) by scheme (19). Since $Lu + f = 0$, we have

$$\psi = \Lambda u + f - Lu + Lu = \Lambda u - Lu,$$

yielding $\psi = \Lambda u - Lu$. From (8) it follows that

$$\psi = \frac{h_1^2}{12}\,\overline{\frac{\partial^4 u}{\partial x_1^4}} + \frac{h_2^2}{12}\,\overline{\frac{\partial^4 u}{\partial x_2^4}} \qquad \text{for} \quad u \in C^{(4)},$$

where the symbol over-bar designates that the values of the arguments are taken at some intermediate points of the intervals $(x_1 - h_1, x_2)$, $(x_1 + h_1, x_2)$ and $(x_1, x_2 - h_2)$, $(x_1, x_2 + h_2)$, respectively.

Within the notation $M_4 = \max\limits_{\bar{G},\alpha} \left| \dfrac{\partial^4 u}{\partial x_\alpha^4} \right|$, we get

$$|\psi| \le M_4 \frac{|h|^2}{12}.$$

The proof of convergence of scheme (19) reduces to the estimation of a solution of problem (21) in terms of the approximation error. In the sequel we obtain such estimates using the maximum principle for domains of arbitrary shape and dimension. In an attempt to fill that gap, a non-equidistant grid

$$\hat{\bar{\omega}}_h = \left\{ x_i = \left(x_1^{(i_1)}, x_2^{(i_2)}\right),\ i_\alpha = 0, 1, \ldots, N_\alpha,\ x_\alpha^{(0)} = 0,\ x_\alpha^{(N_\alpha)} = l_\alpha,\ \alpha = 1, 2 \right\}$$

with steps $h_1^{(i_1)} = x_1^{(i_1)} - x_1^{(i_1-1)}$ and $h_2^{(i_2)} = x_2^{(i_2)} - x_2^{(i_2-1)}$ can be introduced in the rectangle in the usual way for later use of the difference operator (13). Thus emerged instead of (19)–(20) the problem

$$\text{(22)} \quad \Lambda y = -f(x), \qquad \Lambda y = y_{\bar{x}_1 \hat{x}_1} + y_{\bar{x}_2 \hat{x}_2}, \qquad x \in \hat{\omega}_h, \quad y\big|_{\gamma_h} = \mu(x).$$

This scheme provides the local approximation of order 1 since

$$\begin{aligned}\psi_i &= \left(\Lambda u + f(x)\right)_i = \left(u_{\bar{x}_1 \hat{x}_1} + u_{\bar{x}_2 \hat{x}_2} + f(x)\right)_i \\ &= \frac{1}{3} \sum_{\alpha=1}^{2} \left(h_\alpha^{(i_\alpha+1)} - h_\alpha^{(i_\alpha)}\right) \frac{\partial^3 u}{\partial x_\alpha^3} + O(|\hbar|^2) \\ &= O(|\hbar|), \quad |\hbar|^2 = \hbar_1^2 + \hbar_2^2 .\end{aligned}$$

However, by analogy with Chapter 3, Section 4, $\psi_i$ is representable by

$$\psi_i = \sum_{\alpha=1}^{2} \left(\eta_\alpha\right)_{\hat{x},i} + O(|\hbar|^2), \qquad \eta_\alpha = \frac{h_\alpha^2}{6} \frac{\partial^3 u}{\partial x_\alpha^3} = O(h_\alpha^2) .$$

From such reasoning it seems clear that scheme (22) provides an approximation of second total order in the negative norm.

**4. The Dirichlet difference problem in a domain of rather complicated configuration.** If a solution of the Dirichlet problem needs to be determined in a domain $\bar{G}$ with a nonlinear boundary, the grid $\bar{\omega}_h(\bar{G})$ is, generally speaking, non-equidistant near the boundary. We describe below such a grid and give the possible classification of its nodes.

Consider an arbitrary finite domain $G$ with the boundary $\Gamma$ in a $p$-dimensional space, where a point with coordinates $x_1, x_2, \ldots, x_p$ is denoted by $x = (x_1, x_2, \ldots, x_p)$. We proceed to construct a grid in the domain $\bar{G} = G + \Gamma$ and confine ourselves, for the sake of simplicity, to the case of a two-dimensional domain ($p = 2$). Here a constructive supposition about the shape of the domain $\bar{G}$ is taken into account that the intersection of the domain $G$ with any straight line passing through an inner point $x \in G$ in parallel to the coordinate axis $Ox_\alpha$ ($\alpha = 1, 2$) consists only of a finite number of intervals.

If the origin of coordinates is inside the domain $G$, we draw up two families of uniform straight lines

$$x_1^{(i_1)} = i_1 h_1 , \quad i_1 = 0, \pm 1, \pm 2, \ldots , \qquad x_2^{(i_2)} = i_2 h_2 , \quad i_2 = 0, \pm 1, \pm 2 \ldots ,$$

where $h_1 > 0$ and $h_2 > 0$ are fixed numbers. Such straight lines split up the plane $(x_1, x_2)$ into rectangles of sides $h_1$ and $h_2$. The vertices of these rectangles with coordinates $x_1 = i_1 h_1$ and $x_2 = i_2 h_2$ are called **nodes** and the set of all nodes is known as a **grid** in the plane $(x_1, x_2)$. The nodes $x_i = (i_1 h_1, i_2 h_2)$ lying inside the domain $G$ refer to **inner nodes** with the notation $\omega_h = \{x_i \in G\}$ for the set of all such nodes. The points of the intersection of the straight lines $x_\alpha^{(i_\alpha)} = i_\alpha h_\alpha$, $\alpha = 1, 2$ with the boundary $\Gamma$ of the domain $G$ are called **boundary nodes along the direction $x_\alpha$**. We denote by $\gamma_{h,\alpha}$ the set of all boundary nodes along the direction $x_\alpha$. Let $\gamma_h = \gamma_{h,1} + \gamma_{h,2}$ be the set of all **boundary nodes**. It represents such nodes that are boundary at least with respect to one direction $x_\alpha$. The set of all inner and boundary nodes is known as a **grid** $\bar{\omega}_h = \omega_h + \gamma_h$ in the domain $G$ (see Fig. 11).

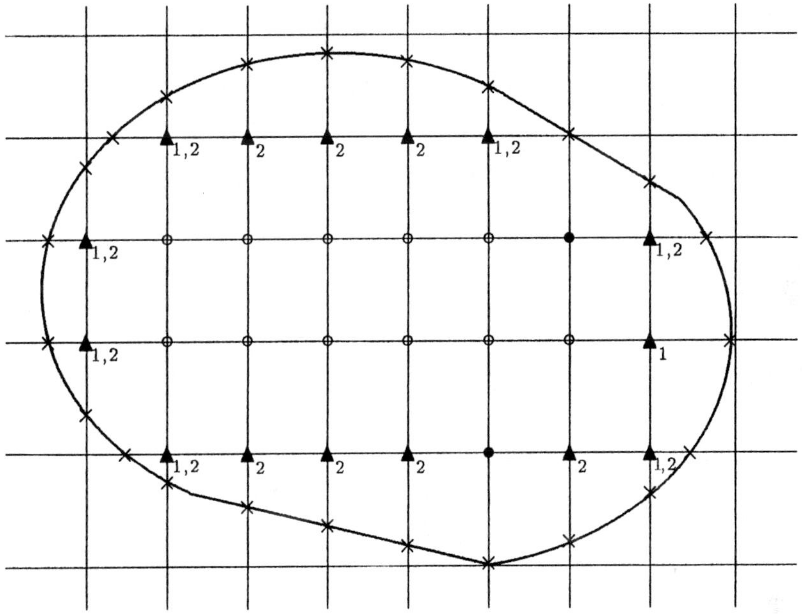


**Figure 11.**

We offer below more a detailed classification of inner nodes. With this aim, let us draw up a straight line parallel to the axis $Ox_\alpha$ through an inner node $x \in \omega_h$. Its intersection with the domain $G$ is an interval (or several intervals), whose ends are boundary nodes along the direction $Ox_\alpha$. The nodes from this interval can be distinguished by the approved rules. We call the nearest to the end of the interval node a **near-boundary node along the direction $Ox_\alpha$** (with respect to $x_\alpha$). If the distance to the boundary $\gamma_h$ is $h_\alpha^* \neq h_\alpha$, any such node is said to be **irregular with respect to $x_\alpha$**. Let

$\omega^*_{h,\alpha}$ be the set of all near-boundary nodes with respect to $x_\alpha$ and $\omega^{**}_{h,\alpha}$ be the set of those near-boundary nodes, which are irregular in the direction $x_\alpha$. Obviously, $\omega^{**}_{h,\alpha} \subseteq \omega^*_{h,\alpha}$. The notation $\omega^*_h$ stands for the set of all **near-boundary nodes** (that is, near-boundary at least in one direction) and $\omega^{**}_h$ stands for the set of all **irregular nodes** (that is, irregular at least in one of the directions $x_1$ or $x_2$). Let $\overset{\circ}{\omega}_h$ be the complement of $\omega^*_h$ to $\omega_h$: $\omega_h = \omega^*_h + \overset{\circ}{\omega}_h$. All the nodes belonging to $\overset{\circ}{\omega}_h$ are called **strictly inner nodes**. The symbol $\overset{\circ}{\omega}_{h,\alpha}$ is in common usage for all strictly inner nodes with respect to $x_\alpha$ (that is, the nodes adjacent to the node $x \in \overset{\circ}{\omega}_{h,\alpha}$ in the direction $Ox_\alpha$ are inner ones).

In Fig. 11 the aforementioned nodes $\overset{\circ}{\omega}_h$ are depicted by the signs $\circ$, the nodes being irregular only with respect to $x_\alpha$ ($\alpha = 1, 2$) — by the signs $\blacktriangle_\alpha$, the nodes being irregular both with respect to $x_1$ and with respect to $x_2$ — by the signs $\blacktriangle_{1,2}$ and, finally, near-boundary nodes regular both with respect to $x_1$ and with respect to $x_2$ — by the signs $\bullet$.

The grid $\bar{\omega}_h$ in view is supposed to be **connected**, it being understood that any two inner nodes can be joined by a polygonal line, the parts of which are parallel to the coordinate axes and vertices coincide with inner nodes of the grid. Then at least one of the four nodes $x^{(\pm\alpha)}$, $\alpha = 1, 2$, of the five-point pattern $(x^{(\pm 1_1)}, x, x^{(\pm 1_2)})$ (regular or irregular) falls within the collection of inner nodes. The assumption on connectedness of the grid entails some limitations both on the choice of spacings $h_1$ and $h_2$ and on the shape of the domain and its position with respect to the grid $\omega_h$ for fixed $h_1$ and $h_2$.

Examples of disconnected and connected grid domains are shown in Fig. 12.a and 12.b, respectively. The assumption on connectedness for a domain with a narrow bridge will be satisfied if we make the step $h_\alpha$ small enough or refine the grid in this part of the domain. Fig. 12.b illustrates the case where the connectedness of the grid is stipulated by the proper choice of its step $h_1$ rather than by successive grid refinements.

The procedure of constructing a grid in the plane domain we have described above can easily be generalized to the case of an arbitrary $p$-dimensional domain. A grid so constructed is a result of the intersection of hyperplanes (planes for $p = 3$ or straight lines for $p = 2$)

$$x_\alpha^{i_\alpha} = i_\alpha h_\alpha , \qquad i_\alpha = 0, \pm 1, \dots , \qquad \alpha = 1, 2, \dots , p ,$$

where $h_\alpha > 0$. The preceding classification of nodes remains unchanged here.

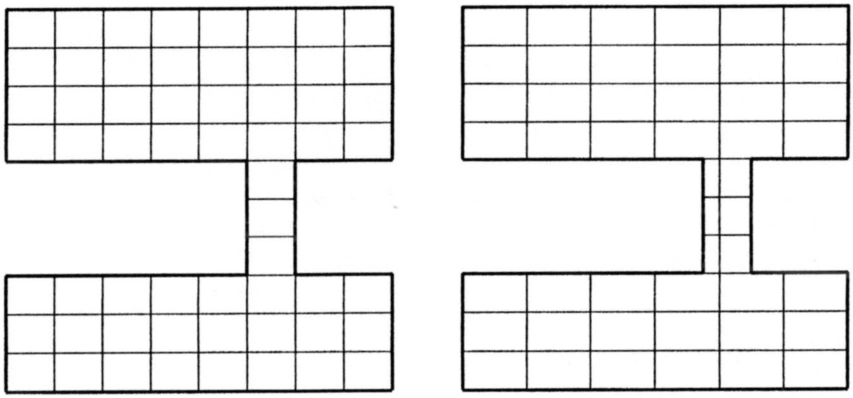


**Figure 12.** a) A disconnected grid. b) A connected grid.

Our purpose here is to construct a difference scheme for solving the Dirichlet problem in the domain $\bar{G} = G + \Gamma$, the complete posing of which is to find an unknown solution to the equation

$$\Delta u = \frac{\partial^2 u}{\partial x_1^2} + \frac{\partial^2 u}{\partial x_2^2} = -f(x), \qquad x = (x_1, x_2) \in G,$$

which is continuous in the closed domain $\bar{G} = G + \Gamma$ and satisfies the boundary condition $u|_{\Gamma} = \mu(x)$.

At each of the inner nodes $x \in \omega_h$ we approximate the difference operator

$$L_\alpha u = \frac{\partial^2 u}{\partial x_\alpha^2}$$

by the three-point difference operator $\Lambda_\alpha$.

If a node $x \in \omega_h$ is regular with respect to $x_\alpha$, then the difference operator $\Lambda_\alpha$ on the regular pattern $\left(x^{(-1_\alpha)}, x, x^{(+1_\alpha)}\right)$ is similar to (11):

$$\Lambda_\alpha y = y_{\bar{x}_\alpha x_\alpha} = \frac{y^{(+1_\alpha)} - 2\,y + y^{(-1_\alpha)}}{h_\alpha^2}\,.$$

But if a node $x \in \omega_{h,\alpha}^{**}$, that is, a node is irregular with respect to $x_\alpha$ on the irregular pattern, then the difference operator $\Lambda_\alpha$ can be rewritten as

$$\text{(23a)} \qquad \Lambda_\alpha^* y = \frac{1}{h_\alpha}\left(\frac{y^{(+1_\alpha)} - y}{h_\alpha} - \frac{y - y^{(-1_\alpha)}}{h_\alpha^*}\right) \quad \text{for} \quad x^{(-1_\alpha)} \in \gamma_{h,\alpha}\,,$$

where $h^*_\alpha$ is the distance between the nodes $x$ and $x^{(-1_\alpha)}$, or

$$(23b)\qquad \Lambda^*_\alpha\, y = \frac{1}{h_\alpha}\left(\frac{y^{(+1_\alpha)}-y}{h^*_\alpha} - \frac{y-y^{(-1_\alpha)}}{h_\alpha}\right) \quad \text{for}\quad x^{(+1_\alpha)} \in \gamma_{h,\alpha}\,,$$

where $h^*_\alpha$ is the distance between the nodes $x$ and $x^{(+1_\alpha)}$.

In this context, it is worth noting that the reader can encounter the case when $x^{(-1_\alpha)} \in \gamma_{h,\alpha}$ and $x^{(+1_\alpha)} \in \gamma_{h,\alpha}$. If this happens, it is recommended to refer to the difference operator

$$(23c)\qquad \Lambda^*_\alpha\, y = \frac{1}{h_\alpha}\left(\frac{y^{(+1_\alpha)}-y}{h^*_{\alpha+}} - \frac{y-y^{(-1_\alpha)}}{h_{\alpha-}}\right) \quad \text{for}\quad x^{(\pm 1_\alpha)} \in \gamma_{h,\alpha}\,,$$

where $h^*_{\alpha\pm} \neq h_\alpha$ is the distance between $x$ and $x^{(\pm 1_\alpha)}$.

Typical situations corresponding to the forms (23a)–(23c) for $p = 2$ are shown in Fig. 13. Using one of the formulae (11) and (23) for approximating $L_\alpha u = \frac{\partial^2 u}{\partial x_\alpha^2}$ by the difference operator, we get instead of (1) the difference equation $\Lambda y + \varphi(x) = 0$ for all $x \in \omega_h$, where $\Lambda = \sum_{\alpha=1}^{p} \Lambda_\alpha$. Here the exact value $y|_{\gamma_h} = \mu(x)$ is taken on the grid boundary $\gamma_h$.

Finally, we arrive at the **Dirichlet difference problem** of determining a grid function $y(x)$ defined for $x \in \bar\omega_h = \omega_h + \gamma_h$, satisfying at the inner nodes the equation

$$(24)\qquad \Lambda\, y + \varphi(x) = 0 \quad \text{at the regular nodes,}$$

$$(25)\qquad \Lambda^* y + \varphi(x) = 0 \quad \text{at the irregular nodes,}$$

and taking the assigned values at the boundary nodes $x \in \gamma_h$:

$$(26)\qquad y = \mu(x), \qquad x \in \gamma_h\,.$$

By analogy with Section 3 we formulate conditions for the accuracy of a scheme under the agreement that $y(x)$ is a solution of the difference problem (24)–(26) and $u = u(x)$ is a solution of the original problem (1). Substitution of $y = z + u$ into (24)–(26) yields

$$(27)\qquad \begin{aligned} \Lambda z &= -\psi && \text{at the regular nodes,}\\ \Lambda^* z &= -\psi^* && \text{at the irregular nodes,}\\ z &= 0 && \text{on}\quad \gamma_h\,, \end{aligned}$$

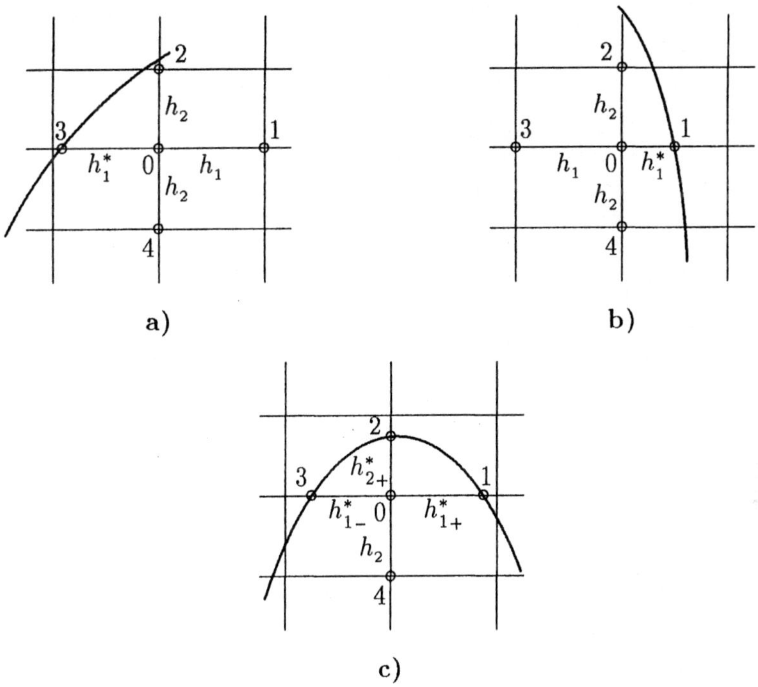


**Figure 13.**

$$\text{a)}\quad \Lambda_1^* y = \frac{1}{\hbar_1}\left(\frac{y_1 - y_0}{h_1} - \frac{y_0 - y_3}{h_1^*}\right), \quad \Lambda_2 y = y_{\bar{x}_2 x_2}, \quad \Lambda^* = \Lambda_1^* + \Lambda_2 .$$

$$\text{b)}\quad \Lambda_1^* y = \frac{1}{\hbar_1}\left(\frac{y_1 - y_0}{h_1^*} - \frac{y_0 - y_3}{h_1}\right), \quad \Lambda_2 y = y_{\bar{x}_2 x_2}, \quad \Lambda^* = \Lambda_1^* + \Lambda_2 .$$

$$\text{c)}\quad \Lambda_1^* y = \frac{1}{\hbar_1}\left(\frac{y_1 - y_0}{h_{1+}^*} - \frac{y_0 - y_3}{h_{1-}^*}\right), \quad \Lambda_2^* y = \frac{1}{\hbar_2}\left(\frac{y_2 - y_0}{h_2^*} - \frac{y_0 - y_4}{h_2}\right),$$

$$\Lambda^* = \Lambda_1^* + \Lambda_2^* .$$

where $\psi$ is the approximation error equal for $\varphi(x) = f(x)$ to

$$
(28) \qquad \begin{aligned} &\psi = \Lambda\, u + \varphi = \Lambda\, u - L\, u && \text{at the regular nodes,} \\ &\psi^* = \Lambda^* u - L\, u && \text{at the irregular nodes.} \end{aligned}
$$

Let $u \in C^{(4)}(\bar{G})$, where $C^{(4)}$ is the class of functions $u(x)$ with four continuous in $\bar{G}$ derivatives with respect to $x_1, \ldots, x_p$. As stated in Section 3, we have at the regular nodes

$$
(29) \qquad |\,\psi\,| \le M_4 \,\frac{|\,h\,|^2}{12}\,, \qquad |\,h\,|^2 = h_1^2 + h_2^2 + \cdots + h_p^2\,.
$$

Furthermore, in giving the approximation error at the irregular nodes as a sum

$$
(30) \qquad \psi^* = \sum_{\alpha=1}^{p} \psi^*_\alpha\,, \qquad \psi^*_\alpha = \Lambda_\alpha\, u - L_\alpha\, u\,,
$$

we apply the results obtained in Section 2 to the current situation:

$$
(31) \qquad \psi^*_\alpha = \frac{h_{\alpha+} - h_{\alpha-}}{2\,h_\alpha}\,\frac{\partial^2 u}{\partial x_\alpha^2} + O(h_\alpha) = O(1)\,, \qquad \psi^* = O(1)\,,
$$

meaning that at the irregular nodes the scheme does not approximate the equation $\Delta u + f(x) = 0$.

Thus, in the $p$-dimensional case a difference scheme such as

$$
\begin{aligned} &\Lambda y = \sum_{\alpha=1}^{p} \Lambda_\alpha\, y = -f(x) && \text{at the regular nodes,} \\ &\Lambda^* y = \sum_{\alpha=1}^{p} \Lambda^*_\alpha\, y = -f(x) && \text{at the irregular nodes,} \end{aligned}
$$

where $\Lambda_\alpha\, y = y_{\bar{x}_\alpha x_\alpha}$ and $\Lambda^*_\alpha$ is specified by the formula

$$
\Lambda^*_\alpha\, y = \frac{1}{h_\alpha}\left(\frac{y^{(+1_\alpha)} - y}{h_{\alpha+}} - \frac{y - y^{(-1_\alpha)}}{h_{\alpha-}}\right),
$$

is associated with problem (1).

**Remark** Quite often, the Dirichlet problem is approximated by the method based on the difference approximation at the near-boundary nodes of the Laplace operator on an irregular pattern, with the use of formulae (14) instead of (16) at the nodes $x \in \omega_h^*$. However, in some cases the difference operator so constructed does not possess several important properties intrinsic to the initial differential equation, namely, the self-adjointness and the property of having fixed sign. For this reason iterative methods are of little use in studying grid equations and will be excluded from further consideration.

**5. The canonical form of a difference equation.** We now consider the $(2p+1)$-point scheme $\Lambda y = -f$ at a regular node

$$\sum_{\alpha=1}^{p} \frac{1}{h_\alpha^2} \left( y^{(+1_\alpha)} - 2\,y + y^{(-1\alpha)} \right) = -f\,,$$

which admits an alternative form of writing

$$\sum_{\alpha=1}^{p} \frac{2}{h_\alpha^2}\, y(x) = \sum_{\alpha=1}^{p} \frac{1}{h_\alpha^2} \left( y^{(+1_\alpha)} + y^{(-1\alpha)} \right) + f(x)\,. \tag{32}$$

To avoid cumbersome calculations, we concentrate primarily on the two-dimensional case. Fig. 6 demonstrates that at a regular node

$$2 \left( \frac{1}{h_1^2} + \frac{1}{h_2^2} \right) y_0 = \frac{1}{h_1^2} (y_1 + y_3) + \frac{1}{h_2^2} (y_2 + y_4) + f_0\,.$$

Let $x \in \omega_{h,1}^*$ be an irregular node. In the case corresponding to Fig. 13.a we obtain

$$\Lambda_1^*\, y_0 = \frac{1}{\hbar_1} \left( \frac{y_1 - y_0}{h_1} - \frac{y_0 - y_3}{h_1^*} \right) = \frac{1}{\hbar_1} \left( \frac{y_1}{h_1} + \frac{y_3}{h_1^*} - \frac{2\,\hbar_1}{h_1\, h_1^*}\, y_0 \right),$$

where $\hbar_1 = \frac{1}{2}(h_1 + h_1^*)$, and

$$\Lambda_2\, y_0 = \frac{1}{h_2^2} (y_2 - 2\,y_0 + y_4)\,.$$

From the equation $\Lambda^* y = \Lambda_1^*\, y + \Lambda_2\, y = -f$ we find that

$$\left( \frac{2\,\hbar_1}{h_1^2\, h_1^*} + \frac{2}{h_2^2} \right) y_0 = \frac{1}{h_1^2}\, y_1 + \frac{1}{h_1\, h_1^*}\, y_3 + \frac{1}{h_2^2} (y_2 + y_4) + f_0\,.$$

In the case corresponding to Fig. 13.c we deduce that

$$\left( \frac{2\,\hbar_1}{h_1\,h^*_{1-}\,h^*_{1+}} + \frac{2\,\hbar_2}{h_2^2\,h_2^*} \right) y_0 = \frac{1}{h_1\,h^*_{1+}}\,y_1 + \frac{1}{h_1\,h^*_{1-}}\,y_3 + \frac{1}{h_2\,h_2^*}\,y_2 + \frac{1}{h_2^2}\,y_4 + f_0\,,$$

where $\hbar_1 = \frac{1}{2}(h^*_{1-} + h^*_{1+})$ and $\hbar_2 = \frac{1}{2}(h_2 + h^*_2)$.

Let $\bar{\omega}_h(\bar{G})$ be a grid in a $p$-dimensional domain and $x \in \omega^*_{h,\alpha}$ be a near-boundary irregular node. Then

$$(33) \qquad \Lambda^*_\alpha\, y = \frac{1}{\hbar_\alpha} \left( \frac{y^{(+1_\alpha)} - y}{h^*_{\alpha+}} - \frac{y - y^{(-1_\alpha)}}{h^*_{\alpha-}} \right)$$

$$= \frac{1}{\hbar_\alpha\, h^*_{\alpha+}}\, y^{(+1_\alpha)} + \frac{1}{\hbar_\alpha\, h^*_{\alpha-}}\, y^{(-1_\alpha)} - \frac{2\,\hbar_\alpha}{\hbar_\alpha\, h^*_{\alpha-}\, h^*_{\alpha+}}\, y\,.$$

Substituting this expression into the equation $\Lambda^* y = -f$ and regarding formally the node $x$ to be irregular in all directions $x_\alpha$, we finally get

$$(34) \qquad \sum_{\alpha=1}^{p} \frac{2\,\hbar_\alpha}{\hbar_\alpha\, h^*_{\alpha-}\, h^*_{\alpha+}}\, y(x) = \sum_{\alpha=1}^{p} \left( \frac{1}{\hbar_\alpha\, h^*_{\alpha+}}\, y^{(+1_\alpha)} + \frac{1}{\hbar_\alpha\, h^*_{\alpha-}}\, y^{(-1_\alpha)} + f(x) \right),$$

$$\hbar_\alpha = \tfrac{1}{2}\left(h^*_{\alpha+} + h^*_{\alpha-}\right).$$

This is acceptable for $h^*_{\beta-} = h^*_{\beta+} = \hbar_\beta = h_\beta$ when $x$ happens to be regular along some direction $x_\beta$. But if $x$ is regular in all directions $x_\alpha$, then $h^*_{\alpha+} = h^*_{\alpha-} = \hbar_\alpha = h_\alpha$ for all $\alpha = 1, 2, \ldots, p$, leading to formula (32). Comparison of (32) and (34) shows that these equations can be represented in the canonical form

$$(35) \qquad A(x)\, y(x) = \sum_{\xi \in Patt'(x)} B(x,\xi)\, y(\xi) + F(x), \qquad x \in \omega_h\,,$$

where $Patt'(x)$ is the set consisting of $2p$ nodes of the $(2p+1)$-point "cross" pattern with center at the point $x$ except for the node $x$ itself, that is, $\xi \neq x$. We call the set $Patt(x)'(x)$ the **neighborhood** of the node $x$. The quantities $A(x)$ and $B(x,\xi)$ are the available coefficients of the equation. From (32) and (34) it is easily seen that

$$(36)\quad A(x) > 0\,,\quad B(x,\xi) > 0\,,\quad \sum_{\xi \in Patt'(x)} B(x,\xi) = A(x) \quad \text{for all } x \in \omega_h\,.$$

Equation (35) is put together with the boundary condition

$$(37)\qquad y\big|_{\gamma_h} = \mu(x)\,.$$

The Dirichlet difference problem is a special case of a more general problem in which it is required to find a grid function $y(x)$ defined on the grid $\bar{\omega}_h = \omega_h + \gamma_h$ and satisfying on $\omega_h$ the equation

$$(38)\qquad \begin{aligned} A(x)\,y(x) &= \sum_{\xi \in Patt'(x)} B(x,\xi)\,y(\xi) + F(x), \qquad x \in \omega_h\,,\\ y(x) &= \mu(x), \qquad x \in \gamma_h\,, \end{aligned}$$

where

$$(39)\quad A(x) > 0,\quad B(x,\xi) > 0,\quad D(x) = A(x) - \sum_{\xi \in Patt'(x)} B(x,\xi) \geq 0$$

for all $x \in \omega_h$.

**Remark** The third difference boundary-value problem for Poisson's equation can always be represented in the form (38), equation (38) being satisfied for all $x \in \bar{\omega}_h$ and conditions (39) being valid. Here, in addition, $D \geq \delta > 0$ on $\gamma_h$.

To prove the existence and uniqueness of a solution of problem (38)–(39), it suffices to check that the homogeneous equation

$$(40)\qquad \begin{aligned} \mathcal{L}\,[y] = A(x)\,y(x) - \sum_{\xi \in Patt'(x)} B(x,\xi)\,y(\xi) &= 0\,, \qquad x \in \omega_h\,,\\ y(x) = 0, \qquad x \in \gamma_h\,, \end{aligned}$$

has only the trivial solution $y(x) = 0$ for $x \in \bar{\omega}_h$. We will show in the sequel that this fact follows immediately from the maximum principle, valid for schemes (38)–(39).

## 4.2 THE MAXIMUM PRINCIPLE

**1. The canonical form of a grid equation of common structure.** The maximum principle is suitable for the solution of difference elliptic and parabolic equations in the space $C$ and is certainly true for grid equations of common structure which will be investigated in this section.

Let $\omega$ be a finite set of nodes (a grid) in some bounded domain of the $n$-dimensional Euclidean space and let $P \in \omega$ be a point of the grid $\omega$. Consider the equation

$$(1) \qquad A(P)\, y(P) = \sum_{Q \in Patt'(P)} B(P,Q)\, y(Q) + F(P), \qquad P \in \omega\,,$$

related to a function $y(P)$ defined on the grid $\omega$. Here the coefficients of the equation $A(P)$ and $B(P,Q)$ and the right-hand side of the equation $F(P)$ are given grid functions; $Patt'(P) \subset \omega$, being the set of all the nodes of the grid $\omega$ except for the node $P$, is the neighborhood of the node $P$. The pattern of the grid equation (1) at the node $P$ consists, evidently, of the node $P$ itself and its neighborhood $Patt'(P)$.

Similar equations do arise in grid approximations of integral equations. In what follows we will suppose that coefficients $A(P)$ and $B(P,Q)$ are subject to the conditions

$$(2) \qquad \begin{cases} A(P) > 0\,, \quad B(P,Q) > 0 \quad \text{for all} \quad P \in \omega, \quad Q \in Patt'(P)\,, \\ D(P) = A(P) - \displaystyle\sum_{Q \in Patt'(P)} B(P,Q) \geq 0\,. \end{cases}$$

A point $P$ is called a **boundary node of the grid** $\omega$ if at this point the value of the function $y(P)$ is known in advance:

$$(3) \qquad y(P) = \mu(P) \quad \text{for} \quad P \in \gamma\,,$$

where $\gamma$ is the set of all boundary nodes.

Comparing (3) and (1) we see that on the boundary $\gamma$ we must set formally $A(P) \equiv 1$, $B(P,Q) \equiv 0$ and $F(P) = \mu(P)$.

We call the nodes, at which equation (1) is valid under conditions (2), **inner nodes of the grid**; $\omega$ is the set of all inner nodes and $\bar{\omega} = \omega + \gamma$ is the set of all grid nodes. The first boundary-value problem completely posed by conditions (1)–(3) plays a special role in the theory of equations (1). For instance, in the case of boundary conditions of the second or third kinds there are no boundary nodes for elliptic equations, that is, $\bar{\omega} = \omega$.

The **grid** $\bar{\omega}$ is taken to be **connected**, that is, for fixed points $\bar{P} \in \bar{\omega}$ and $\bar{\bar{P}} \in \bar{\omega}$ there always exists a sequence of neighborhoods $\{Patt'(P)\}$ such that the passage from $\bar{P}$ to $\bar{\bar{P}}$ can be done using only the nodes of those neighborhoods or, in other words, one can select nodes $P_1, P_2, \dots, P_m$ of the grid $\omega$ such that

$$P_1 \in Patt'(\bar{P}), \ P_2 \in Patt'(P_1), \ \dots, \ P_m \in Patt'(P_{m-1}), \ \bar{\bar{P}} \in Patt'(P_m)$$

with

$$\begin{aligned} &B(P_i, P_{i+1}) \neq 0, \quad i = 1, 2, \dots, m-1, \\ &B(\bar{P}, P_1) \neq 0, \quad B(P_m, \bar{\bar{P}}) \neq 0 . \end{aligned} \tag{4}$$

In the case of the difference scheme for the Dirichlet problem (24)–(26) of Section 1 the definition (4) of connectedness coincides with another definition from Section 1. The very definition implies that the point $\bar{\bar{P}}$ may be boundary and, hence, the connectedness is to be understood that every point of the boundary belongs to the neighborhood $Patt'(P)$ of at least one inner node.

Within the notation

$$\mathcal{L}\, y(P) = A(P)\, y(P) - \sum_{Q \in Patt'(P)} B(P, Q)\, y(Q) , \tag{5}$$

we may attempt equation (1) in the form

$$\mathcal{L}\, y(P) = F(P) . \tag{6}$$

An alternative form of $\mathcal{L}\, y(P)$ may be useful in the further development:

$$\mathcal{L}\, y(P) = D(P)\, y(P) + \sum_{Q \in Patt'(P)} B(P, Q)\, \big(y(P) - y(Q)\big) . \tag{7}$$

In the preceding section the Dirichlet difference problem was set up in the form (1), (3). Consider as one possible example the so-called scheme with weights for the heat conduction equation

$$\frac{\partial u}{\partial t} = \frac{\partial^2 u}{\partial x^2} + f(x, t), \qquad 0 < x < 1, \quad t > 0 ,$$

$$u(x, 0) = u_0(x), \qquad u(0, t) = \mu_1(t), \quad u(1, t) = \mu_2(t) .$$

On the grid $\bar{\omega}_{h\tau} = \{(x_i = ih, t_j = j\tau), i = 0, 1, \ldots, N, hN = 1, j = 0, 1, \ldots\}$ this scheme takes the form

$$\frac{y_i^{j+1} - y_i^j}{\tau} = \Lambda\left(\sigma\, y_i^{j+1} + (1-\sigma)\, y_i^j\right) + \varphi_i^j\,, \tag{8}$$

$$\Lambda\, y = y_{\bar{x}x}\,, \qquad y_i^0 = u_0(x_i), \quad y_0^j = \mu_1(t_j), \quad y_N^j = \mu_2(t_j)\,.$$

These difference equations are convenient to be presented in the canonical form (1), giving $P$ as a node of the grid $\bar{\omega}_{h\tau}$: $P = P(x_i\,, t_{j+1})$, where $Patt'(P)$ consists of the nodes $Q_1 = (x_i\,, t_j)$, $Q_2 = (x_{i-1}\,, t_{j+1})$, $Q_3 = (x_{i+1}\,, t_{j+1})$, $Q_4 = (x_{i-1}\,, t_j)$, $Q_5 = (x_{i+1}\,, t_j)$ and the boundary $\gamma$ consists of the nodes $(x_i\,, 0)$ and $(0, t_j)$, $(1, t_j)$, $i = 0, 1, \ldots, N$, $j = 0, 1, \ldots$. Next, we fix some moment $t = t_{j+1}$ and rewrite (8) as

$$\begin{aligned}\left(\frac{1}{\tau} + \frac{2\sigma}{h^2}\right) y_i^{j+1} &= \frac{\sigma}{h^2}\left(y_{i-1}^{j+1} + y_{i+1}^{j+1}\right)\\ &\quad + \left(\frac{1}{\tau} + \frac{2(\sigma-1)}{h^2}\right) y_i^j\\ &\quad + \frac{1-\sigma}{h^2}\left(y_{i-1}^j + y_{i+1}^j\right) + \varphi_i^j\,.\end{aligned}$$

From here it is easily seen that $B(P,Q) \ge 0$ only if $\tau \le h^2/2(1-\sigma)$ and $0 \le \sigma \le 1$. By the same token, $D(P) = 0$.

**2. The maximum principle.**

**Theorem 1** (the maximum principle) *Let $y(P) \not\equiv const$ be a grid function defined on a connected grid $\bar{\omega}$ and let both conditions (2) and (4) hold. Then the condition $\mathcal{L}\, y(P) \le 0$ ($\mathcal{L}\, y(P) \ge 0$) on the grid $\omega$ implies that $y(P)$ cannot attain the maximal positive (minimal negative) value at the inner nodes $P \in \omega$.*

**Proof** Let $\mathcal{L}\, y(P) \le 0$ at all of the inner nodes $P \in \omega$. On the contrary, let $y(P)$ attain its maximal positive value at an inner node $\bar{P} \in \omega$, so that

$$y(\bar{P}) = \max_{\bar{\omega}} y(P) = M_0 > 0\,.$$

The theorem will be proved if we succeed in showing that there exists an inner point $\widetilde{P}$ at which $\mathcal{L}\, y(\widetilde{P}) > 0$, violating the condition $\mathcal{L}\, y(P) \le 0$.

Since $y(\bar{P}) \geq y(Q)$ for all $Q \in Patt'(\bar{P})$, we find by virtue of the relations $D(\bar{P}) \geq 0$ and $y(\bar{P}) > 0$ that

$$\mathcal{L}\, y(\bar{P}) = D(\bar{P})\, y(\bar{P}) + \sum_{Q \in Patt'(P)} B(\bar{P}, Q)\, \big(y(\bar{P}) - y(Q)\big) \geq D(\bar{P})\, y(\bar{P}) \geq 0\,.$$

From such reasoning it seems clear that only the case $\mathcal{L}\, y(\bar{P}) = 0$ needs investigation. As formula (7) shows, it is possible only if $D(\bar{P}) = 0$ and $y(Q) = y(\bar{P})$ for all $Q \in Patt'(\bar{P})$.

We now take the node $P_1 \in Patt'(\bar{P})$ at which $y(P_1) = y(\bar{P}) = M_0$ and adopt those ideas. Since $y(P) \neq$ const on the grid $\omega$ and the grid is connected, there exists a sequence of nodes $P_1, P_2, \ldots, P_m, \bar{\bar{P}}$, satisfying conditions (4), such that

$$y(P_m) = y(\bar{P}) = M_0\,, \qquad y(\bar{\bar{P}}) < M_0\,,$$

but

$$\bar{\bar{P}} \in Patt'(P_m)\,.$$

Then

$$\mathcal{L}\, y(P_m) \geq D(P_m)\, y(P_m) + B(P_m, \bar{\bar{P}})\, \Big(y(P_m) - y(\bar{\bar{P}})\Big)$$

$$\geq B(P_m, \bar{\bar{P}})\, \big(y(\bar{P}) - y(\bar{\bar{P}})\big) > 0\,,$$

meaning $\widetilde{P} = P_m$ and justifying the first assertion of the theorem. The second assertion will be reduced to the first one once we replace $y(P)$ by $-y(P)$.

**Corollary 1** *Let conditions* (2) *and* (4) *hold and let a grid function $y(P)$ defined on $\omega + \gamma$ be nonnegative on the boundary, that is, $y(P) \geq 0$ for $P \in \gamma$ and $\mathcal{L}\, y(P) \geq 0$ on $\omega$. Then $y(P)$ is nonnegative on $\omega + \gamma$, that is, $y(P) \geq 0$ for all $P \in \omega + \gamma$. But if $y(P) \leq 0$ on $\gamma$ and $\mathcal{L}\, y(P) \leq 0$ on $\omega$, then $y(P) \leq 0$ on $\omega + \gamma$.*

**Proof** Let $\mathcal{L}\, y(P) \geq 0$ on $\omega$ and $y(P) \geq 0$ on $\gamma$. Suppose that $y(P_0) < 0$ at least at one inner node $P_0 \in \omega$. Then $y(P)$ should take its minimal negative value inside the grid $\omega$, what is impossible on account of Theorem 1, since $y(P) \neq$ const on the grid $\omega\big(y(P_0) < 0,\ y|_\gamma \geq 0\big)$. The second assertion of the theorem can be proved in a similar manner.

**Corollary 2** *The homogeneous equation* (1) *subject to the boundary condition*

$$\mathcal{L}\, y(P) = 0 \quad on \quad \omega, \qquad y(P) = 0 \quad on \quad \gamma, \tag{9}$$

*has only the trivial solution* $y(P) \equiv 0$.

It is straightforward to verify that $y(P) \equiv 0$ is a solution of problem (9). Further, assume to the contrary that there exists a solution $y(P) \not\equiv 0$ of problem (9). If $y(P) \neq 0$ at least at one point, then by Corollary 1 the inequalities $y(P) \leq 0$ and $y(P) \geq 0$ must hold simultaneously. But it is possible only if $y(P) \equiv 0$. Thus, we have proved the following assertion.

**Corollary 3** *Problem* (1)–(4) *possesses a unique solution.*

**3. Comparison theorem. The majorant.**

**Theorem 2** *Let* $y(P)$ *be a solution of problem* (1)–(4) *and let* $Y(P)$ *be a solution of the problem*

$$\mathcal{L}\, Y(P) = \bar{F}(P), \quad P \in \omega, \qquad Y(P) = \bar{\mu}(P), \quad P \in \gamma. \tag{10}$$

*Then the conditions*

$$|\, F(P) \,| \leq \bar{F}(P), \quad P \in \omega, \qquad |\, \mu(P) \,| \leq \bar{\mu}(P), \quad P \in \gamma, \tag{11}$$

*provide the validity of the inequality*

$$|\, y(P) \,| \leq Y(P) \quad for \quad P \in \omega + \gamma. \tag{12}$$

**Proof** By Corollary 1 the inequality $Y(P) \geq 0$ is valid on $\omega + \gamma$. The functions $u(P) = Y(P) + y(P)$ and $v(P) = Y(P) - y(P)$ solve the equations $\mathcal{L}\, u = F_u = \bar{F} + F \geq 0$ and $\mathcal{L}\, v = F_v = \bar{F} - F \geq 0$ subject to the boundary conditions $u|_\gamma = (Y + y)|_\gamma = \bar{\mu} + \mu \geq 0$ and $v|_\gamma = (Y - y)|_\gamma = \bar{\mu} - \mu \geq 0$.

Since the conditions of Corollary 1 are satisfied, we have $u \geq 0$ or $y \geq -Y$, $v \geq 0$ or $y \leq Y$. It follows from the foregoing that $-Y \leq y \leq Y$ or $|\, y(P) \,| \leq Y(P)$ on $\omega + \gamma$.

The function $Y(P)$ is called the **majorant** for a solution of problem (1)–(3). Its determination entails immediately the validity of the desired estimate for $\|\, y \,\|_C$.

**Corollary** *For a solution of the problem*

$$\mathcal{L}\, y = 0 \quad \text{on} \quad \omega, \qquad y(P) = \mu(P) \quad \text{on} \quad \gamma, \tag{13}$$

*the estimate*

$$\max_{P \in \omega + \gamma} |\, y(P) \,| = \|\, y \,\|_{\bar{C}} \le \|\, \mu \,\|_{C_\gamma} \tag{14}$$

*is valid with* $\|\, \mu \,\|_{C_\gamma} = \max_{P \in \gamma} |\, \mu(P) \,|$.

Indeed, let us specify the majorant $Y(P)$ by the conditions $\mathcal{L}\, Y = 0$ on $\omega$ and $Y = \|\, \mu \,\|_{C_\gamma}$ on $\gamma$. The function $Y(P)$ is nonnegative on $\omega + \gamma$ and attains its maximum at some node of the grid. This node is none the inner nodes if $Y(P) \ne \text{const}$ and, hence,

$$\|\, Y \,\|_{\bar{C}} = \max_{P \in \omega + \gamma} Y(P) = \max_{P \in \gamma} Y(P) = \|\, \mu \,\|_{C_\gamma} .$$

If $Y(P) = \text{const}$, then $Y(P) = \|\, \mu \,\|_{C_\gamma}$. In both cases $\|\, Y \,\|_{\bar{C}} = \|\, \mu \,\|_{C_\gamma}$. Combination of this relation and the inequality $\|\, y \,\|_{\bar{C}} \le \|\, Y \,\|_{\bar{C}}$ gives estimate (14).

**4. The estimate of a solution to the nonhomogeneous equation.** In the further development a solution of problem (1)–(3) is viewed as a sum

$$y = \bar{y} + v ,$$

where $\bar{y} = \bar{y}(P)$ is a solution to the homogeneous equation

$$\mathcal{L}\, \bar{y} = 0 \quad \text{on} \quad \omega, \qquad \bar{y} = \mu(P) \quad \text{on} \quad \gamma, \tag{15}$$

and $v = v(P)$ is a solution to the nonhomogeneous equation

$$\mathcal{L}\, v(P) = F(P) \quad \text{on} \quad \omega, \qquad v(P) = 0 \quad \text{on} \quad \gamma . \tag{16}$$

We have already obtained estimate (14) for $\bar{y}(P)$ and so it remains to evaluate the function $v(P)$.

**Theorem 3** *If* $D(P) > 0$ *everywhere on the grid* $\omega$, *then a solution of problem* (16) *admits the estimate*

$$\|\, v \,\|_C \le \left\| \frac{F}{D} \right\|_C . \tag{17}$$

**Proof** Let a majorant $Y(P)$ be taken such that

$$\mathcal{L} Y = |\, F(P)\, |, \quad Y|_\gamma = 0, \quad Y(P) \ge 0 \quad \text{for} \quad P \in \omega + \gamma$$

and $Y(P)$ attain its maximum at a node $P_0 \in \omega$. As far as $Y(P_0) = \|\, Y \,\|_C$ is concerned, the equation

$$D(P_0)\, Y(P_0) + \sum_{Q \in Patt'(P_0)} B(P_0\,, Q)\, \big(Y(P_0) - Y(Q)\big) = |\, F(P_0)\, |$$

yields

$$D(P_0)\, Y(P_0) \le |\, F(P_0)\, |, \qquad Y(P_0) \le \frac{|\, F(P_0)\, |}{D(P_0)} \le \Big\| \frac{F}{D} \Big\|_C ,$$

thereby completing the proof.

**Remark** Estimate (17) is still valid for the solution of problem (16) provided that instead of (2) other conditions

$$|\, A(P)\, | \ne 0\,, \quad |\, B(P, Q)\, | \ne 0\,,$$

$$D(P) = |\, A(P)\, | - \sum_{Q \in Patt'(P)} |\, B(P, Q)\, | > 0$$

hold.

Indeed, let $|\, v(P)\, | \ge 0$ take the maximal value at a node $P_0$. Because of this fact,

$$|\, A(P_0)\, | \cdot |\, v(P_0)\, | = \Big| \sum_{Q \in Patt'(P_0)} B(P_0\,, Q)\, v(Q) + F(P_0) \Big|$$

$$\le \sum_{Q \in Patt'(P_0)} |\, B(P_0\,, Q)\, | \cdot |\, v(P_0)\, | + |\, F(P_0)\, |,$$

whence it follows that

$$D(P_0)\ \ |\, v(P_0)\, | \le |\, F(P_0)\, |, \qquad \|\, v \,\|_C = |\, v(P_0)\, | \le \frac{|\, F(P_0)\, |}{D(P_0)} \le \Big\| \frac{F}{D} \Big\|_C .$$

It may happen that $D(P) = 0$ on a subset $\overset{\circ}{\omega}$ of the grid $\omega$ and $D(P) > 0$ on the complement of $\overset{\circ}{\omega}$ to $\omega$: $\overset{\circ}{\omega} + \omega^* = \omega$. This type of situation is covered by the following assertion.

**Theorem 4** *Let the conditions*

$$D(P) = 0 \quad \text{for} \quad P \in \overset{\circ}{\omega}\,, \qquad D(P) > 0 \quad \text{for} \quad P \in \omega^*$$

*hold, where* $\overset{\circ}{\omega}$ *is a connected grid. Then for a solution of problem* (16) *with the right-hand side*

$$F(P) = 0 \quad \text{for} \quad P \in \overset{\circ}{\omega}\,, \qquad F(P) \not\equiv 0 \quad \text{for} \quad P \in \omega^*,$$

*the estimate*

$$\| v \|_C \le \left\| \frac{F}{D} \right\|_{C^*} \tag{18}$$

*is valid in the norm* $\| f \|_{C^*} = \max_{P \in \omega^*} | f(P) |$.

**Proof** Let $Y(P)$ be a majorant and $\mathcal{L}\, Y = | F(P) |$ on the grid $\omega$, $Y|_\gamma = 0$, $Y \ge 0$. The function $Y(P)$ should attain its maximum on a finite set $\omega + \gamma$ at some node, not belonging to the boundary, because $Y|_\gamma = 0$. Also, it does not enter the grid $\overset{\circ}{\omega}$ due to the connectedness of $\overset{\circ}{\omega}$ and the maximum principle. Hence,

$$\max_{P \in \omega} Y(P) = \max_{P \in \omega^*} Y(P) = Y(P_0)\,,$$

where $P_0$ is a node on the set $\omega^*$.

By the initial hypothesis, $D(P_0) > 0$. Arguing as in the proof of Theorem 3 we arrive at (18). An analog of the remark to Theorem 3 is still valid for that case.

## 4.3 STABILITY AND CONVERGENCE OF THE DIRICHLET DIFFERENCE PROBLEM

**1. Estimation of a solution of the Dirichlet difference problem.** We make use of a priori estimates obtained in Section 2 for a grid equation of common structure for constructing a uniform estimate of a solution of the Dirichlet difference problem (24)–(26) arising in Section 1:

$$\begin{aligned} \Lambda\, y &= -\varphi && \text{at the regular nodes,} \\ \Lambda^* y &= -\varphi && \text{at the irregular nodes,} \\ y &= \mu(x) && \text{on the boundary,} \end{aligned} \tag{1}$$

where

$$\Lambda\, y = \sum_{\alpha=1}^{p} \Lambda_\alpha\, y, \qquad \Lambda_\alpha\, y = y_{\bar{x}_\alpha x_\alpha}\,,$$

$$\Lambda^* y = \sum_{\alpha=1}^{p} \Lambda_\alpha^*\, y\,,$$

$$\Lambda_\alpha^*\, y = \frac{1}{\hbar_\alpha} \left( \frac{y^{(+1_\alpha)} - y}{h_{\alpha +}} - \frac{y - y^{(-1_\alpha)}}{h_{\alpha -}} \right).$$

Other ideas are connected with the operator $\bar{\Lambda}$, which coincides with $\Lambda^*$ at the near-boundary nodes and with $\Lambda$ at the remaining inner nodes, leading to an alternative form of writing

(2) $$\bar{\Lambda}\, y = -\varphi, \qquad x \in \omega_h\,, \quad y\big|_{\gamma_h} = \mu(x)\,.$$

In conformity with Section 1, problem (1) can be recast as

(3) $$A(x)\, y(x) = \sum_{\xi \in Patt'(x)} B(x,\xi)\, y(\xi) + F(x), \qquad x \in \omega\,, \quad y\big|_{\gamma_h} = \mu(x)\,,$$

where

$$A(x) > 0, \quad B(x,\xi) > 0, \quad D(x) = A(x) - \sum_{\xi \in Patt'(x)} B(x,\xi) \geq 0\,.$$

We now represent a solution of problem (1) as a sum

$$y = \bar{y} + \tilde{y},$$

where $\bar{y}$ and $\tilde{y}$ are, respectively, solutions of the appropriate problems

(4) $$\bar{\Lambda}\, \bar{y} = 0\,, \qquad x \in \omega_h\,, \quad \bar{y} = \mu \quad \text{for} \quad x \in \gamma_h\,,$$

(5) $$\bar{\Lambda}\, \tilde{y} = -\varphi, \qquad x \in \omega_h\,, \quad \tilde{y} = 0 \quad \text{for} \quad x \in \gamma_h\,.$$

An estimate for a solution of problem (4) such as

(6) $$\| \, \bar{y} \, \|_{\bar{C}} \leq \| \, \mu \, \|_{C_\gamma}$$

has been derived in Section 2 and may be useful in such a setting.

Having decomposed the right-hand side $\varphi$ as

$$\varphi = \overset{\circ}{\varphi} + \varphi^*,$$

where $\overset{\circ}{\varphi} = \varphi$ and $\varphi^* = 0$ at the strictly inner nodes $x \in \overset{\circ}{\omega}_h$ (see Section 1) and $\overset{\circ}{\varphi} = 0$ and $\varphi^* = \varphi$ at the near-boundary nodes $x \in \omega_h^*$, we agree to consider

$$\tilde{y} = v + w$$

with $v$ and $w$ being solutions of the problems

$$\bar{\Lambda}\, v = -\overset{\circ}{\varphi} \qquad \text{for} \quad x \in \omega_h\,, \quad v\big|_{\gamma_h} = 0\,, \tag{7}$$

$$\bar{\Lambda}\, w = -\varphi^* \qquad \text{for} \quad x \in \omega_h\,, \quad w\big|_{\gamma_h} = 0\,. \tag{8}$$

We are going to evaluate separately each of the functions $v(x)$ and $w(x)$. In order to estimate $v(x)$, it is necessary to construct a majorant $Y(x)$. Assuming that the origin is inside the domain $G$, we try to determine a majorant of the type

$$Y(x) = K\,(R^2 - r^2), \qquad r^2 = \sum_{\alpha=1}^{p} x_\alpha^2,$$

where $K > 0$ is a constant and $R$ is the radius of a $p$-dimensional ball (a circle for $p = 2$) with center at the origin containing entirely the domain $G$. The constant K will be chosen a little later.

By virtue of the relations $\Lambda_\alpha\, x_\beta^2 = 0$ for $\alpha \neq \beta$ and

$$\Lambda_\alpha\, x_\alpha^2 = \frac{(x_\alpha + h_\alpha)^2 - 2\,x_\alpha^2 + (x_\alpha - h_\alpha)^2}{h_\alpha^2} = 2\,,$$

$$\Lambda_\alpha^*\, x_\alpha^2 = \frac{1}{\hbar_\alpha}\left[\frac{(x_\alpha + h_{\alpha+})^2 - x_\alpha^2}{h_{\alpha+}} - \frac{x_\alpha^2 - (x_\alpha - h_{\alpha-})^2}{h_{\alpha-}}\right] = 2\,\theta_\alpha\,,$$

$$\theta_\alpha = \frac{h_{\alpha+} + h_{\alpha-}}{2\,\hbar_\alpha}$$

we find that

$$\Lambda Y = \sum_{\alpha=1}^{p} \Lambda_\alpha Y = -2\,p\,K \qquad \text{for} \quad x \in \overset{\circ}{\omega}_h\,,$$

$$\Lambda^* Y = -2\,p\,\theta\,K \qquad \text{for} \quad x \in \omega_h^*\,,$$

where $\theta = p^{-1} \sum_{\alpha=1}^{p} \theta_\alpha$. Here $\theta_\alpha = 1$ if a node $x \in \omega_h^*$ is regular along the direction $x_\alpha$. Thus, $Y$ is a solution of the problem

$$\bar{\Lambda} Y = -\bar{F}(x), \qquad Y|_{\gamma_h} = K\,(R^2 - r^2)|_{\gamma_h} \geq 0,$$

where $\bar{F}(x) = 2\,p\,K$ for $x \in \overset{\circ}{\omega}_h$ and $\bar{F}(x) = 2\,p\,\theta\,K$ for $x \in \omega_h^*$.

Comparison with problem (7), where $F = \overset{\circ}{\varphi}$, that is, $F = 0$ for $x \in \omega_h^*$ and $v|_{\gamma_h} = 0$, shows that $\bar{F}(x) \geq |\,F(x)\,| = |\,\overset{\circ}{\varphi}(x)\,|$ if we accept $K = \frac{1}{2p} \|\,\overset{\circ}{\varphi}\,\|_C$. Here the conditions of the comparison theorem in Section 2 are valid as long as $\bar{F}(x) \geq |\,F(x)\,| = 0$ for $x \in \omega^*$, assuring the relation $\|\,v\,\|_C \leq \|\,Y\,\|_C$.

It is easily seen from the expression for $Y$ that $\|\,Y\,\|_C \leq K R^2$. So, for a solution of problem (7) the estimate

$$\text{(9)} \qquad \|\,v\,\|_C \leq \frac{R^2}{2\,p} \|\,\overset{\circ}{\varphi}\,\|_C = \frac{R^2}{2\,p} \|\,\varphi\,\|_{\overset{\circ}{C}}$$

is valid in the next norm $\|\,\varphi\,\|_{\overset{\circ}{C}} = \max_{x \in \overset{\circ}{\omega}_h} |\,\varphi(x)\,|$.

Our next step is the estimation of the function $w(x)$. First, we are going to show that for problem (8)

$$\text{(10)} \qquad D(x) \geq \frac{1}{h^2}\,, \qquad \text{where} \quad h = \max_\alpha h_\alpha\,, \quad x \in \omega_h^*\,,$$

$$\text{(11)} \qquad D(x) = 0 \qquad \text{for} \quad x \in \overset{\circ}{\omega}_h\,.$$

Assertion (11) is simple to follow.

After that, we look at equation (8) at a near-boundary node $x \in \omega_h^*$

$$\text{(12)} \qquad \begin{aligned} A(x)\,w(x) &= \sum_{\xi \in Patt'(x)} B(x,\xi)\,w(\xi) + F(x), \\ F(x) &= \varphi^*(x), \qquad w|_{\gamma_h} = 0\,. \end{aligned}$$

If one of the nodes $\xi = \xi_0$, say $\xi_0 = x^{(+1_\alpha)}$, happens to be a boundary node, then $w(\xi_0) = 0$ and the neighborhood $Patt'(x)$ contains no point $\xi_0$.

In this case the function $D(x)$ takes on the value

$$D(x) = A(x) - \sum_{\substack{\xi \in Patt'(x) \\ \xi \neq \xi_0}} B(x, \xi)$$

$$= A(x) - \Bigg[ \sum_{\xi \in Patt'(x)} B(x, \xi) - B(x, \xi_0) \Bigg]$$

$$= B(x, \xi_0) \,,$$

since $A(x) = \sum\limits_{\xi \in Patt'(x)} B(x, \xi)$ for the Laplace equation. This provides support for the view that

$$D(x) = B\big(x, x^{(+1_\alpha)}\big) > 0 \,.$$

If a node $x$ is near-boundary not only with respect to $x_\alpha$, but also in other directions, then sum (12) contains no other terms for $\xi = \xi_1, \xi_2, \ldots, \xi_k$, so that

$$D(x) = B(x, \xi_0) + B(x, \xi_1) + \cdots + B(x, \xi_k) > 0 \,.$$

Let $x \in \omega_h^*$ be a near-boundary and irregular node only in some direction $x_\alpha$ and $\xi_0 = x^{(+1_\alpha)} \in \gamma_h$, $x^{(-1_\alpha)} \in \omega_h$. From the equation

$$\Lambda_\alpha^* \, w + \sum_{\substack{\beta=1 \\ \beta \neq \alpha}}^{p} \Lambda_\beta \, w = -\varphi^*(x),$$

where

$$\Lambda_\beta \, y = y_{\bar{x}_\beta x_\beta} \,,$$

$$\Lambda_\alpha^* \, w = \frac{1}{h_\alpha} \left( \frac{w^{(+1_\alpha)} - w}{h_{\alpha+}} - \frac{w - w^{(-1_\alpha)}}{h_\alpha} \right)$$

$$= -\frac{1}{h_\alpha} \left( \frac{w}{h_{\alpha+}} + \frac{w - w^{(-1_\alpha)}}{h_\alpha} \right),$$

we establish the relations

$$A(x) = \frac{1}{h_\alpha\, h_{\alpha+}} + \frac{1}{h_\alpha^2} + \sum_{\substack{\beta=1\\ \beta\neq\alpha}}^{p} \frac{2}{h_\beta^2}\,,$$

$$\sum_{\xi\in Patt'(x)} B(x,\xi) = \frac{1}{h_\alpha^2} + \sum_{\substack{\beta=1\\ \beta\neq\alpha}}^{p} \frac{2}{h_\beta^2}\,,$$

$$D(x) = \frac{1}{h_\alpha\, h_{\alpha+}} \geq \frac{1}{h^2}\,.$$

If $x$ is a regular near-boundary node only with respect to $x_\alpha$, then

$$D(x) = \frac{1}{h_\alpha^2} \geq \frac{1}{h^2}\,.$$

When, in addition, it turns out to be near-boundary in other directions, the function $D(x)$ can only increase. Just for this reason estimate (10) continues to hold on the same footing.

To evaluate a solution of problem (8), we apply Theorem 4 of Section 2 due to which

$$(13) \qquad \|\, w\,\|_C \leq \left\| \frac{\varphi^*}{D} \right\|_{C^*} \leq h^2 \|\, \varphi^*\,\|_C\,.$$

Collecting estimates (6), (9), (13) and then involving the well-known inequality $\|\, y\,\|_{\bar C} \leq \|\, \bar y\,\|_{\bar C} + \|\, v\,\|_C + \|\, w\,\|_C$, we establish the following theorem.

**Theorem 1** *For a solution of the Dirichlet difference problem the estimate*

$$(14) \qquad \|\, y\,\|_{\bar C} \leq \|\, \mu\,\|_{C_\gamma} + \frac{R^2}{2\,p} \|\, \varphi\,\|_{\overset{\circ}{C}} + h^2\, \|\, \varphi\,\|_{C^*}$$

*holds with*

$$\|\, f\,\|_{\bar C} = \max_{x\in\omega_h+\gamma_h} |\, f(x)\,|\,, \qquad \|\, f\,\|_{\overset{\circ}{C}} = \max_{x\in\overset{\circ}{\omega}_h} |\, f(x)\,|\,,$$

$$\|\, f\,\|_{C^*} = \max_{x\in\omega_h^*} |\, f(x)\,|\,, \qquad \|\, f\,\|_{C_\gamma} = \max_{x\in\gamma_h} |\, f(x)\,|\,.$$

This theorem expresses the stability of the Dirichlet difference problem (1) with respect to the boundary data and the right-hand side.

**2. The uniform convergence and the order of accuracy of a difference scheme.** In the study of convergence and accuracy of scheme (2) we begin by placing the problem for the error

$$z = y - u\,,$$

where $y$ is a solution of problem (1) and $u = u(x)$ is a solution of problem (1) arising from Section 1. Substituting $y = z + u$ into (1) or (2) yields

$$\bar{\Lambda}\, z = -\psi(x), \qquad x \in \omega, \quad z|_\gamma = 0\,, \tag{15}$$

where $\psi(x) = \bar{\Lambda}\, u + \varphi(x)$ is the residual.

We stated in Section 1 that

$$\psi(x) = O(|\,h\,|^2) = O(h^2) \qquad \text{at the regular nodes,}$$

$$\psi(x) = O(1) \qquad \text{at the irregular nodes,}$$

or, more specifically,

$$|\,\psi\,| \le \frac{M_4\,|\,h\,|^2}{12} \le p\,\frac{M_4}{12}\,h^2 \qquad \text{at the regular nodes,}$$

$$|\,\psi\,| \le p\,M_2 \qquad \text{at the irregular nodes,}$$

where

$$M_k = \max_{\substack{x\in\bar{G}\\ 1\le\alpha\le p}} \left|\frac{\partial^k u}{\partial x_\alpha^k}\right|, \quad k = 2, 3, 4, \ldots, \quad |\,h\,|^2 = \sum_{\alpha=1}^{p} h_\alpha^2, \quad h = \max_{1\le\alpha\le p} h_\alpha .$$

Also, Theorem 1 of Section 1 asserts that estimate (14) is of the form

$$\|\,z\,\|_C \le \frac{R^2}{2\,p}\,\|\,\psi\,\|_{\overset{\circ}{C}} + h^2\,\|\,\psi\,\|_{C^*}\,.$$

Using the estimates of $|\,\psi\,|$ obtained above we arrive at the relation

$$\|\,z\,\|_C = \|\,y - u\,\|_C \le \left(\frac{R^2}{24}\,M_4 + p\,M_2\right) h^2\,, \tag{16}$$

making it possible to formulate the following statement.

**Theorem 2** *If $u(x) \in C^4(\bar{G})$, that is, a solution possesses continuous derivatives in $\bar{G} = G + \Gamma$ of the first four orders, then the difference scheme converges uniformly with the rate $O(h^2)$, that is, it is of second-order accuracy, so that estimate (16) is valid.*

## 4.4 SOME PROPERTIES OF DIFFERENCE ELLIPTIC OPERATORS

In this section we reveal some properties of difference operators approximating the Laplace operator in a rectangle and derive several estimates for difference approximations to elliptic second-order operators with variable coefficients and mixed derivatives.

**1. Eigenvalue problems for the Laplace difference operator in a rectangle.** Let $\bar{G}_0 = \{0 \le x_\alpha \le l_\alpha\,,\ \alpha = 1, 2\}$ be a rectangle, $\bar{\omega}_h = \{x = (i_1 h_1\,, i_2 h_2),\ i_\alpha = 0, 1, 2, \ldots, N_\alpha\,,\ N_\alpha h_\alpha = l_\alpha\}$ be a grid in $\bar{G}_0$ and $\gamma_h$ be the set of all boundary grid nodes. The grid $\omega_h$ is taken to be equidistant in each direction $x_\alpha$ with step $h_\alpha$.

The eigenvalue problem for the Laplace operator in the rectangle $\bar{G}_0$ subject to the first kind boundary conditions

$$\Delta v + \lambda v = 0, \qquad x \in G_0\,, \quad v|_\Gamma = 0, \quad v(x) \not\equiv 0\,,$$

has an infinite set of the eigenvalues such as

$$\lambda_k = \lambda_{k_1 k_2} = \pi^2 \left( \frac{k_1^2}{l_1^2} + \frac{k_2^2}{l_2^2} \right), \qquad k_\alpha = 1, 2, \ldots, \qquad \alpha = 1, 2\,,$$

associated with the orthonormal system of eigenfunctions

$$v_{k_1 k_2}(x) = \sqrt{\frac{4}{l_1\, l_2}}\ \sin \frac{\pi k_1}{l_1}\, x_1 \ \sin \frac{\pi k_2}{l_2}\, x_2\,,$$

so that

$$\left( v_{k_1 k_2}\,,\ v_{k_1' k_2'} \right) = \delta_{k_1 k_1'}\, \delta_{k_2 k_2'}\,,$$

where

$$(u, v) = \int_0^{l_1} dx_1 \int_0^{l_2} dx_2\, u(x_1\,, x_2)\, v(x_1\,, x_2)\,.$$

This problem can be solved by the method of separation of variables. The eigenvalue problem for the difference Laplace operator $\Lambda y = y_{\bar{x}_1 x_1} + y_{\bar{x}_2 x_2}$ supplied by the first kind boundary conditions may be set up in a quite similar manner as follows: it is required to find the values of the parameter $\lambda$ (eigenvalues) associated with nontrivial solutions of the homogeneous equation subject to the homogeneous boundary conditions

$$\text{(1)} \qquad \Lambda v + \lambda v = 0\,, \qquad x \in \omega_h\,, \quad v\big|_{\gamma_h} = 0, \quad v(x) \not\equiv 0\,.$$

Observe that $\gamma_h$ and, hence, $\bar{\omega}_h$ contain no nodes $(0,0)$, $(0,N_2)$, $(N_1,0)$, $(N_1,N_2)$ being the vertices of the rectangle.

We seek a solution of problem (1) by the method of separation of variables as a product

$$v(x_1,x_2)=\mu(x_1)\,\eta(x_2), \qquad x_1=i_1\,h_1\,, \quad x_2=i_2\,h_2\,.$$

Substitution of this expression into the equation

$$\Lambda\,v+\lambda\,v=v_{\bar{x}_1\,x_1}+v_{\bar{x}_2\,x_2}+\lambda v=0$$

yields

$$\mu_{\bar{x}_1\,x_1}(x_1)\;\eta(x_2)+\mu(x_1)\;\eta_{\bar{x}_2\,x_2}(x_2)+\lambda\,\mu(x_1)\,\eta(x_2)=0\,. \tag{2}$$

Because we are interested only in nontrivial solutions of problem (1), the division of both sides of equation (2) by $\mu(x_1)\,\eta(x_2)\neq 0$ is meaningful. As a final result we get $\mu_{\bar{x}_1\,x_1}/\mu+\eta_{\bar{x}_2\,x_2}/\eta+\lambda=0$ or

$$\frac{\mu_{\bar{x}_1\,x_1}}{\mu}=-\frac{\eta_{\bar{x}_2\,x_2}}{\eta}-\lambda=-\lambda^{(1)} \tag{3}$$

with $\lambda^{(1)}=$ const being independent neither of $x_1$ nor of $x_2$. In view of this, we might set up for $\mu(x_1)$ the eigenvalue problem on the grid

$$\bar{\omega}_h^{(1)}=\{x_1=i_1\,h_1\,,\ i_1=0,1,\ldots,N_1;\quad h_1\,N_1=l_1\}\,,$$

the statement of which is

$$\Lambda_1\,\mu+\lambda^{(1)}\mu=\mu_{\bar{x}_1\,x_1}+\lambda^{(1)}\mu=0\,,$$

$$x_1=i_1\,h_1\,,\quad i_1=1,2,\ldots,N_1-1,\quad \mu_0=0,\quad \mu_{N_1}=0\,.$$

The conditions $\mu_0=\mu_{N_1}=0$ follow immediately from the relations $\mu(0)\,\eta(x_2)=0$, $\mu(l_1)\,\eta(x_2)=0$ and $\eta(x_2)\not\equiv 0$. As we learn from Chapter 2, Section 3, a solution of this problem acquires the form

$$\lambda_{k_1}^{(1)}=\frac{4}{h_1^2}\,\sin^2\frac{\pi\,k_1\,h_1}{2\,l_1}\,,$$

$$\mu_{k_1}(x_1)=\sqrt{\frac{2}{l_1}}\,\sin\frac{\pi\,k_1\,x_1}{l_1},\qquad k=1,2,\ldots,N_1-1\,,$$

where $\{\mu_{k_1}(x_1)\}$ is an orthonormal system of eigenfunctions: $(\mu_{k_1}, \mu_{k_1'})_1 = \delta_{k_1 k_1'}$ under the inner product structure $(y, v)_1 = \sum_{i_1=1}^{N_1-1} y(i_1 h_1)\, v(i_1 h_1)\, h_1$.

From the governing relations (3) a similar problem arises for $\eta(x_2)$ with $\lambda^{(2)} = \lambda - \lambda^{(1)}$:

$$\eta_{\bar{x}_2 x_2} + \lambda^{(2)} \eta(x_2) = 0\,, \qquad x_2 = i_2 h_2\,, \quad i_2 = 1, 2, \dots, N_2 - 1,$$
$$\eta_0 = 0, \quad \eta_{N_2} = 0\,,$$

whose solution is given by

$$\lambda_{k_2}^{(2)} = \frac{4}{h_2^2} \sin^2 \frac{\pi k_2 h_2}{2 l_2}, \quad \eta_{k_2}(x_2) = \sqrt{\frac{2}{l_2}} \sin \frac{\pi k_2 x_2}{l_2},$$
$$k_2 = 1, 2, \dots, N_2 - 1\,.$$

Here $(\eta_{k_2}, \eta_{k_2'})_2 = \delta_{k_2 k_2'}$, where $(y, v)_2 = \sum_{i_2=1}^{N_2-1} y(i_2 h_2)\, v(i_2 h_2)\, h_2$. Thus, problem (1) is completely solved, meaning that to the eigenvalues

$$\lambda_k = \lambda_{k_1 k_2} = \lambda_{k_1}^{(1)} + \lambda_{k_2}^{(2)}$$

or

$$\lambda_{k_1 k_2} = 4 \left( \frac{1}{h_1^2} \sin^2 \frac{\pi k_1 h_1}{2 l_1} + \frac{1}{h_2^2} \sin^2 \frac{\pi k_2 h_2}{2 l_2} \right), \tag{4}$$
$$k_1 = 1, 2, \dots, N_1 - 1\,, \qquad k_2 = 1, 2, \dots, N_2 - 1\,,$$

there correspond the eigenfunctions

$$v_k(x) = v_{k_1 k_2}(x_1, x_2) = \mu_{k_1}(x_1)\, \mu_{k_2}(x_2) \tag{5}$$
$$= \sqrt{\frac{4}{l_1 l_2}} \sin \frac{\pi k_1 x_1}{l_1} \sin \frac{\pi k_2 x_2}{l_2},$$
$$k_1 = 1, 2, \dots, N_1 - 1, \qquad k_2 = 1, 2, \dots, N_2 - 1\,.$$

The total number of the eigenfunctions is equal to $(N_1 - 1)(N_2 - 1) = N$. These constitute an orthonormal system

$$(v_{k_1 k_2}, v_{k_1' k_2'}) = \delta_{k_1 k_1'} \delta_{k_2 k_2'} \tag{6}$$

in the sense of the inner product

$$(7) \qquad (y, v) = \sum_{i_1=1}^{N_1-1} \sum_{i_2=1}^{N_2-1} y(i_1 h_1, i_2 h_2)\, v(i_1 h_1, i_2 h_2)\, h_1 h_2$$

$$= \sum_{x \in \omega_h} y(x)\, v(x)\, h_1 h_2 .$$

Therefore, any function $f(x_1, x_2)$ defined on the grid $\bar{\omega}_h$ and vanishing on its boundary can always be expanded on the system of functions $\{v_{k_1 k_2}(x_1, x_2)\}$:

$$f(x) = \sum_{k=1}^{N} c_k v_k(x), \qquad c_k = (f, v_k) .$$

In the case of the second boundary-value problem with $\partial v/\partial n|_{\Gamma} = 0$, the boundary condition of second-order approximation is imposed on $\gamma_h$ as a first preliminary step. It is not difficult to verify directly that the difference eigenvalue problem of second-order approximation with the second kind boundary conditions is completely posed by

$$\Lambda v + \lambda v = 0 , \qquad x \in \bar{\omega}_h , \quad \Lambda v = \Lambda_1 v + \Lambda_2 v ,$$

$$(8) \qquad \Lambda_\alpha v = \begin{cases} \dfrac{2}{h_\alpha}\, v_{x_\alpha}, & x_\alpha = 0 , \\ v_{\bar{x}_\alpha x_\alpha}, & h_\alpha \le x_\alpha \le l_\alpha - h_\alpha , \\ -\dfrac{2}{h_\alpha}\, v_{\bar{x}_\alpha}, & x_\alpha = l_\alpha, \quad \alpha = 1, 2 . \end{cases}$$

Indeed, assuming that $u(x)$ is a solution to the equation $\Delta u + \lambda u = 0$ subject to the condition $\partial u/\partial n|_{\Gamma} = 0$, the error of approximation for the boundary condition is represented by

$$\nu_1^- = \left(u_{x_1} + \tfrac{1}{2} h_1 \Lambda_2 u + \tfrac{1}{2} h_1 \lambda u\right)\Big|_{x_1=0}$$

$$= \left(\frac{\partial u}{\partial x_1} + \frac{h_1}{2} \frac{\partial^2 u}{\partial x_1^2} + \frac{h_1}{2} \frac{\partial^2 u}{\partial x_2^2} + \frac{h_1 \lambda u}{2}\right)\Bigg|_{x_1=0} + O(h_1^2)$$

$$= \frac{\partial u}{\partial x_1}\Big|_{x_1=0} + \frac{h_1}{2} (\Delta u + \lambda u)\Big|_{x_1=0} + O(h_1^2) = O(h_1^2) .$$

In a similar way we find that

$$\nu_1^+ = \left(-u_{\bar{x}_1} + \tfrac{1}{2}\, h_1\, \Lambda_2\, u + \tfrac{1}{2}\, h_1\, \lambda\, u\right)\Big|_{x_1=l_1} = O(h_1^2)\,,$$

etc. By means of the method of separation of variables problem (8) reduces to the two one-dimensional eigenvalue problems for $\mu(x_1) = \mu(i_1\, h_1)$ and $\eta(x_2) = \eta(i_2\, h_2)$:

$$\mu_{\bar{x}_1 x_1} + \lambda^{(1)}\mu = 0\,, \qquad \left(\mu_{x_1} + \tfrac{1}{2}\, h_1\, \lambda^{(1)}\mu\right)\Big|_{x_1=0} = 0\,,$$

$$\left(-\mu_{\bar{x}_1} + \tfrac{1}{2}\, h_1\, \lambda^{(1)}\mu\right)\Big|_{x_1=l_1} = 0\,;$$

$$\eta_{\bar{x}_2 x_2} + \lambda^{(2)}\eta = 0\,, \qquad \left(\eta_{x_2} + \tfrac{1}{2}\, h_2\, \lambda^{(2)}\eta\right)\Big|_{x_2=0} = 0\,,$$

$$\left(-\eta_{\bar{x}_2} + \tfrac{1}{2}\, h_2\, \lambda^{(2)}\eta\right)\Big|_{x_1=l_2} = 0\,.$$

Their solutions have been already found in Chapter 2, Section 3.

Along these lines, we can immediately write down the eigenvalues and the orthonormal eigenfunctions

(9) $$\lambda_k = \lambda_{k_1 k_2} = \lambda_{k_1}^{(1)} + \lambda_{k_2}^{(2)}\,,$$

(10) $$v_k(x) = v_{k_1 k_2}(x_1\,, x_2) = \mu_{k_1}(x_1)\, \eta_{k_2}(x_2)\,,$$

$$k_\alpha = 0, 1, 2, \ldots, N_\alpha\,, \quad \alpha = 1, 2\,,$$

where

$$\lambda_0^{(1)} = \lambda_0^{(2)} = 0\,,$$

(11) $$\lambda_{k_\alpha}^{(\alpha)} = \frac{4}{h_\alpha^2}\, \sin^2 \frac{\pi\, k_\alpha\, h_\alpha}{2 l_\alpha}\,,$$

$$1 \le k_\alpha \le N_\alpha\,, \quad \alpha = 1, 2\,;$$

$$
(12)\quad \begin{cases}
\mu_0(x_1) = \sqrt{\dfrac{1}{l_1}}\,, \qquad \mu_{N_1}(x_1) = \sqrt{\dfrac{1}{l_1}}\,\cos\dfrac{\pi\, N_1\, x_1}{l_1}\,, \\[2ex]
\mu_{k_1}(x_1) = \sqrt{\dfrac{2}{l_1}}\,\cos\dfrac{\pi\, k_1\, x_1}{l_1}\,, \qquad k_1 = 1, 2, \ldots, N_1 - 1\,, \\[2ex]
\eta_0(x_2) = \sqrt{\dfrac{1}{l_2}}\,, \qquad \eta_{N_2}(x_2) = \sqrt{\dfrac{1}{l_2}}\,\cos\dfrac{\pi\, N_2\, x_2}{l_2}\,, \\[2ex]
\eta_{k_2}(x_2) = \sqrt{\dfrac{2}{l_2}}\,\cos\dfrac{\pi k_2\, x_2}{l_2}\,, \qquad k_2 = 1, 2, \ldots, N_2 - 1\,,
\end{cases}
$$

and

$$
(13)\qquad \left[ v_{k_1 k_2}\,, v_{k_1' k_2'} \right] = \delta_{k_1 k_1'}\,\delta_{k_2 k_2'}
$$

with

$$
\begin{aligned}
(14)\qquad [y, v] = (y, v) &+ \frac{h_1\, h_2}{4}\,\big[ y(0,0)\, v(0,0) + y(0, l_2)\, v(0, l_2) \\
&+ y(l_1\,, 0)\, v(l_1\,, 0) + y(l_1\,, l_2)\, v(l_1\,, l_2) \big] \\
&+ \frac{h_1}{2} \sum_{i_2=2}^{N_2-1} \big[ y(0, i_2\, h_2)\, v(0, i_2\, h_2) \\
&+ y(N_1\, h_1\,, i_2\, h_2)\, v(N_1\, h_1, i_2\, h_2) \big]\, h_2 \\
&+ \frac{h_2}{2} \sum_{i_1=2}^{N_1-1} \big[ y(i_1\, h_1\,, 0)\, v(i_1\, h_1\,, 0) \\
&+ y(i_1\, h_1\,, N_2\, h_2)\, v(i_1\, h_1\,, N_2\, h_2) \big]\, h_1\,,
\end{aligned}
$$

the inner product $(y, v)$ being understood in the sense of (7).

**2. Properties of difference operators.** Introduce the space $\overset{\circ}{\Omega}_h$ of all grid functions defined on the grid $\bar{\omega}_h = \omega_h + \gamma_h$ and vanishing on the boundary $\gamma_h$ and the space $\Omega_h$ comprising all the functions defined on $\omega_h$. The inner product in the space $\Omega_h$ is defined to be

$$
(15)\qquad (y, v) = \sum_{x \in \omega_h} y(x)\, v(x)\, h_1\, h_2\,.
$$

We list below the properties of the difference Laplace operator acting from $\overset{\circ}{\Omega}_h$ into $\Omega_h \subset \overset{\circ}{\Omega}_h$. By definition,

$$\Lambda\, v = \Lambda_1\, v + \Lambda_2\, v, \qquad \Lambda_\alpha\, v = v_{\bar{x}_\alpha\, x_\alpha}, \quad \alpha = 1, 2 .$$

1. The operator $\Lambda$ is self-adjoint:

$$(\Lambda\, y, v) = (y, \Lambda\, v) \quad \text{for all} \quad y, v \in \overset{\circ}{\Omega}_h . \tag{16}$$

To make sure of it, it is straightforward to verify the chain of the relations

$$\begin{aligned}(\Lambda_1\, y, v) &= \sum_{i_2=1}^{N_2-1} h_2 \sum_{i_1=1}^{N_1-1} h_1\, (v\, \Lambda_1\, y)_{i_1 i_2} \\ &= \sum_{i_2=1}^{N_2-1} h_2 \sum_{i_1=1}^{N_1-1} h_1\, (y\, \Lambda_1\, v)_{i_1 i_2} = (y, \Lambda_1\, v)\end{aligned}$$

with a simple observation that the operator $\Lambda_1$ is self-adjoint on the grid

$$\bar{\omega}_h^{(1)} = \left\{ x_1 = i_1\, h_1 \, , \ i_1 = 0, 1, \ldots, N_1 \, , \ N_1\, h_1 = l_1 \right\} .$$

This is due to the fact that the order of summation over $i_1$ and $i_2$ may be interchanged. In a similar way we obtain

$$(\Lambda_2\, y, v) = (y, \Lambda_2\, v) \, ,$$

which leads to (16).

2. The operator $-\Lambda$ is positive definite:

$$(-\Lambda\, y, y) \geq \delta\, \|\, y\, \|^2 \, , \tag{17}$$

where

$$\delta = \frac{4}{h_1^2} \sin^2 \frac{\pi\, h_1}{2\, l_1} + \frac{4}{h_2^2} \sin^2 \frac{\pi\, h_2}{2\, l_2} \geq \frac{8}{l_1^2} + \frac{8}{l_2^2} = \delta_0 \, .$$

This property follows immediately from the relation

$$\lambda_{\min}\, \|\, y\, \|^2 \leq (-\Lambda\, y, y) \, ,$$

which is certainly true with $\delta = \lambda_{\min}$ (see Section 1).

3. The operator $\Lambda$ admits the bilateral estimate

$$\delta \, \| y \|^2 \le (-\Lambda \, y, y) \le \Delta \, \| y \|^2 , \tag{18}$$

where $\delta$ is the same as before and

$$\Delta = \lambda_{\max} = \frac{4}{h_1^2} \cos^2 \frac{\pi \, h_1}{2 \, l_1} + \frac{4}{h_2^2} \cos^2 \frac{\pi \, h_2}{2 \, l_2} < \frac{4}{l_1^2} + \frac{4}{l_2^2} = \Delta_0 \, .$$

From the relations just established and the lower estimate for $\delta$ we deduce that

$$\delta_0 \, \| y \|^2 \le (-\Lambda \, y, y) \le \Delta_0 \, \| y \|^2 . \tag{19}$$

These properties provide support for the view that the operator $A = -\Lambda$ is more convenient for many things than the operator $\Lambda$. The operator $A$ can be treated as an operator either from $H_h = \overset{\circ}{\Omega}_h$ into $\Omega_h \subset H_h$ or from $H_h = \Omega_h$ into $H_h$ by merely setting

$$A y = - \widetilde{\Lambda} y \, , \qquad y \in \Omega_h \, ,$$

where $\widetilde{\Lambda} y = \Lambda y$ for $y \in \overset{\circ}{\Omega}_h$. In accordance with what has been said above, we have in the space $\Omega_h$

$$A = A^* , \qquad \delta \, E \le A \le \Delta \, E$$

with $E$ denoting the identity operator.

Any function $f \in \overset{\circ}{\Omega}_h$, defined on the grid $\bar{\omega}_h$ and vanishing on the boundary $\gamma_h$ or defined only on $\omega_h$, can be expanded into a series of the eigenfunctions of the operator $A$:

$$f(x) = \sum_{k=1}^{N} c_k \, v_k \, , \qquad c_k - (f, v_k) \, ,$$

so that $\sum k = 1^N c_k^2 = \| f \|^2$, where $v_k$ are the eigenvectors of the operator $A$: $A \, v_k = \lambda_k \, v_k$, $N = (N_1 - 1)(N_2 - 1)$.

In the case of the second eigenvalue boundary-value problem (8) the space $H_h = \bar{\Omega}_h$ comprises all the functions defined on the grid $\bar{\omega}_h$; the inner product $(,)$ on $H_h$ is to be understood in the sense (14) and the operator $A$ is defined as a sum

$$A = A_1 + A_2 \, ,$$

where

$$A_\alpha\, y = -\Lambda_\alpha y = -y_{\bar{x}_\alpha x_\alpha} \qquad \text{for} \quad 0 < x_\alpha = i_\alpha\, h_\alpha < l_\alpha\,,$$

$$A_\alpha\, y = -\Lambda_\alpha^- y = -2\, y_{x_\alpha}/h_\alpha \qquad \text{for} \quad x_\alpha = 0\,,$$

$$A_\alpha\, y = -\Lambda_\alpha^+ y = 2\, y_{\bar{x}_\alpha}/h_\alpha \qquad \text{for} \quad x_\alpha = N_\alpha\, h_\alpha = l_\alpha, \quad \alpha = 1, 2\,.$$

With this in mind, problem (8) is convenient to be taken in the operator form $A\,y = \lambda\,y$. The operator $A$ so defined is self-adjoint and nonnegative:

$$0 \le A \le \Delta\, E, \qquad A = A^*,$$

where $\Delta = \lambda_{\max} = 4(h_1^{-2} + h_1^{-2})$.

Observe that the operators $A_1$ and $A_2$ are commuting and self-adjoint both for the first and second boundary-value problems. The methodology of the general theory (see Chapter 2, Section 1) guides the choice of a common system of eigenfunctions coinciding with the system of eigenfunctions for the operator $A = A_1 + A_2$; in so doing, each eigenvalue of the operator $A$ is equal to a sum of the appropriate eigenvalues of the operators $A_1$ and $A_2$:

$$\lambda(A) = \lambda(A_1) + \lambda(A_2)\,.$$

**3. The Laplace operator in a domain consisting of rectangles.** We now consider a domain consisting of a finite number of rectangles, whose sides are parallel to the coordinate axes as shown in Fig. 14.

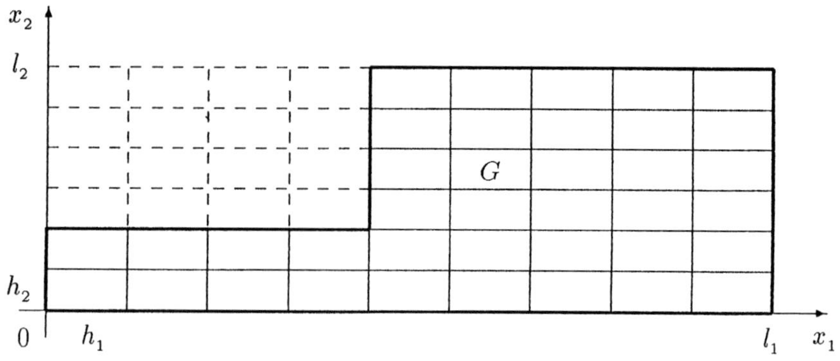


**Figure 14.**

The sides of rectangles, which constitute the domain $G$ are assumed to be commensurable. All this enables us to place in the plane a grid with steps $h_1$ and $h_2$ so that the boundary of the grid domain lies on the boundary of the domain $G$. One trick we have encountered is to complete the domain $G$ to the rectangle and then denote it by $\widetilde{G}$ (see Fig. 14). After that, we construct in $G$ a difference grid $\omega_h$ and extend it to $\widetilde{G}$. The notation $\tilde{\omega}_h$ will be used for the grid in the domain $\widetilde{G}$.

Let $v$ be a grid function defined on the grid $\omega_h$ and satisfying the boundary condition $v|_{\gamma_h} = 0$. Also, it will be sensible to introduce the function

$$\tilde{v}(x) = \begin{cases} v(x), & x \in \omega_h\,, \\ 0, & x \in \tilde{\omega}_h \setminus \omega_h\,. \end{cases}$$

By definition,

$$\|\, \tilde{v} \,\|_{\tilde{\omega}_h} = \|\, v \,\|_{\omega_h} = \|\, v \,\|\,,$$

where

$$\|\, v \,\|^2 = (v, v), \quad (v, y) = \sum_{x \in \omega_h} v(x)\, y(x)\, h_1\, h_2\,,$$

$$\|\, \tilde{v} \,\|^2_{\tilde{\omega}_h} = \sum_{x \in \tilde{\omega}_h} \tilde{v}^2(x)\, h_1\, h_2\,, \qquad \sum_{\alpha=1}^{2} \left\| \tilde{v}_{\bar{x}_\alpha} \right\|^2_{\tilde{\omega}_h} = \sum_{\alpha=1}^{2} \left\| v_{\bar{x}_\alpha} \right\|^2.$$

With these relations established, we deduce that estimate (19) holds for any function $v(x)$ defined on the grid $\omega_h$ in the domain $G$, providing the values

$$\delta_0 = 8\,(l_1^{-2} + l_2^{-2}), \qquad \Delta_0 = 4\,(h_1^{-2} + h_2^{-2})\,,$$

where $l_1$ and $l_2$ are the sides of the rectangle $\widetilde{G}$.

**4. The embedding theorem.** Various a priori estimates for the equation $Ay = \varphi$ can be obtained in the energy norms in light of the properties of elliptic operators. One might expect that the energy estimates imply a uniform estimate, that is, an estimate in the norm

$$\|\, y \,\|_C = \max_{x \in \omega_h} |\, y(x) \,|\,.$$

The following grid theorem gives a definite answer to this hint.

**The embedding theorem.** *For any function $y(x) \in \overset{\circ}{\Omega}_h$ defined on the grid $\tilde{\omega}_h = \{x_i = (i_1 h_1, i_2 h_2),\ i_\alpha = 0, 1, \ldots, N_\alpha,\ h_\alpha = l_\alpha / N_\alpha,\ \alpha = 1, 2\}$ and vanishing on the boundary (for $x_\alpha = 0, l_\alpha$; $\alpha = 1, 2$) the estimate*

$$\|\, y \,\|_C \le M_0 \,\|\, Ay \,\| \tag{20}$$

*is valid with*

$$A\,y = -y_{\bar{x}_1 x_1} - y_{\bar{x}_2 x_2}\,, \qquad M_0 = \frac{l_0^2}{2\sqrt{l_1\, l_2}}\,, \qquad l_0 = \max(l_1\,, l_2)\,.$$

**Proof** Suppose that $v_{k_1k_2}(x)$ and $\lambda_{k_1k_2}(x)$ are the eigenfunctions and eigenvalues of the operator $A$ that we have determined in Section 1 for $k_\alpha = 1, 2, \dots, N_\alpha - 1$ and $\alpha = 1, 2$. The expansion of the function $y(x)$ with respect to the orthonormal functions $\{v_{k_1k_2}(x)\}$

$$y(x) = \sum_{k_1,\,k_2} c_{k_1k_2}\, v_{k_1k_2}(x)$$

yields

$$Ay = \sum_{k_1,\,k_2} c_{k_1k_2}\, \lambda_{k_1k_2}\, v_{k_1k_2}\,,$$

$$\|\, y\,\|^2 = \sum_{k_1,\,k_2} c_{k_1k_2}^2\,,$$

$$\|\, Ay\,\|^2 = \sum_{k_1,\,k_2} c_{k_1k_2}^2\, \lambda_{k_1k_2}^2\,.$$

The function $y(x)$ satisfies the inequality

$$|\, y(x)\,| \le \Big(\sum_{k_1,\,k_2} |\, c_{k_1k_2}\,|\Big) \max_{k_1,\,k_2} |\, v_{k_1k_2}(x)\,|\,.$$

From formula (5) we get $|\, v_{k_1k_2}(x)\,| \le 2/\sqrt{l_1\, l_2}$. Applying the Cauchy–Bunyakovskiĭ inequality we establish the chain of the relations

$$(21) \qquad |\, y(x)\,|^2 \le \frac{4}{l_1\, l_2} \Bigg(\sum_{k_1,\,k_2} |\, c_{k_1k_2}\,|\Bigg)^2$$

$$= \frac{4}{l_1\, l_2} \Bigg(\sum_{k_1,\,k_2} |\, c_{k_1k_2}\,|\, \lambda_{k_1k_2}\, \frac{1}{\lambda_{k_1k_2}}\Bigg)^2$$

$$\le \frac{4}{l_1\, l_2} \sum_{k_1,\,k_2} \lambda_{k_1k_2}^2\, c_{k_1k_2}^2 \sum_{k_1,\,k_2} \frac{1}{\lambda_{k_1k_2}^2}$$

$$= \frac{4}{l_1\, l_2}\, \|\, Ay\,\|^2 \sum_{k_1,\,k_2} \lambda_{k_1k_2}^{-2}\,.$$

With this in mind, it remains only to evaluate the remaining sums

$$S = \sum_{k_1, k_2} \lambda_{k_1 k_2}^{-2} = \lambda_{1,1}^{-2} + \sum_{\substack{k_1=1 \\ k_1+k_2>2}}^{N_1-1} \sum_{k_2=1}^{N_2-1} \lambda_{k_1 k_2}^{-2}.$$

Because the function $\sin \varphi/\varphi$ is monotonically decreasing on the segment $0 \leq \varphi \leq \pi/2$ and, therefore, $\sin \varphi \geq 2\varphi/\pi$, formula (4) reduces in this case to

$$\lambda_{k_1 k_2} = \frac{\pi^2 k_1^2}{l_1^2} \left( \frac{\sin \varphi_1}{\varphi_1} \right)^2 + \frac{\pi^2 k_2^2}{l_2^2} \left( \frac{\sin \varphi_2}{\varphi_2} \right)^2 \geq 4 \left( \frac{k_1^2}{l_1^2} + \frac{k_2^2}{l_2^2} \right),$$

where $\varphi_\alpha = \pi h_\alpha k_\alpha/(2 l_\alpha)$, $\alpha = 1, 2$. Using this estimate behind, we obtain

$$\sum_{\substack{k_1=1 \\ k_1+k_2>2}}^{N_1-1} \sum_{k_2=1}^{N_2-1} \lambda_{k_1 k_2}^{-2} \leq \frac{l_0^4}{16} \sum_{\substack{k_1=1 \\ k_1+k_2>2}}^{N_1-1} \sum_{k_2=1}^{N_2-1} \left( k_1^2 + k_2^2 \right)^{-2} = \frac{l_0^4}{16} J.$$

Majorizing the sum $J$ by the integral

$$J \leq \int_1^\infty \int_0^{\pi/2} \frac{r\, dr\, d\varphi}{r^4} = \frac{\pi}{4}$$

and taking into account that $\lambda_{1,1} \geq \delta_0 = \dfrac{8}{l_1^2} + \dfrac{8}{l_2^2} \geq \dfrac{16}{l_0^4}$, we deduce that

$$S \leq \frac{l_0^4}{16^2} + \frac{\pi\, l_0^4}{4 \cdot 16} = \frac{l_0^4}{64} \left( \pi + \frac{1}{4} \right) \leq \frac{l_0^4}{16}. \tag{22}$$

Substitution of (22) into (21) yields inequality (20).

**5. Equations with variable coefficients.** The Dirichlet problem for the elliptic equation in the domain $G + \Gamma = \bar{G}$ comes next:

$$L u = -f(x), \qquad x \in G, \quad u = \mu(x), \quad x \in \Gamma, \tag{23}$$

with a rectangle $\bar{G} = \{0 \leq x_\alpha \leq l_\alpha,\ \alpha = 1, 2\}$ and

$$Lu = \sum_{\alpha=1}^{2} L_\alpha u, \qquad L_\alpha u = \frac{\partial}{\partial x_\alpha} \left( k_\alpha(x) \frac{\partial u}{\partial x_\alpha} \right),$$

where $k_\alpha(x)$ are sufficiently smooth functions, $0 < c_1 \le k_\alpha(x) \le c_2$, $\alpha = 1, 2$, with constants $c_1 > 0$ and $c_1 > 0$. We compose in just the same way the grid with steps $h_1$ and $h_2$ and denote it by $\bar{\omega}_h = \omega_h + \gamma_h$ as we did in Section 1.

An excellent start in this direction is to approximate the operator $L_\alpha$ by the difference operator

$$\Lambda_\alpha u = \left(a_\alpha u_{\bar{x}_\alpha}\right)_{x_\alpha} \sim L_\alpha u = \frac{\partial}{\partial x_\alpha}\left(k_\alpha(x)\, \frac{\partial u}{\partial x_\alpha}\right),$$

where $a_\alpha(x)$ is so chosen as to satisfy the relations

$$\Lambda_\alpha u - L_\alpha u = O(h_\alpha^2), \qquad 0 < c_1 \le a_\alpha(x) \le c_2 . \tag{24}$$

This means that $\Lambda_\alpha$ provides an approximation of order 2, for instance, for $a_\alpha = k_\alpha^{(-0.5_\alpha)}$ or

$$a_1(x_1, x_2) = k_1(x_1 - \tfrac{1}{2} h_1, x_2), \qquad a_2(x_1, x_2) = k_2(x_1, x_2 - \tfrac{1}{2} h_2) .$$

The next step is to put the operator $\Lambda u = \sum_{\alpha=1}^{2} \Lambda_\alpha u$ in correspondence with the operator $Lu$ due to which the difference Dirichlet problem

$$\begin{aligned} &\Lambda y = -\varphi, \qquad x \in \omega_h, \quad y = \mu(x), \quad x \in \gamma_h, \\ &\Lambda y = (\Lambda_1 + \Lambda_2)\, y, \qquad \Lambda_\alpha y = \left(a_\alpha y_{\bar{x}_\alpha}\right)_{x_\alpha}, \end{aligned} \tag{25}$$

having the approximation error (residual) $\psi = \Lambda u + \varphi = O(|h|^2)$ will be associated with problem (23).

Such an approximation is the result of a natural generalization of homogeneous conservative schemes from Chapter 3 for one-dimensional equations to the multidimensional case. These schemes can be obtained by means of the integro-interpolational method without any difficulties.

We now investigate the properties of the operator $\Lambda$ in the space $\overset{\circ}{\Omega}_h$ of all grid functions with the inner product in the sense of (15). In what follows we accept $A y = -\Lambda y$ for any $y \in \overset{\circ}{\Omega}$.

**Lemma** *The operator $A y = -\Lambda y = -\sum_{\alpha=1}^{2} \Lambda_\alpha y$, $y \in H$, is self-adjoint, positive definite and admits the estimates*

$$c_1 (\overset{\circ}{A} v, v) \le (Av, v) \le c_2 (\overset{\circ}{A} v, v), \tag{26}$$

$$c_1 \delta \| v \|^2 \le (Av, v) \le c_2 \Delta \| v \|^2 \tag{27}$$

*for any* $v \in H$, *where*

$$\overset{\circ}{A}\, y = -\overset{\circ}{\Lambda}\, y = -\sum_{\alpha=1}^{2} y_{\bar{x}_\alpha x_\alpha}$$

*and the constants* $\delta > 0$ *and* $\Delta > 0$ *arise from formulae (18) and (19).*

As far as $v|_{\gamma_h} = 0$, the first Green formula with regard to the operator $\Lambda_\alpha v = \left(a_\alpha v_{\bar{x}_\alpha}\right)_{x_\alpha}$ implies that

$$(\Lambda_\alpha v, v) = -\left(a_\alpha v_{\bar{x}_\alpha}, v_{\bar{x}_\alpha}\right]_\alpha,$$

where

$$(w, z]_1 = \sum_{i_1=1}^{N_1} \sum_{i_2=1}^{N_2-1} w(i_1 h_1, i_2 h_2)\, z(i_1 h_1, i_2 h_2)\, h_1 h_2,$$

$$(w, z]_2 = \sum_{i_1=1}^{N_1-1} \sum_{i_2=1}^{N_2} w(i_1 h_1, i_2 h_2)\, z(i_1 h_1, i_2 h_2)\, h_1 h_2.$$

With these representations in view, we have

$$(Av, v) = \sum_{\alpha=1}^{2} \left(a_\alpha v_{\bar{x}_\alpha}, v_{\bar{x}_\alpha}\right]_\alpha.$$

On the other hand,

$$(\overset{\circ}{A}\, v, v) = \sum_{\alpha=1}^{2} \left(v_{\bar{x}_\alpha}, v_{\bar{x}_\alpha}\right]_\alpha.$$

By condition (24) we are led to (26) and so it remains only to substitute into (26) the bilateral estimate $\delta\, \|v\|^2 \le (\overset{\circ}{A}\, v, v) \le \Delta\, \|v\|^2$, which has been obtained in Section 2. These manipulations permit us to derive (27) with the ingredients

$$\delta = \sum_{\alpha=1}^{2} \frac{4}{h_\alpha^2} \sin^2 \frac{\pi h_\alpha}{2 l_\alpha}, \qquad \Delta = \sum_{\alpha=1}^{2} \frac{4}{h_\alpha^2} \cos^2 \frac{\pi h_\alpha}{2 l_\alpha}.$$

In what follows inequalities (26) and (27) are adopted in operator form and we agree to consider

$$c_1\, \overset{\circ}{A} \le A \le c_2\, \overset{\circ}{A}, \tag{28}$$

$$c_1\, \delta E \le A \le c_2\, \Delta E, \tag{29}$$

where $E$ stands for the identity operator.

The self-adjointness of the operator $A = A_1 + A_2$ can be proved by exactly the same reasoning as in Section 2 with further reference to some properties such as

$$A_\alpha^* = A_\alpha > 0\,, \qquad A_\alpha\, y = -\Lambda_\alpha\, y = -\big(a_\alpha\, y_{\bar{x}_\alpha}\big)_{x_\alpha} \quad \text{for} \quad y \in \mathring{\Omega}_h\,.$$

To make sure of it, it suffices to calculate the row sum with $\alpha = 1$ for fixed $i_2 = 1, 2, \dots, N_2 - 1$. The outcome of this is

$$\sum_{i_1=1}^{N_1-1} (A_1\, y)_{i_1}\, v_{i_1}\, h_1 = - \sum_{i_1=1}^{N_1-1} (a_1\, y_{\bar{x}_1})_{x_1, i_1}\, v_{i_1}\, h_1$$

$$= \sum_{i_1=1}^{N_1-1} y_{i_1}\, (a_1\, v_{\bar{x}_1})_{x_1, i_1}\, h_1$$

$$= \sum_{i_1=1}^{N_1-1} y_{i_1}\, (A_1\, v)_{i_1}\, h_1\,.$$

Multiplying this identity by $h_2$ and summing over $i_2 = 1, 2, \dots, N_2 - 1$, we establish $(A_1\, y, v) = (y, A_1\, v)$ and, in a similar way, $(A_2\, y, v) = (y, A_2\, v)$, yielding

$$(A\, y, v) = \big((A_1 + A_2)\, y, v\big) = (y, Av)\,.$$

By definition, this means $A^* = A$.

**6. Equations with mixed derivatives.** In this section we consider problem (23) involving the elliptic operator $L$ with mixed derivatives

$$Lu = \sum_{\alpha, \beta=1}^{2} L_{\alpha\beta}\, u, \qquad L_{\alpha\beta}\, u = \frac{\partial}{\partial x_\alpha} \Big(k_{\alpha\beta}\, \frac{\partial u}{\partial x_\beta}\Big)\,, \tag{30}$$

assuming the ellipticity conditions

$$c_1 \sum_{\alpha=1}^{2} \xi_\alpha^2 \le \sum_{\alpha, \beta=1}^{2} k_{\alpha\beta}(x)\, \xi_\alpha\, \xi_\beta \le c_2 \sum_{\alpha=1}^{2} \xi_\alpha^2, \qquad x \in G\,, \tag{31}$$

to be valid, where $c_1 > 0$, $c_2 > 0$ are constants and $\boldsymbol{\xi} = (\xi_1\,, \xi_2)$ is an arbitrary vector. Setting first $\xi_1 = 1$, $\xi_2 = 0$ and then $\xi_1 = 0$, $\xi_2 = 1$ it is not difficult to check that

$$0 < c_1 \le k_{\alpha\alpha} \le c_2\,, \qquad \alpha = 1, 2\,.$$

In this line, the operator $L_{\alpha\beta}\, u$ is approximated by the difference operator

$$\Lambda_{\alpha\beta}\, u = \tfrac{1}{2}\left((k_{\alpha\beta}\, u_{\bar{x}\beta})_{x_\alpha} + (k_{\alpha\beta}\, u_{x\beta})_{\bar{x}_\alpha}\right), \tag{32}$$

which is defined for $\alpha \neq \beta$ on the 7-point pattern (see Fig. 15) consisting of the nodes

$$(x_1\,, x_2),\quad (x_1 \pm h_1\,, x_2),\quad (x_1\,, x_2 \pm h_2),\quad (x_1 - h_1\,, x_2 + h_2)\,.$$

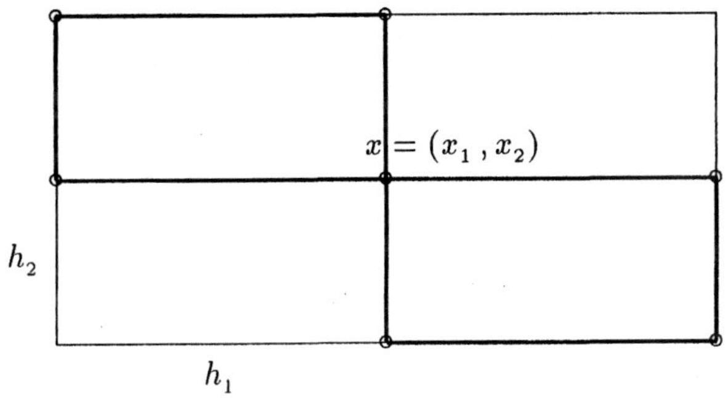


**Figure 15.**

The operator $\Lambda_{\alpha\beta}$ arranges itself as a sum

$$\Lambda_{\alpha\beta} = \tfrac{1}{2}\left(\Lambda^-_{\alpha\beta} + \Lambda^+_{\alpha\beta}\right), \qquad \Lambda^-_{\alpha\beta} u = \left(k_{\alpha\beta}\, u_{\bar{x}_\beta}\right)_{x_\alpha}, \qquad \Lambda^+_{\alpha\beta} u = \left(k_{\alpha\beta}\, u_{x_\beta}\right)_{\bar{x}_\alpha}.$$

We are going to show that $\Lambda^{\pm}_{12}$ and $\Lambda_{12}$ have the first and second order of approximation, respectively:

$$\Lambda^{\pm}_{12}\, u = L_{12}\, u + O(h_1 + h_2), \qquad \Lambda_{12}\, u = L_{12}\, u + O(|\, h\,|^2)\,. \tag{33}$$

Indeed, we refer to the operator $\Lambda^-_{12}\, u = \left(k_{12}\, u_{\bar{x}_2}\right)_{x_1}$. Substituting here the expressions

$$u_{\bar{x}_2} = \frac{\partial u}{\partial x_2} - \frac{h_2}{2}\, \frac{\partial^2 u}{\partial x_2^2} + \frac{h_2^2}{6}\, \frac{\partial^3 u}{\partial x_2^3}(x_1\,, \bar{x}_2), \quad \bar{x}_2 \in [x_2 - h_2\,, x_2]\,,$$

$$v_{x_1} = \frac{\partial v}{\partial x_1} + \frac{h_1}{2}\, \frac{\partial^2 v}{\partial x_1^2} + O(h_1^2)$$

for $v = k_{12}\, u_{\bar{x}_2}$ yields

$$\Lambda_{12}^{-} u = L_{12}\, u + \frac{h_1}{2}\, \frac{\partial}{\partial x_1}\, L_{12}\, u - \frac{h_2}{2}\, L_{12}\, \frac{\partial u}{\partial x_2} + O(|\, h\,|^2)\,.$$

Likewise,

$$\Lambda_{12}^{+} u = L_{12}\, u - \frac{h_1}{2}\, \frac{\partial}{\partial x_1} L_{12}\, u + \frac{h_2}{2}\, L_{12}\, \frac{\partial u}{\partial x_2} + O(|\, h\,|^2)\,,$$

from which estimates (33) follow immediately.

When $\alpha = \beta = 1$ and $\alpha = \beta = 2$ we might have

$$\Lambda_{11}\, u = \tfrac{1}{2} \left( (k_{11}\, u_{\bar{x}_1})_{x_1} + (k_{11}\, u_{x_1})_{\bar{x}_1} \right) = (a_{11}\, u_{\bar{x}_1})_{x_1}\,,$$

$$\Lambda_{22}\, u = \tfrac{1}{2} \left( (k_{22}\, u_{\bar{x}_2})_{x_2} + (k_{22}\, u_{x_2})_{\bar{x}_2} \right) = (a_{22}\, u_{\bar{x}_2})_{x_2}\,,$$

where

$$a_{11} = \tfrac{1}{2} \left[ k_{11} (x_1 - h_1\,, x_2) + k_{11} (x_1\,, x_2) \right],$$

$$a_{22} = \tfrac{1}{2} \left[ k_{22} (x_1\,, x_2 - h_2) + k_{22} (x_1\,, x_2) \right].$$

From the preceding relations it seems clear that $\Lambda_{\alpha\alpha}\, u - L_{\alpha\alpha}\, u = O(h_\alpha^2)$, $\alpha = 1, 2$. Thus, the differential operator (30) is approximated to second order by the difference operator

$$\Lambda\, y = \sum_{\alpha,\, \beta=1}^{2} \Lambda_{\alpha\beta}\, y\,,$$

meaning

$$\Lambda\, u - L\, u = O(|\, h\,|^2)\,.$$

In addition to operator (32), we take into consideration one more operator with the values

$$\Lambda y = \sum_{\alpha,\beta=1}^{2} \widetilde{\Lambda}_{\alpha\beta}\, y, \qquad \widetilde{\Lambda}_{\alpha\beta}\, y = \tfrac{1}{2} \left[ (k_{\alpha\beta}\, y_{x_\beta})_{x_\alpha} + (k_{\alpha\beta}\, y_{\bar{x}_\beta})_{\bar{x}_\alpha} \right],$$

providing an approximation of $O(|\, h\,|^2)$.

Let us investigate the properties of the operator $Ay = -\Lambda y$ for any $y \in \overset{\circ}{\Omega}_h = H$. The operator $Ay = \frac{1}{2}\left(A^- y + A^+ y\right)$ with the members

$$A^- y = - \sum_{\alpha,\beta=1}^{2} \left(k_{\alpha\beta}\, y_{x_\beta}\right)_{\bar{x}_\alpha}, \qquad A^+ y = - \sum_{\alpha,\beta=1}^{2} \left(k_{\alpha\beta}\, y_{\bar{x}_\beta}\right)_{x_\alpha}$$

is self-adjoint: $A = A^*$ for $k_{12} = k_{21}$ and admits both estimates (28) and (29).

The operators $A^-$ and $A^+$ will be the subject of separate investigations. For arbitrary elements $y,\ v \in H$ the summation by parts formula yields

$$(34) \qquad (A^+ y, v) = - \sum_{\alpha,\beta=1}^{2} \left((k_{\alpha\beta}\, y_{\bar{x}_\beta})_{x_\alpha}, v\right) = \sum_{\alpha,\beta=1}^{2} \left(k_{\alpha\beta}\, y_{\bar{x}_\beta}, v_{\bar{x}_\alpha}\right]_\alpha ,$$

$$(35) \qquad (A^- y, v) = - \sum_{\alpha,\beta=1}^{2} \left((k_{\alpha\beta}\, y_{x_\beta})_{\bar{x}_\alpha}, v\right) = \sum_{\alpha,\beta=1}^{2} \left[k_{\alpha\beta}\, y_{x_\beta}, v_{x_\alpha}\right)_\alpha .$$

The interchange of $y$ and $v$ followed by that of $\alpha$ and $\beta$ results in the relations

$$(A^+ v, y) = \sum_{\alpha,\beta=1}^{2} \left(k_{\alpha\beta}\, v_{\bar{x}_\beta}, y_{\bar{x}_\alpha}\right]_\alpha = \sum_{\alpha,\beta=1}^{2} \left(k_{\beta\alpha}\, y_{\bar{x}_\beta}, v_{\bar{x}_\alpha}\right]_\beta .$$

Putting these together with (34) we verify that $(A^+)^* = A^+$ only if $k_{12} = k_{21}$. In a similar manner it is plain to show that in this case $(A^-)^* = A^-$ and, hence, $A^* = A$. In this regard, it is necessary to take into account that $y_{x_1} = 0$ for $x_2 = 0$ and $y_{\bar{x}_1} = 0$ for $x_2 = l_2$, while $y_{x_2} = 0$ for $x_1 = 0$ and $y_{\bar{x}_2} = 0$ for $x_1 = l_1$. Passing now to the expression $(A^+ y, y)$, which is always representable by

$$(A^+ y, y) = \sum_{\alpha,\beta=1}^{2} \left(k_{\alpha\beta}\, y_{\bar{x}_\beta}, y_{\bar{x}_\alpha}\right) + \sum_{i_2=1}^{N_2-1} \left(k_{11}\, (y_{\bar{x}_1})^2\right)_{i_1=N_1} h_1\, h_2$$

$$+ \sum_{i_1=1}^{N_1-1} \left(k_{22}\, (y_{\bar{x}_2})^2\right)_{i_2=N_2} h_1\, h_2 ,$$

we carry the sum over $\alpha, \beta$ under the sign of the inner product in the first summand, leading by (31) to

$$c_1 \sum_{\alpha=1}^{2} \left((y_{\bar{x}_\alpha})^2, 1\right) \leq \left(\sum_{\alpha,\beta=1}^{2} k_{\alpha\beta}\, y_{\bar{x}_\beta}, y_{\bar{x}_\alpha}\right) \leq c_2 \sum_{\alpha=1}^{2} \left((y_{\bar{x}_\alpha})^2, 1\right).$$

The expression for $(A^- y, y)$ can be transformed without difficulties in just the same way. Recalling now that $c_1 \le k_{\alpha\alpha} \le c_2$ and $(\overset{\circ}{A} y, y) = \sum_{\alpha=1}^{2} (y_{\bar{x}_\alpha}, y_{\bar{x}_\alpha}]_\alpha$ we obtain $c_1 \overset{\circ}{A} \le A \le c_2 \overset{\circ}{A}$, yielding inequalities (28). Inequality (29) can be deduced from the above relations by virtue of the estimate $\delta E \le \overset{\circ}{A} \le \Delta E$.

## 4.5 HIGHER–ACCURACY SCHEMES FOR POISSON'S EQUATION

In this section we consider higher-accuracy schemes for the Dirichlet problem (1) of Section 1 in a rectangle.

**1. The statement of the Dirichlet difference problem providing a higher-order approximation.** On the basis of the "cross" scheme it is possible to construct a scheme with the error of approximation $O(|h|^4)$ or $O(h^6)$ on a solution in the case of a square (cube) grid. In order to raise the order of approximation, we exploit the fact that $u = u(x)$ is a solution of Poisson's equation

$$\Delta u = -f(x) . \tag{1}$$

Without loss of generality we may restrict ourselves to the careful analysis of the two-dimensional case ($p = 2$) where

$$\Delta u = L_1 u + L_2 u , \qquad L_\alpha u = \frac{\partial^2 u}{\partial x_\alpha^2} ,$$

by appeal to the difference operator

$$\Lambda u = (\Lambda_1 + \Lambda_2) u , \qquad \Lambda_\alpha u = u_{\bar{x}_\alpha x_\alpha} ,$$

assuming $u = u(x)$ to possess all necessary derivatives. Then

$$\Lambda u - L u = \frac{h_1^2}{12} L_1^2 u + \frac{h_2^2}{12} L_2^2 u + O(|h|^4) . \tag{2}$$

From the equation $L_1 u + L_2 u = -f(x)$ we find that

$$L_1^2 u = -L_1 f - L_1 L_2 u, \qquad L_2^2 u = -L_2 f - L_1 L_2 u ,$$

so that

$$\Lambda u = L u - \frac{h_1^2}{12} L_1 f - \frac{h_2^2}{12} L_2 f - \frac{h_1^2 + h_2^2}{12} L_1 L_2 u + O(|h|^4) . \tag{3}$$

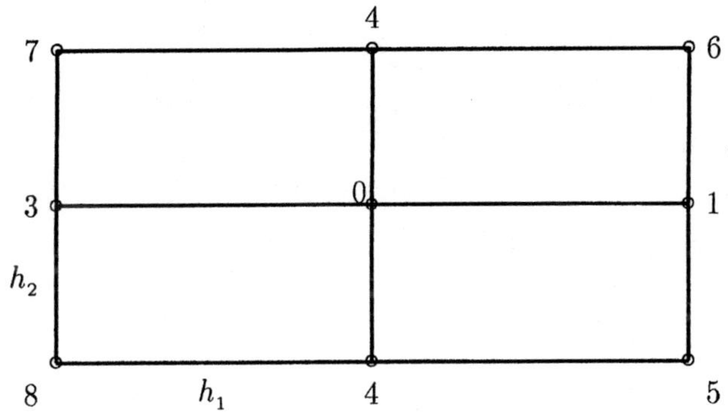


**Figure 16.**

We substitute here $Lu = -f$ and replace $L_1 L_2 u$ by the difference operator. The outcome of this is

$$\Lambda_1 \Lambda_2 u = u_{\bar{x}_1 x_1 \bar{x}_2 x_2} \sim L_1 L_2 u = \frac{\partial^4 u}{\partial x_1^2 \, \partial x_2^2} \; .$$

This operator is defined on the nine-point pattern shown in Fig. 16.

The expression for $\Lambda_1 \Lambda_2 u$ such that

$$\Lambda_1 \Lambda_2 u = \Lambda_1 \left[ \frac{u(x_1 \, , x_2 - h_2) - 2\, u(x_1 \, , x_2) + u(x_1 \, , x_2 + h_2)}{h_2^2} \right]$$

$$= \frac{1}{h_1^2 \, h_2^2} \left\{ u(x_1 - h_1 \, , x_2 - h_2) - 2\, u(x_1 \, , x_2 - h_2) \right.$$

$$+ u(x_1 + h_1 \, , x_2 - h_2) + 4\, u(x_1 \, , x_2)$$

$$- 2\, u(x_1 - h_1 \, , x_2) + u(x_1 - h_1 \, , x_2 + h_2) - 2 u(x_1 \, , x_2 + h_2)$$

$$\left. - 2\, u(x_1 + h_1 \, , x_2) + u(x_1 + h_1 \, , x_2 + h_2) \right\}$$

is needed in the estimation of the error of approximation to $\Lambda_1 \Lambda_2 u - L_1 L_2 u$ by virtue of the well-established expansion

$$\text{(4)} \qquad \Lambda v = v_{\bar{x}x} = \frac{v(x+h) - 2\, v(x) + v(x-h)}{h^2}$$

$$= v''(\xi), \qquad \xi = x + \theta h, \quad |\,\theta\,| \le 1 \, ,$$

assuming that $v(x)$ has a continuous second derivative on the segment $[x-h, x+h]$;

$$(5) \qquad \Lambda\, v = v_{\bar{x}x} = v''(x) + \frac{h^2}{12}\, v^{(4)}(\xi^*), \qquad \xi^* = x + \theta^* h, \quad |\,\theta^*\,| \le 1,$$

$v(x)$ has a continuous fourth derivative on the segment $[x-h, x+h]$. By relating $x_1$ to be fixed we might have

$$\Lambda_2\, u = L_2\, u(x_1\,, x_2) + \frac{h_2^2}{12}\, \frac{\partial^4 u}{\partial x_2^4}(x_1\,, \xi_2), \qquad \xi_2 = x + \theta_2\, h_2\,, \quad |\,\theta_2\,| \le 1\,.$$

Other ideas are connected with the expression

$$\Lambda_1\, \Lambda_2\, u(x_1\,, x_2) = \Lambda_1\, L_2\, u(x_1\,, x_2) + \frac{h_2^2}{12}\; \Lambda_1\; \frac{\partial^4 u}{\partial x_2^4}(x_1\,, \xi_2)\,.$$

Applying formula (5) with $v = L_2\, u$ and $x = x_1$ to the first summand yields

$$\Lambda_1\, L_2\, u(x_1\,, x_2) = L_1\, L_2\, u(x_1\,, x_2) + \frac{h_1^2}{12}\, \frac{\partial^4 u}{\partial x_1^4}(\xi_1^*\,, x_2)$$

$$\xi_1^* = x_1 + \theta_1^*\, h_1\,, \qquad |\,\theta_1^*\,| \le 1\,.$$

The same procedure works for the second summand with respect to formula (4):

$$\frac{h_2^2}{12}\; \Lambda_1\; \frac{\partial^4 u}{\partial x_2^4}(x_1\,, \xi_2) = \frac{h_2^2}{12}\; \frac{\partial^6 u}{\partial x_1^2\, \partial x_2^4}(\xi_1\,, \xi_2), \qquad \xi_1 = x_1 + \theta_1\, h_1\,, \quad |\,\theta_1\,| \le 1\,.$$

What should be done is to bring together the results obtained:

$$(\Lambda_1\, \Lambda_2 - L_1\, L_2)\, u = O(h_1^2) + O(h_2^2) = O(|\,h\,|^2)\,.$$

Substituting into (3) the expression for $\Lambda_1\, \Lambda_2\, u$ in place of $L_1\, L_2\, u$

$$L_1\, L_2\, u = \Lambda_1\, \Lambda_2\, u + O(|\,h\,|^2)\,,$$

and involving the equation $Lu = -f(x)$, we finally get

$$(6) \qquad \Lambda\, u = L\, u - \frac{h_1^2 + h_2^2}{12}\, \Lambda_1\, \Lambda_2\, u - \frac{h_1^2}{12}\, L_1\, f - \frac{h_2^2}{12}\, L_2\, f + O(|\,h\,|^4)$$

$$= -\left( f + \frac{h_1^2}{12}\, L_1\, f + \frac{h_2^2}{12}\, L_2\, f \right) - \frac{h_1^2 + h_2^2}{12}\, \Lambda_1\, \Lambda_2\, u + O(|\,h\,|^4)\,.$$

Because of this, the equation

$$(7)\qquad \begin{cases} \Lambda' y = -\varphi, \quad \Lambda' y = \Lambda\, y + \dfrac{h_1^2 + h_2^2}{12}\, \Lambda_1 \Lambda_2\, y\,, \\ \varphi = f + \dfrac{h_1^2}{12}\, L_1 f + \dfrac{h_2^2}{12}\, L_2 f\,, \end{cases}$$

provides an approximation of order 4 on a solution $u = u(x)$ of Poisson's equation (1). Indeed, formula (6) gives

$$\Lambda' u + \varphi = (\Lambda' u + \varphi) - (L\, u + f) = O(|\, h\,|^4), \qquad L = L_1 + L_2\,.$$

The operator $\Lambda'$ is defined on the nine-point "box" pattern (see Fig. 16) consisting of the nodes $(x_1 + m_1\, h_1\,, x_2 + m_2\, h_2)$; $m_1\,, m_2 = -1, 0, 1$, by means of which scheme (7) is representable by

$$(8)\qquad \frac{5}{3}\left(\frac{1}{h_1^2} + \frac{1}{h_2^2}\right) y = \frac{1}{6}\left(\frac{5}{h_1^2} - \frac{1}{h_2^2}\right)\left(y^{(+1_1)} + y^{(-1_1)}\right)$$

$$+ \frac{1}{6}\left(\frac{5}{h_2^2} - \frac{1}{h_1^2}\right)\left(y^{(+1_2)} + y^{(-1_2)}\right)$$

$$+ \frac{1}{12}\left(\frac{1}{h_1^2} + \frac{1}{h_2^2}\right)\left(y^{(+1_1,+1_2)} + y^{(+1_1,-1_2)}\right.$$

$$\left. + \left(y^{(-1_1,-1_2)} + y^{(-1_1,+1_2)}\right)\right) + \varphi\,,$$

where $y^{(\pm 1_1)} = y(x_1 \pm h_1\,, x_2)$, $y^{(+1_1,-1_2)} = y(x_1 + h_1\,, x_2 - h_2)$, etc. On the square grid ($h_1 = h_2 = h$) the final result is simple to follow:

$$y_0 = \frac{4\,(y_1 + y_2 + y_3 + y_4) + y_5 + y_6 + y_7 + y_8}{20} + \frac{3}{10}\, h^2 \varphi$$

(see Fig. 16).

To avoide cumbersome computations, we substitute $\Lambda_1 f$ for $L_1 f$ and $\Lambda_2 f$ for $L_2 f$ into the formula for $\varphi$ having replaced $\varphi$ by $O(|\, h\,|^4)$, which does not change the order of approximation for $\psi = \Lambda' u + \varphi = O(|\, h\,|^4)$, so that

$$\varphi = f + \frac{h_1^2}{12}\, \Lambda_1 f + \frac{h_2^2}{12}\, \Lambda_2 f\,.$$

**2. Estimation of a solution of the difference boundary-value problem.** Consider now the difference Dirichlet problem for the scheme of accuracy $O(|h|^4)$ in the rectangle $\bar{G} = \{0 \le x_\alpha \le l_\alpha\,,\ \alpha = 1, 2\}$:

$$
(9) \qquad \begin{cases} \Lambda' y = -\varphi\,, & x \in \omega_h\,, \quad y|_{\gamma_h} = \mu(x)\,, \\ \varphi = f + \dfrac{h_1^2}{12}\ \Lambda_1 f + \dfrac{h_2^2}{12}\ \Lambda_2 f\,, & \end{cases}
$$

where $\Lambda' y$ is given by formula (7). Each of the grid nodes is regular, because the nine-point pattern belongs to the rectangle $\bar{G}$ (Fig. 16). The boundary $\gamma_h$ of the grid contains all the nodes on the boundary $\Gamma$ including the vertices of the rectangle. With this in mind, we set up the problem for the error $z = y - u$:

$$
(10) \qquad \Lambda' z = -\psi, \qquad x \in \omega_h\,, \quad z = 0 \quad \text{on} \quad \gamma_h\,,
$$

where $\psi = \Lambda' u + \varphi = O(|h|^4)$ for $x \in \omega_h$ if $u \in C^{(6)}$. To decide for yourself whether the conditions of the maximum principle are satisfied, a first step is to compare (8) with (1) from Section 2. As a final result we get

$$
(11) \qquad B(x, \xi) \ge 0 \qquad \text{for} \quad \frac{1}{\sqrt{5}} \le \frac{h_1}{h_2} \le \sqrt{5}\,.
$$

To evaluate the solution of problem (10), we should have at our disposal the majorant of the type

$$
Y(x) = K\,(l_1^2 - x_1^2 + l_2^2 - x_2^2)\,.
$$

Taking into account that $\Lambda Y = -4K$, $\Lambda_1 \Lambda_2 Y = 0$, $\|Y\| \le K(l_1^2 + l_2^2)$ and accepting $4K = \|\psi\|_C$, we deduce by Theorem 3 for the solution of problem (10) the estimate

$$
\|z\|_C \le \frac{l_1^2 + l_2^2}{4}\ \|\psi\|_C\,,
$$

provided that the condition

$$
\frac{1}{\sqrt{5}} \le \frac{h_1}{h_2} \le \sqrt{5}
$$

holds. This implies that

> *Scheme (9) is of fourth-order accuracy when $u \in C^{(6)}$, $f \in C^{(4)}$ and condition (11) is satisfied.*

It is worth noting here that on the square grid ($h_1 = h_2 = h$) this condition is automatically fulfilled. A proper choice of $\varphi$ guarantees the sixth order of accuracy of scheme (9) on any such grid. Convergence of scheme (9) with the fourth order in the space $C$ can be established without concern of condition (11). An alternative way of covering this is to construct an a priori estimate for $\| \Lambda z \|^2$ and then apply the embedding theorem (see Section 4).

Let $\Omega_h$ be the space of all grid functions defined at the inner nodes $x \in \omega_h$ of the grid $\bar{\omega}_h = \{(i_1 h_1, i_2 h_2)\}$, $0 \le i_\alpha \le N_\alpha$, $h_\alpha N_\alpha = l_\alpha$, $\alpha = 1, 2$, and let $\overset{\circ}{\Omega}_h$ be the space of all grid functions defined on the grid $\bar{\omega}_h$ and vanishing on the boundary $\gamma_h$. The accepted view is that an inner product in the space $\Omega_h$ such as

$$(y, v) = \sum_{i_1=1}^{N_1-1} \sum_{i_2=1}^{N_2-1} y(i_1 h_1, i_2 h_2)\, v(i_1 h_1, i_2 h_2)\, h_1 h_2$$

$$= \sum_{x \in \omega_h} y(x)\, v(x)\, h_1 h_2, \qquad y, v \in \Omega_h, \quad \| y \| = \sqrt{(y, y)}\,,$$

and the operators $A_1$ and $A_2$ specified by the relations

$$A_1 y = -\widetilde{\Lambda}_1 y, \qquad A_2 y = -\widetilde{\Lambda}_2 y\,,$$

where $\widetilde{\Lambda}_\alpha y = \Lambda_\alpha y$ for all $y \subset \overset{\circ}{\Omega}_h$, will complement special investigations. Here $\widetilde{\Lambda}_\alpha$ really acts from $\Omega_h$ into $\Omega_h$ and is identical with $\Lambda_\alpha$ itself in $\overset{\circ}{\Omega}_h$. Therefore, $A_1$ and $A_2$ are linear operators defined on $H_h = \Omega_h$ (they can be treated also as operators from $\overset{\circ}{\Omega}_h$ into $\Omega_h \subset \overset{\circ}{\Omega}_h$). The domain and range of these operators coincide with $\Omega_h = H_h$.

Here the complete posing of problem (10) is concerned with an operator equation

$$(12) \qquad A' z = A_1 z + A_2 z - (\varkappa_1 + \varkappa_2)\, A_1 A_2 z = \psi, \qquad y, \psi \in H_h\,,$$

with $\varkappa_1 = h_1^2/12$ and $\varkappa_2 = h_2^2/12$ incorporated. The operators $A_1$ and $A_2$ are self-adjoint:

$$(A_\alpha\, y, v) = (y, A_\alpha\, v), \qquad \alpha = 1, 2, \quad y, v \in H_h\,,$$

positive definite:

$$(A_\alpha\, y, y) \geq \lambda_1^{(\alpha)} \|\, y \,\|^2, \qquad \lambda_1^{(\alpha)} = \frac{4}{h_\alpha^2} \sin^2 \frac{\pi h_\alpha}{2\, l_\alpha} \geq \frac{8}{l_\alpha^2}\,, \qquad \alpha = 1, 2\,,$$

and commuting: $A_1\, A_2 = A_2\, A_1$. In view of this,

$$A_1\, A_2 = (A_1\, A_2)^* > 0\,.$$

By virtue of the relations

$$A_\alpha \leq \|\, A_\alpha \,\|\, E\,, \qquad \|\, A_\alpha \,\| = \frac{4}{h_\alpha^2} \cos^2 \frac{\pi\, h_\alpha}{2\, l_\alpha} < \frac{4}{h_\alpha^2}\,, \qquad \varkappa_\alpha \,\|\, A_\alpha \,\| < \frac{1}{3}$$

we arrive at

$$A_1\, A_2 \leq \|\, A_1 \,\|\, A_2, \qquad A_1\, A_2 = A_2\, A_1 \leq \|\, A_2 \,\|\, A_1\,,$$

$$\begin{aligned}(\varkappa_1 + \varkappa_2)\, A_1\, A_2 &= \varkappa_1\, A_1\, A_2 + \varkappa_2\, A_2\, A_1 \\ &\leq \varkappa_1 \,\|\, A_1 \,\|\, A_2 + \varkappa_2 \,\|\, A_2 \,\|\, A_1 \\ &< \tfrac{1}{3}\, (A_1 + A_2)\,,\end{aligned}$$

yielding

$$\tfrac{2}{3}\, A \leq A' = A_1 + A_2 - (\varkappa_1 + \varkappa_2)\, A_1\, A_2 \leq A\,,$$

$$A = A_1 + A_2\,,$$

$$\tfrac{2}{3} \,\|\, Az \,\| \leq \|\, A'z \,\| \leq \|\, Az \,\|\,.$$

The operators $A$ and $A'$ are commuting and self-adjoint and, therefore, for the equation $A'z = \psi$ the estimate $\|\, A'z \,\| \leq \frac{3}{2} \,\|\, \psi \,\|$ is certainly true. By the embedding theorem from Section 4,

$$\|\, z \,\|_C \;\leq\; \frac{l_0^2}{2\sqrt{l_1\, l_2}}\, Az \;\leq\; \frac{3 l_0^2}{4\sqrt{l_1\, l_2}} \,\|\, \psi \,\|\,,$$

it being understood that for any steps $h_1$ and $h_2$ the solution of problem (12) can be estimated by

$$\| z \|_C = \| y - u \|_C \leq M \| \psi \| ,$$

where

$$M = \frac{3\, l_0^2}{4 \sqrt{l_1\, l_2}} , \qquad l_0 = \max(l_1 , l_2) .$$

The estimate so constructed implies the uniform convergence of scheme (9) with the rate $O(|h|^4)$ for any ratio $h_1/h_2$.

**3. The multidimensional case.** The method for constructing a scheme of fourth-order accuracy described in Section 1 applies equally well to the case of several variables, making it possible to compose the difference scheme of fourth-order approximation

$$\Lambda' y = -\varphi(x) , \qquad x \in \omega_h , \quad y = \mu(x), \quad x \in \gamma_h , \tag{13}$$

$$\Lambda' y = \Lambda\, y + \sum_{\alpha=1}^{p} \frac{h_\alpha^2}{12} \sum_{\beta \neq \alpha}^{1 \div p} \Lambda_\alpha\, \Lambda_\beta\, y ,$$

$$\Lambda\, y = \sum_{\alpha=1}^{p} \Lambda_\alpha\, y , \qquad \Lambda_\alpha y = y_{\bar{x}_\alpha x_\alpha} , \tag{14}$$

$$\varphi = f + \sum_{\alpha=1}^{p} \frac{h_\alpha^2}{12} \Lambda_\alpha\, f ,$$

associated with problem (1) of Section 1 in the $p$-dimensional parallelepiped $\bar{G}_0 = \{0 \leq x_\alpha \leq l_\alpha , \ \alpha = 1, 2, \ldots , p\}$ on the grid $\bar{\omega}_h = \{x_i = (i_1 h_1 , \ldots , i_p h_p), i_\alpha = 0, 1, 2, \ldots , N_\alpha , h_\alpha N_\alpha = l_\alpha\}$.

By introducing the space $H_h = \Omega_h$ and the operators $A_\alpha$ by analogy with the preceding section we draw the conclusion that the operator

$$A' = A - \sum_{\alpha=1}^{p} \varkappa_\alpha \sum_{\substack{\beta=1 \\ \beta \neq \alpha}}^{p} A_\alpha\, A_\beta , \qquad \varkappa_\alpha = \frac{h_\alpha^2}{12}, \quad A = \sum_{\alpha=1}^{p} A_\alpha , \tag{15}$$

is self-adjoint and possesses the estimates

$$\frac{4-p}{3} A \leq A' \leq A , \qquad p > 1 .$$

These properties are an immediate implication of the chain of the relations

$$0 < \sum_{\alpha=1}^{p} \varkappa_\alpha \sum_{\beta \neq \alpha} A_\alpha A_\beta \leq \frac{1}{3} \sum_{\alpha=1}^{p} \sum_{\beta \neq \alpha} A_\alpha = \frac{1}{3} \sum_{\alpha=1}^{p} (A - A_\alpha) = \frac{p-1}{3} A .$$

For $p \geq 4$ the difference operator (15) lacks the property of having fixed sign (ellipticity). In each such case it is recommended to refer to another operator $A'$, which preserves the ellipticity property for any $p$:

$$\Lambda' y = \sum_{\alpha=1}^{p} \prod_{\substack{\beta=1 \\ \beta \neq \alpha}}^{p} \left( E + \frac{h_\beta^2}{12} \Lambda_\beta \right) \Lambda_\alpha y$$

or

$$A' y = \sum_{\alpha=1}^{p} \prod_{\substack{\beta=1 \\ \beta \neq \alpha}}^{p} (E - \varkappa_\beta A_\beta) A_\alpha y . \tag{16}$$

It is evident that the approximation order remains unchanged and $A'$ is identical with operator (15) up to the terms $O(|h|^4)$.

On the other hand, since $E - \varkappa_\beta A_\beta > \frac{2}{3} E$,

$$A' > \sum_{\alpha=1}^{p} A_\alpha \left(\frac{2}{3}\right)^{p-1} = \left(\frac{2}{3}\right)^{p-1} A$$

and, hence,

$$\left(\frac{2}{3}\right)^{p-1} A < A' < A .$$

With the aid of the above operator inequalities we are able to produce the necessary a priori estimates and justify the convergence with the rate $O(|h|^4)$ for the scheme in hand. Observe that for $p = 2$ operator (16) coincides with operator (15).

We have nothing worthwhile to add to such discussions, so will leave it at this.

# 5

# Difference Schemes for Time-Dependent Equations with Constant Coefficients

In this chapter difference schemes for the simplest time-dependent equations are studied, namely, for the heat conduction equation with one or more spatial variables, the one-dimensional transfer equation and the equation of vibrations of a string. Two-layer and three-layer schemes are designed for the first, second and third boundary-value problems. Stability is investigated by different methods such as the method of separation of variables and the method of energy inequalities as well as by means of the maximum principle. Asymptotic stability of difference schemes is discovered for the heat conduction equation in ascertaining the viability of difference approximations. Finally, stability theory is being used, increasingly, to help us understand a variety of phenomena, so it seems worthwhile to discuss it in full details.

## 5.1 ONE-DIMENSIONAL HEAT CONDUCTION EQUATION WITH CONSTANT COEFFICIENTS

In this section we consider the one-dimensional heat conduction equation with constant coefficients and difference schemes in order to develop various methods for designing the appropriate difference schemes in the case of time-dependent problems.

**1. The original problem.** The heat diffusion process on a straight line is described by the heat conduction equation

$$(1) \qquad c\rho \frac{\partial u}{\partial t} = \frac{\partial}{\partial x}\left(k \frac{\partial u}{\partial x}\right) + \bar{f},$$

where $u = u(x,t)$ is the temperature, $c$ is the specific heat, $\rho$ is the density, $k$ is the thermal conductivity and $\bar{f}$ is the density of heat sources, that is, the amount of heat emitted per a unit of time on a unit of length. Thermal conductivity and specific heat may depend not only on $x$ and $t$, but also on the temperature $u$. In that case the equation is said to be **quasilinear**. If $k$ and $c\rho$ are constant, equation (1) can be written in the form

$$(2) \qquad \frac{\partial u}{\partial t} = a^2 \frac{\partial^2 u}{\partial x^2} + \tilde{f}, \qquad a^2 = \frac{k}{c\rho}, \qquad \tilde{f} = \frac{\bar{f}}{c\rho},$$

where $a^2$ is the thermometric conductivity (Maxwell) or diffusivity (Lord Kelvin).

Without loss of generality we may set $a = 1$ and rewrite equation (2) as

$$(3) \qquad \frac{\partial u}{\partial t} = \frac{\partial^2 u}{\partial x^2} + f.$$

Indeed, by introducing $x' = x/a$ and denoting once again $x'$ by $x$ we obtain (3). Where searching a solution to equation (2) on the segment $0 \le x \le l$, it is sensible to pass to the dimensionless variables

$$x' = \frac{x}{l}, \qquad t' = \frac{a^2 t}{l^2},$$

whose use permits us to rewrite equation (2) in the form (3) with $0 \le x' \le 1$ and $f = l^2 \tilde{f}/a^2$ incorporated.

We concentrate primarily on the first boundary-value problem associated with equation (3) in the rectangle $\overline{D} = \{0 \le x \le 1,\ 0 \le t \le T\}$, in which it is required to find a continuous in $\overline{D}$ solution $u = u(x,t)$ of the problem

$$(\mathrm{I}) \qquad \begin{aligned} &\frac{\partial u}{\partial t} = \frac{\partial^2 u}{\partial x^2} + f(x,t), && 0 < x < 1, \quad 0 < t \le T, \\ &u(x,0) = u_0(x), && 0 \le x \le 1, \\ &u(0,t) = u_1(t), \quad u(1,t) = u_2(t), && 0 \le t \le T. \end{aligned}$$

**2. The family of six-point schemes.** As before, it will be convenient to introduce the grids $\bar{\omega}_h = \{x_i = ih,\ i = 0,1,\dots,N\}$ and $\omega_\tau = \{t_j = j\tau,\ j = 0,1,\dots,j_0\}$. When operating in $\overline{D}$ on the grid

$$\bar{\omega}_{h\tau} = \bar{\omega}_h \times \omega_\tau = \big\{(ih,\ j\tau),\ i = 0,1,\dots,N,\ j = 0,1,\dots,j_0\big\}$$

with steps $h = 1/N$ and $\tau = T/j_0$, we denote by $y_i^j$ the value at the node $(x_i, t_j)$ of the grid function $y$ given on $\bar{\omega}_{h\tau}$. The approximation here consists of replacing the first derivative $\partial u/\partial t$ by the first order difference derivative, the second derivative $\partial^2 u/\partial x^2$ by the second-order difference derivative $u_{\bar{x}x} = \Lambda u$ and introducing an arbitrary real parameter $\sigma$. As a final result we get a one-parameter family of difference schemes

$$(4)\qquad \frac{y_i^{j+1} - y_i^j}{\tau} = \Lambda\Big(\sigma y_i^{j+1} + (1-\sigma) y_i^j\Big) + \varphi_i^j, \quad 0 < i < N, \quad 0 \le j < j_0.$$

Sometimes scheme (4) will be treated as a **scheme with weights**. The supplementary boundary and initial conditions will be explicitly specified with accurate approximations:

$$(5)\qquad y_0^j = u_1^j, \qquad y_N^j = u_2^j,$$

$$(6)\qquad y_i^0 = y(x_i, 0) = u_0(x_i).$$

Here $\varphi_i^j$ is a grid function approximating the right-hand side $f$ of equation (3). With such a variety, we may accept, for example,

$$\varphi_i^j = f(x_i, t_{j+1/2}), \qquad t_{j+1/2} = t_j + 0.5\,\tau,$$

$$\Lambda y_i = y_{\bar{x}x,i} = \big(y_{i-1} - 2y_i + y_{i+1}\big)/h^2.$$

When all conditions (4)–(6) are put together, they are constitute problem (II).

The difference scheme (4) is constructed on the six-point pattern with the nodes

$$(x_{i\pm1}, t_{j+1}), \qquad (x_i, t_{j+1}), \qquad (x_{i\pm1}, t_j), \qquad (x_i, t_j)$$

and center $(x_i, t_{j+1})$ (see Fig. 4.c). The truth of equation (4) is supposed at the nodes $(x_i, t_{j+1})$, $i = 1,2,\dots,N-1$; $j+1 = 1,2,\dots,j_0$, known as the inner nodes. The set of all inner nodes of the grid $\bar{\omega}_{h\tau}$ is denoted by

$$\omega_{h\tau} = \big\{(x_i, t_j),\ 1 \le i \le N-1,\ 1 \le j \le j_0\big\}.$$

The boundary and initial conditions (5) and (6) are specified at the boundary nodes of the grid $\bar{\omega}_{h\tau}$. All of the nodes of the grid $\bar{\omega}_{h\tau}$ lying on the straight line $t = t_j$ constitute what is called a **layer**. Because in scheme (4) knowledge of the values of the sought function $y$ is required on two layers, it can be regarded as a **two-layer scheme**. In the sequel we will show that the accuracy and stability of scheme (4) depend on a proper choice of the parameter $\sigma$. Before proceeding to further careful analysis, it would be prudent to examine some schemes relating to particular values of $\sigma$ in light of those remarks. For $\sigma = 0$ we get the four-point scheme (Fig. 4.a)

$$\frac{y_i^{j+1} - y_i^j}{\tau} = \Lambda\, y_i^j + \varphi_i^j$$

or

$$(7) \qquad y_i^{j+1} = (1 - 2\,\gamma)\, y_i^j + \gamma\,(y_{i-1}^j + y_{i+1}^j) + \tau\,\varphi_i^j\,, \qquad \gamma = \tau/h^2,$$

developed on the pattern with the nodes $(x_i, t_{j+1})$, $(x_i, t_j)$ and $(x_{i\pm1}, t_j)$. The value $y_i^{j+1}$ at every point of the new layer $t = t_{j+1}$ is given by the explicit formula (7) through the values $y_i^j$ on the previous layer $t = t_j$. Since the solution is prescribed by the initial condition $y_i^0 = u_0(x_i)$ at moment $t = t_0$, it is possible to determine all the values of $y$ on any adjacent layer by applying successively formula (7). Because of this, scheme (7) is said to be **explicit**.

For $\sigma \neq 0$ scheme (4) refers to an **implicit two-layer scheme.** When the value $y_i^{j+1}$ is sought on the new layer under the natural premise $\sigma \neq 0$, we obtain the governing system of algebraic equations

$$(8) \qquad \sigma\,\Lambda\, y_i^{j+1} - \frac{1}{\tau}\, y_i^{j+1} = -F_i^j\,, \qquad F_i^j = \frac{1}{\tau}\, y_i^j + (1 - \sigma)\,\Lambda\, y_i^j + \varphi_i^j\,,$$

$$i = 1, 2, \ldots, N - 1\,,$$

subject to the boundary conditions $y_0^{j+1} = u_1^{j+1}$ and $y_N^{j+1} = u_2^{j+1}$. This system can be solved by the right elimination method (see Chapter 1, Section 2.5). In conclusion, it is worth mentioning two other schemes. The first one with $\sigma = 1$ known as the **forward difference scheme** or **pure implicit scheme** will appear as further developments occur:

$$(9) \qquad \frac{y_i^{j+1} - y_i^j}{\tau} = \Lambda\, y_i^{j+1} + \varphi_i^j\,.$$

The **six-point symmetric scheme** with $\sigma = 0.5$ ascribed to Crank and Nicolson is of the form

$$\frac{y_i^{j+1} - y_i^j}{\tau} = \frac{1}{2}\,\Lambda\left(y_i^{j+1} + y_i^j\right) + \varphi_i^j\,. \tag{10}$$

**3. The error of approximation.** In order to evaluate the accuracy of scheme (4)–(6), the solution $y = y_i^j$ of problem (4)–(6) should be compared with the solution $u = u(x,t)$ of problem (I). Since $u = u(x,t)$ is the continuous solution of problem (I), we may set $u_i^j = u(x_i, t_j)$ and deal then with the difference $z_i^j = y_i^j - u_i^j$. For this, the first step in the estimation of the grid function $z_i^j$ on the relevant layer is connected with norms $\|\cdot\|$ of proper form, for example,

$$\|z\| = \|z\|_C = \max_{0\le i\le N} |z_i|\,, \qquad \|z\| = \left(\sum_{i=1}^{N-1} z_i^2\,h\right)^{1/2}.$$

Let us pass, time and again, to the notations without subscripts and superscripts, which are good enough for our purposes:

$$y_i^j = y\,, \qquad y_i^{j+1} = \hat{y}\,, \qquad y_t = (\hat{y} - y)/\tau\,,$$

permitting us to recast the problem we have completely posed by conditions (4)–(6) as

$$\text{(II)}\qquad \begin{aligned} & y_t = \Lambda\left(\sigma\,\hat{y} + (1-\sigma)\,y\right) + \varphi\,, && (x,\,t) \in \omega_{h\tau}\,, \\ & y(0,\,t) = u_1(t)\,, \quad y(1,\,t) = u_2(t)\,, && t \in \omega_\tau\,, \\ & y(x,\,0) = u_0(x)\,, && x \in \bar{\omega}_h\,, \qquad \Lambda\,y = y_{\bar{x}x}\,. \end{aligned}$$

In trying to establish the conditions for determination of $z = y - u$ we substitute $y = z + u$ into (II) and regard $u$ as a known function, making it possible to set up the problem for $z$:

$$\text{(III)}\qquad \begin{aligned} & z_t = \Lambda\left(\sigma\,\hat{z} + (1-\sigma)\,z\right) + \psi\,, && (x,\,t) \in \omega_{h\tau}\,, \\ & z(0,\,t) = z(1,\,t) = 0\,, && t \in \omega_\tau\,, \\ & z(x,\,0) = 0\,, && x \in \bar{\omega}_h\,, \end{aligned}$$

where

$$\psi = \Lambda\left(\sigma\,\hat{u} + (1-\sigma)\,u\right) - u_t + \varphi \tag{11}$$

is the error of approximation (residual) for scheme (II) on the solution $u = u(x,t)$ to equation (I).

Recall the definition of the order of approximation from Chapter 2, Section 1.3 saying that scheme (II) approximates equation (I) with order $(m, n)$ or equation (I) is approximated by scheme (II) to $O(h^m + \tau^n)$ on a solution $u = u(x,t)$ to equation (I) if $\| \psi(x,t) \|_{(2)} = O(h^m + \tau^n)$ or $\| \psi \|_{(2)} \le M\,(h^m + \tau^n)$ for all $t \in \omega_\tau$, where $M$ is a positive constant, not depending on $h$ and $\tau$, and $\| \cdot \|_{(2)}$ is a suitably chosen norm on the grid $\omega_h$.

We proceed to the estimation of the order of approximation for scheme (II) under the agreement that $u = u(x,t)$ possesses a number of derivatives in $x$ and $t$ necessary in this connection for performing current and subsequent manipulations. Within the notations

$$\dot{u} = \frac{\partial u}{\partial t}\,, \qquad u' = \frac{\partial u}{\partial x}\,, \qquad \bar{u} = u(x_i,\, t_{j+1/2})\,, \qquad t_{j+1/2} = t_j + \tfrac{1}{2}\,\tau\,,$$

the development of Taylor's series for the function $u = u(x,t)$ about the node $(x_i, t_{j+1/2})$ leads to the expressions:

$$\hat{u} = 0.5\,(\hat{u} + u) + 0.5\,(\hat{u} - u) = 0.5\,(\hat{u} + u) + 0.5\,\tau\,u_t\,,$$

$$u = 0.5\,(\hat{u} + u) - 0.5\,\tau\,u_t\,,$$

$$\sigma\,\hat{u} + (1-\sigma)\,u = 0.5\,(\hat{u} + u) + (\sigma - 0.5)\,\tau\,u_t\,.$$

All this enables us to reduce $\psi$ to $\psi = 0.5\,\Lambda\,(\hat{u}+u) + (\sigma - 0.5)\,\tau\,\Lambda\,u_t - u_t + \varphi$. Substitution of the set of expansions

$$\Lambda\,u = u'' + \frac{h^2}{12}\,u^{(4)} + O(h^4) = L\,u + \frac{h^2}{12}\,L^2\,u + O(h^4)\,,$$

$$L\,u = \frac{\partial^2 u}{\partial x^2}\,,$$

$$\hat{u} = \bar{u} + \frac{1}{2}\,\tau\,\dot{\bar{u}} + \frac{\tau^2}{8}\,\ddot{\bar{u}} + O(\tau^3)\,,$$

$$u = \bar{u} - \frac{1}{2}\,\tau\,\dot{\bar{u}} + \frac{\tau^2}{8}\,\ddot{\bar{u}} + O(\tau^3)\,,$$

$$\frac{1}{2}\,(\hat{u} + u) = \bar{u} + \frac{\tau^2}{8}\,\ddot{\bar{u}} + O(\tau^3)\,, \qquad u_t = \dot{\bar{u}} + O(\tau^2)$$

yields

$$\psi = (L\,\bar{u} - \bar{\dot{u}} + \varphi) + (\sigma - 0.5)\,\tau\,L\,\bar{\dot{u}} + \frac{h^2}{12}\,L^2\,\bar{u} + O(h^4 + \tau^2)\,, \tag{12}$$

whence it follows that $\psi = (\sigma - 0.5)\,\tau\,L\,\bar{\dot{u}} + O(h^2 + \tau^2)$ for $\varphi = \bar{f} = f(x, t_{j+1/2})$, since $\dot{u} = L\,u + f$. By virtue of the relations $L\,\dot{u} = L^2\,u + L f = u^{(4)} + f''$ and $L^2\,u = L\,\dot{u} - L\,f$ and expression (12) we are led to

$$\psi = (\varphi - \bar{f}) = \Big[(\sigma - \tfrac{1}{2})\,\tau + \tfrac{1}{12}\,h^2\Big]\,L\,\bar{\dot{u}} - \tfrac{1}{12}\,h^2\,L\,\bar{f} + O(h^4 + \tau^2)\,. \tag{13}$$

By equating the expression in the square brackets to zero we find that

$$\sigma = \frac{1}{2} - \frac{h^2}{12\,\tau} = \sigma_*\,. \tag{14}$$

For $\sigma = \sigma_*$ and $\varphi = \bar{f} + \frac{1}{12}\,h^2\,L\bar{f}$ scheme (II) generates an approximation of $O(h^4 + \tau^2)$, that is, $\psi = O(h^4 + \tau^2)$. But the order of approximation of this scheme remains unchanged if $f''$ will be replaced by $f_{\bar{x}x} = \Lambda\,f$, giving $\varphi = \bar{f} + \frac{1}{12}\,(h^2\,\Lambda\,\bar{f})$ or

$$\varphi_i^j = \frac{5}{6}\,f_i^{j+1/2} + \frac{1}{12}\,\Big(f_{i-1}^{j+1/2} + f_{i+1}^{j+1/2}\Big)\,. \tag{15}$$

Observe that the last formula is much more simpler for later use.

Let $C_n^m(\bar{D})$ be the class of functions with $m$ derivatives in $x$ and $n$ derivatives in $t$, all derivatives being continuous in $\bar{D}$. Formulae (13)–(14) justify that scheme (II) provides approximations of

- $O(h^2 + \tau^2)$ for $\sigma = 0.5$, $\varphi = \bar{f}$ or $\varphi = \bar{f} + O(h^2 + \tau^2)$ if $u \in C_3^4$,
- $O(h^2 + \tau)$ for any $\sigma \neq 0.5$, $\varphi = \bar{f} + O(h^2 + \tau)$, for instance, $\varphi = \hat{f}$ or $\varphi = f$ if $u \in C_2^4$,
- $O(h^4 + \tau^2)$ for $\sigma = \sigma_*$ and $\varphi$ specified by formula (15) if $u \in C_3^6$.

Scheme (II) with $\sigma = \sigma_*$ and $\varphi = \bar{f} + \frac{1}{12}\,h^2\,\Lambda\,\bar{f}$ is usually termed a higher-accuracy scheme. The requirement of the retention of the approximation order for a given value $\sigma$ guides a proper choice of the right-hand side $\varphi$. Thus, for $\sigma = 0.5$ it is taken to be $\varphi = 0.5\,(\hat{f} + f)$, $\varphi = \bar{f}$ and more.

It is easily seen from (13) that the error $O(h^2 + \tau^2)$ may be also achieved for $\sigma \neq 0.5$ if we keep $\sigma = 0.5 + h^2\alpha/\tau$, where $\alpha$ is an arbitrary constant independent of $h$ and $\tau$. Certainly, in this case $\sigma$ depends on $h$ and $\tau$. The arbitrariness in the choice of $\alpha$ is limited by the condition of stability of the scheme (a result is ensured under the constraint $\alpha \geq -\frac{1}{4}$, for more detail see Section 4).

**4. Stability with respect to the initial data.** Let us investigate the stability of scheme (II) with homogeneous boundary conditions by the method of separation of variables. For this, we proceed as usual. This amounts to applying to scheme (II) with homogeneous boundary conditions the identities

$$\hat{y} = y + \tau\, y_t\,, \qquad \sigma\,\hat{y} + (1-\sigma)\,y = y + \sigma\,\tau\,y_t$$

and reducing them to the following ones:

$$(16)\qquad \begin{aligned} & y_t - \sigma\,\tau\,\Lambda\,y_t = \Lambda\,y + \varphi\,, && (x,\,t) \in \omega_{h\tau}\,,\\ & y(0,\,t) = y(1,\,t) = 0\,, && t \in \omega_\tau\,,\\ & y(x,\,0) = u_0(x)\,, && x \in \bar{\omega}_h\,. \end{aligned}$$

Scheme (16) is said to be **stable** if for a solution of problem (16) the estimate holds:

$$(17)\qquad \|\,y(t)\,\|_{(1)} \leq M_1\,\|\,u_0\,\|_{(1)} + M_2 \max_{0\leq t'<t} \|\,\varphi(t')\,\|_{(2)}\,, \qquad t \in \omega_\tau\,,$$

where $M_1$, and $M_2$ are positive constants independent of $h$ and $\tau$, $\|\;\cdot\;\|_{(1)}$ and $\|\;\cdot\;\|_{(2)}$ are suitable norms on the grid $\omega_h$.

With $\varphi = 0$ incorporated, the estimate

$$(18)\qquad \|\,y(t)\,\|_{(1)} \leq M_1\,\|\,u_0)\,\|_{(1)}\,, \qquad t \in \omega_\tau\,,$$

expresses the stability of scheme (16) with respect to the initial data. When $y(x,0) = 0$, the meaning of the stability of scheme (16) with respect to the right-hand side is that we should have

$$(19)\qquad \|\,y(t)\,\|_{(1)} \leq M_2 \max_{0\leq t'<t} \|\,\varphi(t')\,\|_{(2)}\,.$$

The stability of scheme (16) with respect to the initial data and the right-hand side is ensured by estimate (17), valid for the solution of problem (16).

Before giving further motivations, let us represent the solution of problem (16) as a sum $y = \bar{y} + \tilde{y}$, where $\bar{y}$ is a solution of the homogeneous equation

$$(16a)\quad y_t - \sigma\,\tau\,\Lambda\,y_t = \Lambda\,y\,, \qquad y(0,\,t) = y(1,\,t) = 0\,, \quad y(x,\,0) = u_0(x)\,,$$

and $\tilde{y}$ is a solution to the nonhomogeneous equation with the initial condition $\tilde{y}(x,\,0) = 0$:

$$(16b)\quad y_t - \sigma\,\tau\,\Lambda\,y_t = \Lambda\,y + \varphi\,, \qquad y(0,\,t) = y(1,\,t) = 0\,, \quad y(x,\,0) = 0\,.$$

To decide for yourself whether scheme (16) is stable with respect to the initial data, a first step is to evaluate the solution of problem (16a). This can be done using the method of separation of variables and deriving estimate (18) in the grid $L_2(\omega_h)$-norm:

$$\| y \|_{(1)} - \| y \| , \quad \text{where} \quad \| y \| = \sqrt{(y, y)} , \quad (y, v) = \sum_{i=1}^{N-1} y_i v_i h .$$

To develop those ideas, we seek the solution to equation (16a) as a product of two functions, one of which $T = T(t_j)$ depends only on $t = t_j$ and the other $X = X(x_i)$ only on $x = x_i$ under the approved decomposition $y(x, t) = X(x)\, T(t)$. Substituting this expression into (16a) and taking into account that

$$\Lambda y = T \Lambda X , \qquad y_t = X\, T_t ,$$

we arrive at the relations

$$\frac{\hat{T} - T}{\tau (\sigma \hat{T} + (1-\sigma) T)} = \frac{\Lambda X}{X} = -\lambda , \qquad \hat{T} = T(t_{j+1}) , \qquad T = T(t_j) ,$$

where $\lambda$ is a separation constant. As a final result we get

$$\hat{T} = q\, T , \qquad q = \frac{1 - (1-\sigma) \tau \lambda}{1 + \sigma \tau \lambda} .$$

The difference eigenvalue problem for $X$ can be viewed as the Sturm–Liouville difference problem:

$$\Lambda X(x) + \lambda X(x) = 0 , \quad 0 < x = i h < 1 , \quad X(0) = X(1) = 0 , \quad X(x) \not\equiv 0 ,$$

which has been under consideration in Chapter 2, Section 3.2. As stated therein, the problem in view has nontrivial solutions identical with the eigenfunctions

$$X^{(k)} F(x) = \sqrt{2} \sin \pi k x , \qquad k = 1, 2, \ldots , N-1 ,$$

wich correspond to the eigenvalues

$$\lambda_k = \frac{4}{h^2} \sin^2 \frac{\pi k h}{2} , \qquad k = 1, 2, \ldots , N-1 , \qquad 0 < \lambda_1 < \cdots < \lambda_{N-1} ,$$

$$\lambda_1 = \frac{4}{h^2} \sin^2 \frac{\pi h}{2} , \qquad \lambda_{N-1} = \frac{4}{h^2} \cos^2 \frac{\pi h}{2} .$$

The eigenfunctions $\{X^{(k)}\}$ constitute an orthonormal system

$$\left(X^{(k_1)},\, X^{(k_2)}\right) = \delta_{k_1,\,k_2}\,,$$

for which Parseval's identity takes place:

$$\|\, f\,\|^2 = \sum_{k=1}^{N-1} f_k^2 . \tag{20}$$

Here $f_k$ are the coefficients of the expansion for an arbitrary grid function $f(x)$ defined on the grid $\bar\omega_h$ and vanishing at the points $x = 0$ and $x = 1$:

$$f(x) = \sum_{k=1}^{N-1} f_k\, X^{(k)}(x)\,, \qquad f_k = \left(f,\, X^{(k)}\right).$$

So, problem (16a) has nontrivial solutions $y_{(k)} = T_k\, X^{(k)} \not\equiv 0$, where $T_k$ can be recovered from the equation

$$\hat T_k = q_k\, T_k \tag{21}$$

or

$$T_k^{j+1} = q_k\, T_k^j = \cdots = q_k^{j+1}\, T_k^0\,, \qquad q_k = \frac{1-(1-\sigma)\,\tau\,\lambda_k}{1+\sigma\,\tau\,\lambda_k}\,.$$

As a matter of fact, the constant $T_k^0$ is free to be chosen in any convenient way.

A solution to equation (16a) having the form $y_{(k)} = T_k\, X^{(k)}$ is called a **harmonic** of the attached number $k$. Clearly, this function satisfies problem (16a) with the initial condition $u_0(x) = T_k^0\, X^{(k)}(x)$. Let us find out the conditions under which every harmonic $y_{(k)}$, $k = 1, 2, \ldots, N-1$, is stable. From the recurrence relations

$$y_{(k)}^{j+1} = X^{(k)}\, T_k^{j+1} = q_k\, X^{(k)}\, T_k^j\,, \qquad y_{(k)}^{j+1} = q_k\, y_{(k)}^j \tag{22}$$

the conclusion can be drawn that for $|\, q_k\,| \ge 1+\varepsilon$, where $\varepsilon = \text{const} > 0$ does not depend on $h$ and $\tau$ both,

$$\|\, y_{(k)}^{j+1}\,\| = |\, q_k\,|\; \|\, y_{(k)}^j\,\| \ge (1+\varepsilon)\; \|\, y_{(k)}^j\,\| \ge (1+\varepsilon)^{j+1}\; \|\, y_{(k)}^0\,\| \longrightarrow \infty$$

as $\tau \to 0$, that is, the problem concerned becomes unstable. If $|\, q_k\,| \le 1$ and $t = j\tau$ is kept fixed, then $\|\, y_{(k)}\,\|$ does not increase along with increasing $j\,(\tau \to 0)$:

$$\|\, y_{(k)}^{j+1}\,\| \le \|\, y_{(k)}^j\,\| \le \cdots \le \|\, y_{(k)}^0\,\|$$

and the harmonic in question is stable. Under the choice $|\, q_k \,| \le 1$, we have $\|\, y^j_{(k)} \,\| \le \|\, y^0_{(k)} \,\|$. When this is the case, the scheme is said to be **stable in every harmonic**.

We now assign the values of $\sigma$ such that either $|\, q_k \,| \le 1$ or $-1 \le q_k \le 1$, which guarantees the stability of the scheme in every harmonic. It is clear from the formula $q_k = 1 - \tau\, \lambda_k\, (1 + \sigma\, \tau\, \lambda_k)^{-1}$ that $q_k < 1$ if $1 + \sigma\, \tau\, \lambda_k > 0$, that is, $\sigma > -(\tau\, \lambda_k)^{-1}$. The bound $q_k \ge -1$ or

$$q_k + 1 = \frac{2 + (2\,\sigma - 1)\,\tau\,\lambda_k}{1 + \sigma\,\tau\,\lambda_k} \ge 0$$

is ensured by the restriction $2 + (2\,\sigma - 1)\,\tau\,\lambda_k \ge 0$ or $\sigma \ge \frac{1}{2} - (\tau\,\lambda_k)^{-1}$. In that case the condition $1 + \sigma\tau\lambda_k > 0$ is automatically fulfilled. Since

$$\lambda_k \le \lambda_{N-1} < \frac{4}{h^2}\,,$$

the following relations occur:

$$-\frac{1}{\tau\,\lambda_k} \le -\frac{1}{\tau\,\lambda_{N-1}} < -\frac{h^2}{4\,\tau}\,.$$

Hence, the condition $|\, q_k \,| \le 1$ holds true for all $k = 1, 2, \ldots, N-1$ under the agreement that

$$\sigma \ge \frac{1}{2} - \frac{h^2}{4\,\tau} = \sigma_0\,. \tag{23}$$

Thus, all of the harmonics $y_{(k)} = T_k\, X^{(k)}$ are stable under one and the same condition $\sigma \ge \sigma_0$. We are going to show that the stability of scheme (16a) in every harmonic known as the spectral stability implies that in the grid $L_2$-norm with respect to the initial data $y(x,\, 0) = u_0(x)$, where $u_0(x)$ is a grid function defined for $0 \le x \le 1$ and vanishing at the points $x = 0$ and $x = 1$. With this aim, the general solution of problem (16a) is sought as the sum of particular solutions having the form (22):

$$\hat{y} = \sum_{k=1}^{N-1} \hat{y}_{(k)} = \sum_{k=1}^{N-1} \hat{T}_k\, X^{(k)}, \qquad \|\, \hat{y} \,\|^2 = \sum_{k=1}^{N-1} \hat{T}_k^2\,.$$

Substituting here $\hat{T}_k = q_k\, T_k$ and taking into account (20), we find that

$$\hat{y} = \sum_{k=1}^{N-1} q_k\, T_k\, X^{(k)},$$

$$\|\, \hat{y} \,\|^2 = \sum_{k=1}^{N-1} q_k^2\, T_k^2 \le \max_k q_k^2 \sum_{k=1}^{N-1} T_k^2 = \max_k q_k^2\, \|\, y \,\|^2\,.$$

Under the constraint $\sigma \geq \sigma_0$ it is not difficult to verify that $\max_k | q_k | \leq 1$ and $\| \hat{y} \| \leq \| y \|$ or

$$\| y^{j+1} \| \leq \| y^j \| \leq \cdots \leq \| y^0 \| = \| u_0 \| .$$

If so, the solution of problem (16a) satisfies the estimate

$$\| y^j \| \leq \| u_0 \| , \qquad j = 1, 2, \ldots , \tag{24}$$

which means that scheme (16) is stable in the grid $L_2(\omega_h)$-norm with respect to the initial data for $\sigma \geq \sigma_0$.

A difference scheme is said to be **conditionally stable** when it is stable only if $\tau$ and $h$ are related in some way, otherwise it becomes **unconditionally stable**. A scheme, stable for arbitrary values of $\tau$ and $h$, is said to be **absolutely stable**. One can encounter schemes stable for sufficiently small $h$ and $\tau$ such that $h \leq h_0$ and $\tau \leq \tau_0$. These schemes are not absolutely stable, but may be unconditionally stable.

Throughout the entire chapter, a special attention is being paid to several types of schemes.

1. The explicit scheme ($\sigma = 0$). Condition (23) gives $0 \geq \frac{1}{2} - h^2 (4\tau)^{-1}$, that is,

$$\frac{\tau}{h^2} \leq \frac{1}{2} . \tag{25}$$

The explicit scheme is stable only under condition (25) relating the grid steps $h$ and $\tau$ (a conditionally stable scheme).

2. For $\sigma \geq \frac{1}{2}$ the implicit scheme is stable for any $h$ and $\tau$, since $\sigma \geq \frac{1}{2} > \sigma_0$. Thus, the forward difference scheme ($\sigma = 1$) and the symmetric scheme ($\sigma = \frac{1}{2}$) are stable for any $h$ and $\tau$, what means, by definition, their absolute stability.

3. The scheme of a higher-order approximation ($\sigma = \sigma_*$, $\sigma_* = \frac{1}{2} - h^2(12\tau)^{-1}$) is absolutely stable. Indeed, for any $h$ and $\tau$

$$\sigma_* - \sigma_0 = - \frac{h^2}{12\,\tau} + \frac{h^2}{4\,\tau} = \frac{h^2}{6\,\tau} > 0 .$$

4. Implicit schemes with $0 < \sigma < \frac{1}{2}$ for $\sigma$ independent of $\gamma = \tau / h^2$ are conditionally stable if one imposes the constraint $\gamma \leq (2 - 4\,\sigma)^{-1}$.

5. Scheme (16) with $\sigma = \frac{1}{2} + h^2\,\alpha\,\tau^{-1}$ providing an approximation of $O(h^2 + t^2)$ is stable for any $h$ and $\tau$ if $\alpha \geq -\frac{1}{4}$.

So, not only the order of approximation, but also the stability of scheme (16) depend on the parameter $\sigma$.

While investigating stability we dealt actually only with two time layers $t_j$, $t_{j+1}$ and the step $\tau = t_{j+1} - t_j$. All the tricks and turns remain unchanged when the grid $\omega_\tau$ becomes non-equidistant, that is, the step $\tau_{j+1} = t_{j+1} - t_j$ depends on the number of the layer. In this situation the parameter $\sigma$ may depend on the number $j+1$ of the layer, say $\sigma = \sigma^{j+1}$, which serves to motivate the presence of another condition $\sigma \geq \sigma_0^{j+1} = \frac{1}{2} - h^2/(4\,\tau_{j+1})$ in place of (23). For instance, for the scheme of accuracy $O(h^4 + \tau_{j+1}^2)$ we may accept $\sigma_*^{j+1} = \frac{1}{2} - h^2/(12\,\tau_{j+1})$. The condition $\sigma \geq \sigma_0^{j+1}$ is sufficient for the stability of the scheme with weights in the case when the grid $\omega_\tau$ is non-equidistant.

**5. Stability with respect to the right-hand side.** As a matter of fact, if $\sigma \geq 0$, condition (23) given by

$$\sigma \geq \sigma_0\,, \qquad \sigma_0 = \frac{1}{2} - \frac{h^2}{4\,\tau}\,,$$

is sufficient for the stability of scheme (16) with respect to the right-hand side as well. It is interesting to consider problem (16b) and look for its solution in the form

$$(26) \qquad \hat{y} = \sum_{k=1}^{N-1} \hat{T}_k\, X^{(k)}\,, \qquad \|\,\hat{y}\,\|^2 = \sum_{k=1}^{N-1} \hat{T}_k^2\,,$$

by representing the right-hand side $\varphi$ in terms of the eigenfunctions $\{X^{(k)}\}$:

$$(27) \qquad \varphi = \sum_{k=1}^{N-1} \varphi_k\, X^{(k)}, \qquad \|\,\varphi\,\|^2 = \sum_{k=1}^{N-1} \varphi_k^2\,.$$

Substituting (26) and (27) into (16b) and recalling that $\Lambda X^{(k)} = -\lambda_k X^{(k)}$, we derive the series

$$\sum_{k=1}^{N-1} \{\, T_{kt}\,(1 + \sigma\tau\lambda_k) + \lambda_k\, T_k - \varphi_k \,\}\, X^{(k)} = 0\,,$$

from which it follows that the expression in the curly brackets vanishes as far as an orthogonal system of eigenfunctions is concerned:

$$(28) \qquad \hat{T}_k = q_k\, T_k + \frac{\tau\,\varphi_k}{1 + \sigma\tau\lambda_k}\,, \qquad q_k = \frac{1 - (1-\sigma)\,\tau\,\lambda_k}{1 + \sigma\tau\lambda_k}\,.$$

Putting these together with (26) we find that

$$\hat{y} = \sum_{k=1}^{N-1} \hat{T}_k \, X^{(k)} = \sum_{k=1}^{N-1} q_k \, T_k \, X^{(k)} + \tau \sum_{k=1}^{N-1} \frac{\varphi_k}{1+\sigma\tau\lambda_k} \, X^{(k)} .$$

With the aid of the well-known triangle inequality $\| v+w \| \leq \| v \| + \| w \|$ we establish

$$\| \hat{y} \| \leq \max_k | q_k | \left( \sum_{k=1}^{N-1} T_k^2 \right)^{1/2} + \max_k \frac{\tau}{| 1+\sigma\tau\lambda_k |} \left( \sum_{k=1}^{N-1} \varphi_k^2 \right)^{1/2}$$

or

$$\| \hat{y} \| \leq \max_k | q_k | \cdot \| y \| + \max_k \frac{\tau}{| 1+\sigma\tau\lambda_k |} \, \| \varphi \| . \tag{29}$$

Further, let

$$\sigma \geq \sigma_0 , \qquad \sigma_0 = \frac{1}{2} - \frac{h^2}{4\tau} , \qquad \sigma \geq 0 . \tag{30}$$

When this is the case, $| q_k | \leq 1$, $1+\sigma\tau\lambda_k \geq 1$ and $\| \hat{y} \| \leq \| y \| + \tau \| \varphi \|$ or $\| y^{j'+1} \| \leq \| y^{j'} \| + \tau \, \| \varphi^{j'} \|$. Summing over $j' = 0, 1, 2, \ldots, j$ and exploiting the fact that $\| y^0 \| = 0$ for the solution of problem (16b), we derive the estimate

$$\| y^{j+1} \| \leq \sum_{j'=0}^{j} \tau \, \| \varphi^{j'} \| . \tag{31}$$

Let us stress here that estimate (31) was obtained under condition (30). In subsequent reasonings we get rid of the bound $\sigma \geq 0$ and impose then the constraint

$$\sigma \geq \sigma_\varepsilon , \qquad \sigma_\varepsilon = \frac{1}{2} - \frac{1-\varepsilon}{4\,\tau} \, h^2 , \qquad 0 < \varepsilon < 1 , \tag{32}$$

where $\varepsilon = \text{const} > 0$ does not depend on $h$ and $\tau$. True, it is to be shown that

$$| q_k | \leq 1 , \qquad 1+\sigma\tau\lambda_k = 1 + (\sigma - \sigma_\varepsilon) \, \tau \, \lambda_k + \sigma_\varepsilon \, \tau \, \lambda_k$$

$$\geq 1 + \sigma_\varepsilon \, \tau \, \lambda_k = 1 + \tfrac{1}{2} \, \tau \, \lambda_k - \frac{(1-\varepsilon) \, h^2 \, \lambda_k}{4}$$

$$> 1 - \frac{(1-\varepsilon) \, h^2 \, \lambda_{N-1}}{4} > 1 - \frac{(1-\varepsilon) \, h^2}{4} \, \frac{4}{h^2} = \varepsilon ,$$

implying that $1 + \sigma\tau\lambda_k > \varepsilon$ for all $k = 1, 2, \ldots, N-1$ and providing in combination with (29) the validity of the following estimates:

$$\|\hat{y}\| \le \|y\| + \tau\|\varphi\|\varepsilon^{-1}$$

and

(33)
$$\|y^{j+1}\| \le \frac{1}{\varepsilon}\sum_{j'=0}^{j} \tau\,\|\varphi^{j'}\|.$$

For $\sigma = \sigma_*$ the condition $\sigma_* < 0$ ensures $\tau < \frac{1}{6}h^2$ and the value $\varepsilon = \frac{2}{3}$ can easily be found from the equation $\frac{1}{4}(1-\varepsilon) = \frac{1}{12}$. Collecting estimates (24), (31) and (33) and summarizing the preceding results, we deduce the following assertion.

If the conditions $\sigma \ge \frac{1}{2} - \frac{1}{4}h^2\tau^{-1} = \sigma_0$ and $\sigma \ge 0$ hold simultaneously, then scheme (16) is stable with respect to the initial data and the right-hand side, so that the solution of problem (16) admits the estimate

$$\|y^{j+1}\| \le \|u_0\| + \sum_{j'=0}^{j} \tau\,\|\varphi^{j'}\|.$$

When $\sigma < 0$, for the stability of scheme (16) with respect to the right-hand side

$$\sigma \ge \frac{1}{2} - \frac{(1-\varepsilon)\,h^2}{4\tau} = \sigma_\varepsilon\,, \qquad 0 < \varepsilon < 1\,,$$

is a sufficient condition. Here $\varepsilon \in (0, 1)$ is an arbitrary constant independent of $h$ and $\tau$. For such a choice, the solution of problem (16) satisfies the estimate

$$\|y^{j+1}\| \le \|u_0\| + \frac{1}{\varepsilon}\sum_{j'=0}^{j} \tau\,\|\varphi^{j'}\|.$$

For the scheme of accuracy $O(h^4 + \tau^2)$ we thus have $\varepsilon = \frac{2}{3}$ and $\sigma_* < 0$ if $\tau < \frac{1}{6}h^2$.

**6. Convergence and accuracy in the space $\mathbf{L_2(\omega_h)}$.** We state here that the convergence of scheme (II) follows from its stability and approximation. The error $z = y - u$ is just the solution of problem (III). Using a priori estimate (31) behind we deduce that

(34)
$$\|z^{j+1}\| \le \sum_{j'=0}^{j} \tau\,\|\psi^{j'}\| \qquad \text{as} \qquad \sigma \ge \sigma_0\,, \qquad \sigma \ge 0\,.$$

This type of situation is covered by the following assertion.

*If scheme* (II) *is stable with respect to the right-hand side and approximates problem* (I), *then it converges and the order of accuracy coincides with the order of approximation.*

Upon substituting the estimates for the approximation error obtained in Section 3 into (34) we find that

$$\| y^j - u^j \| = \begin{cases} O(h^2 + \tau^2), & \sigma = \frac{1}{2}, \quad u \in C_3^4, \\ O(h^4 + \tau^2), & \sigma = \sigma_*, \quad u \in C_3^6, \\ O(h^2 + \tau), & \sigma \neq \frac{1}{2}, \quad \sigma \neq \sigma_*, \quad u \in C_2^4. \end{cases} \tag{35}$$

So far we have investigated stability and convergence in mean, that is, in the grid norm of the space $L_2(\omega_h)$. Meanwhile, many situations exist in which a uniform estimate, that is, an estimate for the error of a solution will be of practical significance with regard to the norm

$$\| y^j - u^j \|_C := \max_{x \in \omega_h} | y^j - u^j | .$$

**7. Stability and convergence in the space C.** In this section we expound certain devices for obtaining uniform estimates for the problem (16) solution:

- the maximum principle;
- the energy method which makes it possible to establish stability in the space $C$ with respect to the right-hand side on account of embedding theorems;
- the representation of a solution in integral form in terms of the grid function of the impulse point source (Green's function).

The maximum principle and, in particular, Theorem 3 in Chapter 1, Section 2 will be quite applicable once we rearrange problem (II) supplied by homogeneous boundary conditions (scheme (16)) with obvious modifications and minor changes. The traditional tool for carrying out this work is connected with

$$\begin{gathered} \sigma \tau \Lambda \hat{y} - \hat{y} = -y - (1-\sigma) \tau \Lambda y - \tau \varphi = -F , \\ \hat{y}_0 = 0 , \qquad \hat{y}_N = 0 , \qquad y_i^0 = u_0(x_i) , \end{gathered} \tag{36}$$

or

$$\sigma\gamma\,\hat{y}_{i-1} - (2\,\sigma\,\gamma + 1)\,\hat{y}_i + \sigma\gamma\,\hat{y}_{i+1} = -F_i\,, \qquad i = 1,\,2,\,\dots\,, N-1\,,$$

$$(37)\quad F_i = (\,1-\sigma)\,\gamma\;y_{i-1} + (1 - 2\,(1-\sigma)\,\gamma)\;y_i + (1-\sigma)\,\gamma\;y_{i+1} + \tau\,\varphi_i\,,$$

$$\gamma = \frac{\tau}{h^2}\,, \qquad i = 1,\,2,\,\dots\,, N-1\,.$$

The theorem we have mentioned above asserts that for a solution to the difference equation

$$A_i\,y_{i-1} - C_i\,y_i + B_i\,y_{i+1} = -F_i\,, \qquad i = 1,\,2,\,\dots\,, N-1\,,$$

$$y_0 = 0\,, \qquad y_N = 0\,, \qquad |\,A_i\,| \neq 0\,, \qquad |\,B_i\,| \neq 0\,,$$

the estimate

$$\|\,y\,\|_C \leq \left\|\,\frac{F}{D}\,\right\|_C$$

is valid if and only if $D_i = |\,C_i\,| - |\,A_i\,| - |\,B_i\,| > 0$. For problem (36) these conditions ($|\,A_i\,| \neq 0$, $|\,B_i\,| \neq 0$, $D_i > 0$) are satisfied for $\sigma > 0$ and $D_i = 1$, so that a solution to equation (37) can be most readily evaluated as follows:

$$\|\,\hat{y}\,\|_C \leq \|\,F\,\|_C \quad \text{for} \quad \sigma \geq 0\,,$$

($\hat{y}_i = F_i$ for $\sigma = 0$). A simple observation that

$$\|\,F\,\|_C \leq \|\,y\,\|_C + \tau\,\|\,\varphi\,\|_C \quad \text{if} \quad 1 - 2\,(1-\sigma)\,\gamma \geq 0\,, \qquad 1-\sigma \geq 0\,,$$

may be useful in the further establishment of the inequality

$$(38)\qquad \|\,y^{j+1}\,\|_C \leq \|\,y^j\,\|_C + \tau\,\|\,\varphi^j\,\|_C\,,$$

which is valid for scheme (36) under the constraint

$$(39)\qquad \tau \leq \frac{h^2}{2\,(1-\sigma)}\,.$$

Summation of (38) over $j' = 0, 1, 2, \dots, j$ leads to the estimate for the problem (16) solution

$$(40)\qquad \|\,y^{j+1}\,\|_C \leq \|\,y^0\,\|_C + \sum_{j'=0}^{j} \tau\,\|\,\varphi^{j'}\,\|_C\,.$$

Summarizing, under condition (39) scheme (16) is stable with respect to the initial data and the right-hand side.

Applying a priori estimate (40) to problem (III) yields

$$\| z^{j+1} \|_C \le \sum_{j'=0}^{j} \tau \, \| \psi^{j'} \|_C \, .$$

We give a brief survey afforded by the above results: scheme (II) converges uniformly with the same rate as in the grid $L_2(\omega_h)$-norm (see (35)) if and only if condition (39) holds. The stability condition (39) in the space $C$ for the explicit scheme with $\sigma = 0$, namely $\tau \le \frac{1}{2} h^2$, coincides with the stability condition (25) in the space $L_2(\omega_h)$ that we have established for the case $\sigma < \frac{1}{2}$. The forward difference scheme with $\sigma = 1$ is absolutely stable in the space $C$. The symmetric difference scheme with $\sigma = \frac{1}{2}$ is stable in the space $C$ under the constraint $\tau \le h^2$.

**8. The method of energy inequalities.** The well-developed method of energy inequalities from Chapter 2 seems to offer more advantages in investigating the stability of scheme (II) with weights.

We first illustrate its employment for a differential equation in tackling problem (I) with homogeneous boundary conditions

$$(41) \qquad \begin{aligned} &\frac{\partial u}{\partial t} = \frac{\partial^2 u}{\partial x^2} + f(x, t) \, , \qquad 0 < x < 1 \, , \qquad 0 < t \le T \, , \\ &u(0, t) = u(1, t) = 0 \, , \qquad u(x, 0) = u_0(x) \, . \end{aligned}$$

Here the inner product and associated norm are defined by

$$(u, v) = \int_0^1 u(x) \, v(x) \, dx \, , \qquad \| u \| = \sqrt{(u, u)} \, ,$$

where $u(x)$ and $v(x)$ are functions defined on the segment $0 \le x \le 1$ and vanishing at the points $x = 0$ and $x = 1$. Let us multiply the equation by $\partial u / \partial t$ and integrate it with respect to $x$ from 0 to 1. The outcome of this is

$$\left\| \frac{\partial u}{\partial t} \right\|^2 + \left( - \frac{\partial u}{\partial t} \, , \, \frac{\partial^2 u}{\partial x^2} \right) = \left( f \, , \, \frac{\partial u}{\partial t} \right) .$$

Integrating the second term by parts and taking into account the equality

$$\left. \frac{\partial u}{\partial t} \, \frac{\partial u}{\partial x} \right|_0^1 = 0 \, ,$$

we find that

$$\left\| \frac{\partial u}{\partial t} \right\|^2 + \frac{1}{2} \frac{\partial}{\partial t} \left\| \frac{\partial u}{\partial x} \right\|^2 = \left( f, \frac{\partial u}{\partial t} \right).$$

To majorize the right-hand side, we make use of the Cauchy–Bunyakovskiĭ inequality and the $\varepsilon$-inequality $|ab| \le \varepsilon a^2 + \frac{1}{4}\varepsilon^{-1} b^2$ with $\varepsilon = 1$, permitting us to arrive at the relations

$$\left( f, \frac{\partial u}{\partial t} \right) \le \left| \left( f, \frac{\partial u}{\partial t} \right) \right| \le \left\| \frac{\partial u}{\partial t} \right\| \cdot \| f \| \le \left\| \frac{\partial u}{\partial t} \right\|^2 + \frac{1}{4} \| f \|^2,$$

implying that

$$\frac{\partial}{\partial t} \left\| \frac{\partial u}{\partial x} \right\|^2 \le \frac{1}{2} \| f(x, t) \|^2.$$

By integrating with respect to $t$ we thus have

$$\left\| \frac{\partial u(t)}{\partial x} \right\| \le \left\| \frac{\partial u(0)}{\partial x} \right\| + \frac{1}{\sqrt{2}} \left[ \int_0^t \| f(t') \|^2 \, dt' \right]^{1/2}$$

$$\le \left\| \frac{\partial u(0)}{\partial x} \right\| + \sqrt{\frac{T}{2}} \max_{0<t\le T} \| f(t) \|.$$

With the aid of the relation

$$\| u \|_C = \max_{0\le x\le 1} | u(x) | \le \frac{1}{2} \left\| \frac{\partial u}{\partial x} \right\|$$

we finally get

$$\| u(t) \|_C \le \frac{1}{2} \left\| \frac{\partial u(0)}{\partial x} \right\| + \frac{1}{2} \sqrt{\frac{T}{2}} \max_{0<t\le T} \| f(t) \|.$$

At the next stage a similar estimate is needed in this connection for the difference problem (16). We proceed as usual. This amounts to introducing the inner products and associated norms:

$$(y, v) = \sum_{i=1}^{N-1} y_i \, v_i \, h, \qquad \| y \| = \sqrt{(y, y)},$$

$$(y, v] = \sum_{i=1}^{N} y_i \, v_i \, h, \qquad \| y ]| = \sqrt{(y, y]},$$

and taking into consideration the trivial identities

$$y = \tfrac{1}{2}(\hat{y}+y) - \tfrac{1}{2}\tau y_t, \qquad \hat{y} = \tfrac{1}{2}(\hat{y}+y) + \tfrac{1}{2}\tau y_t,$$

$$\sigma\hat{y} + (1-\sigma)y = (\sigma - \tfrac{1}{2})\tau y_t + \tfrac{1}{2}(\hat{y}+y).$$

Putting these together with (16b) we recast the problem concerned as

$$(42) \qquad \begin{gathered} y_t - (\sigma - \tfrac{1}{2})\tau\Lambda y_t - \tfrac{1}{2}\Lambda(\hat{y}+y) = \varphi, \\ y(x, 0) = 0, \qquad y(0, t) = y(1, t) = 0, \end{gathered}$$

which reduces by multiplying equation (42) by $2\tau y_t h = 2(\hat{y}-y)h$ and summing the resulting equality over the inner nodes $x = ih$ of the grid $\omega_h$ to

$$(43) \quad 2\tau\|y_t\|^2 - 2(\sigma - \tfrac{1}{2})\tau^2(\Lambda y_t, y_t) - (\Lambda(\hat{y}+y), \hat{y}-y) = 2\tau(\varphi, y_t).$$

By appeal to the Green difference formula derived in Chapter 2, Section 3.1

$$(\Lambda v, w) = (v_{\bar{x}x}, w) = -(v_{\bar{x}}, w_{\bar{x}}], \qquad v_0 = w_0 = 0, \qquad v_N = w_N = 0,$$

with $v = y_t$, $w = y_t$ and $v = \hat{y}+y$, $w = \hat{y}-y$ incorporated, we find in view of $y_0 = y_N = 0$ that

$$(\Lambda y_t, y_t) = -\|y_{t\bar{x}}]|^2,$$

$$(\Lambda(\hat{y}+y), \hat{y}-y) = -(\hat{y}_{\bar{x}} + y_{\bar{x}}, \hat{y}_{\bar{x}} - y_{\bar{x}}] = -(\|\hat{y}_{\bar{x}}]|^2 - \|y_{\bar{x}}]|^2).$$

Substitution of these expressions into (43) leads to the **energy identity**

$$(44) \quad 2\tau(\|y_t\|^2 + (\sigma - \tfrac{1}{2})\tau\|y_{t\bar{x}}]|^2) + \|\hat{y}_{\bar{x}}]|^2 = \|y_{\bar{x}}]|^2 + 2\tau(\varphi, y_t),$$

which is valid for any $\sigma$. Let $\sigma \geq \sigma_0$. The accepted view is that a reasonable form is

$$J = \|v\|^2 + (\sigma - \tfrac{1}{2})\tau\,\|v_{\bar{x}}]|^2 \quad \text{with} \quad v = y_t.$$

By virtue of the estimate emerged in Chapter 2, Section 3

$$(45) \qquad \|v_{\bar{x}}]|^2 \leq \frac{4}{h^2}\|v\|^2$$

one succeeds in showing that $J \geq 0$ for $\sigma \geq \sigma_0$. Because of (45),

$$J = \| v \|^2 + (\sigma - \sigma_0)\, \tau\, \| v_{\bar{x}} ]|^2 + (\sigma_0 - \tfrac{1}{2})\, \tau\, \| v_{\bar{x}} ]|^2$$
$$\geq \| v \|^2 - \tfrac{1}{4} h^2 \| v_{\bar{x}} ]|^2 \geq 0$$

for $\sigma \geq \sigma_0 = \frac{1}{2} - h^2 (4\tau)^{-1}$. Combination of the preceding relations reveals a profound result known as the **energy inequality**:

$$\| \hat{y}_{\bar{x}} ]|^2 \leq \| y_{\bar{x}} ]|^2 + 2\tau (\varphi, y_t), \qquad \sigma \geq \sigma_0 .$$

If $\varphi = 0$ and $y$ is a solution of problem (16a), then $\| y_{\bar{x}}^{i+1} ]| \leq \cdots \leq \| y_{\bar{x}}^{0} ]|$, meaning that for $\sigma \geq \sigma_0$ scheme (16) is stable with respect to the initial data in the norm $\| y \|_{(1)} = \| y_{\bar{x}} ]|$, which is a grid analog of the $\overset{\circ}{W}{}_2^1$-norm. No doubt, it sounds interesting, but the discovery of stability with respect to the right-hand is of special merit in this matter. Letting

$$(46) \qquad \sigma \geq \sigma_\varepsilon , \qquad \sigma_\varepsilon = \frac{1}{2} - \frac{(1-\varepsilon)\, h^2}{4\tau}, \qquad 0 < \varepsilon \leq 1,$$

we claim that

$$(47) \qquad J \geq \varepsilon \| v \|^2 \quad \text{for} \quad \sigma \geq \sigma_\varepsilon .$$

Indeed,

$$J = \| v \|^2 + (\sigma - \sigma_\varepsilon)\, \tau \| v_{\bar{x}} ]|^2 + (\sigma_\varepsilon - \tfrac{1}{2})\, \tau \| v_{\bar{x}} ]|^2$$
$$\geq \| v \|^2 - \frac{(1-\varepsilon)\, h^2}{4} \| \dot{v}_{\bar{x}} ]|^2$$
$$\geq \| v \|^2 - (1-\varepsilon) \| v \|^2 = \varepsilon \| v \|^2 .$$

Substituting (47) into (44) yields the energy inequality

$$(48) \qquad 2\tau\varepsilon \| y_t \|^2 + \| \hat{y}_{\bar{x}} ]|^2 \leq \| y_{\bar{x}} ]|^2 + 2\tau (\varphi, y_t), \qquad \sigma \geq \sigma_\varepsilon .$$

By applying successively the Cauchy–Bunyakovskiĭ inequality and the $\varepsilon$-inequality we arrive at the chain of the relations

$$(49) \qquad 2\tau (\varphi, y_t) \leq 2\tau \| \varphi \| \cdot \| y_t \| \leq 2\tau\varepsilon \| y_t \|^2 + \frac{\tau}{2\varepsilon} \| \varphi \|^2 .$$

Substitution of (49) into (48) gives

$$\| y_{\bar{x}}^{j'+1} ]|^2 \le \| y_{\bar{x}}^{j'} ]|^2 + \frac{\tau}{2\,\varepsilon} \, \| \varphi^{j'} \|^2 .$$

Summing over $j' = 0, 1, \ldots, j$ and keeping $y^0 = 0$, we deduce that

$$\| y_{\bar{x}}^{j+1} ]|^2 \le \frac{1}{2\,\varepsilon} \sum_{j'=0}^{j} \tau \, \| \varphi^{j'} \|^2 .$$

In agreement with Lemma 1 in Chapter 2, Section 3, we might have

$$\| y \|_C \le \frac{1}{2} \, \| y_{\bar{x}} ]| ,$$

thereby justifying that

$$\| y^{j+1} \|_C \le \frac{1}{2\sqrt{2\,\varepsilon}} \left[ \sum_{j'=0}^{j} \tau \, \| \varphi^{j'} \|^2 \right]^{1/2} , \qquad \sigma \ge \sigma_\varepsilon .$$

Applying this estimate to problem (III) we get

$$\| z^{j+1} \|_C \le M_2 \max_{0 \le j' \le j} \| \psi^{j'} \| , \qquad M_2 = \frac{\sqrt{T}}{2\sqrt{2\,\varepsilon}} , \qquad \sigma \ge \sigma_\varepsilon ,$$

thus demonstrating that scheme (III) converges uniformly, so that

$$\| y^j - u^j \|_C \le \begin{cases} M\,(h^2 + \tau) , & \sigma > \frac{1}{2} , \quad u \in C_2^4 , \\ M\,(h^2 + \tau^2) , & \sigma = \frac{1}{2} , \quad u \in C_3^4 , \\ M\,(h^4 + \tau^2) , & \sigma = \sigma_* , \quad u \in C_3^6 . \end{cases}$$

A case in point is that for the explicit scheme with $\sigma = 0$ the uniform convergence does not follow from (46) under the constraint $\tau \le \frac{1}{2} h^2$. But the a priori estimate emerged in Section 5.7, namely

$$\| y^j \|_C \le \| y^0 \|_C + \sum_{j'=0}^{j-1} \tau \, \| \varphi^{j'} \|_C , \qquad \tau \le \tfrac{1}{2} h^2 ,$$

provides support for the view that

$$\| z^{j+1} \|_C \le \sum_{j'=0}^{j} \tau \, \| \psi^{j'} \|_C .$$

Whence it follows immediately that the explicit scheme converges uniformly with the rate $O(\tau + h^2)$.

**9. The third kind boundary conditions.** The first kind boundary conditions we have considered so far are satisfied on a grid exactly. In Chapter 2 we have suggested one effective method, by means of which it is possible to approximate the third kind boundary condition for the forward difference scheme ($\sigma = 1$) and the explicit scheme ($\sigma = 0$) and generate an approximation of $O(\tau + h^2)$. Here we will handle scheme (II) with weights, where $\sigma$ is kept fixed. In preparation for this, the third kind boundary condition

$$\frac{\partial u(0,\, t)}{\partial x} = \beta_1\, u(0,\, t) = \mu_1(t)\,, \qquad \beta_1 = \text{const} > 0\,, \tag{50}$$

will be imposed at the point $x = 0$, providing later the difference boundary condition on the four-point pattern with the nodes $(0, t_{j+1})$, $(h, t_{j+1})$, $(0, t_j)$ and $(h, t_j)$. Once supplied by the difference condition

$$\sigma\,(\hat{y}_x - \beta_1\, \hat{y})_0 + (1-\sigma)\,(y_x - \beta_1\, y)_0 = \tfrac{1}{2}\, h\, y_{t,0} - \tilde{\mu}_1\,, \qquad \tilde{\mu}_1 = \bar{\mu}_1 + \tfrac{1}{2}\, h \bar{f}_0\,, \tag{51}$$

where $\bar{f}_0 = f(0, t_{j+1/2})$ and $\bar{\mu}_1 = \mu_1(t_{j+1/2})$, we will show that it approximates condition (50) on a solution $u = u(x,t)$ to equation (3) subject to condition (50) and the order of approximation is the same as we obtained in approximating equation (3) by scheme (II) for a given value of $\sigma$.

Upon substituting $y = z + u$ into (51) we find that

$$\sigma\,(\hat{z}_x - \beta_1\, \hat{z})_0 + (1-\sigma)\,(z_x - \beta_1\, z)_0 = \tfrac{1}{2}\, h\, z_{t,0} - \tilde{\nu}_1\,, \tag{52}$$

where $\tilde{\nu}_1 = \sigma\,(\hat{u}_x - \beta_1\, \hat{u})_0 + (1-\sigma)\,(u_x - \beta_1\, u)_0 - \frac{1}{2}\, h\, u_{t,0} + \tilde{\mu}_1$ is the error of approximation of condition (50) by the difference condition (51) on a solution $u$. Developing Taylor's series for $u$ about the node $(0,\, t_j + \frac{1}{2}\tau)$ and denoting by $\bar{v}_0$ the value of the function $v$ at this node, we can write down in the preceding notations $u' = \partial u/\partial x$ and $\dot{u} = \partial u/\partial t$

$$\tilde{\nu}_1 = (\bar{u}'_0 - \beta_1\, \bar{u}_0 + \tilde{\mu}_1)_0 + (\sigma - \tfrac{1}{2})\,\tau\,(\bar{\dot{u}}' - \beta_1\, \bar{\dot{u}})_0$$
$$- \tfrac{1}{2}\, h\, \bar{\dot{u}}_0 + \tfrac{1}{2}\, h\, \bar{u}''_0 + O(h^2 + \tau^2)\,.$$

Inserting here $\bar{u}'_0 = \beta_1\, \bar{u}_0 - \bar{\mu}_1$ and $\bar{u}''_0 = \bar{\dot{u}}_0 - \bar{f}_0$ both recovered from the equation

$$\tilde{\nu}_1 = (\sigma - \tfrac{1}{2})\,\tau\,(\bar{\dot{u}}'_0 - \beta_1\, \bar{\dot{u}}_0) + O(h^2 + \tau^2)\,,$$

we deduce that

$$\begin{aligned} &\tilde{\nu}_1 = O(h^2 + \tau)\,, && \sigma \neq \tfrac{1}{2}\,, \\ &\tilde{\nu}_1 = O(h^2 + \tau^2)\,, && \sigma = \tfrac{1}{2}\,. \end{aligned} \tag{53}$$

It is straightforward to verify that at the point $x = 1$ the boundary condition

$$(54) \qquad -\frac{\partial u(1,t)}{\partial x} = \beta_2\, u(1,t) - \mu_2(t)\,, \qquad \beta_2 = \text{const} > 0\,,$$

is approximated to the same order by the difference condition

$$(55) \qquad -\big[\sigma\,(\hat{y}_{\bar{x}} + \beta_2\,\hat{y})_N + (1-\sigma)\,(y_{\bar{x}} + \beta_2\,y)_N\big] = \tfrac{1}{2}\,h\,y_{t,N} - \tilde{\mu}_2\,,$$

where $\tilde{\mu}_2 = \bar{\mu}_2 + \frac{1}{2}\,h\,\bar{f}_N$, $\bar{\mu}_2 = \mu_2(t_{j+1/2})$ and $\bar{f}_N = \bar{f}(1, t_{j+1/2})$. By setting

$$\tilde{\mu}_1 = \bar{\mu}_1 + \frac{h^2}{12}\,\dot{\bar{\mu}}_1 + \frac{h^2}{6}\,(\bar{f}_0' - \beta_1\,\bar{f}_0) + \frac{h^2}{2}\,\rho_1\,\varphi_0\,,$$

$$\tilde{\mu}_2 = \bar{\mu}_2 + \frac{h^2}{12}\,\dot{\bar{\mu}}_2 - \frac{h^2}{6}\,(\bar{f}_N' + \beta_2\,\bar{f}_N) + \frac{h^2}{2}\,\rho_2\,\varphi_N\,,$$

$$\varphi = \bar{f} + \frac{h^2}{12}\,\bar{f}''$$

and replacing $\frac{1}{2}\,h\,y_{t,0}$ and $\frac{1}{2}\,h\,y_{t,N}$ in (51) and (55) by $\frac{1}{2}\,h\,\rho_1 y_{t,0}$ and $\frac{1}{2}\,h\,\rho_2 y_{t,N}$, respectively, with $\rho_k = 1 + \frac{1}{3}\,h\,\beta_k$, $k = 1, 2$, we establish the difference boundary conditions providing an approximation of $O(h^4 + \tau^2)$ for $\sigma = \sigma_* = \frac{1}{2} - \frac{1}{12}\,h^2\,\tau^{-1}$.

To avoid cumbersome calculations, it will be sensible to introduce more compact notations

$$\Lambda^- y = \frac{y_x - \beta_1\,y}{\frac{1}{2}\,h}\,, \qquad \Lambda^+ y = -\frac{y_{\bar{x}} + \beta_2\,y}{\frac{1}{2}\,h}$$

and rewrite the difference boundary conditions (51) and (55) in the same forms as approved before for scheme (II):

$$(56) \qquad \begin{aligned} y_t &= \Lambda^-\big(\sigma\,\hat{y} + (1-\sigma)\,y\big) + \varphi^-\,, \qquad && x = 0\,,\\ y_t &= \Lambda^+\big(\sigma\,\hat{y} + (1-\sigma)\,y\big) + \varphi^+\,, \qquad && x = 1\,, \end{aligned}$$

with $\varphi^- = 2\,\tilde{\mu}_1/h$ and $\varphi^+ = 2\,\tilde{\mu}_2/h$, showing the new notations to be sensible ones. For $\beta_1 = \beta_2 = 0$ these can be viewed as the difference approximations of the second kind boundary conditions. Also, the order of approximation happens to be the same as we obtained for the boundary-value problem of the third kind.

Let us reduce condition (51) to an alternative form convenient for current manipulations. The outcome of solving (51) with respect to $\hat{y}_0 = y_0^{j+1}$ is

$$(57) \qquad \begin{aligned} &\hat{y}_0 = \varkappa_1 \hat{y}_1 + \nu_1 , \qquad \varkappa_1 = \frac{\sigma}{\Delta_1} , \qquad \Delta_1 = \sigma (1 + \beta_1 h) + \frac{h^2}{2\tau} , \\ &\nu_1 = \frac{1}{\Delta_1} \left\{ (1-\sigma) y_1 + \left[ \frac{h^2}{2\tau} - (1-\sigma)(1+\beta_1 h) \right] y_0 + h \tilde{\mu}_1 \right\} . \end{aligned}$$

Along these lines, condition (55) becomes

$$(58) \qquad \begin{aligned} &\hat{y}_N = \varkappa_2 \hat{y}_{N-1} + \nu_2 , \qquad \varkappa_2 = \frac{\sigma}{\Delta_2} , \qquad \Delta_2 = \sigma (1 + \beta_2 h) + \frac{h^2}{2\tau} , \\ &\nu_2 = \frac{1}{\Delta_2} \left\{ (1-\sigma) y_{N-1} + \left[ \frac{h^2}{2\tau} - (1-\sigma)(1+\beta_2 h) \right] y_N + h \tilde{\mu}_2 \right\} , \end{aligned}$$

so that $0 < \varkappa_\alpha \leq 1$ for $\beta_\alpha \geq 0$, $\sigma > 0$ and $\alpha = 1, 2$. In the determination of $\hat{y}$ on a new layer we obtain the difference equation (8) with boundary conditions (57)–(58). This problem can be solved by the right elimination method (see Chapter 1, Section 2.5).

Stability of scheme (II) with the third kind boundary conditions can be discovered following established practice either by the method of separation of variables or on account of the maximum principle.

**10. Three-layer schemes for the heat conduction equation.** One of the first schemes arising in numerical solution of the heat conduction equation $\partial u / \partial t = \partial^2 u / \partial x^2$ was the **Richardson** explicit three-layer scheme such as

$$(59) \qquad \frac{y^{j+1} - y^{j-1}}{2\tau} = \Lambda y^j \qquad \text{or} \qquad y_{\overset{\circ}{t}} = \Lambda y ,$$

where

$$y_{\overset{\circ}{t}} = \frac{\hat{y} - \check{y}}{2\tau} , \qquad \hat{y} = y^{j+1} , \qquad \check{y} = y^{j-1} , \qquad y = y^j , \qquad \Lambda y = y_{\bar{x}x} .$$

By exactly the same reasoning as before, it is not difficult to verify that this scheme is of order 2 with respect to $\tau$ and $h$ both: $\psi = \Lambda u - u_{\overset{\circ}{t}} = O(h^2 + \tau^2)$.

But it is absolutely unstable: this scheme becomes unstable for any way of tending $\tau$ and $h$ to zero.

As further developments occur, we rewrite equation (59) as

$$\frac{y_i^{j+1} - y_i^{j-1}}{2\,\tau} = \frac{y_{i-1}^j - 2\,y_i^j + y_{i+1}^j}{h^2}\,. \tag{60}$$

Upon substituting the sum $y_i^{j+1} + y_i^{j-1}$ for $2\,y_i^j$ in the right-hand side of equation (60) the resulting three-layer **"rhombus" scheme** known as the **Du-Fort–Frankel scheme**

$$\frac{y_i^{j+1} - y_i^{j-1}}{2\,\tau} = \frac{y_{i-1}^j - y_i^{j+1} - y_i^{j-1} + y_{i+1}^j}{h^2} \tag{61}$$

becomes explicit with respect to $y_i^{j+1}$ and appears to be absolutely stable for any $h$ and $\tau$. In giving an alternative form of the "rhombus" scheme

$$y_{\overset{\circ}{t}} + \frac{\tau^2}{h^2}\, y_{\bar{t}t} = \Lambda\, y \tag{62}$$

with $y_{\bar{t}t} = \left(y_i^{j+1} - 2\,y_i^j + y_i^{j-1}\right)\tau^{-2}$ we may attempt the right-hand side of equation (61) in the form

$$\begin{aligned} \frac{y_{i-1} - \hat{y}_i - \check{y}_i + y_{i+1}}{h^2} &= \frac{y_{i-1} - 2\,y_i + y_{i+1}}{h^2} - \frac{\hat{y}_i - 2\,y_i + \check{y}_i}{h^2} \\ &= y_{\bar{x}x} - \frac{\tau^2}{h^2}\, y_{\bar{t}t}\,. \end{aligned}$$

Substitution of the last expression into (61) yields (62). As a matter of fact, the "rhombus" scheme is some modification of the Richardson scheme with the extra member on the left-hand side of (61) $\tau^2\,h^{-2}\,y_{\bar{t}t}$, which assures us of its stability. The proof of stability of scheme (62) is an immediate implication of the general theory of Chapter 6 and is omitted here. The error of approximation for scheme (62) is

$$\psi = \Lambda\,u - u_{\overset{\circ}{t}} - \frac{\tau^2}{h^2}\,u_{\bar{t}t} = u'' - \dot{u} - \frac{\tau^2}{h^2}\,\ddot{u} + O(h^2+\tau^2) = -\frac{\tau^2}{h^2}\,\ddot{u} + O(h^2+\tau^2)\,.$$

From such reasoning it seems clear that the "rhombus" scheme provides the conditional approximation

$$\psi = O\left(\tau^2 + h^2 + \tau^2\,h^{-2}\right) = O(h^2) \quad \text{as} \quad \tau = O(h^2)\,.$$

Keeping $\tau = \alpha h \, (1 + O(h))$ with $\alpha = \text{const}$ we draw the conclusion that scheme (62) approximates any equation of the form

$$\frac{\partial u}{\partial t} + \alpha^2 \frac{\partial^2 u}{\partial t^2} = \frac{\partial^2 u}{\partial x^2} \,.$$

The following implicit three-layer schemes with weights are quite applicable in solving equation (3):

- symmetric schemes

$$y_{\overset{\circ}{t}} = \Lambda \left( \sigma \, \hat{y} + (1 - 2\,\sigma) \, y + \sigma \, \check{y} \right) + \varphi \, ; \tag{63}$$

- nonsymmetric schemes

$$y_t + \sigma \, \tau \, y_{\bar{t}t} = \Lambda \, \check{y} + \varphi \, . \tag{64}$$

Equations (63) and (64) contain the three layers $(t_{j-1}, \, t_j, \, t_{j+1})$, so that, in what follows, $t_j \geq \tau$, $j \geq 1$. The value $y(x, 0) = u_0(x)$ is known in advance and the value $y(x, \tau)$ should be preassigned in addition to the available information. For example, this value can be determined either by

$$y_t(x, \, 0) = \bar{u}_0(x) \qquad \text{or} \qquad y(x, \, \tau) = y(x, \, 0) + \tau \, \bar{u}_0(x)$$

with $\bar{u}_0(x) = u_0''(x) + f(x, 0)$ still subject to the condition $y(x, \, \tau) - u(x, \, \tau) = O(\tau^2)$ (see Chapter 2, Section 1).

Sometimes two-layer schemes are effectively used to assign the value $y(x, \tau)$.

By virtue of the asymptotic relation

$$\sigma \, \hat{u} + (1 - 2\,\sigma) \, u + \sigma \, \check{u} = u + \sigma \, \tau^2 \, u_{\bar{t}t} = u + O(\tau^2)$$

the symmetric scheme (63) is of order 2 in $\tau$ and $h$ for any $\sigma$. The error of approximation for scheme (64) is representable by

$$\begin{aligned}\psi &= \Lambda \, \hat{u} + \varphi - (\, u_t + \sigma \, \tau \, u_{\bar{t}t}) \\ &= L \, \hat{u} + \varphi - (\, \dot{\hat{u}} - \tfrac{1}{2} \, \tau \, \ddot{\hat{u}} + \sigma \, \tau \, \ddot{\hat{u}} \,) + O(h^2 + \tau^2) \\ &= (\, L \, \hat{u} + \hat{f} - \dot{\hat{u}} \,) + (\, \varphi - \hat{f} \,) - (\, \sigma - \tfrac{1}{2} \,) \, \tau \, \ddot{\hat{u}} + O(h^2 + \tau^2) \, ,\end{aligned} \tag{65}$$

which serves to motivate the accuracies which interest us in the following cases:

$$\psi = O(h^2 + \tau^2) \qquad \text{for} \qquad \sigma = \tfrac{1}{2} \, , \qquad \varphi = \hat{f},$$

$$\psi = O(h^2 + \tau) \qquad \text{for} \qquad \sigma \neq \tfrac{1}{2} \, , \qquad \varphi = \hat{f}.$$

Omitting the terms $O(h^2)$ in (65) and involving the equation $\dot{u} = u'' + f$ insight, we deduce that scheme (64) provides an approximation of $O(h^4+\tau^2)$ with the members

$$\sigma = \tfrac{1}{2} + \tfrac{1}{12} h^2 \tau^{-1}, \qquad \varphi = \hat{f} + \tfrac{1}{12} h^2 \left( \hat{f}'' + \hat{\dot{f}} \right).$$

In trying to recover $\hat{y}$ from (63)–(64) we obtain the three-point equations

$$A\, \hat{y}_{i-1} - C\, \hat{y}_i + B\, \hat{y}_{i+1} = -F_i \tag{66}$$

with the right-hand sides $-F_i$ depending on $y$, $\check{y}$, $\varphi$ and the usual boundary conditions for $i = 0$ and $i = N$. This problem can be solved by the right elimination method. In the process of computing the values of $y$ and $\check{y}$ on two previous layers should be placed in the storage. In the case of two-layer schemes it suffices to save merely one preceding layer.

Stability of three-layer schemes will be established in Chapter 6. For the purposes of current section we confine ourselves to sufficient stability conditions for the symmetric scheme (63) and scheme (64), respectively: $\sigma > \frac{1}{4}$ and $\sigma \geq -\frac{1}{2}$. In just the same way as we did for two-layer schemes the difference boundary conditions with a highly accurate approximation can be developed for the third kind boundary conditions (50) and (54). For the symmetric scheme (63) the boundary conditions providing an approximation of $O(h^2 + \tau^2)$ reduce to

$$y_{\overset{\circ}{t}} = \Lambda^- \left( \sigma\, \hat{y} + (1 - 2\,\sigma)\, y + \sigma\, \check{y} \right) + \varphi^-, \qquad i = 0,$$

$$y_{\overset{\circ}{t}} = \Lambda^+ \left( \sigma\, \hat{y} + (1 - 2\,\sigma)\, y + \sigma\, \check{y} \right) + \varphi^+, \qquad i = N,$$

with

$$\Lambda^- y = \frac{y_x - \beta_1\, y}{\frac{1}{2} h}, \qquad \Lambda^+ y = -\frac{y_{\bar{x}} + \beta_2\, y}{\frac{1}{2} h},$$

$$\varphi^- = f(0, t) + \frac{\mu_1(t)}{\frac{1}{2} h}, \qquad \varphi^+ = f(1, t) + \frac{\mu_2(t)}{\frac{1}{2} h},$$

showing the new notations to be sensible ones.

On the same grounds as before, the third kind difference boundary conditions for the nonsymmetric scheme (64) are specified by the relations

$$\begin{aligned} &\rho_1 \left( y_t + \sigma\, \tau\, y_{\bar{t}t} \right) = \Lambda^- \hat{y} + \varphi^-, \qquad i = 0, \\ &\rho_2 \left( y_t + \sigma\, \tau\, y_{\bar{t}t} \right) = \Lambda^+ \hat{y} + \varphi^+, \qquad i = N. \end{aligned} \tag{67}$$

These boundary conditions generate an approximation of $O(h^2+\tau)$ for any $\sigma$ if we agree to consider $\rho_1 = \rho_2 = 1$ and

$$\varphi^- = f(0,\, t+\tau) + \frac{\mu_1(t+\tau)}{0.5\,h}\,, \qquad \varphi^+ = f(1,\, t+\tau) + \frac{\mu_2(t+\tau)}{0.5\,h}\,.$$

But the same procedure works for $\sigma = \frac{1}{2}$ and the boundary conditions (67), making it possible to achieve $O(h^2+\tau^2)$. Eventually, by successive use of the members

$$\sigma = \frac{1}{2} + \frac{h^2}{12\,\tau}\,, \qquad \rho_1 = 1 + \frac{h\,\beta_1}{3}\,, \qquad \rho_2 = 1 + \frac{h\,\beta_2}{3}\,,$$

$$\varphi^- = \rho_1\,\varphi + \frac{\nu_1}{0.5\,h}\,, \qquad \varphi^+ = \rho_2\,\varphi + \frac{\nu_2}{0.5\,h}\,,$$

$$\varphi = f(x,\, t+\tau) + \frac{h^2}{12}\left(f''(x,\, t+\tau) + \dot{f}(x,\, t+\tau)\right),$$

$$\nu_1 = \mu_1(t+\tau) + \frac{h^2}{6}\left(\dot{\mu}_1(t+\tau) + f'(0,\, t+\tau) - \beta_1 f(0,\, t+\tau)\right),$$

$$\nu_2 = \mu_2(t+\tau) + \frac{h^2}{6}\left(\dot{\mu}_2(t+\tau) - f'(1,\, t+\tau) - \beta_2 f(1,\, t+\tau)\right)$$

we arrive at scheme (64) of accuracy $O(h^4+\tau^2)$ and the error of approximation $O((h^4+\tau^2)/h)$ at the nodes $x=0$ and $x=1$.

## 5.2 ASYMPTOTIC STABILITY

**1. $\rho$-stability.** In studying various schemes for the heat conduction equation we have already mentioned that the simplest way of relaxation of the stability condition connected with upper bounds on the step in time $\tau$ is the transition to implicit schemes. All of the implicit schemes with a weight of the upper layer $\sigma \geq \sigma_0$, $\sigma_0 = \frac{1}{2} - \frac{1}{4}\,h^2\,\tau^{-1}$, are stable. In particular, the schemes with $\sigma = \frac{1}{2}$ and $\sigma = 1$ are absolutely stable. At the final stage of research the scientists are interested in the accuracy with which an approximate solution to a differential equation could be found. It is natural to try to reach a prescribed accuracy by carrying out the minimal possible number of computer operations (arithmetic and logic), that is, with the minimal expenses of execution time. A number of operations may be minimized by improvements of difference schemes as a result of modeling the basic properties of a differential equation in the space of grid functions as well as possible. One of such typical properties is the true asymptotic behavior of its solution as $t \to \infty$.

To illustrate our approach, let us consider, for example, the heat conduction equation

$$(1) \qquad \frac{\partial u}{\partial t} = \frac{\partial^2 u}{\partial x^2}\,, \qquad 0 < x < 1\,, \qquad t > 0\,,$$

$$u(x,0) = u_0(x)\,, \qquad 0 \le x \le 1\,, \qquad u(0,t) = u(1,t) = 0\,, \qquad t \ge 0\,.$$

The problem concerned can be solved by the method of separation of variables, within the framework of which we seek its solution as a sum

$$u(x,\,t) = \sum_{k=1}^{\infty} c_k\, e^{-\lambda_k t}\, X_k(x)\,, \qquad X_k(x) = \sqrt{2}\,\sin \pi k x\,,$$

where

$$\lambda_k = k^2\,\pi^2, \qquad c_k = \big(u_0,\, X_k\big) = \int_0^1 u_0(x)\, X_k(x)\, dx\,,$$

$$\|\, u(x,\,t)\,\|^2 = \sum_{k=1}^{\infty} c_k^2\, e^{-2\lambda_k t}\, \|\, X_k\,\|^2 = \sum_{k=1}^{\infty} c_k^2\, e^{-2\lambda_k t},$$

since $\|\, X_k \,\| = 1$. As $\lambda_k$ increases along with the subscript $k$, we thus have $\lambda_k > \lambda_1$ and

$$\|\, u(t)\,\|^2 \le e^{-2\lambda_1 t} \sum_{k=1}^{\infty} c_k^2 = e^{-2\lambda_1 t}\, \|\, u_0\,\|^2,$$

which means that the solution of problem (1) admits the estimate

$$(2) \qquad \|\, u(t)\,\| \le e^{-\lambda_1 t}\, \|\, u(0)\,\| \quad \text{for any} \quad t > 0\,.$$

With increasing $t$ the harmonics $u_{(k)} = c_k\, e^{-\lambda_k t}\, X_k(x)$, $k > 1$, damp more rapidly than the first one, so that for sufficiently large $t$

$$(3) \qquad u(x,\,t) \approx c_1\, e^{-\lambda_1 t}\, X_1(x)\,, \qquad c_1 \ne 0\,.$$

This stage of the process refers to the **regular behavior**. When the solution of a difference scheme for problem (1) also possesses the properties similar to (2) and (3), the scheme is said to be **asymptotically stable**.

We now deal with the scheme with weights

$$
\begin{aligned}
& y_t = \Lambda\left(\sigma\,\hat{y} + (1-\sigma)\,y\right), && x \in \omega_h, \qquad t = j\tau > 0, \\
(4) \qquad & y_0 = y_N = 0, \qquad y(x, 0) = u_0(x), && x \in \bar{\omega}_h, \\
& \Lambda\, y = y_{\bar{x}x} .
\end{aligned}
$$

The solution of the problem under consideration has been obtained in Section 1.4 by the method of separation of variables and can be expressed by the following series

$$
y_i^j = y(x_i, t_j) = \sum_{k=1}^{N-1} c_k\, q_k^j\, X_k(x_i),
$$

where

$$
q_k = \frac{1-(1-\sigma)\,\tau\lambda_k^h}{1+\sigma\,\tau\,\lambda_k^h}, \qquad \lambda_k^h = \frac{4}{h^2}\sin^2\frac{\pi h k}{2},
$$

$$
X_k(x_i) = \sqrt{2}\,\sin \pi k x_i, \qquad c_k = (u_0, X_k), \qquad (y, v) = \sum_{i=1}^{N-1} y_i\, v_i\, h .
$$

The orthonormality of the system $\{X_k\}$ gives

$$
\|\, y^j \,\|^2 = \sum_{k=1}^{N-1} c_k^2\, q_k^{2j} \leq \rho^{2j} \sum_{k=1}^{N-1} c_k^2 = \rho^{2j}\, \|\, y^0 \,\|^2,
$$

which assures us of the validity of the estimate

$$
(5) \qquad \|\, y^j \,\| \leq \rho^j\, \|\, y^0 \,\|, \qquad \rho = \max_{1\leq k\leq N-1} |\, q_k \,| .
$$

Scheme (4) is said to be $\rho$-stable with respect to the initial data if estimate (5) is valid. An estimate of the form (3) is admissible for the solution of problem (4) when $\rho < 1$.

**2. Asymptotic stability.** Let us describe rather mild conditions under which the behavior of the solution $y^j$ of the difference problem (4) becomes regular if one makes $t_j = j\tau$ large enough:

$$y_i^j = y(x_i\,,\, t_j) \approx c_1\, q_1^j\, X_1(x_i)\,. \tag{6}$$

Clearly, it is possible only in the case when the first harmonic dominates, that is, $\max\limits_k |\, q_k \,|$ is attained for $k = 1$, so that

$$\max_{1\le k\le N-1} q_k = \rho = \left| \frac{1-(1-\sigma)\,\tau\,\lambda_1^h}{1+\sigma\,\tau\,\lambda_1^h} \right| . \tag{7}$$

What is available to find out the conditions ensuring the validity of formula (7) is the function

$$f(\mu) = \left( \frac{1-(1-\sigma)\,\mu_1}{1+\sigma\mu_1} \right)^2 - \left( \frac{1-(1-\sigma)\,\mu}{1+\sigma\mu} \right)^2 ,$$

where $\mu_1 = \tau\,\lambda_1^h$ and $\mu = \tau\,\lambda_k^h \ge \mu_1$ under the natural premise $f(\mu) \ge 0$, $\mu \ge \mu_1$. Alternative forms of the function $f(\mu)$ may be chosen in a number of different ways. In subsequent reasonings we prefer one of them:

$$\begin{aligned}
f(\mu) &= f_1(\mu)\; f_2(\mu)\; K^{-1}, \qquad\qquad K = (1+\sigma\,\mu_1)^2\,(1+\sigma\,\mu)^2 > 0\,,\\
f_1(\mu) &= \big(1-(1-\sigma)\,\mu_1\big)\;(1+\sigma\,\mu) - \big(1-(1-\sigma)\,\mu\big)\;(1+\sigma\,\mu_1)\\
&= \mu - \mu_1 \ge 0\,,\\
f_2(\mu) &= \big(1-(1-\sigma)\,\mu_1\big)\;(1+\sigma\,\mu) + \big(1-(1-\sigma)\,\mu\big)\;(1+\sigma\,\mu_1)\\
&= 2 + (2\sigma - 1)(\mu + \mu_1) - 2\,(1-\sigma)\,\sigma\,\mu\,\mu_1\,,
\end{aligned}$$

implying for $\tau \le \bar\tau_0$ that

$$f_2(\mu) = \begin{cases} 2 + \mu + \mu_1 > 0 & \text{if } \sigma = 1\,,\\ 2 - (\mu + \mu_1) \ge 2 - \tau\,(\Delta + \delta) \ge 0 & \text{if } \sigma = 0\,, \end{cases}$$

where $\bar\tau_0 = 2\,(\delta + \Delta)^{-1}$, $\delta = \lambda_1^h$ and $\Delta = \lambda_{N-1}^h$. In particular, we have

$$f_2(\mu) = 2 - \tfrac{1}{2}\,\mu\,\mu_1 \ge 2 - \tfrac{1}{2}\,\mu_{N-1}\,\mu_1 = 2 - \tfrac{1}{2}\,\tau^2\,\Delta\,\delta \ge 0$$

for the symmetric scheme with $\sigma = \frac{1}{2}$ under the constraint $\tau \le 2\,(\delta\,\Delta)^{-1/2}$. We bring together the results obtained and state that

$$\rho = \begin{cases} \dfrac{1}{1+\tau\,\delta} & \text{for} \quad \sigma = 1 \quad \text{and any} \quad \tau > 0\,, \\ 1 - \tau\,\delta & \text{for} \quad \sigma = 0 \quad \text{and} \quad \tau \le \bar{\tau}_0 = \dfrac{2}{\delta + \Delta}\,, \\ \dfrac{1 - \frac{1}{2}\,\tau\,\delta}{1 + \frac{1}{2}\,\tau\,\delta} & \text{for} \quad \sigma = \frac{1}{2} \quad \text{and} \quad \tau \le \tau_0 = \dfrac{2}{\sqrt{\delta\,\Delta}}\,. \end{cases}$$

In each such case the estimate $\|\, y^j \,\| \le \rho^j \,\|\, y^0 \,\|$ is certainly true. Moreover, $\rho^j \to 0$ occurs as $t_j = j\,\tau \to \infty$. As we will see a little later, it will be convenient to represent $\rho$ by

$$\rho = e^{-\mu_1 + \tau\,\varphi}\,, \qquad \tau\,\varphi = \mu_1 - \log\,\frac{1}{\rho}\,, \qquad \mu_1 = \tau\,\delta\,.$$

It is straightforward to verify that for $\sigma = \frac{1}{2}$

$$\tau\,\varphi = \mu_1 - \log\,\frac{1 + \frac{1}{2}\,\mu_1}{1 - \frac{1}{2}\,\mu_1} = -\left(\frac{\mu_1^3}{12} + \frac{\mu_1^5}{80} + \cdots\right) = -\tau\left(\frac{\tau^2\,\delta^3}{12} + \frac{\tau^4\,\delta^5}{80} + \cdots\right).$$

It follows from the foregoing that

$$\rho^j = \exp\left\{ -\delta\,t_j - \left(\frac{\tau^2\,\delta^3}{12} + \frac{\tau^4\,\delta^5}{80} + \cdots\right) t_j \right\} \; < \; e^{-\delta\,t_j}\,.$$

The regular behavior for the symmetric scheme with $\sigma = \frac{1}{2}$ can be described as follows:

$$y(x_i\,,\,t_j) \;\approx\; c_1\,e^{-\delta\,t_j + \beta\,t_j}\,X_1(x_i)\,,$$

where

$$\beta = -\left(\frac{\tau^2\,\delta^3}{12} + \frac{\tau^4\,\delta^5}{80} + \cdots\right) = O(\tau^2)\,, \qquad\qquad \delta = \lambda_1^h\,.$$

In dealing with the pure implicit scheme we might have

$$\rho = \frac{1}{1+\tau\delta} \; > \; e^{-\tau\delta}\,, \qquad\qquad \rho^j = e^{-\delta\,t_j + \beta\,t_j}$$

with the asymptotic expansion $\beta = \frac{1}{2}\,\tau\,\delta^2 - \frac{1}{3}\,\tau^2\,\delta^3 + \cdots = O(\tau)$. It is evident that $\rho^j$ approximates $e^{-\delta\, t_j}$ poorly as compared with another scheme we have mentioned above.

Our approach is especially clear in the forthcoming example with

$$\tau = \tau_0 = \frac{2}{\sqrt{\delta\,\Delta}}\,, \qquad \tau\,\delta = 2\sqrt{\frac{\delta}{\Delta}}\,.$$

For the value $\delta\,/\,\Delta = 0.01$ the quantities

$$\rho|_{\sigma=0.5} = \rho_{0.5} = \frac{0.9}{1.1} = 0.8182\,, \qquad \rho|_{\sigma=1} = \rho_1 = \frac{1}{1.2} = 0.8333$$

need to be considered with more care than usual. Their comparison with $e^{-\tau\delta} = e^{-0.2} = 0.8187$ gives

$$\rho_{0.5} = (1 - 0.0006)\, e^{-\tau\delta}, \qquad \rho_1 = (1 + 0.018)\, e^{-\tau\delta}\,.$$

To be more specific, for $\delta\, t_j = 1$ $(j = 5)$ we have

$$\rho_{0.5}^j \approx 0.997\, e^{-\delta\, t_j}\,, \qquad \rho_1^j \approx 1.09\, e^{-\delta\, t_j}\,,$$

thereby clarifying that for $\delta\, t_j = 1$ the quantity $\rho_{0.5}^j$ differs from $e^{-\delta\, t_j}$ by 0.3% and $\rho_1^j$ – by 9%.

The above example shows that although the scheme with $\sigma = 1$ is absolutely stable and, in principle, may be used for any $\tau$, it is not accurate enough at the stage of the regular behavior when $\tau$ grows. In order to retain a prescribed accuracy (here it is meaninful to speak only about the relative accuracy), we should refine successively the step $\tau = \tau_j$ with increasing $t_j$. So, we are much disappointed by the main advantage of the scheme with $\sigma = 1$ stipulated by its stability for any $\tau > 0$.

The symmetric scheme with $\sigma = \frac{1}{2}$, which is absolutely stable in the usual sense: $\|\, y^j \,\| \le \|\, y^0 \,\|$, is conditionally asymptotically stable for $\tau \le \tau_0$. Being concerned with the explicit scheme for $\sigma = 0$, we observe that the condition of asymptotic stability $\tau \le 2\,(\delta + \Delta)^{-1}$ practically coincides with the usual stability condition $\tau \le 2\,\Delta^{-1}$ for small values of $\delta/\Delta$.

For the heat conduction equation

$$\delta = \frac{4}{h^2}\sin^2\frac{\pi h}{2}\,, \qquad \Delta = \frac{4}{h^2}\cos^2\frac{\pi h}{2}\,, \qquad \delta + \Delta = \frac{4}{h^2}$$

and the stability condition $\tau \le h^2/2$ also implies that $\tau \le 2\,(\delta + \Delta)^{-1}$. In the case of a symmetric scheme

$$\tau_0 = \frac{2}{\sqrt{\delta\,\Delta}} = \frac{h^2}{\sin\,\pi h} \approx \frac{h}{\pi}$$

and the condition $\tau < h/\pi$ is not burdensome. Thus emerged the typical relation $\gamma = \tau/h^2 < 1/(\pi h)$, so that $\gamma < 10/\pi$ for $h = 0.1$ and $\gamma < 100/\pi$ for $h = 0.01$. The first eigenvalue $\lambda_1^h = \delta$ of the difference problem concerned is

$$\lambda_1^h = \delta = \pi^2 \left( \frac{\sin\,\xi}{\xi} \right)^2 = \pi^2 \left( 1 - \frac{\xi^2}{3} + \frac{2}{45}\,\xi^4 + \cdots \right)^2,$$

where $\xi = \pi h/2$. Plain calculations show that

$$\lambda_1^h = \begin{cases} 0.97\,\lambda_1 & \text{for} \quad h = 0.1\,, \\ 0.9997\,\lambda_1 & \text{for} \quad h = 0.01\,, \end{cases} \qquad \lambda_1 = \pi^2 .$$

It is interesting to learn what may happen in dealing with the symmetric scheme with $\sigma = \frac{1}{2}$ and one of the possible steps $\tau > \tau_0$, say $\tau = m\,\tau_0$, $m > 1$? Then $\max_k\,|\,q_k\,|$ is attained for $k = N - 1$:

$$\rho = |\,q_{N-1}\,| = \frac{\frac{1}{2}\,\tau\,\Delta - 1}{\frac{1}{2}\,\tau\,\Delta + 1} = \frac{1 - 2\,\tau^{-1}\,\Delta^{-1}}{1 + 2\,\tau^{-1}\,\Delta^{-1}}\,.$$

Upon substituting here

$$\frac{1}{\tau\,\Delta} = \frac{1}{m\,\tau_0\,\Delta} = \frac{1}{2\,m}\sqrt{\frac{\delta}{\Delta}} = \frac{\delta}{4\,m}\,\frac{2}{\sqrt{\delta\,\Delta}} = \frac{\delta\,\tau_0}{4\,m} = \frac{\tau\,\delta}{4\,m^2} = \tau\,\bar{\delta}$$

we get $\rho = e^{-\tau\,\bar{\delta} + \tau\,\beta(\tau,\,\delta)}$ with

$$\bar{\delta} = \frac{\delta}{4\,m^2}\,, \qquad \beta = O(\tau^2)\,.$$

Thus, in the symmetric scheme with $\sigma = \frac{1}{2}$ the improper asymptotics is revealed for $m > 1$:

$$y(x_i\,,\,t_j) \approx c_{N-1}\,\exp\left\{ -\,\frac{\delta}{4\,m^2}\,t_j \right\} \widetilde{X}_{N-1}(x_i)\,,$$

where

$$\widetilde{X}_{N-1}(x_i) = \sin \pi(N-1)\, x_i = \sin \pi(N-1)\, ih$$
$$= \sin \pi(1-h)\, i = (-1)^{i-1} \sin \pi x_i$$
$$= (-1)^{i-1}\, \widetilde{X}_1(x_i)\,,$$

so that

$$y(x_i, t_j) \approx c_{N-1} \exp\left\{ -\frac{\delta}{4\, m^2}\, t_j \right\} (-1)^{i-1}\, \widetilde{X}_1(x_i)\,, \quad \widetilde{X}_1(x_i) = \sin \pi x_i\,.$$

This provides support for the view that the solution is completely distorted. From such reasoning it seems clear that asymptotic stability of a given scheme is intimately connected with its accuracy. When asymptotic stability is disturbed, accuracy losses may occur for large values of time. On the other hand, the forward difference scheme with $\sigma = 1$ is asymptotically stable for any $\tau$ and its accuracy becomes worse with increasing $t_j$, because its order in $t$ is equal to 1. In practical implementations the further retention of a prescribed accuracy is possible to the same value for which the explicit scheme is applicable. Hence, it is not expedient to use the forward difference scheme for solving problem (1) on the large time intervals.

**3. The scheme of second-order accuracy (unconditionally stable in the asymptotic sense).** Before taking up the general case, our starting point is the existing scheme of order 2 for the heat conduction equation possessing the unconditional asymptotic stability and having the form

$$(8) \qquad \begin{aligned} \left(E - \tfrac{1}{2}\, \sigma\, \tau\, \Lambda\right) \frac{y^{j+1/2} - y^j}{\tau} &= \left(1 - \tfrac{1}{2}\, \sigma\right) \Lambda\, y^j\,, \\ \frac{y^{j+1} - y^{j+1/2}}{\tau} &= \tfrac{1}{2}\, \sigma\, \Lambda\, y^{j+1}\,, \end{aligned}$$

where $\Lambda\, y = y_{\bar{x}x}$, $E$ is the identity operator, $y^{j+1/2}$ is the intermediate value and $\sigma = 2 - \sqrt{2}$. For $i = 0$ and $i = N$ the homogeneous boundary conditions are specified by

$$(9) \qquad y_0^{j+1/2} = y_N^{j+1/2} = 0\,, \qquad y_0^j = y_N^j = 0\,.$$

The usual trick amounts to eliminating $y^{j+1/2}$ from (8) so as to reduce the governing difference equations to the following ones:

$$\left( E - \frac{\sigma\,\tau}{2}\,\Lambda \right) y^{j+1/2} = \left( E + \tau\,(1-\sigma)\,\Lambda \right) y^j ,$$

$$\left( E - \frac{\sigma\,\tau}{2}\,\Lambda \right) y^{j+1} = y^{j+1/2} .$$

As a corollary to this fact, we see that

$$\left( E - \frac{\sigma\,\tau}{2}\,\Lambda \right)^2 y^{j+1} = \left( E + \tau\,(1-\sigma)\,\Lambda \right) y^j . \tag{10}$$

Having substituted into (10) the identity of artificial character

$$y^{j+1} = y^j + \tau\,\frac{y^{j+1} - y^j}{\tau}$$

or $\hat{y} = y + \tau\,y_t$, we recast the scheme at hand as

$$\left( E - \frac{\sigma\,\tau}{2}\,\Lambda \right)^2 y_t = \left( \Lambda + \frac{\sigma^2\,\tau}{4}\,\Lambda^2 \right) y . \tag{11}$$

The accurate account of the approximation error of scheme (11) on a solution $u = u(x,\,t)$ to equation (1) can be done using

$$\psi = \Lambda\,u - \frac{\sigma^2\,\tau}{4}\,\Lambda^2\,u - u_t + \sigma\,\tau\,\Lambda\,u_t - \frac{\sigma^2\,\tau^2}{4}\,\Lambda^2\,u_t . \tag{12}$$

Upon substituting into (12) the expansions

$$u = \frac{\hat{u} + u}{2} - \frac{\tau\,u_t}{2} = \bar{u} - \frac{\tau}{2}\,\frac{\partial \bar{u}}{\partial t} + O(\tau^2) ,$$

$$\Lambda\,u = L\,u + \frac{h^2}{12}\,L^2\,u + O(h^4) ,$$

$$\Lambda^2\,u = \Lambda\,(\Lambda\,u) = \Lambda \left( L\,u + \frac{h^2}{12}\,L^2\,u(\xi,t) \right) = \Lambda \left( \frac{\partial u}{\partial t} + \frac{h^2}{12}\,\frac{\partial^2 u}{\partial t^2}\,(\xi,t) \right)$$

$$= L^2\,u + O(h^2) ,$$

where

$$L\,u = \frac{\partial^2 u}{\partial x^2} = \frac{\partial u}{\partial t}\,, \qquad \bar{u} = u|_{t=t_{j+1/2}}\,,$$

we find that

$$\psi = L\,\bar{u} - \frac{\partial \bar{u}}{\partial t} - \tau\left(\frac{\sigma^2}{4} - \sigma + \frac{1}{2}\right) L^2\,\bar{u} + O(h^2+\tau^2)\,.$$

Hence, $\psi = O(h^2+\tau^2)$ for $\sigma = 2-\sqrt{2}$, satisfying the equation $\frac{1}{4}\,\sigma^2 - \sigma + \frac{1}{2} = 0$. Other ideas are connected with the method of separation of variables, whose use permits us to look for a solution to equation (10) with zero boundary conditions $y_0 = y_N = 0$ as a sum

$$y^j = \sum_{k=1}^{N-1} T_k^j\, X_k(x)\,, \qquad X_k(x) = \sqrt{2}\,\sin \pi k x\,.$$

By inserting this expression in (10) we obtain the equation related to $T_k^j$:

$$(1 + \tfrac{1}{2}\,\sigma\,\tau\,\lambda_k^h)^2\; T_k^{j+1} = (1 - (1-\sigma)\,\tau\,\lambda_k^h)\; T_k^j$$

or, what amounts to the same, $T_k^{j+1} = q_k\,T_k^j$, where

$$q_k = \frac{1 - (1-\sigma)\,\tau\,\lambda_k^h}{\left(1 + \frac{1}{2}\,\sigma\,\tau\,\lambda_k^h\right)^2}\,.$$

Furthermore, adopting arguments similar to those being used for the scheme with weights we arrive at the chain of the inequalities

$$\|\,y^{j+1}\,\| \le \rho\,\|\,y^j\,\| \le \cdots \le \rho^{j+1}\,\|\,y^0\,\|$$

with $\rho = \max_k\,|\,q_k\,|$. In the further analysis we shall need yet, among other things, the formula

$$\rho = \frac{1 - (1-\sigma)\,\tau\lambda_1^h}{\left(1 + \frac{1}{2}\,\sigma\,\tau\lambda_1^-h\right)^2}\,,$$

where $\lambda_1^h = 4\,h^{-2}\sin^2(\pi h/2)$. In giving it we will show that $\max_k |\,q_k\,|$ is attained for $k = 1$. With this aim, we have occasion to use the function

$$f(\mu) = \frac{1 - (1-\sigma)\,\mu}{\left(1 + \frac{1}{2}\,\sigma\,\mu\right)^2}\,,$$

for which

$$|f(\mu)| - f(\mu_1) < 0 \qquad \text{for} \qquad \mu > \mu_1 \qquad \text{if} \qquad \mu_1 < 1 .$$

It is necessary to consider merely the case $(1-\sigma)\,\mu > 1$, since otherwise $|f(\mu)| = f(\mu) < f(\mu_1)$ for $\mu > \mu_1$. In that case the difference in question becomes

$$\begin{aligned} f(\mu_1) - |f(\mu)| &= \frac{1-(1-\sigma)\,\mu_1}{\left(1+\frac{1}{2}\sigma\,\mu_1\right)^2} - \frac{(1-\sigma)\,\mu - 1}{\left(1+\frac{1}{2}\sigma\,\mu\right)^2} \\ &= \frac{1}{D}\Big\{2 + (2\,\sigma - 1)\,\mu_1 + \frac{1}{4}\,\sigma^2\,\mu_1^2 \\ &\quad - \Big((1-2\,\sigma) + 2\,\sigma\,(1-\sigma)\,\mu_1 + \frac{1}{4}\,\sigma^2\,(1-\sigma)\,\mu_1^2\Big)\,\mu \\ &\quad + \frac{1}{4}\,\sigma^2\,(1-(1-\sigma)\,\mu_1)\,\mu^2\Big\}, \end{aligned}$$

where $D = \left(1+\frac{1}{2}\sigma\,\mu\right)^2\left(1+\frac{1}{2}\sigma\,\mu_1\right)^2$.

Recalling that $\sigma = 2-\sqrt{2}$, $\sigma^2 = 4\,\sigma - 2$ and $\mu_1 < 1$ we get the lower estimate which will be the subject of special investigations in the near future:

$$\begin{aligned} & f(\mu_1) - |f(\mu)| \\ &\quad > \frac{1}{D}\Big\{2 - \Big(2\,\sigma\,(1-\sigma) + \frac{1}{4}\,\sigma^2\,(1-\sigma) - (2\,\sigma - 1)\Big)\,\mu + \frac{1}{4}\,\sigma^3\,\mu^2\Big\} \\ &\quad = \frac{1}{D}\{\,2 - (6.5 - 10.5\,\sigma)\,\mu + (3.5\,\sigma - 2)\,\mu^2\,\}. \end{aligned}$$

The square trinomial in the curly brackets equals $0.05\,\mu^2 - 0.349\,\mu + 2$ and has the negative discriminant $d = 0.349^2 - 0.402 < 0$. Because of this, we deduce that $f(\mu_1) - |f(\mu)| > 0$, meaning

$$\rho = \frac{1-(1-\sigma)\,\mu_1}{\left(1+\frac{1}{2}\sigma\,\mu_1\right)^2} = \frac{1-(1-\sigma)\,\tau\,\delta}{\left(1+\frac{1}{2}\sigma\,\tau\,\delta\right)^2}$$

under the constraint $\mu_1 = \tau\,\delta < 1$, which is valid if being replaced by the condition $\tau\,\lambda_1 = \tau\,\pi^2 < 1$.

With this relation in view, it is not difficult to derive the asymptotic expansion

$$\log \rho = -\mu_1 - \frac{\mu_1^3}{24.6} - \frac{\mu_1^4}{271.7} + \cdots$$

with the members

$$\rho = e^{-\tau\,\delta + \tau\,\beta}\,, \qquad \beta = -\frac{\tau^2\,\delta^3}{24.6} - \frac{\tau^3\,\delta^4}{271.7} + \cdots\,,$$

which serve to motivate the double inequality $\rho^j < e^{-\delta\,t_j + \beta\,t_j} < e^{-\delta\,t_j}$ and the a priori estimate $\|\,y^j\,\| \le e^{-\delta\,t_j}\,\|\,y^0\,\|$ for scheme (8) under the constraint $\tau\,\delta < 1$.

**4. Asymptotic stability of the three-layer scheme.** The object of investigation here is the three-layer scheme

$$(13) \qquad \tfrac{3}{2}\,y_t - \tfrac{1}{2}\,y_{\bar t} + A\,\hat y = 0\,, \qquad A = A^*\,, \qquad A \ge \delta E\,, \qquad \delta > 0\,,$$

which is unconditionally asymptotically stable for $\tau < 1/(2\delta)$. To make sure of it, an alternative form $y_{\overset{\circ}{t}} + \tau\,y_{\bar t t} + A\,\hat y = 0$, which provides an approximation of order 2 for the heat conduction equation with constant coefficients (1), will appear in the further development:

$$\begin{aligned}\psi &= u_{\overset{\circ}{t}} + \tau\,u_{\bar t t} + A\,\hat u = u_{\overset{\circ}{t}} + \tau\,u_{\bar t t} - \Lambda\,u - \tau\,\Lambda\,u_{\overset{\circ}{t}} - \tfrac{1}{2}\,\tau^2\,\Lambda\,u_{\bar t t}\\ &= \dot u + \tau\,\ddot u - L\,u - \tau\,L\,\dot u - \tfrac{1}{2}\,\tau^2\,L\,\ddot u + O(h^2 + \tau^2)\\ &= (\dot u - L\,u) + \tau\,(\ddot u - L\,\dot u) + O(h^2 + \tau^2) = O(h^2 + \tau^2)\,.\end{aligned}$$

Here we used also the initial equation $\dot u = L\,u$ and its corollary $\ddot u = L\,\breve u$.

In conformity with the method of separation of variables,

$$y_k^j = T_k(t_j)\,X_k\,,$$

where $X_k$ refers to an eigenfunction of the operator $A$. Recall that, by definition,

$$A\,X_k = \lambda_k^h\,X_k\,, \qquad k = 1, 2, \ldots, N\,,$$

With this in mind, we initiate the derivation of the difference equation

$$\tfrac{3}{2}\,T_k^{j+1} - 2\,T_k^j + \tfrac{1}{2}\,T_k^{j-1} + \lambda_k^h\,\tau\,T_k^{j+1} = 0$$

or

$$(3 + 2\tau\lambda_k^h)\, T_k^{j+1} - 4\, T_k^j + T_k^{j-1} = 0\,,$$

whose solution is sought in the form $T_k^j = q_k^j$. Omitting the subscript $k$ for a while, we investigate the quadratic equation related to $q_k$:

$$(3 + 2\mu)\, q^2 - 4\, q + 1 = 0\,, \qquad \mu = \tau\lambda\,,$$

with discriminant $D - 4\,(1 - 2\mu)$. For $D < 0$ its roots $q^{(1,2)}$ are complex-conjugate and $q^{(1)} q^{(2)} = |\, q\,|^2 = (3 + 2\mu)^{-1}$, that is,

$$\left| q_k^{(1,2)} \right| = \frac{1}{\sqrt{3 + 2\,\tau\,\lambda_k}} < \frac{1}{\sqrt{3 + 2\,\tau\,\lambda_1}} = \left| q_1^{(1,2)} \right| .$$

Let now $D > 0$. In that case

$$q^{(1,2)} = \frac{2 \pm \sqrt{1 - 2\mu}}{3 + 2\mu}\,, \quad q^{(1)} > 0\,, \quad q^{(2)} > 0\,, \quad q^{(2)} = \frac{2 - \sqrt{1 - 2\mu}}{3 + 2\mu} < 1\,.$$

A new function

$$\varphi(\mu) = \frac{2 + \sqrt{1 - 2\mu}}{3 + 2\mu}\,, \qquad \mu_1 \le \mu \le \mu_N\,,$$

with the derivative

$$\frac{1}{\varphi} \cdot \frac{\partial \varphi}{\partial \mu} = - \frac{1}{(2 + \sqrt{1 - 2\mu})\,\sqrt{1 - 2\mu}} - \frac{2}{3 + 2\mu} < 0$$

is aimed at establishing that $|q_{k_1}^{(1)}| < |q_{k_2}^{(1)}|$ for $k_1 > k_2$. Hence, $\max\limits_{\mu_1 \le \mu \le \mu_N} \varphi(\mu)$ is attained for $\mu = \mu_1 = \tau\lambda_1$, giving

$$\max_{\mu_1 \le \mu \le \mu_N} \varphi(\mu) = \varphi(\mu_1) = \frac{2 + \sqrt{1 - 2\mu_1}}{3 + 2\mu_1} \qquad \text{if} \qquad 1 - 2\mu_1 > 0\,.$$

It may happen that $\frac{1}{4} D_k = 1 - 2\mu_k < 0$ for some $k > 1$, thus causing the occurrence of the event that

$$|\, q_k^{(1,2)}|^2 = \frac{1}{3 + 2\,\mu_k}\,.$$

Since $(3 + 2\,\mu_k)^{-1} < \varphi^2(\mu_1)$, the maximum value $\max\limits_k \left| q_k^{(1,2)} \right|$, which can always be attainable for $k = 1$, is equal to

$$\rho = \frac{2 + \sqrt{1 - 2\,\mu_1}}{3 + 2\,\mu_1} \qquad \text{if} \qquad \mu_1 < \frac{1}{2}\,.$$

For large $j$ the first harmonic $y^j \approx \alpha_1 \rho^j X_1$ dominates in the solution

$$y^j = \sum_{k=1}^{N-1} \left( \alpha_k \, (q_k^{(1)})^j + \beta_k \, (q_k^{(2)})^j \right) X_k ,$$

which has been constructed before.

By means of the quantity $\rho^j = e^{-j \log(1/\rho)}$ with

$$\rho = \frac{1 - \dfrac{2}{3} \dfrac{\mu_1}{1 + \sqrt{1 - 2\mu_1}}}{1 + \frac{2}{3} \mu_1}$$

it is plain to calculate

$$\log \frac{1}{\rho} = \log \left( 1 + \frac{2}{3} \mu_1 \right) - \log \left( 1 - \frac{2}{3} \frac{\mu_1}{1 + \sqrt{1 - 2\mu_1}} \right) .$$

The expansion in powers of $\mu_1$ yields $\log(1/\rho) = \mu_1 + \frac{1}{3} \mu_1^3 + \cdots$, so that

$$\rho^j = \exp \left\{ -\delta \, t_j - \tfrac{1}{3} \tau^2 \delta^3 \, t_j \right\} .$$

Thus, the three-layer scheme (13) of accuracy $O(h^2 + \tau^2)$ possesses the proper asymptotics as $t_j \to \infty$ under the unique restriction $\tau \delta < 1/2$, which is not burdensome. Comparison of the final results with the two-layer scheme (4) reveals some formal advantage of the three-layer scheme over the symmetric two-layer scheme with $\sigma = \frac{1}{2}$, which is conditionally asymptotically stable if we imposed the extra constraint $\tau \leq \tau_0 = 1/\sqrt{\delta \, \Delta}$, $\tau_0 = \tau_0(h)$, in addition to the usual one $\tau \delta < 1$. However, this restriction is sufficiently weak in real-life situations and, therefore, it is meaningless to speak about any practical advantage of the three-layer scheme. For this reason the two-layer scheme is quite applicable and more efficient in practical implementations.

## 5.3 SCHEMES FOR THE HEAT CONDUCTION EQUATION WITH SEVERAL SPATIAL VARIABLES

**1. The explicit difference scheme.** The schemes considered in Section 1 may be generalized to the case of the heat conduction equation with several spatial variables.

In a common setting, let $\bar{G} = G + \Gamma$ be a $p$-dimensional domain with the boundary $\Gamma$ and let $\mathbf{x} = (x_1, x_2, \ldots, x_p)$ denote an arbitrary point in it. It is required to find in the cylinder $\bar{Q}_T = \bar{G} \times [0 \le t \le T]$ a continuous function $u(\mathbf{x}, t)$, $\mathbf{x} \in \bar{G}$, satisfying the governing equation

$$(1) \qquad \frac{\partial u}{\partial t} = \Delta u + f(\mathbf{x}, t), \qquad \mathbf{x} \in G, \qquad t > 0,$$

and the supplementary conditions

$$u(\mathbf{x}, 0) = u_0(\mathbf{x}), \quad \mathbf{x} \in \bar{G}, \qquad u(\mathbf{x}, t) = \mu(\mathbf{x}, t), \quad \mathbf{x} \in \Gamma, \quad t \ge 0,$$

where

$$\Delta u = \sum_{\alpha=1}^{p} \frac{\partial^2 u}{\partial x_\alpha^2}$$

is the well-known Laplace operator.

For the purposes of the present section, let us introduce the grid $\bar{\omega}_h = \{ x_i \in \bar{G} \}$ in $\bar{G}$ and denote by $\gamma_h$ the set of all nodes of $\bar{\omega}_h$ belonging to $\Gamma$ and by $\omega_h$ the set of all inner nodes $x_i \in G$, so that $\bar{\omega}_h = \omega_h + \gamma_h$. The starting point in the further development of the difference scheme is the approximation of the elliptic operator $\Delta u$. We learn from Section 1 of Chapter 4 that at all of the inner nodes $\Delta u \sim \Lambda u$ for $\mathbf{x} \in \omega_h$.

By inserting in (1) the difference operator $\Lambda$ in place of the Laplace operator we are led to the system of differential-difference equations

$$(2) \qquad \frac{\partial v}{\partial t} = \Lambda v + \varphi(\mathbf{x}, t), \qquad \mathbf{x} \in \omega_h,$$

where $v(\mathbf{x}, t)$ is defined on the grid $\bar{\omega}_h$ for every $t \ge 0$. The size of the system (2) is equal to the total number of the inner nodes $N$ of the grid $\omega_h$. Here the function $\varphi(\mathbf{x}, t)$ approximates $f(\mathbf{x}, t)$ on the grid $\omega_h$.

At the next stage we introduce the grid in $t$

$$\bar{\omega}_\tau = \{ t_j = j\tau, \ j = 0, 1, 2, \ldots, j_0, \ j_0 \tau = T \}$$

with step $\tau$. In order to pass to the difference scheme for a new function $y(\mathbf{x}, t)$ defined on the grid $\bar{\omega}_{h\tau} = \bar{\omega}_h \times \bar{\omega}_\tau = \{ (x_i, t_j), \ \mathbf{x} \in \bar{\omega}_h, \ t \in \bar{\omega}_\tau \}$, it is necessary to replace the system of differential equations (2) by some difference scheme written in terms of $t$. Choosing, for instance, Euler's scheme we obtain the explicit scheme

$$(3) \qquad \begin{aligned} &\frac{y^{j+1} - y^j}{\tau} = \Lambda y^j + \varphi^j, && j = 0, 1, 2, \ldots, \\ &y^0 = y(\mathbf{x}, 0) = u_0(\mathbf{x}), \quad y^j \big|_{\gamma_h} = \mu^j, && j = 0, 1, 2, \ldots . \end{aligned}$$

The value $y^{j+1}$ on the relevant layer is to be determined by the recursion

$$y^{j+1} = y^j + \tau\,(\Lambda\, y^j + \varphi^j)\,, \qquad j = 0,\, 1,\, \ldots .$$

For the sake of simplicity we are working in a parallelepiped $\bar{G}$ such that $\{\,0 \leq x_\alpha \leq l_\alpha,\ \alpha = 1, 2, \ldots, p\,\}$ and on a certain grid $\bar{\omega}_h = \{\, x_i = (i_1\, h_1, \ldots, i_p\, h_p),\ i_\alpha = 0, 1, \ldots, N_\alpha,\ h_\alpha = l_\alpha\, N_\alpha^{-1}\}$ equidistant in each of the directions $x_\alpha$. We refer to the $(2p+1)$-point operator $\Lambda$ of second-order approximation with the values

$$(4)\qquad \Lambda\, y = \sum_{\alpha=1}^{p} \Lambda_\alpha\, y\,, \qquad \Lambda_\alpha\, y = y_{\bar{x}_\alpha x_\alpha} = \frac{y^{(+1_\alpha)} - 2\,y + y^{(-1_\alpha)}}{h_\alpha^2}\,,$$

where $y$ is the value of the function $y(\mathbf{x}, t)$ at a fixed node $\mathbf{x} = (i_1\, h_1, \ldots, i_p\, h_p)$, $y^{(\pm 1_\alpha)} = y(\mathbf{x}^{(\pm 1_\alpha)}, t)$, $\mathbf{x}^{(\pm 1_\alpha)} = (i_1\, h_1, \ldots, i_{\alpha-1}\, h_{\alpha-1}, (i_\alpha \pm 1)\, h_\alpha, i_{\alpha+1}\, h_{\alpha+1}, \ldots, i_p\, h_p)$ is the node adjacent to $\mathbf{x}$ ($\mathbf{x}^{(+1_\alpha)}$ stands on the right from $\mathbf{x}$ and $\mathbf{x}^{(-1_\alpha)}$ stands on the left from $\mathbf{x}$). With the aid of the well-established relations

$$\Lambda\, u = \Delta\, u + \sum_{\alpha=1}^{p} \frac{h_\alpha^2}{12}\, L_\alpha^2\, u + O(\,|\,\mathbf{h}\,|^4)\,,$$

$$L_\alpha\, u = \frac{\partial^2 u}{\partial x_\alpha^2}\,, \qquad |\,\mathbf{h}\,|^2 = h_1^2 + h_2^2 + \cdots + h_p^2\,,$$

we calculate the residual

$$\psi = \Lambda\, u^j + \varphi^j - \frac{u^{j+1} - u^j}{\tau} = O(\,|\,\mathbf{h}\,|^2 + \tau)\,,$$

provided the asymptotic expansion $\varphi^j = f(\mathbf{x}, t_j) + O(\,|\,\mathbf{h}\,|^2 + \tau)$ holds.

Scheme (3) is conditionally stable in the space $C$ with respect to the initial data, the right-hand side and the boundary data. The maximum principle for the difference problem (3) may be of help in establishing the indicated properties with further reference to the canonical form

$$(5)\qquad y^{j+1} = \left(1 - 2 \sum_{\alpha=1}^{p} \frac{\tau}{h_\alpha^2}\right) y^j + \sum_{\alpha=1}^{p} \frac{\tau}{h_\alpha^2} \left(y^j\,(\mathbf{x}^{(+1_\alpha)}) + y^j\,(\mathbf{x}^{(-1_\alpha)})\right) + \tau\,\varphi^j .$$

Comparison with the general difference equation

$$(6) \qquad A(P)\, y(P) = \sum_{Q \in Patt'(P)} B(P,\, Q)\, y(Q) + F(P)\,,$$

which has been considered in Section 2 of Chapter 4, shows that in the case of interest

$$P = (\mathbf{x}, t_{j+1}), \quad Q = (\mathbf{x}, t_j), \left(\mathbf{x}^{(+1_\alpha)}, t_j\right), \left(\mathbf{x}^{(-1_\alpha)}, t_j\right), \quad \alpha = 1, 2, \ldots, p,$$

$$A(P) = 1\,, \qquad B(P,\, Q) = 1 - 2 \sum_{\alpha=1}^{p} \frac{\tau}{h_\alpha^2}\,.$$

The boundary $\gamma$ of the grid $\bar{\omega}$ for equation (5) consists of the nodes $(\mathbf{x}, 0)$, $\mathbf{x} \in \bar{\omega}_h$, and $(\mathbf{x}, t_{j'})$, $\mathbf{x} \in \gamma_h$, $t_{j'} \leq t_{j+1}$. It is straightforward to check that

$$D(P) = A(P) - \sum_{Q \in Patt'(P)} B(P,\, Q) = 0\,, \qquad B(P,\, Q) \geq 0$$

if

$$(7) \qquad 1 - 2 \sum_{\alpha=1}^{p} \frac{\tau}{h_\alpha^2} \geq 0\,.$$

Common practice involves the decomposition for the solution of the problem concerned as a sum $y = \bar{y} + \tilde{y}$, where $\bar{y}$ is a solution to the homogeneous equation

$$\bar{y}_t = \Lambda\, \bar{y}\,, \qquad \bar{y}\big|_{\gamma_h} = \mu\,, \qquad \bar{y}(\mathbf{x},\, 0) = u_0(\mathbf{x})\,,$$

and $\tilde{y}$ is a solution to the nonhomogeneous equation

$$\tilde{y}_t = \Lambda\, \tilde{y} + \varphi\,, \qquad \tilde{y}\big|_{\gamma_h} = 0\,, \qquad \tilde{y}(\mathbf{x},\, 0) = 0\,.$$

As a matter of fact, we obtain for $\bar{y}$ equation (6) with $F = 0$. On the strength of the maximum principle (see corollaries to Theorem 2, Chapter 4, Section 2.3) we have

$$\|\, \bar{y}(\mathbf{x},\, t_j)\, \|_C \leq \max\Big( \max_{\mathbf{x} \in \gamma_h} \max_{t_{j'} \leq t_j} |\, \mu(\mathbf{x},\, t_{j'})\, |,\, \|\, u_0(\mathbf{x})\, \|_C \Big)\,,$$

provided condition (7) holds. With this relation in view, we derive the inequality

$$(8)\qquad \|\bar{y}^j\|_C \le \max_{0\le j'\le j} \|\mu^{j'}\|_{C_\gamma} + \|u_0\|_C,$$

where $\|\mu\|_{C_\gamma} := \max_{\mathbf{x}\in\gamma_h} |\mu(\mathbf{x},t)|$, which expresses the stability of the explicit scheme (3) with respect to boundary conditions and initial data for

$$(9)\qquad \tau \le \tau_0, \qquad \tau_0 = \frac{1}{2}\left(\sum_{\alpha=1}^{p}\frac{1}{h_\alpha^2}\right)^{-1}.$$

By analogy with the one-dimensional case in the estimation of $\tilde{y}$ (see Section 1.7), let us recast the difference equation as $\tilde{y}^{j+1} = F^j$, where

$$F^j = \left(1 - 2\sum_{\alpha=1}^{p}\frac{\tau}{h_\alpha^2}\right)\tilde{y} + \sum_{\alpha=1}^{p}\frac{\tau}{h_\alpha^2}\left(\tilde{y}^{(+1_\alpha)} + \tilde{y}^{(-1_\alpha)}\right) + \tau\varphi$$

and $\|F^j\|_C \le \|\tilde{y}\|_C + \tau\|\varphi\|_C$ for $\tau \le \tau_0$, thereby justifying that

$$\|\tilde{y}^{j'+1}\|_C = \|F^{j'}\|_C \le \|\tilde{y}^{j'}\|_C + \tau\|\varphi^{j'}\|_C$$

for the same values of $\tau \le \tau_0$. Summation over $j' = 0, 1, \dots, j$ yields the inequality

$$(10)\qquad \|\tilde{y}^{j+1}\|_C \le \sum_{j'=0}^{j}\tau\|\varphi^{j'}\|_C, \qquad \tau \le \tau_0,$$

characterizing the stability of the explicit scheme with respect to the right-hand side. For $\tau \le \tau_0$ relations (8) and (10) together imply the estimate

$$(11)\qquad \|y^{j+1}\|_C \le \|u_0\|_C + \max_{0\le j'\le j+1}\|\mu^{j'}\|_{C_\gamma} + \sum_{j'=0}^{j}\tau\|\varphi^{j'}\|_C.$$

With the notation $h = \min h_\alpha$, the stability condition (9) takes the form

$$\tau \le \frac{h^2}{2p},$$

thus demonstrating that the admissible step $\tau$ of the explicit scheme decreases with increasing the dimension $p$.

**2. The explicit three-layer scheme.** We now turn to the simplest explicit three-layer scheme known as the Richardson scheme and being an analog of scheme (59) from Section 1:

$$\frac{y^{j+1}-y^{j-1}}{2\tau} = \Lambda y^j + \varphi^j . \tag{12}$$

However, it is absolutely unstable. Substitution of the half-sum

$$\tfrac{1}{2}\left(y_i^{j-1}+y_i^{j+1}\right) = \tfrac{1}{2}\left(y_i^{j-1}-2y_i^j+y_i^{j+1}\right)+y_i^j = \tfrac{1}{2}\tau^2 y_{\bar t t}^j + y_i^j$$

for $y_i^j$, where

$$y_{\bar t t} = \frac{y^{j+1}-2y^j+y^{j-1}}{\tau^2},$$

into the right-hand side yields the $p$-dimensional analog of the Du-Fort–Frankel scheme. The forthcoming substitution

$$\Lambda y^j = \sum_{\alpha=1}^{p} \frac{y^j\left(\mathbf{x}^{(+1_\alpha)}\right)+y^j\left(\mathbf{x}^{(-1_\alpha)}\right)}{h_\alpha^2} - 2\sum_{\alpha=1}^{p}\frac{1}{h_\alpha^2}\, y^j(\mathbf{x})$$

leads to an alternative form of writing

$$\Lambda y = \tau^2 \sum_{\alpha=1}^{p} \frac{1}{h_\alpha^2}\, y_{\bar t t},$$

which will be involved further in the explicit scheme

$$y_{\overset{\circ}{t}} + \frac{\tau^2\Delta}{4}\, y_{\bar t t} = \Lambda y + \varphi, \tag{13}$$

where

$$\Delta = 4\sum_{\alpha=1}^{p}\frac{1}{h_\alpha^2}.$$

In the case of a cube grid with $h_1 = h_2 = \cdots = h_p = h$ one succeeds in designing one more scheme instead of (13):

$$y_{\overset{\circ}{t}} + \frac{p\,\tau^2}{h^2}\, y_{\bar t t} = \Lambda y + \varphi \tag{13'}$$

and in establishing as its immediate implication the useful formula for determining $\hat y = y^{j+1}$:

$$(1+2\gamma)\,\hat y = (1-2\gamma)\,\check y + 4\gamma\, y + 2\tau\Lambda y + 2\tau\varphi,$$

where $y = y^j$, $\check{y} = y^{j-1}$, $\gamma = p\tau/h^2$ and $\Delta = 4ph^{-2}$ Also, we point out here without proving that scheme (13) is stable for any $\tau$ and $h$. This fact follows from the general stability theory developed in Chapter 6. Also, this scheme generates conditional approximation, since its residual behaves for $\tau = O(h^2)$ like

$$\psi = \Lambda u + \varphi - u_{\overset{\circ}{t}} - \frac{p\tau^2}{h^2}\, u_{\bar{t}t} = O(h^2)$$

if we accept $\varphi = f$ and $h_1 = h_2 = \cdots = h_p = h$.

**3. Schemes with weights.** When discretizing equation (2) in $t$, the scheme with weights arises naturally in one or another form:

$$\frac{y^{j+1} - y^j}{\tau} = \Lambda\left(\sigma y^{j+1} + (1-\sigma)\, y^j\right) + \varphi_j\,,$$

or

$$(14)\qquad \begin{aligned} &y_t = \Lambda\left(\sigma\hat{y} + (1-\sigma)\, y\right) + \varphi\,, && \mathbf{x} \in \omega_h\,, \quad t = j\tau > 0\,,\\ &y\big|_{\gamma_h} = \mu\,, \qquad y(\mathbf{x},\, 0) = u_0(\mathbf{x})\,, && \mathbf{x} \in \bar{\omega}_h\,. \end{aligned}$$

In preparation for this, we agree to consider $\varphi = \bar{f} = f(\mathbf{x}, t_{j+1/2})$. As before, we suppose once again that $\bar{G}$ is a parallelepiped and $\Lambda$ is specified by formula (4). We investigate the order of approximation by appeal to the expression

$$\sigma\hat{u} + (1-\sigma)\, u = \frac{\hat{u}+u}{2} + \left(\sigma - \frac{1}{2}\right)\tau\, u_t$$

for $\sigma\hat{u} + (1-\sigma)\, u$ and, after this, touch upon the residual

$$\begin{aligned} \psi &= \Lambda\left(\sigma\hat{u} + (1-\sigma)\, u\right) + \varphi - u_t\\ &= \Lambda\,\frac{\hat{u}+u}{2} + \left(\sigma - \frac{1}{2}\right)\tau\,\Lambda\, u_t + \varphi - u_t\\ &= L\bar{u} + \left(\sigma - \frac{1}{2}\right)\tau\, L\dot{u} + \bar{f} - \dot{u} + (\varphi - \bar{f}) + O(|\, h\,|^2 + \tau^2)\\ &= \left(\sigma - \frac{1}{2}\right)\tau\, L\dot{u} + O(|\, h\,|^2 + \tau^2)\,, \end{aligned}$$

where $Lu = \Delta u$, $\bar{u} = u(\mathbf{x}, t_{j+1/2})$, $\varphi = \bar{f}$ and $\dot{u} = \partial u/\partial t$. It is therefore concluded that

$$\psi = \begin{cases} O(|\, h\,|^2 + \tau^2) & \text{if}\quad \sigma = \frac{1}{2}\,,\\ O(|\, h\,|^2 + \tau) & \text{if}\quad \sigma \neq \frac{1}{2}\,. \end{cases}$$

The maximum principle can be applied to any such scheme with weights under the constraint $\tau \le \tau_0$, where

$$\tau_0 = \frac{1}{2\,(1-\sigma)} \left( \sum_{\alpha=1}^{p} \frac{1}{h_\alpha^2} \right)^{-1}, \qquad 0 \le \sigma \le 1,$$

by means of which it is not hard to show that for $\sigma = 1$ there is no limitation on $\tau$. Because of this, a priori estimate (11) for the problem solution is certainly true. Indeed, it is possible to reduce (14) to (6), by means of which

$$A(P) = 1 + 2\,\sigma \sum_{\alpha=1}^{p} \frac{\tau}{h_\alpha^2}, \qquad \sigma \ge 0,$$

$$B(P,\,Q) = \sigma\, \frac{\tau}{h_\alpha^2}, \quad (1-\sigma)\,\frac{\tau}{h_\alpha^2}, \quad 1 - 2\,(1-\sigma) \sum_{\alpha=1}^{p} \frac{\tau}{h_\alpha^2}.$$

The restriction $\tau \le \tau_0$ follows from the initial assumption concerning the nonnegativity of the coefficients $B(P,Q)$. It is clear that $D(P) = 0$.

Instead of (14) it is reasonable to deal with the scheme with different weights $\sigma_\alpha$ related to the directions $x_\alpha$:

$$y_t = \sum_{\alpha=1}^{p} \Lambda_\alpha \left( \sigma_\alpha\, \hat{y} + (1-\sigma_\alpha)\, y \right),$$

for which

$$\tau \le \tau_0, \qquad \tau_0 = \frac{1}{2} \left( \sum_{\alpha=1}^{p} \frac{1-\upsilon_\alpha}{h_\alpha^2} \right)^{-1}.$$

**4. The scheme with increased accuracy.** For the stationary problem

$$\Delta\, u = -f(\mathbf{x}), \qquad u|_\Gamma = \mu(\mathbf{x}),$$

in a parallelepiped the scheme of accuracy $O(|\,h\,|^4)$ has been constructed in Chapter 4. In the two-dimensional case $p = 2$ this scheme reduces to

$$\Lambda'\, y = -\varphi,$$

where

$$\Lambda'\, y = \Lambda_1\, y + \Lambda_2\, y + \frac{h_1^2 + h_2^2}{12}\, \Lambda_1\, \Lambda_2\, y, \qquad \varphi = f + \frac{h_1^2}{12}\, \Lambda_1\, f + \frac{h_2^2}{12}\, \Lambda_2\, f.$$

We are going to show that the scheme

$$(15) \qquad y_t = \sum_{\alpha=1}^{2} \Lambda_\alpha \big(\sigma_\alpha \hat{y} + (1-\sigma_\alpha) y\big) + \frac{h_1^2+h_2^2}{12} \Lambda_1 \Lambda_2 \, y + \varphi ,$$

$$y|_{\gamma_h} = \mu , \qquad y(\mathbf{x}, 0) = u_0(\mathbf{x}) ,$$

with the ingredients

$$(16) \qquad \begin{aligned} &\sigma_\alpha = \frac{1}{2} - \frac{h_\alpha^2}{12\tau} , \qquad \alpha = 1, 2 , \\ &\varphi = \bar{f} + \frac{h_1^2}{12} \Lambda_1 \bar{f} + \frac{h_2^2}{12} \Lambda_2 \bar{f} , \qquad \bar{f} = f^{j+1/2} , \end{aligned}$$

provides an approximation of $O(|\,h\,|^4 + \tau^2)$ in light of the representation of the residual

$$\psi = \sum_{\alpha=1}^{2} \Big( \Lambda_\alpha \frac{\hat{u}+u}{2} + \Big(\sigma_\alpha - \frac{1}{2}\Big) \tau \Lambda_\alpha u_t \Big) + \frac{h_1^2+h_2^2}{12} \Lambda_1 \Lambda_2 \, u + \varphi - u_t .$$

This can be done by inserting the well-established expansions

$$\frac{\hat{u}+u}{2} = \bar{u} + O(\tau^2) , \qquad u_t = \bar{\dot{u}} + O(\tau^2) ,$$

$$\Lambda_\alpha \bar{u} = L_\alpha \bar{u} + \frac{h_\alpha^2}{12} L_\alpha^2 \bar{u} , \qquad \Lambda_\alpha u_t = L_\alpha \bar{\dot{u}} + O(\tau^2 + h_\alpha^2) ,$$

$$L_\alpha^2 \bar{u} = L_\alpha \bar{\dot{u}} - L_1 L_2 \bar{u} - L_\alpha \bar{f} , \qquad \alpha = 1, 2 .$$

The outcome of this is

$$\psi = \sum_{\alpha=1}^{2} \Big[ \frac{h_\alpha^2}{12} + \Big(\sigma_\alpha - \frac{1}{2}\Big) \tau \Big] L_\alpha \bar{\dot{u}} + O(|\,h\,|^4 + \tau^2) = O(|\,h\,|^4 + \tau^2)$$

with $\sigma_\alpha$ still subject to the first condition (16). In order to prove the convergence of this scheme with the rate $O(|\,h\,|^4 + \tau^2)$, it is necessary to obtain an a priori estimate for a solution of the problem with zero initial and boundary conditions. Having no opportunity to touch upon this topic we address the readers to Chapter 6 and Concluding remarks therein. We note in concluding that scheme (15) is stable for any $\tau$ and $h_\alpha$ and do not pursue here analysis of this: the ideas needed to do so have been covered.

The second, no less importance, question is: how to solve the system of difference equations

$$\sum_{\alpha=1}^{2} \tau\, \sigma_\alpha\, \Lambda_\alpha\, \hat{y} - \hat{y} = -F\,.$$

At first glance, the matrix elimination method suits us perfectly. But $O(N^2)$ operations are required in its application, where $N$ is the total number of the nodes of the grid $\omega_h$. Just for this reason scheme (15) is unacceptable in practical implementations. We will show later that it may be replaced by some schemes of the same order $O(|\,h\,|^4 + \tau^2)$ and $O(N)$ arithmetic operations are required in this connection for determination of $\hat{y}$ by applying successively the scalar elimination for a three-point equation. The resulting schemes are said to be **economical**, so there is some reason to be concerned about this. One needs to exercise good judgment in deciding which to consider. The scheme we have constructed above is aimed at designing other schemes with the indicated property. Here we will not elaborate on these matters. If you wish to explore this more deeply, you might find it helpful to refer to Concluding Remarks at the very end of the present chapter and references therein.

## 5.4 SCHRÖDINGER TIME-DEPENDENT EQUATION

**1. Two-layer scheme with weights.** We are now interested in various difference schemes for the **Schrödinger equation**

$$(1)\qquad \begin{aligned} & i\,\frac{\partial u}{\partial t} = \frac{\partial^2 u}{\partial x^2}\,, \qquad 0 < x < 1\,, \qquad t > 0\,, \\ & u(x,\,0) = u_0(x)\,, \qquad u(0,\,t) = u(1,\,t) = 0\,, \qquad i = \sqrt{-1}\,. \end{aligned}$$

By analogy with the heat conduction equation we employ the method of separation of variables, in the framework of which a solution of this problem is sought as the series

$$u(x,\,t) = \sum_{k=1}^{\infty} c_k\, e^{i\lambda_k t}\, X_k(x)\,,$$

where

$$c_k = (u_0, X_k) = \int_0^1 u_0(x)\, X_k(x)\, dx\,,$$

$$\lambda_k = k^2\,\pi^2, \qquad X_k(x) = \sqrt{2}\,\sin\,\pi k x.$$

Granted the grid $\bar{\omega}_{h\tau} = \bar{\omega}_h \times \bar{\omega}_\tau$ with $\bar{\omega}_h = \{x_s = s\,h,\ s = 0, 1, \ldots, N,\ hN = 1\}$ and $\bar{\omega}_\tau = \{t_j = j\,\tau,\ j = 0, 1, \ldots\}$, the difference scheme with weights

$$(2) \quad i\,y_t = \Lambda\left(\sigma\,\hat{y} + (1-\sigma)\,y\right), \qquad y(x, 0) = u_0(x)\,, \qquad y_0 = y_N = 0\,,$$

where $\Lambda\,y = y_{\bar{x}x}$ and $\sigma = \sigma_0 + i\,\sigma_1$ is a complex number, comes first.

Of our initial concern is the residual

$$\begin{aligned}
\psi &= \Lambda\left(\sigma\,\hat{u} + (1-\sigma)\,u\right) - i\,u_t \\
&= \Lambda\,\frac{\hat{u}+u}{2} + \tau\left(\sigma - \frac{1}{2}\right)\Lambda\,u_t - u_t \\
&= \Lambda\,\bar{u} + \left(\sigma - \frac{1}{2}\right)\tau\,\Lambda\,\bar{\dot{u}} - i\,\bar{\dot{u}} + O(\tau^2) \\
&= (L\,\bar{u} - i\,\bar{\dot{u}}) + \frac{h^2}{12}\,L^2\,\bar{u} + (\sigma - \tfrac{1}{2})\,\tau\,L\,\bar{\dot{u}} + O(\,h^4 + \tau^2) \\
&= \left(\frac{h^2}{12}\,i + \left(\sigma - \frac{1}{2}\right)\tau\right)L\,\bar{\dot{u}} + O(\,h^4 + \tau^2)\,,
\end{aligned}$$

where $\bar{u} = u(x, t_j + \frac{1}{2}\,\tau)$, $\dot{u} = \partial u/\partial t$ and $L\,u = \partial^2 u/\partial x^2$. It follows from the foregoing that

$$\psi = \begin{cases} O(h^2 + \tau^2) & \text{if} \quad \sigma = \dfrac{1}{2}\,, \\ O(h^4 + \tau^2) & \text{if} \quad \sigma = \dfrac{1}{2} - \dfrac{i\,h^2}{12\,\tau} = \sigma_*\,, \\ O(h^2 + \tau) & \text{if} \quad \sigma \neq \dfrac{1}{2}\,, \quad \sigma \neq \sigma_*\,. \end{cases}$$

The scheme so constructed is of increased accuracy for the parameters

$$\sigma = \frac{1}{2} - \frac{i\,h^2}{12\,\tau}\,, \qquad \sigma_0 = \frac{1}{2}\,, \qquad \sigma_1 = -\,\frac{h^2}{12\,\tau}\,.$$

The metodology of the method of separation of variables guides the choice of $y(x,t)$:

$$y^j(x_s) = \sum_{k=1}^{N-1} c_k^j \, X_k(x_s),$$

where $X_k(x_s)$ refer to the eigenfunctions of the operator $\Lambda$, meaning, by definition,

$$\Lambda X_k + \lambda_k X_k = 0, \qquad k = 1, 2, \dots, N-1, \qquad X_k(0) = X_k(1) = 0,$$

$$X_k(x) = \sqrt{2} \sin \pi k x, \qquad \lambda_k = \frac{4}{h^2} \sin^2 \frac{\pi k h}{2}.$$

On the strength of the preceding decomposition we establish the recurrence relation for the coefficients

$$c_k^{j+1} = q_k \, c_k^j, \qquad k = 1, 2, \dots, N-1,$$

where

$$q_k = \frac{i - (1-\sigma)\tau\lambda_k}{i + \sigma\tau\lambda_k} = \frac{-(1-\sigma_0)\tau\lambda_k + i(1+\sigma_1\tau\lambda_k)}{\sigma_0\tau\lambda_k + i(1+\sigma_1\tau\lambda_k)},$$

and the useful relations for $y^{j+1}$:

$$y^{j+1} = \sum_{k=1}^{N-1} q_k \, c_k^j \, X_k,$$

$$\| y^{j+1} \|^2 = \sum_{k=1}^{N-1} | q_k |^2 \, | c_k^j |^2 \le \max_k | q_k |^2 \cdot \| y^j \|^2.$$

Simple algebra gives

$$| q_k |^2 = \frac{(1+\sigma_1\tau\lambda_k)^2 + (1-\sigma_0)^2\tau^2\lambda_k^2}{(1+\sigma_1\tau\lambda_k)^2 + \sigma_0^2\tau^2\lambda_k^2} = 1 - \frac{(2\sigma_0 - 1)\tau^2\lambda_k^2}{(1+\sigma_1\tau\lambda_k)^2 + \sigma_0^2\tau^2\lambda_k^2},$$

implying that $| q_k | \le 1$ and thereby justifying that this scheme is stable:

$$\| y^{j+1} \| \le \| y^j \| \le \dots \le \| y^0 \| \qquad \text{for} \qquad \sigma_0 \ge \tfrac{1}{2}.$$

If $\sigma_0 < \frac{1}{2}$, then $| q_k | > 1$ and the scheme becomes unstable, thus causing some difficulties. In particular, the explicit scheme with $\sigma = 0$ is unstable for any $\tau / h^2 = \text{const}$. In contrast to the approved schemes for the heat conduction equation, there are no conditionally stable schemes among ones with $\tau / h^2 = \text{const}$: all the schemes with $\sigma_0 \geq \frac{1}{2}$ are stable, while others turn out to be unstable.

We have found earlier for the scheme of accuracy $O(h^4 + \tau^2)$ that

$$\sigma = \frac{1}{2} - \frac{i\, h^2}{12\, \tau} .$$

The same is still valid for the scheme at hand. From what has been said above it is clear that for $\sigma_0 = \frac{1}{2}$ this scheme is unconditionally stable.

The meaning of stability $\| y^j \| \leq M \| y^0 \|$ with constant $M > 1$ is that we should have

$$\max_k | q_k | \leq 1 + c_0 \tau < e^{c_0 \tau}, \qquad c_0 > 0 .$$

A similar estimate for $| q_k |$ can be derived also for $\sigma_0 < \frac{1}{2}$ by merely setting $\tau = O(h^4)$ or

$$\sigma_0 = \frac{1}{2} - \frac{c_0}{\tau \Delta^2}, \qquad \Delta = \frac{4}{h^2} .$$

Indeed, in the case of interest

$$K = \frac{(1 - 2\sigma_0)\, \tau^2 \lambda_k^2}{(1 + \sigma_1 \tau \lambda_k)^2 + \sigma_0^2 \tau^2 \lambda_k^2}$$

$$\leq \frac{(1 - 2\sigma_0)\, \tau^2 \Delta^2}{(1 + \sigma_1 \tau \lambda_k)^2 + \sigma_0^2 \tau^2 \lambda_k^2}$$

$$\leq (1 - 2\sigma_0)(\tau \Delta^2)\, \tau \leq c_0 \tau$$

if

$$\tau \leq \frac{c_0}{(1 - 2\sigma_0)\, \Delta^2} = \frac{c_0 h^4}{(1 - 2\sigma_0)\, 16}$$

or, what amounts to the same,

$$K \leq \frac{c_0 \tau}{(1 + \sigma_1 \tau \Delta)^2 + \sigma_0^2 \tau^2 \Delta^2} < c_0 \tau \qquad \text{for} \qquad \sigma = \frac{1}{2} - \frac{c_0}{\tau \Delta^2} .$$

This provides support for the view that the explicit scheme with $\sigma = 0$ is stable:

$$\| y^j \| \leq e^{c_0 t_j} \| y^0 \|$$

if $\tau \leq c_0 \Delta^{-2} = \frac{1}{16} c_0 h^4$, where $c_0 > 0$ is an arbitrary number. The condition $\tau \leq \frac{1}{16} c_0 h^4$ is very tough and unnatural. Here $\tau / h^2$ (but not $\tau / h^4$) is a dimensionless quantity. Because of this, the explicit scheme is unacceptable for the **Schrödinger equation.**

**2. Three-layer schemes.** Of special interest is the three-layer scheme with weights

$$
(3) \qquad \begin{aligned} & i\, y_{\overset{\circ}{t}} = \Lambda\left(\sigma\,\hat{y} + (1 - 2\,\sigma)\,y + \sigma\,\check{y}\right), \\ & y_0 = y_N = 0\,, \qquad y^0(x) = u_0(x)\,, \qquad y^1(x) = \bar{u}_0(x)\,, \end{aligned}
$$

where $\hat{y} = y^{j+1}$, $\check{y} = y^{j-1}$, $y = y^j$ and $\sigma = \sigma_0$ is a real number. For any $\sigma$ the scheme is of second-approximation order with $\psi = O(h^2 + \tau^2)$. For later use, we seek a particular solution in the form $y_k^j(x_s) = c_k\, q_k^j\, X_k(x_s)$. Substituting this expression into equation (3), recalling the definition of eigenfunctions $\Lambda\, X_k = -\lambda_k\, X_k$ and omitting the subscript $k$ of $q_k$ and $\lambda_k$, we eventually get the quadratic equation for $q$:

$$
(\,i + 2\,\mu\,\sigma)\,q^2 - 2\,\mu\,(2\,\sigma - 1)\,q + 2\,\mu\,q - i = 0\,, \qquad \mu = \tau\,\lambda\,,
$$

with discriminant

$$
\frac{D}{4} = \mu^2\,(2\,\sigma - 1)^2 - 1 - 4\,\mu^2\,\sigma^2 = (1 - 4\,\sigma)\,\mu^2 - 1\,.
$$

Plain calculations show that

$$
D < 0 \qquad \text{for} \qquad \sigma \geq \frac{1}{4} - \frac{1}{4\,\tau^2\,\Delta^2}
$$

and give the roots of the quadratic equation

$$
q^{(1,\,2)} = \frac{(2\,\sigma - 1)\,\mu \pm i\,\sqrt{1 + 4\,\mu^2\,\sigma^2 - (2\,\sigma - 1)^2\,\mu^2}}{i + 2\,\mu\,\sigma}\,, \qquad \left|\,q^{(1,\,2)}\,\right| = 1\,.
$$

Under the constraint

$$
\sigma \geq \frac{1}{4} - \frac{1}{4\,\tau^2\,\Delta^2}\,, \qquad \Delta = \frac{4}{h^2}\,,
$$

the particular solutions $y_k^j$ do not increase with increasing $j$:

$$
\|\,y_k^{j+1}\,\| \leq \|\,y_k^j\,\|\,.
$$

Being concerned with $q_k^{(1,\,2)} = e^{\pm i \varphi_k}$, we look for the general solution of the problem in view as a sum

$$
y^j = \sum_{k=1}^{N-1} \left(\alpha_k \cos j\varphi_k + \beta_k \sin j\varphi_k\,\right) X_k(x)
$$

with $\alpha_k$ and $\beta_k$ arising from the initial conditions

$$y^0 = u_0 , \qquad y^1 = \bar{u}_0$$

or

$$y_t^0 = \frac{y^1 - y^0}{\tau} = u_1 ,$$

$$\alpha_k = (u_0, X_k) , \qquad \beta_k = \left( (u_1 , X_k) + \frac{2 \alpha_k \sin^2 (\varphi_k/2)}{\tau} \right) \frac{\tau}{\sin \varphi_k} .$$

No wishing to load the book down with more a detailed derivation of possible estimates we quote here only the final result

$$\| y^j \| \leq M ( \| y^0 \| + \| y_t^0 \| ) \qquad \text{for} \qquad \sigma \geq \frac{1}{4} - \frac{h^4}{16 \tau^2} .$$

Careful explorations of this sort will appear in Section 6 in trying to establish stability of schemes for the equation of vibrations of a string.

## 5.5 THE TRANSFER EQUATION

**1. Explicit schemes for the Cauchy problem.** The first-order equation

$$\frac{\partial u}{\partial t} + a \frac{\partial u}{\partial x} = 0 ,$$

known as the **transfer equation**, describes, for instance, the behavior of the density $\rho = \rho(x,t)$ of incompressible liquid moving along the $Ox$-axis with velocity $v$:

$$\frac{\partial \rho}{\partial t} + v \frac{\partial \rho}{\partial x} = 0 .$$

Here we treat it as a model one. However, the arguments about this matter can result in the design of interesting experiments, whose aims and scope are to test and improve admisssible schemes for rather complicated equations of acoustics, kinematic integrodifferential equations of neutron transfer, nonlinear equations of gas dynamics, etc. Because of the enormous range and variety of problems dealt with by mathematical physics, the contents of this section would be of the methodological merit.

The problem we must solve is the Cauchy problem

$$(1)\quad \frac{\partial u}{\partial t}+a\,\frac{\partial u}{\partial x}=0\,,\qquad -\infty<x<\infty\,,\qquad t>0\,,\qquad u(x,0)=u_0(x)\,,$$

under the natural premise $a=\text{const}\neq 0$. The solution of problem (1) is a "travelling wave":

$$u(x\,,\,t)=u_0(x-at)\,,$$

where $a$ is the wave velocity and $u_0(\xi)$ is a differentiable function.

For the purposes of the present section we introduce in the plane $(x,t)$ the grid

$$\bar{\omega}_{h\tau}=\omega_h\times\omega_\tau\,,$$

$$\omega_h=\{\,x_i=ih\,,\quad i=0,\,\pm 1,\,\pm 2,\,\ldots\}\,,$$

$$\omega_\tau=\{\,t_j=j\tau\,,\quad j=0,\,1,\,2,\,\ldots\}$$

with steps $h$ and $\tau$ in $x$ and $t$, respectively, and begin by constructing one of the well-known explicit schemes for the Cauchy problem

$$(2)\qquad \frac{y_i^{j+1}-y_i^j}{\tau}+a\,\frac{y_{i+1}^j-y_i^j}{h}=0\,,\qquad y_i^0=u_0(x_i)\,,$$

or $y_t+a\,y_x=0$. The pattern of this scheme consists of the three nodes $(x_i\,,\,t_j)$, $(x_{i+1}\,,\,t_j)$ and $(x_i\,,\,t_{j+1})$ (see Fig. 17a).

Obviously, scheme (2) is of first-approximation order with respect to $\tau$ and $h$, since its residual behaves as follows:

$$\psi=u_t+a\,u_x=(\dot{u}+a\,u')+\tfrac{1}{2}\tau\,\ddot{u}+\tfrac{1}{2}ah\,u''+O(h^2+\tau^2)=O(\tau+h)\,.$$

For $a>0$ the scheme at hand turns out to be absolutely unstable. Now what we must do is to discover that some particular solution becomes un stable albeit with obvious modifications of equation (2):

$$(3)\qquad y_k^{j+1}=-\gamma\,y_{k+1}^j+(\,1+\gamma)\,y_k^j\,,\qquad \gamma=\frac{a\,\tau}{h}\,.$$

With this aim, we seek a particular solution to this equation in harmonic form. This amounts to

$$(4)\qquad y_k^j=q^j\,e^{ik\varphi},\qquad i=\sqrt{-1}\,,\qquad \varphi\neq 0\,.$$

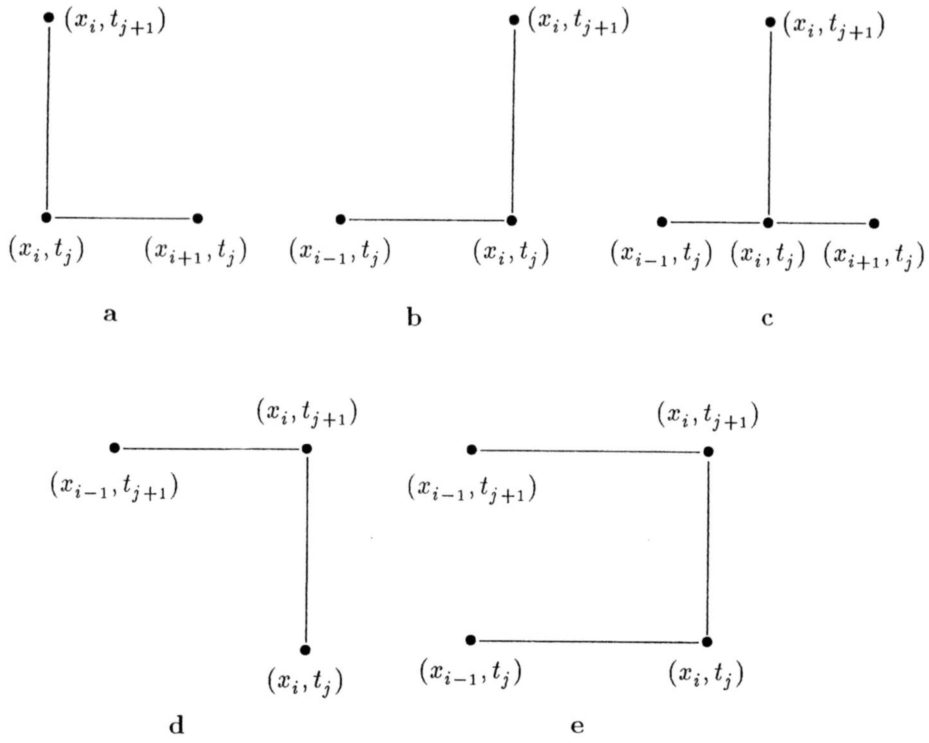


**Figure 17.**

Having substituted (4) into (3), we obtain

$$q = -\gamma\, e^{i\varphi} + \gamma + 1 = 1 + (1 - \cos\varphi)\,\gamma - i\,\gamma \sin\varphi$$

and find, by simple algebra, that

$$|\,q\,|^2 = \big(1 + (1 - \cos\varphi)\,\gamma\big)^2 + \gamma^2 \sin^2\varphi = 1 + 4\,\gamma\,(\gamma + 1)\,\sin^2(\varphi/2)\,.$$

Hence, $|\,q\,| > 1$ for any fixed $\gamma$, bounded from below as $\tau \to \infty$, if $\sin\varphi/2 \neq 0$. It is worth noting here that the case where $\sin\varphi/2 = 0$ corresponds to the values $y_k^0 = 1 = \text{const}$. Then

$$|\,y_k^j\,| = |\,q\,|^j \longrightarrow \infty \quad \text{as} \quad j \to \infty\,.$$

**Remark** Keeping $\gamma = a\tau/h = O(h)$, that is, $\tau = O(h^2)$, we might have $|q| \le 1 + c_0\,\tau$, where $c_0 > 0$ does not depend on $\tau$ and $h$, and the harmonic in question remains bounded:

$$|y_k^j| \le e^{c_0\,\tau j} = e^{c_0\,t_j} \le e^{c_0\,T} = M \qquad \text{for} \qquad 0 < t_j \le T\,.$$

This means that

$$|y_k^j| \le M\,|y_k^0|$$

with the constraint $a\,\tau/h^2 \le c_1$ (in our example $|y_k^0| = 1$), where $c_1 =$ const $> 0$ does not depend on $h$ and $\tau$.

We may recommend one more explicit scheme on the pattern consisting of the three nodes $(x_i, t_j)$, $(x_i, t_{j+1})$ and $(x_{i-1}, t_j)$ (see Fig. 17b):

$$\text{(5)} \qquad \frac{y_i^{j+1} - y_i^j}{\tau} + a\,\frac{y_i^j - y_{i-1}^j}{h} = 0\,, \qquad a > 0\,,$$

or $y_t + a\,y_{\bar{x}} = 0$. This scheme also generates an approximation of order 1, since $\psi = u_t + a\,u_{\bar{x}} = O(h+\tau)$. After scrutinising the available information on the difference equation

$$y_i^{j+1} = (1-\gamma)\,y_i^j + \gamma\,y_{i-1}^j\,, \qquad \gamma = \frac{a\,\tau}{h} > 0\,,$$

we deduce for $\gamma \le 1$ that

$$\|\,y^{j+1}\,\|_C \le (1-\gamma)\,\|\,y^j\,\|_C + \gamma\,\|\,y^j\,\|_C = \|\,y^j\,\|_C\,,$$

meaning the stability of the scheme in the space $C$:

$$\text{(6)} \qquad \|\,y^{j+1}\,\|_C \le \|\,y^0\,\|_C \quad \text{for} \quad 0 < \gamma \le 1\,.$$

By analogy with scheme (2) for $a > 0$, it is straightforward to verify that scheme (5) is unstable for $a < 0$, while scheme (2) is stable under the constraint $|\gamma| = |a|\,\tau\,h^{-1} \le 1$, so that the inequality

$$\|\,y^{j+1}\,\|_C \le \|\,y^j\,\|_C$$

holds true for all $j = 0, 1, 2, \ldots$.

**2. Explicit schemes of a higher-order approximation.** Of major importance is the explicit scheme of accuracy $O(h^2 + \tau)$ having the form

$$\frac{y_i^{j+1} - y_i^j}{\tau} + a\,\frac{y_{i+1}^j - y_{i-1}^j}{2\,h} = 0 \tag{7}$$

or $y_t + a\,y_{\overset{\circ}{x}} = 0$, where $y_{\overset{\circ}{x},\,i} = (y_{i+1} - y_{i-1})/(2\,h)$. This scheme is constructed on the four-point pattern (Fig. 17c) consisting of the four nodes $(x_i, t_{j+1})$, $(x_i, t_j)$, $(x_{i-1}, t_j)$ and $(x_{i+1}, t_j)$. Obviously, scheme (7) is unstable for every fixed $\gamma = a\,\tau\,h^{-1}$ and arbitrary sign of the coefficient $a$. Indeed, upon substituting harmonic (4) into equation (7) we get the equation for $q$:

$$q - 1 + \gamma\,\frac{e^{i\varphi} - e^{-i\varphi}}{2} = 0\,, \qquad i = \sqrt{-1}, \qquad \gamma = \frac{a\,\tau}{h}\,, \qquad q = 1 - i\,\gamma\,\sin\varphi\,.$$

Whence it follows that $|\,q\,|^2 = 1 + \gamma^2 \sin^2\varphi > 1$, so that $|\,y_k^j\,| = |q|^j \to \infty$ as $j \to \infty$. Adopting the arguments similar to those used for the scheme of Section 1.10 and replacing $y_k^j$ by the half-sum $\frac{1}{2}(y_{k+1}^j + y_{k-1}^j)$, we obtain the stable scheme on the same pattern:

$$\frac{y_k^{j+1} - \frac{1}{2}(y_{k+1}^j + y_{k-1}^j)}{\tau} + a\,\frac{y_{k+1}^j - y_{k-1}^j}{2\,h} = 0\,, \tag{8}$$

for which

$$y_k^{j+1} = \tfrac{1}{2}\,(1+\gamma)\,y_{k-1}^j + \tfrac{1}{2}\,(1-\gamma)\,y_{k+1}^j\,,$$

$$\|\,y^{j+1}\,\|_C \le \|\,y^j\,\|_C \le \cdots \le \|\,y^0\,\|_C$$

for any $|\,\gamma\,| \le 1$ and arbitrary sign of the coefficient $a$. More specifically, for $a < 0$ we thus have $1 + \gamma = 1 - |\,\gamma\,| \ge 0$ and $1 - \gamma = 1 + |\,\gamma\,| > 0$.

In the estimation of the residual for scheme (8) it is possible to rewrite it as

$$y_t - \frac{0.5\,h^2}{\tau}\,y_{\bar{x}x} + a\,y_{\overset{\circ}{x}} = 0$$

in light of the trivial relations

$$\tfrac{1}{2}\,(y_{k+1} + y_{k-1}) = \tfrac{1}{2}\,(y_{k+1} + y_{k-1} - 2y_k) + y_k = y_k + \tfrac{1}{2}\,h^2\,y_{\bar{x}x,\,k}\,.$$

The residual on a solution $u(x,t)$ is

$$\psi = u_t + a\,u_{\overset{\circ}{x}} - \frac{0.5\,h^2}{\tau}\,u_{\bar{x}x} = \frac{1}{2}\,\tau\,\ddot{u} - \frac{1}{2}\,\frac{h^2}{\tau}\,u'' + O(h^2+\tau^2)\,.$$

With the members recovered from the equation $\ddot{u} = a^2\,u''$, the preceding becomes

$$\psi = \left(\frac{a^2\tau}{2} - \frac{h^2}{2\tau}\right)\,u'' + O(h^2+\tau^2)\,,$$

which provides support for the view that scheme (8) generates a conditional approximation, because it approximates the equation only if $h^2/\tau \to 0$ as $h \to 0$ and $\tau \to 0$. In my view, two important things are that for $\tau = O(h)$ we might achieve $\psi = O(\tau + h)$ and the second order of approximation $\psi = O(h^2)$ is attained for $\tau = h/|\,a\,|$.

To make our exposition more complete, we involve the four-point scheme of second-order accuracy

$$y_t + a\,y_{\overset{\circ}{x}} - \tfrac{1}{2}\,\tau\,a^2 y_{\bar{x}x} = 0 \qquad (9)$$

on the same pattern as was done for scheme (7) (see Fig. 17c). Plain calculations of the residual with $\dot{u} + a\,u' = 0$ and $\ddot{u} = -a\,\dot{u}' = a^2\,u''$ incorporated give

$$\begin{aligned}\psi &= u_t + a\,u_{\overset{\circ}{x}} - \frac{1}{2}\,\tau\,a^2\,u_{\bar{x}x}\\ &= \dot{u} + \frac{1}{2}\,\tau\,\ddot{u} + O(\tau^2) + a\,u' + O(h^2) - \frac{1}{2}\,a^2\,\tau\,u'' + O(h^2+\tau)\\ &= (\dot{u} + a\,u') + \frac{1}{2}\,\tau\,(\,\ddot{u} - a^2\,u'') + O(h^2+\tau^2) = O(h^2+\tau^2)\,,\end{aligned}$$

so that $\psi = O(h^2+\tau^2)$.

Let us investigate the stability of scheme (9) by the spectral method having represented it beforehand in the form

$$y_k^{j+1} = (1-\gamma^2)\,y_k^j + \frac{1}{2}\,\gamma\,(\gamma-1)\,y_{k+1}^j + \frac{1}{2}\,\gamma\,(\gamma+1)\,y_{k-1}^j\,.$$

Via transform $y_k^j = q^j\,e^{ik\varphi}$ we derive the useful expression for $q$:

$$q = 1 - \gamma^2\,(1-\cos\,\varphi) - i\,\gamma\,\sin\,\varphi\,, \qquad |\,q\,|^2 = 1 - \gamma^2\,(1-\gamma^2)\,(1-\cos\,\varphi)^2,$$

by means of which we establish that $|\,\gamma\,| \le 1$ is a necessary stability condition, whereas $|\gamma| > 1$ is a sufficient unstability condition in the situations when $|\gamma|$ is fixed and $|\gamma| = \text{const}$ with varying parameters $\tau$ and $h$.

**3. The boundary-value problem.** We are interested in learning more about the boundary-value problem when the boundary value $\mu(t)$ is specified at the point $x = 0$ and a solution is sought for $x > 0$ and $t > 0$:

$$(10)\qquad \begin{aligned} &\frac{\partial u}{\partial t} + a\,\frac{\partial u}{\partial x} = 0\,, \qquad t > 0\,, \qquad 0 < x < \infty\,, \qquad a = \text{const} > 0\,, \\ &u(x,\,0) = u_0(x)\,, \qquad x \geq 0\,, \qquad u(0,\,t) = \mu(t)\,, \qquad t \geq 0\,, \end{aligned}$$

under the natural premise $u_0(0) = \mu(0)$. In the case of the differentiable members $u_0(x)$ and $\mu(t)$ the function

$$u(x,\,t) = \begin{cases} u_0(x - at) & \text{for } \ t \leq x/a\,, \\ \mu(t - x/a) & \text{for } \ t \geq x/a\,, \end{cases}$$

is just the solution of the problem under consideration.

In working on the grid $\omega_h = \{\,x_i = ih,\ i = 0, 1, 2, \ldots\}$ with spacing $h$ and the grid $\omega_\tau = \{\,t_j = j\tau,\ j = 0, 1, \ldots\}$ with spacing $\tau$ we construct the implicit scheme on the pattern depicted in Fig. 17d in the usual way:

$$(11)\qquad \frac{y_k^{j+1} - y_k^j}{\tau} + a\,\frac{y_k^{j+1} - y_{k-1}^{j+1}}{h} = 0\,.$$

Alternative forms of this scheme

$$y_t + a\,\hat{y}_{\bar{x}} = 0$$

and

$$y_k^{j+1} = \frac{\gamma}{\gamma+1}\,y_{k-1}^{j+1} + \frac{1}{\gamma+1}\,y_k^j\,, \qquad \gamma = \frac{a\,\tau}{h} > 0\,,$$

are more convenient for later use. With these, the computational procedure may be carried out with the starting point $k = 1$, $j = 0$. Then

$$y_1^1 = \frac{\gamma}{\gamma+1}\,y_0^1 + \frac{1}{\gamma+1}\,y_1^0 = \frac{\gamma}{\gamma+1}\,\mu(t_1) + \frac{1}{\gamma+1}\,u_0(x_1)\,.$$

With knowledge of $y_1^1$ we are able to calculate the remaining values $y_1^j$ up to some $j = j_c$. After that, we will find $y_2^j$ for $0 < j \leq j_0$, etc. by setting $k = 2$.

We now deal with the family of schemes on the four-point pattern (see Fig. 17e)

$$(12)\qquad \begin{aligned} &y_t + \sigma\,\hat{y}_{\bar{x}} + (1-\sigma)\,y_{\bar{x}} = 0\,, \qquad x \in \omega_h\,, \qquad t \in \omega_\tau\,, \\ &y(x,\,0) = u_0(x)\,, \qquad y(0,\,t_j) = 0\,, \qquad a \equiv 1\,. \end{aligned}$$

Observe that scheme (11) belongs to this family, corresponds to the case $\sigma = 1$ and has the residual $\psi = u_t + \sigma\,\hat{u}_{\bar{x}} + (1-\sigma)\,u_{\bar{x}}$. By inserting here the asymptotic expansions

$$u_t = \bar{\dot{u}} + O(\tau^2)\,, \qquad u_{\bar{x}} = u' - \tfrac{1}{2}\,h\,u'' + O(h^2)\,,$$

$$\hat{v} = \bar{v} + \tfrac{1}{2}\,\tau\,\bar{\dot{v}} + O(\tau^2)\,, \qquad v = \bar{v} - \tfrac{1}{2}\,\tau\,\bar{\dot{v}} + O(\tau^2)\,,$$

where $\bar{v} = v|_{t_j+\tau/2}$ and $v = u_{\bar{x}}$, we arrive at

$$\begin{aligned}\psi &= \tfrac{1}{2}\,\tau\,(2\,\sigma - 1)\,\bar{\dot{u}}' - \tfrac{1}{2}\,h\,u'' + O(h^2+\tau^2)\\ &= \tfrac{1}{2}\,(2\,\sigma\,\tau + h - \tau)\,\bar{\dot{u}}' + O(h^2+\tau^2)\,.\end{aligned}$$

which serves to motivate that the scheme with weights is of order 2: $\psi = O(h^2+\tau^2)$ if $\sigma = \frac{1}{2} - \frac{1}{2}\,h\,\tau^{-1} = \sigma_0$. In all other cases ($\sigma \neq \sigma_0$) it is of order 1, since $\psi = O(\tau + h)$.

In order to demonstrate that scheme (12) is stable with respect to initial data for

$$\sigma \geq \frac{1}{2} - \frac{h}{2\,\tau} = \sigma_0\,,$$

we proceed as usual. This amounts to forming the grid $\bar{\omega}_h = \{x_i = i\,h,\ i = 0, 1, \ldots, N,\ hN = l\}$ on the segment $0 \leq x \leq l$ and introducing the inner product and associated norm by

$$(\,y,\,v\,] := \sum_{i=1}^{N} y_i\,v_i\,h\,, \qquad \|\,y\,\| := \sqrt{(\,y,\,y\,]}\,.$$

Other ideas are connected with the well-established expressions

$$\hat{v} = \frac{1}{2}\,(\hat{v} + v) + \frac{1}{2}\,(\hat{v} - v)$$

and

$$v = \frac{1}{2}\,(\hat{v} + v) - \frac{1}{2}\,(\hat{v} - v)$$

and forthcoming substitution $v = y_{\bar{x}}$. Upon receipt of the available information we reduce the scheme concerned to

$$y_t + (\,\sigma - \tfrac{1}{2})\,\tau\,y_{\bar{x}t} + \tfrac{1}{2}\,(\hat{y}_{\bar{x}} + y_{\bar{x}}) = 0\,, \qquad y(0,\,t) \equiv 0\,.$$

The next step is to multiply the resulting equation by $2\tau\, y_{\bar{x}t} = 2\,(\hat{y}_{\bar{x}} - y_{\bar{x}})$ and take into account that

$$2\, y_t\, y_{\bar{x}t} = 2\, v\, v_{\bar{x}} = (v^2)_{\bar{x}} + h\,(v_{\bar{x}})^2 = (y_t^2)_{\bar{x}} + h\,(y_{\bar{x}t})^2$$

with $v = y_t$. The outcome of this is

$$\tau\,(y_t^2)_{\bar{x}} + 2\,\big((\sigma - \tfrac{1}{2})\,\tau + \tfrac{1}{2}\,h\big)\,(y_{\bar{x}t})^2 + (\hat{y}_{\bar{x}})^2 - (y_{\bar{x}})^2 = 0\,.$$

With the aid of the relations

$$\sum_{i=1}^{N} (y_t^2)_{\bar{x},\,i}\, h = \sum_{i=1}^{N} \big[(y_t)_i^2 - (y_t)_{i-1}^2\big] = (y_t)_N^2 - (y_t)_0^2$$

and $y_{t,0} = y_t(0,t) = 0$, since $y(0,t) \equiv 0$, we get

$$\tau\,(y_t)_N^2 + 2\,\big((\sigma - \tfrac{1}{2})\,\tau + \tfrac{1}{2}h\big)\,\|\, y_{\bar{x}t}\,\|^2 + \|\,\hat{y}_{\bar{x}}\,\|^2 = \|\, y_{\bar{x}}\,\|^2 \tag{13}$$

as a final result of multiplying once again by $h$ and summing over all grid nodes $x = ih$, $i = 1, 2, \dots, N$. It follows from the foregoing that

$$\|\, y_{\bar{x}}^{j+1}\,\| \le \|\, y_{\bar{x}}^{j}\,\| \le \cdots \le \|\, y_{\bar{x}}^{0}\,\|$$

if $(\sigma - \frac{1}{2})\,\tau + \frac{1}{2}\,h \ge 0$, giving $\sigma \ge \sigma_0$. This scheme is stable in the energy norm $\|\, y\,\|_{(1)} = \|\, y_{\bar{x}}\,\|$.

In Chapter 6 we will deal with the two-layer scheme of general form

$$B\, y_t + A\, y = 0\,, \tag{14}$$

where both operators $A$ and $B$ really act in a prescribed Euclidean space $H$, $A = A^* > 0$ and $B > 0$. A necessary and sufficient stability condition will be established in the form

$$B \ \ge\ \tfrac{1}{2}\,\tau\, A\,. \tag{15}$$

Moreover, $\|\, y^{j+1}\,\|_A \le \|\, y^0\,\|_A$ in the norm $\|\, y\,\|_A = \sqrt{(Ay,\, y)}$.

In studying problem (12) we refer to the space $H$ of all grid functions defined on the grid $\bar{\omega}_h$ and vanishing for $i = 0$ and the operator $\widetilde{A}\, y = y_{\bar{x}}$, for which the estimate

$$(\widetilde{A}\, y,\, y) = \tfrac{1}{2}\, y_N^2 + \tfrac{1}{2}\, h\,\|\, y_{\bar{x}}\,\|^2 > 0$$

holds true due to the relation $y\,y_{\bar{x}x} = \frac{1}{2}\,(y^2)_{\bar{x}} + \frac{1}{2}\,h\,(y_{\bar{x}})^2$. An alternative form of scheme (12) may be useful in the further development:

$$(E + \sigma\,\tau\,\widetilde{A}\,)\,y_t + \widetilde{A}\,y = 0\,. \tag{16}$$

Since $\widetilde{A} > 0$, there always exists the inverse $\widetilde{A}^{-1} > 0$, by means of which we recast (16) as

$$(\,\widetilde{A}^{-1} + \sigma\,\tau\,E\,)\,y_t + y = 0\,. \tag{17}$$

Comparison of (17) with (14) shows that $B = \widetilde{A}^{-1} + \sigma\,\tau\,E$, $A = E$ and condition (15) signifies that

$$\Big(\big(\,\widetilde{A}^{-1} + \sigma\,\tau\,E\,\big)\,x,\;x\Big) - \tfrac{1}{2}\,\tau\,(x,\,x)$$
$$= (\,\widetilde{A}^{-1}x,\,x\,) + (\sigma - \frac{1}{2})\;\tau\;(x,\,x) \geq 0\,,$$

yielding $(\,\widetilde{A}\,y,\,y\,) + (\sigma - \frac{1}{2})\,\tau\,\|\,\widetilde{A}\,y\,\|^2 \geq 0$ for the substitution $\widetilde{A}^{-1}x = y$ and providing the relation $(\,\widetilde{A}\,y,\,y\,) = \frac{1}{2}\,y_N^2 + \frac{1}{2}\,h\,\|\,\widetilde{A}\,y\,\|^2$ in the case of interest. The stability condition (15) taking now the form

$$\tfrac{1}{2}\,y_N^2 + (\,\tfrac{1}{2}\,h + (\sigma - \tfrac{1}{2})\,\tau\,)\,\|\,A\,y\,\|^2 \geq 0 \tag{18}$$

is valid for $\sigma \geq \sigma_0$. This provides enough reason to conclude that the a priori estimate $\|\,y^j\,\| \leq \|\,y^0\,\|$ is certainly true. Likewise, a scheme with weights may be designed for the system

$$\frac{\partial u}{\partial t} = \frac{\partial v}{\partial x}\,, \qquad \frac{\partial v}{\partial t} = \frac{\partial u}{\partial x}\,,$$

which is equivalent to the equation of vibrations of a string

$$\frac{\partial^2 u}{\partial t^2} = \frac{\partial^2 u}{\partial x^2}$$

and the stability conditions for such a scheme may be established in a similar way as we did before.

## 5.6 DIFFERENCE SCHEMES FOR THE EQUATION OF VIBRATIONS OF A STRING

**1. The statement of the difference problem and calculations of the approximation error.** In this section we study the equation of vibrations of a string

$$\frac{\partial^2 u}{\partial t_1^2} = a^2 \frac{\partial^2 u}{\partial x_1^2} + f(x_1, t_1), \qquad 0 < x_1 < l, \qquad t_1 > 0.$$

In this view, it seems reasonable to pass to the dimensionless variables $x = x_1/l$ and $t = at_1/l$, due to which the initial equation is representable by

$$(1) \qquad \frac{\partial^2 u}{\partial t^2} = \frac{\partial^2 u}{\partial x^2} + f(x, t), \qquad 0 < x < 1, \qquad 0 < t \leq T.$$

At the initial moment the supplementary conditions are specified by

$$(2) \qquad u(x, 0) = u_0(x), \qquad \frac{\partial u(x, 0)}{\partial t} = \bar{u}_0(x),$$

where $u_0(x)$ is the initial deviation and $\bar{u}_0(x)$ is the initial velocity. The string's ends move in accordance with the known laws

$$(3) \qquad u(0, t) = \mu_1(t), \qquad u(1, t) = \mu_2(t).$$

By analogy with Section 1.2 we introduce in the domain $\overline{D} = \{0 \leq x \leq 1, 0 \leq t \leq T\}$ the rectangular grid $\bar{\omega}_{h\tau}$. As equation (1) contains the second derivative in $t$, the number of layers cannot be any smaller than 3. Retaining the preceding notations, we have

$$y = y^j, \qquad \hat{y} = y^{j+1}, \qquad \check{y} = y^{j-1}, \qquad y_t = \frac{\hat{y} - y}{\tau}, \qquad y_{\bar{t}} = \frac{y - \check{y}}{\tau},$$

$$\Lambda y = y_{\bar{x}x},$$

$$y_{\bar{t}t} = \frac{y_t - y_{\bar{t}}}{\tau} = \frac{\hat{y} - 2y + \check{y}}{\tau^2}, \qquad y_{\overset{\circ}{t}} = \frac{y_t + y_{\bar{t}}}{2} = \frac{\hat{y} - \check{y}}{2\tau}.$$

As before, we replace the derivatives built into equation (1) by the formulae

$$\frac{\partial^2 u}{\partial t^2} \sim u_{\bar{t}t}, \qquad \frac{\partial^2 u}{\partial x^2} \sim \Lambda u = u_{\bar{x}x}, \qquad f \sim \varphi.$$

A key role in subsequent discussions is played by a family of schemes with weights

$$(4)\qquad \begin{aligned} &y_{\bar{t}t} = \Lambda\left(\sigma\,\hat{y} + (1-2\,\sigma)\,y + \sigma\,\check{y}\right) + \varphi\,, \qquad\qquad \varphi = f(x,\,t_j)\,,\\ &y_0 = \mu_1(t),\quad y_N = \mu_2(t),\quad y(x,0) = u_0(x),\quad y_t(x,0) = \tilde{u}_0(x)\,, \end{aligned}$$

where $\tilde{u}_0(x)$ will be specified below.

The boundary conditions and the first initial condition $u(x,\,0) = u_0(x)$ on the grid $\bar{\omega}_{h\tau}$ are satisfied exactly. A choice of $\tilde{u}_0(x)$ is stipulated by the wishes that the approximation error $\tilde{u}(x) - \partial u(x,0)/\partial t = \tilde{u}(x) - \bar{u}_0(x)$ would be a quantity of order $O(\tau^2)$. From the chain of the relations

$$\begin{aligned} u_t(x,\,0) &= \dot{u}(x,\,0) + \frac{1}{2}\,\tau\,\ddot{u}(x,\,0) + O(\tau^2)\\ &= \bar{u}_0(x) + \frac{1}{2}\,\tau\left(u''(x,\,0) + f(x,\,0)\right) + O(\tau^2)\\ &= \bar{u}_0(x) + \frac{1}{2}\,\tau\left(u_0''(x) + f(x,\,0)\right) + O(\tau^2) \end{aligned}$$

it is readily seen that $\tilde{u}(x) - u_t(x,0) = O(\tau^2)$ if we accept

$$(5)\qquad \tilde{u}_0(x) = \bar{u}_0(x) + \tfrac{1}{2}\,\tau\left(u_0''(x) + f(x,\,0)\right).$$

Thus, the difference problem (4)–(5) is completely posed. With regard to $\hat{y} = y^{j+1}$ we set up on the basis of (4) the boundary-value problem

$$\begin{gathered} \sigma\,\gamma^2\left(y_{i+1}^{j+1} + y_{i-1}^{j+1}\right) - (1 + 2\,\sigma\,\gamma^2)\,y_i^{j+1} = -F_i\,, \qquad 0 < i < N\,,\\ y_0 = \mu_1\,, \qquad y_N = \mu_2\,, \qquad \gamma = \frac{\tau}{h}\,,\\ F_i = \left(2\,y_i^j - y_i^{j-1}\right) + \tau^2\,(1-2\,\sigma)\,\Lambda\,y^j + \sigma\,\tau^2\,\Lambda\,y^{j-1} + \tau^2\,\varphi\,, \end{gathered}$$

which can be solved by the right elimination method, stable for any $\sigma > 0$ (see Chapter 1, Section 2.6).

The next step is to calculate the approximation error of (4) in the case $\varphi = f(x,t_j)$ by investigating the difference between a solution $y$ of problem (4)–(5) and a solution $u = u(x,t)$ of problem (1)–(3). Substituting $y = z+u$ into (4) yields

$$(6)\qquad \begin{aligned} &z_{\bar{t}t} = \Lambda\left(\sigma\,\hat{z} + (1-2\,\sigma)\,z + \sigma\,\check{z}\right) + \psi\,,\\ &z_0 = z_N = 0\,,\qquad z(x,\,0) = 0\,,\qquad z_t(x,\,0) = \nu(x)\,, \end{aligned}$$

where $\psi = \Lambda(\sigma\hat{u} + (1-2\sigma)u + \sigma\check{u}) + \varphi - u_{\bar{t}t}$ is the approximation error of scheme (4) on the solution $u = u(x,t)$ and $\nu = \tilde{u}_0(x) - u_t(x,0)$ is the approximation error for the second initial condition $y_t = \tilde{u}_0(x)$. In accordance with what has been said above, $\nu = O(\tau^2)$. The well-established expressions $\hat{u} = u + \tau u_t$ and $\check{u} = u - \tau u_{\bar{t}}$ allow us to deduce that

$$\sigma\hat{u} + (1-2\sigma)u + \sigma\check{u} = u + \sigma\tau^2 u_{\bar{t}t},$$

implying that

$$\begin{aligned}\psi &= \Lambda u + \sigma\tau^2\Lambda u_{\bar{t}t} + \varphi - u_{\bar{t}t}\\ &= Lu + \sigma\tau^2 L\ddot{u} + f - \ddot{u} + O(h^2+\tau^2).\end{aligned} \tag{7}$$

Moreover, $\psi = O(h^2+\tau^2)$ for an arbitrary constant $\sigma$, which does not depend on $\tau$ and $h$.

Let $\sigma = \bar{\sigma} - h^2/(12\tau^2)$, where the constant $\bar{\sigma}$ independent of $h$ and $\tau$ is so chosen as to provide the stability of scheme (4). As can readily be observed, it suffices to take $\bar{\sigma} \geq \frac{1}{4}(1-\varepsilon)^{-1}$, since scheme (4) is stable for $\sigma \geq \frac{1}{4}(1-\varepsilon)^{-1} - \frac{1}{4}\gamma^{-2}$, $\gamma = \tau/h$, $\varepsilon > 0$. Scheme (4) with the member

$$\varphi = f + \frac{h^2}{12}f'' \tag{8}$$

permits us to make the order of approximation more higher and achieve $\psi = O(h^4+\tau^2)$.

The boundary conditions of the third kind

$$\frac{\partial u(0,t)}{\partial x} = \beta_1 u(0,t) - \mu_1(t), \qquad -\frac{\partial u(1,t)}{\partial x} = \beta_2 u(1,t) - \mu_2(t)$$

are approximated by the following difference equations:

$$\begin{aligned}\rho_1 y_{\bar{t}t} &= \Lambda^-(\sigma\hat{y} + (1-2\sigma)y + \sigma\check{y}) + \varphi^-, && i=0,\\ \rho_2 y_{\bar{t}t} &= \Lambda^+(\sigma\hat{y} + (1-2\sigma)y + \sigma\check{y}) + \varphi^+, && i=N,\end{aligned}$$

where

$$\Lambda^- y = \frac{y_x - \beta_1 y}{h/2}, \qquad \Lambda^+ y = -\frac{y_{\bar{x}} + \beta_2 y}{h/2},$$

$$\varphi^- = \rho_1\varphi + \frac{\nu_1}{h/2}, \qquad \varphi^+ = \rho_2\varphi + \frac{\nu_2}{h/2}.$$

Here the approximation error for the boundary conditions is a quantity $O(h^2 + \tau^2)$ if

$$\varphi = f(x,\, t)\,, \qquad \rho_1 = \rho_2 = 1\,, \qquad \nu_1 = \mu_1(t)\,, \qquad \nu_2 = \mu_2(t)\,.$$

The forthcoming substitutions

$$\sigma = -\frac{h^2}{12\,\tau^2} + \bar{\sigma}\,, \qquad \bar{\sigma} = \text{const}\,,$$

$$\rho_1 = 1 + \frac{h\,\beta_1}{3}\,, \qquad \rho_2 = 1 + \frac{h\,\beta_2}{3}\,,$$

$$\varphi = f(x,\, t) + \frac{h^2}{12}\, f''\,,$$

$$\nu_1(t) = \mu_1(t) + \frac{h^2}{6}\left(\frac{\ddot{\mu}_1(t)}{2} + f'(0,\, t) - \beta_1\, f(0,\, t)\right),$$

$$\nu_2(t) = \mu_2(t) + \frac{h^2}{6}\left(\frac{\ddot{\mu}_2(t)}{2} - f'(1,\, t) - \beta_2\, f(1,\, t)\right)$$

are best suited for the design of the scheme of accuracy $O(h^4 + \tau^2)$ approximating the initial equation at the nodes $x = 0$ and $x = 1$ to $O((h^4 + \tau^2)/h)$ and the boundary conditions to $O(h^4 + \tau^2)$.

**2. Stability analysis.** We now investigate the stability of scheme (4) with respect to initial data in the case of homogeneous boundary conditions and zero right-hand side of the equation. A reasonable statement of the problem is

$$\text{(4a)} \qquad \begin{gathered} y_{\bar{t}t} = \Lambda\left(\sigma\,\hat{y} + (1 - 2\,\sigma)\,y + \sigma\,\check{y}\right) = \Lambda\, y^{(\sigma)}, \\ y_0 = y_N = 0\,, \qquad y(x,\, 0) = u_0(x)\,, \qquad y_t(x,\, 0) = \tilde{u}_0(x)\,. \end{gathered}$$

On the same grounds as before, we look for its solution by the method of separation of variables, still using the framework of Section 1.4 where particular solutions are sought as a product $y(x,t) = X(x)\,T(t) \not\equiv 0$. Substitution of $y = X\,T$ into equation (4a) gives

$$\text{(9)} \qquad \frac{\Lambda\, X}{X} = \frac{T_{\bar{t}t}}{T^{(\sigma)}} = -\,\lambda\,.$$

Putting these together with the boundary conditions $y_0 = y_N = 0$ we set up the eigenvalue problem for $X(x)$:

$$\Lambda\, X + \lambda\, X = 0\,, \qquad x \in \omega_h\,, \qquad X(0) = X(1) = 0\,, \qquad X(x) \not\equiv 0\,,$$

whose solutions are

$$\lambda_k = \frac{4}{h^2} \sin^2 \frac{\pi k h}{2} , \qquad X^{(k)}(x) = \sqrt{2} \sin \pi k x .$$

From (9) we get the difference equation of second order for $T_k(t)$:

$$(T_k)_{\bar{t}t} + \lambda_k T_k^{(\sigma)} = 0$$

or the equation

$$(1 + \sigma \tau^2 \lambda_k) \hat{T}_k - 2 \left( 1 + (\sigma - \tfrac{1}{2}) \tau^2 \lambda_k \right) T_k + (1 + \sigma \tau^2 \lambda_k) \check{T}_k = 0 ,$$

which can be rewritten as

$$(10) \qquad \hat{T}_k - 2 (1 - \alpha_k) T_k + \check{T}_k = 0 , \qquad \alpha_k = \frac{\frac{1}{2} \tau^2 \lambda_k}{1 + \sigma \tau^2 \lambda_k} .$$

We may attempt a solution of the preceding equation in the form $T_k = T_k(t_j) = q_k^j$. Thus the quadratic equation for $q$ arises from (10): $q^2 - 2(1 - \alpha) q + 1 = 0$ (the subscript $k$ is omitted for a while). Careful analysis of its roots $q_{1,2} = 1 - \alpha \pm \sqrt{\alpha^2 - 2\alpha}$ shows that for $0 < \alpha < 2$, the values $q_{1,2} = 1 - \alpha \pm i \sqrt{\alpha (2 - \alpha)}$ are complex with $| q_{1,2} | = 1$. It is sensible to pass to a new variable $\varphi_k$ for which

$$\cos \varphi_k = 1 - \alpha_k , \qquad \sin \varphi_k = \sqrt{\alpha_k (2 - \alpha_k)} ,$$

making it possible to get $q_1^{(k)} = e^{i \varphi_k}$ and $q_2^{(k)} = e^{-i \varphi_k}$, due to which the general solution to equation (10) is representable by

$$T_k(t_j) = C_k \left( q_1^{(k)} \right)^j + D_k \left( q_2^{(k)} \right)^j = A_k \cos j\varphi_k + B_k \sin j\varphi_k ,$$

where $A_k$ and $B_k$ stand for arbitrary constants.

After that, we look for a solution of problem (4a) as a sum of particular solutions

$$(11) \qquad y^j = \sum_{k=1}^{N-1} \left( A_k \cos j\varphi_k + B_k \sin j\varphi_k \right) X^{(k)}(x) .$$

Let $u_{0k}$ and $\tilde{u}_{0k}$ be the appropriate coefficients in the relevant expansions of $u_0(x)$ and $\tilde{u}_0(x)$:

$$(12) \qquad u_0(x) = \sum_{k=1}^{N-1} u_{0k} X^{(k)}(x) , \qquad \tilde{u}_0(x) = \sum_{k=1}^{N-1} \tilde{u}_{0k} X^{(k)}(x) .$$

By relating the initial conditions $y^0 = u_0$ and $y_t^0 = (y^1 - y^0)/\tau = \tilde{u}_0(x)$ to hold for sum (11) we establish the relationships for determining the coefficients $A_k$ and $B_k$:

$$A_k = u_{0k}\,, \qquad A_k \,\frac{\cos\varphi_k - 1}{\tau} + B_k \,\frac{\sin\varphi_k}{\tau} = \tilde{u}_{0k}\,,$$

making it possible to find that

$$(13) \qquad A_k = u_{0k}\,, \qquad B_k = \frac{1-\cos\varphi_k}{\sin\varphi_k}\, u_{0k} + \frac{\tau}{\sin\varphi_k}\,\tilde{u}_{0k}\,.$$

Having substituted $A_k$ and $B_k$ into (11), we are led by minor changes to

$$(14) \qquad y^j = \sum_{k=1}^{N-1} \left( \frac{\cos\,(j-\frac{1}{2})\,\varphi_k}{\cos\,\frac{1}{2}\varphi_k}\, u_{0k} + \frac{\tau\,\sin\,j\varphi_k}{\sin\,\varphi_k}\,\tilde{u}_{0k} \right) X^{(k)}(x)\,.$$

After the first stage the estimate of $\|\, y^j \,\|$ for scheme (4a) will be derived for $\sigma = 0$ relating to the scheme

$$(15) \qquad y_{\bar{t}t} = \Lambda y\,, \qquad y_0 = y_N = 0, \quad y(x,0) = u_0(x), \quad y_t(x,0) = \tilde{u}_0(x)\,.$$

In that case

$$\alpha_k = \tfrac{1}{2}\,\tau^2\,\lambda_k = \mu_k\,, \qquad \cos\varphi_k = 1 - \mu_k\,, \qquad \sin\varphi_k = \sqrt{\mu_k\,(2-\mu_k)}\,.$$

When the steps of the grid $\omega_{h\tau}$ are related by

$$(16) \qquad \frac{\tau^2}{h^2} \le \frac{1}{1+\varepsilon}\,,$$

where $\varepsilon > 0$ is an arbitrary number, we find that

$$(17) \qquad \mu_k \le \frac{2}{1+\varepsilon}\,, \qquad k = 1, 2, \ldots, N-1\,,$$

yielding

$$(18) \qquad \cos\,\tfrac{1}{2}\,\varphi_k = \sqrt{1 - \frac{\mu_k}{2}} \ge \sqrt{\frac{\varepsilon}{1+\varepsilon}}\,.$$

By virtue of the relations

$$\frac{\sin\varphi_k}{\tau} = \frac{2\,\sin\,\frac{1}{2}\varphi_k}{\tau}\,\cos\,\tfrac{1}{2}\varphi_k \ge \frac{2\,\sin\,\frac{1}{2}\,\varphi_k}{\tau}\,\sqrt{\frac{\varepsilon}{1+\varepsilon}}$$

and

$$\frac{2\sin\frac{1}{2}\varphi_k}{\tau} = -\frac{\sqrt{2(1-\cos\varphi_k)}}{\tau} = \frac{\sqrt{2\mu_k}}{\tau} = \sqrt{\lambda_k}$$

the estimate occurs:

$$\frac{\sin\varphi_k}{\tau} \geq \sqrt{\lambda_k \frac{\varepsilon}{1+\varepsilon}}\,. \tag{19}$$

Observe that expansion (14) implies that

$$\| y^j \| \leq \left\| \sum_{k=1}^{N-1} \frac{\cos(j-\frac{1}{2})\varphi_k}{\cos\frac{1}{2}\varphi_k} u_{0k} X^{(k)} \right\| + \left\| \sum_{k=1}^{N-1} \frac{\tau\sin j\varphi_k}{\sin\varphi_k} \tilde{u}_{0k} X^{(k)} \right\|.$$

Using estimates (18) and (19) behind we finally establish

$$\| y^j \| \leq \sqrt{\frac{1+\varepsilon}{\varepsilon}} \left( \| u_0 \| + \left( \sum_{k=1}^{N-1} \frac{(\tilde{u}_{0k})^2}{\lambda_k} \right)^{1/2} \right).$$

We note in passing that the expression

$$\left( \sum_{k=1}^{N-1} \frac{(\tilde{u}_{0k})^2}{\lambda_k} \right)^{1/2}$$

is nothing than one admissible norm on the space $H_{A^{-1}}$ known as the negative norm:

$$\| \tilde{u}_0 \|_{A^{-1}} = (A^{-1}\tilde{u}_0, \tilde{u}_0)^{1/2},$$

where $Ay = -\Lambda y = -y_{\bar{x}x}$ in the space of all functions $y$ defined on the grid $\bar{\omega}_h$ and vanishing at the points $x=0$ and $x=1$. Indeed,

$$A^{-1}\tilde{u}_0 = \sum_{k=1}^{N-1} \tilde{u}_{0k} A^{-1} X^{(k)} = \sum_{k=1}^{N-1} \frac{\tilde{u}_{0k}}{\lambda_k} X^{(k)},$$

implying that

$$(A^{-1}\tilde{u}_0, \tilde{u}_0) = \sum_{k=1}^{N-1} \frac{(\tilde{u}_{0k})^2}{\lambda_k}.$$

Thus, we have proved that under condition (16) scheme (15) admits the estimate

$$\| y^j \| \leq \sqrt{\frac{1+\varepsilon}{\varepsilon}} \left( \| u_0 \| + \| \tilde{u}_0 \|_{A^{-1}} \right). \tag{20}$$

This estimate remains valid for scheme (4a) under the constraint

$$\sigma \geq \frac{1+\varepsilon}{4} - \frac{h^2}{4\,\tau^2}\,, \tag{21}$$

where $\varepsilon > 0$ is an arbitrary number. To make sure of it, we should replace everywhere in the above proof $\mu_k$ by $\alpha_k = (\frac{1}{2}\,\tau^2\,\lambda_k)\,(1+\sigma\,\tau^2\,\lambda_k)^{-1}$.

The superposition principle unveils its potential in investigating the stability of scheme (4) with respect to the right-hand side by considering the problem

$$y_{\bar{t}t} = \Lambda\, y^{(\sigma)} + \varphi\,, \quad y_0 = y_N = 0\,, \quad y(x,\,0) = 0\,, \quad y_t(x,\,0) = 0\,, \tag{4b}$$

whose solution is sought as a sum

$$y^j = \sum_{j'=0}^{j} \tau\, Y^{j,j'}, \tag{22}$$

where $Y^{j,j'}$ as a function of $j$ solves for fixed $j'$ the homogeneous equation

$$Y_{\bar{t}t}^{j,j'} = \Lambda\left(\sigma\, Y^{j+1,j'} + (1-2\,\sigma)\, Y^{j,j'} + \sigma\, Y^{j-1,j'}\right)_{\tau}, \qquad 0 \leq j' < j\,, \tag{23}$$

supplied by the boundary conditions

$$Y_0^{j,j'} = Y_N^{j,j'} = 0 \tag{24}$$

and the initial conditions

$$Y^{j',j'} = 0\,, \qquad Y_t^{j',j'} = \frac{Y^{j'+1,j'} - Y^{j',j'}}{\tau} = \frac{Y^{j'+1,j'}}{\tau} = \Phi^{j'}, \tag{25}$$

where $\Phi^{j'}$ is so chosen as to satisfy the nonhomogeneous equation (4b).

A similar problem arises naturally for the function $\Phi^j$. By the definition of $Y^{j,j'}$,

$$y_{\bar{t}t}^j = \frac{1}{\tau}\, Y^{j+1,j} + \sum_{j'=0}^{j-1} \tau\, Y_{\bar{t}t}^{j,j'},$$

$$\Lambda\, y^{(\sigma)} = \sigma\,\tau\,\Lambda\, Y^{j+1,j} + \sum_{j'=0}^{j-1} \tau\,\Lambda\,(Y^{j,j'})^{(\sigma)}.$$

Putting these together with (4b) we find that

$$Y^{j+1,j} - \sigma \tau^2 \Lambda Y^{j+1,j} = \tau \varphi^j, \tag{26}$$

which allows us to derive the equation for $\Phi = \Phi^j = Y^{j+1,j} \tau^{-1}$:

$$\Phi - \sigma \tau^2 \Lambda \Phi = \varphi, \qquad \Phi_0 = \Phi_N = 0. \tag{27}$$

For the purposes of the present section, let us estimate a solution $y^j$ of problem (4b) in terms of the right-hand side $\varphi$, provided the stability condition (21) holds. Having stipulated this condition, estimate (20) is certainly true for a solution of problem (23) and takes for now the form

$$\left\| Y^{j,j'} \right\| \le \sqrt{\frac{1+\varepsilon}{\varepsilon}} \left\| Y_t^{j',j'} \right\|_{A^{-1}} = \frac{1}{\tau} \sqrt{\frac{1+\varepsilon}{\varepsilon}} \left\| Y^{j'+1,j'} \right\|_{A^{-1}}.$$

By the triangle inequality we deduce from (22) that

$$\| y^j \| \le \sqrt{\frac{1+\varepsilon}{\varepsilon}} \sum_{j'=0}^{j-1} \left\| Y^{j'+1,j'} \right\|_{A^{-1}}.$$

Equation (27) is needed in obtaining a bound of $\left\| Y^{j'+1,j'} \right\|_{A^{-1}}$, in which $\Phi$ and $\varphi$ should be expanded in the series with respect to the eigenfunctions of the system $\{X^{(k)}\}$:

$$\Phi = \sum_{k=1}^{N-1} \Phi_k X^{(k)}, \qquad \varphi = \sum_{k=1}^{N-1} \varphi_k X^{(k)}. \tag{28}$$

Substituting (28) into (27) gives $\Phi_k = \varphi_k (1 + \sigma\tau^2 \lambda_k)^{-1}$. With these relations in view, we establish for $\sigma \ge 0$

$$\| \Phi \|_{A^{-1}}^2 = \sum_{k=1}^{N-1} \frac{\Phi_k^2}{\lambda_k} = \sum_{k=1}^{N-1} \frac{\varphi_k^2}{(1 + \sigma \tau^2 \lambda_k)^2 \lambda_k} \le \sum_{k=1}^{N-1} \frac{\varphi_k^2}{\lambda_k} = \| \varphi \|_{A^{-1}}^2,$$

meaning

$$\left\| Y^{j'+1,j'} \right\|_{A^{-1}} \le \tau \, \| \varphi^{j'} \|_{A^{-1}}.$$

Thus, if $\sigma \ge 0$ and condition (21) is fulfilled, the estimate holds for scheme (4b):

$$\| y^j \| \le \sqrt{\frac{1+\varepsilon}{\varepsilon}} \sum_{j'=0}^{j-1} \tau \, \| \varphi^{j'} \|_{A^{-1}}.$$

By exactly the same reasoning as before we deduce for problem (4) with zero boundary conditions $y_0 = y_N = 0$ that

$$\| y^j \| \le \sqrt{\frac{1+\varepsilon}{\varepsilon}} \left( \| y^0 \| + \| y_t^0 \|_{A^{-1}} + \sum_{j'=0}^{j-1} \tau \, \| \varphi^{j'} \|_{A^{-1}} \right)$$

under condition (21) and the restriction $\sigma \ge 0$. Intuition suggests and in this case it does not deceive us that for a special choice $y^1 = y(\tau)$ the stability of scheme (15) under the constraint $\tau \le h$ may be proved in the space $L_2$.

We learn from Section 5.1 that the difference scheme

$$(29) \qquad y_{\bar{t}t}^j = y_{\bar{x}x}^j \,, \qquad j = 1, 2, \ldots,$$

$$(30) \qquad y^0 = u_0 \,, \qquad y_t^0 = \bar{u}_0 + \tfrac{1}{2} \tau \, y_{\bar{x}x}^0 \,.$$

provides an approximation of $O(h^2 + \tau^2)$ to equation (1)–(2). Other ideas are connected with expressions of the solution $y^j$ in terms of $y^0 = u_0$ and $\bar{u}_0$. Following established practice, we find that

$$(31) \qquad y^j = \sum_{k=1}^{N-1} \left( u_{0k} \cos j\varphi_k + \frac{\tau \, \bar{u}_{0k}}{\sin \varphi_k} \sin j\varphi_k \right) X^{(k)} \,,$$

where $\bar{u}_{0k}$ are Fourier coefficients of $\bar{u}_0(x)$, the sense of the quantities $u_{0k}$ and $\varphi_k$ being reserved. Squaring (31) and applying the estimate

$$2 \left( \frac{\tau \, \bar{u}_{0k}}{\sin \varphi_k} \cos j\varphi_k \right) u_{0k} \sin j\varphi_k \le \tau^2 \frac{\bar{u}_{0k}^2}{\sin^2 \varphi_k} \cos^2 j\varphi_k + u_{0k}^2 \sin^2 j\varphi_k \,,$$

we obtain

$$\| y^j \|^2 \le \| u_0 \|^2 + \tau^2 \sum_{k=1}^{N-1} \frac{\bar{u}_{0k}^2}{\sin^2 \varphi_k} \,.$$

To evaluate the lower bound of the expression

$$\frac{\tau^2}{\sin^2 \varphi_k} = \frac{1}{\lambda_k \, (1 - \frac{1}{4} \tau^2 \lambda_k)} \,,$$

we take $\gamma = \tau / h \le 1$. Along these lines, it is evident that

$$\begin{aligned} \lambda_k \left( 1 - \frac{\tau^2 \lambda_k}{4} \right) &= \frac{4}{h^2} \sin^2 \frac{\pi k h}{2} \left( 1 - \gamma^2 \sin^2 \frac{\pi k h}{2} \right) \\ &\ge \frac{4}{h^2} \sin^2 \frac{\pi k h}{2} \left( 1 - \sin^2 \frac{\pi k h}{2} \right) \\ &= \frac{4}{h^2} \sin^2 \frac{\pi k h}{2} \cos^2 \frac{\pi k h}{2} \\ &\ge \frac{4}{h^2} \sin^2 \frac{\pi h}{2} \cos^2 \frac{\pi h}{2} = \frac{\sin^2 \pi h}{h^2} \,. \end{aligned}$$

In Chapter 2 it has been already shown that $4\,h_1^{-2}\sin^2(\frac{1}{2}\pi h_1) \geq 8$ if $h_1 \leq \frac{1}{2}$. Therefore, keeping $h_1 = 2h$ we find that

$$\frac{\sin^2 \pi h}{h^2} = \frac{4}{h_1^2}\sin^2\frac{\pi h_1}{2} \geq 8\,, \qquad h \leq \frac{1}{4}\,.$$

Thus, if $\tau \leq h$ and $h \leq 1/4$ the proper estimate for the solution of problem (29)–(30) is

$$\| y^j \|^2 \leq \| u_0 \|^2 + \tfrac{1}{8}\| \bar{u}_0 \|^2.$$

**3. The energy inequality method.** An investigation of difference schemes for the string vibration equation may be carried out by means of the energy inequality method (see Section 1). Here we restrict ourselves to stability with respect to the initial data with regard to the problem

$$\begin{aligned} &y_{\bar{t}t} = \Lambda\,(\sigma\hat{y} + (1-2\sigma)\,y + \sigma\check{y})\,, \\ &y_0 = y_N = 0\,, \qquad y(x,\,0) = u_0(x)\,, \qquad y_t(x,\,0) = \tilde{u}_0(x)\,. \end{aligned} \tag{32}$$

Bearing in mind that $\sigma\,\hat{y} + (1-2\,\sigma)\,y + \sigma\,\check{y} = y + \sigma\,\tau^2\,y_{\bar{t}t}$ and attempting equation (32) in the form

$$(E - \sigma\,\tau^2\,\Lambda)\,y_{\bar{t}t} = \Lambda\,y \tag{33}$$

where $E$ is the identity operator, we take the inner product of the resulting equation (33) and the quantity $y_{\overset{\circ}{t}} = (y_t + y_{\bar{t}})/2$. The outcome of this is

$$((E - \sigma\tau^2\Lambda)\,y_{\bar{t}t}\,,\, y_{\overset{\circ}{t}}) = (\Lambda y,\, y_{\overset{\circ}{t}})\,. \tag{34}$$

The trivial identities

$$(y_{\bar{t}t}\,,\, y_{\overset{\circ}{t}}) = \frac{1}{2}\,(\| y_{\bar{t}} \|^2)_t\,, \qquad -(\Lambda y_{\bar{t}t},\, y_{\overset{\circ}{t}}) = (y_{\bar{x}\bar{t}t}\,,\, y_{\bar{x}\overset{\circ}{t}}\,] = \frac{1}{2}\,(\| y_{\bar{x}\bar{t}} ]|^2)_t$$

help rearrange the left-hand side of equality (34) as

$$((E - \sigma\,\tau^2\,\Lambda)\,y_{\bar{t}t}\,,\, y_{\overset{\circ}{t}}) = \frac{1}{2}\,(\| y_{\bar{t}} \|^2 + \sigma\,\tau^2\,\| y_{\bar{t}\bar{x}} ]|^2)_t\,. \tag{35}$$

Further, we will show that for any function $y = y(x,\,t_n)$ vanishing at the points $x = 0$ and $x = 1$

$$-(\Lambda y,\, y_{\overset{\circ}{t}}) = \frac{1}{8}\,(\| y_{\bar{x}} + \check{y}_{\bar{x}} ]|^2)_t - \frac{\tau^2}{8}\,(\| y_{\bar{t}\bar{x}} ]|^2)_t\,. \tag{36}$$

Indeed, from the first Green formula (see Chapter 2, Section 3) it follows that

$$-(\Lambda y,\, y_{\overset{\circ}{t}}) = (v,\, v_{\overset{\circ}{t}}\,]$$

with $v = y_{\bar{x}}$. Having involved in the further development

$$v \cdot v_{\overset{\circ}{t}} = \frac{1}{8}\left((v+\check{v})^2\right)_t - \frac{\tau^2}{8}\left((v_{\bar{t}})^2\right)_t ,$$

we arrive at identity (36).

Substituting (35) and (36) into (34) yields the energy identity

$$\text{(37)} \qquad \left(\|\, y_{\bar{t}}\,\|^2 + \left(\sigma - \tfrac{1}{4}\right)\tau^2 \,\|\, y_{\bar{t}\bar{x}}\,]|^2 + \tfrac{1}{4}\,\|\, y_{\bar{x}} + \check{y}_{\bar{x}}\,]|^2\right)_t = 0$$

or $\mathcal{E}^{j+1} = \mathcal{E}^{j}$, where

$$\text{(38)} \qquad \mathcal{E}^{j} := \|\, y_{\bar{t}}^{j}\,\|^2 + \left(\sigma - \tfrac{1}{4}\right)\tau^2 \,\|\, y_{\bar{t}\bar{x}}^{j}\,]|^2 + \tfrac{1}{4}\,\|\, y_{\bar{x}}^{j} + \check{y}_{\bar{x}}^{j-1}\,]|^2 .$$

Let us find the values of $\sigma$, for which the quantity $\mathcal{E}^{j}$ is nonnegative when $y^j$ and $y^{j-1}$ are arbitrarily taken, by simple observations (see Chapter 2, Section 3) that

$$\|\, y_{\bar{t}\bar{x}}\,]|^2 \le \frac{4}{h^2}\,\|\, y_{\bar{t}}\,\|^2$$

and

$$\|\, y_{\bar{t}}\,\|^2 + \left(\sigma - \tfrac{1}{4}\right)\tau^2 \,\|\, y_{\bar{t}\bar{x}}\,]|^2 \ge \left(\tfrac{1}{4}\,h^2 + \left(\sigma - \tfrac{1}{4}\right)\tau^2\right)\|\, y_{\bar{t}\bar{x}}\,]|^2 .$$

With this in mind, we conclude that the right-hand side of (38) is nonnegative if

$$\text{(39)} \qquad \sigma \ge \frac{1}{4} - \frac{1}{4\gamma^2}\,, \qquad \gamma = \frac{\tau}{h}\,.$$

Here the expression $\left(\mathcal{E}^{j}\right)^{1/2} = \|\, y^j\,\|_*$ can be viewed as a norm (or, more exactly, as a seminorm), permitting us to write down

$$\text{(40)} \qquad \mathcal{E}^{j} = \|\, y^j\,\|_*^2 = \|\, y_{\bar{t}}^{j}\,\|^2 + \left(\sigma - \tfrac{1}{4}\right)\tau^2 \,\|\, y_{\bar{t}\bar{x}}^{j}\,]|^2 + \tfrac{1}{4}\,\|\, y_{\bar{x}}^{j} + y_{\bar{x}}^{j-1}\,]|^2 .$$

Note that such combined norms depending on values of $y$ on several layers are typical for multilayer schemes. This is especially true for three-layer schemes.

The identity (37) implies the stability with respect to the initial data in the norm (40): $\| y^{j+1} \|_* = \| y^0 \|_*$ for all $j = 0, 1, \ldots$.

So, condition (39) is sufficient for the stability of scheme (32) with respect to the initial data in the norm (40). In particular, scheme (32) with $\sigma = 0$ is stable with respect to the initial data under the condition

$$\tau \le h \,. \tag{41}$$

Often this stability condition is named the **Courant condition** because it has been proved for the first time by R. Courant, C. Friedrichs and G. Levy in 1928.

For the equation in the dimensional variables

$$\frac{\partial^2 u}{\partial t^2} = a^2 \, \frac{\partial^2 u}{\partial x^2} \tag{42}$$

condition (41) takes the form $\tau \le h/a$, where $a$ is the sound velocity.

**4. Determination of nonsmooth solutions by the difference method.** Numerous problems of mathematical physics describing shock processes in gases, liquids and solids lead to the problem of determining nonsmooth solutions of second-order hyperbolic equations, the simplest of which is equation (42) of vibrations of a string. Since those solutions do not possess the second-order derivatives involved in the equation, the words "a solution satisfies the equation" should be understood in some generalized sense. One of the possible definitions of generalized solutions is due to the fact that the differential equation follows from an integral conservation law if continuous derivatives emerged in equation (42) exist. In this case the generalized solution is meant as a function $u(x,t)$ having in the domain $\bar{G} = \{0 \le x \le l,\, 0 \le t \le T\}$ the bounded piecewise continuous derivatives $\partial u/\partial x$, $\partial u/\partial t$ and satisfying the integral equation

$$\oint_C \left( \frac{\partial u}{\partial t} \, dx + a^2 \, \frac{\partial u}{\partial x} \, dt \right) = 0 \,,$$

where $C$ is an arbitrary closed curve in the domain $\bar{G}$. If the first derivatives are discontinuous, then for the characteristics $x \pm at = \text{const}$ the jump conditions should be satisfied:

$$\left[ \frac{\partial u}{\partial t} \right] = \pm a \left[ \frac{\partial u}{\partial x} \right] , \qquad [f] := f(\xi + 0) - f(\xi - 0) \quad \text{as} \quad \xi = x \pm at \,.$$

In giving a generalized solution of the problem

$$
(43)\qquad
\begin{aligned}
&\frac{\partial^2 u}{\partial t^2}dx = a^2\,\frac{\partial^2 u}{\partial x^2}0\,, \qquad 0 < x < l\,, \quad t > 0\,,\\
&u(x,0) = u_0(x), \quad \frac{\partial u}{\partial t}(x,0) = u_1(x), \quad u(0,t) = 0, \quad u(l,t) = 0\,,
\end{aligned}
$$

we rely on the scheme with weights

$$
(44)\qquad y_{\bar{t}t} = \Lambda(\sigma\,\hat{y} + (1-2\sigma)\,y + \sigma\,\check{y})\,, \qquad \Lambda\,y = a^2\,y_{\bar{x}x}\,,
$$

with the appropriate supplementary conditions. When studying the convergence of the scheme with weights, it is supposed that a solution of problem (43) exists and is smooth enough. It is possible under certain conditions relating to the smoothness of the initial data.

Does the same scheme converge if $u = u(x,t)$ is a generalized solution? The grid solution of problem (44) turns out to converge to the generalized solution with the rate $O(\sqrt{\sqrt{h}+\tau})$. We do not dwell on confirming this statement.

In trying to find a generalized solution of problem (43) one can come across oscillations of the grid solution and its derivatives ("ripple"), which essentially reduce the accuracy of a scheme. What is more, the lines of discontinuity of derivatives spread over several grid intervals, thus causing the difficulties in determination of proper discontinuity propagation velocity. This is the result of introducing the fictitious friction (dissipation) in the difference approximation.

The ripple is stipulated by the fact that difference harmonics reveal the dispersion, that is, determination of a harmonic velocity depends on its number, whereas for the differential equation all harmonics have the same velocity $a$. In order to improve the quality of a scheme, one needs to minimize the dispersion. Among various schemes (44) with weights the scheme relating to

$$
\sigma = \sigma_* = \frac{1}{12}\left(1 - \frac{1}{\gamma^2}\right), \qquad \gamma = \frac{a\tau}{h}\,,
$$

has the minimal dispersion and allows us to overcome the obstacles mentioned above. It is of order 4, that is, $O(\tau^4 + h^4)$ on sufficiently smooth solutions $u = u(x,t)$. On nonsmooth generalized solutions the approximation error for scheme (44) with the weight $\sigma = \sigma_*$ is as large as for schemes with any weight $\sigma \neq \sigma_*$. However, due to reduced dispersion the scheme with the weight $\sigma = \sigma_*$ is more precise and reproduces much more better the characteristic features of generalized solutions.

A stability condition established for scheme (44) such as

$$\sigma \geq \frac{1}{4} - \frac{1}{4\,\gamma^2} = \frac{1}{4}\left(1 - \frac{1}{\gamma^2}\right)$$

is certainly true for the scheme with the weight $\sigma = \sigma_*$ if $\gamma \leq 1$ or $\tau \leq h/a$, that is, under the same condition as for the explicit scheme. In this context,

$$\sigma_* = \frac{1}{12}\left(1 - \frac{1}{\gamma^2}\right) \leq 0\,.$$

Let us stress here, that in an attempt to relax the "ripple" by introducing the viscosity, the distortion of the solution profile and accuracy losses occurred.

## 5.7 SELECTED PROBLEMS

**1.** For the heat conduction equation the difference scheme is suggested:

$$y^j_{t,\,i} = \frac{1}{2}\,(y^{j+1}_{\bar{x}x,\,i} + y^j_{\bar{x}x,\,i})\,, \qquad i = 1, 2, \ldots, N-1\,,$$

$$y^j_0 = y^0_N = 0\,, \qquad y^0_i = u_0(x_i)\,.$$

Prove its absolute stability, find the order of approximation and point out the method for solving the problem.

**2.** Find the order of approximation for the difference scheme

$$\frac{y^{j+1}_i - y^j_i}{\tau} = \frac{1}{2}\left(y^{j+1}_{\bar{x}x,\,i} + y^j_{\bar{x}x,\,i}\right) - \frac{h^2}{12}\,y^j_{t\,\bar{x}\,x,\,i}\,,$$

$$i = 1, 2, \ldots, N-1\,, \qquad y^j_0 = y^j_N = 0\,, \qquad y^0_i = u_0(x_i)\,,$$

prove its absolute stability and investigate the stability of the elimination method being used for determination of

$$y^{j+1}_i\,, \qquad i = 1, 2, \ldots, N-1\,.$$

**3.** The statement of the problem is

$$\frac{\partial u}{\partial t} = \frac{\partial^2 u}{\partial x^2}, \qquad 0 < x < 1, \qquad t > 0,$$

$$\frac{\partial u}{\partial x}(0, t) = \mu_1(t), \qquad u(1, t) = \mu_2(t),$$

$$u(x, 0) = u_0(x).$$

Construct a perfect difference scheme of accuracy $O(\tau^2 + h^2)$.

**4.** Find stability conditions for the difference scheme

$$i\, y_{t,j} = \sigma\, y^{n+1}_{\bar{x}x,j} + (1-\sigma)\, y^n_{\bar{x}x,j} - q\,(\mu\, y^{n+1}_j + (1-\mu)\, y^n_j),$$

$$j = 1, 2, \ldots, N-1, \qquad y^n_0 = y^n_N = 0, \qquad y^0_j = u_0(x_j),$$

$$\sigma = \sigma_0 + i\,\sigma_1, \qquad \mu = \mu_0 + i\,\mu_1,$$

approximating the equation

$$i\,\frac{\partial u}{\partial t} = \frac{\partial^2 u}{\partial x^2} - q\,u, \qquad q = \text{const} > 0,$$

$$u(0, t) = u(1, t) = 0,$$

$$u(x, 0) = u_0(x).$$

**5.** Determine the exact solutions of the boundary-value problem

$$\frac{\partial u}{\partial t} + \frac{\partial u}{\partial x} = 0, \qquad 0 < x \le X, \qquad 0 < t \le T,$$

$$u(x, 0) = u_0(x), \qquad u(0, t) = u_1(t),$$

and of the difference scheme approximating this problem:

$$y_{t,i} + y^n_{\bar{x},i} = 0, \qquad i = 1, 2, \ldots, \qquad n = 0, 1, 2, \ldots,$$

$$y^0_i = u_0(x_i), \qquad y^n_0 = u_1(t_n),$$

$$x_i = i\,h, \qquad t_n = n\,\tau.$$

Proceeding from the explicit representation of solutions, show that if the condition $\gamma = \tau/h = \text{const}$ is fulfilled as $\tau \to 0$, $h \to 0$, then convergence occurs only for $\gamma \le 1$ and the difference problem solution coincides with the differential problem solution for $\gamma = 1$.

**6.** The problem

$$\frac{\partial u}{\partial t} = \frac{\partial^2 u}{\partial x_1^2} + \frac{\partial^2 u}{\partial x_2^2}, \qquad 0 < x_1 < l_1, \quad 0 < x_2 < l_2,$$

$$u(\mathbf{x}, 0) = 0, \qquad \mathbf{x} \in G = \{0 < x_1 < l_1, 0 < x_2 < l_2\},$$

$$u(\mathbf{x}, t) = 0, \qquad \mathbf{x} \in \partial G, \quad t \geq 0,$$

is approximated by the explicit difference scheme on the grid

$$\omega_{h\tau} = \omega_h \times \omega_\tau, \qquad \omega_\tau = \{t_n = n\tau, \ \tau > 0\},$$
$$\omega_h = \{x_1^{(i)}, x_2^{(j)}\}, \qquad x_1^{(i)} = i h_1, \qquad x_2^{(j)} = j h_2,$$
$$i = 1, 2, \dots, N_1 - 1, \quad j = 1, 2, \dots, N_2 - 1, \qquad h_1 N_1 = l_1, \ h_2 N_2 = l_2.$$

What maximal step $\tau$ in time should be taken to provide stability of the scheme when $l_1 = 1$, $l_2 = 10$, $N_1 = 10$, $N_2 = 100$?

**7.** Prove for the problem

$$\frac{\partial u}{\partial t} = \frac{\partial u}{\partial x}, \qquad t > 0, \qquad -\infty < x < 0,$$

$$u(x, 0) = u_0(x), \qquad x \leq 0,$$

$$u(0, t) = 0, \qquad t \geq 0,$$

that the difference scheme

$$\frac{y_i^{j+1} - y_i^j}{\tau} = \frac{y_i^j - y_{i-1}^j}{h}$$

is absolutely unstable and the scheme

$$\frac{y_i^{j+1} - y_i^j}{\tau} = \frac{y_{i+1}^j - y_i^j}{h}$$

is stable under the condition $\tau \leq h$.

**8.** Show that for any $\tau$ and $h$ a pure implicit difference scheme (a forward difference scheme) approximating the problem

$$\frac{\partial u}{\partial t} = \frac{\partial^2 u}{\partial x^2}, \qquad 0 < x < 1, \qquad t > 0,$$

$$u(x, 0) = u_0(x),$$

$$\frac{\partial u}{\partial x}(0, t) = \frac{\partial u}{\partial x}(1, t) = 0,$$

is not asymptotically stable.

**9.** For the three-layer scheme

$$y_{\bar{t}t} = \sigma_1 \, y_{\bar{x}x}^{n+1} + (1 - \sigma_1 - \sigma_2) \, y_{\bar{x}x}^{n} + \sigma_2 \, y_{\bar{x}x}^{n-1}$$

a) investigate the approximation error, b) find a stability condition, c) impose the initial conditions $y_i^0$, $y_i^1$, $i = 0, 1, \ldots, N$, not changing the order of approximation for this proposal.

**10.** What differential equation is approximated by the difference equation

$$(1 + 2\gamma) \, y_i^{n+1} = (1 - 2\gamma) \, y_i^{n-1} + 2\gamma \, (y_{i+1}^{n} + y_{i-1}^{n}) \, ,$$

where $\gamma = \tau / h^2$. Find the order of approximation for this.

# 6

# Stability Theory of Difference Schemes

In this chapter we study the stability with respect to the initial data and the right-hand side of two-layer and three-layer difference schemes that are treated as operator-difference schemes with operators in Hilbert space. Necessary and sufficient stability conditions are discovered and then the corresponding a priori estimates are obtained through such an analysis by means of the energy inequality method. A regularization method for the further development of various difference schemes of a desired quality (in accuracy and economy) in the class of stability schemes is well-established. Numerous concrete schemes for equations of parabolic and hyperbolic types are available as possible applications, bring out the indisputable merit of these methods and unveil their potential.

## 6.1 OPERATOR–DIFFERENCE SCHEMES

**1. Introduction.** In Section 4 of Chapter 2 the boundary-value problems for the differential equations $Lu = -f(x)$ have been treated as the operator equations $Au = f$, where $A$ is a linear operator in a Banach space $\mathcal{B}$.

In the study of nonstationary processes described by partial differential equations of parabolic and hyperbolic types

$$\frac{\partial u}{\partial t} = L u + f(x,t)\,, \qquad \frac{\partial^2 u}{\partial t^2} = L u + f(x,t)\,, \qquad 0 < t \le t_0\,,$$

the variable $t$ (time) plays a key role and, following established practice, should be marked out throughout the entire chapter. Here $L$ is a differential operator acting on $u(x,t)$ as a function of a point $x = (x_1, x_2, \dots, x_p)$ in some $p$-dimensional domain $G$. For any fixed $t$, the function $u(x,t)$ is an element of the Banach space $\mathcal{B}$. Therefore, instead of $u(x,t)$ we obtain an abstract function $u(t)$ of the variable $t$, $0 \le t \le t_0$, with the values in the space $\mathcal{B}$; meaning $u(t) \in \mathcal{B}$ for all $t \in [0, t_0]$. The operator $L$ acting on $u(x,t)$ as a function of the variable $x$ is replaced by an operator $\mathcal{A}$ acting in the space $\mathcal{B}$. The operator $\mathcal{A}$ generally acts from a space $\mathcal{B}_1$ into a space $\mathcal{B}_2$ (its domain $\mathcal{D}(\mathcal{A}) \subset \mathcal{B}_1$ is everywhere dense in $\mathcal{B}_1$, while its range $\mathcal{R}(\mathcal{A}) \subseteq \mathcal{B}_2$). In this regard, we take for granted that $\mathcal{B}_1 = \mathcal{B}_2 = \mathcal{B}$, making it possible to set up the abstract Cauchy problem

$$\frac{du}{dt} + \mathcal{A}\, u = f(t), \qquad 0 \le t \le t_0, \qquad u(0) = u_0,$$

where $u_0$ is a given element from the domain $\mathcal{D}(\mathcal{A})$.

The above reasoning is of a heuristic nature and is aimed at carrying out some analogy between the methods of the general theory of differential equations and those of the theory of difference schemes, the framework and methodology of which are outlined in this chapter.

A Cauchy problem is said to be **stable with respect to the initial data and right-hand side** if

$$\| u(t) \| \le M_1 \| u_0 \| + M_2 \int_0^t \| f(t') \| \, dt',$$

where $M_1 = \text{const} > 0$ and $M_2 = \text{const} > 0$.

In conformity with the superposition principle ($\mathcal{A}$ is a linear operator), the stability of the Cauchy problem with respect to the right-hand side follows from the uniform stability with respect to the initial data

$$\| u(t) \| \le M_1 \| u(t') \|, \qquad t \ge t' \ge 0,$$

where $u(t)$ is a solution of the homogeneous equation.

**2. Operator-difference schemes.** We now consider a linear system $\mathcal{B}_h$ depending on a parameter $h$ as a vector of some normed space equipped with the norm $|h|$. With regard to the linear system $\mathcal{B}_h$, it is reasonable to introduce a collection of norms $\| \cdot \|_h$, $\| \cdot \|_{(1_h)}$, $\| \cdot \|_{(2_h)}$, ..., thus causing linear normed spaces $\mathcal{B}_h$, $\mathcal{B}_h^{(1)}$, $\mathcal{B}_h^{(2)}$, .... For the sake of simplicity, we involve in further development the norms $\|\cdot\|_{(1_h)}$, $\|\cdot\|_{(2_h)}$, ... on the space $\mathcal{B}_h$, assuming $\|\cdot\|_h$ to be a basic one on $\mathcal{B}_h$.

At the next stage we introduce on the interval $0 \le t \le t_0$ a $\tau$-step grid $\bar{\omega}_\tau = \{t_j = j\tau,\ j = 0, 1, \dots, j_0,\ \tau = t_0/j_0\}$, $\omega_\tau = \{t_j = j\tau,\ 0 < j \le j_0\}$ and pass to abstract $\mathcal{B}_h$-valued functions $y_{h\tau}^{(t)}$, $\varphi_{h\tau}^{(t)}$, etc. of one discrete argument $t = j\tau \in \bar{\omega}_\tau$, so that $y_{h\tau}^{(t)} \in \mathcal{B}_h$ for all $t = j\tau \in \bar{\omega}_\tau$. Let $A_{h\tau}^{(t)}$, $B_{h\tau}^{(t)}$, $C_{h\tau}^{(t)}$, ... be linear operators dependent on the parameters $h$ and $\tau$ and acting from $\mathcal{B}_h$ into $\mathcal{B}_h$ for every $t \in \bar{\omega}_\tau$. For the moment, omitting the subscripts $h$ and $\tau$ we can write down $y_n = y(n\tau) = y(t_n) = y$, $A(t)$, $B(t)$ and $C(t)$. This should cause no confusion.

We call a family of difference equations of order $r-1$

$$B_0(t_n)\, y_{n+1} = \sum_{s=1}^{r-1} C_s(t_n)\, y_{n+1-s} + f_n\,, \qquad n = r-2,\ r-1,\ r,\ \dots,$$

depending on the parameters $h$ and $\tau$ with operator coefficients $B_0$, $C_1$, ..., $C_{r-1}$, which are linear operators acting in the space $\mathcal{B}_h$ and dependent on $h$ and $\tau$, an $r$-layer operator-difference scheme or simply **r-layer scheme**. If the inverse operator $B_0^{-1}$ exists, a solution $y_{n+1}$ of this problem can be expressed in terms of the initial vectors $y_0$, $y_1$, ... and the right-hand side $f$. As usual, we assume all the vectors to be given and consider only two-layer and three-layer schemes

$$(1) \qquad B_0\, y_{n+1} + B_1\, y_n = \tau\, \varphi_n\,, \qquad n = 0,\ 1,\ \dots, \qquad y_0 \ \text{ given},$$

$$(2) \qquad B_0\, y_{n+1} + B_1\, y_n + B_2\, y_{n-1} = \tau\, \varphi_n\,, \quad y_0 \ \text{ and } \ y_1 \ \text{ given}.$$

**3. The canonical form of two-layer schemes.** Any two-layer scheme (1) can be written in the form

$$(3) \qquad B(t_n)\, \frac{y_{n+1} - y_n}{\tau} + A(t_n)\, y_n = \varphi_n\,,$$

$$n = 0,\ 1,\ \dots, \qquad y_0 \in \mathcal{B}_h \ \text{ given}.$$

Indeed, by comparing (1) with (3) we see that $B = B_0$ and $A = \big(B_0 + B_1\big)/\tau$. Within more compact notations

$$y = y_n = y(t_n)\,, \qquad \hat{y} = y_{n+1} = y(t_{n+1}) = y(t_n + \tau)\,,$$

$$\check{y} = y_{n-1}\,, \quad y_t = \frac{\hat{y} - y}{\tau}\,, \quad y_{\bar{t}} = \frac{y - \check{y}}{\tau}\,,$$

equation (3) can be recast as

(4) $B\, y_t + A\, y = \varphi(t)\,, \qquad t = t_n = n\,\tau \in \omega_\tau\,, \qquad y(0) = y_0 \in \mathcal{B}_h\,.$

We call both equations (3) and (4) the **canonical form of two-layer schemes**. Equation (4) is similar to the differential equation

$$\mathcal{B}\,\frac{d\,u}{d\,t} + \mathcal{A}\,u = f(t)\,.$$

**Example 1** For the heat conduction equation

$$\frac{\partial\,u}{\partial\,t} = L\,u + f\,, \qquad L\,u = \frac{\partial}{\partial\,x}\left(k(x)\,\frac{\partial\,u}{\partial\,x}\right),$$

we made in Chapter 5 the design of the two-layer scheme with weights

$$y_t = \Lambda\left(\sigma\,\hat{y} + (1-\sigma)\,y\right) + \varphi\,, \qquad \Lambda\,v = \left(a(x)\,v_{\bar{x}}\right)_x\,.$$

With the aid of the the identity

$$\hat{y} = y + \tau\,\frac{\hat{y} - y}{\tau} = y + \tau\,y_t$$

we reduce it to

$$y_t - \sigma\,\tau\,\Lambda\,y_t - \Lambda\,y = \varphi\,.$$

Comparison of this equation with equation (4) gives

$$B = E + \sigma\,\tau\,A\,, \qquad A = -\Lambda\,.$$

This form of writing reveals what is meant by the canonical form of a weighted two-layer scheme.

At the next stage we proceed to solve equation (4) with respect to $\hat{y} = y_{n+1}$. If the inverse operator $B^{-1}$ exists, one can write down

(5) $\hat{y} = S\,y + \tau\,\tilde{\varphi}\,, \qquad S = E - \tau\,B^{-1}A\,, \qquad \tilde{\varphi} = B^{-1}\varphi\,.$

The operator $S$ is called the **transition operator** (from one layer to another). In addition to the canonical form (4), alternative forms of writing will appear in the sequel for two-layer schemes: $B\,\hat{y} = C\,y + \tau\,\varphi$ or $B\,y_{n+1} = C\,y_n + \tau\,\varphi_n$, where $C = E - \tau\,A$, $E$ being the identity operator. In the case $B = E$, scheme (3) is called an **explicit two-layer scheme**:

$$\frac{y_{n+1} - y_n}{\tau} + A\,y_n = \varphi_n\,,$$

permitting us to determine the value $y_{n+1}$ on the upper layer by the formula $y_{n+1} = y_n - \tau\,A\,y_n + \tau\,\varphi_n$. If $B \neq E$, then scheme (3) is called an **implicit two-layer scheme**.

**4. The canonical form of three-layer schemes.** We may attempt the three-layer scheme (2) in the canonical form

$$B\,\frac{y_{n+1}-y_{n-1}}{2\,\tau}+R\,(\,y_{n+1}-2\,y_n+y_{n-1}\,)+A\,y_n\;=\varphi\,. \tag{6}$$

By comparing (6) with (2) we see that such a writing is always possible if we agree to consider

$$B=B_0-B_2\,,\qquad R=\frac{1}{2\,\tau}\,(\,B_0+B_2\,)\,,\qquad A=\frac{1}{\tau}\,(\,B_0+B_1+B_2\,)\,. \tag{7}$$

Also, it will be sensible to introduce the notations

$$y_{\overset{\circ}{t}}\;=\;\frac{\hat y-\check y}{2\,\tau}\,,\qquad\qquad y_{\bar t\,t}\;=\;\frac{y_t-y_{\bar t}}{\tau}\;=\;\frac{\hat y-2\,y+\check y}{\tau^2}$$

and regard the equation

$$\begin{aligned}&B\,y_{\overset{\circ}{t}}\;+\tau^2\,R\,y_{\bar t\,t}+A\,y\;=\;\varphi(t)\,,\\&0<t=n\,\tau\in\omega_\tau\,,\qquad y(0)=y_0\,,\qquad y(\tau)=y_1\,,\end{aligned} \tag{8}$$

to the canonical form of a three-layer scheme together with (6).

**Example 2** We now turn to the weighted three-layer scheme

$$y_{\overset{\circ}{t}}\;+A\,\Big(\,\sigma_1\,\hat y+(\,1-\sigma_1-\sigma_2\,)\,y+\sigma_2\,\check y\,\Big)\;=\;\varphi \tag{9}$$

and try to reduce it to the canonical form. With this aim, we make use of the formulae

$$\hat y\;=\;y+\frac{\hat y-\check y}{2}+\frac{\hat y-2\,y+\check y}{2}\;=\;y+\tau\,y_{\overset{\circ}{t}}\;+\;\frac{\tau^2}{2}\;y_{\bar t\,t}\,,$$

$$\check y\;=\;y-\frac{\hat y-\check y}{2}+\frac{\hat y-2\,y+\check y}{2}\;=\;y-\tau\,y_{\overset{\circ}{t}}\;+\;\frac{\tau^2}{2}\;y_{\bar t\,t}\,,$$

$$\sigma_1\,\hat y+(\,1-\sigma_1-\sigma_2\,)\,y+\sigma_2\,\check y\;=\;y+(\,\sigma_1-\sigma_2)\,\tau\,y_{\overset{\circ}{t}}\;+\;\frac{(\sigma_1+\sigma_2)\,\tau^2}{2}\;y_{\bar t\,t}\,.$$

Upon substituting these expressions into (9) we write the weighted scheme in the canonical form (8), where

$$B\;=\;E+\tau\,(\,\sigma_1-\sigma_2\,)\,A\,,\qquad\qquad R\;=\;\tfrac{1}{2}\,(\,\sigma_1+\sigma_2\,)\,A\,. \tag{10}$$

The passage from (6) or (8) to (2) leads to the equation for determining the value $y_{n+1}$ on the upper layer $t = t_{n+1}$:

$$(B + 2\tau R)\, y_{n+1} = 2\tau\,(2R - A)\, y_n + (B - 2\tau R)\, y_{n-1} + 2\tau \varphi_n ,$$

thereby clarifying that problem (8) is solvable if the inverse operator $(B + 2\tau R)^{-1}$ exists. Moreover, the value $y_{n+1}$ can be expressed in terms of the values $y_n$ and $y_{n-1}$ on the preceding two layers. Therefore, two initial vectors $y_0$ and $y_1$ (or $y_0 = y(0)$ and $\bar{y}_0 = y_t(0)$) are required to be given in such matters. If $B_0 = B + 2\tau R = E$ is the identity operator, then the three-layer scheme (8) is said to be **explicit**; we thus have for it

$$y_{n+1} = 2\tau\,(2R - A)\, y_n + (B - 2\tau R)\, y_{n-1} + 2\tau \varphi_n .$$

In the case when $B + 2\tau R \neq E$, scheme (8) is said to be **implicit**.

Sometimes it may be useful to have at our disposal together with the canonical form (8) the three-layer scheme in the form (2) or

$$B\, y_{\overset{\circ}{t}} + (E + \tau^2 R)\, y_{\bar{t}t} + A y = \varphi . \tag{11}$$

This equation is obtained from equation (8) by the formal substitution of $E/\tau^2 + R$ for $R$.

**5. The notion of stability.** The notion of stability for three-layer schemes is of our initial concern. By a two-layer scheme we mean a set of operator-difference equations (4) depending on the parameters $h$ and $\tau$. We pre-assumed here that the operators $A$ and $B$ are given on the entire space $\mathcal{B}_h$.

As a matter of fact, we will consider the set of solutions $\{y_{h\tau}(t)\}$ of Cauchy problem (4) dependent on the input data $\{\varphi_{h\tau}(t)\}$ and $\{y_{0h}\}$. Scheme (4) is said to be **well-posed** if for all sufficiently small $\tau \le \tau_0$ and $|h| \le h_0$

(1) a solution of problem (4) exists and is unique for any initial data $y_{0h} \in \mathcal{B}_h$ and right-hand sides $\varphi_{h\tau}(t) \in \mathcal{B}_h$ for all $t \in \bar{\omega}_\tau$;

(2) there are positive constants $M_1$ and $M_2$ independent of $h$ and $\tau$ and disregarding to the choice of $y_{0h} \in \mathcal{B}_h$ and $\varphi_{h\tau}(t) \in \mathcal{B}_h$ such that for any $t \in \bar{\omega}_\tau$ for a solution of problem (4) the estimate holds:

$$\| y_{h\tau}(t+\tau) \|_{(1_h)} \le M_1 \| y_{0h} \|_{(1^\circ_h)} + M_2 \max_{0 \le t' \le t} \| \varphi_{h\tau}(t') \|_{(2_h)} , \tag{12}$$

where $\|\cdot\|_{(1_h)}$, $\|\cdot\|_{(1^\circ_h)}$ and $\|\cdot\|_{(2_h)}$ are suitable norms on the space $\mathcal{B}_h$.

Inequality (12) expresses the property of continuous dependence which is uniform in $h$ and $\tau$ of the Cauchy problem (4) upon the input data. Here and below the meaning of this property is **stability**. A difference scheme is said to be **absolutely stable** if it is stable for any $\tau$ and $h$ (not only for all sufficiently small ones). It is fairly common to distinguish the notion of stability with respect to the initial data and that with respect to the right-hand side. Scheme (4) is said to be **stable with respect to the initial data** if a solution to the homogeneous equation

$$ (4a) \qquad B\,y_t + A\,y = 0\,, \qquad t = n\,\tau > 0\,, \qquad y(0) = y_0\,, $$

satisfies the estimate

$$ (12a) \qquad \|\,y_{h\tau}(t+\tau)\,\|_{(1_h)} \leq M_1\,\|\,y_{0h}\,\|_{(1_h^0)}\,. $$

Scheme (4) is said to be **stable with respect to the right-hand side** if for a solution to equation (4) with the zero initial conditon $y(0) = 0$

$$ (4b) \qquad B\,y_t + A\,y = \varphi\,, \qquad y(0) = 0\,, $$

the inequality holds:

$$ (12b) \qquad \|\,y_{h\tau}(t+\tau)\,\|_{(1_h)} \leq M_2 \max_{0 \leq t' \leq t} \|\,\varphi_{h\tau}(t')\,\|_{(2_h)}\,. $$

For later use, we approve for a solution $y$ of problem (4) the decomposition $y = y^{(1)} + y^{(2)}$, where $y^{(1)}$ is a solution of problem (4a) and $y^{(2)}$ is a solution of problem (4b). On the strength of the triangular inequality

$$ \|\,y_{h\tau}(t+\tau)\,\|_{(1_h)} \leq \|\,y_{h\tau}^{(1)}(t+\tau)\,\|_{(1_h)} + \|\,y_{h\tau}^{(2)}(t+\tau)\,\|_{(1_h)} $$

the combination of estimates (12a) and (12b) gives estimate (12). In a similar way it makes sense to introduce the notion of stability for three-layer schemes. However, in this case we are to consider the pair of vectors $Y_{n+1} = \{\,y_n\,,\,y_{n+1}\,\}$ with the norm of the special type

$$ (13) \qquad \|\,Y_{n+1}\,\|_{(1_h)}^2 = \|\,y_n + y_{n+1}\,\|_{(1_h^*)}^2 + \|\,y_{n+1} - y_n\,\|_{(1_h^{**})}^2\,, $$

where $\|\cdot\|_{(1_h^*)}$ and $\|\cdot\|_{(1_h^{**})}$ are suitable norms on the space $\mathcal{B}_h$. Various norms of the type (13) appear, time and again, in the stability analysis of three-layer schemes by means of the energy inequality method (for more detail see Chapter 5, Section 6).

Thus, the three-layer scheme (8) is said to be **stable** if for any initial data $y_0$, $y_1$ and for any right-hand side $\varphi(t)$ its solution satisfies the estimate

$$(14)\quad \| Y_{h\tau}(t+\tau)\|_{(1_h)} \le M_1 \| Y_{h\tau}(\tau)\|_{(1_h^0)} + M_2 \max_{0<t'\le t} \| \varphi_{h\tau}(t')\|_{(2_h)} ,$$

where $M_1$ and $M_2$ are positive constants independent of $h$ and $\tau$ and disregarding to the choice of $y_0$, $y_1$ and $\varphi(t)$.

The main problem under consideration amounts to the following one. Suppose that equation (4) is uniquely solvable with respect to $y_{n+1}$ for any $y_n$ and $\varphi(t)$. What are conditions on the operators $A$ and $B$ for a scheme to be stable in the sense of the above definition? In other words, we wish to establish sufficient stability conditions for scheme (4) and obtain a priori estimates of the form (12). Moreover, sufficient conditions should be convenient for practical verifications in the case of concrete difference schemes associated with equations of mathematical physics.

In what follows stability of differential schemes will be given special investigation irrelevant to approximation and convergence.

**6. Sufficient stability conditions for two-layer schemes in linear normed spaces.** We now raise the question concerning sufficient stability conditions for two-layer schemes in linear normed spaces. In full details these investigations will be carried out in Section 2 for the case when $\mathcal{B}_h = H_h$ is a real Hilbert space.

In what follows the Cauchy problem (1) is supposed to be solvable, that is, the inverse operator $B^{-1}$ exists. Therefore, scheme (4) can be written in the form

$$(15)\quad y_{n+1} = S_n\, y_n + \tau\, f_n\,, \quad f_n = B_n^{-1}\, \varphi_n\,, \quad n = 0, 1, \ldots, \quad y_0 \in \mathcal{B}_h\,,$$

where

$$(16)\qquad S_n = E - \tau\, B_n^{-1}\, A_n$$

is the transition operator. The operator $S_n$ depends on $t_n = n\,\tau$, $h$ and $\tau$, however, neither for $S_n$ nor for $B_n$, $A_n$, $y_n$, $\varphi_n$ and $y_0$ the dependence on $h$ and $\tau$ is explicitly indicated.

By virtue of the recurrence relation (15) we find that

$$(17)\qquad y_{n+1} = T_{n+1,\,0}\, y_0 + \sum_{j=0}^{n} \tau\, T_{n+1,\,j+1}\, f_j\,,$$

where

$$T_{n+1,j} = S_n\, S_{n-1}\, \dots\, S_{j+1}\, S_j\,,$$

$$T_{n+1,0} = S_n\, S_{n-1}\, \dots\, S_1\, S_0\,, \quad T_{n+1,n+1} = E\,.$$

The operator $T_{n+1,j}$ is called the **transition operator** from the layer $j$ to the layer $n+1$, while the operator $T_{n+1,0}$ refers to the **resolving operator**. The triangular inequality yields

$$(18) \quad \| y_{n+1} \|_{(1)} \le \| T_{n+1,0} \| \cdot \| y_0 \|_{(1)} + \sum_{j=0}^{n} \tau \, \| T_{n+1,j+1} \| \cdot \| f_j \|_{(1)}\,,$$

where $\| \cdot \|_{(1)}$ is any suitable norm on the space $\mathcal{B}_h$, making it possible to arrive at the following assertion.

**Theorem 1** *For the stability of scheme* (15) *it is sufficient that for any* $0 \le j \le n \le n_0$

$$(19) \qquad \| T_{n,j} \| \le M_1\,.$$

*Moreover, having stipulated this condition, the solution of the difference problem* (3) *satisfies the a priori estimate*

$$(20) \qquad \| y_{n+1} \|_{(1)} \le M_1 \left( \| y_0 \|_{(1)} + \sum_{j=0}^{n} \tau \, \| B_j^{-1} \varphi_j \|_{(1)} \right)$$

*for all* $0 \le n \le n_0$.

Note that estimate (20) implies (12) with constant $M_2 = M_1\, t_0$ and $\| \varphi_j \|_{(2)} = \| B_j^{-1} \varphi_j \|_{(1)}$ incorporated.

**Theorem 2** *For the stability of scheme* (3) *it is sufficient that for the norm of its transition operator* $S_j$ *the estimate*

$$(21) \qquad \| S_j \| \le 1 + c_0 \tau$$

*is valid for all* $j = 0, 1, \dots, n_0 - 1$, *where* $c_0 \ge 0$ *is a constant independent of* $\tau$ *and* $h$. *Moreover, under condition* (21) *a priori estimate* (20) *holds with constant* $M_1 = \exp\{c_0 t_0\}$.

To prove this assertion, it suffices to verify that estimate (21) implies estimate (19):

$$\begin{aligned}\| T_{n,j} \| &= \| S_{n-1} S_{n-2} \cdots S_{j+1} S_j \| \\ &\le \| S_{n-1} \| \cdot \| S_{n-2} \| \cdots \| S_{j+1} \| \cdot \| S_j \| \le (1 + c_0 \tau)^{n-j} \\ &\le (1 + c_0 \tau)^n \le (1 + c_0 \tau)^{n_0} \le e^{c_0 n_0 \tau} = e^{c_0 t_0} = M_1 .\end{aligned}$$

Quite often, the reader can encounter the statement that "stability with respect to the initial data implies stability with respect to the right-hand side". How is one to understand the nature of this assertion?

We say that scheme (4) is **uniformly stable with respect to the initial data** if the Cauchy problem is stable:

$$y_{n+1} = S_n y_n \,, n = j, j+1, \ldots , \quad y_j \,, j = 0, 1, \ldots , n \quad \text{given}, \tag{22}$$

for any $j = 0, 1, \ldots , n_0 - 1$, that is,

$$\|y_n\|_{(1)} \le M_1 \, \|y_j\|_{(1)} \quad \text{for all} \quad 0 \le j < n \le n_0 \,, \tag{23}$$

where $M_1$ is a constant independent of $\tau$ and $h$ both.

If the condition of uniform stability is satisfied, then estimate (19) holds true for the resolving operator $T_{n,j}$. Therefore, Theorem 1 asserts that estimate (20) is valid for a solution of problem (4). This type of situation is covered by the following results.

**Theorem 3** *If scheme* (4) *is uniformly stable with respect to the initial data, then it is also stable with respect to the right-hand side under the condition of the norm concordance*

$$\|\varphi\|_{(2)} = \|B^{-1}\varphi\|_{(1)} \,. \tag{24}$$

*Moreover, condition* (24) *assures us of the validity of a priori estimate* (20).

It is worth noting here that condition (21) is sufficient for uniform stability with respect to the initial data.

The object of investigation is the two-layer scheme with constant operators $A$ and $B$, not depending on $t_n = n\tau$:

$$\begin{aligned} & y_{n+1} = S y_n + \tau f_n \,, \quad f_n = B^{-1} \varphi_n \,, \quad n = 0, 1, \ldots , \\ & S = E - \tau B^{-1} A \,, \end{aligned} \tag{25}$$

where $y_0$ is given. If a scheme with a constant transition operator $S$ is stable with respect to the initial data, then it is uniformly stable with respect to the initial data, since

$$T_{n,j} = T_{n-j,0} = S^{n-j}. \tag{26}$$

**Theorem 4** *The stability with respect to the initial data of scheme* (25) *with constant operators is necessary and sufficient for the stability with respect to the right-hand side, provided condition* (24) *of the norm of concordance holds. Moreover, in that case a priori estimate* (20) *is valid.*

*Sufficiency.* The stability with respect to the initial data means the boundedness of the resolving operator

$$\| T_{n,0} \| \leq M_1 . \tag{27}$$

In accordance with what has been said above, this implies condition (19) and so it remains to use Theorem 1.

*Necessity.* Let scheme (25) be stable with respect to the right-hand side, meaning the validity of the inequality for the solution of problem (4b)

$$\|y_{n+1}\|_{(1)} \leq M_1 \sum_{j=0}^{n} \tau \, \|B^{-1}\varphi_j\|_{(1)}, \quad n = 0, 1, \ldots . \tag{28}$$

This estimate holds true for any right-hand side $f_j = B^{-1}\varphi_j$. This provides enough reason to conclude from (17) that

$$y_{n+1} = \sum_{j=0}^{n} \tau \, T_{n+1,j+1} \, f_j . \tag{29}$$

The assertion will be proved if we succeed in showing that property (27) is true. Choosing $\tau f_j = \delta_{j,0}$ we deduce from (29) that

$$y_{n+1} = T_{n+1,1} f = T_{n,0} f$$

and

$$\| y_{n+1} \|_{(1)} \leq \| T_{n,0} \| \cdot \| f \|_{(1)} .$$

On the other hand, (28) yields $\| y_{n+1} \|_{(1)} \leq M_1 \| f \|_{(1)}$. Comparison of these inequalities gives (27), from which the stability with respect to the initial data immediately follows.

Summarizing, we draw some conclusions which will be needed in the sequel.

1. When the transition operator happens to be constant, the stability analysis with respect to the initial data is mostly based on estimates of the norms of the transition operator.

2. The condition of concordance between the norms of the right-hand side and the solution such as

$$\|\varphi\|_{(2)} = \|B^{-1}\varphi\|_{(1)}$$

is a severe constraint. If $\| B^{-1} \| \leq c_1$, where $c_1 > 0$ is a constant independent of $h$ and $\tau$, then $\| \varphi \|_{(2)} \leq c_1 \| \varphi \|_{(1)}$ and instead of (20) we obtain the estimate

$$(30) \quad \|y(t+\tau)\|_{(1)} \leq M_1 \|y(0)\|_{(1)} + M_2 \sum_{t'=0}^{t} \tau \, \|\varphi(t')\|_{(1)} \, , M_2 = M_1 \, c_1 \, .$$

In Section 2 a priori estimates will be obtained for which condition (24) of the norm concordance is not required.

3. Scheme (4) is stable if $\| S_j \| \leq 1 + c_0 \tau$ for all $j = 0, 1, \ldots, n_0 - 1$. In practical applications of this sufficient stability criterion one needs to reveal some properties of the operators $A$ and $B$ ensuring condition (21). Such conditions are established in Section 2 of the present chapter. They asquire the form of linear operator inequalities for the operators $A$ and $B$ acting in the Hilbert space $H_n = \mathcal{B}_h$.

**7. Approximation and convergence.** The notions of approximation, convergence and accuracy for operator-difference schemes are introduced by analogy with the corresponding notions for the operator schemes $A_n y_n = \varphi_n$ arising earlier in Chapter 2, Section 4. Only a few editorial changes will appear in this matter. For instance, together with the norms $\| \cdot \|_{(1_h)}$ and $\| \cdot \|_{(2_h)}$ it will be sensible to introduce the additional norms

$$\|y_{h\tau}(t_n)\|_{(1_{h\tau})} = \max_{0 \leq t = n'\tau \leq t_n} \|y_{h\tau}(t_{n'})\|_{(1_h)} \, ,$$

$$\|\varphi_{h\tau}(t_n)\|_{(2_{h\tau})} = \max_{0 \leq t = n'\tau \leq t_n} \|\varphi_{h\tau}(t_{n'})\|_{(2_h)} \, ,$$

which complement further stability analysis.

Thus, let $\mathcal{B}_h$ be a linear space with norms $\| \cdot \|_{(1_h)}$ and $\| \cdot \|_{(2_h)}$ on it. We denote by $\mathcal{B}_h^{(1)}$ and $\mathcal{B}_h^{(2)}$, respectively, the resulting normed spaces and assume in the sequel that

$$y_{h\tau}(t_n) \in \mathcal{B}_h^{(1)} \, , \quad \varphi_{h\tau}(t_n) \in \mathcal{B}_h^{(2)}$$

for all $t_n = n\tau$ and that the operators of scheme (4) carry out $\mathcal{B}_h^{(1)}$ into $\mathcal{B}_h^{(2)}$:

$$A_h\,, B_h \,:\, \mathcal{B}_h^{(1)} \mapsto \mathcal{B}_h^{(2)}\,.$$

Let $\mathcal{B}_0^{(1)}$ and $\mathcal{B}_0^{(2)}$ be normed spaces with norms $\|\cdot\|_{(1_0)}$ and $\|\cdot\|_{(2_0)}$, $u(t)$ be an abstract $\mathcal{B}_0^{(1)}$-valued function of the argument $t \in [0,T]$ and let $f(t)$ be an abstract $\mathcal{B}_0^{(2)}$ valued function of the argument $t \in [0,T]$. Furthermore, we refer to linear operators $\mathcal{P}_h^{(\alpha)}$, $\alpha = 1, 2$, projecting $\mathcal{B}_0^{(\alpha)}$ onto $\mathcal{B}_h^{(\alpha)}$:

$$u_h = \mathcal{P}_h^{(1)}\, u \in \mathcal{B}_h^{(1)} \quad \text{if} \quad u \in \mathcal{B}_0^{(1)}\,,$$

$$f_h = \mathcal{P}_h^{(2)}\, f \in \mathcal{B}_h^{(2)} \quad \text{if} \quad f \in \mathcal{B}_0^{(2)}\,.$$

We take for granted the condition of the norm concordance

$$\lim_{|h|\to 0} \|\psi_h^{(\alpha)}\|_{(\alpha_h)} = \|\psi^{(\alpha)}\|_{(\alpha_0)}\,, \qquad \text{where} \quad \psi^{(\alpha)} \in \mathcal{B}_0^{(\alpha)}, \alpha = 1, 2\,.$$

Let $y_{h\tau}(t_j)$ be a solution of problem (4) and $u(t)$ be a continuous function of the argument $t$, so that

$$u(t)|_{t=t_j} = u(t_j) = u^j\,.$$

In such a setting the error $z_{h\tau}^j = y_{h\tau}^j - u_h^j$ needs investigation. We say that scheme (4) converges on an abstract function $u(t) \in \mathcal{B}_0^{(1)}$ if

$$\lim_{|h|\to 0, \tau \to 0} \ \max_{0 \le j \le j_0} \|y_{h\tau}^j - u_h^j\|_{(1_h)} = 0\,.$$

Scheme (4) converges with the rate $O\big(|h|^m + \tau^k\big)$ or is of accuracy $O\big(|h|^m + \tau^k\big)$ on an abstract function $u(t) \in \mathcal{B}_0^{(1)}$ if

$$\max_{0 \le j \le j_0} \|y_{h\tau}^j - u_h^j\|_{(1_h)} \le M\big(|\,h\,|^m + \tau^k\big)\,,$$

where $M = \text{const} > 0$ is independent of $h$ and $\tau$ both.

An a priori characteristic of a scheme is the error of approximation. The approximation error on a function $u(t)$ for scheme (4) is known as the residual

$$\psi_{h\tau}^j = B_{h\tau}\, \frac{u_h^{j+1} - u_h^j}{\tau} + A_{h\tau}\, u_h^j - \varphi_{h\tau}^j\,, \qquad \psi_{h\tau}^j = \psi(t_j\,, h, \tau)\,.$$

Scheme (4) provides an approximation on a function $u(t)$ if

$$\max_{0\le j\le j_0} \|\psi^j\|_{(2_h)} \longrightarrow 0 \quad \text{as} \quad |h|\to 0, \quad \tau\to 0. \tag{31}$$

Scheme (4) provides an approximation of $O\big(|h|^m+\tau^k\big)$ on a function $u(t)\in \mathcal{B}_0^{(1)}$ if

$$\max_{0\le j\le j_0} \|\psi^j\|_{(2_h)} \le M\big(|h|^m+\tau^k\big),$$

where $M = \text{const} > 0$ is independent of $h$ and $\tau$ both.

Along these lines, we may set up the problem for the error $z^j = y^j - u_h^j$:

$$B\,\frac{z^{j+1}-z^j}{\tau} + A\,z^j = \psi^j, \quad j=0,1,\dots, \quad z^0 = y_0 - u_h^0. \tag{32}$$

If scheme (4) is stable, then for a solution of problem (32) the estimate holds:

$$\|z^j\|_{(1_h)} \le M_1\,\|y_0-u_h^0\|_{(1_h)} + M_2 \max_{0\le j'<j} \|\psi^{j'}\|_{(2_h)},$$

thereby justifying the next assertions.

> *Scheme* (4) *converges on a function* $u(t)$ *if it is stable, generates an approximation on* $u(t)$ *and the initial value* $y_0$ *approximates the element* $u(0)$:
>
> $$\|y_{0h\tau} - u_h^0\|_{(1_h)} \longrightarrow 0 \quad \text{as} \quad |h|\to 0, \quad \tau\to 0.$$

> *Scheme* (4) *is of accuracy* $O\big(|h|^m+\tau^k\big)$ *on a function* $u(t)$ *if it is stable, provides an approximation of* $O\big(|h|^m+\tau^k\big)$ *on* $u(t)$ *and*
>
> $$\|\,y_0 - u_h^0\,\|_{(1_h)} = O\big(|h|^m+\tau^k\big).$$

In particular, $u(t)$ may be a solution of a certain differential equation. In that case we say that the difference scheme approximates the difference equation, provided condition (31) holds, etc.

We note in passing that one is to understand the statement "if a scheme is stable and provides an approximation, then it is convergent" given in Chapter 2, Section 2 as follows: both the difference equation and the initial value generate an approximation (if we accept $y_0 = \mathcal{P}_h^{(1)}u(0)$, then $\|\,y_{0h_\tau} - u_h(0)\,\|_{(1_h)} = 0$).

## 6.2 CLASSES OF STABLE TWO-LAYER SCHEMES

**1. The problem statement.** We pursue the stability analysis of two-layer schemes by having recourse to their canonical form

$$(1) \qquad Bu_t + Ay - \varphi(t), \qquad t = n\tau \in \omega_\tau, \quad y(0) = y_0$$

Let $\mathcal{B}_h = H_h$ be a finite-dimensional real space, ( , ) be an inner product and $\| x \| = \sqrt{(x, x)}$ be the associated norm in the space $H_h$. The operators $A$ and $B$ of scheme (1) depend, in general, on $h$, $\tau$ and $t$. Here and below, we agree not to indicate explicitly the dependence on $t$.

The main goal of our studies is to find out sufficient conditions for the stability of scheme (1) and obtain a priori estimates for a solution of problem (1) expressing the stability of this scheme with respect to the right-hand side and the initial data. In preparation for this, a solution of problem (1) can be written as a sum $y = \bar{y} + \tilde{y}$, where $\bar{y}$ is a solution to the homogeneous equation with the initial condition $\bar{y}(0) = y(0) = y_0$:

$$Bu_t + Ay = 0, \qquad t = \in \omega_\tau, \quad y(0) = y_0,$$

and $\tilde{y}$ is a solution to the nonhomogeneous equation with the zero initial condition:

$$Bu_t + Ay = \varphi(t), \qquad t = \in \omega_\tau, \quad y(0) = 0.$$

The meaning of the estimate

$$(2) \qquad \|y(t+\tau)\|_{(1)} \le M_1 \, \|y(0)\|_{(1)}$$

of a solution of problem (1a) is that scheme (1) is stable with respect to the initial data, while the estimate

$$(3) \qquad \|y(t+\tau)\|_{(1)} \le M_2 \max_{0 \le t' \le t} \|\varphi(t')\|_{(2)}$$

of a solution of problem (1b) expresses the stability of scheme (1) with respect to the right-hand side.

We will also use an alternative definition of stability of a scheme with respect to the right-hand side:

$$(4) \qquad \|y(t+\tau)\|_{(1)} \le M_2 \max_{0 \le t' \le t} \left( \|\varphi(t')\|_{(2)} + \|\varphi_{\bar{t}}(t')\|_{(2^*)} \right),$$

where

$$\varphi_{\bar{t}}(t') = \left(\varphi(t') - \varphi(t' - \tau)\right)/\tau.$$

When the triangle inequality

$$\| y \|_{(1)} \le \| \bar{y} \|_{(1)} + \| \tilde{y} \|_{(1)}$$

is put together with estimates (2), (3) and (4), we derive the a priori estimates

$$(5) \qquad \|y(t+\tau)\|_{(1)} \le M_1 \|y_0\|_{(1)} + M_2 \max_{0 \le t' \le t} \|\varphi(t')\|_{(2)}$$

and

$$(6) \quad \|y(t+\tau)\|_{(1)} \le M_1 \|y_0\|_{(1)} + M_2 \max_{0 \le t' \le t} \left( \|\varphi(t')\|_{(2)} + \|\varphi_{\bar{t}}(t')\|_{(2^*)} \right) .$$

The forthcoming energy norms will be taken as the norm $\| \cdot \|_{(1)}$ in addition to the approved basic norm $\| y \| = \sqrt{(y, y)}$:

$$(7) \qquad \|y\|_A = \sqrt{(Ay, y)} \qquad \text{for} \quad A = A^* > 0 ,$$

$$(8) \qquad \|y\|_B = \sqrt{(By, y)} \qquad \text{for} \quad B = B^* > 0 .$$

Scheme (1) is said to be stable in the space $H_A$ (or in $H_B$) if estimate (5) is valid in the norm $\| \cdot \|_{(1)} = \| \cdot \|_A$ (or $\| \cdot \|_1 = \| \cdot \|_B$).

**2. A primary family of schemes.** We will discover stability in a certain primary family of difference schemes. Before going further, we regard operators $A$ and $B$ to be bounded linear operators defined on the entire space $H_h$, $\mathcal{D}(A) = \mathcal{D}(B) = H_h$. In what follows the difference problem (1) is presupposed to be solvable for any input data $y_0$ and $\varphi(t)$, that is, there exists a bounded operator $B^{-1}$ with the domain $\mathcal{D}(B^{-1}) = H_h$. For the sake of simplicity, we take for granted in the detailed account below that

(1) the operators $A$ and $B$ are independent of $t$, that is, are constant operators;
(2) $B$ is a positive operator: $B > 0$;
(3) $A$ is a self-adjoint positive operator: $A = A^* > 0$.

Conditions (1)–(3) in combination with the solvability requirement single out a family of admissible schemes known as a primary family from the set of all possible schemes (1). Observe that condition 1) can be weakened in a number of different ways. Sometimes we will deal with operators $A$ and $B$, which are dependent on $t$, that is, with$A = A(t)$ and $B = B(t)$.

In the weighted scheme for the heat conduction equation in Example 1 of Section 3 the operators $A$ and $B$ are specified by $A = -\Lambda$ and $B = E+$

$\sigma\tau A$. Since $A \le \|A\| E$ and $E \ge A/\|A\|$, we have $B \ge (1/\|A\| + \sigma\tau)A > 0$ under the restriction $\sigma > -1/(\tau\|A\|)$. The operator $A = A^* > 0$ is independent of $t$, so that conditions (1)–(3) are satisfied and the scheme concerned belongs to the primary family for $\sigma \ge -h^2/(4c_2\tau)$, where $c_2 = \max_{x\in\omega_h} a(x)$.

**3. The energy identity.** We study the problem of stability of scheme (1) by the method of the energy inequalities involving as the necessary manipulations the inner product of both sides of equation (1) with $2\tau y_t = 2(\hat{y} - y)$:

$$2\tau(By_t\,, y) + 2\tau(Ay\,, y_t) = 2\tau(\varphi\,, y_t)\,. \tag{9}$$

Using the formula

$$y = \frac{\hat{y}+y}{2} - \frac{\hat{y}-y}{2} = \frac{1}{2}\,(\hat{y}+y) - \frac{\tau}{2}\, y_t\,, \tag{10}$$

we rewrite (9) in the form

$$2\tau\big((B - 0.5\tau A)y_t\,, y_t\big) + \big(A(\hat{y}+y), \hat{y}-y\big) = 2\tau(\varphi\,, y_t)\,. \tag{11}$$

**Lemma 1** *Let $A$ be a self-adjoint operator, then*

$$\big(A(\hat{y}+y), \hat{y}-y\big) = (A\hat{y}, \hat{y}) - (Ay, y)\,. \tag{12}$$

Indeed, the chain of the relations

$$(Ay, \hat{y}) = (y, A\hat{y}) = (A\hat{y}, y)$$

occurs due to the fact that $A$ is self-adjoint operator and, therefore,

$$\begin{aligned}\big(A(\hat{y}+y), \hat{y}-y\big) &= (A\hat{y}, \hat{y}) + (Ay, \hat{y}) - (A\hat{y}, y) - (Ay, y)\\ &= (A\hat{y}, \hat{y}) - (Ay, y)\,.\end{aligned}$$

Substituting (12) into (11) we obtain the energy identity for scheme (1):

$$2\tau\big((B - 0.5\tau A)y_t\,, y_t\big) + (A\hat{y}, y) = (Ay, y) + 2\tau(\varphi\,, y_t)\,. \tag{13}$$

**4. Stability with respect to the initial data in $\mathbf{H_A}$.** Stability of scheme (1) with respect to the initial data is investigated with further estimation of a solution of problem (1a).

**Theorem 1** *The condition*

$$B \geq \tfrac{\tau}{2} A \tag{14}$$

*is necessary and sufficient for the stability in $H_A$ of scheme (1) from the primary family with respect to the initial data with constant $M_1 = 1$, that is, condition (14) is necessary and sufficient for the validity of the estimate*

$$\| y_n \|_A \leq \| y_0 \|_A, \qquad n = 1, 2, \ldots, \tag{15}$$

*where $y_n$ is a solution of problem* (1a).

*Sufficiency.* Granted condition (14), the energy identity for problem (1a) (with $\varphi = 0$)

$$2\tau((B - 0.5\tau A)y_t, y_t) + (A\hat{y}, y) = (Ay, y) \tag{16}$$

implies the inequality $(A\hat{y}, \hat{y}) \leq (Ay, y)$ or $\| \hat{y} \|_A^2 \leq \| y \|_A^2$, yielding

$$\| y_{n+1} \|_A \leq \| y_n \|_A \leq \cdots \leq \| y_0 \|_A .$$

*Necessity.* Suppose that scheme (1a) is stable and estimate (15) is satisfied. We are going to show that this leads to the operator inequality (14), that is,

$$(Bv, v) \geq 0.5\tau (Av, v) \qquad \text{for any} \quad v \in H . \tag{17}$$

We begin by placing identity (16) on the first layer ($n = 0$):

$$2\tau((B - 0.5\tau A)y_t(0), y_t(0)) + (Ay_1, y_1) = (Ay_0, y_0) .$$

By virtue of (15) this identity can be satisfied only for

$$2\tau((B - 0.5\tau A)y_t(0), y_t(0)) = (Ay_0, y_0) - (Ay_1, y_1) \geq 0 ,$$

meaning

$$((B - 0.5\tau A)y_t(0), y_t(0)) \geq 0 .$$

If $y_0 \in H$ is an arbitrary element, then so is the element $v = y_t(0) = -B^{-1}Ay_0 \in H$. Indeed, taking any element $v = y_t(0) \in H$ we find $y_0 = -A^{-1}Bv \in H$ for $A^{-1}$ does exist. Thus, the inequality is met for any $v = y_t(0) \in H$, that is, the operator inequality (14) takes place, thereby completing the proof.

**Remark 1** Condition (14) is sufficient for stability in the sense of (15) of scheme (1) if $B = B(t) > 0$ is a non-self-adjoint variable operator.

**Remark 2** The fact that condition (14) is, in a certain sense, natural can be clarified a little bit by considering the simplest example relating to the difference scheme

$$b\,\frac{y_{n+1}-y_n}{\tau} + a\,y_n = 0\,, \qquad n = 0, 1, \dots\,, \qquad y_0 = u_0\,,$$

where $a$ and $b$ are positive numbers corresponding to the differential equation

$$b\,\frac{du}{dt} + a\,u = 0\,, \qquad t > 0, \qquad u(0) = u_0\,,$$

The difference equation yields

$$y_{n+1} = \left(1 - \frac{\tau\,a}{b}\right) y_n\,, \qquad |\,y_{n+1}\,| \le \left|1 - \frac{\tau\,a}{b}\right| \cdot |\,y_n\,|\,.$$

The stability requirement

$$|\,y_{n+1}\,| \le |\,y_0\,|$$

is obviously satisfied if $|1 - \tau a/b| \le 1$ or $-1 \le 1 - \tau a/b \le 1$, that is, for $b \ge \frac{1}{2}\tau a$. The similarity with the operator equation $B \ge \frac{1}{2}\tau a$ is clear at the first glance.

**Example** We now consider the weighted scheme

$$y_t + A\left(\sigma\hat{y} + (1-\sigma)\,y\right) = 0\,,$$

whose use permits us to illustrate the effectiveness of the stability condition (14). For this, we write down this scheme in canonical form accepted in Section 1:

$$(E + \sigma\,\tau\,A)\,y_t + A\,y = 0\,, \qquad B = E + \sigma\,\tau\,A\,. \tag{18}$$

If $A = A^* > 0$ is independent of $t$ and $\sigma > -1/(\tau\|\,A\,\|)$, then the weighted scheme belongs to the primary family (see Section 2). The necessary and sufficient stability condition (14) is of the form

$$B - 0.5\,\tau\,A = E + (\sigma - 0.5)\,\tau\,A \ge 0\,.$$

With the relations $A \le \| A \| E$ and $E \ge A/\| A \|$ in view, we obtain

$$B - 0.5\,\tau A \ge \left(1/\| A \| + (\sigma - 0.5)\,\tau\right) A .$$

The preceding inequality implies that condition (14) is equivalent to the constraint

$$\sigma \ge \sigma_0 , \qquad \sigma_0 = \frac{1}{2} - \frac{1}{\tau \| A \|} .$$

This condition is necessary and sufficient for the stability of the weighted scheme, which interests us.

In the case of a model heat conduction scheme

$$y_t + \Lambda\left(\sigma \hat{y} + (1-\sigma)\, y\right) = 0 , \quad \Lambda\, y = y_{\bar{x}x} , \quad 0 < x = ih < 1 , \quad hN = 1 ,$$

$$y(0, t_n) = 0 , \quad y(1, t_n) = 0 , \quad y(x, 0) = u_0(x) , \quad x = ih \in [0, 1] ,$$

corresponding to the first boundary-value problem for the heat conduction equation

$$\frac{\partial u}{\partial t} = \frac{\partial^2 u}{\partial x^2} , \qquad 0 < x < 1 , \qquad t > 0 ,$$

$$u(0,t) = u(1,t) = 0 , \quad t \ge 0 , \quad u(x,0) = u_0(x) ,$$

we deduce that (see Chapter 5, Section 1)

$$A = -\widetilde{\Lambda} , \qquad \widetilde{\Lambda}\, y = \Lambda\, y \quad \text{for } y \in \overset{\circ}{\Omega}_h = H_h ,$$

$$\| A \| = \frac{4}{h^2} \cos^2 \frac{\pi h}{2} \le \frac{4}{h^2} , \qquad \sigma_0 = \frac{1}{2} - \frac{h^2}{4\,\tau \cos^2 \frac{\pi h}{2}} .$$

The condition $\sigma \ge \bar{\sigma}_0$, $\bar{\sigma}_0 = h^2/(4\tau) > \sigma_0$, has been found in Chapter 5 by the method of separation of variables.

Suppose now the operator $A > 0$ not to be self-adjoint. Then scheme (18) does not belong to the primary family. However, it can be replaced by an equivalent scheme from the primary family. Since $A > 0$, there exists an inverse operator $A^{-1} > 0$, whose use with regard to equation (18) permits us to confine ourselves to

$$\widetilde{B}\, y_t + \widetilde{A}\, y = 0 , \qquad \widetilde{B} = A^{-1} + \sigma \tau E , \qquad \widetilde{A} = E .$$

The operator $\widetilde{A} = E = \widetilde{A}^* > 0$ is independent of $t$ and the operator $\widetilde{B}$ is positive for $\sigma \geq 0$. Stability condition (14) in the space $H_{\widetilde{A}} = H$ takes for now the form $\widetilde{B} - \frac{1}{2}\tau\widetilde{A} = A^{-1} + (\sigma - \frac{1}{2})\tau E \geq 0$, valid with $\sigma \geq \frac{1}{2}$. The condition $\sigma \geq \frac{1}{2}$ is sufficient for the occurrence of the relation

$$\| y_n \| \leq \| y_0 \| \qquad \text{for } \sigma \geq 0.5\,, \quad A \neq A^*, \quad A > 0\,.$$

**5. Stability with respect to the initial data in $\mathbf{H_B}$.** Let us write down the second energy identity for scheme (1a) assuming $B$ also to be a self-adjoint operator: $B = B^* > 0$. At the first stage we take the inner product of (1a) and $2\tau\hat{y}$:

$$2\,\tau\,(B\,y_t, \hat{y}) + 2\,\tau\,(A\,y, \hat{y}) = 0\,. \tag{19}$$

Relying on the formulae

$$\hat{y} = \frac{1}{2}\,(\hat{y} + y) + \frac{\tau}{2}\,y_t\,, \qquad y = \frac{1}{2}\,(\hat{y} + y) - \frac{\tau}{2}\,y_t$$

and making use of Lemma 1, we find that

$$\begin{aligned} 2\,\tau\,(B\,y_t, \hat{y}) &= \big(B\,(\hat{y} - y), \hat{y} + y\big) + \tau^2\,(B\,y_t, y_t) \\ &= \|\,\hat{y}\,\|_B^2 - \|\,y\,\|_B^2 + \tau^2\,\|\,y_t\,\|_B^2\,, \\ 2\,\tau\,(A\,y, \hat{y}) &= \frac{\tau}{2}\,\big(A\,(\hat{y} + y - \tau\,y_t), \hat{y} + y + \tau\,y_t\big) \\ &= \frac{\tau}{2}\,\|\,\hat{y} + y\,\|_A^2 - \frac{\tau^3}{2}\,\|\,y_t\,\|_A^2\,. \end{aligned}$$

Substitution of these expressions into (13) leads to

$$\|\,\hat{y}\,\|_B^2 + \tau^2\,(\|\,y_t\,\|_B^2 - 0.5\,\tau\,\|\,y_t\,\|_A^2) + 0.5\,\tau\,\|\,\hat{y} + y\,\|_A^2 = \|\,y\,\|_B^2\,. \tag{20}$$

**Theorem 2** *Let the operators $A$ and $B$ involved in scheme (1) be independent of $t$, $A^* = A > 0$ and $B^* = B > 0$. Condition (14) is sufficient for scheme (1) to be stable with respect to the initial data in the space $H_B$ with constant $M_1 = 1$.*

Indeed, let $B \geq \frac{1}{2}\tau A$, then

$$\|\,y_t\,\|_B^2 - 0.5\,\tau\,\|\,y_t\,\|_A^2 = ((B - 0.5\,\tau\,A)\,y_t, y_t) \geq 0$$

and relation (20) yields $\| \hat{y} \|_B \leq \| y \|_B$, meaning

$$\| y(t) \|_B \leq \| y(0) \|_B \, .$$

**Remark** If the operators $A$ and $B$ are commuting, then condition (14) is necessary and sufficient for the stability of scheme (1a) in the space $H_D$:

$$\| y_n \|_D \leq \| y_0 \|_D$$

where $D = D^* > 0$ is any operator commuting with $A$ and $B$. This is acceptable if we involve, for instance, $D = E$, $D = A^2$ or $D = B^2$ for $B = B^*$, so that $\| y_n \| \leq \| y_0 \|$, $A\| y_n \| \leq A\| y_0 \|$, $B\| y_n \| \leq B\| y_0 \|$, etc.

**6. Estimates of the norm of the transition operator.** Stability considerations are connected with the use of a new method based on the estimation of the norm for the operator of transition from one layer to another. This method actually falls within the category of energy methods.

We may attempt the difference scheme (1a) in the form

$$\hat{y} = S\,y \,, \qquad S = E - \tau\, B^{-1} A \,, \tag{21}$$

where $S$ is the transition operator. Let $D = D^*$ be an arbitrary constant operator in the space $H$. With these, it is plain to show that

$$\| \hat{y} \|_D = \| S\,y \|_D \leq \| S \|_D \cdot \| y \|_D \,, \tag{22}$$

where $\| S \|_D^2$ is adopted as the smallest constant $M$ subject to the inequality

$$(DSy, Sy) \leq M(Dy, y) \, .$$

As can readily be observed from (22), scheme (21) is stable in the space $H_D$, that is, $\| y_n \|_D \leq \| y_0 \|_D$ if the norm of the transition operator does not exceed unity: $\| S \|_D \leq 1$. This condition is equivalent to being nonnegative of the functional ($M = 1$)

$$J_D[y] = (D\,y, y) - (D\,S\,y, S\,y) \geq 0 \, .$$

Of special interest is the case $D = A$, for which $\| S \|_A \leq 1$ for $B \geq \frac{1}{2}\tau A$. Indeed, upon substituting the expression for $S$ into the functional $J_A$ we obtain

$$\begin{aligned} J_A[y] &= (A\,y, y) - (A\,S\,y, S\,y) \\ &= (A\,y, y) - \big(A\,(E - \tau\,B^{-1}\,A)\,y, (E - \tau\,B^{-1}\,A)\,y\big) \\ &= 2\,\tau\,(A\,y, B^{-1}\,A\,y) - \tau^2\,(A\,B^{-1}\,A\,y, B^{-1}\,A\,y) \, . \end{aligned}$$

The outcome of further substitutions $x = B^{-1}Ay$ or $y = A^{-1}Bx$ is the same:

(23) $$J_A[y] = \widetilde{J}_A[x] = 2\,\tau\left[(B\,x, x) - \tfrac{1}{2}\,\tau\,(A\,x, x)\right].$$

This implies that the conditions $\|S\|_A \le 1$ and $B \ge \frac{1}{2}\tau A$ are equivalent, providing support for the view that Theorem 1 holds.

Since $A = A^* > 0$, there exists a square root $A^{1/2} = \left(A^{1/2}\right)^* > 0$, by means of which the expression for $J_A[y]$ can be recast as

$$\begin{aligned} J_A[y] &= (A^{1/2}y, A^{1/2}y) - (A^{1/2}\,S\,y, A^{1/2}S\,y) \\ &= \|A^{1/2}y\|^2 - \|A^{1/2}S\,y\|^2 \\ &= \|A^{1/2}y\|^2 - \|(E - \tau\,C)\,A^{1/2}\,y\|^2, \end{aligned}$$

where $C = A^{1/2}B^{-1}A^{1/2}$. Substitution $u = A^{1/2}y$ gives

$$J_A[y] = \|u\|^2 - \|(E - \tau\,C)\,u\|^2.$$

This implies that the condition

(24) $$\|E - \tau\,C\| \le 1$$

is equivalent to the relation $B \ge \frac{1}{2}\tau A$. No assumption is made here that $B$ is a self-adjoint operator.

It seems worthwhile giving the special case when

$$B = B^* > 0, \qquad A = A^* > 0,$$

for which the equivalence of the conditions

(25) $$J_B[y] \ge 0 \qquad \text{and} \qquad B \ge 0.5\,\tau A$$

is certainly true. Since $B = B^* > 0$, there exists $B^{1/2} = \left(B^{1/2}\right)^* > 0$ and this property is valid for the operator

$$C = B^{1/2}\,A^{-1}\,B^{1/2}, \qquad C = C^* > 0.$$

By applying successively the substitutions $B^{-1/2}Ay = x$, $C^{1/2}x = u$, $C = B^{1/2}A^{-1}B^{1/2}$ and $B^{-1/2}u = v$ we arrive at the chain of the relations

$$\begin{aligned} J_B[y] &= 2\,\tau\,(A\,y, y) - \tau^2\,(A\,y, B^{-1}\,A\,y) \\ &= 2\,\tau\,(B^{1/2}x, A^{-1}\,B^{1/2}) - \tau^2\,(x, x) \\ &= 2\,\tau\,(Cx, x) - \tau^2\,(x, x) = 2\,\tau\,(u, u) - \tau^2\,(C^{-1}\,u, u) \\ &= 2\,\tau\,(u, u) - \tau^2\,(B^{-1/2}\,A\,B^{-1/2}u, u) \\ &= 2\,\tau\left((B\,v, v) - \frac{\tau}{2}\,(A\,v, v)\right). \end{aligned}$$

Thus, the functional $J_B[y]$ is transformed into the same form (23) as was done before for $J_A[y]$:

$$J_B[y] = 2\tau\left((B\,v, v) - \frac{\tau}{2}(A\,v, v)\right),$$

where

$$v = B^{-1/2}\,C^{1/2}\,B^{-1/2}\,A\,y\,, \qquad y = A^{-1}\,B^{1/2}\,C^{-1/2}\,B^{1/2}\,v\,.$$

This implies the equivalence of conditions (25), which assures us of the validity of Theorem 2.

Therefore, the method of estimation of the transition operator norm permits us to prove that condition (14) is necessary and sufficient for the stability of scheme (1) with respect to the initial data in the space $H_A$ (for $B \neq B^*$) and in the space $H_B$ (for $B = B^* > 0$) with constant $M_1 = 1$.

**7. The method of separation of variables.** If the operators $A$ and $B$ are self-adjoint

$$A = A^* > 0\,, \qquad B = B^* > 0\,, \tag{26}$$

then the stability of scheme (1a) in the spaces $H_A$ and $H_B$ for

$$B \geq \tfrac{\tau}{2} A$$

can be proved by means of the method of separation of variables following established practice (for more detail see Chapter 5).

Let $N$ be the dimension of a finite-dimensional space $H$, $\lambda_k$ be eigenvalues and $\xi_k$ be orthonormal eigenfunctions of the problem (see Chapter 1, Section 1 and Chapter 2, Section 1)

$$A\,\xi_k = \lambda_k\,B\,\xi_k\,, \qquad k = 1, 2, \dots, N\,, \tag{27}$$

where, in addition, $(B\xi_k, \xi_m) = \delta_{km}$ ($\delta_{kk} = 1$ and $\delta_{km} = 0$ for $k \neq m$). All the eigenvalues $\lambda_k$ of problem (27) are positive, because $A > 0$.

In such a setting a solution of problem (1a) is sought in the form

$$y(t) = \sum_{k=1}^{N} c_k(t)\,\xi_k\,. \tag{28}$$

By virtue of the relations

$$Ay = \sum_{k=1}^{N} c_k\,A\,\xi_k = \sum_{k=1}^{N} \lambda_k\,c_k\,B\,\xi_k$$

we find upon substituting (28) into (1a) that

$$\sum_{k=1}^{N}\left\{\frac{c_k(t+\tau)-c_k(t)}{\tau}+\lambda_k\, c_k(t)\right\} B\,\xi_k = 0\,.$$

Putting this together with the orthogonality of the system $\{\xi_k\}$ we arrive at

$$\frac{c_k(t+\tau)-c_k(t)}{\tau}+\lambda_k\, c_k(t)=0\,,\quad c_k(t+\tau)=(1-\tau\,\lambda_k)\,c_k(t)\,,$$

$$y(t+\tau)=\sum_{k=1}^{N} c_k(t+\tau)\,\xi_k=\sum_{k=1}^{N}(1-\tau\,\lambda_k)\,c_k(t)\,\xi_k\,.$$

The norm $\|\, y(t+\tau)\,\|_A^2 = \big(Ay(t+\tau),\, y(t+\tau)\big)$ can be most readily evaluated by observing that

$$\|\, y(t)\,\|_A^2 = \Big(\sum_{k=1}^{N} c_k(t)\,\lambda_k\, B\,\xi_k,\, \sum_{k=1}^{N} c_k(t)\,\xi_k\Big) = \sum_{k=1}^{N} \lambda_k\, c_k^2(t)\,.$$

All this enables us to deduce that

$$\|\, y(t+\tau)\,\|_A^2 = \sum_{k=1}^{N} \lambda_k\, c_k^2(t+\tau) \le \max_{1\le k\le N}\,(1-\tau\,\lambda_k)^2 \sum_{k=1}^{N} \lambda_k\, c_k^2(t)$$

yielding

$$\|\, y(t+\tau)\,\|_A \le \max_{1\le k\le N}\, |1-\tau\,\lambda_k|\, \|\, y(t)\,\|_A\,.$$

Whence it follows that

(29) $$\|\, y(t+\tau)\,\|_A \le \|\, y(t)\,\|_A \le \|\, y(0)\,\|_A$$

if

$$|1-\tau\,\lambda_k| \le 1\,,\qquad k=1,2,\ldots N\,.$$

This condition provides support for the view that $-1 \le 1-\tau\lambda_k \le 1$ or

(30) $$0 < \lambda_k \le 2/\tau\,,\qquad k=1,2,\ldots N\,.$$

Let us show that inequalities (14) and (30) are equivalent. To this end, we operate with the expression

$$B\,y - 0.5\,\tau\,A\,y = \sum_{k=1}^{N} c_k(t)\,(B\,\xi_k - 0.5\,\tau\,A\,\xi_k)$$

$$= \sum_{k=1}^{N} c_k(t)\,(1 - 0.5\,\tau\,\lambda_k)\,B\xi_k\,,$$

by means of which it is plain to calculate the functional

$$(By, y) - 0.5\,\tau\,(Ay, y) = \sum_{k=1}^{N} c_k^2(t)\,(1 - 0.5\,\tau\,\lambda_k)\,.$$

Because of this form, the equivalence of (14) and (30) is obvious.

We thus have shown that under conditions (26) inequality (14) is sufficient for the stability of scheme (1a) in the space $H_A$, that is, relation (29) occurs. Let us stress that the requirement of self-adjointness of the operator $B$ is necessary here, while the energy method demands only the positivity of $B$ and no more.

One can prove in a similar way the stability of scheme (1a) in the space $H_B$, provided condition (30) holds.

**8. The $\rho$-stability condition.** A more general definition of stability with respect to the initial data became rather urgent and extremely important.

Let $D = D^* > 0$ be a constant operator. Scheme (1) is said to be $\rho$-stable with respect to the initial data if for a solution of problem (1a) the inequality holds for any $y_0 \in H$:

$$(31) \qquad \|\, y_n \,\|_D \le \rho^n \,\|\, y_0 \,\|_D\,,$$

where $\rho = \exp\{c_0 \tau\}$, $c_0$ is a constant independent of $h$, $\tau$ and irrelevant to the choice of $y_0$. If scheme (1a) is $\rho$-stable in the space $H_D$, then it is stable in the same space $H_D$:

$$\|\, y_n \,\|_D \le M_1 \,\|\, y_0 \,\|_D\,, \qquad t_n = n\tau \le \bar{t}_0$$

with constants $M_1 = \exp\{c_0 \bar{t}_0\}$ for $c_0 > 0$ and $M_1 = 1$ for $c_0 \le 0$.

The two-layer scheme (1a) with constant operators $A$ and $B$ can be reduced to the explicit scheme

$$(32) \qquad \frac{x_{n+1} - x_n}{\tau} + C\,x_n = 0 \qquad \text{or} \qquad x_{n+1} = (E - \tau\,C) - x_n\,,$$

provided that the forthcoming substitutions are carried out before going further:

1) $x_n = B^{1/2} y_n$ for $B = B^* > 0$, where $C = C_1 = B^{-1/2} A B^{-1/2}$;

2) $x_n = A^{1/2} y_n$ for $A = A^* > 0$, where $C = C_2 = A^{1/2} B^{-1} A^{1/2}$.

Definitions 1) and 2) together imply that

$$\| x_n \| = \| y_n \|_B \quad \text{for } C = C_1 , \quad x_n = B^{1/2} y_n ,$$

$$\| x_n \| = \| y_n \|_A \quad \text{for } C = C_2 , \quad x_n = A^{1/2} y_n .$$

The condition of $\rho$-stability of the implicit scheme (1a) in the space $H_D$ for the choices $D = B$ and $D = A$ is equivalent to the condition of $\rho$-stability of the explicit scheme (32) in the space $H$: $\| x_n \| \le \rho^n \| x_0 \|$, $n = 1, 2, \ldots$.

**Lemma 2** *Let scheme* (32) *be given with a constant operator* $C$. *The condition of* $\rho$*-stability of this scheme is equivalent to the boundedness of the norm of the transition operator*

$$\| S \| = \| E - \tau C \| \le \rho .$$

Indeed, we have $x_1 = S x_0$ and $\| x_1 \| \le \| S \| \cdot \| x_0 \|$ for $n = 1$. By comparing this inequality with (31) for $n = 1$ we finish the proof of Lemma 2.

**Lemma 3** *If* $A = A^* > 0$ *and* $B = B^* > 0$, *then the inequalities*

$$\gamma_1 B \le A \gamma_2 \le B \qquad \gamma_1 E \le C \gamma_2 \le E \tag{33}$$

*are equivalent for the operators* $C = B^{-1/2} A B^{-1/2}$ *and* $C = A^{1/2} B^{-1} A^{1/2}$.

**Proof** Let $C = B^{-1/2} A B^{-1/2}$ and $\gamma$ be an arbitrarily taken number. The difference

$$(Cx, x) - \gamma (x, x) = (B^{-1/2} A B^{-1/2} x, x) - \gamma (x, x) = (Ay, y) - \gamma (By, y)$$

with regard to $y = B^{-1/2} x$ shows that the signs of the operators $C - \gamma E$ and $A - \gamma B$ coincide. No property of positiveness of the operator $A$ is required here.

Let now $C = A^{1/2} B^{-1} A^{1/2}$. First, we are going to establish the equivalence of the inequalities $C \ge \gamma E$ ($C \le \gamma E$) and $E \ge \gamma C^{-1}$ ($E \le \gamma C^{-1}$). Upon substituting $y$ for $C^{1/2} x$ we obtain

$$(Cx, x) - \gamma (x, x) = (C^{1/2} x, C^{1/2} x) - \gamma (x, x) = (y, y) - \gamma (C^{-1} y, y) ,$$

thereby justifying coinciding of the signs of the operators $C - \gamma E$ and $E - \gamma C^{-1}$. We now insert $A^{-1/2}BA^{-1/2}$ in place of the inverse $C^{-1}$ and accept $v = A^{-1/2}y$. The outcome of this is

$$(Cx, x) - \gamma (x, x) = (y, y) - \gamma (A^{-1/2}BA^{-1/2}y, y) = (Av, v) - \gamma (Bv, v),$$

meaning that the operators $C - \gamma E$ and $A - \gamma B$ possess the same signs. By merely setting $\gamma = \gamma_1$ and $\gamma = \gamma_2$ we draw the conclusion that inequalities (33) are equivalent.

**Lemma 4** *If the operator $C = C^* > 0$ and $\tau > 0$, then the conditions*

$$\| S \| = \| E - \tau C \| \leq \rho, \tag{34}$$

$$\frac{1-\rho}{\tau} E \leq C \leq \frac{1+\rho}{\tau} E \tag{35}$$

*are equivalent.*

Indeed, since the operator $S = E - \tau C = S^*$ is self-adjoint,

$$\| S \| = \sup_{\| x \| = 1} | (Sx, x) | = \sup_{\| x \| = 1} | ((E - \tau C) x, x) |,$$

so that

$$-\| S \| E \leq S \leq \| S \| E$$

or

$$-\rho E \leq S \leq \rho E$$

and

$$-\rho E \leq E - \tau C \leq \rho E,$$

we deduce that

$$\frac{1-\rho}{\tau} E \leq C \leq \frac{1+\rho}{\tau} E.$$

In view of this, condition (34) implies condition (35). The converse implication is simple to follow.

**Theorem 3** *Let $A$ and $B$ be positive operators and $A = A^*$, $B = B^* > 0$. Then the conditions*

$$\frac{1-\rho}{\tau} B \leq A \leq \frac{1+\rho}{\tau} B \tag{36}$$

*are necessary and sufficient for the $\rho$-stability in the space $H_B$ of scheme* (1a)*:*

$$\| y_n \|_B \le \rho^n \| y_0 \|_B \, .$$

*When, in addition, $A$ is a positive operator, these are necessary and sufficient for the $\rho$-stability in the space $H_A$:*

$$\| y_n \|_A \le \rho^n \| y_0 \|_A \, .$$

In order to prove this theorem, we beforehand reduce the implicit scheme (1a) to the explicit scheme (32) with the operator $C = B^{-1/2} A B^{-1/2}$ (or $C = A^{1/2} B^{-1} A^{1/2}$ for $A > 0$) and then collect the results of Lemmas 3, 2 and 4.

**Remark** One fails to prove Theorem 3 on account of the energy identity (20). However, the method developed in Section 6 may be of assistance in achieving this aim. In the case $D = B$ conditions (36) are equivalent to the condition $\| E - \tau B^{-1} A \|_B \le \rho$ or to being nonnegative of the functional

$$J_B[y] = \rho^2 \, (By, y) - \big( B \, (E - \tau \, B^{-1} A) \, y, (B - \tau \, B^{-1} A) \, y \big) \, .$$

Suppose now that $H$ is a finite-dimensional space of the dimension $N$. Substituting $y = \sum_{k=1}^{N} c_k \, \xi_k$, where $\xi_k$ is an eigenelement of problem (27), into the expression for the fuctional $J_B[y]$ yields

$$J_B[y] = \sum_{k=1}^{N} c_k^2 \, (\rho^2 - (1 - \tau \, \lambda_k)^2) \ge 0 \quad \text{for } \frac{1-\rho}{\tau} \le \lambda_k \le \frac{1+\rho}{\tau} \, ,$$

which is equivalent to condition (36).

**9. Stability with respect to the right-hand side.** Recall that in Section 1 we have established a Theorem 3. This is a way of saying that the stability in the norm $\| \cdot \|_{(1)}$ with respect to the initial data implies the stability with respect to the right-hand side taken in the norm $\| \varphi \|_{(2)} = \| B^{-1} \varphi \|_{(1)}$. An immediate implication of this is covered by the following assertion.

**Theorem 4** *If condition* (14) *is satisfied, then scheme* (1) *from the primary family of schemes is stable with respect to the right-hand side and for a solution of problem* (1) *the a priori estimate holds:*

$$\| y_{n+1} \|_A \le \| y_0 \|_A + \sum_{k=0}^{n} \tau \, \| B^{-1} \varphi_k \|_A \, .$$

*If, in addition, the operator $B$ is self-adjoint, then*

$$\| y_{n+1} \|_B \le \| y_0 \|_B + \sum_{k=0}^{n} \tau \| \varphi_k \|_{B^{-1}} \,.$$

What is more, a priori estimates (3) and (4) are valid with the members $\| \varphi \|_{(2)} = \| \varphi \|_{A^{-1}}$ and $\| \varphi \|_{(2)} = \| \varphi \|$.

**Theorem 5** *Under the condition $B \ge \frac{1}{2} \tau A$, scheme (1) from the primary family of schemes is stable with respect to the right-hand side and for a solution of problem (1) the a priori estimate holds:*

$$(37) \qquad \| y_{n+1} \|_A \le \| y_0 \|_A + \| \varphi_0 \|_{A^{-1}} + \sum_{k=1}^{n} \tau \| \varphi_{\bar{t},k} \|_{A^{-1}} \,.$$

**Proof** In preparation for this, a solution of problem (1) can be arranged as a sum

$$(38) \qquad y_n = v_n + w_n \,,$$

where $w_n$ is a solution of the equation (of the so-called "stationary" problem)

$$(39) \qquad A\, w_n = \varphi_{n-1} \,, \qquad n = 1, 2, \dots \,, \qquad w_0 = w_1 \,.$$

Upon substituting (38) and (39) into equation (1) a new problem arises for $v_n$:

$$(40) \qquad B\, v_t + A\, v = \tilde{\varphi} \,, \qquad v_0 = y_0 - w_0 \,,$$

where $\tilde{\varphi}_n = -(B - \tau A) w_{t,n}$ and $\tilde{\varphi}_0 = 0$. In the estimation of $v$ Theorem 4 gives

$$(41) \qquad \| v_{n+1} \|_A \le \| v_0 \|_A + \sum_{k=0}^{n} \tau \| B^{-1} \tilde{\varphi}_k \|_A \,.$$

A simple observation that

$$w_t = A^{-1} \varphi_{\bar{t}} \,, \qquad \| A^{1/2} B^{-1} \tilde{\varphi} \| = \| (E - \tau C)\, A^{-1/2} \varphi_{\bar{t}} \|$$

with $C = A^{1/2}B^{-1}A^{1/2}$ enables us to evaluate the summand

$$\| B^{-1}\tilde{\varphi}_k \|_A = \| A^{1/2}B^{-1}\tilde{\varphi}_k \|$$

in inequality (41). Stability condition (14) has been used in Section 2 to obtain through such an analysis estimate (24)

$$\| E - \tau C \| \le 1 ,$$

implying that

$$\| A^{1/2}B^{-1}\tilde{\varphi} \| \le \| A^{-1/2}\varphi_{\bar{t}} \| \le \| \varphi_{\bar{t}} \|_{A^{-1}} .$$

where the norm $\|\varphi_{\bar{t}}\|_{A^{-1}} = \sqrt{(A^{-1}\varphi_{\bar{t}}, \varphi_{\bar{t}})}$. Consequently, a solution of problem (40) satisfies the estimate

$$\| v_{n+1} \|_A \le \| v_0 \|_A + \sum_{k=1}^{n} \tau \| \varphi_{\bar{t},k} |_{A^{-1}} ,$$

which in combination with the inequality

$$\| v_0 \|_A \le \| y_0 \|_A + \| w_0 \|_A = \| y_0 \|_A + \| \varphi_0 \|_{A^{-1}}$$

gives

$$\| v_{n+1} \|_A \le \| y_0 \|_A + \| \varphi_0 |_{A^{-1}} + \sum_{k=1}^{n} \tau \| \varphi_{\bar{t},k} |_{A^{-1}} .$$

Finally, by appeal to the triangular inequality from (38) we obtain (37). Thus, Theorem 5 is completely proved.

**Theorem 6** *If the conditions*

$$B \ge \frac{1+\varepsilon}{2} \tau A , \qquad B = B^* \tag{42}$$

*are satisfied, then for scheme* (1) *from the primary family of schemes the a priori estimate holds:*

$$\| y_{n+1} \|_A^2 \le \| y_0 \|_A^2 + \frac{1+\varepsilon}{2\varepsilon} \sum_{k=1}^{n} \tau \| \varphi_k \|_{B^{-1}}^2 , \tag{43}$$

*where* $\varepsilon > 0$ *is a constant independent of* $h$ *and* $\tau$ *both.*

**Proof** The energy identity (13) is involved at the first stage. Estimation of its right-hand side $2\tau(\varphi, y_t)$ is stipulated by successive use of the generalized Cauchy–Bunyakovskiĭ inequality and the $\varepsilon$-inequality

$$2\tau\,(\varphi, y_t) \le 2\tau\,\|\,\varphi\,\|_{B^{-1}} \cdot \|\,y_t\,\|_B \le 2\tau\,\varepsilon_1\,\|\,y_t\,\|_B^2 + \frac{\tau}{2\,\varepsilon_1}\,\|\,\varphi\,\|_{B^{-1}}^2 \,.$$

Upon substituting this estimate into (13) we obtain

$$2\tau\,\Big(\big((1-\varepsilon_1)\,B - \tfrac{\tau}{2}\,A\big)\,y_t\,,y_t\Big) + (A\hat{y},\hat{y}) \le (Ay,y) + \frac{\tau}{2\,\varepsilon_1}\,\|\,\varphi\,|_{B^{-1}} \,.$$

If condition (42) is satisfied, the number $\varepsilon_1$ may be chosen so that $1/(1-\varepsilon_1) = 1+\varepsilon$, that is, $\varepsilon_1 = \varepsilon/(1+\varepsilon)$. Under such an approach we arrive at

$$(1-\varepsilon_1)\,B - \frac{\tau}{2}\,A = (1-\varepsilon_1)\Big(B - \frac{1+\varepsilon}{2}\,\tau A\Big) \ge 0\,,$$

$$(Ay_{k+1}\,,y_{k+1}) \le (Ay_k\,,y_k) + \frac{1+\varepsilon}{2\,\varepsilon}\,\tau\,\|\,\varphi_k\,\|_{B^{-1}}^2 \,.$$

Summing up the latter inequality over $k = 0, 1, 2, \dots$ , leads to estimate (43).

What are the conditions under which the stability in the norm $\|\,\varphi\,\|_{(2)} = \|\,\varphi\,\|$ reveals itself? The following assertion answers this question.

**Theorem 7** *Let the condition*

$$B \ge \varepsilon\,E + 0.5\,\tau\,A \tag{44}$$

*be satisfied, where $\varepsilon$ is a positive number, and scheme (1) belongs to the primary family of schemes. Then for a solution of problem (1) the a priori estimate holds:*

$$\|\,y_{n+1}\,\|_A^2 \le \|\,y_0\,\|_A^2 + \frac{1}{2\,\varepsilon}\sum_{k=0}^{n} \tau\,\|\,\varphi_k\,\|^2. \tag{45}$$

**Proof** With regard to identity (13) the Cauchy–Bunyakovskiĭ inequality and the $\varepsilon$-inequality together give

$$2\tau\,(\varphi, y_t) \le 2\tau\,\|\,\varphi\,\| \cdot \|\,y_t\,\| \le 2\tau\,\varepsilon\,\|\,y_t\,\|^2 + \frac{\tau}{2\,\varepsilon}\,\|\,\varphi\,\|^2.$$

Furthermore, we substitute this estimate into (13). Having stipulated condition (44), it follows from the foregoing that

$$\| \hat{y} \|_A^2 \le \| y \|_A^2 + \frac{\tau}{2\varepsilon} \| \varphi \|^2$$

or

$$\| y_{j+1} \|_A^2 \le \| y_j \|_A^2 + \frac{\tau}{2\varepsilon} \| \varphi_j \|^2 .$$

Summing up then over $j = 0, 1, \dots, n$ leads to (45). Thus, Theorem 7 is completely proved.

**Remark 1** Theorem 7 remains valid in the case of a variable operator $B = B(t)$ and Theorem 2 continues to hold for a variable operator $A = A(t)$. The reader is invited to verify these facts on his/her own with the aid of the proofs of the aforementioned theorems.

**Remark 2** If $A = A^*$, $B = B^* > 0$ and conditions (36) are satisfied, then scheme (1) is $\rho$-stable in the space $H_B$ with respect to the initial data, that is, a solution of problem (1a) satisfies the inequality

$$\| y_n \|_B \le \rho^n \| y_0 \|_B .$$

On the strength of Theorem 3 in Section 1 this implies the following estimate for a solution of problem (1):

$$\| y_{n+1} \|_B \le \rho^{n+1} \| y_0 \|_B + \sum_{k=0}^{n} \tau \rho^{n-k} \|\varphi_k\|_{B^{-1}} .$$

It is worth noting here that the positiveness of the operator $A$ is unnecessary in this matter. Apparently, this remark needs certain clarification. Let, for instance, $A \ge -c_* E$ and $c_* > 0$. When this is the case, the condition

$$A \ge \frac{1-\rho}{\tau} B \qquad \text{or} \qquad A + \frac{\rho - 1}{\tau} B \ge 0$$

can be met only for $\rho = \exp\{c_0 \tau\} > 1$, that is, for $c_0 \ge 0$. Under the agreement $B \ge \varepsilon E$, $\varepsilon > 0$, we find that

$$A + \frac{\rho - 1}{\tau} B \ge A + c_0 B \ge (-c_* + c_0 \varepsilon) E$$

and, therefore,

$$A \ge \frac{1-\rho}{\tau} B$$

if we accept $c_0 \ge c_*/\varepsilon$. To make sure of it, we may choose as $B$, for example, the operator

$$B = E + \tau R , \qquad R = 0.5 A' , \qquad A' = A + c_* E \ge 0 .$$

Then $\varepsilon = 1$, $c_0 = c_*$ and $\rho = \exp\{c_* \tau\}$.

**10. Stability of the weighted scheme.** As an example of applying the theorems just established the weighted scheme comes first

$$(46) \qquad y_t + A\left(\sigma \hat{y} + (1-\sigma)\, y\right) = \varphi\,, \qquad y(0) = y_0\,.$$

Recall that in Section 1 scheme (46) has been already reduced to the canonical form

$$(47) \qquad (E + \sigma\,\tau\, A)\, y_t + Ay = \varphi\,, \qquad y(0) = y_0\,.$$

By comparing (46) with (1) we see that $B = E + \sigma\tau A$. Suppose there exists an inverse operator $A^{-1}$. Applying $A^{-1}$ to (47) reveals the second canonical form of the weighted scheme:

$$(48) \qquad \begin{gathered} \widetilde{B} y_t + \widetilde{A} y = \widetilde{\varphi}\,, \qquad y(0) = y_0\,, \\ \widetilde{B} = A^{-1} + \sigma\,\tau\, E\,, \qquad \widetilde{A} = E\,, \qquad \widetilde{\varphi} = A^{-1}\varphi\,. \end{gathered}$$

In the sequel the form (47) is more convenient for the case of self-adjoint operators $A$ and the form (48) – for the case of non-self-adjoint positive definite operators $A = A(t)$. We will elaborate on this later.

The first analysis is connected with the case when $A$ is a constant self-adjoint positive operator $A = A^* > 0$. As we have shown in Section 2, a necessary and sufficient condition for the stability of the weighted scheme (47) with respect to the initial data is

$$\sigma \geq \sigma_0\,, \qquad \sigma_0 = \frac{1}{2} - \frac{1}{\tau \|\, A\,\|}\,.$$

Under this condition estimate (15) holds true for a solution of problem (47). In particular, for an explicit scheme (for $\sigma = 0$) the condition $\sigma \geq \sigma_0$ implies $\tau \leq 2/\|\, A\,\|$, that is, an explicit scheme is stable in the space $H_A$ for $\tau \leq 2/\|\, A\,\|$. A scheme with $\tau \geq \frac{1}{2}$ is unconditionally stable, that is, for any $\tau$. In Section 2.4 a model example with $Ay = -\Lambda y = -y_{\bar{x}x}$ for $y \in \overset{\circ}{\Omega}_h = H_h$ has been considered in full details, in which $\|\, A\,\| < 4/h^2$ and the appropriate explicit scheme is stable for $\tau \leq \frac{1}{2} h^2$.

For the heat conduction equation with a variable coefficient $k(x)$ we might have

$$L\,u = \frac{\partial}{\partial x}\left(k(x)\,\frac{\partial u}{\partial x}\right), \qquad 0 < k \leq c_2\,,$$

$$\Lambda\, y = \left(a(x)\, y_{\bar{x}}\right)_x\,, \qquad 0 < a \leq c_2\,,$$

and $\sigma_0 < \frac{1}{2} - h^2/(4c_2\tau)$, while an explicit scheme is stable for $\tau \leq \frac{1}{2} h^2/c_2$.

**Theorem 8** *Let $A$ be a self-adjoint positive operator independent of $t = n\tau$; $A = A^* > 0$. Then for the weighted scheme (47) estimate (37) is valid for $\sigma \geq \sigma_0$, estimate (43) for $\sigma \geq \frac{1+\varepsilon}{2} - \frac{1}{\tau \| A \|}$, $\varepsilon > 0$, and estimate (45) for $\sigma \geq \sigma_\varepsilon$, $\sigma_\varepsilon = \frac{1}{2} - (1-\varepsilon)/(\tau \| A \|)$, $0 < \varepsilon \leq 1$, where a number $\varepsilon$ is independent of $h$ and $\tau$ both.*

The above statements will be proved if we succeed in verifying the fulfilment of the conditions of Theorems 5, 6, 7 with the aid of the inequality

$$B - \frac{1}{2}\tau A = E + \left(\sigma - \frac{1}{2}\right)\tau A \geq \left(\frac{1}{\| A \|} + \left(\sigma - \frac{1}{2}\right)\tau\right) A .$$

Suppose now that $A = A(t) > 0$ and $A^* \neq A$, that is, $A = A(t)$ is a positive non-self-adjoint variable operator.

**Theorem 9** *Let $A = A(t) > 0$ be a positive non-self-adjoint variable operator. If $\sigma \geq \frac{1}{2}$, then for scheme (46) the estimate holds:*

$$(49) \quad \| y_{n+1} \| \leq \| y_0 \| + \| (A^{-1}\varphi)_0 \| + \| (A^{-1}\varphi)_n \| + \sum_{k=1}^{n} \tau \| (A^{-1}\varphi)_{\bar{t},\,k} \| .$$

**Proof** Consider scheme (48) with the right-hand side $\tilde{\varphi} = A^{-1}\varphi$. When $\sigma \geq \frac{1}{2}$, the conditions of Theorem 5 are satisfied for this scheme and inequality (37) applies here together with the established relations: $\|y\|_{\tilde{A}} = \| y \|$ and

$$\|\tilde{\varphi}_{\bar{t},k}\|_{\tilde{A}^{-1}} = \| \tilde{\varphi}_{\bar{t},k} \| = \| (A^{-1}\varphi)_{\bar{t},k} \| .$$

As a final result we obtain (49).

**Remark** If $A^* = A^*(t)$ is a self-adjoint operator, then estimate (49) holds true for $\sigma \geq \sigma_0$.

**Theorem 10** *Let $A(t) = A^*(t) > 0$ and $\sigma \geq \sigma_0$. Then the following estimate is valid:*

$$(50) \qquad \| y_{n+1} \|^2 \leq \| y_0 \|^2 + + \frac{1}{2\,\varepsilon} \sum_{k=1}^{N} \tau \| \varphi_k \|^2_{A_k^{-1}} .$$

**Proof** The energy identity (13) for scheme (48) takes for now the form

$$(51) \quad 2\tau \left( (A^{-1} + (\sigma - 0.5)\,\tau\, E)\, y_t, y_t \right) + \| \hat{y} \|^2 = \| y \|^2 + 2\tau\,(A^{-1}\varphi, y_t) .$$

The condition $\sigma \geq \sigma_\varepsilon$ implies that

$$\widetilde{B} - \frac{\tau}{2}\,\widetilde{A} \geq \varepsilon\, A^{-1}\,. \tag{52}$$

Indeed, for $\sigma \geq \sigma_\varepsilon$

$$\begin{aligned}
\widetilde{B} - \frac{\tau}{2}\,\widetilde{A} &= A^{-1} + (\sigma - 0.5)\,\tau E \\
&= \varepsilon\, A^{-1} + (1-\varepsilon)\, A^{-1} + (\sigma - 0.5)\,\tau E \\
&\geq \varepsilon\, A^{-1} + \frac{1-\varepsilon}{\|\, A\,\|}\, E + (\sigma - 0.5)\,\tau E \\
&= \varepsilon\, A^{-1} + (\sigma - \sigma_\varepsilon)\,\tau E \geq \varepsilon\, A^{-1}\,.
\end{aligned}$$

We have taken into account here that $A^{-1} \geq E/\|\,A\,\|$. Substituting (52) into (51) we obtain

$$2\,\tau\,\varepsilon\, \|y_t\|^2_{A^{-1}} + \|\,\hat{y}\,\|^2 \leq \|\,y\,\|^2 + 2\tau\,(A^{-1}\varphi, y_t)\,. \tag{53}$$

The generalized Cauchy–Bunyakovskiĭ inequality and the $\varepsilon$-inequality together yield

$$2\,\tau\,(A^{-1}\varphi, y_t) \leq 2\,\tau\,\|\varphi\|_{A^{-1}} \cdot \|y_t\|_{A^{-1}} \leq 2\,\tau\varepsilon\,\|y_t\|^2_{A^{-1}} + \frac{\tau}{2\,\varepsilon}\,\|\varphi\|^2_{A^{-1}}\,. \tag{54}$$

Substituting (54) into (53) we arrive at the relation

$$\|\,\hat{y}\,\|^2 \leq \|\,y\,\|^2 + \frac{\tau}{2\,\varepsilon}\,\|\varphi\|^2_{A^{-1}}$$

or

$$\|\,y_{k+1}\,\|^2 \leq \|\,y_k\,\|^2 + \frac{\tau}{2\,\varepsilon}\,\|\varphi_k\|^2_{A_k^{-1}}\,.$$

Summing the preceding over $k = 0, 1, 2, \ldots$ leads to estimate (50).

**Lemma 5** *Let $A$ be a positive operator for which the inequality*

$$\|\,Ax\,\|^2 \leq \Delta\,(Ax, x)\,, \tag{55}$$

*is valid with $\Delta = const > 0$. Then*

$$A^{-1} \geq \frac{1}{\Delta}\,E \qquad \text{and} \qquad A \leq \Delta\, E\,. \tag{56}$$

Indeed, putting $x = A^{-1}y$ we obtain from (55) the inequality

$$(y, y) \le \Delta(A^{-1}y, y),$$

meaning $E \le \Delta A^{-1}$ or $A^{-1} \ge E/\Delta$. By virtue of the inequalities

$$(Ax, x)^2 \le \| Ax \|^2 \cdot \| x \|^2 \le \Delta(Ax, x) \| x \|^2$$

we deduce that $(Ax, x) < \Delta \| x \|^2$, giving $A \le \Delta E$. Lemma 5 implies that

$$\widetilde{B} = A^{-1} + \sigma\tau E \ge (1/\Delta + \sigma\tau) E, \qquad \widetilde{B} - 0.5\tau\widetilde{A} \ge 0 \quad \text{for } \sigma \ge \bar{\sigma}_0,$$

where $\widetilde{A} = E$, $\bar{\sigma}_0 = \frac{1}{2} - 1/(\tau\Delta)$.

**Lemma 6** *Let $A$ be a positive definite operator and inequality* (55) *hold. Then*

$$(57) \qquad \| (E + \sigma\tau A)^{-1} (E - (1 - \sigma)\tau A) \| \le 1 \quad \text{for } \sigma \ge \bar{\sigma}_0 = \frac{1}{2} - \frac{1}{\tau\Delta},$$

$$(58) \qquad \| (E + \sigma\tau A)^{-1} \| \le 1 \quad \text{for } \sigma \ge 0,$$

$$(59) \qquad \| (E + \sigma\tau A)^{-1} \| \le \frac{1}{\varepsilon} \quad \text{for } \sigma \ge \bar{\sigma}_\varepsilon = \frac{1}{2} - \frac{1-\varepsilon}{\tau\Delta},$$

$$0 < \varepsilon \le 1.$$

**Proof** 1) Since $\widetilde{B} \ge \frac{1}{2}\tau\widetilde{A}$ for $\sigma \ge \bar{\sigma}_0$, by applying Theorem 1 to scheme (48) we deduce that for a solution of problem (48) the estimate is valid for any $y_n \in H$ and $\varphi = 0$:

$$(60) \qquad \| y_{n+1} \| \le \| y_n \| .$$

As can readily be observed, scheme (47) for $\varphi = 0$ can be rewritten as

$$y_{n+1} = S y_n, \qquad S = (E + \sigma\tau A)^{-1} (E - (1 - \sigma)\tau A).$$

From here and relation (60) we obtain estimate (57).

2) In order to estimate the norm $\| B^{-1} \|$, where $B = E + \sigma\tau A$, it suffices to establish an inequality of the form $B \ge \sigma E$, $\sigma > 0$. Then

$$\delta \| x \|^2 \le (Bx, x) \le \| Bx \| \cdot \| x \|, \qquad \| Bx \| \ge \delta \| x \|,$$

and, consequently, $\| B^{-1} \| \le 1/\delta$. If $\sigma \ge 0$, then $B \ge E$ and $\| B^{-1} \| \le 1$. If $\sigma \ge \bar{\sigma}_\varepsilon$, then

$$B \ge E + \bar{\sigma}_\varepsilon \tau A = E + 0.5\,\tau A - \frac{1-\varepsilon}{\Delta} A.$$

In conformity with (56), $A \le \Delta E$, making it possible to conclude that

$$B \ge E + 0.5\,\tau A - (1 - \varepsilon) E \ge \varepsilon E + 0.5\,\tau A > \varepsilon E$$

and, therefore, $\| B^{-1} \| \le 1/\varepsilon$. Observe that estimate (58) holds true also for any non-self-adjoint operator $A > 0$, thereby completing the proof of the lemma.

Joint use of Lemma 6 and inequality (60) permits us to state the following.

**Theorem 11** *Let $A = A(t)$ be a positive operator and condition* (55) *hold. Then for scheme* (46) *with $\sigma \geq \sigma_\varepsilon$ the a priori estimate*

$$\| y(t+\tau) \| \leq \| y(0) \| + \frac{1}{\varepsilon} \sum_{t'=0}^{t} \tau \| \varphi_{t'} \| \tag{61}$$

*is true. If the following conditions*

$$\sigma \geq 0, \qquad \sigma \geq \bar{\sigma}_0, \qquad \bar{\sigma}_0 = \frac{1}{2} - \frac{1}{\tau \Delta}$$

*are simultaneously satisfied, then estimate* (61) *is attained for $\varepsilon = 1$.*

To prove this theorem, let us write down scheme (46) in the form

$$y_{n+1} = S\, y_n + \tau B^{-1} \varphi_n ,$$

where $S = (E + \sigma\tau A)^{-1}(E - (1-\sigma)\tau A)$ and $B = E + \sigma\tau A$. Using the triangular inequality and estimates (57)–(59) behind, we obtain the relation

$$\| y_{n+1} \| \leq \| y_n \| + \frac{\tau}{\varepsilon} \| \varphi_n \| ,$$

from which estimate (61) immediately follows.

**11. A priori estimates in the case of a variable operator A.** So far we have established stability in $H_A$ under the agreement that operator $A$ is constant, that is, independent of $t$. In the case when $A(t) = A^*(t) > 0$ depends on $t$, this obstacle necessitates imposing the **Lipschitz continuity of the operator** $A(t)$ in the variable $t$

$$|((A(t) - A(t-\tau))\, x, x| \leq \tau\, c_3 \big(A(t-\tau)\, x, x\big) \tag{62}$$

for all $x \in H$, $0 < t < n_0\tau$, where $c_3$ is a positive constant independent of $h$ and $\tau$ both.

A primary family of describing schemes is specified by the following restrictions:

$$\begin{aligned} & A(t) = A^*(t) > 0 \qquad \text{for all} \quad t \in \bar{\omega}_\tau , \\ & A(t) \qquad \text{is Lipschitz continuous in} \quad t , \\ & B(t) > 0 \qquad \text{for all} \quad t \in \bar{\omega}_\tau . \end{aligned} \tag{63}$$

As before we assume the existence of an inverse operator $B^{-1}(t)$, which assures us of solvability of problem (1) for any input data $y_0$ and $\varphi(t)$. This family obviously contains the primary family introduced in Section 2.2.

The method of energy inequalities shows in such a setting that the conditions

$$B(t) \geq 0.5\,\tau\,A(t) \qquad \text{for all} \quad t \in \omega_\tau\,, \tag{64}$$

$$B(t) \geq \varepsilon\,E + 0.5\,\tau\,A(t) \qquad \text{for all} \quad t \in \omega_\tau\,, \quad 0 < \varepsilon \leq 1\,, \tag{65}$$

turn out to be sufficient for the stability of scheme (1) with variable operators $A(t)$ and $B(t)$. The norms $\|\cdot\|_A$, $\|\cdot\|_{A^{-1}}$ themselves happen to be dependent on the variable $t$:

$$\|\,y\,\|_A = \|y\|_{A(t)} = \sqrt{(A(t)\,y,y)}\,, \qquad \|\varphi\|_{A^{-1}(t)} = \sqrt{(A^{-1}(t)\,\varphi,\varphi)}\,.$$

Therefore, it makes sense to speak about stability in the space $H_{A(t)}$ (instead of $H_A$) and $H_{B(t)}$.

The energy identity (13) with $A = A(t)$ is the starting point in special investigations. To obtain a recurrence inequality, we should modify the expression

$$\begin{aligned}(Ay,y) = \big(A(t)\,y(t),y(t)\big) &= \big(A(t-\tau)\,y(t),y(t)\big) \\ &+ \big((A(t) - A(t-\tau))\,y(t),y(t)\big)\end{aligned}$$

and estimate the second summand on the right-hand side with the aid of inequality (62):

$$\big(A(t)\,y(t),y(t)\big) \leq (1+\tau\,c_3)\big(A(t-\tau)\,y(t),y(t)\big)\,.$$

Upon substituting this estimate into (13) we obtain the energy inequality

$$\begin{aligned}2\tau\,\big((B(t) - 0.5\,\tau\,A(t))\,y(t),y(t)\big) &+ \mathcal{E}(t+\tau) \\ &\leq (1+\tau\,c_3)\mathcal{E}(t) + 2\tau\,\big(\varphi(t),y(t)\big)\,,\end{aligned} \tag{66}$$

where

$$\mathcal{E}(t+\tau) = \big(A(t)\,y(t+\tau),y(t+\tau)\big) = \|y(t+\tau)\|^2_{A(t)}\,.$$

Inequality (66) with the member $\varphi = 0$ implies that

$$\mathcal{E}(t+\tau) \leq (1+\tau\,c_3)\,\mathcal{E}(t) \leq e^{c_3\,t}\mathcal{E}(\tau) \quad \text{for} \quad t \geq \tau\,, \tag{67}$$

provided condition (64) holds. The energy identity for $t = 0$ admits the form

$$2\tau\left((B(0) - 0.5\,\tau A(0))\, y(t), y(t)\right) + \mathcal{E}(\tau) = \|y(0)\|^2_{A(0)} + 2\tau\left(\varphi(0), y_t(0)\right).$$

With conditions (64) and $\varphi = 0$ in view, we deduce from here that

$$\mathcal{E}(\tau) \le \|y(0)\|^2_{A(0)}\,. \tag{68}$$

Collecting (67) and (68) with regard to problem (1) with $\varphi = 0$ we establish

$$\|y(t+\tau)\|^2_{A(t)} \le M_1\,\|y(0)\|_{A(0)}\,, \qquad M_1 = e^{0.5\,c_3 t_0}.$$

Such a reasoning discovers actually only one essential difference between the cases of variable and constant operators.

We now summarize the above results in some aspects as the analogs of Theorems 5 and 7 that furnish the justification for what we wish to do.

**Theorem 12** *Let operators $A = A(t)$ and $B = B(t)$ be dependent on $t$ and conditions* (63) *and* (64) *be satisfied. Then for scheme* (1) *we have the estimate*

$$\|y_{n+1}\|_{A_n} \le M_1 \times \left\{\|y_0\|_{A_0} + \|\varphi_0\|_{A_0^{-1}} + \|\varphi_n\|_{A_n^{-1}} + \sum_{k=1}^{n} \tau\,\|(A^{-1}\varphi)_{\bar t,\,k}\|_{A_k}\right\}, \tag{69}$$

*where $M_1 = \exp\left\{\frac{1}{2}c_3 t_0\right\}$ and $A_n = A(t_n)$.*

If $c_3 = 0$, then the operator $A$ is independent of the variable $t$ and (69) becomes (37) on account of the equality

$$\|\,A^{-1}\varphi_{\bar t}\,\|_A = \|\varphi_{\bar t}\|_{A^{-1}}\,.$$

**Theorem 13** *Let operators $A = A(t)$ and $B = B(t)$ be dependent on $t$ and conditions* (63) *and* (65) *be satisfied. Then for scheme* (1) *we have the estimate*

$$\|y_{n+1}\|^2_{A_n} \le M_1^2\left\{\|y_0\|^2_{A_0} + \frac{1}{2\,\varepsilon}\sum_{k=0}^{n} \tau\,\|\,\varphi_k\,\|^2\right\}, \qquad M_1 = e^{0.5\,c_3 t_0}. \tag{70}$$

Estimates (37) and (45) are obtained for the case of constant operators $A$ by merely setting $M_1 = 1$ or $c_3 = 0$.

**12. Example.** In order to apply the general stability theory for particular difference schemes, one needs to perform several manipulations:

- to reduce a two-layer scheme to the canonical form (1), that is, to specify the operators $A$ and $B$ involved;
- to introduce the space $H_h$ of all grid functions and reveal the desirable properties of the operators $A$ and $B$ such as positiveness, self-adjointness, etc. as operators acting in the space $H_h$;
- to verify whether or not the scheme belongs to the primary family of schemes as well as the fulfilment of sufficient stability conditions (64) or (65);
- if the preceding conditions are satisfied, the scheme at hand is stable and a priori estimates hold for it such as, for example, (69) and (70).

The first step within this framework is to reduce a scheme to the canonical form, but the above sufficient conditions provide a real possibility of writing stable difference schemes immediately in canonical form.

We cite here only one possible example which helps motivate what is done. Those ideas are connected with the heat conduction equation

$$\frac{\partial u}{\partial t} = \frac{\partial^2 u}{\partial x^2}\,, \qquad 0 < x < 1\,, \qquad t > 0\,,$$

$$u(x,0) = u_0(x)\,, \qquad u(0,t) = u(1,t) = 0\,,$$

and associated asymmetric scheme which is given on the grid

$$\bar{\omega}_{h\tau} = \bar{\omega}_h \times \bar{\omega}_\tau\,,$$

$$\bar{\omega}_h = \{x_i = ih,\ 0 \le i \le N\}\,, \quad \bar{\omega}_\tau = \{t_j = j\tau,\ 0 \le j \le j_0\}\,,$$

and asquires the form

$$(71)\quad y_{i,j+1} = \frac{1}{\omega + \alpha}\left(\alpha\, y_{i-1,j+1} + (1-\alpha)\, y_{i-1,j} + y_{i+1,j} - (2-\omega-\alpha)\, y_{i,j}\right),$$

where $\omega = h^2/\tau$ and $\alpha$ is a parameter.

We do follow established practice in a step-by-step fashion. 1) Reduction of the scheme which interests us to the canonical form. Denoting $y_{i,j} = y_i$ and $y_{i,j+1} = \hat{y}_i$ we first rewrite (71) as

$$(72)\qquad (\omega + \alpha)\,\hat{y}_i = \alpha\,\hat{y}_{i-1} + (1-\alpha)\,y_{i-1} + y_{i+1} - (2-\omega-\alpha)\,y_i\,.$$

After that, taking into account that

$$y_{i-1} = -h\, y_{\bar{x},i} + y_i\,, \quad y_{i+1} = h\, y_{x,i} + y_i\,, \quad h\, y_{x,i} - h\, y_{\bar{x},i} = h^2 y_{\bar{x}x,i}\,,$$

$$\hat{y}_{i-1} = y_{i-1} + \tau\, y_{t,i-1} = y_{i-1} - h\, y_{\bar{x},i} + \tau\, y_{t,i} - h\,\tau\, y_{\bar{x}t,i}\,,$$

substituting these expressions into (72) and omitting the subscript $i$, we come to

$$\omega\,\tau\, y_t = h^2\, y_{\bar{x}x} - \alpha\, h\,\tau\, y_{\bar{x}t}\,. \tag{73}$$

Dividing (73) by $h^2$ generates

$$y_t + \frac{\alpha\,\tau}{h}\, y_{\bar{x}t} = y_{\bar{x}x}\,. \tag{74}$$

2) Let $H_h$ be the space of grid functions $\overset{\circ}{\Omega}_h$ (see Examples 1 and 2 in Chapter 2, Section 4.1) defined on the grid $\omega_h = \{x_i = ih, 0 < i < N\}$ under the inner product structure $(y, v) = \sum_{i=1}^{N-1} y_i\, v_i\, h$. In conformity with the results obtained in Section 4.1 of Chapter 2, the operators $Ay = -y_{\bar{x}x}$ and $R_1 y = \frac{1}{h} y_{\bar{x}}$ involved in the scheme are positive definite: $(R_1 y, y) = \frac{1}{2}(Ay, y)$. The operator $A$ is self-adjoint, $\|A\| \le 4/h^2$.

3) The operators $A$ and $R_1$ are constant. Because of this, it will be convenient to write scheme (74) in the form

$$(E + \alpha\,\tau\, R_1)\, y_t + Ay = 0\,, \tag{75}$$

so that $B = E + \alpha\tau R_1$.

The condition $B \ge \frac{1}{2}\tau A$ is satisfied for $\alpha \ge 1 - 2/(\tau\|A\|)$. Indeed, for any $x \in H$

$$\big((B - 0.5\,\tau\, A)\, x, x\big) = \big((E + \alpha\,\tau\, R_1 - 0.5\,\tau\, A)\, x, x\big)$$
$$= \big((E + 0.5\,\tau\,(\alpha - 1)\, A)\, x, x\big)\,,$$

yielding

$$B - 0.5\,\tau\, A \ge E + 0.5\,\tau\,(\alpha - 1)\, A \ge \Big(\frac{1}{\|A\|} + 0.5\,\tau\,(\alpha - 1)\Big) A \ge 0\,.$$

4) Since $\|A\| \le 4/h^2$, scheme (71) is stable in the space $H_A$ (in the grid norm of the space $W_2^1$) for

$$\alpha \ge 1 - \frac{h^2}{2\,\tau}\,. \tag{76}$$

Together with scheme (71) one can write down one more asymmetric scheme which after reducing it to the canonical form becomes

$$(E + \alpha\,\tau\,R_2)\,y_t + A\,y = 0\,, \qquad R_2\,y = -\frac{1}{h}\,y_x\,.$$

Since $(R_1 y, y) = (R_2 y, y)$, this scheme is stable under the same condition (76). Condition (76) shows that the asymmetric schemes are unconditionally stable for $\alpha \ge 1$.

**13. The case of a skew-symmetric operator A.** The main results of stability theory for two-layer schemes

$$B\,y_t + A\,y = \varphi$$

have been obtained under the agreement that $A = A^* > 0$ is a positive self-adjoint operator, while the operator $B > 0$ may be, generally speaking, non-self-adjoint. An exception is a weighted scheme in which the operator $B$ is of the special type $B = E + \sigma\tau A$ and $A^* \ne A$.

Assume that $A$ is a non-self-adjoint operator with the approved decomposition $A = A_0 + A_1$, where $A_0 = \frac{1}{2}(A + A^*)$, $A_1 = \frac{1}{2}(A - A^*)$, $A_0^* = A_0$, $A_1^* = -A_1$, that is, $A_1$ is a skew-symmetric operator,

$$(A_1 y, v) = -(y, A_1 v)$$

and

$$(A_1 y, y) = -(y, A_1 y) = 0\,.$$

To avoid generality, for which we have no real need, we restrict ourselves here to the case when $A = A_1$ is a skew-symmetric operator involved in the weighted scheme

$$y_t + A\,y^{(\sigma)} = 0\,,$$

where $A^* = -A$, $(Ay, y) = 0$, $y^{(\sigma)} = \sigma\hat{y} + (1-\sigma)y$, and attempt it in the form

$$B\,y_t + A\,y = 0\,, \qquad B = E + \sigma\,\tau\,A\,.$$

Since $A$ and $B$ are non-self-adjoint operators, we cannot use the obtained results of the general theory, so there is some reason to be concerned about this. With this aim, we proceed to develop a new approach in such matters.

Observe, first of all, that

$$(By, y) = (y + \sigma\,\tau\,A\,y, y) = (y, y) + \sigma\,\tau\,(A\,y, y) = \|\,y\,\|^2\,,$$

that is, there exists an inverse operator $B^{-1}$. By inserting $y = B^{-1}x$ it is plain to establish the relations $\| B^{-1}x \|^2 = (B^{-1}x, x) \le \| B^{-1}x \| \cdot \| x \|$, giving, $\| B^{-1}x \| \le \| x \|$ and $\| B^{-1} \| \le 1$. The next step is to rewrite the scheme in the form

$$B\hat{y} = By - \tau A y\,, \qquad \hat{y} = y - \tau B^{-1} A y\,,$$

and calculate the inner product

$$(B\hat{y}, \hat{y}) = (By, y) - \tau (Ay, y) - \tau (By, B^{-1}Ay) + \tau^2 (Ay, B^{-1}Ay)\,.$$

For later use, it will be sensible to represent the operator $B$ in the form

$$B = E + \sigma\tau A = (E - \sigma\tau A) + 2\sigma\tau A = B^* + 2\sigma\tau A$$

and make the obvious transformations

$$\begin{aligned}(By, B^{-1}Ay) &= (B^*y, B^{-1}Ay) + 2\sigma\tau (Ay, B^{-1}Ay) \\ &= (y, Ay) + 2\sigma\tau (Ay, B^{-1}Ay) \\ &= 2\sigma\tau (Ay, B^{-1}Ay)\,.\end{aligned}$$

As a final result we obtain

$$\| \hat{y} \|^2 = \| y \|^2 - (2\sigma - 1)\,\tau^2 (B^{-1}Ay, Ay)\,,$$

yielding

$$\| \hat{y} \| \le \| y \| \qquad \text{if } \sigma \ge 0.5\,,$$

$$\| \hat{y} \|^2 \le \| y \|^2 + (1 - 2\sigma)\,\tau^2 \| A \|^2 \cdot \| y \|^2 \le (1 + c_0\tau) \| y \|^2$$

or

$$\| \hat{y} \| \le \rho \| y \|\,, \qquad \rho = e^{0.5 c_0 \tau}\,, \qquad c_0 = (1 - 2\sigma)\,c_2\,,$$

if $\sigma < \frac{1}{2}$ and $\tau \| A \|^2 \le c_2$.

We have obtained through such an analysis the following estimates for the norm of the transition operator:

$$\| S \| \le 1 \quad \text{for} \quad \sigma \ge 0.5\,,$$

$$\| S \| \le \rho \quad \text{for} \quad \sigma < 0.5 \quad \text{and} \quad \tau \| A \|^2 \le c_2\,.$$

This implies the estimate for a solution of the nonhomogeneous equation $By_t + Ay = \varphi$:

$$\| y_{n+1} \| \le \rho \| y_n \| + \tau \| \varphi_n \| \le \rho^{n+1} \| y_0 \| + \sum_{k=0}^{n} \tau \rho^{n-k} \| \varphi_k \| .$$

Let now $H$ be a complex space and $B$ be a non-self-adjoint operator. Then a necessary and sufficient condition for the stability in the space $H_A$ with respect to the initial data of the scheme

$$B y_t + A y = 0$$

is of the form

$$B_0 = \operatorname{Re} B \ge 0.5 \tau A , \qquad A = A^* > 0 ,$$

where $B_0 = \frac{1}{2} (B + B^*) = B_0^*$. In particular, the scheme

$$i y_t + A y = 0 , \qquad A = A^* > 0$$

is unstable in the space $H_A$, since $B = iE$ and $\operatorname{Re} B = 0$. However, $A' = -iA$ is a skew-symmetric operator $(A')^* = -A'$ and, as stated before, this scheme is conditionally stable in the space $H$:

$$\| y_{n+1} \| \le \rho^{n+1} \| y_0 \| , \quad \rho = e^{c_0 \tau} , \quad c_0 = c_2 \quad \text{if } \tau \| A \|^2 \le c_2 .$$

In the case of the Schrödinger equation $Ay = -y_{\bar{x}x}$, $y \in \overset{\circ}{\Omega}_h[0,1]$, we have $\| A \| < 4/h^2$ and the restriction on $\tau$ takes the form

$$\tau \le \frac{h^4}{16} c_2 ,$$

which is improper for parabolic equations. However, the weighted scheme

$$i y_t + A y^{(\sigma)} = 0 , \qquad A = A^* > 0 ,$$

will be stable in both spaces $H_A$ and $H$ for $\sigma \ge \frac{1}{2}$. If so,

$$B_0 = \operatorname{Re} B = \sigma \tau A > \tfrac{1}{2} \tau A$$

and $\| y_n \|_A \le \| y_0 \|_A$. On the other hand, in dealing with the skew-symmetric operator $A' = -iA$ we get the estimate $\| y_n \| \le \| y_0 \|$ for $\sigma \ge \frac{1}{2}$.

## 6.3 CLASSES OF STABLE THREE-LAYER SCHEMES

**1. The problem statement.** In this section we establish sufficient stability conditions and a priori estimates for three-layer schemes on the basis of their canonical form

$$(1)\qquad \begin{gathered} B\, y_{\overset{\circ}{t}} + \tau^2 R\, y_{\bar{t}t} + A\, y = \varphi(t)\,, \qquad y(0) = y_0\,, \qquad y(\tau) = y_1\,, \\ 0 < t = n\tau < t_0\,, \qquad n = 1, 2, \ldots, n_0 - 1\,, \qquad t_0 = n_0 \tau\,. \end{gathered}$$

Here and below, $y_0$ and $y_1$ are arbitrary given vectors of a finite-dimensional real space $H$, $\varphi(t)$ is a given arbitrary abstract $H$-valued function of the variable $t \in \omega_\tau$; $A$, $B$ and $R$ are linear operators in the space $H$. The dependence of $y(t) = y_{h\tau}(t)$, $\varphi(t) = \varphi_{h\tau}(t)$, $A(t) = A_{h\tau}$, $B$, $R$, $y_0$ and $y_1$ on $h$ and $\tau$ is not explicitly indicated. We proceed as usual. This amounts to introducing more compact notations:

$$y = y(t_n) = y_n\,, \qquad \hat{y} = y(t_n + \tau) = y_{n+1}\,, \qquad \check{y} = y(t_n - \tau) = y_{n-1}\,,$$

$$y_t = (\hat{y} - y)/\tau\,, \qquad y_{\bar{t}} = (y - \check{y})/\tau\,, \qquad y_{\overset{\circ}{t}} = (\hat{y} - \check{y})/(2\,\tau)\,,$$

$$y_{\bar{t}t} = (\hat{y} - 2y + \check{y})/\tau^2\,,$$

allowing a simpler writing of the ensuing formulae.

All the tricks and turns remain unchanged by analogy with Section 2: first, a solution of problem (1) can be arranged as a sum $y = \bar{y} + \tilde{y}$, where $\bar{y}$ is a solution of the homogeneous equation:

$$(1a)\qquad \begin{gathered} B\, y_{\overset{\circ}{t}} + \tau^2 R\, y_{\bar{t}t} + A\, y = 0\,, \qquad y(0) = y_0\,, \qquad y(\tau) = y_1\,, \\ 0 < t = n\tau < t_0\,, \end{gathered}$$

and $\tilde{y}$ is a solution of the nonhomogeneous equation with the zero initial data:

$$(1b)\qquad \begin{gathered} B\, y_{\overset{\circ}{t}} + \tau^2 R\, y_{\bar{t}t} + A\, y = \varphi(t)\,, \qquad y(0) = y(\tau) = 0\,, \\ 0 < t = n\tau < t_0\,, \end{gathered}$$

Second, an alternative form of writing appears useful:

$$(2)\qquad \begin{gathered} (B + 2\,\tau R)\, y_{n+1} = \Phi_n\,, \\ \Phi_n = 2\,(2R - A)\,\tau y_n + (B - 2\tau R)\, y_{n-1} + 2\tau\varphi_n \end{gathered}$$

($A$, $B$ and $R$ are, in general, variables, that is, they depend on $t_n$). From such reasoning it seems clear that problem (1) is solvable if an inverse operator $(B+2\tau R)^{-1}$ exists. In the sequel this condition is supposed to be satisfied. Moreover, we take for granted that

(3) the operator $B + 2\tau R$ is positive definite.

Further development of the three-layer scheme (3) is connected with the functional known as the compound-norm:

(4) $$\| Y_{n+1} \|^2 = \tfrac{1}{4} \|y_n + y_{n+1}\|^2_{(1_1)} + \|y_{n+1} - y_n\|^2_{(1_2)},$$

where $\|\cdot\|_{(1_1)}$ and $\|\cdot\|_{(1_2)}$ are suitable norms on the linear system $H$. In order to understand the structure of this norm a little better, it is appropriate to introduce the space $H^2 = H \oplus H$ being the direct sum of two copies of $H$. The space $H^2$ is defined as the set of all vectors of the form

$$Y = \{Y^{(1)} + Y^{(2)}\} \in H^2, \qquad Y^{(\alpha)} \in H\,, \qquad \alpha = 1, 2\,,$$

where the operations of addition of vectors and multiplication of a vector by a number are carried out in a coordinate-wise fashion:

$$Y + \bar{Y} = \{Y^{(1)} + \bar{Y}^{(1)},\ Y^{(2)} + \bar{Y}^{(2)}\}, \qquad \alpha Y = \{\alpha\, Y^{(1)}, \alpha Y^{(2)}\}.$$

The norm on the space $H^2$ is natural to be defined by

$$\| Y \|^2 = \|Y^{(1)}\|^2_{(1_1)} + \|Y^{(2)}\|^2_{(1_2)}.$$

In our case the vector

$$Y_{n+1} = \left\{ \tfrac{1}{2}\,(y_{n+1} + y_n), y_{n+1} - y_n \right\} \in H^2$$

possesses the coordinates $Y^{(1)}_{n+1} = \frac{1}{2}(y_{n+1} + y_n)$ and $Y^{(2)}_{n+1} = y_{n+1} - y_n$. It is easy to see that functional (4) satisfies all the axioms of the norm, namely $\| \alpha Y_{n+1} \| = |\alpha| \| Y_{n+1} \|$, $\| Y_{n+1} \| \geq 0$ for any $y_n \in H$, $y_{n+1} \in H$ and $\| Y_{n+1} \| = 0$ only for $y_n = y_{n+1} = 0$; $\| \bar{Y}_{n+1} + Y_{n+1} \| \leq \| \bar{Y}_{n+1} \| + \| Y_{n+1} \|$.

We now in a position to define the notion of stability of scheme (1). The three-layer scheme (1) is said to be **stable** if there exists the norm (4) and for all sufficiently small $\tau \leq \tau_0$ and $|h| \leq h_0$ one can point out positive constants $M_1$ and $M_2$ independent of $\tau$, $h$ and disregarding to the choice of $y_0$, $y_1$, $\varphi(t)$ such that for any $y_0$, $y_1$, $\varphi(t)$ and all $t = \tau, 2\tau, \ldots, (n_0 - 1)\tau$ a solution of problem (1) satisfies one of the following estimates:

$$\|Y(t+\tau)\|_{(1)} \le M_1 \|Y(\tau)\|_{(1^0)} + M_2 \max_{0<t'\le t} \|\varphi(t')\|_{(2)}, \tag{5}$$

$$\|Y(t+\tau)\|_{(1)} \le M_1 \|Y(\tau)\|_{(1^0)} + M_2 \max_{\tau<t'\le t} \left(\|\varphi(t')\|_{(2)} + \|\varphi_{\bar t}(t')\|_{(2)}\right), \tag{6}$$

where $\|\cdot\|_2$ is some suitable norm on the space $H$, $\|Y(t+\tau)\|_{(1)}$ and $\|Y(\tau)\|_{(1^0)}$ are defined by the formula of the form (4), so that

$$\|Y(t+\tau)\|^2_{(1)} = \frac{1}{4}\|y(t+\tau) + y(t)\|^2_{(1_1)} + \|y(t+\tau) - y(t)\|^2_{(1_2)}, \tag{7}$$

$$\|Y(\tau)\|^2_{(1^0)} = \frac{1}{4}\|y_1 + y_0\|^2_{(1^0_1)} + \|y_1 - y_0\|^2_{(1^0_2)}, \tag{8}$$

where $\|\cdot\|_{(1^0_1)}$ and $\|\cdot\|_{(1^0_2)}$ are suitable norms on the space $H$.

As far as constant operators $A$ and $R$ are concerned, the norms $\|Y\|_{(1)}$ and $\|Y\|_{(1^0)}$ normally coincide. In the general case $\|Y(t+\tau)\|_{(1)}$ and $\|\varphi(t)\|_{(2)}$ depend on $t = n\tau$, so that one should write $\|Y(t+\tau)\|_{(1,t)}$ instead of $\|Y(t+\tau)\|_{(1)}$ and $\|\varphi(t)\|_{(2,t)}$ instead of $\|\varphi(t)\|_{(2)}$.

As we will see later, the norms $\|\cdot\|_{(1_1)}$ and $\|\cdot\|_{(1_2)}$ are energy norms constructed for the operators $A$ and $R$. For this reason we will assume that these operators are

$$\text{self-adjoint:} \quad A = A^*, \qquad R = R^*, \tag{9}$$

$$\text{positive:} \quad A > 0, \qquad R > 0, \tag{10}$$

if $H$ is a Hilbert space.

**2. The basic energy identity.** We will carry out the derivation of the energy identity for the three-layer scheme (1) with variable operators $A = A(t)$, $B = B(t)$ and $R = R(t)$. This identity is aimed at achieving a priori estimates expressing the stability of a scheme with respect to the initial data and right-hand side.

By virtue of the relations

$$y = \frac{1}{2}(\hat y + \check y) - \frac{1}{2}(\hat y - 2y + \check y) = \frac{1}{2}(\hat y + \check y) - \frac{\tau^2}{2}\, y_{\bar t t}$$

we represent scheme (1) in the form

$$
B\, y_{\overset{\circ}{t}} + \tau^2 \left(R - \tfrac{1}{2}\, A\right) y_{\bar{t}t} + \tfrac{1}{2}\, A\,(\hat{y} + \check{y}) = \varphi\,, \qquad y(0) = y_0\,, \qquad y(\tau) = y_1\,, \tag{11}
$$

where $A = A(t_n) = A_n$, $B = B(t_n) = B_n$ and $R = R(t_n) = R_n$. Taking the inner product of (11) and $2\tau y_{\overset{\circ}{t}} = \tau(y_t + y_{\bar{t}})$, we find that

$$
2\tau \left(B\, y_{\overset{\circ}{t}}\,, y_{\overset{\circ}{t}}\right) + \tau^2 \left(\left(R - \tfrac{1}{2}\, A\right)(y_t - y_{\bar{t}}), y_t + y_{\bar{t}}\right) + \tfrac{1}{2}\left(A\,(\hat{y} + \check{y}), \hat{y} - \check{y}\right) = 2\tau\left(\varphi, y_{\overset{\circ}{t}}\right). \tag{12}
$$

Let $A$ and $R$ be self-adjoint operators. Then

$$
R - 0.5\, A = (R - 0.5\, A)^*.
$$

By Lemma 1 of Section 2 we thus have

$$
\left(\left(R - \tfrac{1}{2}\, A\right)(y_t - y_{\bar{t}}), y_t + y_{\bar{t}}\right) = \left(\left(R - \tfrac{1}{2}\, A\right)(y_t\,, y_t\right) - \left(\left(R - \tfrac{1}{2}\, A\right)(y_{\bar{t}}, y_{\bar{t}}\right), \tag{13}
$$

$$
\left(A\,(\hat{y} + \check{y}), \hat{y} - \check{y}\right) = (A\,\hat{y}, \hat{y}) - (A\,\check{y}, \check{y})\,. \tag{14}
$$

At the final stage we add and subtract $(Ay, y)$ on the right-hand side of (14). The outcome of this is

$$
\left(A\,(\hat{y} + \check{y}), \hat{y} - \check{y}\right) = \left[(A\hat{y}, \hat{y}) + (Ay, y)\right] - \left[(Ay, y) + (A\check{y}, \check{y})\right]. \tag{15}
$$

**Lemma 1** *Let $A = A^*$ be a self-adjoint operator. Then*

$$
(Av, v) + (Az, z) = \tfrac{1}{2}\left(A(v + z), v + z\right) + \tfrac{1}{2}\left(A(v - z), v - z\right) \tag{16}
$$

*for any vectors $v$ and $z$ of the space $H$.*

**Proof** Since $A = A^*$, we have $(Av, z) = (v, Az) = (Az, v)$ and

$$
\begin{aligned}
&((A(v + z), v + z)) + ((A(v - z), v - z)) \\
&\quad = \left[(Av, v) + 2(Av, z) + (Az, z)\right] + \left[(Av, v) - 2(Av, z) + (Az, z)\right] \\
&\qquad = 2\left[(Av, v) + (Az, z)\right].
\end{aligned}
$$

By inserting in (16) $v = \hat{y}$ and $z = y$ we transform (15) into

$$(17)\quad (A(\hat{y} + \check{y}), \hat{y} - \check{y}) = 0.5\left[(A(\hat{y} + y), \hat{y} + y) + (A(\hat{y} - y), \hat{y} - y)\right] - 0.5\left[(A(y + \check{y}), y + \check{y}) + (A(y - \check{y}), y - \check{y})\right].$$

We now substitute (17) and (13) into (12) and take into account that

$$(A(\hat{y} - y), \hat{y} - y) = \tau^2 (Ay_t, y_t), \quad y_{\bar{t}} = (y - \check{y})/\tau = \check{y}_t,$$
$$(A(y - \check{y}), y - \check{y}) = \tau^2 (Ay_{\bar{t}}, y_{\bar{t}}),$$

making it possible to establish the basic energy identity for the three-layer scheme (1):

$$(18)\quad 2\tau\left(By_{\overset{\circ}{t}}, y_{\overset{\circ}{t}}\right) + \left[\tfrac{1}{4}\left(A(\hat{y} + y), \hat{y} + y\right) + \tau^2\left(\left(R - \tfrac{1}{4}A\right) y_t, y_t\right)\right] = \left[\tfrac{1}{4}\left(A(y + \check{y}), y + \check{y}\right) + \tau^2\left(\left(R - \tfrac{1}{4}A\right) y_{\bar{t}}, y_{\bar{t}}\right)\right] + 2\tau\left(\varphi, y_{\overset{\circ}{t}}\right).$$

In giving it we preassumed only property (9) concerning the self-adjointness of the operators $A$ and $R$ and no more.

**3. Stability with respect to the initial data.** Recall the definition of stability with respect to the initial data and the right-hand side. Scheme (1) is said to be stable with respect to the initial data if for problem (1a) the a priori estimate holds:

$$(19)\qquad \|Y(t + \tau)\|_{(1)} \le M_1 \, \|Y(\tau)\|_{(1^\circ)}.$$

Scheme (1) is said to be stable with respect to the right-hand side if for problem (1b) the estimate

$$(20)\qquad \|Y(t + \tau)\|_{(1)} \le M_2 \max_{0 < t' \le t} \|\varphi(t')\|_{(2)}$$

or the estimate

$$(21)\qquad \|Y(t + \tau)\|_{(1)} \le M_2 \max_{\tau < t' \le t} \left(\|\varphi(t')\|_{(2)} + \|\varphi_{\bar{t}}(t')\|_{(2)}\right)$$

holds.

Making use of the triangle inequality and collecting (19) and (20) or (21), we obtain estimate (5) or (6).

In our basic account $A$ and $R$ are taken to be constant self-adjoint positive operators and $B$ refers to a non-self-adjoint nonnegative operator:

$$A = A^* > 0, \qquad R = R^* > 0\,, \qquad B \geq 0\,. \tag{22}$$

These restrictions permit us to specify and extract a primary family of schemes.

With the regard to problem (1a) identity (18) takes the form

$$2\tau\,(By_{\overset{\circ}{t}}\,, y_{\overset{\circ}{t}}) + \|\,Y(t+\tau)\,\|^2 = \|\,Y(t)\,\|^2, \qquad t = n\,\tau\,, \tag{23}$$

where

$$\begin{aligned}\|\,Y(t+\tau)\,\|^2 &= \tfrac{1}{4}\,(A(y(t+\tau)+y(t)), y(t+\tau)+y(t)) \\ &\quad + \tau^2\left(\left(R - \tfrac{1}{4}\,A\right) y_t\,, y_t\right),\end{aligned} \tag{24}$$

$$\begin{aligned}\|\,Y(t)\,\|^2 &= \tfrac{1}{4}\,(A(y(t)+y(t-\tau)), y(t)+y(t-\tau)) \\ &\quad + \tau^2\left(\left(R - \tfrac{1}{4}\,A\right) y_{\bar{t}}\,, y_{\bar{t}}\right).\end{aligned} \tag{25}$$

To avoid cumbersome calculations, we will also use index denotations by setting $Y(t+\tau) = Y_{n+1}$ and

$$\begin{aligned}\|\,Y_{n+1}\,\|^2 &= \tfrac{1}{4}\,(A(y_{n+1}+y_n), y_{n+1}+y_n) + \tau^2\left(\left(R - \tfrac{1}{4}\,A\right) y_{t,n}\,, y_{t,n}\right) \\ &= \tfrac{1}{4}\,\|\,y_{n+1}+y_n\,\|_A^2 + \|y_{n+1}-y_n\|_{R-\frac{1}{4}A}^2\,.\end{aligned}$$

Observe that relation (24) implies that $\|\,Y(t+\tau)\,\|^2 > 0$ for any $y(t) \neq 0$, $y(t+\tau) \neq 0$, provided that the operators $A$ and $R - A/4$ are positive, $A > 0$ and $R > A/4$.

**Theorem 1** *Let $A = A^* > 0$ and $R = R^* > 0$ be positive operators. Then the conditions*

$$B = B(t) \geq 0 \qquad \text{for all} \quad t \in \omega_\tau\,, \tag{26}$$

$$R > \tfrac{1}{4}\,A \tag{27}$$

*are sufficient for the stability of scheme* (1) *with respect to the initial data. Under conditions* (26) *and* (27) *for problem* (1a) *the estimate holds:*

$$\|\,Y(t+\tau)\,\| \leq \|\,Y(\tau)\,\|\,, \tag{28}$$

*where* $\| Y \|$ *is defined in accordance with rule* (24).

Indeed, for $B \geq 0$ identity (23) implies that

$$\| Y(t+\tau) \|^2 \leq \| Y(t) \|^2, \qquad \| Y(t+\tau) \| \leq \| Y(t) \| \leq \cdots \leq \| Y(\tau) \| .$$

**Remark 1** Theorem 1 holds true under the constraints

$$R > \tfrac{1}{4} A, \qquad A \geq 0 .$$

However, in that case $\| Y \|$ may fail to be a norm, but it is always a seminorm.

**Remark 2** The three-layer scheme (1a) can be reduced to the two-layer scheme

$$\mathcal{B} Y_t + \mathcal{A} Y = 0, \qquad Y_0 \in H^2, \tag{29}$$

where $Y = Y_n \in H^2$, $Y_t = (Y_{n+1} - Y_n)/\tau$, $\mathcal{A}$ and $\mathcal{B}$ are operators in the space $H^2$. This can be done using the vector

$$Y_n = \left\{ \tfrac{1}{2} (y_n + y_{n-1}), y_n - y_{n-1} \right\}$$

and treating the operators $\mathcal{A}$ and $\mathcal{B}$ as operator matrices with elements being operators in the space $H$:

$$\mathcal{A} = \begin{pmatrix} A & 0 \\ 0 & R - \frac{1}{4} A \end{pmatrix} \qquad \mathcal{B} = \begin{pmatrix} B + 0.5\,\tau A & \tau\left(R - \frac{1}{4} A\right) \\ -\tau\left(R - \frac{1}{4} A\right) & \frac{\tau}{2}\left(R - \frac{1}{4} A\right) \end{pmatrix}$$

If $A = A^*$, $R = R^*$, then the operator $\mathcal{A} : H^2 \longrightarrow H^2$ is self-adjoint: $\mathcal{A} = \mathcal{A}^*$, while the operator $\mathcal{B}$ is non-self-adjoint. All this enables us to conclude that

$$\|Y\|_{\mathcal{A}}^2 = (\mathcal{A} Y, Y) = (A\, Y^{(1)}, Y^{(1)}) + \left( (R - \tfrac{1}{4} A) Y^{(2)}, Y^{(2)} \right) .$$

In addition, we have $\mathcal{A} > 0$, provided that $A > 0$ and $R > \frac{1}{4} A$.

It is easy to verify that the stability condition in the space $H_{\mathcal{A}}^2$ of the two-layer scheme (29)

$$\mathcal{B} \geq \frac{\tau}{2} \mathcal{A} \tag{30}$$

is equivalent to the requirements

$$B = B(t) \geq 0 \qquad \text{for all } t \in \omega_\tau\,, \tag{26*}$$

$$R \geq \frac{1}{4}\,A\,, \tag{27*}$$

which assure us of the validity of the estimate

$$\|Y_{n+1}\|_{\mathcal{A}} \leq \|Y_1\|_{\mathcal{A}}\,, \tag{28*}$$

where

$$\|Y_{n+1}\|^2_{\mathcal{A}} = \frac{1}{4}\,\|\,y_{n+1} + y_n\,\|^2_A + \|y_{n+1} - y_n\|^2_{R-\frac{1}{4}A}\,.$$

In concluding this section, let us stress that condition (26*) is not only sufficient, but also necessary for the validity of estimate (28*).

**4. Stability with respect to the right-hand side.** We now consider problem (1b) under conditions (9), (10) and (27), which are put together for later use. Since $A$ and $R$ are constant operators, identity (18) for (1b) is of the form

$$2\,\tau\,(By_{\overset{\circ}{t}}\,, y_{\overset{\circ}{t}}) + \|\,Y(t+\tau)\,\|^2 = \|\,Y(t)\,\|^2 + 2\tau\,(\varphi, y_{\overset{\circ}{t}}\,)\,, \qquad t = n\,\tau\,. \tag{31}$$

In the further development of a priori estimates of the form (20) or (21) a key role is played by the estimation of the functional $2\tau(\varphi, y_{\overset{\circ}{t}})$. First of all, we give below the obvious inequality

$$2\,\tau\,(\varphi, y_{\overset{\circ}{t}}\,) \leq \tau\,\varepsilon_0\|\,y_{\overset{\circ}{t}}\,\|^2 + \frac{\tau}{\varepsilon_0}\,\|\,\varphi\,\|^2\,, \tag{32}$$

where $\varepsilon_0 = \text{const} > 0$ is independent of $\tau$ and $h$ both.

**Theorem 2** *Let $A = A^* > 0$ and $R = R^* > 0$ be positive operators. Then under the conditions $B \geq \varepsilon E$, $R \geq A/4$, $\varepsilon = const > 0$, a solution of problem* (1) *satisfies the a priori estimate*

$$\|\,Y(t+\tau)\,\| \leq \|\,Y(\tau)\,\| + \frac{1}{\sqrt{2\,\varepsilon}}\left[\,\sum_{t'=\tau}^{t}\,\tau\,\|\,\varphi(t')\,\|^2\,\right]^{1/2}. \tag{33}$$

It suffices to estimate only a solution of problem (1b), since Theorem 1 remains valid for $B \geq \varepsilon E$. By merely setting in (32) $\varepsilon_0 = 2\varepsilon$ we deduce from (31) that

$$\|\,Y(t+\tau)\,\|^2 \leq \|\,Y(t)\,\|^2 + \frac{\tau}{2\,\varepsilon}\,\|\,\varphi(t)\,\|^2. \tag{34}$$

To finish the proof of the theorem, it remains to sum up this inequality with respect to the variable $t = \tau, 2\tau, \ldots, n\tau$, exploit the fact that $\|\,Y(\tau)\,\| = 0$ and apply Theorem 1.

We cite here without proving the following assertion.

**Theorem 3** *Let the conditions of Theorem 1 be satisfied. Then scheme (1) is stable with respect to the right-hand side and for it the estimate holds for $t > \tau$:*

$$(35)\qquad \| Y(t+\tau) \| \le \| Y(\tau) \| + M_2 \max_{\tau < t' \le t} \left( \|\varphi(t')\|_{A^{-1}} + \|\varphi_{\bar t}(t')\|_{A^{-1}} \right),$$

*where $M_2 = const > 0$ depends only on $t_0$.*

**Theorem 4** *Let $A = A^* \ge 0$ and $R = R^* > 0$ be constant nonnegative operators and $B = B(t)$ be a variable non-self-adjoint positive definite operator*

$$(36)\qquad B \ge \varepsilon E, \qquad \varepsilon = const > 0,$$

*where a number $\varepsilon$ is independent of $h$ and $\tau$, and let the condition*

$$(37)\qquad R \ge \frac{1}{4} A$$

*be satisfied. Then a solution of problem (1b) admits a priori estimate*

$$(38)\qquad \| y(t+\tau) \| \le \frac{2\sqrt{t}}{\varepsilon} \left[ \sum_{t'=\tau}^{t} \tau \| \varphi(t') \|^2 \right]^{1/2}.$$

Look at identity (31). It follows from (36) and (32) for $\varepsilon_0 = \varepsilon$ that

$$(39)\qquad \tau\varepsilon \| y_{\bar t} \|^2 + \| Y(t+\tau) \|^2 \le \| Y(t) \|^2 + \frac{\tau}{\varepsilon} \| \varphi(t) \|^2.$$

With the relations $\| Y(\tau) \| = 0$ and $\| Y(t+\tau) \|^2 \ge 0$ in view, summing up over $t = \tau, 2\tau, \dots, n\tau$ yields

$$(40)\qquad \varepsilon \sum_{t'=\tau}^{t} \tau \| y_{\bar t}(t') \|^2 + \| Y(t+\tau) \|^2 \le \frac{1}{\varepsilon} \sum_{t'=\tau}^{t} \tau \| \varphi(t') \|^2$$

or

$$(41)\qquad \sum_{t'=\tau}^{t} \tau \| y_{\bar t}(t') \|^2 \le \frac{1}{\varepsilon^2} \sum_{t'=\tau}^{t} \tau \| \varphi(t') \|^2.$$

Further development of estimate (38) from (41) is based on the following assertion of auxiliary character.

**Lemma 2** *If* $y(0) = y(\tau) = 0$, *then*

$$(42)\qquad \| y(t) \|^2 + \| y(t+\tau) \|^2 \le 4t \sum_{t'=\tau}^{t} \tau \| y_{\bar{t}}(t') \|^2 .$$

Indeed,

$$(43)\qquad \begin{aligned} y(t+\tau) + y(t) &= 2 \sum_{t'=\tau}^{t} \tau\, y_{\bar{t}}(t') , \\ \| y(t+\tau) + y(t) \|^2 &\le 4t \sum_{t'=\tau}^{t} \tau \| y_{\bar{t}}(t') \|^2 . \end{aligned}$$

In the notation $w_n = y_n - y_{n-1}$ we obtain

$$w_{n+1} = 2\tau\, y_{\bar{t},n} - w_n , \qquad w_1 = 0 ,$$

which implies the inequality

$$\| w_{n'+1} \| \le 2\tau \| y_{\bar{t},n'} \| + \| w_{n'} \| , \qquad n' = 1, 2, \dots , n .$$

Summing up the preceding over $n'$ from 1 to $n$ we find that

$$\| w_{n+1} \| \le 2 \sum_{n'=1}^{n} \tau \| y_{\bar{t},n'} \|$$

or

$$(44)\qquad \begin{aligned} \| y(t+\tau) - y(t) \| &\le 2 \sum_{t'=\tau}^{t} \tau \| y_{\bar{t}}(t') \| , \\ \| y(t+\tau) - y(t) \|^2 &\le 4t \sum_{t'=\tau}^{t} \tau \| y_{t}(t') \|^2 . \end{aligned}$$

Putting inequalities (43) and (44) together with the obvious identity

$$(45)\quad \| y(t+\tau) + y(t) \|^2 + \| y(t+\tau) - y(t) \|^2 = 2 \left( \| y(t) \|^2 + \| y(t+\tau) \|^2 \right) ,$$

we arrive at (42). Substituting estimate (42) into (41) yields (38).

**Theorem 5** *Let $A = A^* > 0$, $R = R^* > 0$ and $B = B^* > 0$, $A$ and $R$ be constant operators and $R > \frac{1}{4}A$. Then for scheme (1) the estimate holds:*

$$\| Y_{n+1} \|^2 \leq \| Y_1 \|^2 + \frac{1}{2} \sum_{k=1}^{n} \tau \, \|\varphi_k\|^2_{B^{-1}} . \tag{46}$$

The fact that the operator $B$ is self-adjoint is kept in mind in the estimation of the expression

$$2\tau\,(\varphi, y_{\bar t}) \leq \| y_{\bar t} \|_B \cdot \|\varphi\|_{B^{-1}} \leq \tau\,\varepsilon \, \| y_{\bar t} \|^2_B + \frac{\tau}{\varepsilon} \, \|\varphi\|^2_{B^{-1}} , \tag{47}$$

which is involved in identity (31). Putting $\varepsilon = 2$ and substituting this estimate into (31), we obtain the inequality

$$\| Y_{n+1} \|^2 \leq \| Y_n \|^2 + \frac{\tau}{2} \, \|\varphi_n\|^2_{B^{-1}} , \tag{48}$$

implying the desired estimate. Note that the operator $B$ may depend on $t$: $B = B(t)$. In that case we must write $\|\varphi_k\|^2_{B_k^{-1}}$, where $B_k^{-1} = B^{-1}(t_k)$.

**5. Schemes with variable operators.** If operators $A$ and $R$ depend on the variable $t$, the extra property of the Lipschitz continuity of $A$ and $R$ with respect to the variable $t$ is needed in this connection:

$$\big|((A(t) - A(t-\tau))\,x, x)\big| \leq \tau\, c_3 \big(A(t-\tau)\,x, x\big) \tag{49}$$

for all $x \in H$ and $t = 2\tau, \ldots, (n_0 - 1)\tau$, where $c_3 = \text{const} > 0$ is independent of $h$ and $\tau$; the condition imposed on the operator $R$ is analogous. In this case the compound norm $\| Y(t+\tau) \| = \|Y(t+\tau)\|_{(t)}$ depends on the variable $t$:

$$\begin{aligned} \|Y(t+\tau)\|^2_{(t)} &= \frac{1}{4}\,(A(t)\,(y(t+\tau) + y(t)), y(t+\tau) + y(t)) \\ &\quad + \tau^2 \Big( \big(R(t) - \tfrac{1}{4}\,A(t)\big)\, y_t(t), y_t(t) \Big) , \end{aligned} \tag{50}$$

$$\begin{aligned} \|Y(t)\|^2_{(t-\tau)} &= \frac{1}{4}\,(A(t-\tau)\,(y(t) + y(t-\tau)), y(t) + y(t-\tau)) \\ &\quad + \tau^2 \Big( \big(R(t-\tau) - \tfrac{1}{4}\,A(t-\tau)\big)\, y_{\bar t}(t), y_{\bar t}(t) \Big) . \end{aligned} \tag{51}$$

We modify the expression in the brackets on the right-hand side of identity (18) by a simple observation that

$$\big(A(y+\hat{y}),y+\hat{y}\big) = \big(\check{A}(y+\hat{y}),y+\hat{y}\big) + \tau\,\big(A_{\bar{t}}(y+\hat{y}),y+\hat{y}\big)\,,$$

$$\Big(\big(R-\tfrac{1}{4}A\big)\,y_{\bar{t}},y_{\bar{t}}\Big) = \Big(\big(\check{R}-\tfrac{1}{4}\check{A}\big)\,y_{\bar{t}},y_{\bar{t}}\Big) + \tau\,\Big(\big(R-\tfrac{1}{4}A\big)_{\bar{t}}\,y_{\bar{t}},y_{\bar{t}}\Big)$$

$$\check{A} = A(t-\tau)\,, \qquad A_{\bar{t}} = (A-\check{A})/\tau\,.$$

Also, it will be convenient to introduce more compact notations

$$\hat{J} = J(t+\tau) + \|Y(t+\tau)\|^2_{(t)}\,, \qquad J = J(t) = \|Y(t)\|^2_{(t-\tau)}\,. \tag{52}$$

With these, identity (18) is recast as

$$2\tau\,\big(By_{\overset{\circ}{t}}\,,y_{\overset{\circ}{t}}\big) + \hat{J} = J + 2\tau\,\big(\varphi,y_{\overset{\circ}{t}}\big) + \tau\,F\,, \tag{53}$$

$$F = \frac{1}{4}\,\big(A_{\bar{t}}(y+\hat{y}),y+\hat{y}\big) + \tau^2\,\Big(\big(R-\tfrac{1}{4}A\big)_{\bar{t}}\,y_{\bar{t}},y_{\bar{t}}\Big)\,, \tag{54}$$

showing the new members to be sensible ones. If the operators $R-A/4$ and $A$ satisfy condition (49), then

$$\|\,F\,\| \le \frac{c_3}{4}\,\big(\check{A}(y+\hat{y}),y+\hat{y}\big) + \tau^2\,c_3\,\Big(\big(\check{R}-\tfrac{1}{4}\check{A}\big)_{\bar{t}}\,y_{\bar{t}},y_{\bar{t}}\Big) = c_3\,J$$

and from identity (53) for $R \ge A/4$ it follows that

$$2\tau\,\big(By_{\overset{\circ}{t}}\,,y_{\overset{\circ}{t}}\big) + \hat{J} \le (1+\tau\,c_3)\,J + 2\tau\,\big(\varphi,y_{\overset{\circ}{t}}\big)\,. \tag{55}$$

In the general case when each of the operators $R(t)$ and $A(t)$ satisfies condition (49), we obtain

$$\begin{aligned} \|\,F\,\| &\le \frac{c_3}{4}\,\big(\check{A}(y+\hat{y}),y+\hat{y}\big) + \tau^2\,c_3\,\Big[(\check{R}y_t,y_t)) + \frac{1}{4}\,\big(\check{A}y_{\bar{t}},y_{\bar{t}}\big)\Big] \\ &\le c_3\,\Big(1+\frac{2}{\varepsilon}\Big)\,J \end{aligned} \tag{56}$$

under the constraint $R \ge \frac{1+\varepsilon}{4}A$, where $\varepsilon = \text{const} > 0$ is independent of $h$ and $\tau$, since

$$J > \frac{\varepsilon\,\tau^2}{4}\,\big(\check{A}y_{\bar{t}},y_{\bar{t}}\big) \qquad \text{for} \quad R \ge \frac{1+\varepsilon}{4}\,A\,.$$

Identity (53) yields

$$(57) \qquad 2\tau\left(By_{\overset{\circ}{t}}, y_{\overset{\circ}{t}}\right) + J(t+\tau) \le \left(1 + \frac{(2+\varepsilon)\,c_3}{\varepsilon}\,\tau\right) J(t) + 2\tau\left(\varphi, y_{\overset{\circ}{t}}\right).$$

Having at our disposal the energy inequality (57), we can derive the a priori estimates in just the same way as was done for constant operators $A$ and $R$. For example, for $B \ge 0$ inequality (57) for problem (1a) implies the estimate

$$(58) \qquad \|Y(t+\tau)\|_{(t)} \le M_1\,\|Y(\tau)\|_{(\tau)} \qquad \text{if} \quad R \ge \frac{1+\varepsilon}{4}\,A\,,$$

where $M_1$ depends only on $\varepsilon$, $c_3$ and $t_0$.

We bring together the basic facts in the following assertion.

**Theorem 6** *Let variable operators $A = A(t) = A^*(t) > 0$ and $R = R(t) = R^*(t) > 0$ be Lipschitz continuous in $t$ and let*

$$(59) \qquad R(t) \ge \frac{1+\varepsilon}{4}\,A(t) \qquad \text{for all} \quad 0 < t = n\,\tau < t_0\,,$$

*where $\varepsilon = \text{const} > 0$ is independent of $\tau$ and $h$. Then for scheme (1) the following estimates hold:*

$$(60) \qquad \|Y(t+\tau)\|_{(t)} \le M_1\,\|Y(\tau)\|_{(\tau)} + M_2 \max_{\tau < t' \le t} \left[\|\varphi(t')\|_{A^{-1}(t')} + \|\varphi_{\bar t}(t')\|_{A^{-1}(t')}\right]$$

*for $B(t) \ge 0$, $0 < t = n\tau < t_0$,*

$$(61) \qquad \|Y(t+\tau)\|_{(t)} \le M_1\,\|Y(\tau)\|_{(\tau)} + M_2 \max_{\tau < t' \le t} \|\varphi(t')\|$$

*for $B(t) \ge \bar\varepsilon E$, where $\bar\varepsilon = \text{const} > 0$, $M_1 > 0$ and $M_2 > 0$ are independent of $\tau$ and $h$ both.*

To avoid needless repetitions, we omit here the proof of the theorem.

**Remark** Some requirements of Theorem 6 can be relaxed. For instance, the condition $B \ge 0$ can be replaced by the following one:

$$(62) \qquad B \ \ge\ -c_4\,\tau^2\,A\,,$$

where $c_4 = \text{const} > 0$ is independent of $\tau$ and $h$ both. Under condition (62) estimate (60) holds true for $\tau \le \tau_0$, $\tau_0 = 1/(4\,c_4)$.

**6. Weighted schemes.** In practice the reader frequently encountered the weighted schemes

$$(63)\qquad \begin{aligned} & y_{\overset{\circ}{t}} + A\, y^{(\sigma_1,\sigma_2)} = \varphi(t)\,, \qquad \tau \le t = n\,\tau < t_0\,, \\ & y(0) = y_0\,, \qquad y(\tau) = y_1\,, \end{aligned}$$

where $y^{(\sigma_1,\sigma_2)} = \sigma_1 \hat{y} + (1-\sigma_1-\sigma_2) y + \sigma_2 \check{y}$. The accepted view is that stability and accuracy of the scheme are governed by selection rules for real numbers $\sigma_1$ and $\sigma_2$. In Section 1 scheme (63) was written in the canonical form (1), making it possible to recover the operators

$$(64)\qquad B = E + \tau\,(\sigma_1-\sigma_2)\,A\,, \qquad R = \frac{\sigma_1+\sigma_2}{2}\,A\,.$$

Assumming that there exists an inverse operator $A^{-1}$ and applying it, on the same grounds, to both sides of (1) with operators (64), we obtain

$$(65)\qquad \begin{aligned} & \widetilde{B}\, y_{\overset{\circ}{t}} + \tau^2\, \widetilde{R}\, y_{\bar{t}t} + \widetilde{A}\, y = \varphi\,, \qquad \tau \le t = n\,\tau < t_0\,, \\ & y(0) = y_0\,, \qquad y(\tau) = y_1\,, \end{aligned}$$

where $\widetilde{B} = A^{-1} + (\sigma_1-\sigma_2)\,\tau\,E\,,\quad \widetilde{R} = \dfrac{\sigma_1+\sigma_2}{2}\,\varepsilon\,,\quad \widetilde{A} = E\,,\quad \tilde{\varphi} = A^{-1}\varphi\,.$

This implies that constant operators $\widetilde{A}$ and $\widetilde{R}$ are self-adjoint.

On account of Theorem 1 with regard to (65) the operator inequalities hold:

$$(66)\qquad \widetilde{R} - \frac{1}{4}\,\widetilde{A} = \Big(\frac{\sigma_1+\sigma_2}{2} - \frac{1}{4}\Big)\,E > 0 \quad \text{for} \quad \sigma_1+\sigma_2 > 0.5\,,$$

$$(67)\qquad \widetilde{B} = A^{-1} + (\sigma_1-\sigma_2)\,\tau\,E \ge 0 \quad \text{for} \quad \sigma_1 \ge \sigma_2 \quad \text{and any } A(t) > 0\,.$$

**Theorem 7** *If $A(t)$ is a variable positive operator and the conditions*

$$(68)\qquad \sigma_1 \ge \sigma_2\,, \qquad \sigma_1+\sigma_2 > 0.5$$

*are fulfilled, then scheme (63) is stable and the estimate holds:*

$$(69)\qquad \|\,Y(t+\tau)\,\| \le \|\,Y(\tau)\,\| + \sqrt{2\,(\sigma_1+\sigma_2)}\,\sum_{t'=\tau}^{t} \tau\,\|\,\varphi(t')\,\|\,,$$

*where*

$$(70)\quad \|Y(t+\tau)\|^2 = \tfrac{1}{4}\|y(t+\tau)+y(t)\|^2 + \tfrac{1}{2}\left(\sigma_1+\sigma_2-\tfrac{1}{2}\right)\|y(t+\tau)-y(t)\|^2.$$

**Proof** In what follows we distinguish two separate tasks.

1) *Stability with respect to the initial data.* Since the conditions $\widetilde{R} > \widetilde{A}/4$ and $\widetilde{B} \geq 0$ of Theorem 1 are fulfilled, we have for a solution of problem (65) with $\varphi = 0$ the estimate

$$\|Y(t+\tau)\| \leq \|Y(t')\|, \qquad t' < t,$$

and, in particular,

$$(71)\qquad \|Y(t+\tau)\| \leq \|Y(\tau)\|,$$

where $\|Y(t+\tau)\|$ is specified by formula (70) being a particular case of formula (24) with $\widetilde{A} = E$ and $\widetilde{R} = \frac{1}{2}(\sigma_1+\sigma_2)E$ incorporated .

2) *Stability with respect to the right-hand side.* Consider problem (63) for $y(0) = y(\tau) = 0$ and seek its solution in the form

$$(72)\qquad y_{n+1} = \sum_{s=1}^{n} \tau\, g_{n+1,s}, \qquad y_0 = 0,$$

where $g_{n+1,s}$ as a function of $n$ for fixed $s = 1, 2, \dots, n$ satisfies equation (63) with $\varphi = 0$ for $n > s+1$ and the initial data

$$(73)\qquad g_{s+1,s} + 2\sigma_1 \tau A\, g_{s+1,s} = 2\varphi_s, \qquad g_{s,s} = 0.$$

Substituting (72) into (63) and taking into account (78), we conclude that (72) is just the solution of problem (63). As we stated in (71), on the strength of stability with respect to the initial data we have for $g_{n,s}$

$$(74)\qquad \|G_{n+1,s}\| \leq \|G_{s+1,s}\| \quad \text{for fixed} \quad s = 1, 2, \dots,$$

where $\|G_{n+1,s}\|$ is expressed through $g_{n,s}$ and $g_{n+1}$ by relation (70). We find from (73) that $g_{s+1,s} = 2(E+2\sigma_1\tau A)^{-1}\varphi$ and establish the relations $\|E+2\sigma_1\tau A)^{-1}\| \leq 1$ and $\|g_{s+1,s}\| \leq 2\|\varphi_s\|$. This is due to the fact that $E + 2\sigma_1\tau A \geq E$ for $\sigma_1 \geq 0$

By assumption, $g_{s,s} = 0$. This provides support for the view that

$$\|G_{s+1,s}\|^2 = \frac{1}{4}\|g_{s+1,s}\|^2 + \frac{1}{2}\left(\sigma_1+\sigma_2-\frac{1}{2}\right)\|g_{s+1,s}\|^2$$

$$= \frac{1}{2}(\sigma_1+\sigma_2)\|g_{s+1,s}\|^2 \leq 2(\sigma_1+\sigma_2)\|\varphi_s\|^2,$$

giving

$$\|G_{n+1,s}\| \le \|G_{s+1,s}\| \le \sqrt{2(\sigma_1+\sigma_2)}\,\|\varphi_s\|. \tag{75}$$

Substituting (75) into the right-hand side of the inequality

$$\|Y_{n+1}\| \le \sum_{s=1}^{n} \tau\,\|G_{n+1,s}\|$$

we obtain for a solution of problem (63) with $y(0)=y(\tau)=0$ the estimate

$$\|Y(t+\tau)\| \le \sqrt{2(\sigma_1+\sigma_2)}\,\sum_{t'=\tau}^{t} \tau\,\|\varphi(t')\|, \tag{76}$$

which in combination with (71) implies (69).

**Theorem 8** *If $A(t)=A^*(t)>0$ is a positive operator and conditions* (68) *are satisfied, then for a solution of problem* (63) *the inequality*

$$\|Y(t+\tau)\| \le \|Y(\tau)\| + \frac{1}{\sqrt{2}}\left[\sum_{t'=\tau}^{t} \tau\,\|\varphi(t')\|_{A^{-1}(t')}\right]^{1/2} \tag{77}$$

*holds, where $\|Y(t+\tau)\|$ is given by relation* (70).

To prove this assertion the estimates

$$2\tau\,(\tilde{\varphi}, y_{\overset{\circ}{t}}) = 2\tau\,(A^{-1}\varphi, y_{\overset{\circ}{t}}) \le 2\tau\,\|y_{\overset{\circ}{t}}\|_{A^{-1}}\cdot\|\varphi\|_{A^{-1}}$$

$$\le 2\tau\,\|y_{\overset{\circ}{t}}\|^2_{A^{-1}} + \frac{\tau}{2}\,\|\varphi\|_{A^{-1}},$$

$$2\tau\,(\widetilde{B}\,y_{\overset{\circ}{t}}, y_{\overset{\circ}{t}}) \ge 2\tau\,(A^{-1}y_{\overset{\circ}{t}}, y_{\overset{\circ}{t}}) = 2\tau\,\|y_{\overset{\circ}{t}}\|^2_{A^{-1}}$$

are incorporatred in identity (18) for scheme (65).

Applying Theorem 3 to scheme (65) with a constant positive operator $A$, it is plain to derive under conditions (68) the estimate

$$\|Y(t+\tau)\| \le \|Y(\tau)\| + +M_2 \max_{\tau<t'\le t}\left(\|A^{-1}\varphi(t')\| + \|A^{-1}\varphi_{\bar{t}}(t')\|\right). \tag{78}$$

Note that estimate (60) holds true for scheme (63) if $A(t)=A^*(t)>0$, $A(t)$ being Lipschitz continuous in $t$, and

$$\sigma_1 \ge \sigma_2 - \frac{1}{\tau\|A\|},$$

while the estimate of the form (61) is valid for

$$\sigma_1 \geq \sigma_2 - \frac{1-\varepsilon}{\tau \| A \|}, \qquad 0 < \varepsilon \leq 1 .$$

**7. Examples.** It seems worthwhile giving several schemes of particular forms.

1) Scheme (1) with the operators $R = \varkappa E$ and $B = E$

(79) $$y_{\bar{t}} + \varkappa \tau^2 y_{\bar{t}t} + A\, y = \varphi$$

is stable for $\varkappa E > A/4$, that is, for

(80) $$\varkappa > \tfrac{1}{4} \| A \| .$$

A particular case of scheme (79) is the Du Fort–Frankel scheme known as the "rhombus" scheme for the heat conduction equation

$$\frac{\partial u}{\partial t} = \frac{\partial^2 u}{\partial x^2}, \qquad 0 < x < 1, \quad t > 0, \quad u(x,0) = u_0(x),$$
$$u(0,t) = u(1,t) = 0,$$

emerging from the explicit unstable scheme

$$y_{\bar{t}} + A\, y = 0, \qquad Ay = -y_{\bar{x}x} = \frac{y(x+h,t) - 2y(x,t) + y(x-h,t)}{h^2},$$

upon replacing $y(x,t) = y_i^j$ by the half-sum $\frac{1}{2}(y_i^{j+1} + y_i^{j-1}) = \frac{1}{2}(\hat{y}_i + \check{y}_i)$. The outcome of this is

(81) $$\frac{\hat{y}_i - \check{y}_i}{2\tau} = \frac{y_{i+1} - (\hat{y}_i + \check{y}_i) + y_{i-1}}{h^2} .$$

Also, it will be sensible to write (81) in canonical form. Since

$$\hat{y} + \check{y} = 2y + \tau^2 y_{\bar{t}t},$$

the right-hand side of (81) equals $y_{\bar{x}x} - \dfrac{\tau^2}{h^2} y_{\bar{t}t}$. Therefore,

$$y_{\bar{t}} + \frac{\tau^2}{h^2}\, y_{\bar{t}t} + A\, y = 0, \quad Ay = -y_{\bar{x}x}, \quad \| A \| = \frac{4}{h^2} \cos^2 \frac{\pi h}{2} < \frac{4}{h^2} .$$

Comparison of this equation with (79) gives $\varkappa = 1/h^2 > \|A\|/4$, that is, the Du Fort–Frankel scheme is stable for any $\tau$ and $h$. This is a way of producing an analog of this scheme for the case when $L$ is any elliptic operator with $\varkappa$ still subject to condition (80).

2) The asymmetric three-layer scheme

$$\frac{3}{2}\, y_t - \frac{1}{2}\, y_{\bar t} + A\, \hat y = \varphi \quad \text{or} \quad \frac{3\hat y - 4y + \check y}{2\,\tau} + A\, \hat y = \varphi$$

suits us perfectly for solving the heat conduction equation. Using the formulae

$$y_t = y_{\overset{\circ}{t}} + \frac{\tau}{2}\, y_{\bar t t}\,, \quad y_{\bar t} = y_{\overset{\circ}{t}} - \frac{\tau}{2}\, y_{\bar t t}\,, \quad \hat y = y + \tau\, y_{\overset{\circ}{t}} + \frac{\tau^2}{2}\, y_{\bar t t}\,,$$

it is plain to reduce it to the canonical form

$$(E + \tau\, A)\, y_{\overset{\circ}{t}} + \tau^2 \left(\frac{E}{\tau} + \frac{1}{2}\, A\right) y_{\bar t t} + A\, y = \varphi\,,$$

giving $B = E + \tau A$ and $R = E/\tau + \frac{1}{2}A$. These assure us of the validity of the conditions

$$R \ge \frac{1+\varepsilon}{4}\, A\,, \qquad 0 < \varepsilon \le 1\,, \qquad B \ge E$$

for any $A > 0$. If $A = A^* > 0$, the scheme concerned is stable in the norm

$$\|Y(t+\tau)\|^2_{(1)} = \tfrac{1}{2}\left(\|y(t+\tau)\|^2_A + \|y(t)\|^2_A\right) + \tau\, \|y_t\|^2_A\,.$$

**8. Three-layer schemes with non-self-adjoint operators.** The three-layer explicit scheme with a self-adjoint operator $A$

$$y_{\overset{\circ}{t}} + A\, y = 0\,, \qquad A = A^* > 0\,,$$

is absolutely unstable; the necessary stability condition $R \ge \frac{1}{4}A$ is violated for this case, since $R = 0$. This scheme is unstable in any norm $\|\cdot\|_D$ and refers to a generalization of the well-known Richardson scheme for the heat conduction equation

$$\frac{\hat y_i - \check y_i}{2\,\tau} = \frac{y_{i-1} - 2y_i + y_{i+1}}{h^2}\,.$$

The implicit scheme $By_{\overset{\circ}{t}} + Ay = 0$ with any operator $B = B^* > 0$ is also absolutely unstable.

We now consider the scheme with a skew-symmetric operator $A$:

(82) $$y_{\overset{\circ}{t}} + A\,y = 0\,, \qquad A = -A^*\,.$$

We are going to show that this scheme is stable for $\tau \|\,A\,\| < 1$ and for it the energy identity holds true:

$$\mathcal{E}_{n+1} = \mathcal{E}_1\,,$$

where $\mathcal{E}_{n+1} = \|\,y_{n+1}\,\|^2 + 2\,\tau(\,y_{n+1}, Ay_n) + \|\,y_n\,\|^2 > 0$.

All the tricks and turns remain unchanged: first, write the difference equation in the form

$$\hat{y} + \tau\,A\,y = \check{y} - \tau\,A\,y$$

and compute the squared norms on the left and right-hand sides

$$\|\,\hat{y}\,\|^2 + 2\,\tau\,(\hat{y}, Ay) + \tau^2 \|\,Ay\,\|^2 = \|\,\check{y}\,\|^2 - 2\,\tau\,(Ay, \check{y}) + \tau^2 \|\,Ay\,\|^2,$$

then add $\|\,y\,\|^2$ to both sides and take into account that $A$ is a skew-symmetric operator, that is, $(Ay, \check{y}) = -(y, A\check{y})$. As a final result we obtain $\mathcal{E}_{n+1} = \mathcal{E}_n = \ldots = \mathcal{E}_1$. What is more, we claim that $\mathcal{E}_{n+1} > 0$. Indeed,

$$\begin{aligned}\mathcal{E}_{n+1} &\geq \|\,y_{n+1}\,\|^2 - 2\,\tau\,\|\,y_{n+1}\,\| \cdot \|\,Ay_n\,\| + \|\,y_n\,\|^2 \\ &\geq \|\,y_n\,\|^2 - \tau^2\,\|\,Ay_n\,\|^2 > 0\end{aligned}$$

under the restriction $\tau \|\,A\,\| > 1$. Here $H$ is a real space. In the case of a complex space $H$ we have the quantity

$$\mathcal{E}_{n+1} = \|\,y_{n+1}\,\|^2 + 2\,\tau\,\mathrm{Re}\,(y_{n+1}, Ay_n) + \|\,y_n\,\|^2\,,$$

with respect to which both the above reasoning and results are still valid.

**Example** Of special interest is the Schrödinger equation

$$i\,\frac{\partial u}{\partial t} = \frac{\partial^2 u}{\partial x^2} - q\,u\,, \qquad q = \text{const} > 0\,, \quad 0 < x < 1\,,$$

$$u(0,t) = u(1,t) = 0\,, \qquad u(x,0) = u_0(x)\,.$$

We introduce, as usual, on the segment $0 \leq x \leq 1$ a uniform grid

$$\bar{\omega}_h = \{x_i = i\,h\,, \quad i = 0, 1, \ldots, N\,; \quad h\,N = 1\}\,.$$

Let $H = \overset{\circ}{\Omega}_h$ be the space of all complex functions defined on the grid $\bar{\omega}_h$ and vanishing at the points $x = 0$ and $x = 1$. Also, it will be sensible to introduce

$$(y, v) = \sum_{k=1}^{N-1} y_k \, \bar{v}_k \, h \, ,$$

where $\bar{v}_k$ is the complex conjugate function of $v_k$. The next step is to write the initial equation in the form

$$\frac{\partial u}{\partial t} = -i \, \frac{\partial^2 u}{\partial x^2} + i \, q \, u$$

and introduce in the space $H$ the operator

$$A \, y = i \, y_{\bar{x}x} - i \, q \, u \, .$$

The operator $A$ is skew-symmetric: $(Ay, v) = -(y, Av)$, since

$$(Ay, v) = \sum_{k=1}^{N-1} (i \, y_{\bar{x}x,k}) \, \bar{v}_k \, h + \sum_{k=1}^{N-1} (-i \, q \, y)_k \, \bar{v}_k \, h$$

$$= - \sum_{k=1}^{N-1} y_k \, \overline{(i \, v_{\bar{x}x,k})} \, h - \sum_{k=1}^{N-1} y_k \, \overline{(-i \, q \, v)_k} \, h = -(y, Av)$$

The norm of the operator $A$ admits the estimate

$$\| A \| \le \frac{4}{h^2} + q \, .$$

The explicit scheme (82) for the Schrödinger equation takes the form

$$y_{\overset{\circ}{t}} + i \, y_{\bar{x}x} - i \, q \, y = 0 \, , \qquad y \in \overset{\circ}{\Lambda}_h \, .$$

It is stable for $\tau \| A \| < 1$, that is, under the constraints

$$\left( \frac{4}{h^2} + q \right) \tau < 1 \qquad \text{or} \qquad \tau < \frac{h^2}{4 + q \, h^2} \, .$$

As the second possible example we look at the scheme

$$y_{\overset{\circ}{t}} + y_{\overset{\circ}{x}} = 0$$

associated with the transfer equation

$$\frac{\partial u}{\partial t} + \frac{\partial u}{\partial x} = 0\,,$$

which is stable for $\tau/h \le 1$.

Observe that the explicit three-layer scheme

$$y_{\bar{t}t} + A\,y = 0$$

with a skew-symmetric operator $A = -A^*$ is absolutely unstable, while the scheme with a self-adjoint operator $A = A^* > 0$, as stated before, is conditionally stable for

$$E > \frac{\tau^2}{4}\,A \qquad \text{or} \qquad \tau < \frac{2}{\sqrt{\|A\|}}$$

(for $Ay = -y_{\bar{x}x}$ we obtain $\tau < h$).

**9. Other a priori estimates.** Together with scheme (1) the reader may be confronted with schemes written in the form

$$(83) \qquad (E + \tau^2 R)\, y_{\bar{t}t} + B\, y_{\overset{\circ}{t}} + A\, y = \varphi\,, \quad y(0) = y_0\,, \quad y(\tau) = y_1\,.$$

Such schemes formally can be obtained from (1) by replacing $R$ by $\widetilde{R} = R + E/\tau^2$. With this substitution in mind, one can easily conclude that scheme (83) is stable for $R \ge A/4$ and write down the appropriate estimates.

Compound norms $\|Y\|$ arise naturally in connection with the energy balance equation. Their structure seems to be rather complicated. It is desirable to possess a priori estimates for solutions of problems (1) and (83) in the usual energy norms of the spaces $H_A$ and $H_R$. We proceed to the derivation of such estimates. This amounts to setting any three-layer scheme in the form

$$(84) \qquad \begin{aligned} &D\, y_{\bar{t}t} + B\, y_{\overset{\circ}{t}} + A\, y = \varphi(t)\,, \qquad 0 < t \in \omega_\tau\,, \\ &y(0) = y_0\,, \qquad y_t(0) = \bar{y}_0\,, \end{aligned}$$

where $D = D(t)$, $A = A(t)$ and $B = B(t)$ are linear operators. In particular, $D = \tau^2 R$ for scheme (1) and $D = E + \tau^2 R$ for scheme (83).

Together with (84) we consider the problems

$$(84a) \qquad D\, y_{\bar{t}t} + B\, y_{\overset{\circ}{t}} + A\, y = 0\,, \quad y(0) = y_0\,, \quad y_t(0) = \bar{y}_0\,,$$

$$(84b) \qquad D\, y_{\bar{t}t} + B\, y_{\overset{\circ}{t}} + A\, y = \varphi(t)\,, \quad y(0) = y_t(0) = 0\,,$$

assuming that

$$A(t) = A^*(t) > 0\,, \quad D(t) = D^*(t) > 0\,, \quad B(t) \ge 0\,, \tag{85}$$

(86) $A(t)$ and $D(t)$ are Lipschitz continuous in $t$ with constant $c_3$.

Theorem 6 implies that scheme (84) under conditions (85)–(86) and the extra restriction

$$D \ge \frac{1+\varepsilon}{4}\,\tau^2 A\,, \tag{87}$$

where a number $\varepsilon > 0$ is independent of $\tau$ and $h$, is stable with respect to the initial data and a solution of problem (84a) admits the estimate

$$\|Y_{n+1}\|_{(n)} \le M_1\,\|Y_1\|_{(1)}\,, \tag{88}$$

where $M_1 = M_1(c_3, \varepsilon, t_0) > 0$ is independent of $\tau$, $h$, $n$ and

$$\|Y_{n+1}\|^2_{(n)} = \frac{1}{4}\,\|y_n + y_{n+1}\|^2_{A(t_n)} + \Big(\Big(D(t_n) - \frac{\tau^2}{4}\,A(t_n)\Big)\,y_{t,n}, y_{t,n}\Big)\,, \tag{89}$$

$$\|Y_1\|^2_{(1)} = \frac{1}{4}\,\|y_0 + y_1\|^2_{A(\tau)} + \Big(\Big(D(\tau) - \frac{\tau^2}{4}\,A(\tau)\Big)\,y_t(0), y_t(0)\Big)\,. \tag{90}$$

The constant $M_1$ equals 1 if the operators $A$ and $D$ are independent of $t$.

In order to turn from (88) to the estimates in the spaces $H_A$ and $H_D$, we shall need yet some bilateral estimates of the functional $\|\,Y_{n+1}\,\|$.

**Lemma 3** *Under conditions* (85) *and* (87) *one has*

$$\|Y_{n+1}\|_{(n)} \le \|y_n\|_{A(t_n)} + \|y_t(t_n)\|_{D(t_n)}\,, \tag{91}$$

$$\|Y_{n+1}\|_{(n)} \ge \sqrt{\frac{\varepsilon}{1+\varepsilon}}\,\|y_{n+1}\|_{A(t_n)}\,, \tag{92}$$

$$\|Y_{n+1}\|_{(n)} \ge \frac{1}{2}\sqrt{\frac{\varepsilon}{1+\varepsilon}}\,\Big(\|y_{n+1}\|_{A(t_n)} + \|y_t(t_n)\|_{D(t_n)}\Big)\,. \tag{93}$$

**Proof** To make our exposition more transparent, it is convenient to deal with

$$J = \frac{1}{4} \| \hat{y} + y \|_A^2 + \left( \left( D - \frac{\tau^2}{4} A \right) y_t, y_t \right).$$

To prove inequality (91), we concentrate mainly on

$$J = \frac{1}{4} \left( \| y \|_A^2 + 2 (Ay, \hat{y}) + \| \hat{y} \|_A^2 \right)$$

$$- \frac{1}{4} \left( \| y \|_A^2 - 2 (A\hat{y}, y) + \| \hat{y} \|_A^2 \right) + (Dy_t, y_t)$$

$$= (Ay, \hat{y}) + \|y_t\|_D^2$$

with respect to a new variable $\hat{y} = y + \tau y_t$. With this relation established, we arrive at

$$J = \| y \|_A^2 + \tau (Ay, y_t) + \|y_t\|_D^2 \le \| y \|_A^2 + \tau \| y \|_A \cdot \| y_t \|_A + \|y_t\|_D^2 .$$

Condition (87) yields the estimate

$$\| y_t \|_A \le \frac{2}{\tau \sqrt{1 + \varepsilon}} \|y_t\|_D ,$$

so that

$$J \le \| y \|_A^2 + \frac{2}{\sqrt{1 + \varepsilon}} \| y \|_A \cdot \|y_t\|_D + \|y_t\|_D^2 < \left( \| y \|_A + \|y_t\|_D \right)^2 .$$

This implies the first inequality of the lemma.

A simple observation that $J = (A\hat{y}, y) + \|y_t\|_D^2$ justifies the forthcoming substitution $y = \hat{y} - \tau y_t$, leaving us with

$$J = (A\hat{y}, \hat{y}) - \tau (A\hat{y}, y_t) + \|y_t\|_D^2 .$$

Making use of the generalized Cauchy–Bunyakovskiĭ inequality

$$(A\hat{y}, y_t) \le \| \hat{y} \|_A \cdot \| y_t \|_A$$

and taking into account (87), we obtain

$$J \ge \| \hat{y} \|_A^2 - \tau \| \hat{y} \|_A \cdot \| y_t \|_A + \|y_t\|_D^2$$

$$\ge \| \hat{y} \|_A^2 - \frac{2}{\sqrt{1 + \varepsilon}} \| \hat{y} \|_A \cdot \|y_t\|_D + \|y_t\|_D^2 .$$

Applying now the inequality $ab \le \delta a^2 + b^2/(4\delta)$ yields

$$J \ge (1-\delta)\,\|\,\hat{y}\,\|_A^2 + \Big(1 - \frac{1}{\delta\,(1+\varepsilon)}\Big)\,\|y_t\|_D^2\,. \tag{94}$$

Assuming the second coefficient to vanish, we deduce that $\delta = 1/(1+\varepsilon)$ and

$$J \ge \frac{\varepsilon}{1+\varepsilon}\|\,\hat{y}\,\|_A^2\,.$$

The second estimate of the lemma is proved.

In order to obtain the third estimate, we require the equality of the coefficients at the members $\|\,\hat{y}\,\|_A^2$ and $\|y_t\|_D^2$ in (94); this yields

$$\delta = \frac{1}{\sqrt{1+\varepsilon}}\,, \qquad 1-\delta = \frac{\sqrt{1+\varepsilon}-1}{\sqrt{1+\varepsilon}} = \frac{\varepsilon}{1+\varepsilon+\sqrt{1+\varepsilon}}\,.$$

Since $\sqrt{1+\varepsilon} \le 1+\varepsilon$ for any $\varepsilon > 0$, it follows from the foregoing that $1-\delta > \frac{\varepsilon}{2(1+\varepsilon)}$ and

$$J \ge \frac{\varepsilon}{2\,(1+\varepsilon)}\,\big(\|\,\hat{y}\,\|_A^2 + \|y_t\|_D^2\big) \ge \frac{\varepsilon}{4\,(1+\varepsilon)}\,\big(\|\,\hat{y}\,\|_A + \|y_t\|_D\big)^2\,.$$

Thus, the lemma is completely proved.

Upon substituting (91)–(93) into (88) we obtain estimates for problem (84a)

$$\|y_{n+1}\|_{A(t_n)} \le M_1\,\sqrt{\frac{1+\varepsilon}{\varepsilon}}\,\Big(\|y_0\|_{A(\tau)} + \|y_t(0)\|_{D(\tau)}\Big)\,, \tag{95}$$

$$\|y_{n+1}\|_{A(t_n)} + \|y_t\|_{D(t_n)} \le 2\,M_1\,\sqrt{\frac{1+\varepsilon}{\varepsilon}}\,\Big(\|y_0\|_{A(\tau)} + \|y_t(0)\|_{D(\tau)}\Big)\,. \tag{96}$$

In order to prove the stability of scheme (84) with respect to the right-hand side, let us employ the superposition principle and seek a solution of problem (84b) as a sum

$$y_n = \sum_{s=1}^{n} \tau\, g_{n,s}\,, \qquad n = 1,2,\dots\,, \qquad y_0 = 0\,, \qquad s = 1,2,\dots\,, \tag{97}$$

where $g_{n,s}$ as a function of $n$ for any fixed $s$ satisfies equation (84a) and the initial conditions

$$(0.5\,\tau\,B(t_s) + D(t_s))\,\frac{g_{s+1,s} - g_{s,s}}{\tau} = \varphi_s\,, \qquad g_{s,s} = 0\,.$$

Since $D > 0$ and the space $H$ is finite-dimensional, $D \geq \delta E$ and the inverse operator $D^{-1}$ exists ($\delta > 0$). As far as $B \geq 0$ and $D = D^* \geq \delta E$, we have for a solution of the equation $(0.5\,\tau B + D)\,w = \varphi$ the estimate $\|w\|_D \leq \|\varphi\|_{D^{-1}}$, so that

$$\|(g_t)_{s,s}\|_{D(t_s)} \leq \|\varphi_s\|_{D^{-1}(t_s)} \,.$$

By virtue of estimate (95) we are led to

$$\|g_{n+1,s}\|_{A(t_n)} \leq M_1 \sqrt{\frac{1+\varepsilon}{\varepsilon}}\;\|(g_t)_{s,s}\|_{D(t_s)}$$

$$\leq M_1 \sqrt{\frac{1+\varepsilon}{\varepsilon}}\;\|\varphi_s\|_{D^{-1}(t_s)} \,.$$

Making use of (97) and the triangle inequality, we derive the estimate for a solution of problem (84b)

$$\|y_{n+1}\|_{A(t_n)} \leq M_1 \sqrt{\frac{1+\varepsilon}{\varepsilon}}\;\sum_{s=1}^{n} \tau\,\|\varphi_s\|_{D^{-1}(t_s)} \,. \tag{98}$$

We summarize all the results obtained in the following assertion.

**Theorem 9** *Let conditions* (85)–(87) *be satisfied. Then scheme* (84) *is stable with respect to the initial data and right-hand side and a solution of problem* (84) *satisfies the a priori estimate*

$$\|y_{n+1}\|_{A(t_n)} \tag{99}$$

$$\leq M_1 \sqrt{\frac{1+\varepsilon}{\varepsilon}} \left( \|y(0)\|_{A(\tau)} + \|y_t(0)\|_{D(\tau)} + \sum_{s=1}^{n} \tau\,\|\varphi_s\|_{D^{-1}(t_s)} \right) .$$

**Corollary** *If* $D = E + \tau^2 R > E$ *and* $D^{-1} = E$, *then* $\|\varphi_s\|_{D^{-1}} \leq \|\,\varphi\,\|$ *and for a solution of pròblem* (84b) *the following estimate is valid:*

$$\|y_{n+1}\|_{A(t_n)} \leq M_1 \sqrt{\frac{1+\varepsilon}{\varepsilon}}\;\sum_{s=1}^{n} \tau\,\|\varphi_s\| \,. \tag{100}$$

Finer estimates that are similar to estimates established for the string vibration equation (for more detail see Chapter 5) hold true in a more narrower class of schemes

$$D\,y_{\bar{t}t} + A\,y = \varphi\,, \quad 0 < t = n\tau < t_0\,, \quad y(0) = y_0\,, \quad y_t(0) = \bar{y}_0\,. \tag{101}$$

In order to understand the nature of this a little better, we take for granted that

$$(102)\qquad \begin{aligned} &A \quad \text{and} \quad D \quad \text{are constant operators,} \\ &A \quad \text{and} \quad D \quad \text{are self-adjoint positive operators.} \end{aligned}$$

Then under condition (87) for any scheme of the form (101) estimate (99) is valid with $M_1 = 1$. By merely setting $x = D^{1/2} y$ and $C = D^{-1/2} A D^{-1/2}$ we reduce (101) to the form

$$(103)\qquad x_{\bar{t}t} + C\, x = \tilde{\varphi}\,, \qquad x(0) = x_0\,, \qquad x_t(0) = \bar{x}_0\,.$$

Applying the inverse operator $C^{-1}$ to scheme (103) yields

$$(104)\qquad C^{-1} x_{\bar{t}t} + x = C^{-1} \tilde{\varphi}\,, \qquad x(0) = x_0\,, \qquad x_t(0) = \bar{x}_0\,.$$

Comparison with scheme (84) reveals the correspondences

$$C^{-1} \sim D\,, \qquad E \sim a\,, \qquad C^{-1} \tilde{\varphi} \sim \varphi\,.$$

With this in mind, condition (87) acquires the form

$$C^{-1} \geq \frac{1+\varepsilon}{4}\, \tau^2 E \quad \text{or} \quad E \geq \frac{1+\varepsilon}{\varepsilon}\, \tau^2 C\,.$$

Involving estimate (99) with $M_1 = 1$ and taking into account that $C$ is a constant operator, we arrive at

$$(105)\qquad \|x_{n+1}\| \leq \sqrt{\frac{1+\varepsilon}{\varepsilon}} \left( \|x(0)\| + \|x_t(0)\|_{C^{-1}} + \sum_{s=1}^{n} \tau\, \|C^{-1} \tilde{\varphi}\|_C \right)$$

with $x = D^{1/2} y$ and $\tilde{\varphi} = D^{-1/2} \varphi$ incorporated. What has been done is to derive the desirable expressions

$$\begin{aligned} \|x_t(0)\|^2_{C^{-1}} &= (C^{-1} x_t(0), x_t(0)) \\ &= (D^{1/2} A^{-1} D^{1/2} D^{1/2} y_t(0), D^{1/2} y_t(0)) \\ &= \|D y_t(0)\|^2_{A^{-1}}\,, \\ \|C^{-1} \tilde{\varphi}\|^2_C &= (C^{-1} \tilde{\varphi}, \tilde{\varphi}) \\ &= (D^{1/2} A^{-1} D^{1/2} D^{-1/2} \varphi, D^{-1/2} \varphi) \\ &= (A^{-1} \varphi, \varphi) = \|\varphi\|^2_{A^{-1}}\,, \end{aligned}$$

so there is some reason to be concerned about estimate (105) in the original variables:

$$(106)\quad \|y_{n+1}\|_D \leq \sqrt{\frac{1+\varepsilon}{\varepsilon}} \left( \|y(0)\|_D + \|D y_t(0)\|_{A^{-1}} + \sum_{s=1}^{n} \tau\, \|\varphi_s\|_{A^{-1}} \right).$$

Thus, we have proved the useful assertion.

**Theorem 10** *If conditions (87) and (102) are satisfied, then for scheme (101) a priori estimate (106) takes place. In particular, for scheme (101) with $D = E$ and $y_0 = \bar{y}_0 = 0$ we have*

$$\|y_{n+1}\| \le \sqrt{\frac{1+\varepsilon}{\varepsilon}} \sum_{s=1}^{n} \tau \, \|\varphi_s\|_{A^{-1}}$$

(see Chapter 5, Section 6.2).

As an example consider the weighted scheme

$$y_{\bar{t}t} + A\left(\sigma\,\hat{y} + (1-2\sigma)\,y + \sigma\,\check{y}\right) = \varphi, \quad A^* = A \ge \delta\,E, \quad \delta > 0\,.$$

Substituting $y + \sigma\tau^2 y_{\bar{t}t}$ for $\sigma\hat{y} + (1-2\sigma)y + \sigma\check{y}$ we obtain

$$(E + \sigma\,\tau^2\,A)\,y_{\bar{t}t} + A\,y = \varphi\,, \tag{107}$$

yielding $D = E + \sigma\tau^2 A$. The stability condition $D \ge \frac{1+\varepsilon}{4}\tau^2 A$ or $E \ge \left((1+\varepsilon)/4 - \sigma\right)\tau^2 A$ is ensured by

$$\sigma \ge \frac{1+\varepsilon}{4} - \frac{1}{\tau^2 \|\,A\,\|}\,.$$

For the explicit scheme with $\sigma = 0$ this implies that

$$\tau \le \frac{2}{\sqrt{(1+\varepsilon)\,\|\,A\,\|}}\,.$$

Having stipulated this condition, the explicit scheme ($y_{\bar{t}t} = y_{\bar{x}x}$) for the string vibration equation is stable for $\tau/h \le 1/\sqrt{(1+\varepsilon)}$ (see Chapter 5, Section 6).

**10. On regularization of difference schemes.** Stability theory of difference schemes outlined in this chapter may be useful for the statement of a general principle (the regularization principle) providing with schemes of a desired quality, that is, stable, generating an approximation and satisfying the extra economy requirement of minimizing the arithmetic operations necessary in computer implementations of resulting difference equations.

The economy requirement in the case of nonstationary problems in mathematical physics generally means that the number of arithmetic operations needed in connection with solving difference equations in passing from one layer to another is proportional to the total number of grid nodes.

In studying the canonical form of two-layer and three-layer schemes

$$(108)\qquad (E+\tau R)\, y_t + A\, y = \varphi, \quad y(0)=y_0, \quad B = E + \tau R,$$

$$(109)\qquad B\, y_{\overset{\circ}{t}} + \tau^2 R\, y_{\bar{t}t} + A\, y = \varphi, \quad y(0)=y_0, \quad y(\tau)=y_1,$$

it has been disclosed that the operator $R$ (**regularizator**) is responsible for stability.

Sufficient stability conditions have now simplified forms

$$(110)\qquad \begin{aligned} &R \geq \sigma_0\, A\,, \quad \sigma_0 = \frac{1}{2} - \frac{1}{\tau\, \| A \|} \quad \text{for two-layer schemes,} \\ &R > \frac{1}{4}\, A \quad \text{or} \quad R \geq \frac{1+\varepsilon}{4}\, A \quad \text{for three-layer schemes.} \end{aligned}$$

Stability or instability of a scheme from the primary family depends only on selection rules for the operator $R$. From the point of view of stability theory the arbitrariness in the choice of the operator $R$ is restricted by the following requirements:

- a scheme should belong to the primary family, that is, $B = E + \tau R$ for (108) and $R = R^* > 0$ for (109);
- conditions (110) should be valid.

To obtain a stable scheme of a desired quality, one is to construct a scheme generating an approximation of the attainable order and being economical, that is, it is required to solve the equations $(E+\tau R)\hat{y} = F$ for (108) or $(B + 2\tau R)\hat{y} = F$ for (109) in a minimal number of operations (in a certain sense).

First of all observe that if scheme (108) or (109) with an operator $R$ is stable, then so is a scheme with an operator $\widetilde{R} \geq R$. Common practice in designing difference schemes involves the development of a scheme which generates an approximation of the attainable order and is economical. With the indicated properties, its stability will be given special investigation.

The main idea behind regularization of difference schemes is that the schemes of a desired quality should be sought in the class of stable schemes starting from an original scheme and replacing it, by changing the operator $R$, by another scheme of a desired quality belonging to the class of stable schemes.

Many modes of constructing schemes of a particular form can be treated as simplest regularization modes. The canonical form of a scheme is convenient not only for practical tests of stability, but also for proper evaluation of the order of approximation. The factor at the member $R$ in (108) is $\tau$, while in (109) is $\tau^2$. Therefore, if in the case of two-layer schemes the condition $\| Ru_t \| = O(1)$ with such a variety of $R$ continues to hold ($u$ is a solution of the initial differential equation), then the approximation error changes by a quantity $O(\tau)$ when $R$ changes. In the case of three-layer schemes the condition $\| Ru_{\bar{t}t} \| = O(1)$ ensures that the regularization leads to schemes with the approximation error differing by a quantity $O(\tau^2)$. For this reason three-layer schemes are good enough for the purposes of the present chapter in designing stable schemes of second-order approximation in $\tau$.

The main problem here is connected with selecting the regularizator $R$. Since regularity conditions became operator inequalities, it seems reasonable to choose as $R$ operators of the most simplest structures which are energetically equivalent to the operator $A$. Let, for instance, $A$ and $A_0$ be energetically equivalent operators with constants $\gamma_1$ and $\gamma_2$, so that

$$(111) \qquad \gamma_1 A_0 \le A \le \gamma_2 A_0 , \qquad \gamma_1 > 0 , \qquad \gamma_2 > 0 .$$

Keeping then $R = \sigma A_0$ we have at our disposal stable schemes: for $\sigma \ge \sigma_0 \gamma_2$ (or $\sigma \ge \frac{1}{2}\gamma_2$) in the case (108) and for $\sigma > \gamma_2/4$ in the case (109).

A simplest form of $R$ is the operator $R = \sigma E$ ($A_0 = E$). Stability conditions are satisfied if $\sigma \ge \sigma_0 \| A \|$ for scheme (108) and $\sigma > \frac{1}{4}\| A \|$ for scheme (109).

**Example 1** The explicit three-layer Du Fort–Frankel scheme for the heat conduction equation from Section 3.7 belongs to the family

$$(112) \qquad y_{\overset{\circ}{t}} + \sigma \tau^2 y_{\bar{t}t} + A y = 0 , \qquad \sigma > \tfrac{1}{4} \| A \| , \qquad A = A^* > 0 .$$

Indeed,

$$Ay = -\Lambda y , \quad \Lambda y = y_{\bar{x}x} , \quad \| A \| = \frac{4}{h^2} \cos^2 \frac{\pi h}{2} < \frac{4}{h^2} , \quad \sigma = \frac{1}{h^2} ,$$

that is, the condition $\sigma > \frac{1}{4}\| A \|$ is satisfied. This scheme provides a conditional approximation of $O(h^2)$ for $\tau = O(h^2)$.

It is not difficult to write down an explicit stable scheme for the heat conduction equation with variable coefficients

$$(113) \qquad \begin{gathered} \frac{\partial u}{\partial t} = \frac{\partial}{\partial x} \left( k(x,t) \frac{\partial u}{\partial x} \right) , \qquad 0 < c_1 \le k(x,t) \le c_2 , \\ t > 0 , \qquad 0 < x < 1 , \\ u(0,t) = u(1,t) = 0 , \qquad u(x,0) = u_0(x) . \end{gathered}$$

In this case we agree to consider in (112) $\sigma = c_2/h^2$, $Ay = -\Lambda y$ and $\Lambda y = (ay_{\bar{x}})_x$).

For the multiple heat conduction equation

$$(114)\quad \frac{\partial u}{\partial t} = \sum_{\alpha=1}^{p} \frac{\partial}{\partial x_\alpha} \left( k_\alpha(x,t) \frac{\partial u}{\partial x_\alpha} \right), \qquad x = (x_1, \ldots, x_p) \in G,$$

$$u|_\Gamma = 0, \qquad u(x,0) = u_0(x),$$

$G$ being a parallelepiped ($0 \le x_\alpha \le l_\alpha$, $\alpha = 1, 2, \ldots, p$), $0 < c_1 \le k_\alpha \le c_2$, other ideas are connected with

$$\sigma = c_2 \sum_{\alpha=1}^{p} \frac{1}{h_\alpha^2}, \qquad A\,y = -\sum_{\alpha=1}^{p} \left( a_\alpha\, y_{\bar{x}_\alpha} \right)_{x_\alpha},$$

where $h_\alpha$ is the step of a grid $\omega_h = \{x = (x_1, \ldots, x_p) \in \bar{G}\}$ along the direction $x_\alpha$.

Let $A_0 = A_1 + A_2$, where $A_1$ and $A_2$ are adjoint or "triangular" (with a triangular matrix) operators, so that

$$(A_0 y, y) = 2\,(A_1 y, y) = 2\,(A_2 y, y)\,.$$

Setting $R = \sigma A_1$ or $R = \sigma A_2$ we arrive at scheme (108), which is stable for $\sigma \ge 2\gamma_2 2\sigma_0$ ($\gamma_2$ is a constant involved in (111)).

**Example 2** The asymmetric scheme for the heat conduction equation

$$\frac{\partial u}{\partial t} = \frac{\partial^2 u}{\partial x^2}$$

belongs to the family of "triangle" schemes having the form

$$(115)\qquad y_t + \frac{\sigma\,\tau}{h}\, y_{\bar{x}t} = \Lambda\, y\,, \qquad \Lambda\, y = y_{\bar{x}x}\,.$$

Here

$$Ay = -\Lambda y\,, \quad Ry = \frac{\sigma}{h}\, y_{\bar{x}}\,, \quad A_0 = A\,, \quad \gamma_1 = \gamma_2 = 1\,.$$

Scheme (115) is stable for $\sigma \ge 1 - \frac{h^2}{2\tau}$ and conditionally approximates the heat conduction equation to $O(h)$ as $\tau = O(h^2)$.

For equation (113) we accept

$$\gamma_2 = c_2 \,, \quad Ay = -(a y_{\bar{x}}) \,, \quad Ry = \frac{\sigma}{h} y_{\bar{x}}$$

and take

$$\sigma \geq c_2 \big(1 - \frac{h^2}{2\, c_2 \tau}\big) \,.$$

With regard to problem (114) we thus have

$$Ry = \sigma \sum_{\alpha=1}^{p} \frac{1}{h_\alpha} y_{\bar{x}_\alpha} \quad \text{or} \quad Ry = -\sigma \sum_{\alpha=1}^{p} \frac{1}{h_\alpha} y_{x_\alpha} \,, \quad \gamma_2 = c_2 \,,$$

$$\sigma \geq c_2 \left(1 - \frac{1}{2\, \tau\, \Delta}\right), \qquad \Delta = c_2 \sum_{\alpha=1}^{p} \frac{1}{h_\alpha^2} \,.$$

It is important to note that while constructing the Du Fort–Frankel scheme as an original scheme the explicit unstable scheme $y_{\overset{\circ}{t}} = \Lambda y$ generating an approximation of $O(\tau^2 + h^2)$ has been taken with further modification corresponding to the regularizator of the simplest type ($R = \frac{1}{h^2} E$, $\sigma = \frac{1}{h^2}$ in (112)).

Afterwards when the sweep formulae became customary, one began to analyze in full details two-layer implicit schemes (weighted schemes) for which $R = \sigma A$. These schemes obviously represent a particular case of the scheme with $R = \sigma A_0$.

We point out one more selection rule for $R$. Let $A_0 = A_1 + A_2$, $A_2 = A_1^* > 0$. Choose $R$ in such a way that the two-layer scheme possesses the factorized operator

$$B = (E + \sigma\, \tau\, A_1)\,(E + \sigma\, \tau\, A_2) = E + \tau\, (\sigma\, A_0 + \sigma^2\, \tau\, A_1\, A_2) \,,$$

so that

$$R = \sigma\, A_0 + \gamma^2 \tau\, A_1\, A_2 \,.$$

Since $(A_1\, A_2\, y, y) = (A_2 y, A_2 y) = \|\, A_2 y\,\|^2 > 0$, this scheme is stable if $\sigma \geq \gamma_2 \sigma_0$.

Various schemes with the factorized operator

$$B = (E + \omega\, R_1)\,(E + \omega\, R_2) \,, \qquad R_2 = R_1^* \,,$$

are in common usage as iteration schemes for solving equations of the form $Ay = \varphi$.

# 7

# Homogeneous Difference Schemes for Time-Dependent Equations of Mathematical Physics with Variable Coefficients

## 7.1 HOMOGENEOUS DIFFERENCE SCHEMES FOR THE HEAT CONDUCTION EQUATION WITH VARIABLE COEFFICIENTS

In the present chapter the objects of investigation are various homogeneous difference schemes for the heat conduction equation and a second-order equation of hyperbolic type with variable coefficients in several settings: on nonequidistant grids, with the boundary conditions of the third kind, etc. The results obtained in the preceding chapters find a wide range of applications in designing homogeneous difference schemes and establishing their stability. Especial attention here is being paid to one-dimensional problems.

**1. The original problem.** We begin by placing the first boundary-value problem for for the heat conduction equation in which it is required to find a continuous in the rectangle $\bar{D}_T = \{0 \le x \le 1,\ 0 \le t \le T\}$ solution to the equation

$$(1) \qquad \frac{\partial u}{\partial t} = L\,u + f(x,t)\,, \qquad L\,u = \frac{\partial}{\partial x}\Big(k(x,t)\,\frac{\partial u}{\partial x}\Big)\,,$$

satisfying the initial condition

$$(2) \qquad u(x,0) = u_0(x)\,, \qquad 0 \le x \le 1\,,$$

and the boundary conditions

$$u(0,t) = u_1(t)\,, \quad u(1,t) = u_2(t)\,, \quad 0 \le t \le T\,, \tag{3}$$

under the assumptions that the coefficient $k(x,t)$ is bounded from below and from above:

$$0 < c_1 \le k(x,t) \le c_2\,, \quad (x,t) \in \bar{D}_T\,, \tag{4}$$

where $c_1$ and $c_2$ are constants.

Also, we take for granted that problem (1)–(3) possesses a unique solution with all necessary derivatives.

**2. Homogeneous difference schemes with weights.** In a common setting it seems natural to expect that a difference scheme capable of describing this or that nonstationary process would be suitable for the relevant stationary process, that is, for $\partial u/\partial t \equiv 0$ we should have at our disposal a difference scheme from a family of homogeneous conservative schemes, whose use permits us to solve the equation $Lu + f = 0$.

One way of covering this for the heat conduction equation is to construct a homogeneous conservative scheme by means of the integro-interpolation method. To make our exposition more transparent, we may assume that the coefficient of heat conductivity $k = k(x)$ is independent of $t$. The general case $k = k(x,t)$ will appear on this basis in Section 8 without any difficulties.

We proceed as usual. This amounts to introducing the following grids: an equidistant grid on the segment $0 \le x \le 1$ with step $h$

$$\bar{\omega}_h = \{x_i = ih\,, \ i = 0, 1, \ldots, N\,, \ h = 1/N\};$$

a grid on the segment $0 \le t \le T$ with step $\tau$

$$\bar{\omega}_\tau = \{t_j = j\tau\,, \ j = 0, 1, \ldots, N_0\,, \ \tau = T/N_0\};$$

a grid in the rectangle $\bar{D}_T$

$$\bar{\omega}_{h\tau} = \bar{\omega}_h \times \bar{\omega}_\tau = \{(x_i, t_j)\,, \ x_i \in \bar{\omega}_h\,, \ t_j \in \bar{\omega}_\tau\}$$

and to forming

$$\omega_{h\tau} = \omega_h \times \omega_\tau = \{(x_i, t_j)\,, \ x_i = ih\,, \ 0 < i < N\,, \ t_j = j\tau\,, \ 0 < i \le N_0\}\,.$$

The starting point is the balance equation written in the rectangle $\{x_{i-1/2} \le x \le x_{i+1/2}\}$, $t_j \le t \le t_{j+1}$ with regard to the governing equation

$$\frac{\partial u}{\partial t} = L\,u + f(x,t)\,, \qquad L\,u = \frac{\partial}{\partial x}\Big(k(x)\,\frac{\partial u}{\partial x}\Big)\,.$$

The outcome of this is

$$(5) \quad \int\limits_{x_{i-0.5}}^{x_{i+0.5}} \big[u(x,t_{j+1}) - u(x,t_j)\big]\,dx = \int\limits_{t_j}^{t_{j+1}} \big[w(x_{i+1/2},t) - w(x_{i-1/2},t)\big]\,dt$$

$$+ \int\limits_{t_j}^{t_{j+1}} dt \int\limits_{x_{i-0.5}}^{x_{i+0.5}} f(x,t)\,dx\,, \quad w(x,t) = k\,\frac{\partial u}{\partial x}\,,$$

where the integrals and derivatives are yet to be replaced by

$$\int\limits_{x_{i-0.5}}^{x_{i+0.5}} u(x,t)\,dx \sim h\,u(x_i,t)\,, \quad w_{i-1/2} \sim a_i\,u_{\bar{x},i}\,,$$

$$\frac{1}{\tau}\int\limits_{t_j}^{t_{j+1}} w(x_{i-1/2},t)\,dt \sim \sigma\,w_{i-1/2}^{j+1} + (1-\sigma)\,w_{i-1/2}^{j}\,,$$

$$\int\limits_{t_j}^{t_{j+1}} dt \int\limits_{x_{i-0.5}}^{x_{i+0.5}} f(x,t)\,dx \sim h\,\tau\,\varphi_i^j\,,$$

where $\approx$ designates approximation, $\sigma$ is a numerical parameter and the coefficients $a_i$ are expressed through the values of $k(x)$ for $x_{i-1} \le x \le x_i$ by means of pattern functionals $A[\bar{k}(s)]$, $-1 \le s \le 1$, so that

$$a(x_i) = a_i = A\,\big[k(x_i + sh)\big] \quad \text{or} \quad \frac{1}{a_i} = A\left[\frac{1}{k(x_i + sh)}\right].$$

Here $A[\bar{k}(s)]$ is linear nondecreasing functional, for which the conditions $A[1] = 1$ and $A[s] = -0.5$ hold.

From such reasoning it seems clear that the operator

$$L\,u = \frac{\partial}{\partial x}\left(k\,\frac{\partial u}{\partial x}\right)$$

is approximated to second order by the difference operator $\Lambda u = (a\,u_{\bar{x}})_x$, meaning

$$\Lambda\,u - L\,u = O(h^2) \quad \text{if} \quad k(x) \in C^{(3)}\,.$$

In so doing $0 < c_1 \le a(x) \le c_2$.

Upon substituting the resulting expressions into (5) and replacing $u$ by $y$ we obtain the difference scheme for the grid function $y(x_i, t_j)$:

$$\text{(6)} \qquad \frac{y_i^{j+1} - y_i^j}{\tau} = \Lambda\left(\sigma\,y_i^{j+1} + (1-\sigma)\,y_i^j\right) + \varphi_i^j\,,$$

$$i = 1, 2, \ldots, N-1\,, \quad j \ge 0\,,$$

where $\Lambda y = (a y_{\bar{x}})_x$.

The simplest formulas

$$a_i = k_{i-1/2}\,, \qquad a_i = 0.5\,(k_{i-1} + k_i)\,,$$

$$\varphi_i^j = f_i^{j+1/2}, \qquad \varphi_i^j = 0.5\,(f_i^j + f_i^{j+1/2})$$

suit us perfectly for determination of $\varphi_i^j$ and $a_i$.

When the difference equation (6) is put together with the supplementary conditions $y_i^0 = u_0(x_i)$, $y_0^j = \mu_1(t_j)$, $y_N^j = \mu_2(t_j)$, there arises naturally the difference boundary-value problem

$$\text{(7)} \qquad y_t = \Lambda\left(\sigma\,\hat{y} + (1-\sigma)\,y\right) + \varphi\,, \quad x \in \omega_h\,, \quad t = j\tau \ge 0\,,$$

$$y(x, 0) = u_0(x)\,, \quad x \in \bar{\omega}_h\,,$$

$$y(0, t) = \mu_1(t)\,, \quad y(1, t) = \mu_2(t)\,, \quad t \in \bar{\omega}_t\,,$$

$$\Lambda\,y = (a\,y_{\bar{x}})_x\,, \quad 0 < c_1 \le a \le c_2\,,$$

within the usual notations

$$y = y_i^j = y(x_i, t_j)\,, \quad \hat{y} = y_i^{j+1} = y(x_i, t_{j+1})\,.$$

The next step is to reduce the resulting scheme to more convenient forms in trying to avoid cumbersome calculations. The case $\sigma = 0$, relating to the explicit scheme, is simple to follow:

$$\begin{aligned} y_i^{j+1} &= y_i^j + \tau\,(\Lambda\, y_i^j + \varphi_i^j) \\ &= \left(1 - \frac{\tau\,(a_i + a_{i+1})}{h^2}\right) y_i^j \\ &\quad + \frac{\tau}{h^2}\,(a_i y_{i-1}^j + a_{i+1} y_{i+1}^j) + \tau\,\varphi_i^j\,, \end{aligned}$$

permitting us to find a solution on every new layer.

The case $\sigma \neq 0$, relating to the implicit scheme, is connected with the equation related to the unknown $\hat{y} = y^{j+1}$:

$$\sigma\,\tau\,\Lambda\,\hat{y} - \hat{y} = -F\,, \qquad F = y + \tau\,(1-\sigma)\,\Lambda\,y + \tau\,\varphi\,,$$

which can be written in the augmented form

$$\text{(8)} \qquad A_i \hat{y}_{i-1} - C_i \hat{y}_i + A_{i+1} \hat{y}_{i+1} = -F_i\,, \quad i = 1, 2, \ldots, N-1\,,$$

$$A_i = \sigma\,\tau\,a_i/h^2\,, \quad C_i = A_i + A_{i+1} + 1\,,$$

$$\begin{aligned} F_i &= \left(1 - \frac{\tau}{h^2}\,(1-\sigma)\,(a_i + a_{i+1})\right) y_i \\ &\quad + \frac{\tau}{h^2}\,(1-\sigma)\,(a_i y_{i-1} + a_{i+1} y_{i+1}) + \tau\,\varphi_i \end{aligned}$$

with the supplementary boundary conditions for $i = 0$ and $i = N$

$$\hat{y}_0 - \mu_1(t_{j+1})\,, \qquad \hat{y}_N - \mu_2(t_{j+1})\,.$$

The difference boundary-value problem associated with the difference equation (7) of second order can be solved by the standard elimination method, whose computational algorithm is stable, since the conditions $A_i \neq 0$, $|C_i| \geq |A_i| + |A_{i+1}|$ are certainly true for $\sigma > 0$.

The recurrence formula

$$F_i^j = \frac{1}{\sigma}\,y_i^j - \frac{1-\sigma}{\sigma}\,F_i^{j-1} + \tau\,\varphi_i^j$$

is aimed at finding the right-hand side $F_i = F_i^j$ of equation (8) with a smaller volume of computations.

The following implicit schemes are frequently encountered in the theory and practice:

a) the symmetric scheme ($\sigma = 0.5$)

$$y_t = 0.5\,\Lambda\,(\hat{y} + y) + \varphi\,;$$

b) the forward scheme or purely implicit scheme ($\sigma = 1$)

$$y_t = \Lambda\,y + \varphi\,.$$

**3. Stability and convergence.** The general stability theory for two-layer schemes applies equally well to the stability analysis of the weighted scheme (7). With this aim, the appropriate difference scheme with the homogeneous boundary conditions comes first:

(9)
$$y_t = \Lambda\,\big(\sigma\,\hat{y} + (1-\sigma)\,y\big) + \varphi\,, \quad x \in \omega_h\,, \quad t \geq 0\,,$$
$$y(x,0) = u_0(x)\,, \quad y = 0 \quad \text{for} \quad x = 0\,, x = 1\,.$$

All the tricks and turns remain unchanged: we, first, introduce the space $\overset{\circ}{\Omega}$ of all grid functions given on the grid $\bar{\omega}_h$ and vanishing on the boundary for $x = 0$ and $x = 1$ and then define in that space $H = \overset{\circ}{\Omega}$ an inner product

$$(y, v) = \sum_{i=1}^{N-1} y_i\,v_i\,h$$

and associated norm $\|\,y\,\| = \sqrt{(y,y)}$. By means of a linear operator $A$ acting in accordance with the rule

$$A\,y = -\Lambda\,y = -(a\,y_{\bar{x}})_x \quad \text{for} \quad y \in H$$

the preceding scheme (9) can be rewritten as

(10)
$$y_t + A\,\big(\sigma\,\hat{y} + (1-\sigma)\,y\big) = \varphi\,, \quad t = j\tau \geq 0\,, \quad y(0) = u_0\,.$$

Observe the operator $A$ so defined is self-adjoint and positive definite:

$$A^* = A > 0\,, \quad \delta\,E \leq A \leq \Delta\,E\,, \quad \delta > 0\,,$$

where

$$\delta = \min_{1 \leq k \leq N-1} \lambda_k(A)\,, \qquad \Delta = \|\,A\,\| = \max_{1 \leq k \leq N-1} \lambda_k(A)$$

and $\lambda_k(A)$ is the $k$th eigenvalue of the operator $A$. It follows from Green's formula

$$(Ay, y) = -((a\, y_{\bar{x}})_x, y) = (a, (y_{\bar{x}})^2]$$

that

$$c_1 \overset{\circ}{A} \le A \le c_2 \overset{\circ}{A}\,,$$

where $\overset{\circ}{A}\, y = -y_{\bar{x}x}$ for $y \in H$, with further reference to the relation

$$(\overset{\circ}{A}\, y, y) = ((y_{\bar{x}})^2, 1]\,.$$

The smallest and greatest eigenvalues of the operator $\overset{\circ}{A}$ are given by the formulas

$$\overset{\circ}{\delta} = \frac{4}{h^2} \sin^2 \frac{\pi h}{2}\,, \qquad \overset{\circ}{\Delta} = \frac{4}{h^2} \cos^2 \frac{\pi h}{2}\,,$$

thereby justifying the estimates $\delta \ge c_1 \overset{\circ}{\delta}$ and $\Delta \ge c_2 \overset{\circ}{\Delta}$.

We know from the general stability theory that scheme (10) is stable in the space $H_A$ with respect to the initial data, that is,

$$\|y^j\|_A \le \|y^0\|_A \quad \text{if} \quad \varphi^j \equiv 0$$

under the constraint

$$\sigma \ge \sigma_0\,, \qquad \sigma_0 = \frac{1}{2} - \frac{1}{\tau\Delta}\,.$$

In the case where

$$\sigma \ge \sigma_\varepsilon\,, \qquad \sigma_\varepsilon = \frac{1}{2} - \frac{1-\varepsilon}{\tau\Delta}\,, \qquad 0 < \varepsilon < 1\,,$$

a solution of problem (10) satisfies the estimate

$$\|y^{j+1}\|_A \le \|y^0\|_A + \frac{1}{\sqrt{2\varepsilon}} \left( \sum_{j'=0}^{j} \tau \|\varphi^{j'}\|^2 \right)^{1/2}. \tag{11}$$

For the explicit scheme ($\sigma = 0$) we might have

$$y_i^{j+1} = \Big(1 - \frac{\tau}{h^2}(a_i + a_{i+1})\Big) y_i^j + \frac{\tau}{h^2}\left(a_i y_{i-1}^j + a_{i+1} y_i^j\right) + \tau \varphi_i^j\,, \tag{12}$$

where the coefficient at the member $y_i^j$ is non-negative, provided the condition

$$1 - \frac{2\tau}{h^2} c_2 \ge 0$$

holds. Under this condition the relation

$$\|y^{j+1}\|_C \le \|y^j\|_C + \|\varphi^j\|_C$$

takes place. Summing up the preceding other $j = 0, 1, 2, \dots$, we obtain the inequality

$$(13) \qquad \|y^{j+1}\|_C \le \|y^0\|_C + \sum_{j'=0}^{j} \tau \, \|\varphi^{j'}\|_C ,$$

which expresses the stability of scheme (10) in the space $C$ if

$$(14) \qquad \tau \le \frac{h^2}{2\, c_2} .$$

This condition is only sufficient for the indicated property. A necessary and sufficient condition for the stability of the explicit scheme with respect to the initial data in the space $H_A$ is

$$(15) \qquad \tau \le \frac{2}{\Delta}, \qquad \text{where} \qquad \Delta \le \frac{4\, c_2}{h^2} .$$

But it may happen that the coefficient $k(x)$ varies very fastly. In that case the estimate $\Delta \le 4c_2/h^2$ is too rough and condition (14) gives a severe restriction. In mastering the difficulties involved, we are forced to apply the maximum principle to the weighted scheme (9) written in the canonical form for any $\sigma$:

$$(16) \qquad \Big(1 + \frac{\sigma\,\tau}{h^2}\,(a_i + a_{i+1})\Big)\, y_i^{j+1} = \frac{\sigma\,\tau}{h^2}\,(a_i y_{i-1}^{j+1} + a_{i+1} y_{i+1}^{j+1}) + F_i^j ,$$

$$F_i^j = \Big(1 - \frac{(1-\sigma)\,\tau}{h^2}\,(a_i + a_{i+1})\Big)\, y_i^j$$

$$+ \frac{(1-\sigma)\,\tau}{h^2}\,(a_i y_{i-1}^j + a_{i+1} y_{i+1}^j) + \tau\, \varphi_i^j .$$

The boundary conditions $y(0) = y(1) = 0$ together with Theorem 3 in Chapter 4, Section 2 for equation (16) give us the estimate

$$\|y^{j+1}\|_C \le \|F^j\|_C .$$

The coefficient at the member $y_i^j$ becomes non-negative for

$$\tau \le \frac{h^2}{2\,(1-\sigma)\,c_2}\,, \tag{17}$$

which assures us of the validity of the relations

$$\|F^j\|_C \le \|y^j\|_C + \tau\,\|\varphi^j\|_C$$

and

$$\|y^{j+1}\|_C \le \|y^j\|_C + \tau\,\|\varphi^j\|_C \le \|y^0\|_C + \sum_{j'=0}^{j} \tau\,\|\varphi^{j'}\|_C\,. \tag{18}$$

Summarizing, the weighted scheme (9) is stable in the space $C$, provided condition (17) holds. For the purely implicit scheme with $\sigma = 1$ estimate (18) is valid for any value of $\tau$.

The accurate account of the accuracy is stipulated by more a detailed exploration of the residual

$$\psi = \Lambda\,(\sigma\,\hat{u} + (1-\sigma)\,u) + \varphi - u_t$$

on the solution $u$ of the original problem (1)–(3). Upon substituting the expansions

$$\hat{u} = \frac{\hat{u}+u}{2} + \frac{\tau}{2}\,u_t = \bar{u} + \frac{\tau}{2}\,\dot{\bar{u}} + O(\tau^2)\,,$$

$$u = \frac{\hat{u}+u}{2} - \frac{\tau}{2}\,u_t = \bar{u} - \frac{\tau}{2}\,\dot{\bar{u}} + O(\tau^2)\,,$$

$$u_t = \dot{\bar{u}} + O(\tau^2)$$

with the members

$$\bar{u} = u^{j+1/2}\,, \qquad \dot{u} = \frac{\partial u}{\partial t}$$

it is plain to calculate the residual

$$\begin{aligned}\psi &= \Lambda\,\frac{\hat{u}+u}{2} + (\sigma - 0.5)\,\tau\,\Lambda\,u_t + \varphi - u_t \\ &= (\Lambda\,\bar{u} + \bar{f} - \dot{\bar{u}}) \\ &\quad + (\Lambda\,\bar{u} - L\,\bar{u}) + (\varphi - \bar{f}) + (\sigma - 0.5)\,\tau\,\Lambda\,\dot{\bar{u}} + O(\tau^2)\,.\end{aligned}$$

Taking into account the relations $Lu + f - \dot{u} = 0$ and

(19) $$\Lambda u = L u + O(h^2), \qquad \varphi = \bar{f} + O(\tau^2 + h^2),$$

we arrive at

(20) $$\psi = (\sigma - 0.5)\, \tau\, L \bar{\dot{u}} + O(\tau^2 + h^2).$$

This supports the view that the order of approximation for a given value of $\sigma$ coincides with the order of approximation established before for the unit constant coefficient $k(x) \equiv 1$:

$$\psi = O(\tau^2 + h^2) \quad \text{for} \quad \sigma = 0.5,$$

$$\psi = O(\tau + h^2) \quad \text{for} \quad \sigma \neq 0.5.$$

Along these lines, the error of approximation $z^j = y^j - u^j$, where $y^j$ is the solution of problem (7) and $u = u(x, t)$ is the solution of the governing problem (1)–(3), can be most readily evaluated with the aid of the equation

$$z_t = \Lambda\,(\sigma \hat{z} + (1 - \sigma)\, z) + \psi, \quad x \in \omega_h, \quad t = j\tau \geq 0,$$

$$z(x, 0) = 0, \quad z(0, t) = z(1, t) = 0.$$

Some progress can be achieved by having recourse to relation (11), leaving us with the estimate

(21) $$\|z^{j+1}\|_A \leq \frac{1}{\sqrt{2\,\varepsilon}} \left( \sum_{j'=0}^{j} \tau \, \| \psi^{j'} \|^2 \right)^{1/2}.$$

Using the results obtained in Chapter 2, Section 3, namely the relations

$$\|z\|_C \leq \tfrac{1}{2}\, (1, (z_{\bar{x}})^2]^{1/2} = \tfrac{1}{2} \sqrt{(\overset{\circ}{A}\, z, z)}$$

and the estimate $c_1 \, \|z\|^2_{\overset{\circ}{A}} \leq \|z\|^2_A$, we derive the inequality

(22) $$\|z^{j+1}\|_C \leq \frac{1}{2\,\sqrt{2\,\varepsilon\, c_1}} \left( \sum_{j'=0}^{j} \tau \, \| \psi^{j'} \|^2 \right)^{1/2},$$

thereby justifying by virtue of representation (20) that scheme (7) converges uniformly with the rate $O(h^2 + \tau^{m_\sigma})$, where

$$m_\sigma = \begin{cases} 2 & \text{for} \quad \sigma = 0.5, \\ 1 & \text{for} \quad \sigma \neq 0.5, \end{cases}$$

provided that the conditions hold under which the describing scheme is of accuracy $O(h^2 + \tau^{m_\sigma})$ and $\sigma \geq \sigma_\varepsilon$ for $0 < \varepsilon \leq 1$. A similar conclusion was drawn in Chapter 5 for the relevant equation with a constant coefficient $k(x)$.

**4. The equation with a discontinuous coefficient of heat conductivity.** In the general setting the convergence of the weighted homogeneous scheme (7) is of our initial concern. A case in point is that the coefficient $k(x)$ has a discontinuity of the first kind on the straight line $x = \xi$ in the plane $(x, t)$. Some consensus of opinion is that the usual condition of conjugation

$$[u] = 0\,, \quad \left[k\,\frac{\partial u}{\partial x}\right] = 0 \quad \text{for} \quad x = \xi\,, \quad t \geq 0\,, \tag{23}$$

is fulfilled on every discontinuity line. In the physical language, this is a way of saying that the temperature $u(x, t)$ and the heat flow $(-ku')$ are continuous.

By relating the functions $k(x)$, $f(x, t)$ and a solution $u(x, t)$ to be sufficiently smooth everywhere except this discontinuous line we proceed to evaluate the residual (the error of approximation)

$$\psi = \Lambda\,(\sigma\,\hat{u} + (1-\sigma)\,u) + \varphi - u_t$$

$$= \Lambda\,u^{(\sigma)} + \varphi - u_t\,,$$

$$u^{(\sigma)} = \sigma\,\hat{u} + (1-\sigma)\,u\,.$$

Let now $\xi = x_n + \theta h$, $x_n = nh$, $0 \leq \theta \leq 1$, $n \geq 1$. As far as a three-point operator $L$ is concerned, we might have

$$\psi_i = O(h^2 + \tau^{m_\sigma}) \quad \text{for all} \quad i \neq n\,, \quad i \neq n+1\,. \tag{24}$$

On this basis it remains to calculate $\psi_i$ for $i = n$ and $i = n+1$. Also, it will be sensible to introduce

$$h\,\psi_n = \tilde{w}^{(\sigma)}_{n+1} - \tilde{w}^{(\sigma)}_n + h\,\varphi_n - h\,u_{t,n}\,, \quad \tilde{w}_i = a\,u_{\bar{x},i}\,. \tag{25}$$

Since

$$(a\,u_{\bar{x}})_i = (k\,u')_{i-1/2} + O(h^2)$$

for $k \in C^{(2)}[x_{i-1}, x_i]$ and $u \in C^{(3)}[x_{i-1}, x_i]$, we obtain

$$\tilde{w}_n = (k\,u')_{n-1/2} + O(h^2) \tag{26}$$

$$= (k\,u')_{\text{left}} - (\theta + 0.5)\,h\,(k\,u')'_{\text{left}} + O(h^2)\,,$$

where $v_{\text{left}} = v(\xi - 0, t)$. By the same token,

$$h\,\psi_{n+1} = \tilde{w}^{(\sigma)}_{n+2} - \tilde{w}^{(\sigma)}_{n+1} + h\,\varphi_{n+1} - h\,u_{t,n+1}\,, \tag{27}$$

$$\tilde{w}_{n+2} = (a\,u_{\bar{x}})_{n+2} = (k\,u')_{\text{right}} + (1.5 - \theta)\,h\,(k\,u')'_{\text{right}} + O(h^2)\,, \tag{28}$$

where $v_{\text{right}} = v(\xi + 0, t)$.

Furthermore, using the expansions behind

$$u_n = u(\xi) - \theta\, h\, u'_{\text{left}} + 0.5\, \theta^2\, h^2\, u''_{\text{left}} + O(h^3)\,,$$

$$u_{n+1} = u(\xi) + (1-\theta)\, h\, u'_{\text{right}} + 0.5\,(1-\theta)^2\, h^2\, u''_{\text{right}} + O(h^3)\,,$$

we find that

$$\tilde{w}_{n+1} = a_{n+1}\, u_{\bar{x},n+1} = a_{n+1}\left(\theta\, u'_{\text{left}} + (1-\theta)\, u'_{\text{right}}\right) + O(h)\,.$$

As a corollary to the condition of conjugation $[ku'] = 0$, we might have

$$(k\, u')_{\text{left}} = (k\, u')_{\text{right}} = w(\xi, t)\,, \quad u'_{\text{left}} = \frac{w(\xi,t)}{k_{\text{left}}}\,, \quad u'_{\text{right}} = \frac{w(\xi,t)}{k_{\text{right}}}\,,$$

so that

$$\tilde{w}_{n+1} = a_{n+1}\left(\frac{\theta}{k_{\text{left}}} + \frac{1-\theta}{k_{\text{right}}}\right) w(\xi,t) + O(h)\,. \tag{29}$$

Upon substituting (26), (28) and (29) into (25) and (27) we finally get

$$h\,\psi_n = Q_n^{(\sigma)} + O(h)\,,$$

$$Q_n = \left[a_{n+1}\left(\frac{\theta}{k_{\text{left}}} + \frac{1-\theta}{k_{\text{right}}}\right) - 1\right] w(\xi,t)\,,$$

$$\begin{aligned} h\,(\psi_n + \psi_{n+1}) &= \tilde{w}^{(\sigma)}_{n+2} - \tilde{w}^{(\sigma)}_n + h\,(\varphi_n + \varphi_{n+1}) - h\,(u_{t,n+1} + u_{t,n}) \\ &= (1.5-\theta)\, h\left((k\,u')'_{\text{right}}\right)^{(\sigma)} + (0.5+\theta)\, h\left((k\,u')'_{\text{left}}\right)^{(\sigma)} \\ &\quad + h\,(\varphi_n + \varphi_{n+1}) - h\,(u_{t,n+1} + u_{t,n}) + O(h^2)\,, \end{aligned}$$

showing the new members to be sensible ones. The limiting values emerging from equation (1) are taken to be

$$(k\,u')'_{\text{left}} = (\dot{u} - f)_{x=\xi}\,, \qquad (k\,u')'_{\text{right}} = (\dot{u} - f)_{x=\xi}\,,$$

since $[\dot{u}] = [f] = 0$ for $x = \xi$. The next step in this direction is to insert the assigned values in the preceding expansions with further reference to

$$\varphi_n = \bar{f}_n + O(h^2 + \tau^2)\,, \qquad \tilde{w}^{(\sigma)} = \frac{\hat{\tilde{w}} + \tilde{w}}{2} + (\sigma - 0.5)\,\tau\,\tilde{w}_t\,,$$

making it possible to find that

$$h\,(\psi_n + \psi_{n+1}) = 2\,h\,(\dot{u} - f)^{(\sigma)}_{x=\xi} + h\,(\varphi_n + \varphi_{n+1} - u_{t,n} - u_{t,n+1}) + O(h^2)$$

$$= O((\sigma - 0.5)\,\tau\,h + \tau^2\,h + h^2)$$

and deduce for any weighted scheme of the form (7) that

$$h\,\psi_n = O(1)\,, \tag{30}$$

$$h\,(\psi_n + \psi_{n+1}) = (\sigma - 0.5)\,O(\tau\,h) + O(\tau^2\,h + h^2)\,. \tag{31}$$

Under the special choice

$$a(x) = \overset{\circ}{a}(x) = \left( \int_{-1}^{0} \frac{ds}{k\,(x + s\,h)} \right)^{-1} \tag{32}$$

it is straightforward to verify by analogy with the available procedures (for more detail see Chapter 3, Section 3) that

$$Q_n = O(h)\,, \qquad h\,\psi_n = O(h)\,.$$

From such reasoning it seems clear that the following estimates are valid:

$$h\,\psi_{\bar{t},n} = O(1)\,, \tag{33}$$

$$h\,(\psi_{\bar{t},n} + \psi_{\bar{t},n+1}) = (\sigma - 0.5)\,O(\tau\,h) + O(\tau^2\,h + h^2)\,, \tag{34}$$

thereby justifying the representations

$$\begin{aligned} \psi &= \overset{\circ}{\psi} + \psi^*\,, & \overset{\circ}{\psi}_i &= \psi_i\,(\delta_{i,n} + \delta_{i,n+1})\,, \\ \psi^* &= O(h^2 + \tau^{m_\sigma})\,, & \psi_i^* &= 0 \quad \text{for} \quad i \neq n\,,\ i \neq n+1\,, \end{aligned} \tag{35}$$

where $\delta_{i,n}$ is, as usual, Kronecker's delta.

The intervention of a new grid function

$$\eta_i = \sum_{k=1}^{i-1} h\,\overset{\circ}{\psi}_k\,, \quad \eta_1 = 0\,, \quad i = 2, 3, \dots, N\,,$$

arising from the representation

$$(36) \qquad \psi = \eta_x + \psi^* , \quad \text{that is,} \quad \overset{\circ}{\psi} = \eta_x ,$$

gives rise to a perfect model for a more simpler estimation:

$$(37) \qquad \eta_i = \begin{cases} 0 & \text{for} \quad i < n+1 , \quad \eta_{n+1} = h\,\psi_n , \\ h\,(\psi_n + \psi_{n+1}) & \text{for} \quad i > n+1 . \end{cases}$$

The order of accuracy of scheme (7) can be most readily evaluated with the aid of the representation for the error $z = y - u$ as

$$z = v + z^* ,$$

where $v$ and $z^*$ are solutions of the related problems:

$$v_t = \Lambda\,(\sigma\,\hat{v} + (1-\sigma)\,v) + \overset{\circ}{\psi} , \qquad v_0 = v_N = 0 , \quad v(x,0) = 0 ,$$

$$z_t^* = \Lambda\,(\sigma\,\hat{z}^* + (1-\sigma)\,z^*) + \psi^*, \quad z_0^* = z_N^* = 0 , \quad z^*(x,0) = 0 .$$

In the accurate account of $v$ and $z^*$ we apply the results of the general stability theory (Theorems 9 and 11 from Section2, Chapter 6):

$$(38) \qquad \| v^{j+1} \| \leq \| A^{-1} \overset{\circ}{\psi}{}^0 \| + \| A^{-1} \overset{\circ}{\psi}{}^j \|$$

$$+ \sum_{j'=1}^{j} \tau \, \| A^{-1} \overset{\circ}{\psi}{}_{\bar t}^{j'} \| \quad \text{for} \quad \sigma \geq \sigma_0 ,$$

$$(39) \qquad \| z^{*j+1} \| \leq \sum_{j'=0}^{j} \tau \, \| \psi^{*j'} \| \quad \text{for} \quad \sigma \geq \sigma_0 , \quad \sigma \geq 0 .$$

Since $A^{-1} \overset{\circ}{\psi} = \zeta$ is a solution to the equation $A\zeta = \overset{\circ}{\psi} = \eta_x$, we have occasion to use the relation

$$(40) \qquad \| A^{-1} \overset{\circ}{\psi} \| \leq \frac{1}{c_1} \, (1, |\eta|] ,$$

the right-hand side of which can be modified on account of (37) into

$$(1, |\eta|] = \sum_{i=n+1}^{N} h \, |\eta_i| = h^2 \, |\psi_n| + h \, |\psi_n + \psi_{n+1}| \, (1 - x_{n+1}) .$$

By substituting estimates (30) and (31) into inequality (40) and, in turn, estimates (33) and (34) into a similar inequality for $\| A^{-1} \overset{\circ}{\psi} \|$ and then applying estimates (35), (38), (39) and the inequality $\| z \| \leq \| z^* \| + \| v \|$ to the resulting expressions we establish the convergence of scheme (7) in the class of discontinuous coefficients.

*Let $k(x)$ possess a discontinuity of the first kind for $x = \xi$ and conditions (23), (24), (30) and (31) hold simultaneously. Then for $\sigma \geq \sigma_0$, $\sigma \geq 0$, scheme (7) converges in the norm of the grid space $L_2$ with the rate $O(h^2 + \tau^{m+\sigma})$, while the best scheme with coefficient (32) does the same with the rate $O(h^2 + \tau^{m+\sigma})$; $m_\sigma = 2$ for $\sigma = 0.5$ and $m_\sigma = 1$ for $\sigma \neq 0.5$.*

A priori estimates obtained in Chapter 6, Section 2 provides the sufficient background for a uniform estimate of accuracy in the norm of the grid space $C$ such as

$$\text{(41)} \qquad \| v^{j+1} \|_A \leq \| \overset{\circ}{\psi}{}^0 \|_{A^{-1}} + \| \overset{\circ}{\psi}{}^j \|_{A^{-1}} + \sum_{j'=1}^{j} \tau \, \| \overset{\circ}{\psi}{}_{\bar t}^{j'} \|_{A^{-1}} \quad \text{for} \quad \sigma \geq \sigma_0 \,,$$

$$\text{(42)} \qquad \| z^{*j+1} \|_A \leq \frac{1}{\sqrt{2\,\varepsilon}} \left( \sum_{j'=0}^{j} \tau \, \| \psi^{*j'} \|^2 \right)^{1/2} \quad \text{for} \quad \sigma \geq \sigma_\varepsilon \,,$$

where $\| z \|_A^2 = (Az, z) = (a, (z_{\bar x})^2]$ and the inequalities (see Chapter 2, Section 3) are taken into account:

$$\|z\|_C \leq \tfrac{1}{2} \, (1, (z_{\bar x})^2]^{1/2}, \qquad \|z\|_A^2 = (a, (z_{\bar x})^2] \geq c_1 \, (1, (z_{\bar x})^2] \, .$$

In conformity with Chapter 2, Section 4 we deduce that

$$\| \overset{\circ}{\psi} \|_{A^{-1}} = \|\eta_x\|_{A^{-1}} \leq \frac{1}{\sqrt{c}} \, \|\eta]| = \frac{1}{\sqrt{c_1}} \left( \sum_{i=1}^{N} h \, \eta_i^2 \right)^{1/2} ,$$

which in combination with (37) leads to

$$\|\eta\|]^2 = h\,\eta_{n+1}^2 + (1 - x_{n+1})\,\eta_{n+2}^2$$

$$= h\,(h\,\psi_n)^2 + (1 - x_{n+1})\,\big(h\,(\psi_n + \psi_{n+1})\big)^2,$$

thereby justifying that

$$\|\eta\|] = O(\sqrt{h} + \tau^{m_\sigma}) \quad \text{for any scheme (7)},$$

$$\|\eta\|] = O(h^{3/2} + \tau^{m_\sigma}) \quad \text{for the best scheme with coefficient (32)}.$$

Having stipulated the same conditions (23), (24), (30) and (31) as before, scheme (7) converges uniformly with the rate $O(\sqrt{h} + \tau^{m_\sigma})$:

$$\|z^j\|_C = \|y^j - u^j\|_C = O(\sqrt{h} + \tau^{m_\sigma}).$$

This estimate can be improved for the forward difference scheme with $\sigma = 1$ by means of the maximum principle and the method of extraction of "stationary nonhomogeneities", what amounts to setting

$$z^j = \bar{v}^j + w^j\,,$$

where $w^j$ is a solution to the equation

$$A\,\hat{w} = \overset{\circ}{\psi} = \eta_x\,, \tag{43}$$

so that

$$\|\hat{w}\|_C \le \frac{1}{c_1}\,(1, |\eta|]\,, \qquad \|w_{\bar{t}}\|_C \le \frac{1}{c_1}\,(1, |\eta_{\bar{t}}|]\,.$$

Here $\bar{v}^j$ needs to be recovered from the relations

$$\bar{v}_t + A\,\hat{\bar{v}} = \bar{\psi}\,, \quad \bar{\psi} = \psi^* - w_t\,, \quad \bar{v}(x,0) = -w(x,0)\,. \tag{44}$$

In this regard, the maximum principle with regard to equation (44) yields

$$\|\bar{v}^{j+1}\|_C \le \|\bar{v}^0\|_C + \sum_{j'=1}^{j} \tau\,(\|\psi^{*j'}\|_C + \|w_t^{j'}\|_C)\,.$$

By inserting here the estimates for $\|w\|_C$, $\|w_t\|_C$ and $\|\psi^*\|_C$ we arrive at $\|\bar{v}^j\|_C = O(\tau + h)$ and, hence, $\|z^j\|_C = O(\tau + h)$. For the scheme with the coefficient specified by (32), it retains the order of accuracy in the class of discontinuous coefficients as occurred before in the stationary case: $\|z^j\|_C = O(h^2 + \tau)$.

**5. Homogeneous schemes on nonequidistant grids.** There is no doubt that the users come across nonequidistant grids in both $x$ and $t$ variables in practical implementations of some or other problems. All the preceding results and estimates remain valid for the two-layer scheme being used on a nonequidistant in $t$ grid. That is to say, the nonuniformity in $t$ has no considerable impact in such matters and it should be taken into account only in the selection rule for the step $\tau = \tau_j$ depending now on the subscript $j$. The order of approximation in $t$ remains unchanged, but the symbol $O(\tau^m)$ will stand either for $O(\tau_j^m)$ or for $O((\max_{1\le j\le j_0} \tau_j)^m)$, where there is no danger of confusion. The situation in which the grid is nonequidistant in $x$ needs investigation by exactly the same reasoning as before.

Let $\hat{\bar{\omega}}_h = \{x_i\,,\ i = 0, 1, , \ldots , N,\ x_0 = 0,\ x_N = 1\}$ be an arbitrary grid on the segment $0 \le x \le 1$ with steps $h_i = x_i - x_{i-1}$, $i = 1, 2, \ldots , N$. In line with established priorities from Chapter 3, Section 4 the operator

$$Lu = \frac{\partial}{\partial x}\left(k\,\frac{\partial u}{\partial x}\right)$$

is approximated by the difference operator

$$\Lambda\, y = (a\, y_{\bar{x}})_{\hat{x},i} = \frac{1}{\hbar_i}\left[\frac{a_{i+1}(y_{+1} - y_i)}{h_{i+1}} - \frac{a_i(y_i - y_{i-1})}{h_i}\right]$$

with the same coefficients $a_i$ as was done on equidistant grids. This is acceptable if we take, for example, under the constraint $-1 \le s \le 0$ (see Section 2)

$$(45)\qquad a_i = A\left[k\,(x_i + sh_i)\right] \qquad \text{or} \qquad \frac{1}{a_i} = A\left[\frac{1}{k\,(x_i + sh_i)}\right].$$

The right-hand side is calculated by the simple formula

$$(46)\qquad \varphi = \varphi_i^j = f(x_i, t_{j+0.5}) = f(x_i, \bar{t}),$$

assuming $f(x,t)$ to be a continuous function of the argument $x$. When some such function may have discontinuities of the first kind at the nodal points, the obstacles involved can be avoided by setting either

$$(47)\qquad \varphi_i = \frac{h_i\, f_{i-0} + h_{i+1}\, f_{i+0}}{2\,\hbar_i},$$

where $f_{i\pm 0} = f(x_{i\pm 0}, \bar{t})$, or

$$\varphi_i = \overset{\circ}{\varphi}_i = \frac{1}{\hbar_i} \int\limits_{x_{i-0.5}}^{x_{i+0.5}} f(x, \bar{t})\, dx \,. \tag{48}$$

In order to clarify the main idea behind this approach, we turn to the scheme with weights

$$y_t = \Lambda\left(\sigma\, \hat{y} + (1-\sigma)\, y\right) + \varphi\,, \quad x \in \omega_h\,, \quad 0 \le t = j\tau < T\,, \tag{49}$$

$$y(x,0) = y_0(x)\,, \quad x \in \bar{\omega}_h\,, \quad y(0,t) = u_1(t)\,, \quad y(1,t) = u_2(t)\,,$$

$$\Lambda\, y = \left(a(x)\, y_{\bar{x}}\right)_{\hat{x}}\,, \quad 0 < c_1 \le a \le c_2\,,$$

whose augmented form is suitable for subsequent calculations. The elimination method unveils its potential once again and permits us to find $y^{j+1}$ on every new layer $t = t_{j+1}$ in terms of a known value $y^j$ on the current layer $t = t_j$:

$$A_i\, \hat{y}_{i-1} - C_i\, \hat{y}_i + B_i\, \hat{y}_{i+1} = -F_i\,, \quad i = 1, 2, \ldots, N-1\,,$$

$$\hat{y}_0 = u_1(t_{j+1})\,, \quad \hat{y}_N = u_2(t_{j+1})\,,$$

$$A_i = \frac{\sigma\, a_i\, \tau}{h_i\, \hbar_i}\,, \quad B_i = \frac{\sigma\, a_{i+1}\, \tau}{h_{i+1}\, \hbar_i}\,, \quad C_i = A_i + B_i + 1\,,$$

$$F_i = \left[1 - \frac{(1-\sigma)\,\tau}{\hbar_i}\left(\frac{a_i}{h_i} + \frac{a_{i+1}}{h_{i+1}}\right)\right] y_i$$

$$+ \frac{(1-\sigma)\,\tau}{\hbar_i}\left[\frac{a_i\, y_{i-1}}{h_i} + \frac{a_{i+1}\, y_{i+1}}{h_{i+1}}\right] + \varphi_i\,.$$

Under such a choice of the computational algorithm the accuracy of scheme (49) will be given special investigation. We are going to show that it converges uniformly with the rate $O(h^2 + \tau^{m_\sigma})$ in the case of smooth functions $k(x)$ and $f(x,t)$.

For this, we proceed as usual. This amounts to evaluating the error $z_i^j = y_i^j - u_i^j$, where $y_i^j$ is a solution of the original problem (1)–(3) and

$u(x,t)$ is a solution of the preceding difference problem. Upon substituting $y = z + u$ into (49) the complete posing of the problem for the error $z$ is

$$z(x,0) = 0\,, \quad x \in \omega_h\,, \quad z(0,t) = z(1,t) = 0\,, \quad 0 \le t = j\tau < T\,,$$
(50)
$$z(x,0) = 0\,, \quad x \in \hat{\omega}_h\,, \quad z(0,t) = z(1,t) = 0\,, \quad 0 \le t = j\tau < T\,,$$

where

(51) $$\psi(x,t) = \Lambda\left(\sigma\,\acute{u} + (1-\sigma)\,u\right) + \varphi - u$$

is the error of approximation of problem (1)–(3) by scheme (49). In such a setting the balance equation on the segment $x_{i-0.5} \le x \le x_{i+0.5}$ with the ends $x_{i-0.5} = x_i - 0.5\,h_i$ and $x_{i+0.5} = x_i + 0.5\,h_{i+1}$ gives

(52) $$\int\limits_{x_{i-0.5}}^{x_{i+0.5}} \frac{\partial u}{\partial t}(x,\bar{t})\,dx = w(x_{i+0.5},\bar{t}) - w(x_{i-0.5},\bar{t}) + \int\limits_{x_{i-0.5}}^{x_{i+0.5}} f(x,\bar{t})\,dx\,.$$

Here
$$w(x,\bar{t}) = k(x)\,\frac{\partial u}{\partial t}(x,\bar{t})\,.$$

Some modification of the residual $\psi$ is possible with minor changes. Dividing both sides of identity (52) by $\hbar_i = 0.5\,(h_i + h_{i+1})$ and then substracting the resulting expressions from (51) reveal

$$\psi = \left(a\,u_{\bar{x}}^{(\sigma)} - \overline{k\,u'}\right)_{\hat{x},i} + \tilde{\psi}\,,$$

where $\bar{v} = v(x_{i-0.5},\bar{t})$ and

$$\tilde{\psi}_i = \varphi_i - \overset{\circ}{\varphi}_i + \frac{1}{\hbar_i}\int\limits_{x_{i-0.5}}^{x_{i+0.5}} \frac{\partial u}{\partial t}(x,\bar{t})\,dx - u_{t,i}\,.$$

Using an expansion arising from Chapter 3, Section 4 such as

$$\overset{\circ}{\varphi}_i = f_i + \tfrac{1}{8}\left(h_i^2\,f'_{i-0.5}\right)_{\hat{x},i} + O(\hbar_i^2)$$

and allowing an alternative form of writing

(53) $$\psi = \eta_{\hat{x}} + \psi^*\,,$$

(54) $$\eta_i^j = a_i\,u_{\bar{x},i}^{(\sigma),j} - (ku')_{i-1/2}^{j+1/2} + \frac{h_i^2}{8}\left(\acute{u}' - f'\right)_{i-1/2}^{j+1/2}$$

with $\dot{u}' = \partial^2 u/\partial t \partial x$ and $f' = \partial f/\partial x$ incorporated, we arrive at

$$\psi_i^* = O(\hbar_i^2 + \tau^2).$$

It seems clear that representation (53) for the residual $\psi$ is an immediate implication of the balance equation (52) by observing that

$$(k\,u')_{i+1/2} = (k\,u')_i + 0.5\,h_{i+1}(k\,u')_i' + \frac{1}{8}\,h_{i+1}^2\,(k\,u')_i'' + O(h_{i+1}^3),$$

$$(k\,u')_{i-1/2} = (k\,u')_i - 0.5\,h_i(k\,u')_i' + \frac{1}{8}\,h_i^2(k\,u')_i'' + O(h_i^3)$$

and, hence,

$$(k\,u')_i' = [(k\,u')_{i-1/2}]_{\hat{x},i} - \frac{1}{8\,\hbar_i}\,[h_{i+1}^2(k\,u')_i'' - h_i^2(k\,u')_i''] + O(\hbar_i^2).$$

By virtue of the relations

$$v_i = v_{i-1/2} + O(h_i), \qquad v_i = v_{i+1/2} + O(h_{i+1}),$$

$$(k\,u')' = \dot{u} - f, \qquad (k\,u')'' = \dot{u}' - f'$$

the desired result will be substantiated if we succeed in showing that

$$\frac{1}{\hbar_i}\,(h_{i+1}^2\,v_i - h_i^2\,v_i) = (h_i^2\,v_{i-1/2})_{\hat{x},i} + O(\hbar_i^2).$$

An alternative form of the residual

$$\psi = \Lambda\,u^{(\sigma)} + \varphi - u_t - \left((k\,u')' + f - \dot{u}\right)_{t=\bar{t}}$$

is best suited for our purposes in the applications of formula (54), thereby justifying the final results

$$\eta = 0.5\,a\,(\hat{u}_{\bar{x}} + u_{\bar{x}}) - \overline{k\,u'} + \tau\,(\sigma - 0.5)\,a\,u_{\bar{x}t} + O(h^2) = O(h^2 + \tau^{m_\sigma}).$$

The next question we have raised above is the accuracy of scheme (49), the accurate account of which can be done using a priori estimates of the problem (50) solution with further reference to the special structure (50) of the right-hand side of $\psi$. Following established practice, we introduce the space $\overset{\circ}{\Omega} = H$ of all grid functions given on the grid $\hat{\omega}_h$ and vanishing at the points $x = 0$ and $x = 1$ and then define several inner products by

$$(y, v)_* = \sum_{i=1}^{N-1} y_i\, v_i\, \hbar_i\,, \quad (y, v) = \sum_{i=1}^{N-1} y_i\, v_i\, h_i\,, \quad (y, v] = \sum_{i=1}^{N} y_i\, v_i\, h_i\,.$$

In this direction we refer to the operator $A : H \mapsto H$ with the values

$$A\, y = -\Lambda\, y = -(a\, y_{\bar{x}})_{\hat{x}} \quad \text{for any} \quad y \in H\,.$$

We know that the operator $A$ so defined is self-adjoint and positive definite:

$$A = A^* > 0\,,$$

since $(Ay, v) = (ay_{\bar{x}}, v_{\bar{x}}] = (y, Av)$. By the same token,

$$(Ay, y) = (a\, y_{\bar{x}}, y_{\bar{x}}] \geq c_1\, (y_{\bar{x}}, y_{\bar{x}}]\,,$$

$$\|y\|_C = \max_{0<i<N} |y_i| \leq \frac{1}{2}\, (y_{\bar{x}}, y_{\bar{x}}]^{1/2} \leq \frac{1}{2\,\sqrt{c_1}}\, (Ay, y)^{1/2}\,.$$

By obvious rearranging of the problem (50) solution as

$$z = v + w\,,$$

where $v$ is a solution of the same problem with another right-hand side $\psi = \eta_{\hat{x}}$ and $w$ is a solution of problem (50) with the right-hand side $\psi = \psi^*$.

As can readily be observed, scheme (50) is stable under the constraints

$$\sigma \geq \sigma_0\,, \qquad \sigma_0 = \frac{1}{2} - \frac{1}{\tau\, \|A\|}\,, \qquad \|A\| \leq \frac{4\, c_2}{h_0^2}\,,$$

where $h_0 = \min\limits_{1 \leq i \leq N} h_i$.

Other ideas are connected with a priori estimate (41) for $v$ and estimate (42) for $w$. When providing such manipulations and establishing the relation

$$\|\eta_{\hat{x}}\|_{A^{-1}} \leq \frac{1}{\sqrt{c_1}}\, \|\eta]|\,,$$

where $\|\eta]| = \sqrt{(1, \eta^2]}$ and $\| z \| \le \| v \| + \| w \|$, we get for $\sigma \ge \sigma_\varepsilon$, $\sigma_\varepsilon = 0.5 - (1-\varepsilon)/(\tau \| A \|)$, the estimate

$$\|z^{j+1}\|_C \le \frac{1}{2\,c_1} \left\{ \|\eta^0]| + \|\eta^j]| + \sum_{j'=1}^{j} \tau \, \|\eta_{\bar t}^{j'}]| \right\} + \frac{1}{2\sqrt{2\,\varepsilon\, c_1}} \left( \sum_{j'=0}^{j} \tau \, \| \psi^{*j'} \|^2 \right)^{1/2} . \tag{55}$$

In what follows it is supposed that $k(x)$, $f(x,t)$, $u_0(x)$, $u_1(t)$ and $u_2(t)$ are smooth functions and the conditions under which

$$\eta = O(|h|^2 + \tau^{m_\sigma}), \quad \eta_t = O(|h|^2 + \tau^{m_\sigma}), \quad \psi^* = O(|h|^2 + \tau^2) \tag{56}$$

hold. Then scheme (49) converges uniformly on any sequence of nonuniform grids $\{\hat{\bar\omega}_h\}$ with the rate $O(\tau^{m_\sigma} + |h|^2)$, where $|h| = \max\limits_{1 \le i \le N} h_i$, if $\sigma \ge \sigma_\varepsilon$. This fact follows immediately from the combination of estimates (55) and (56).

We touch upon briefly the convergence of scheme (49) in the class of discontinuous coefficients and will pursue some analogy with the stationary case which has been considered on the same footing in Chapter 3, Section 4 under the following assumptions:

(a) the functions $k(x)$ and $f(x,t)$ may have only a finite number of discontinuities of the first kind on straight lines parallel to the coordinate axis $Ot$;

(b) the grid $\hat{\bar\omega}_h = \hat{\bar\omega}_h(K)$ is chosen in such a way that all of the discontinuity lines of the functions $k(x)$ and $f(x,t)$ will pass through the nodal points;

(c) the functions $k(x)$, $f(x,t)$ and $u(x,t)$ have all necessary derivatives in the regions lying between the discontinuity lines so that formulas (53)–(54) and estimate (56) are still valid at all the nodes of the grid $\hat{\bar\omega}_h(K)$.

We note in passing that $\varphi_i$ is determined by formula (47), due to which scheme (49) converges uniformly with the rate $O(\tau^{m_\sigma} + |h|^2)$ under the aforementioned conditions on sequences of special grids.

**Remark 1** Convergence in the norm of the grid space $L_2$ occurs with the same rate if condition c) is relaxed:

$$\eta = O(|h|^2 + \tau^{m_\sigma}), \qquad \psi^* = O(|h|^2 + \tau^2). \tag{57}$$

This fact follows immediately from a priori estimate (42):

$$\|z^{j+1}\| \leq \frac{1}{\sqrt{2\varepsilon}} \left( \sum_{j'=1}^{j} \tau \, \|\psi^{j'}\|^2_{A^{-1}} \right)^{1/2} \quad \text{for} \quad \sigma \geq \sigma_\varepsilon \, .$$

**Remark 2** Uniform convergence with the rate $O(|h|^2 + \tau)$ of the forward difference scheme with $\sigma = 1$ can be established by means of the maximum principle and the reader is invited to carry out the necessary manipulations on his/her own.

**6. An one-point heat source.** Of special interest is the nonstationary heat conduction problem in the situation when a heat source is located only at a single point $x = \xi$ under the agreement that at this point the solution of problem (1)–(3) satisfies the condition of conjugation

$$[u] = 0 \, , \qquad \left[ k \, \frac{\partial u}{\partial x} \right] = -Q \quad \text{for} \quad x = \xi \, , \tag{58}$$

where $Q = Q(t)$ is a power of the heat source.

The discontinuity condition of the heat flow

$$\left[ k \, \frac{\partial u}{\partial x} \right] = -Q$$

is to be understood as the discontinuity property of the first derivatives $k \, \dfrac{\partial u}{\partial x}$. That is to say, the solution $u = u(x,t)$ has a week discontinuity on the straight line $x = \xi$ by relating at the same time the coefficient $k(x)$ and the function $f(x,t)$ to be smooth enough.

By means of the integro-interpolation method it is possible to construct a homogeneous difference scheme, whose design reproduces the availability of the heat source $Q$ of this sort at the point $x = \xi$. This can be done using an equidistant grid $\omega_h$ and accepting $\xi = x_n + \theta h$, $0 \leq \theta \leq 0.5$. Under such an approach the difference equation takes the standard form at all the nodes $x_i \neq x_n$ $(i \neq n)$. In this line we write down the balance equation on the segment $x_{n-1/2} \leq x \leq x_{n+1/2}$ for fixed $t = \bar{t} = t_{j+0.5}$. With the aid of the relations

$$\int\limits_{x_{n-0.5}}^{x_{n+0.5}} (k \, u')' \, dx = \int\limits_{x_{n-0.5}}^{\xi} (k \, u')' \, dx + \int\limits_{\xi}^{x_{n+0.5}} (k \, u')' \, dx$$

$$= k \, u' \Big|_{x_{n+0.5}}^{x_{n-0.5}} + [k \, u'] = w_{n+1/2} - w_{n-1/2} - Q \, ,$$

$$w = k \, u' \, ,$$

we arrive at

$$(59)\qquad \int\limits_{x_{n-1/2}}^{x_{n+1/2}} \frac{\partial u}{\partial t}(x,\bar t)\,dx = w(x_{n+1/2},\bar t) - w(x_{n-1/2},\bar t) - Q(\bar t) + h\,\overset{\circ}{\varphi}_n\,.$$

The usual transition to a difference equation leads to

$$y_t = \Lambda\,y^{(\sigma)} + \varphi + \frac{Q(\bar t)}{h} \quad \text{for} \quad x = x_n\,,$$

where $\Lambda y = (a\,y_{\bar x})_x$.

Summarizing, the difference scheme for the problem described by (1)–(3), (58) is of the form

$$(60)\quad y_t = \Lambda\,y^{(\sigma)} + \varphi + \frac{1}{h}\,Q(\bar t)\,\delta_{i,n}\,,\quad 0 < x = x_i < 1\,,\quad t = t_j \ge 0\,,$$

$$y_0 = \mu_1\,,\quad y_N = \mu_2\,,\quad y(x,0) = u_0(x)\,,$$

where $\delta_{i,n}$ is, as usual, Kronecker's delta.

For the error $z = y - u$ we have equation with the right-hand side

$$(61)\qquad \psi = \Lambda\,u^{(\sigma)} + \varphi + \frac{1}{h}\,Q(\bar t)\,\delta_{i,n} - u_t$$

and the boundary conditions

$$(62)\qquad z_0 = 0\,,\qquad z_N = 0\,,\qquad z(x,0) = 0\,.$$

On account of the balance equation (59) the residual is representable by

$$(63)\qquad \psi = \eta_x + \psi^*\,,\qquad \eta = a\,u_{\bar x}^{(\sigma)} - \overline{k\,u'}$$

$$(64)\qquad \psi_i^* = \varphi_i - \overset{\circ}{\varphi}_i - \left( u_{t,i} - \frac{1}{h}\int\limits_{x_{i-0.5}}^{x_{i+0.5}} \frac{\partial u}{\partial t}(x,\bar t)\,dx \right).$$

Here we adopt $\bar v_i = v(x_{i-0.5},\bar t)$ as was done before. For the sake of simplicity we take $\theta \le 0.5$, that is, $x_n \le \xi \le x_{n+1/2}$, permitting us to deduce that

$$\eta_i = O(h^2 + \tau^{m_\sigma}) \quad \text{for all} \quad i \ne n+1\,,$$

$$\psi_i^* = O(h^2 + \tau^2) \quad \text{for all} \quad i \ne n\,.$$

By analogy with the case of discontinuous coefficients we find that

$$u_{\bar{x},n+1} = \theta\, u'_{\text{left}} + (1-\theta)\, u'_{\text{right}} + 0.5\, h\, \left[(1-\theta)^2\, u''_{\text{right}} - \theta^2\, u''_{\text{left}}\right] + O(h^2)\,.$$

Since $a_{n+1} = k(\xi) + (0.5-\theta)\, h\, k'(\xi) + O(h^2)$, we might have

$$a_{n+1}\, u_{\bar{x},n+1} = \theta\, w_{\text{left}} + (1-\theta)\, w_{\text{right}} + O(h)\,.$$

With the aid of the relations

$$w_{\text{right}} - w_{\text{left}} = -Q\,, \quad (k\, u')_{n+1/2} = w_{\text{right}} + h\,(0.5-\theta)\,(k\, u')'_{\text{right}} + O(h^2)$$

we establish

$$\eta_{n+1} = \theta\, Q + O(h) = O(1) \tag{65}$$

by observing further that

$$\psi^*_n = O(h)\,. \tag{66}$$

The accurate account of the error $z$ can be done as in Section 4, leading to the same rate of convergence. No progress is achieved for $a = \overset{\circ}{a}$ in line with approved rules, because the choice of the coefficient should not cause the emergence of a higher-order accuracy. From the formula $\eta_{n+1} = \theta\, Q + O(h)$ it is easily seen that $\eta_{n+1} = O(h)$ and, hence, $\|\, z\, \| = O(h^2 + \tau^{m_\sigma})$ if $\theta = 0$, meaning that the heat source is located at one of the nodal points.

This guides a proper choice of the nonequidistant grid $\omega_h(Q)$ so that the heat source will appear at one of the nodal points. When this is the case, scheme (60) converges uniformly, on the same grounds as before, with the rate $O(|h|^2 + \tau^{m_\sigma})$. But a special choice of the coefficients $a_i$ given by the formulas of the truncated scheme with second-order accuracy (see Chapter 3, Section 7) improves our chances of constructing the difference scheme of accuracy $O(|h|^2 + \tau^{m_\sigma})$ for any $\theta \in [0, 1]$, that is, disregarding to the possible locations of the heat source.

**7. A concentrated heat capacity.** We now consider the boundary-value problem for the heat conduction equation with some unusual condition placing the concentrated heat capacity $C_0$ on the boundary, say at a single point $x = 0$. The traditional way of covering this is to impose at the point $x = 0$ an unusual boundary condition such as

$$C_0\, \frac{\partial u}{\partial t} = k\, \frac{\partial u}{\partial x}\,, \qquad x = 0\,, \qquad C_0 = \text{const} > 0\,, \tag{67}$$

thereby completing the statement of the problem under consideration:

$$\frac{\partial u}{\partial t} = L\,u + f(x,t)\,, \quad 0 < x < 1\,, \quad t > 0\,, \tag{68}$$

$$L\,u = \frac{\partial}{\partial x}\left(k\,\frac{\partial u}{\partial x}\right),$$

$$C_0\,\frac{\partial u}{\partial t} = k\,\frac{\partial u}{\partial x} \quad \text{for} \quad x = 0\,, \quad u(1,t) = 0\,, \quad u(x,0) = u_0(x)\,.$$

The design of a homogeneous difference scheme necessitates approximating the boundary condition at the point $x = 0$.

The first step during the course of the integro-interpolation method is to rely on the balance equation, say in the rectangle $\{0 \le x \le x_{1/2} = 0.5\,h,$ $t_j \le t \le t_{j+1}\}$, leading to

$$\int\limits_0^{x_{1/2}} \big[u(x,t_{j+1}) - u(x,t_j)\big]\,dx$$

$$= \int\limits_{t_j}^{t_{j+1}} \big[w(x_{1/2},t) - w(0,t)\big]\,dt + \int\limits_{t_j}^{t_{j+1}} \int\limits_0^{x_{1/2}} f(x,t)\,dx\,dt\,,$$

where $w(x,t) = k\,\dfrac{\partial u}{\partial x}$. The next step is to substitute here

$$w(0,t) = \left(k\,\frac{\partial u}{\partial x}\right)_{x=0} = C_0\,\frac{\partial u}{\partial t}(0,t)$$

and take into account that

$$\int\limits_{t_j}^{t_{j+1}} C_0\,\frac{\partial u}{\partial t}(0,t)\,dt = C_0\,\big(u(0,t_{j+1}) - u(0,t_j)\big) = C_0\,\tau\,u_{t,0}\,.$$

Then upon replacing the integrals in $x$ by the simplest expressions $0.5\,h\,u_0$ and $0.5\,h\,f_0$ and the integrals in $t$ by the expressions $\tau w_{1/2}^{(\sigma)}$ and $\tau f_0^{(\sigma)}$, the difference boundary condition is taken to be

$$C\,y_{t,0} = a_1\,y_{x,0}^{(\sigma)} + 0.5\,h\,f_0^{(\sigma)}\,, \qquad C = C_0 + 0.5\,h\,. \tag{69}$$

In this way, the difference scheme

$$\text{(70)} \qquad y_t = \Lambda\, y^{(\sigma)} + \varphi\,, \quad 0 < x = ih < 1\,, \quad t = j\tau \geq 0\,,$$

$$C\, y_{t,0} = a_1\, y_{x,0}^{(\sigma)} + 0.5\, h\, f_0^{(\sigma)}\,, \quad t \geq 0\,, \quad y_N = \mu_2\,, \quad y(x,0) = u_0(x)\,,$$

is responsible for problem (68). It is plain to recover from the condition at the point $x = 0$ that

$$\text{(71)} \qquad \hat{y}_0 = \varkappa_1\, \hat{y}_1 + \nu_1\,, \qquad \varkappa_1 = \frac{\sigma\, a_1\, \tau}{\sigma\, a_1\, \tau + C\, h}\,,$$

$$\nu_1 = h\, \big[0.5\, h\, \tau\, f_0^{(\sigma)} + a_1\, (1-\sigma)\, \tau\, y_{x,0} + C\, y\big] / (C\, h + a_1\, \sigma\, \tau)\,,$$

yielding $0 \leq \varkappa_1 < 1$ for $\sigma \geq 0$. In turn, the boundary condition of the first kind is imposed for $i = N$:

$$\text{(72)} \qquad y_N = \mu_2\,.$$

With these, for determination of $\hat{y}_i = y_i^{j+1}$ we obtain a second-order difference equation supplied by the boundary conditions (71)–(72) that can be solved by the standard elimination method.

The intuition suggests that in such a setting the governing difference equation and the boundary condition at the point $x = 0$ have one and the same order of approximation $O(\tau^{m_\sigma} + h^2)$. To make sure of it, it suffices only to evaluate the residual

$$\psi_\nu = C\, u_{t,0} - a_1\, u_{x,0}^{(\sigma)} - 0.5\, h\, f_0^{(\sigma)}\,.$$

Substituting here the expressions

$$a_1\, u_{x,0} = (a\, u_{\bar{x}})_1 = (k\, u')_{1/2} + O(h^2)$$

$$= (k\, u')_0 + 0.5\, h\, (k\, u')_0' + O(h^2)\,,$$

$$(k\, u')_0 = C_0\, \dot{u}_0\,, \quad (k\, u')_0' = \dot{u}_0 - f_0\,,$$

yields

$$\psi_\nu = C\, (u_{t,0} - u_0^{(\sigma)}) + O(h^2) = O(\tau^{m_\sigma} + h^2)$$

as required.

For the error $z = y - u$ we may set up the related problem

$$z_t = \Lambda\, y^{(\sigma)} + \psi\,, \quad \psi = \Lambda\, u^{(\sigma)} + \varphi - u_t\,, \quad 0 < x < 1\,, \tag{73}$$

$$C\, z_{t,0} = a_1\, z_{x,0}^{(\sigma)} + \psi_\nu\,, \quad z_N = 0\,, \quad z(x,0) = 0\,.$$

One obvious way of proceeding is to introduce the space $H$ of all grid functions defined on $\bar{\omega}_h$ and vanishing for $i = N$. Under the inner product structure

$$[y, v) = \sum_{i=1}^{N-1} y_i\, v_i\, h + 0.5\, h\, y_0\, v_0$$

we refer to operators $A$ and $D$ acting in accordance with the rules

$$(A\, y)_i = -(\Lambda\, y)_i \quad \text{for} \quad 0 < i < N\,, \quad (A\, y)_0 = -\frac{a_1\, y_{x,0}}{0.5\, h}\,,$$

$$(D\, y)_i = y_i \qquad \text{for} \quad 0 < i < N\,, \quad (D\, y)_0 = \frac{1}{0.5\, h}\, C\, y_0\,.$$

With the detailed forms in mind, problem (73) can be recast as

$$D\, z_t + A\, z^{(\sigma)} = \psi\,, \quad t = j\tau \geq 0\,, \quad z(0) = 0\,, \tag{74}$$

showing the new members to be sensible ones. All this enables us too write $B = D + \sigma\, \tau\, A$.

From the general operator theory outlined in Chapter 2, Section 4 it seems clear that the operator $D$ is self-adjoint and positive definite:

$$D \geq c_*\, E\,, \quad \text{where} \quad c_* = \min\,(1,\, 2\, C/h)\,.$$

However, this is certainly so with the operator $A$:

$$A = A^* > 0\,.$$

What is more, the operators $A$ and $D$ are commuting: $AD = DA$. Due to these properties the stability condition for scheme (74) is expressed by

$$B - 0.5\, \tau\, A = D^+ + (\sigma - 0.5)\, \tau\, A \geq 0\,,$$

which is valid only for

$$\sigma \geq 0.5 - \frac{c_*}{\tau\, \|\, A\, \|} = \sigma_0\,.$$

More a detailed proof of convergence of this scheme is concerned with the form (74) and a priori estimates obtained in Chapter 6, Section 2 and so it is omitted here. As a final result we deduce that scheme (70) converges uniformly with the rate $O(\tau^{m_\sigma} + h^2)$.

**8. The case when the coefficient of conductivity $k$ depends on $t$, $k = k(x,t)$.** So far we have preassumed for the sake of clarity that the coefficient of conductivity $k$ depends only on one variable $x$.

In a common setting the governing equation (1) of the general form

$$\frac{\partial u}{\partial t} = \frac{\partial}{\partial x}\left(k(x,t)\,\frac{\partial u}{\partial x}\right) + f(x,t)\,, \quad 0 < c_1 \le k(x,t) \le c_2\,,$$

is put together with the boundary and initial conditions (2)–(3). In preparation for this, the intention is to use instead of (6) another difference scheme

$$y_t = \Lambda(\bar{t})\, y^{(\sigma)} + \varphi\,, \tag{75}$$

where $\Lambda(\bar{t})\, v = \big(a(x,\bar{t})\, v_{\bar{x}}\big)_x$, $\bar{t} = t_{j+0.5}$ and the coefficient $a(x,t)$ for fixed $t$ appears by exactly the same reasoning as before (see Sections 2 or 4 of the present chapter). The error of approximation $\psi$ of the scheme concerned is of $O(\tau^{m_\sigma} + h^2)$ if $k(x,t_*) \in C^{(3)}[0,1]$ for every fixed $t = t_*$ and this is also consistent with the results obtained.

In mastering the difficulties involved, we refer to a variable operator $A(t)$ in the space $\overset{\circ}{\Omega} = H$ of all grid functions with the values

$$A\,y = -\Lambda(\bar{t})\, y\,, \qquad y \in \overset{\circ}{\Omega}\,.$$

In connection with its dependence on $t$ the usual practice is to impose, in addition, the Lipshitz condition with respect to the variable $t$:

$$\big|((A(t) - A(t-\tau))\, y, y)\big| \le \tau\, c_3\, (A(t-\tau)\, y, y)\,,$$

making it possible to apply the general stability theory. As can readily be observed, the Lipshitz condition is ensured if

$$|k(x,t) - k(x,t-\tau)| \le \tau\, c_3\, k(x,t-\tau)\,,$$

thereby confirming the validity of the preceding results obtained in Sections 2–4 for scheme (75).

Further development of this trend of research is devoted to a more complex problem in which the governing equation acquires the form

$$c(x,t)\,\frac{\partial u}{\partial t} = \frac{\partial}{\partial x}\left(k(x,t)\,\frac{\partial u}{\partial x}\right) + f(x,t)$$

with $c(x,t) \geq c_1 > 0$ and $0 < c_1 \leq k(x,t) \leq c_2$. In view of this, the corresponding homogeneous difference scheme becomes

$$\rho(x,\bar{t})\, y_t = \Lambda(\bar{t})\, y^{(\sigma)} + \varphi(x,\bar{t}) , \tag{76}$$

where $\rho$ and $\varphi$ are calculated by means of the same pattern functional $\rho(x_i,\bar{t}) = c(x_i,\bar{t})$ or $\rho(x_i,\bar{t}) = \frac{1}{2}\left(c(x_i-0,\bar{t})+c(x_i+0,\bar{t})\right)$ in the case where $c(x,t)$ is discontinuous at the node $x = x_i$.

With the detailed forms of these functionals, scheme (76) is stable under the constraints

$$\sigma \geq \sigma_0(\bar{t}) , \qquad \sigma_0 = \frac{1}{2} - \frac{c_1}{\tau\, \| A(\bar{t}) \|} , \qquad \| A \| = \| \Lambda \| .$$

Because of this fact, it is unconditionally stable for $\sigma \geq 0.5$.

**Remark** So far we have considered only equidistant grids in $t$. But it is not difficult to show that the preceding estimates for two-layer schemes remain valid on nonequidistant grids with step $\tau_j = t_j - t_{j-1}$ being a function of the subscript $j$. It is obvious that the grid in $t$ is rather flexible in comparison with the grid $\omega_h$ as a result of refining the step $\tau_j$ in the regions of the widely varying right-hand side $f(x,t)$, boundary values $\mu_1(t)$, $\mu_2(t)$ and the coefficient $k = k(x,t)$ in $t$. With knowledge of the behaviour of the problem solution on a sparse grid, successive grid refinement will be caused by the necessity of diminishing the step $\tau_j$ in some intervals during which the solution varies very fastly in $t$. Any changes in the composition of the grid $\omega_h$ with changes of $t_j$ in the process of calculations are connected with indeterminate values of the function $y_i^j$ at new (additional) nodes of the grid $\omega_h$ in view.

For example, a greater gain in accuracy in $\tau$ will be achieved once we perform parallel calculations on several grids $\omega_{\tau_1}$ and $\omega_{\tau_2}$ by the rules approved in Chapter 3, Section 4.

The forthcoming procedures serve to motivate what is done on an equidistant grid $\omega_{h\tau}$, on which the representation takes place:

$$y_i^j = u_i^j + \alpha_{ij}\, h^{m_1} + \beta_{ij}\, \tau^{n_1} + O(h^{m_2} + \tau^{n_2}) , \tag{77}$$

$$m_2 > m_1 > 0 , \quad n_2 > n_1 > 0 ,$$

where $\alpha_{ij}$ and $\beta_{ij}$ are independent of $h$ and $\tau$ both.

Let $y_{h\tau_1}(x_i, t_j)$ and $y_{h\tau_2}(x_i, t_j)$ be solutions of the same difference problem with different steps $\tau_1 = \tau$ and $\tau_2 = \tau$. Via the linear transform

$$\tilde{y}_{h\tau}(x,t) = c_1\, y_{h\tau_1}(x,t) + c_2\, y_{h\tau_2}(x,t)\,, \quad t = j\tau\,, \quad j = 0, 1, \ldots\,,$$

with expression (77) standing for $y_{h\tau_1}$ and $y_{h\tau_2}$ both, it is necessary to reduce to zero the coefficient at the member $\tau^{n_1}$. All this enables us to find with reasonable accuracy

$$\tilde{y} = u + O(h^{m_2} + \tau^{n_2})$$

the values

$$c_2 = 1 - c_1\,, \qquad c_1 = -1/(2^{n_1} - 1)\,.$$

A similar procedure works on two different grids $\omega_h$ and $\omega_{0.5\,h}$ for fixed $\tau$, thus causing a grid solution $\tilde{\tilde{y}}$ such that

$$\tilde{\tilde{y}} = u + O(h^{m_2} + \tau^{n_2})\,.$$

When providing current manipulations, the solution $u = u(x,t)$ and the available data of the original problem are preassumed to be smooth enough and sufficient for the existence of the asymptotic expansion

$$y_{n\tau} = u + \alpha\, h^{m_1} + \beta\, \tau^{n_1} + O(h^{m_2} + \tau^{n_2})\,.$$

**9. The third boundary-value problem.** For the moment, the statement of the problem is

$$\frac{\partial u}{\partial t} = L\,u + f(x,t)\,, \quad L\,u = \frac{\partial}{\partial x}\left(k(x,t)\,\frac{\partial u}{\partial x}\right), \quad 0 < x < 1\,,$$

$$k(0,t)\,\frac{\partial u(0,t)}{\partial x} = \beta_1(t)\,u(0,t) - \mu_1(t)\,, \quad \beta_1 \geq 0\,,$$

$$-k(1,t)\,\frac{\partial u(1,t)}{\partial x} = \beta_2(t)\,u(1,t) - \mu_2(t)\,, \quad \beta_2 \geq 0\,.$$

In Chapter 3, Section 5 we have formed a difference equation of the third kind for the stationary equation $Lu + f = 0$. The formal passage from the stationary equation to the nonstationary one is provided by the replacement of $f$ by $f - \partial u/\partial t$. This trick has been already encountered in specifying the difference conditions approximating the boundary conditions of the third kind. The outcome of this is

$$y_t = \Lambda(\bar{t})\,(\sigma\,\hat{y} + (1-\sigma)\,y) + \varphi\,, \quad 0 < x_i = ih < 1\,, \quad t_j = j\tau > 0\,,$$

$$a_1(\bar{t})\,(\sigma\,\hat{y}_{\bar{x},1} + (1-\sigma)\,y_{\bar{x},1}) = \beta_1(\bar{t})\,(\sigma\,\hat{y}_0 + (1-\sigma)\,y_0) + \bar{\mu}_1(\bar{t})$$

$$+ 0.5\,h\,y_{t,0} - a_N(\bar{t})\,(\sigma\,\hat{y}_{\bar{x},N} + (1-\sigma)\,y_{\bar{x},N})$$

$$= \beta_2(\bar{t})\,(\sigma\,\hat{y}_N + (1-\sigma)\,y_N) + \bar{\mu}_2(\bar{t}) + 0.5\,h\,y_{t,N}\,.$$

Here $\bar{\mu}_1 = \mu_1(\bar{t}) - 0.5\,h\,f(0,\bar{t})$ and $\bar{\mu}_2 = \mu_2(\bar{t}) - 0.5\,h\,f(1,\bar{t})$ for $\bar{t} = t_j + 0.5\,\tau$. The resulting scheme is of accuracy $O(\tau^2 + h^2)$ for $\sigma = 0.5$ and it is of accuracy $O(\tau + h^2)$ for $\sigma > 0.5$.

The best possible choice for later use of the elimination method is due to

$$\Lambda(\bar{t})\,\hat{y} - \frac{\hat{y}}{\sigma\,\tau} = -F\,, \quad \hat{y}_0 = \varkappa_1\,\hat{y}_1 + \nu_1\,, \quad \hat{y}_N = \varkappa_2\,\hat{y}_{N-1} + \nu_2\,,$$

where

$$\varkappa_1 = \frac{a_1(\bar{t})}{a_1(\bar{t}) + h\,\beta_1(\bar{t}) + h^2/(2\,\sigma\,\tau)}\,,$$

$$\varkappa_2 = \frac{a_N(\bar{t})}{a_N(\bar{t}) + h\,\beta_2(\bar{t}) + h^2/(2\,\sigma\,\tau)}\,,$$

$$\nu_1 = \frac{(1-\sigma)(a_1(\bar{t})\,y_{\bar{x},1} - \beta_1(\bar{t})\,y_0) + 0.5\,h\,y_0/\tau - \bar{\mu}_1}{\sigma\,(a_1(\bar{t})/h + \beta_1(\bar{t}) + h/(2\,\sigma\,\tau))}\,,$$

$$\nu_2 = \frac{(1-\sigma)(-a_N(\bar{t})\,y_{\bar{x},N} - \beta_2(\bar{t})\,y_N) + 0.5\,h\,y_N/\tau - \bar{\mu}_2}{\sigma\,(a_N(\bar{t})/h + \beta_2(\bar{t}) + h/(2\,\sigma\,\tau))}\,,$$

$$F = ((1-\sigma)\,\Lambda(\bar{t})\,y + y/\tau + \varphi(t))\,\sigma^{-1}\,.$$

Having completed the elimination, we observe that the computational procedure is stable if $\sigma > 0$, because $0 < \varkappa_1 < 1$ and $0 < \varkappa_2 < 1$.

**10. Monotone schemes for parabolic equations of general form.** It is required to find a solution of the following problem foor a parabolic equation of the general form in the rectangle $\bar{D}_T = \{0 \le x \le 1,\ 0 \le t \le T\}$:

$$c(x,t)\,\frac{\partial u}{\partial t} = L\,u + f(x,t)\,, \tag{78}$$

$$u(0,t) = u_1(t)\,, \quad u(1,t) = u_2(t)\,, \quad u(x,0) = u_0(x)\,,$$

$$L\,u = \frac{\partial}{\partial x}\Big(k(x,t)\,\frac{\partial u}{\partial x}\Big) + r(x,t)\,\frac{\partial u}{\partial x} - q(x,t)\,u\,,$$

$$0 < c_1 \le k(x,t) \le c_2\,, \quad c(x,t) \ge c_1 > 0\,, \quad q \ge 0\,.$$

A similar problem has been solved in Chapter 3, Section 5 for the stationary equation $Lu + f = 0$ through the use of monotone schemes of second-order accuracy attainable for any step $h$ and the function $r(x)$.

In order to construct a monotone scheme for problem (78) for which the maximum principle would be valid for any $h$ and $\tau$, we involve in subsequent considerations the equation of the same type, but with the perturbed operator $\tilde{L}$:

$$c(x,t)\,\frac{\partial u}{\partial t} = \tilde{L}\,u + f\,, \quad \tilde{L}\,u = \varkappa\,\frac{\partial}{\partial x}\Big(k\,\frac{\partial u}{\partial x}\Big) + r\,\frac{\partial u}{\partial x} - q\,u\,, \tag{79}$$

$$\varkappa = (1+R)^{-1}\,, \qquad R = 0.5\,h\,|r|/k\,.$$

As usual, the operator $\tilde{L}$ is approximated for fixed $t = \bar{t} = t_{j+1/2}$ by the difference operator

$$\tilde{\Lambda}\,y = \varkappa(a\,y_{\bar{x}})_x + b^+ a^{(+1)} y_x + b^- a\,y_{\bar{x}} - d\,y\,,$$

where (for more detail see Chapter 3, Section 5)

$$a = A\,\big[k(x+sh,\bar{t})\big]\,, \quad d = F\big[q(x+sh,\bar{t})\big]\,, \quad b^{\pm} = F\big[\tilde{r}^{\pm}(x+sh,\bar{t})\big]\,,$$

$$\tilde{r}^{\pm} = r^{\pm}/k\,, \quad r^+ = 0.5\,(r+|r|) \ge 0\,, \quad r^- = 0.5\,(r-|r|) \le 0\,.$$

Here the same pattern functionals $A$ and $F$ are adopted in achieving much progress as was done in Chapter 3, Section 2, making it possible to generate an approximation of order 2.

To equation (79) there corresponds the purely implicit (four-point) homogeneous scheme

$$\rho(x, \bar{t})\, y_t = \tilde{\Lambda}(\bar{t})\, \hat{y} + \varphi\,, \tag{80}$$

$$y(x,0) = u_0(x)\,, \quad y(0,t) = u_1(t)\,, \quad y(1,t) = u_2(t)\,,$$

where the coefficients $\rho$ and $\varphi$ are calculated by the same formulas as before for $d$ and $b^{\pm}$. A special structure of the operator $\tilde{\Lambda}$ built into this scheme results in the error of approximation $\psi = O(\tau + h^2)$ and may be of help in achieving this aim.

The maximum principle applies equally well to the estimation of the problem (80) solution with zero boundary conditions $y_0 = y_N = 0$ in tackling the governing equation in the canonical form

$$(\rho_i/\tau + \alpha_i + \beta_i + d_i)\, \hat{y}_i = \alpha_i\, \hat{y}_{i-1} + \beta_i\, \hat{y}_{i+1} + F_i\,,$$

$$F_i = \rho_i\, y_i/\tau + \varphi_i\,, \quad \alpha_i = a_i(\varkappa_i - h\, b_i^-)/h^2, \quad \beta_i = a_{i+1}(\varkappa_i + h\, b_i^+)/h^2.$$

The conditions of Theorem 3 in Chapter 4, Section 2 are easily verified for this equation, due to which we might have

$$\|\hat{y}\|_C = \max_{0<i<N} |\hat{y}_i| \le \left\| \frac{F}{D} \right\|_C, \qquad D_i = \rho_i/\tau + d_i\,.$$

Substituting here the appropriate expressions for $F_i$ and $D_i$ yields

$$\|y^{j+1}\|_C \le \|y^j\|_C + \tau \left\| \frac{\varphi^j}{\rho} \right\|_C \le \|y^j\|_C + \frac{\tau}{c_1}\, \|\varphi^j\|\,,$$

which assures us of the validity of the estimate

$$\|y^{j+1}\|_C \le \|y^0\|_C + \frac{1}{c_1} \sum_{j'=0}^{j} \tau\, \|\varphi^{j'}\|_C\,.$$

This provides enough reason to conclude that scheme (80) converges uniformly with the rate $O(\tau + h^2)$.

**11. Cylindrically symmetric and spherically symmetric heat conduction problems.** In explorations of many physical processes such as diffusion or heat conduction it may happen that the shape of available bodies is cylindrical. In this view, it seems reasonable to introduce a cylindrical system of coordinates $(r, \varphi, z)$ and write down the heat conduction equation with respect to these variables (here $x = r$):

$$\frac{\partial u}{\partial t} = \frac{1}{x} \frac{\partial}{\partial x} \Big( k(x,t)\, x\, \frac{\partial u}{\partial x} \Big) + f(x,t)\,. \tag{81}$$

In the physical language, this is a way of saying that the temperature is independent of $\varphi$ and $z$.

In the case of a spherical symmetry the heat conduction equation acquires the form

$$\frac{\partial u}{\partial t} = \frac{1}{x^2} \frac{\partial}{\partial x} \Big( k(x,t)\, x^2\, \frac{\partial u}{\partial x} \Big) + f(x,t)\,. \tag{81'}$$

Homogeneous difference schemes for stationary equations in spherical and cylindrical coordinate systems have been designed in Chapter 3.

Still using its framework, the starting point correcting that situation is to impose at the point $x = 1$ the usual condition of the first or third kind

$$u(1,t) = \mu_2(t) \tag{82}$$

and then require the boundedness of a solution at the point $x = 0$:

$$\lim_{x\to 0} k\, x\, \frac{\partial u}{\partial x} = 0 \quad \text{for (81)}, \qquad \lim_{x\to 0} k\, x^2\, \frac{\partial u}{\partial x} = 0 \quad \text{for (81')}\,.$$

The cylindrically symmetric heat conduction problem is reproduced by

$$\frac{\partial u}{\partial t} = L\, u + f(x,t)\,, \qquad L\, u = \frac{1}{x} \frac{\partial}{\partial x} \Big( x\, k(x,t)\, \frac{\partial u}{\partial x} \Big)\,, \tag{83}$$

$$t > 0\,, \qquad 0 < x < 1\,,$$

$$u(x,0) = u_0(x)\,, \quad 0 \le x \le 1\,,$$

$$x\, k\, \frac{\partial u}{\partial x}\Big|_{x=0} = 0\,, \quad u(1,t) = \mu_2(t)\,, \quad t \ge 0\,.$$

In line with the usual practice we introduce on the segment $0 \le x \le 1$ an equidistant grid

$$\omega_h = \{x_i = ih\,,\ i = 0, 1, \ldots, N\,,\ hN = 1\}$$

and a time grid $\bar{\omega}_\tau = \{t_j = j\tau\,,\ j = 0, 1, \dots\}$ on the segment $0 \le t \le T$ to expound exploratory devices for obtaining difference schemes by analogy with Chapter 3.

The operator $L$ is approximated by the difference operator

$$\Lambda(\bar{t})\, u_i = \frac{1}{x_i}\left(x_{i-1/2}\, a_i\, u_{\bar{x},i}\right)_{x,i} \sim L\,u\,, \quad \text{where} \quad a_i = a(x_i, \bar{t})\,,$$

thereby establishing a correspondence between equation (83) and one of the weighted schemes

$$y_t = \Lambda(\bar{t})\, y^{(\sigma)} + \varphi\,, \qquad \varphi = f(x, \bar{t})\,.$$

The complete posing of this includes the difference boundary condition at the point $x = 0$. The methodology of Chapter 3 furnishes the justification of the forthcoming substitutions into the stationary equation: we first replace $y_r$ by $y_x^{(\sigma)}$ and $f(0)$ by $\left(f - \dfrac{\partial u}{\partial t}\right)_{x=0}$ and then $\dfrac{\partial u}{\partial t}$ by $u_t$ and $u$ by $y$. The outcome of this is

$$a_1(\bar{t})\, y_{x,0}^{(\sigma)} = \frac{h}{4}\, y_{t,0} - \frac{h}{4}\, f_0^{(\sigma)}\,,$$

which admits an alternative form of writing

$$y_{t,0} = \frac{4}{h}\, a_1(\bar{t})\, y_{x,0}^{(\sigma)} + \varphi_0\,, \qquad \varphi_0 = f_0^{(\sigma)}\,.$$

In this regard, we observe that it is possible to insert $f(0, \bar{t})$ in place of $f_0^{(\sigma)}$.

When the conditions for $x = 1$ and $t = 0$ are put together with the preceding equation, their collection constitutes what is called the difference boundary-value problem

$$y_t = \Lambda(\bar{t})\, y^{(\sigma)} + \varphi\,, \quad 0 \le x = ih < 1\,, \quad t = j\tau \ge 0\,,$$

$$y_N = \mu_2\,, \quad y(x, 0) = u_0(x)\,, \quad x \in \bar{\omega}_h\,,$$

where

$$\Lambda(\bar{t})\, y = \frac{4}{h}\, a_1(\bar{t})\, y_x \qquad \text{for} \quad x = 0\,,$$

$$\Lambda(\bar{t})\, y = \frac{1}{x}\left(\bar{x}\, a(x, \bar{t})\, y_{\bar{x}}\right)_x \quad \text{for} \quad 0 < x = ih < 1\,, \quad \bar{x} = x_i - 0.5\,h\,.$$

If $\sigma \neq 0$, then the difference equation related to $\hat{y}_i$ can be solved by the standard elimination method.

The statement of the problem for the error $z = y - u$ amounts to

$$z_t = \Lambda(\bar{t})\, z^{(\sigma)} + \psi\,, \quad 0 \le x = ih < 1\,, \quad t_j \ge 0\,, \quad z_0 = 0\,, \quad z(x,0) = 0\,,$$

where $\psi = \Lambda(\bar{t})\, u^{(\sigma)} + \varphi - u_t$. By exactly the same reasoning as in Chapter 3, Section 5 the residual $\psi$ is representable by

$$\psi = \frac{1}{x}\,(\bar{x}\,\eta)_x + \psi^* + \psi^{**}\,, \quad \eta = a\,u_{\bar{x}}^{(\sigma)} - (k\,u')_{x=\bar{x}\,,\,t=\bar{t}}\,,$$

with the members

$$\eta = O(h^2 + \tau^{m_\sigma})\,, \quad \psi^* = O(h^2/x)\,, \quad \psi^{**} = O(h^2 + \tau^2)\,.$$

Having no opportunity to touch upon this topic, we refer the readers to the aforementioned chapters of the manograph “The Theory oof Difference Schemes”, in which the method of extraction of “stationary nonhomogeneities” was employed with further reference to a priori estimates of $z$. The forward difference scheme with $\sigma = 1$ converges uniformly with the rate $O(h^2 + \tau)$ due to the maximum principle.

Being concerned with the heat conduction problem in the case of a spherical symmetry, we are now in a position to produce on the same grounds the difference scheme associated with problem (81′)–(82):

$$y_t = \Lambda(\bar{t})\, y^{(\sigma)} + \varphi \quad \text{for} \quad 0 \le x = ih < 1\,, \quad t_j \ge 0\,,$$

$$y_N^j = \mu_2^j\,, \quad j \ge 0\,, \quad y(x_i, 0) = u_0(x_i)\,, \quad x_i \in \bar{\omega}_h\,,$$

where

$$\Lambda(\bar{t})\, y_i = \frac{1}{x_i^2}\left(\bar{x}_i^2\, a(x_i, \bar{t})\, y_{\bar{x},i}\right)_{x,i} \quad \text{for} \quad i > 0\,,$$

$$\Lambda(\bar{t})\, y_0 = \frac{6}{h}\, a_1\, y_{x,0}\,,$$

$$\varphi_i = f(x_i, \bar{t}) \quad \text{or} \quad \varphi_i = f_i^{(\sigma)} \quad \text{for} \quad 0 \le i < N\,.$$

This is also consistent with the results expounded in Chapter 3, Section 5. The standard elimination method applies equally well to such a setting. We omit here more a detailed exploration of the residual and the accurate account of accuracy of the describing scheme. The reader is invited to do this on his/her own in line with established priorities for difference schemes on “flowing” grids $\bar{\omega}_h$.

**12. A periodic problem.** We are now interested in the problem of the heat distribution over a uniform thin circlic ring $0 \le \varphi \le 2\pi$ of radius $r_0$:

$$\frac{\partial u}{\partial t'} = \frac{a^2}{r_0^2}\,\frac{\partial^2 u}{\partial \varphi^2}\,, \qquad 0 \le \varphi \le 2\,\pi\,, \quad t' > 0\,, \quad u(\varphi, 0) = u_0(\varphi)\,.$$

A unique determination of a solution $u(\varphi, t')$ necessitates imposing the condition of periodicity

$$u(\varphi + 2\pi, t') = u(\varphi, t') \quad \text{for any} \quad \varphi \in [0, 2\,\pi]\,,$$

which, in turn, can be replaced by the condition of conjugation at the point $\varphi = 0$:

$$u(0+0, t') = u(2\,\pi - 0, t')\,, \qquad \left.\frac{\partial u}{\partial \varphi}\right|_{\varphi=0+0} = \left.\frac{\partial u}{\partial \varphi}\right|_{\varphi=2\,\pi-0}\,.$$

By interchanging the variables

$$x = \varphi/(2\,\pi)\,, \qquad t = a^2 t'/(2\,\pi\, r_0^2)$$

the segment $0 \le \varphi \le 2\pi$ is carried into the segment $0 \le x \le 1$. In view of this, the governing equation is modified into

$$\frac{\partial u}{\partial t} = \frac{\partial^2 u}{\partial x^2}\,, \quad 0 \le x \le 1\,, \quad t > 0\,, \quad u(x, 0) = u_0(x)\,,$$

$$u(0+0, t) = u(1-0, t)\,, \qquad \frac{\partial u(0+0, t)}{\partial x} = \frac{\partial u(1-0, t)}{\partial x}\,,$$

which is not surprising. On the grid

$$\omega_h = \{x_i = ih,\ i = 0, 1, \dots, N,\ h = 1/N\}$$

we have occasion to use the simplest implicit scheme

$$y_t = \hat{y}_{\bar{x}x}\,, \quad 0 < x = i\,h < 1\,, \quad t = j\,\tau > 0\,, \quad y(x, 0) = u_0(x)\,,$$

which is supplemented by the condition

$$y_0 = y_N$$

as a corollary of the condition of conjugation $u(0+0,t) = u(1-0,t)$. By analogy with Chapter 3, Section 5 the second condition of conjugation is approximated by the equation $y_{0,t} = \hat{y}_{\bar{x}x,0}$. By identifying the endpoints $x = 0$ and $x = 1$ it is supposed that

$$y_{N+1} = y_1 \,.$$

In accordance with what has been said above, the difference scheme in question is constructed at all the nodes $i = 1, 2, \dots, N$ of the grid $\bar{\omega}_h$ under the periodicity condition $y_{N+1} = y_N$ imposed at the node $i = N$.

The same procedures are workable in constructing the appropriate difference scheme associated with the equation with variable periodic coefficients

$$\frac{\partial u}{\partial t} = \frac{\partial}{\partial x}\left(k(x,t)\,\frac{\partial u}{\partial x}\right) + f(x,t)\,, \quad 0 < x < 1\,, \quad t > 0\,,$$

$$u(x,0) = u_0(x)\,, \quad 0 \leq x \leq 1\,,$$

and the boundary conditions of periodicity that are known to us as the conjugation ones:

$$u(0+0,t) = u(1-0,t)\,, \qquad k\,\frac{\partial u}{\partial x}\bigg|_{x=0+0} = \frac{\partial u}{\partial x}\bigg|_{x=1-0}\,.$$

Here all the functions $k(x,t)$, $f(x,t)$ and $u_0(x)$ are periodic of period 1 so that

$$u_0(x+1) = u_0(x)\,, \quad f(x+1,t) = f(x,t)\,, \quad k(x+1,t) = k(x,t)\,.$$

Let us stress here that the available coefficients $k(0+0,t)$ and $k(1-0,t)$ may be different: $k(0+0,t) \neq k(1-0,t)$. When this is the case, the derivatives $\partial u/\partial x$ involved happen to be discontinuous:

$$\frac{\partial u(0+0,t)}{\partial x} \neq \frac{\partial u(1-0,t)}{\partial x}\,.$$

By identifying the endpoints $x = 0$ and $x = 1$ on the same grounds as before, the condition of periodicity is to be understood as the condition of conjugation at a discontinuity point of the coefficient $k(x,t)$. From such reasoning it seems clear that, having stipulated the condition $y_{N+1} = y_1$, the design of the scheme in question includes all the nodes $i = 1, 2, \dots, N$

of the grid at hand. As a final result we get the homogeneous difference scheme with weights

$$y_t = \Lambda(\bar{t})\, y^{(\sigma)} + \varphi(x, \bar{t}), \quad x = ih, \quad i = 1, 2, \dots, N, \quad \bar{t} = (j + 0.5)\,\tau,$$

$$y(x, 0) = u_0(x), \quad y_{N+1} = y_1, \quad y_0 = y_N,$$

where $\Lambda y = \big(a(x, \bar{t})\, y_{\bar{x}}\big)_x$ and the coefficients $a$ and $\varphi$ are given by the usual formulas. For example, it is fairly common to deal with

$$a_i = k_{i-1/2}, \qquad \varphi_i = 0.5\,(f_{i-0} + f_{i+0}).$$

The solution can uniquely be found from the above conditions. This scheme has the approximation order $O((\sigma - 0.5)\tau + \tau^2 + h^2)$.

In these concerns, there arises the problem for determination of $\hat{y} = y^{j+1}$:

$$A_i \hat{y}_{i-1} - B_i \hat{y}_i + A_{i+1} \hat{y}_{i+1} = -F_i, \quad i = 1, 2, \dots, N,$$

$$\hat{y}_{N+1} = \hat{y}_1, \quad \hat{y}_0 = \hat{y}_N, \quad A_i = \sigma\,\tau\, a_i / h^2, \quad C_i = A_i + A_{i+1} + 1,$$

which can be solved by the cyclic elimination method established in Chapter 1, Section 2.

In an attempt to cover all the issues, we should raise the questions of stability and accuracy for the approximation just established. With this aim, we introduce the space $H$ of all grid functions $y(x_i)$ given for $i = 1, 2, \dots, N, N+1$ and satisfying the condition of periodicity $y_{N+1} = y_1$, $y_N = y_0$. An inner product and associated norm in that space are defined by $(v, w) = \sum_{i=1}^{N} v_i w_i h$ and $\| v \| = \sqrt{(u, v)}$.

The operator $A$ is specified by the relation

$$A\, y = -\Lambda\, y \quad \text{for} \quad y \in H.$$

It seems clear that Green's formulas are certainly true in the case when the operator $A$ is defined in such a way. Moreover, $A = A^* > 0$. All this provides the sufficient background for the possible applications of the general stability theory outlined in Chapter 6, within the framework of which the scheme concerned is unconditionally stable for $\sigma \geq 0.5$.

For $\sigma = 1$ the maximum principle is in full force for any $\tau$ and $h$, due to which the resulting scheme is uniformly stable with respect to the initial data and the right-hand side. What is more, the uniform convergence occurs with the rate $O(\tau^2 + h^2)$.

## 7.2 HOMOGENEOUS DIFFERENCE SCHEMES FOR HYPERBOLIC EQUATIONS

**1. The original problem.** In a common setting it is required to find in the rectangle

$$\bar{D}_T = [0 \le x \le 1] \times [0 \le t \le T]$$

a solution of the first boundary-value problem for a second-order equation of hyperbolic type

$$(1) \qquad \frac{\partial^2 u}{\partial t^2} = L\,u + f(x,t)\,, \quad L\,u = \frac{\partial}{\partial x}\left(k(x,t)\,\frac{\partial u}{\partial x}\right), \quad (x,t) \in D_T\,,$$

$$(2) \qquad u(x,0) = u_0(x)\,, \quad \frac{\partial u}{\partial t}(x,0) = \bar{u}_0(x)\,,$$

$$(3) \qquad u(0,t) = u_1(t)\,, \quad u(1,t) = u_2(t)\,,$$

$$0 < c_1 \le k(x,t) \le c_2\,,$$

where $D_T = (0 < x < 1) \times (0 < t \le T]$, under the following assumptions: the problem is uniquely solvable, its solution is continuous in the closed domain $\bar{D}_T$ and possesses all necessary derivatives which do arise in the further development, the coefficient $k(x,t)$ and the right-hand $f(x,t)$ may have discontinuities of the first kind on a finite number of straight lines parallel to the axis $Ot$ ("immovable discontinuities"), on every discontinuity line $x = \xi_s$, $s = 1, 2, \ldots, s_0$, the conditions of conjugation relating to the continuity of the functions $u$ and $k\,\partial u/\partial x$ for $x = \xi_s$, $s = 1, 2, \ldots, s_0$, must hold:

$$(4) \qquad [u] = u(\xi_s + 0, t) - u(\xi_s - 0, t) = 0\,, \quad [k\,\partial u/\partial x] = 0\,.$$

**2. Homogeneous difference schemes.** In preparation for designing a homogeneous weighted scheme associated with problem (1)–(3), let

$$\hat{\bar{\omega}}_h = \{x_i,\ i = 0, 1, \ldots, N\,,\ x_0 = 0\,,\ x_N = 1\}$$

be a nonequidistant grid on the segment $0 \le x \le 1$, $\bar{\omega}_\tau = \{t_j = i\tau,\ j = 0, 1, 2, \ldots, j_0\}$ be an equidistant grid on the segment $0 \le t \le T$ and let a suitable grid in the rectangle $D_T$ be made up by $\bar{\omega}_{h\tau} = \hat{\omega}_h \times \bar{\omega}_\tau$. A homogeneous difference scheme for solving problem (1)–(3) can be obtained for fixed $t \in \omega_\tau$ through the approximation $\Lambda\, u + \varphi = \big(a(x,t)\, u_{\bar{x}}\big)_{\hat{x}} + \varphi$ to $Lu + f$. The forthcoming replacements $\partial^2 u/\partial t^2|_{t=t_j} \sim u_{\bar{t}t}$ and $Lu + f \sim \Lambda(t_j)\, u^{(\sigma_1,\sigma_2)} + \varphi$, where

$$u^{(\sigma_1,\sigma_2)} = \sigma_1 \hat{u} + (1 - \sigma_1 - \sigma_2)\, u + \sigma_2 \check{u}\,,$$

$$\Lambda(t_j)\, u = \big(a(x,t_j)\, u_{\bar{x}}\big)_{\hat{x}}\,,$$

$$u = u^j\,, \qquad \check{u} = u^{j-1}, \qquad \hat{u} = u^{j+1},$$

complement subsequent constructions. As a final result we obtain the homogeneous weighted three-layer scheme

$$y_{\bar{t}t} = \Lambda(t_j)\, y^{(\sigma_1,\sigma_2)} + \varphi\,. \tag{5}$$

The coefficient $a$ can be taken on the middle layer $t = t_j$.

Substituting $\hat{y} = y + \tau y_{\overset{\circ}{t}} + 0.5\tau^2 y_{\bar{t}t}$ and $\check{y} = y - \tau y_{\overset{\circ}{t}} + 0.5\tau^2 y_{\bar{t}t}$, where $y_{\overset{\circ}{t}} = (\hat{y} - \check{y})/(2\tau)$ and $y_{\bar{t}t} = (\hat{y} - 2y + \check{y})/\tau^2$, into (5) yields

$$y^{(\sigma_1,\sigma_2)} = y + (\sigma_1 - \sigma_2)\, \tau\, y_{\overset{\circ}{t}} + 0.5(\sigma_1 + \sigma_2)\tau^2 y_{\bar{t}t}\,,$$

which admits an alternative form

$$\big(E - 0.5\,(\sigma_1 + \sigma_2)\, \tau^2 \Lambda\big)\, y_{\bar{t}t} - (\sigma_1 - \sigma_2)\, \tau\, \Lambda\, y_{\overset{\circ}{t}} = \Lambda\, y + \varphi\,, \tag{6}$$

where $E$ is the identity operator. For $\sigma_1 = \sigma_2 = \sigma$ it becomes the symmetric scheme

$$(E - \sigma\, \tau^2 \Lambda)\, y_{\bar{t}t} = \Lambda\, y + \varphi(x,t)\,, \quad 0 < t = j\tau\,, \tag{7}$$

which will be given special investigation in the sequel.

Before going further, we must append to (7) the initial and boundary conditions. These depend on the range of variables; thus, the boundary conditions and the first initial condition are specified exactly:

$$y(0,t) = u_1(t)\,, \quad y(1,t) = u_2(t)\,, \quad y(x,0) = u_0(x)\,. \tag{8}$$

There are two ways of approximating the second initial condition $\partial u/\partial t|_{t=0} = \bar{u}_0(x)$, one of which generates an approximation of order 2 in $\tau$:

$$y_t(x,0) = \tilde{u}_0(x)\,, \quad \text{where} \quad \tilde{u}_0(x) = \bar{u}_0(x) + 0.5\, \tau\, (L\, u_0 + f)_{t=0}\,. \tag{9}$$

The second one is connected with the difference equation for determination of $y(\tau)$:

$$(9') \qquad (E - \sigma\tau^2\Lambda(0))\, y_t(x,0) = \bar{u}_0(x) + 0.5\,\tau\left(\Lambda\, u_0 + f(x,0)\right).$$

Summarizing, the homogeneous difference scheme (7)–(9) or (7), (8), (9′) is put in correspondence with the original problem (1)–(3) under consideration.

The computational procedures for three-layer schemes were established before. For their successful realization we need to know the values $y^j$ and $y^{j-1}$ on two preceding layers for searching $\hat{y} = y^{j+1}$ by the elimination method being used on every new layer $t = t_{j+1}$ in solving the boundary-value problem with respect to $\hat{y} = y^{j+1}$:

$$(10) \quad (E - \sigma\tau^2\Lambda)\,\hat{y} = F\,, \quad 0 < x = ih < 1\,, \quad \hat{y}_0 = \hat{u}_1\,, \quad \hat{y}_N = \hat{u}_2\,,$$

$$F(t) = 2\,y - \check{y} - \tau^2\Lambda\left((2\,\sigma - 1)\,y - \sigma\,\check{y}\right) + \tau^2\varphi\,, \quad t \geq \tau\,,$$

$$F(0) = u_0 + \tau^2(0.5 - \sigma)\,\Lambda(0)\,u_0 + \tau\,\bar{u}_0(x) + 0.5\,\tau^2 f(x,0)\,.$$

**3. The error of approximation.** Let $u(x,t)$ be a solution of the original problem (1)–(3) and $y(x_i,t_j) = y_i^j$ be a solution of the difference problem (7)–(9). The next step is to set up the difference problem for the error $z_i^j = y_i^j - u_i^j$, where $u_i^j = u(x_i,t_j)$, by inserting in (7)–(9) the sum $y = z + u$:

$$(11) \qquad (E - \sigma\tau^2\Lambda)\, z_{\bar{t}t} = \Lambda\, z + \psi(x,t)\,, \quad 0 < x < 1\,, \quad t > 0\,,$$

$$z(x,0) = 0\,, \quad z(0,t) = z(1,t) = 0\,, \quad z_t(x,0) = \nu(x)\,,$$

where

$$(12) \qquad \psi(x,t) = \Lambda(t)\,u - u_{\bar{t}t} + \sigma\tau^2\Lambda\, u_{\bar{t}t}\,,$$

$$\nu(x) = 0.5\,\tau\,(L\,u_0 + f)_{t=0} + \bar{u}_0(x) - u_t(x,0)$$

are the errors of approximation on the solution of problem (1)–(3) associated with equation (1) and the second initial condition (2), respectively.

If the coefficient $k(x,t)$ and the right-hand side $f(x,t)$ possess only a finite number of immovable discontinuities, the grid $\hat{\omega}_h = \hat{\omega}_h(K)$ will be so chosen that all discontinuity lines will pass through the nodal points (the main idea behind this approach was explained earlier in Chapter 3, Section 4 and Section 1 of the present chapter).

For convenience in analysis, the residual $\psi$ is representable by

$$\psi = \eta_{\hat{x}} + \psi^* , \tag{13}$$

where

$$\eta_i = a_i u_{\bar{x},i} - \left(k \frac{\partial u}{\partial x}\right)_{i-1/2} + \sigma \tau^2 a_i u_{\bar{t}t\bar{x},i} + \frac{h_i^2}{8} \left(\frac{\partial f}{\partial x} - \frac{\partial^3 u}{\partial t^2 \, \partial x}\right)_{i-1/2}, \tag{14}$$

$$\psi^* = O(\tau^2 + h^2) . \tag{15}$$

Apparently, the current situation needs certain clarification. Having integrated equation (1) with respect to $x$ at a fixed moment $t = t_j$ from $x_{i-1/2}$ to $x_{i+1/2}$

$$\left(k \frac{\partial u}{\partial x}\right)_{x=x_i+0.5} - \left(k \frac{\partial u}{\partial x}\right)_{x=x_i-0.5} + \int\limits_{x_{i-1/2}}^{x_{i+1/2}} f(x, t_j) \, dx - \int\limits_{x_{i-1/2}}^{x_{i+1/2}} \frac{\partial^2 u(x, t_j)}{\partial t^2} \, dx = 0 , \tag{16}$$

we then divide this identity by $\hbar_i$, subtract the resulting expressions from the right-hand side of representation (12) for the residual $\psi$ and, finally, get

$$\psi_i = \left( a_i u_{\bar{x},i} + \sigma \tau^2 a_i u_{\bar{t}t\bar{x},i} - \left(k \frac{\partial u}{\partial x}\right)_{i-1/2} \right)_{\hat{x},t} - u_{\bar{t}t,i} + \varphi_i - \frac{1}{\hbar_i} \int\limits_{x_{i-1/2}}^{x_{i+1/2}} \left( f(x, t_j) - \frac{\partial^2 u(x, t_j)}{\partial t^2} \right) dx , \tag{17}$$

where the coefficients $a_i$ and $\varphi_i$ are given for fixed $t = t_j$ by the same formulas as stated in Chapter 3, Section 4.

Let $x_i$ be a discontinuity point of both functions $k$ and $f$. To avoid cumbersome calculations, the usual practice involves the simplest formulas for finding $a_i$ and $\varphi_i$:

$$a_i = k_{i-1/2}, \quad \varphi_i = \frac{h_i f_i^- + h_{i+1} f_i^+}{2 \, \hbar_i} , \quad f_i^{\pm} = f(x_i \pm 0) . \tag{18}$$

Following the procedures of Chapter 3, Section 4 and taking into account that the second partial derivative $\partial^2 u/\partial t^2$ is continuous on the line of discontinuity $x = \xi$ of the functions $k(x,t)$ and $f(x,t)$, we deduce through such an analysis that

$$\frac{1}{\hbar_i} \int\limits_{x_{i-1/2}}^{x_{i+1/2}} \left( f(x,t_j) - \frac{\partial^2 u(x,t_j)}{\partial t^2} \right) dx = \frac{h_i f_i^- + h_{i+1} f_i^+}{2\,\hbar_i} - \left( \frac{\partial^2 u}{\partial t^2} \right)_i$$

$$+ \frac{1}{8} \left( h_i^2 \left( \frac{\partial f}{\partial x} - \frac{\partial^3 u}{\partial t^2\, \partial x} \right)_{i-1/2} \right)_{\hat{x},i} + O(\hbar_i^2)\,.$$

Because of this, formulas (13)–(15) are an immediate implication of formula (17) with the preceding expansion involved. From representation (14) it seems clear that

$$\eta_i = O(h_i^2 + \tau^2)\,, \qquad \eta_{t,i} = O(h_i^2 + \tau^2)\,. \tag{19}$$

**4. Stability and convergence.** No restrictions are made regarding the smoothness of the coefficients and the solution in the further estimation of the accuracy of scheme (7)–(9). This can be done using various a priori estimates for the operator-difference three-layer scheme

$$D\, z_{\bar{t}t} + A\, z = \psi(t)\,, \quad t = i\tau > 0\,, \tag{20}$$

$$z(0) = 0\,, \quad z_t(0) = \nu\,.$$

Here $D$ and $A$ stand for linear operators in a Hilbert space $H$, $z(t)$ and $\psi(t)$ refer to abstract functions of the argument $t \in \omega_\tau$ with the values in the space $H$ and $\nu$ is an element of the space $H$ (for more detail see Chapter 6). In preparation for this, $H = \overset{\circ}{\Omega}$ is the space of all grid functions given on the grid $\hat{\omega}_h$ and vanishing on the boundary at the points $x = 0$ and $x = 1$. The usual inner products are defined to be

$$(z,v)_* = \sum_{i=1}^{N-1} z_i v_i \hbar_i\,, \quad (z,v) = \sum_{i=1}^{N-1} z_i v_i h_i\,, \quad (z,v] = \sum_{i=1}^{N} z_i v_i h_i\,.$$

In the general setting three types of suitable norms are in common usage:

$$\|z\|_C = \max_{x \in \hat{\omega}_h} |z(x)|\,, \quad \|\, z\,\| = \sqrt{(z,z)_*}\,, \quad \|z\|_A = \sqrt{(Az,z)}\,.$$

A comparison of (7) and (20) provides enough reason to conclude that these operators $A$ and $D$ are identical:

$$A = -\Lambda\,, \qquad D = E - \sigma\,\tau^2 \Lambda = E + \sigma\,\tau^2 A\,.$$

The operator $A$ is self-adjoint, positive definite and satisfies the estimate (see Chapter 2, Section 4)

$$\|\,A\,\| \le 4\,c_2/h_{\min}^2\,, \qquad h_{\min} = \min_{1\le i\le N} h_i\,.$$

The general stability theory outlined in Chapter 6 asserts that scheme (20) is stable under the condition

$$D \ge \frac{1+\varepsilon}{4}\,\tau^2 A \quad \text{or} \quad (Dy, y) \ge \frac{1+\varepsilon}{4}\,\tau^2\,(Ay, y)\,,$$

where $\varepsilon > 0$ is an arbitrary number independent of $h$. We will pursue the further stability analysis of this with

$$D - \frac{1+\varepsilon}{4}\,\tau^2 A = E + \left(\sigma - \frac{1+\varepsilon}{4}\right)\tau^2 A$$

$$\ge \left(\frac{1}{\|\,A\,\|} + \left(\sigma - \frac{1+\varepsilon}{4}\right)\tau^2\right) A \ge 0\,,$$

which is certainly true for

$$\sigma \ge \sigma_\varepsilon = \frac{1+\varepsilon}{4} - \frac{h_{\min}^2}{4\,\tau^2\,c_2}\,.$$

If you wish to explore this more deeply, you might find it helpful to refer to Chapter 6, Section 3 of the monograph "The Theory of Difference Schemes", in which the following estimates were derived for problem (20):

$$(21) \qquad \|z^{j+1}\|_{A(t_j)} \le M\,\sqrt{\frac{1+\varepsilon}{\varepsilon}}\,\left(\|z_t(0)\|_{D(\tau)} + \sum_{k=1}^{j} \tau\,\|\psi^k\|_{D^{-1}(t_k)}\right),$$

$$(22) \qquad \|z^{j+1}\|_{A(t_j)} \le M\,\sqrt{\frac{1+\varepsilon}{\varepsilon}}\,\Bigg(\|z_t(0)\|_{D(\tau)}$$

$$+ \max_{0<k\le j}\left(\|\psi^k\|_{A^{-1}(t_k)} + \|\psi_t^k\|_{A^{-1}(t_k)}\right)\Bigg).$$

For the weighted scheme (20) these estimates are ensured by $\sigma \geq \sigma_\varepsilon$ and $|a_i| \leq c_3 a$.

We shall need yet, among other things, some modification of the well-known estimates on an equidistant grid (see Chapter 2, Section 4), taking on an arbitrary nonequidistant grid the form

$$\|z\|_A \geq \sqrt{c_1}\, \|z_{\bar{x}}]| > \sqrt{c_1}\, \|z\|_C\,,$$

$$\|\psi\|_{A^{-1}} = \|\eta_{\hat{x}}\|_{A^{-1}} \leq \frac{2}{\sqrt{c_1}}\, \|\eta]|\,, \tag{23}$$

$$\|\psi_t\|_{A^{-1}} \leq \frac{2}{\sqrt{c_1}}\, \|\eta_t]| \quad \text{for} \quad \psi = \eta_{\hat{x}}\,.$$

Since $D$ is a self-adjoint operator and

$$D = E + \sigma\,\tau^2 A = E + (\sigma - \sigma_\varepsilon)\,\tau^2 A + \sigma_\varepsilon\,\tau^2 A$$

$$\geq E + 0.5\,\tau^2 A - \frac{1-\varepsilon}{\|A\|} A \geq \varepsilon\, E\,,$$

we obtain

$$D^{-1} \leq \frac{1}{\varepsilon} E \quad \text{and} \quad \|\psi\|_{D^{-1}} \leq \frac{1}{\sqrt{\varepsilon}}\, \|\psi\|\,.$$

As usual, we may attempt the solution $z$ of problem (11)–(13) as a sum $z = v + w$ with the members $v$ and $w$ satisfying the conditions

$$v_{\bar{t}t} = \Lambda\, v^{(\sigma)} + \eta_{\hat{x}}\,, \quad v(x,0) = v_t(x,0) = 0\,, \quad v_0 = v_N = 0\,,$$

$$w_{\bar{t}t} = \Lambda\, w^{(\sigma)} + \psi^*\,, \quad w(x,0) = 0\,, \tag{24}$$

$$w_t(x,0) = \nu(x)\,, \quad w_0 = w_N = 0\,.$$

Putting these together with (21)–(23) we deduce for $v$ and $w$ that

$$\|v^{j+1}\|_C \leq \frac{M}{\sqrt{\varepsilon}} \max_{0<k\leq j} \left(\|\eta^k]| + \|\eta_t^k]|\right), \tag{21'}$$

$$\|w^{j+1}\|_C \leq \frac{M}{\sqrt{\varepsilon}} \left(\|\nu\|_D + \sum_{k=1}^{j} \tau\, \|\psi^{*k}\|\right), \tag{22'}$$

where $\|\nu\|_D^2 = \left((E + \sigma\,\tau^2 A)\,\nu, \nu\right)_* = \|\nu\|^2 + \sigma\,\tau^2 (a, \nu_{\bar{x}}^2]$.

**Theorem** *Let the functions $k(x,t)$ and $f(x,t)$ have discontinuities of the first kind on a finite number of the straight lines $x = \xi_s$, $s = 1, 2, \dots, s_0$ parallel to the axis $Ot$ and in the regions of the special configurations*

$$\Delta_s = (\xi_s < x < \xi_{s+1},\ 0 \le t \le t_0), \quad s = 0, 1, \dots, s_0, \quad \xi_0 = 0, \quad \xi_{s_0+1} = 1,$$

*the coefficients $k(x,t)$ and $f(x,t)$ and the solution $u(x,t)$ are smooth enough so that both conditions* (19) *and* (15) *hold true. Then under condition* (22) *scheme* (7)–(9) *converges uniformly with the rate $O(\tau^2 + h^2)$ on special sequences of nonequidistant grids $\hat{\omega}_h(K)$ and the solution of problem* (11) *satisfies the estimate*

$$(25) \qquad \|z^j\|_C = \|y^j - u^j\|_C \le M\,(\tau^2 + h_0^2)\,, \quad \text{where} \quad h_0 = \max_{1 \le i \le N} h_i\,.$$

To prove this assertion, it suffices to bring together a priori estimates (21)–(22) with relations (15) and (19).

**Remark** The theorem is still valid upon replacing (7) by the scheme

$$(26) \qquad (E - \sigma\,\tau^2\,\overset{\circ}{\Lambda})\,y_{\bar{t}t} = \Lambda\,y + \varphi\,,$$

where the constant operator $\overset{\circ}{\Lambda}\,y = y_{\bar{x}\hat{x}}$ is adopted as a regularizer. In that case $A = -\Lambda$, $R = -\sigma\,\overset{\circ}{\Lambda}$ and $D = E + \tau^2 R$. The sufficient stability condition (21) is ensured if we agree to consider

$$(27) \qquad \sigma = (1 + \varepsilon)\,c_2/4\,.$$

In evaluating the error of approximation all the tricks and turns remain unchanged except for formula (14) for $\eta$, in which the member $\sigma\tau^2 a u_{\bar{t}t\bar{x}}$ should be replaced by $\sigma\tau^2 u_{\bar{t}t\bar{x}}$, where a constant $\sigma$ is specified by (27).

# 8

# Difference Methods for Solving Nonlinear Equations of Mathematical Physics

In this chapter the new difference schemes are constructed for the quasilinear heat conduction equation and equations of gas dynamics with placing a special emphasis on iterative methods available for solving nonlinear difference equations. Among other things, the convergence of Newton's method is established for implicit schemes of gas dynamics.

## 8.1 DIFFERENCE METHODS FOR SOLVING THE QUASILINEAR HEAT CONDUCTION EQUATION

**1. The stationary problem.** To avoid misunderstanding, we concentrate primarily on the simplest problem, the statement of which is related to the stationary heat conduction problem with nonlinear sources:

(1) $$u'' = -f(u)\,, \quad 0 < x < 1\,, \quad u(0) = 0\,, \quad u(1) = 0\,.$$

An excellent start in this direction is to introduce on the segment $0 \le x \le 1$ an equidistant grid $\bar{\omega}_h = \{x_i = ih,\, i = 0, 1, \ldots, N,\, hN = 1\}$ and proceed to design the difference scheme

(2) $$y_{\bar{x}x} = -f(y)\,, \qquad x = ih\,, \qquad i = 1, 2, \ldots, N-1\,,$$

$$y_0 = y_N = 0\,,$$

making it possible to set up the difference problem for the error $z = y - u$:

$$(3) \qquad z_{\bar{x}x} + f'(\bar{y})\, z = -\psi\,, \qquad x = ih\,, \qquad i = 1, 2, \ldots, N-1\,,$$

$$z_0 = z_N = 0\,,$$

where $\bar{y} = u + \theta z$, $0 \le \theta \le 1$, and $\psi = u_{\bar{x}x} + f(u)$ is the residual.

It seems clear that scheme (2) generates an approximation of order 2:

$$\psi = O(h^2)\,.$$

If $f'(y) \le 0$, then a solution of the difference problem (3) satisfies the estimate

$$(4) \qquad \|z\|_C \le \|\psi\|_C\,,$$

which serves to motivate the uniform convergence of scheme (2) with the rate $O(h^2)$:

$$\|z\|_C = \|y - u\|_C = O(h^2)\,.$$

What is more, a solution of the difference problem (2) is bounded, so that

$$(5) \qquad \|y\|_C \le |f(0)| = c_0$$

under the constraint $f'(y) \le 0$. Indeed, simple algebra gives

$$f(y) = f(0) + \big(f(y) - f(0)\big) = f(0) + f'(\bar{y})\, y\,,$$

where $\bar{y} = \theta\, y\,, 0 \le \theta \le 1$, yielding

$$y_{\bar{x}x} + f'(\bar{y})\, y = -f(0)\,, \quad y_0 = y_N = 0\,.$$

Whence estimate (5) follows on account of (4).

In this regard, Newton's method suits us perfectly in connection with solving the nonlinear difference equation (2). It is worth recalling here its algorithm:

$$\overset{k+1}{y}_{\bar{x}x} + f'(\overset{k}{y})\,(\overset{k+1}{y} - \overset{k}{y}) = -f(\overset{k}{y})\,,$$

where $k$ is the iteration number, $k = 0, 1, 2, \ldots$, leaving us with the three-point linear difference equation related to $\overset{k+1}{y}$:

$$(6) \qquad \overset{k+1}{y}_{\bar{x}x} + f'(\overset{k}{y})\,\overset{k+1}{y} = -\big(f(\overset{k}{y}) - f'(\overset{k}{y})\,\overset{k}{y}\big)\,, \quad \overset{k+1}{y}_0 = \overset{k+1}{y}_N = 0\,.$$

This can be solved by the standard elimination method, whose computational procedure is stable under the condition $f'(y) \le 0$.

The convergence rate of such iterations needs investigation with regard to the error

$$\overset{k+1}{v} = \overset{k+1}{y} - y\,,$$

where $y$ is the exact solution of problem (2). Upon substituting two subsequent iterations $\overset{k}{y} = y + \overset{k}{v}$ and $\overset{k+1}{y} = y + \overset{k+1}{v}$ into equation (6) we may set up the problem for the error $\overset{k+1}{v}$:

$$(7) \qquad \overset{k+1}{v}_{\bar{x}x} + f'(\overset{k}{y})\,\overset{k+1}{v} = -\overset{k}{F}\,, \qquad \overset{k+1}{v}_0 = \overset{k+1}{v}_N = 0\,,$$

$$(8) \qquad \overset{k}{F} = f(\overset{k}{y}) - f(y) + (y - \overset{k}{y})\,f'(\overset{k}{y})\,.$$

Taking into account the well-established decomposition

$$f(y) = f(\overset{k}{y}) + f'(\overset{k}{y})\,(y - \overset{k}{y}) + \tfrac{1}{2}\,f''(\overset{k}{\tilde{y}})\,(\overset{k}{y} - y)^2,$$

where $\overset{k}{\tilde{y}} = \overset{k}{y} + \theta(y - \overset{k}{y})$, $0 \le \theta \le 1$, we find that

$$\overset{k}{F} = -\tfrac{1}{2}\,f''(\overset{k}{\tilde{y}})\,\overset{k}{v}{}^2\,.$$

Thus, it is required to evaluate a solution of the problem

$$(9) \qquad \overset{k+1}{v}_{\bar{x}x} + f'(\overset{k}{y})\,\overset{k+1}{v} = \tfrac{1}{2}\,f''(\overset{k}{\tilde{y}})\overset{k}{v}{}^2\,, \qquad \overset{k+1}{v}_0 = \overset{k+1}{v}_N = 0\,.$$

If $f(y)$ is a concave function, that is, $f''(y) \ge 0$, then due to the maximum principle we might have

$$\overset{k+1}{v} = \overset{k+1}{y} - y \le 0\,, \qquad \overset{k+1}{y} \le y\,,$$

thereby clarifying that the iterations approach the exact solution of problem (2) from below. It is plain to show that for a solution of problem (9) either of the following estimates

$$(10) \qquad \|\overset{k+1}{v}\|_C = \tfrac{1}{16}\,\|f''(\overset{k}{y})\|_C \cdot \|\overset{k}{v}\|_C^2$$

or

$$\|\overset{k+1}{v}\|_C \le q\,\|\overset{k}{v}\|_C^2$$

is valid under the condition $\|f''(y)\|_C = 16q$.

Indeed, in conformity with the maximum principle (for more detail see Chapter 4, Section 2) problem (9) has for $f'(y) \le 0$ the majorant

$$V(x) = K\,x\,(1-x)\,, \quad \|V(x)\|_C \le \tfrac{1}{4}\,K\,,$$

where $K = \frac{1}{4}\,\|f''(y)\|_C \cdot \|\overset{k}{v}\|_C^2$, so that

$$\|\overset{k+1}{v}\|_C \le \|V\|_C \le \tfrac{1}{16}\,\|f''(y)\|_C \cdot \|\overset{k}{v}\|_C^2 \le q\,\|\overset{k}{v}\|_C^2\,.$$

With the chain of the relations $\|q\overset{k+1}{v}\|_C \le \|q\overset{k}{v}\|_C^2 \le \cdots \le \|q\overset{0}{v}\|_C^{2^{k+1}}$ in view, we deduce that

$$\|\overset{k+1}{v}\|_C \le \frac{1}{q}\,\|q\overset{0}{v}\|_C^{2^{k+1}}\,,$$

thereby confirming the quadratic law of the convergence of iterations with the initial approximation $\overset{0}{y}$ subject to the condition

$$q\,\|\overset{0}{y} - y\|_C < 1 \quad \text{or} \quad q\,\|\overset{0}{v}\|_C < 1\,.$$

When $f'(y) > -c_1$, $c_1 > 0$, an alternative estimate in the grid norm of the space $L$ instead of (10) is such that

$$\|\overset{k+1}{v}\| \le \frac{\|f''(y)\|_C}{2\,(\delta + c_1)}\,\|\overset{k}{v}\|^2 = \frac{q_1}{\delta + c_1}\,\|\overset{k}{v}\|^2\,, \tag{11}$$

where

$$\delta = \frac{4}{h^2}\,\sin^2\frac{\pi h}{2} \quad \text{and} \quad q_1 = \frac{1}{2}\,\|f''(y)\|_C\,.$$

Other iterative methods apply equally well to problem (2). Among them the method with the recurrence relation

$$\overset{k+1}{y}_{\bar{x}x} = \theta\,\overset{k}{y}_{\bar{x}x} - (1-\theta)\,f(\overset{k}{y})$$

will be appreciated. Here the parameter $\theta$ is given by the formula

$$\theta = \frac{\lambda}{\lambda + 1}\,, \qquad \lambda = \frac{c^*}{2\,\delta}\,, \qquad c^* = \max_y\,|f'(y)|\,;$$

in so doing the iterations converge with the rate of a geometric progression with denominator $q_2 = \theta$, so that

$$\|\overset{k+1}{v}\| \le q_2\,\|\overset{k}{v}\| \le q_2^{k+1}\,\|\overset{0}{v}\|\,.$$

It is worth noting here that the iterations converge no matter how the initial approximation $\overset{0}{y}$ is chosen, because $\theta < 1$.

**2. The quasilinear heat conduction equation.** So far we have considered merely the linear heat conduction equation in spite of the fact that in plenty of real physical processes the coefficient of heat conductivity is, generally speaking, a nonlinear function of temperature (and density). In some problems it gives, in addition, a function of the temperature gradient. High-temperature processes in plasma physics are in line with these statements. Being the right-hand sides of the heat conduction equation, heat sources may depend on the temperature when, for example, the heat transfer is caused by a chemical reaction. Such processes are described by the nonlinear heat conduction equation

$$\frac{\partial \varepsilon(x,t,u)}{\partial t} = -\frac{\partial w}{\partial x} + f(x,t,u)\,, \tag{12}$$

where the heat flow

$$w = w\Big(x,t,u,\frac{\partial u}{\partial x}\Big)$$

is a nonlinear function of temperature $u$ and its derivative. If the heat flow is linearly dependent on the derivative $\partial u/\partial x$ and it is governed by the Fourier law

$$w = -k(x,t,u)\,\frac{\partial u}{\partial x}\,,$$

we obtain as a final result the quasilinear heat conduction equation

$$c(x,t,u)\,\frac{\partial u}{\partial t} = \frac{\partial}{\partial x}\Big(k(x,t,u)\,\frac{\partial u}{\partial x}\Big) + f(x,t,u)\,, \tag{13}$$

$$c(x,t,u) > 0\,, \qquad k(x,t,u) > 0\,.$$

When this is the case, the heat capacity $c$, the coefficient of conductivity $k$ and the right-hand side $f$ depend on the temperature $u(x,t)$. In inhomogeneous media $k$, $c$ and $f$ may have discontinuities of various kinds and this dependence upon the temperature $u$ may be different and depends on the range of situations to be considered.

In this view, it seems reasonable in a typical situation when the functions $k = k(u)$, $c = c(u)$ and $f = f(u)$ depend only on the temperature $u$ and give rise to the govering equation

$$c(u)\,\frac{\partial u}{\partial t} = \frac{\partial}{\partial x}\Big(k(u)\,\frac{\partial u}{\partial x}\Big) + f(u)\,, \tag{14}$$

to introduce a new variable $v = \int_0^u k(\xi)\,d\xi$ for later use in equation (14) with further simplification of the ensuing formulas. The outcome of this is

$$\frac{\partial \varphi(v)}{\partial t} = \frac{\partial^2 v}{\partial x^2} + \tilde{f}(v), \quad \text{where} \quad \varphi(v) = \int_0^u c(\xi)\,d\xi\,,$$

By merely setting $v = \int_0^u c(\xi)\,d\xi$ we are led to an alternative form of equation (14):

$$\frac{\partial v}{\partial t} = \frac{\partial}{\partial x}\left(\varkappa(v)\,\frac{\partial v}{\partial x}\right) + \tilde{f}(v)\,,$$

so that

$$\int_0^v \varkappa(v)\,dv = \int_0^u k(u)\,du\,.$$

When $c(u)$ and $k(u)$ can be expressed through the power functions of temperature $u$, that is,

$$c(u) = c_0 u^\alpha\,, \qquad k(u) = k_0 u^\beta\,,$$

it makes sense to introduce one more variable

$$v = \int_0^u c(\xi)\,d\xi = c_0\,\frac{u^{\alpha+1}}{\alpha+1}$$

and take into account that

$$k\,\frac{\partial u}{\partial x} = \frac{k}{c_0\,u^\alpha}\,\frac{\partial v}{\partial x} = \frac{k_0}{c_0}\,u^{\beta-\alpha}\,\frac{\partial v}{\partial x} = \frac{k_0}{c_0}\left(\frac{\alpha+1}{c_0}\right)^{\frac{\beta-\alpha}{\alpha+1}} v^{\frac{\beta-\alpha}{\alpha+1}}\,\frac{\partial v}{\partial x}\,,$$

permitting us to recast equation (14) as

$$\frac{\partial v}{\partial t} = \frac{\partial}{\partial x}\left(\varkappa_0\,v^\sigma\,\frac{\partial v}{\partial x}\right) + \tilde{f}(v)\,,$$

$$\sigma = \frac{\beta-\alpha}{\alpha+1}\,, \qquad \varkappa_0 = \frac{k_0}{c_0}\left(\frac{\alpha+1}{c_0}\right)^{\frac{\beta-\alpha}{\alpha+1}}.$$

**3. Some analytical solutions to the quasilinear heat conduction equation.** Nonlinearity of the coefficient of heat conductivity results in the new physical effects, the main of which is a final velocity of heat conducting. In what follows we expound some exploratory devices for obtaining the simplest particular solutions to the equation

$$\frac{\partial u}{\partial t} = \frac{\partial}{\partial x}\left(\varkappa_0 u^\sigma \frac{\partial u}{\partial x}\right), \quad \varkappa_0 > 0, \quad \sigma > 0, \quad x > 0. \tag{15}$$

Here the subsidiary information is the temperature at the point $x = 0$:

$$u = u_0 t^n. \tag{16}$$

With these, it is required to find a solution to equation (15) in the domain $\{x > 0,\ t > 0\}$ with the zero initial temperature

$$u(x, 0) = 0. \tag{17}$$

We may attempt a solution of this problem in the form of a "travelling" wave

$$u(x,t) = U(Dt - x), \qquad D = \text{const},$$

where $U(\xi)$ is the unknown function which is sought. Inserting this expression in (15) and taking into account that

$$\frac{\partial u}{\partial t} = D\frac{dU}{d\xi}, \qquad \frac{\partial u}{\partial x} = -\frac{dU}{d\xi},$$

we derive the ordinary differential equation for the function $U(\xi)$:

$$D U' = (\varkappa_0 U^\sigma U')',$$

yielding

$$\varkappa_0 U^\sigma U' = D U + \text{const}.$$

In the case const $= 0$, we obtain

$$\varkappa_0 U^\sigma U' = D U \qquad \text{or} \qquad \frac{\varkappa_0 (U^\sigma)'}{D\sigma} = 1,$$

which upon integrating once again becomes

$$\frac{\varkappa_0}{D\sigma} U^\sigma = \xi + c_0.$$

Since $U = 0$ for $t = x = 0$ ($\xi = 0$), we find that $c_0 = 0$ and

$$u = U(\xi) = \left(\frac{D\sigma}{\varkappa_0}\xi\right)^{1/\sigma} = \left(\frac{D^2\sigma}{\varkappa_0}\right)^{1/\sigma} t^{1/\sigma}\left(1 - \frac{x}{Dt}\right)^{1/\sigma},$$

whence it follows that at the point $x = 0$

$$u(0, t) = \left( \frac{D^2 \sigma}{\varkappa_0} \right)^{1/\sigma} t^{1/\sigma} .$$

A comparison with (16) gives

$$n = \frac{1}{\sigma} , \qquad \frac{D^2 \sigma}{\varkappa_0} = u_0^\sigma .$$

Thus, we have proved by having recourse to a "travelling" wave that problem (15)–(17) is solvable and its solution admits the form

$$(18)\quad u(x, t) = \begin{cases} u_0\, t^{1/\sigma} \left( 1 - \dfrac{x}{Dt} \right)^{1/\sigma} = \dfrac{u_0}{D^{1/\sigma}} (Dt - x)^{1/\sigma} , & 0 \le x \le Dt, \\ 0 , & x \ge Dt, \end{cases}$$

provided the condition $n = 1/\sigma$ holds. Any solution of the form (18) is called a **"temperature wave"** with a finite velocity. What is more, it depends on three parameters $\varkappa_0$, $\sigma$ and $u_0$ in accordance with the law

$$D = \sqrt{\varkappa_0\, u_0^\sigma / \sigma} .$$

At the next stage we focus our attention on the heat flow

$$w = -\varkappa_0\, u^\sigma \frac{\partial u}{\partial x} = \varkappa_0 \frac{u_0^{\sigma+1}}{\sigma\, D^{1+1/\sigma}} (Dt - x)^{1/\sigma} = \varkappa_0 \frac{u_0^\sigma}{\sigma\, D} u(x, t) .$$

In view of this, on the front of the temperature wave $x = Dt$ the temperature and heat flow vanish for $\sigma > 0$ and the partial derivative

$$\frac{\partial u}{\partial x} = - \frac{u_0}{\sigma\, D^{1/\sigma}} \frac{1}{(Dt - x)^{1-1/\sigma}}$$

tends to $\infty$ for $\sigma > 1$, it is finite for $\sigma = 1$ and becomes zero for $0 < \sigma < 1$. Therefore, it is meaningful to speak for $\sigma > 1$ only about a generalized solution to the heat conduction equation (15).

A case in point is a nonlinear dependence of the coefficient of heat conductivity upon the temperature. From the formula for $D$ it is easily seen that we formally have $D = \infty$ for the linear heat conductivity when $\sigma = 0$; meaning that the velocity of heat conducting turns out to be infinite.

It may happen that the temperature front is held fixed, that is, $D = 0$. Such a solution always exists in a special boundary regime such as

$$u(0,t) = u_0 \frac{1}{(t_0 - t)^m}, \tag{19}$$

where $t_0$ is an arbitrary constant, under the agreement that the initial condition was imposed for $t = -\infty$:

$$u(x, -\infty) = 0\,. \tag{20}$$

Still using the framework of the method of separation of variables, a solution to equation (15) is sought in the form

$$u(x,t) = v(x)\,T(t)\,.$$

Upon substituting this product in (15) and separating the variables we obtain

$$\frac{1}{v}\frac{d}{dx}\left(\varkappa_0\, v^\sigma \frac{dv}{dx}\right) = \frac{1}{T^{\sigma+1}}\frac{dT}{dt} = \lambda\,,$$

where $\lambda$ is a separation parameter. Whence it follows that

$$\frac{d}{dx}\left(\varkappa_0\, v^\sigma \frac{dv}{dx}\right) - \lambda\, v = 0\,, \tag{21}$$

$$\frac{dT}{dt} = \lambda\, T^{\sigma+1}\,. \tag{22}$$

Along these lines, we may attempt a solution to equation (21) in the form

$$v^\sigma = \alpha\,(x_1 - x)^\beta\,,$$

where the numbers $\alpha$ and $\beta$ are free to be chosen and $x_1$ is an arbitrary number. Substituting $v^\sigma$ into (21) yields

$$\varkappa_0 \alpha^{1+1/\sigma}\, \frac{\beta}{\sigma}\left(\frac{\beta}{\sigma} + \beta - 1\right)(x_1 - x)^{\beta/\sigma+\beta-2} - \lambda\, \alpha^{1/\sigma}(x_1 - x)^{\beta/\sigma} = 0$$

and reveals

$$\beta = 2\,, \qquad \alpha = \frac{\lambda\,\sigma^2}{2\,\varkappa_0(2+\sigma)}\,.$$

Having integrated the equation related to $T$:

$$T(t) = \left[\sigma\,\lambda\,(c_0 - t)\right]^{-1/\sigma},$$

where $c_0 = \text{const}$, we find the function in question

$$u(x,t) = v(x)\,T(t) = \left(\frac{\alpha}{\sigma\,\lambda}\right)^{1/\sigma} \left(\frac{(x_1 - x)^2}{c_0 - t}\right)^{1/\sigma},$$

but minor changes are needed in complying it with the special boundary regime (19):

$$m = 1/\sigma, \quad c_0 = t_0, \quad u_0 = \left(\frac{x_1^2\alpha}{\sigma\,\lambda}\right)^{1/\sigma} = \left(\frac{x_1^2\sigma}{2\,\varkappa_0(\sigma+2)}\right)^{1/\sigma}$$

and, therefore,

$$x_1^2 = 2\,\varkappa_0(\sigma+2)\,u_0^\sigma/\sigma\,.$$

Thus, equation (15) with the special boundary regime (19) possesses the solution

$$u(x,t) = \begin{cases} u_0\left(\dfrac{1 - x/x_1}{\sqrt{t_0 - t}}\right)^{2/\sigma}, & 0 \le x \le x_1, \\ 0, & x > x_1, \end{cases} \tag{23}$$

where $x_1$ is the width of the region of the heat distribution.

As a matter of fact, the front of the temperature wave becomes immovable, since $x_1 = \text{const}$ is independent of $t$ and depends only on the parameters $\varkappa_0$, $\sigma$, $u_0$ of the problem concerned. Moreover, at the front the heat flow and temperature vanish for any $\sigma > 0$, while the partial derivative becomes $\partial u/\partial x = \infty$ for $\sigma > 2$ (at the front of the "travelling" wave $\partial u/\partial x = \infty$ for $\sigma > 1$).

A solution known as a "staying wave" exists during the interval of time $t < t_0$. This is stipulated by the special boundary regime (19) relating to the regimes with "breaking down".

For the heat conduction equation with a heat source depending on the temperature in accordance with the law

$$\frac{\partial u}{\partial t} = \frac{\partial}{\partial x}\left(\varkappa_0\,u^\sigma\,\frac{\partial u}{\partial x}\right) + q_0 u^\beta, \tag{24}$$

there exist both types of the aforementioned solutions which fall within the category of travelling waves for $\beta < \sigma + 1$ and the category of staying waves for $\beta = \sigma + 1$.

Numerical solutions of such problems cause some difficulties during the course of many methods in connection with nonlinearity and tendency to infinity of the partial derivatives at the front of the temperature way. It is hoped that the exact solutions obtained in such a way help motivate what is done and could serve in practical implementations as "goodness-of-fit" tests.

**4. A difference scheme. Newton's method.** We now proceed to constructing difference schemes for the quasilinear heat conduction equation.

For this, it seems unreasonable to employ explicit schemes with fastly varying ingredients $k(u)$, $c(u)$ and $f(u)$. The power functions of temperature reflect in full measure the difficulties involved in such a case. For any implicit scheme one possible stability condition

$$\frac{\tau}{h^2} \le \frac{1}{2} \frac{\min c(u)}{\max k(u)}$$

is connected with successive step refinement in $t$ to a considerable extent. Quite often, it depends on the values of $k$ and $c$ in a smaller number of nodal points. This supports the conclusion that explicit schemes are useless for our purposes. In an attempt to fill that gap, a considerable amount of effort has been expended in designing unconditionally stable implicit schemes.

In order to understand some things a little better, the governing equation is put together with the boundary conditions

$$(25) \qquad \frac{\partial \varphi(u)}{\partial t} = \frac{\partial^2 u}{\partial x^2}, \qquad 0 < x < 1,$$

$$u(x,0) = u_0(x), \quad u(0,t) = \mu_1(t), \quad u(1,t) = \mu_2(t).$$

A nonlinear difference scheme with respect to $y^{j+1}$

$$(26) \qquad \frac{\varphi(y^{j+1}) - \varphi(y^j)}{\tau} = y_{\bar{x}x}^{j+1},$$

$$x = x_i = ih, \qquad 0 < i < N, \qquad hN = 1,$$

may be employed in such a setting under the conditions $\varphi'(y) \ge c_1 > 0$ and $|\varphi''(y)| \le c_2$, providing its stability and convergence in the space $C$ with the rate $O(\tau + h^2)$. The proof of these assertions is somewhat lengthy and cumbersome and so it is omitted here.

The nonlinear equation

$$\varphi(y^{j+1}) - \tau\, y_{\bar{x}x}^{j+1} = \varphi(y^j)$$

is aimed at determining $y^{j+1}$ on every new layer by making several iterations of Newton's method

$$(27) \qquad \varphi(\overset{k}{y}) + \varphi'(\overset{k}{y})(\overset{k+1}{y} - \overset{k}{y}) - \tau\, y_{\bar{x}x}^{j+1} = \varphi(y^j).$$

Under the boundary conditions

$$(28) \qquad \overset{k+1}{y_0} = \mu_1(t_{j+1}), \qquad \overset{k+1}{y_N} = \mu_2(t_{j+1})$$

the elimination method is quite applicable in giving $\overset{k+1}{y}$. There is no doubt that its stability is ensured by the condition

$$\varphi'(\overset{k}{y}) \geq 0.$$

Indeed, this fact is an immediate implication of an alternative form of writing the governing equation

$$\frac{\tau}{h^2} \overset{k+1}{y}_{i-1} - \left(\varphi'(\overset{k}{y_i}) + \frac{2\tau}{h^2}\right) \overset{k+1}{y_i} + \frac{\tau}{h^2} \overset{k+1}{y}_{i+1} = \overset{k}{F_i},$$

$$i = 1, 2, \ldots, N-1,$$

where $\overset{k}{F_i} = \varphi(\overset{k}{y}) - \varphi(y) - \varphi'(\overset{k}{y})\overset{k}{y}$ and $y = y^j$.

The rate of convergence of iterations can be evaluated by means of the difference

$$\overset{k+1}{v_i} = \overset{k+1}{y_i} - \hat{y}_i, \qquad \hat{y}_i = y_i^{j+1}.$$

This can be done by inserting $\overset{k}{y_i} = \hat{y}_i + \overset{k}{v_i}$ and $\overset{k+1}{y_i} = \hat{y}_i + \overset{k+1}{v_i}$ in (27)

$$\begin{aligned} \varphi'(\overset{k}{y}) \overset{k+1}{v} - \tau \overset{k+1}{v}_{\bar{x}x} &= \varphi(y) - \varphi(\overset{k}{y}) + \varphi'(\overset{k}{y}) \overset{k}{v} + \tau \hat{y}_{\bar{x}x} \\ &= \varphi(\hat{y}) - \varphi(\overset{k}{y}) + \varphi'(\overset{k}{y})(\overset{k}{y} - \hat{y}) \end{aligned}$$

and taking into account the basic relation

$$\varphi(\hat{y}) = \varphi(\overset{k}{y}) + \varphi'(\overset{k}{y})(\hat{y} - \overset{k}{y}) + \tfrac{1}{2}\varphi''(\tilde{y})(\hat{y} - \overset{k}{y})^2$$

and its corollary

$$\varphi(\hat{y}) - \varphi(\overset{k}{y}) + \varphi'(\overset{k}{y})(\overset{k}{y} - \hat{y}) = \tfrac{1}{2}\varphi''(\tilde{y}) \overset{k}{v}{}^2,$$

where $\tilde{y} = \overset{k}{y} + \theta \overset{k}{v}$, $0 \leq \theta \leq 1$. The outcome of this is the equation

$$(29) \qquad \varphi'(\overset{k}{y}) \overset{k+1}{v} - \tau \overset{k+1}{v}_{\bar{x}x} = \tfrac{1}{2}\varphi''(\tilde{y}) \overset{k}{v}{}^2 = \overset{k}{F}, \quad x = ih, \quad 0 < i < N,$$

with the homogeneous boundary conditions

$$\overset{k+1}{v}_0 = 0\,, \qquad \overset{k+1}{v}_N = 0\,. \tag{30}$$

In this regard, the maximum principle states that

$$\|\overset{k+1}{v}\|_C \le 0.5\,\|\varphi''(\tilde{y})/\varphi'(\overset{k}{y})\|_C \cdot \|\overset{k}{v}\|_C^2 \le q\,\|\overset{k}{v}\|_C^2\,, \tag{31}$$

where $q = 0.5\,\|\varphi''(\tilde{y})/\varphi'(\overset{k}{y})\|_C \le 0.5\,\|\varphi''(\tilde{y})\|_C/c_1 \le 0.5\,c_2/c_1 = q_0$, since $\varphi'(y) \ge c_1 > 0$ and $|\varphi''(y)| \le c_2$.

Using this estimate behind, it is not difficult to establish that for the convergence of iterations in accordance with a quadratic law, it suffices to choose the initial approximation so as to satisfy the condition

$$\|\overset{0}{y} - \hat{y}\|_C < 2\,c_1/c_2\,. \tag{32}$$

The meaning of this is that we should have for the choice $\overset{0}{y} = y = y^j$

$$\tau\,\|y_t\|_C < 2\,c_1/c_2\,,$$

which is always valid for sufficiently small $\tau$.

In practical implementations Newton's method converges with any prescribed accuracy $\varepsilon$ only if

$$\varphi(y) = y^\alpha\,, \quad \alpha < 1\,, \quad \varphi'(y) = \alpha\,y^{\alpha-1}\,,$$

$$\varphi''(y) = -\alpha\,(1-\alpha)\,y^{\alpha-2}\,.$$

If $y \ge 0$, then $c_2 = \infty$ and unfortunately the preceding estimates are meaningless.

However, due to the maximum principle a solution of the boundary-value problem (29)–(30) is non-negative:

$$\overset{k+1}{v} = \overset{k+1}{y} - \hat{y} \le 0\,, \qquad \overset{k+1}{y} \le \hat{y}\,,$$

provided the condition $\varphi''(y) < 0$ holds. The preceding justifies that the iterations approach the solution from below. Because of this tendency, the first iteration is such that $\overset{1}{y} < \hat{y}$ if the initial approximation was taken so that $\overset{0}{y} < \hat{y}$. But it may happen that $\overset{1}{y} < 0$, thus terminating subsequent computations.

**5. Various implicit schemes for the quasilinear heat conduction equation.** Other ideas are connected with two types of purely implicit difference schemes (the forward ones with $\sigma = 1$) available for the simplest quasi-linear heat conduction equation

$$(33) \qquad \frac{\partial u}{\partial t} = \frac{\partial}{\partial x}\left(k(u)\,\frac{\partial u}{\partial x}\right) + f(u)\,, \quad 0 < x < 1\,, \quad 0 < t \le T\,,$$

$$u(x,0) = u_0(x)\,, \quad u(0,t) = u_1(t)\,, \quad u(1,t) = u_2(t)\,,$$

where $k(u) > 0$.

The structures of both schemes are well-characterized by

$$(34) \qquad \frac{\hat{y}_i - y_i}{\tau} = \frac{1}{h}\left[a_{i+1}(y)\,\frac{\hat{y}_{i+1} - \hat{y}_i}{h} - a_i(y)\,\frac{\hat{y}_i - \hat{y}_{i-1}}{h}\right] + f(y_i)$$

for the scheme a) and by

$$(35) \qquad \frac{\hat{y}_i - y_i}{\tau} = \frac{1}{h}\left[a_{i+1}(\hat{y})\,\frac{\hat{y}_{i+1} - \hat{y}_i}{h} - a_i(\hat{y})\,\frac{\hat{y}_i - \hat{y}_{i-1}}{h}\right] + f(\hat{y}_i)$$

for the scheme b), where $\hat{y}_i = y_i^{j+1}$, $y_i = y_i^j$, $a_i(v) = a(v_{i-1}, v_i)$. For example, we might agree to consider

$$(36) \qquad a_i(v) = 0.5\,\big(k(v_{i-1} + k(v_i)\big)\,,$$

$$(37) \qquad a_i(v) = k\left(\frac{v_{i-1} + v_i}{2}\right),$$

$$(38) \qquad a_i(v) = \frac{2\,k(v_{i-1})\,k(v_i)}{k(v_{i-1}) + k(v_i)}\,.$$

A greater gain in accuracy in connection with the temperature wave depends significantly on how well we calculate the coefficients $a_i(v)$. In the case where $k = k_0 u^\sigma$ is a power function of temperature, numerical experiments showed that formula (38) is useless and formula (36) is much more flexible than (37), so there is some reason to be concerned about this. Further comparison of schemes (34) and (35) should cause some difficulties. Both schemes are absolutely stable and have the same error of approximation $O(\tau + h^2)$. The scheme a) is linear with respect to the value of the function $y^{j+1}$ on the layer $t_{j+1}$ and so the value $y^{j+1}$ on every new layer

$t_{j+1}$ can be found, for example, by the elimination method in terms of the values of the function $y^j$ on the current layer $t_j$. Because the scheme is absolute stable, the well-founded choice of the step $\tau$ is stipulated by accuracy reasoning only. Unlike the preceding scheme, the scheme b) is nonlinear with respect to the value of the function $y^{j+1}$, so there is a real need for employing the iterative method. Still using its framework, the iteration process is governed by the rule

$$\frac{\overset{(s+1)}{y}_i - y_i}{\tau} = \frac{1}{h} \left[ a_{i+1}(\overset{(s)}{y}) \frac{\overset{(s+1)}{y}_{i+1} - \overset{(s+1)}{y}_i}{h} - a_i(\overset{(s)}{y}) \frac{\overset{(s+1)}{y}_i - \overset{(s+1)}{y}_{i-1}}{h} \right] + f(\overset{(s)}{y}_i) . \tag{39}$$

As a final result of such operations a revised scheme becomes linear with respect to the value of the function $\overset{s+1}{y}$ and only the initial approximation $\overset{0}{y}$ remains as yet unknown. One way of avoiding this obstacle is to accept $\overset{0}{y} = y^j$. We note in passing that most of the iterative methods converge in practical implementations for rather broader classes of coefficients $k$ and $f$ after two-three iterations performed. Even if the process in view is divergent, two iterations can result in improved accuracy of the describing scheme. In trying to adapt the iteration scheme (35), (39) the usual practice is connected with specifying the condition

$$\max_i \left| \overset{(s+1)}{y}_i - \overset{(s)}{y}_i \right| \le \varepsilon ,$$

where either the total number of iterations is known in advance or a desirable accuracy $\varepsilon$ is beforehand prescribed.

Let us stress here that the iteration scheme (35), (39) requires the double storage in comparison with the scheme a). This is caused by the necessity of calculating and saving the values of the function $\overset{s+1}{y}$ in terms of the values of two functions $\overset{s}{y}$ and $y$.

One more difference lies in the fact that the transition from the value $\overset{j}{y}$ to the value $\overset{j+1}{y}$ is possible only after several iterations made in scheme (35), (39), while the value $\overset{j+1}{y}$ immediately follows in the algorithm of the scheme a).

But it would be erroneous to think in light of the same properties of the indicated schemes such as their absolute stability and the same order of approximation that the scheme a) offers in such matters more advantages than the scheme b).

Practical implementations showed that the computational procedures of the scheme b) can work with a larger step in time, thus reducing essentially the total volume of computations and the time complexity despite the extra iterations required in this connection.

However, a preference relation between such schemes is some consensus of opinion. The reader can encounter in the theory and practice various schemes generating approximations of order 2 in time and space:

$$y_t = \frac{1}{2} \left[ (a(\hat{y})\, \hat{y}_{\bar{x}})_x + (a(y)\, y_{\bar{x}})_x \right] + f\Big(\frac{\hat{y} + y}{2}\Big) .$$

As can readily be observed, they are not monotone, thus causing some "ripple". This obstacle can be avoided by refining some suitable grids in time. When solving equations of the form (13) with a weak quasilinearity for the coefficients $k = k(x,t)$, $f = f(u)$ and $c = c(x,t)$, common practice involves **predictor-corrector** schemes of accuracy $O(\tau^2 + h^2)$. Such a scheme for the choice $c = k = 1$, $f = f(u)$ is available now:

(40)
$$\frac{\bar{y} - y}{0.5\,\tau} = \bar{y}_{\bar{x}x} + f(y) , \qquad \bar{y} = y(t_{j+1/2}) ,$$
$$\frac{\breve{y} - y}{\tau} = \frac{1}{2} \left( \bar{y}_{\bar{x}x} + y_{\bar{x}x} \right) + f(y) ,$$

whose composition is depicted in Figure 1.

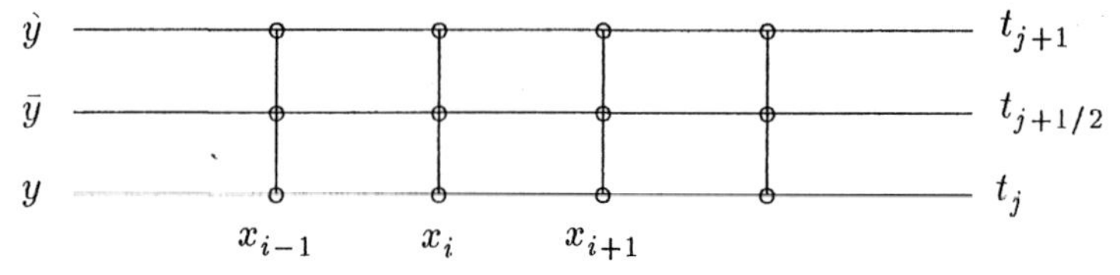


**Figure 1.**

We omit here theoretical investigations of the preceding schemes relying on cumbersome calculations and leading to unsatisfactory and rough estimates that can result in wrong reasoning. Such difficulties are, generally speaking, typical for nonlinear problems in many branches of science, engineering and technology. In tackling nonlinear problems some preliminary tests may be of help in verifying the quality of numerical methods. The traditional way of covering this is to compare numerical solutions of a simple specimen problem with known analytical solutions of the same problem.

It is worth noting here that Newton's method is quite applicable for solving problem (35) in addition to the well-established method of iterations.

**6. Calculations of the temperature waves.** Of special interest is the case where the coefficient $k(u)$ is a function of temperature such that

$$\frac{\partial u}{\partial t} = \frac{\partial}{\partial x}\left(\varkappa_0\, u^\sigma\, \frac{\partial u}{\partial x}\right). \tag{41}$$

As we have mentioned above, the process of heat conducting emerged in that case with a finite velocity and the derivative $\partial u/\partial x$ tends to $\infty$ behind the front for $\sigma > 1$.

The temperature waves can be found through the use of the scheme b) relating to "continuous through execution" ones. No fixation of the front applies here. Under the guidelines of the preceding section with regard to problem (15)–(17) having the exact solution given by formula (18), the calculations permit us to discover that almost everywhere except several near-front nodes the approximate solution deviates very slightly from the exact one, not exceeding 0.002 for $\varkappa_0 = 0.5$, $\sigma = 2$, $D = 5$ and the total number of nodal points $N = 50$. In so doing the number of the necessary iterations is no greater than 3 and $t < 0.2$. When the temperature wave is moving from the left to the right along the zero temperature background, more and more grid intervals are captured in a step-by-step fashion in the process of numerical solutions in a number of different ways in connection with possible computations of the coefficient $a_i(y)$.

Apparently, the main idea behind this approach needs certain clarification. For example, formula (38) necessitates imposing a nonzero background temperature prior to the front. In spite of this fact, there are some delays in introducing new intervals, thus causing large deviations of a solution in a vicinity of the front. Formula (37) is useless for very large values of the index $\sigma$ ($\sigma \geq 20$). Formula (36) has the best accuracy and reproduces rather accurately without concern of the background temperature.

**7. The Stephan problem (problem of the phase transition).** Subsequent considerations include two phases with the coefficients of heat conductivity $k_1(u)$, $k_2(u)$ and of heat capacity $c_1(u)$, $c_2(u)$, in either of which it is supposed that the temperature satisfies the equation

$$c_s(u)\,\frac{\partial u}{\partial t} = \frac{\partial}{\partial x}\left(k_s(u)\,\frac{\partial u}{\partial x}\right), \qquad s = 1,2. \tag{42}$$

At the same time, on the boundary of these phases the temperature is constant and coincides with the temperature of the phase transition: $u(x,t) = u^*$. The velocity of the boundary $\xi$ of the phase transition is subject to the equation

$$k_1\,\frac{\partial u}{\partial x}\bigg|_{x=\xi+0} - k_2\,\frac{\partial u}{\partial x}\bigg|_{x=\xi-0} = -\lambda\,\frac{\partial \xi}{\partial t}\,,$$

if $u < u^*$ in the first phase and $u > u^*$ in the second one.

With the boundary condition for the phase transition in view, we rewrite equation (42) by means of the $\delta$-function as

$$\big(c(u) + \lambda\,\delta\,(u - u^*)\big)\,\frac{\partial u}{\partial t} = \frac{\partial}{\partial x}\left(k(u)\,\frac{\partial u}{\partial x}\right),$$

$$c(u) = \begin{cases} c_1(u)\,, & u < u^*\,, \\ c_2(u)\,, & u > u^*\,, \end{cases} \qquad k(u) = \begin{cases} k_1(u)\,, & u < u^*\,, \\ k_2(u)\,, & u > u^*\,. \end{cases}$$

The method of smoothing is available for solving the Stephan problem. As a matter of experience, this amounts to replacing the $\delta$-function by a nonzero $\delta$-type function $\delta(u - u^*, \Delta)$, not equal to zero only on the interval $(u^* - \Delta, u^* + \Delta)$ and must satisfy the normalization condition

$$\int\limits_{u^*-\Delta}^{u^*+\Delta} \delta(u - u^*, \Delta)\,du = 1\,.$$

The quasilinear equation

$$\tilde{c}(u)\,\frac{\partial u}{\partial t} = \frac{\partial}{\partial x}\left(\tilde{k}(u)\,\frac{\partial u}{\partial x}\right)$$

arises in the process of smoothing the functions $k_1(u)$, $k_2(u)$, $c_1(u)$ and $c_2(u)$ on the interval $(u^* - \Delta, u^* + \Delta)$. All the schemes we have mentioned above find a wide range of applications for its numerical solution. However, there are other numerical methods for solving the Stephan problem concerned.

## 8.2 CONSERVATIVE DIFFERENCE SCHEMES OF NONSTATIONARY GAS DYNAMICS

**1. One-dimensional equations of nonstationary gas dynamics in Lagrangian variables.** Plenty of problems arising in mechanics and physics such as the harnessing of nuclear energy, the creation of nuclear reactors, aircrafr (planes and space-vehicles) design, the dynamics of space flights, plasma physics (governed thermo-nuclear synthesis) led to equations of gas dynamics that are, generally speaking, nonlinear and can be solved by the universal difference methods.

In spite of the fact that problems in gas dynamics began to spread to more and more branches of science, engineering and technology as they gradually took on an important place in real-life situations, for the time being there are no rigorous mathematical results regarding convergence of schemes even in the simplest and typical situations. The desirable quality of schemes are verified with the aid of linear models in the acoustic approximation by applying numerical tests for specimen problems, whose analytical solutions are known to the users in explicit form.

As a rule, equations of gas dynamics are discontinuous. From a physical point of view it is fairly common to distinguish weak discontinuities relating to "cutting waves" and strong discontinuities relating to "shock waves". For these reasons successive grid refinement can be made with caution when the accurate account of accuracy of numerical methods is performed.

In this section we initiate the design of difference methods for numerical solutions of the simplest problems in gas dynamics. Of our initial concern is the problem about one-dimensional nonstationary gas flow in a plane with the following ingredients: velocity $v$, density $\rho$, temperature $T$, pressure $p$, internal energy $\varepsilon$.

In preparation for this, the equations of gas dynamics will reproduce the conservation laws of impuls, mass and energy that can be written in a number of different ways with respect to Eulerian $(x, t)$ or Lagrangian $(s, t)$ variables, where $x$ is the coordinate of a particle and $s$ is the initial coordinate of a particle or the quantity

$$s = \int_0^x \rho(\xi, 0)\, d\xi\,,$$

that is, the value of mass in the volume $0 \le \xi \le x$. The usual practice involves the system of gas dynamics equations in Lagrangian variables $(s, t)$:

(1) $$\frac{\partial v}{\partial t} = -\frac{\partial p}{\partial s} \quad \text{(the law of the impuls conservation)},$$

(2) $$\frac{dx}{dt} = v \,,$$

(3) $$\frac{1}{\rho} = \frac{\partial x}{\partial s} \quad \text{(the law of the mass conservation)},$$

(4) $$\frac{\partial}{\partial t}\Big(\varepsilon + \frac{v^2}{2}\Big) = -\frac{\partial}{\partial s}(p\,v) - \frac{\partial w}{\partial s} \quad \text{(the law of the energy conservation)},$$

(5) $$p = p(\rho, T)\,, \quad \varepsilon = \varepsilon(\rho, T) \quad \text{(the state equation)},$$

where $w$ is the heat flow.

A combination of the second and third equations we have mentioned above gives

(6) $$\frac{\partial}{\partial t}\Big(\frac{1}{\rho}\Big) = \frac{\partial v}{\partial s}$$

by observing that equation (2) can be excluded from the governing system through the possible separate integration.

The expression for the heat flow

(7) $$w = -\varkappa(\rho, T)\,\rho\,\frac{\partial T}{\partial s}\,,$$

where $\varkappa = \varkappa(\rho, T)$ is a coefficient of heat conductivity, is intended to complete the above system of equations. It should be noted here that $\varkappa$ can usually be expressed through the power function of $T$ and $\rho$.

The functions $p(\rho, T)$, $\varepsilon(\rho, T)$, $\varkappa(\rho, T)$ for this system of equations must be given.

For example, the state equations of the ideal gas are of the form $p = R\,\rho\,T$ and $\varepsilon = \varepsilon(T)$. The readers can encounter $\varepsilon = c_0 T$, where $R$ and $c_0$ are constants, $R/c_0 = \gamma - 1$, $\gamma$ is constant, so that

(8) $$\varepsilon = p/\big((\gamma - 1)\,\rho\big)\,.$$

In this context, two limiting cases of interest are as follows:

a) The adiabatic flow of the ideal gas when $w = 0$, that is, $\varkappa = 0$.

Because of this, equations of gas dynamics (1), (6), (4) for the adiabatic flow of the ideal gas can be represented by

$$\frac{\partial v}{\partial t} = \frac{\partial p}{\partial s}, \qquad \frac{\partial}{\partial t}\left(\frac{1}{\rho}\right) = \frac{\partial v}{\partial s}, \tag{9}$$

$$\frac{\partial}{\partial t}\left(\varepsilon + \frac{v^2}{2}\right) = -\frac{\partial}{\partial s}(p\,v), \tag{10}$$

which will be put together with equation (8)

$$\varepsilon = \frac{p}{(\gamma - 1)\,\rho}. \tag{11}$$

Thus, the resultant system of equations comprises 4 equations with respect to four unknowns $v$, $p$, $\rho$ and $\varepsilon$.

We will use below the volume $\eta = 1/\rho$ instead of density $\rho$. In such a setting the preceding equations can be represented by

$$\frac{\partial \eta}{\partial t} = \frac{\partial v}{\partial s}, \tag{12}$$

$$p\,\eta = (\gamma - 1)\,\varepsilon. \tag{13}$$

Equation (10) capable of describing the tootal energy can be replaced by one of the newly formed equations

$$\frac{\partial \varepsilon}{\partial t} = -p\,\frac{\partial v}{\partial s}, \tag{14}$$

$$\frac{\partial \varepsilon}{\partial t} = -p\,\frac{\partial \eta}{\partial t}. \tag{15}$$

Indeed, taking into account the first equation (9) and (12), we obtain

$$0 = \frac{\partial}{\partial t}\left(\varepsilon + \frac{v^2}{2}\right) + \frac{\partial}{\partial s}(p\,v) = \frac{\partial \varepsilon}{\partial t} + \left(v\,\frac{\partial v}{\partial t} + v\,\frac{\partial p}{\partial s}\right) + p\,\frac{\partial v}{\partial s}$$

$$= \frac{\partial \varepsilon}{\partial t} + p\,\frac{\partial v}{\partial s} = \frac{\partial \varepsilon}{\partial t} + p\,\frac{\partial \eta}{\partial t}.$$

b) The isotermic flow of the ideal gas when the temperature of gas $T = \text{const}$ and the equation of energy is missing. The condition $T = \text{const}$

corresponds to the case when $\varkappa \to \infty$. The system of equations of gas dynamics for the isotermic flow of the ideal gas comprises the equations

$$(16) \qquad \frac{\partial v}{\partial t} = -\frac{\partial p}{\partial s}\,, \qquad \frac{\partial}{\partial t}\Big(\frac{1}{\rho}\Big) = \frac{\partial v}{\partial s}\,, \qquad p = c^2 \rho\,,$$

where $c = \text{const} > 0$ is the velocity of sound, or

$$(17) \qquad \frac{\partial v}{\partial t} = -\frac{\partial p}{\partial s}\,, \qquad \frac{\partial \eta}{\partial t} = \frac{\partial v}{\partial s}\,, \qquad p\,\eta = c^2\,.$$

which are consistent with the adiabatic case indicated above. In the sequel our exposition is mostly based on more a detailed exploration of three equations (9), (10) and (11) capable of describing the gas dynamics.

The complete posing of this necessitates specifying the boundary and initial conditions in addition to equations (9)–(10). Knowing

$$(18) \qquad v(x,0)\,, \quad \rho(x,0)\,, \quad p(x,0)\,,$$

we may attempt, for example, the boundary conditions in the form

$$(19) \qquad p(0,t) = p_0(t) \quad \text{for} \quad s = 0\,, \qquad p(M,t) = p_1(t) \quad \text{for} \quad s = M\,,$$

or

$$(19') \qquad v(0,t) = v_0(t) \quad \text{for} \quad s = 0\,, \qquad p(M,t) = p_1(t) \quad \text{for} \quad s = M\,.$$

Summarizing, it is required to construct difference schemes for equations of gas dynamics (9)–(10) in the complex closed domain $\{0 \le s \le M, t \ge 0\}$.

**2. Equations with psevdoviscosity.** In practical implementations there is a real need for forming homogeneous and conservative difference equations of gas dynamics. The meaning of homogeneity here is that difference equations are written at all the nodes of the grid in just the same way regardless of the possible continuity or discontinuity of a solution so that subsequent calculations should be carried out by the same ensuing formulas. Homogeneous schemes or "through execution" schemes of gas dynamics contain the extra members with psevdoviscosity, a key role of which is to spread the front of shock waves over several intervals of the grid. From a formal point of view, the viscosity $\omega$ arises as the additional member to the pressure $p$, so that equations (9)–(10) contain for now instead of $p$ the sum

$$g = p + \omega\,,$$

where the "viscous" pressure $\omega = \omega(\rho, v'_s, h)$ depends on $\rho$, $v'_s$ and the step $h$. The reader can encounter two types of viscosity:

a) a linear viscosity

$$\omega = -\frac{\nu_0\, h\, \rho}{2}\left(\frac{\partial v}{\partial s} - \left|\frac{\partial v}{\partial s}\right|\right), \tag{20}$$

b) a quadratic viscosity or Neuman's viscosity

$$\omega = \frac{\nu_0}{2}\, h^2\, \rho\, \frac{\partial v}{\partial s}\left(\frac{\partial v}{\partial s} - \left|\frac{\partial v}{\partial s}\right|\right) \tag{21}$$

where $\nu_0$ is the coefficient of viscosity. It follows from the foregoing that the function $\omega = 0$ for $\partial v/\partial s > 0$ and $\omega \neq 0$ for $\partial v/\partial s < 0$, that is, only within the zone of the shock wave.

Thus, the psevdoviscosity may emerge only within the zone of the shock wave. The accepted view is to use

$$\omega = -\frac{\nu}{\eta}\,\frac{\partial v}{\partial s}$$

or

$$\omega = \frac{\nu}{\eta}\left(\frac{\partial v}{\partial s}\right)^2,$$

assuming that the coefficient of viscosity depends on the sign of the partial derivative $\partial v/\partial s$, so that $\nu \equiv 0$ for $\partial v/\partial s > 0$.

**3. Conservative homogeneous schemes.** The presence of psevdoviscosity makes it possible to design homogeneous difference schemes or "through execution" ones, permitting us to reveal the gas distribution caused by shock waves. Since the equations of gas dynamics express the conservation laws of impulse, mass and energy, the scientists wish to have at their disposal conservative difference schemes for which the difference equations involved reproduce the analogs of these conservation laws on the grid.

Equations of gas dynamics in integral form are aimed at designing conservative difference schemes by means of the integro-interpolation method:

$$\oint (v\, ds - p\, dt) = 0\,, \tag{22}$$

$$\oint (\eta\, ds + v\, dt) = 0\,, \tag{23}$$

$$\oint ((\varepsilon + 0.5\, v^2)\, ds - p\, v\, dt) = 0\,, \tag{24}$$

where integration is accomplished along a closed curve in the plane $(s,t)$.

To make our exposition more transparent, we introduce the grids

$$\bar{\omega}_h = \{s_i = ih\,,\ i = 0, 1, \dots, N\,,\ hN = M\}\,,$$

$$\bar{\omega}_\tau = \{t_j = j\tau\,,\ j = 0, 1, \dots, j_0\,,\ j_0\tau = t_0\}\,,$$

$$\bar{\omega}_{h\tau} = \bar{\omega}_h \times \bar{\omega}_\tau\,,$$

retaining the notations $v$, $\eta$, $p$, $\varepsilon$ with respect to difference equations and regarding the function $v$ to integer points $s = s_i$ on the grid $\omega_h$ and $p$, $\eta$, $\varepsilon$ – to half-integer points $s = s_{i+1/2}$ on the same grid.

Other ideas are connected with setting equation (22) in the rectangle $\{s_{i-1/2} \le s \le s_{i+1/2},\ t_j \le t \le t_{j+1}\}$:

$$\int\limits_{s_{i-1/2}}^{s_{i+1/2}} (v^{j+1} - v^j)\, ds + \int\limits_{t_j}^{t_{j+1}} (p_{j+1/2} - p_{i-1/2})\, dt = 0$$

and writing equations (23)–(24) in another rectangle $\{s_i \le s \le s_{i+1},\ t_j \le t \le t_{j+1}\}$:

$$\int\limits_{s_i}^{s_{i+1}} (\eta^{j+1} - \eta^j)\, ds - \int\limits_{t_j}^{t_{j+1}} (v_{i+1} - v_i)\, dt = 0\,,$$

$$\int\limits_{s_i}^{s_{i+1}} \left[(\varepsilon + 0.5\, v^2)^{j+1} - (\varepsilon + 0.5\, v^2)^j\right] ds + \int\limits_{t_j}^{t_{j+1}} \left[((p\,v)_{i+1} - (p\,v)_i)\right] dt = 0\,.$$

The integrals built into these identities are replaced by the newly formed expressions

$$\int\limits_{t_j}^{t_{j+1}} p\, dt \approx p^{(\sigma_1)}\tau\,, \qquad \int\limits_{t_j}^{t_{j+1}} v\, dt \approx v^{(\sigma_2)}\tau\,, \qquad \int\limits_{t_j}^{t_{j+1}} (p\,v)_i\, dt = p_{*i}^{(\sigma_3)}\, v_i^{(\sigma_4)}\tau\,,$$

where

$$f^{(\sigma_\alpha)} = \sigma_\alpha f^{j+1} + (1 - \sigma_\alpha) f^j\,, \quad \sigma_\alpha \quad \text{is arbitrary parameter}\,,$$

$$p_{*i} = 0.5\,(p_{i-1/2} + p_{i+1/2})\,, \qquad \alpha = 1, 2, 3, 4\,,$$

and

$$\int_{s_{i-1/2}}^{s_{i+1/2}} v\,ds \approx v_i\,h\,, \qquad \int_{s_i}^{s_{i+1}} \eta\,ds \approx \eta_{i+1/2}\,h\,, \qquad \text{etc.}$$

As a final result we get the difference scheme

$$\frac{v_i^{j+1}-v_i^j}{\tau} + \left(\frac{p_{i+1/2}-p_{i-1/2}}{h}\right)^{(\sigma_1)} = 0\,, \tag{25}$$

$$\frac{\eta_{i+1/2}^{j+1}-\eta_{i+1/2}^{j}}{\tau} = \left(\frac{v_{i+1}-v_i}{h}\right)^{(\sigma_2)}, \tag{26}$$

$$\frac{1}{\tau}\left[\left(\varepsilon_{i+1/2}+\frac{v_i^2+v_{i+1}^2}{4}\right)^{j+1} - \left(\varepsilon_{i+1/2}+\frac{v_i^2+v_{i+1}^2}{4}\right)^{j}\right] = \frac{p_{*i+1}^{(\sigma_3)}\,v_{i+1}^{(\sigma_4)} - p_{*i}^{(\sigma_3)}\,v_i^{(\sigma_4)}}{h}\,, \tag{27}$$

which falls within the category of conservative schemes for any admissible values of parameters $\sigma_1, \sigma_2, \sigma_3, \sigma_4$. In particular, for $\sigma_1 = 0$, $\sigma_2 = 1$, $\sigma_3 = 1$, $\sigma_4 = 1$ we constitute a system of difference equations, whose solution is found successively by the explicit formulas: at the first stage – $v_i^{j+1}$, at the second stage – $\eta_{i+1/2}^{j+1}$ and at the third stage – $p_{i+1/2}$, $i = 0, 1, \ldots, N-1$ by the elimination method from the energy equation and the state equation $p\eta = (\gamma-1)\varepsilon$ being supplemented with the suitably chosen boundary conditions for $i = 0$ and $i = N-1$. For example, we might agree to impose conditions (19).

In this context, it should be noted that conservative difference schemes may be good enough for the equation of the total energy, but approximate poorly the equation of the internal energy (14)

$$\frac{\partial \varepsilon}{\partial t} = -p\,\frac{\partial v}{\partial s}\,.$$

This can result in improper choices of computational procedures in giving the temperature. The lack of energetic balances cannot be avoided by refining the grid in a spatial variable $s$. A presence of energetic disbalances in a scheme can be interpreted as a presence of energetic sources of a purely difference nature connected with some "lack of agreement" between separate difference equations of a scheme being inconsistent each to other.

The emerging disbalances are characterized by a typical solution with a wide range of values. For example, they are sufficiently small on smooth functions, but could grow on solutions varying fastly in time and space prior to the total energy of the system concerned.

**4. Fully conservative schemes.** Other ideas are connected with successive use of conservation laws and more detailed balances of the kinetic and internal energy.

All the schemes with these properties are called **fully conservative schemes**. As a matter of fact, the requirement of the full conservatism is equivalent to being approximated of both equations (14) and (15) in addition to the usual requirements of approximation:

$$\frac{\partial \varepsilon}{\partial t} = -p\,\frac{\partial v}{\partial s}\,, \qquad \frac{\partial \varepsilon}{\partial t} = -p\,\frac{\partial \eta}{\partial t}\,.$$

Before going further, it will be sensible to introduce more compact notations

$$\bar{p}_i = p_{i+1/2}\,, \quad \bar{\eta}_i = \eta_{i+1/2}\,, \quad \bar{\varepsilon}_i = \varepsilon_{i+1/2}\,, \quad \bar{p} = \bar{p}_i^j\,, \quad v = v_i^j\,, \quad \text{etc},$$

$$\frac{1}{h}\,(p_{i+1/2} - p_{i-1/2}) = \bar{p}_{\bar{s}}\,, \qquad \frac{v_{i+1} - v_i}{h} = v_s\,.$$

Where there is no danger of confusion, we will omit the symbol "bar" over $p$, $\eta$, $\varepsilon$. Within these notations, equations (25)–(26) can be reduced to

$$v_t = -p_{\bar{s}}^{(\sigma_1)}\,, \qquad \eta_t = v_s^{(\sigma_2)}\,. \tag{28}$$

Instead of (27), let us consider the scheme generating an approximation to equation (14) capable of describing the law of the internal energy

$$\varepsilon_t = -p^{(\sigma_3)}\,v_s^{(\sigma_4)}\,. \tag{29}$$

In this connection there arises a four-parameter family of schemes, from which a fully conservative scheme needs to be selected through the approximations to equations (15) and (10) by appeal to scheme (28)–(29).

We will use the obvious relationship

$$f^{(\beta)} = f^{(\alpha)} + \tau\,(\beta - \alpha)\,f_t\,, \tag{30}$$

where $\alpha$ and $\beta$ are arbitrary numbers and $f^{(\alpha)} = \alpha\hat{f} + (1-\alpha)f$. The quantity of interest $\varepsilon_t$ can be discovered from

$$\varepsilon_t = -p^{(\sigma_3)}\,\eta_t + \delta_1 E\,, \quad \delta_1 E = -\tau\,(\sigma_4 - \sigma_2)\,p^{(\sigma_3)}\,v_{st}\,, \tag{31}$$

which, in turn, can be obtained from the equations $\eta_t = v_s^{(\sigma_2)}$ and $v_s^{(\sigma_4)} = v_s^{(\sigma_2)} - \tau(\sigma_4 - \sigma_2)\, v_{st} = \eta_t + \tau(\sigma_4 - \sigma_2) v_{st}$. From here it seems clear that (31) is consistent with equation (15) only for $\sigma_4 = \sigma_2$.

At the next stage scheme (28)–(29) is obliged to be conservative. This can always be done by multiplying the equation $v_t = -p_{\bar{s}}^{(\sigma_1)}$ by $v^{(0.5)} = 0.5(v + \hat{v})$, adding the resulting equation

$$\tfrac{1}{2}\,(v^2)_t = -v^{(0.5)}\, p_{\bar{s}}^{(\sigma_1)} \tag{32}$$

to equation (29):

$$(\varepsilon + 0.5\, v^2)_t = -p^{(\sigma_3)}\, v_s^{(\sigma_4)} - v^{(0.5)}\, p_{\bar{s}}^{(\sigma_1)} \tag{33}$$

and rearranging the right-hand side of (33) by means of formula (30) as

$$\begin{aligned} p^{(\sigma_3)}\, v_s^{(\sigma_4)} + v^{(0.5)}\, p_{\bar{s}}^{(\sigma_1)} &= \left(p^{(\sigma_1)} + \tau\,(\sigma_3 - \sigma_1)\, p_t\right) \\ &\quad \times \left(v_s^{(0.5)} + \tau\,(\sigma_4 - 0.5)\, v_{st}\right) + v^{(0.5)}\, p_{\bar{s}}^{(\sigma_1)} \\ &= \left(p_{(-1)}^{(\sigma_1)}\, v^{(0.5)}\right)_s + \delta_2\, E\,, \end{aligned} \tag{34}$$

where

$$p_{(-1)} = \bar{p}_{i-1} = p_{i-1/2}\,,$$

$$\begin{aligned} \delta_2\, E &= \tau\,(\sigma_3 - \sigma_1)\, v_s^{(0.5)}\, p_t + \tau\,(\sigma_4 - 0.5)\, p^{(\sigma_1)}\, v_{st} \\ &\quad + \tau^2 (\sigma_3 - \sigma_1)\,(\sigma_4 - 0.5) p_t\, v_{st}\,. \end{aligned}$$

The outcome of rearranging equation (33) is

$$(\varepsilon + 0.5\, v^2)_t = -\left(p_{(-1)}^{(\sigma_1)}\, v^{(0.5)}\right)_s - \delta_2\, E\,. \tag{35}$$

In the preceding the quantity $\delta_2 E$ means the disbalance of the internal energy. By equating $\delta_2\, E$ to 0 for any $p$ and $v$ we find that $\sigma_3 = \sigma_1$ and $\sigma_4 = 0.5$ and, hence, $\sigma_2 = 0.5$.

Thus, under such an approach a one-parameter family of fully conservative schemes is given by

$$v_t = -p_{\bar{s}}^{(\sigma_1)}\,, \quad \eta_t = v_s^{(0.5)}\,, \quad \varepsilon_t = -p^{(\sigma_1)}\, v_s^{(0.5)}\,, \tag{36}$$

within which the third equation can be replaced by one of the following equations:

$$\varepsilon_t = -p^{(\sigma_1)}\,\eta_t\,, \tag{37}$$

$$(\varepsilon + 0.5\,v^2)_t = -\big(p^{(\sigma_1)}_{(-1)}\,v^{(0.5)}\big)_s\,. \tag{38}$$

Observe that instead of the latter equation there seem to be at least two alternatives

$$(\varepsilon + 0.5\,v^2_{(+1)})_t = -\big(p^{(\sigma_1)}\,v^{(0.5)}\big)_s\,, \tag{39}$$

$$\big(\varepsilon + (v^2 + v^2_{(+1)})/4\big)_t = -\big(p^{(\sigma_1)}_*\,v^{(0.5)}\big)_s\,. \tag{40}$$

Equation (39) can be derived from the foregoing by involving equation (32) at the $(i+1)$th node

$$0.5\,(v^2_{i+1})_t = -v^{(0.5)}_{i+1}\,p^{(\sigma_1)}_{s,i} \quad \text{or} \quad 0.5\,(v^2_{(+1)})_t = -v^{(0.5)}_{(+1)}\,p^{(\sigma_1)}_s. \tag{41}$$

Combination of (38) and (39) gives immediately (40).

Further comparison of (40) with (27) shows that a new family of fully conservative schemes is contained in family (25)–(27) of describing conservative schemes with four parameters as a result of employing the integro-interpolation method.

Another conclusion can be drawn from the preceding that for any $\sigma_1$ scheme (36) generates an approximation of $O(\tau + h^2)$ and for $\sigma_1 = 0.5$ it provides an approximation of $O(\tau^2 + h^2)$, that is, only one scheme of the form (36) can guarantee a second-order approximation in $\tau$:

$$v_t = -p^{(0.5)}_{\bar{s}}\,, \quad \eta_t = v^{(0.5)}_s\,, \quad \varepsilon_t = -p^{(0.5)}\,v^{(0.5)}_s\,. \tag{42}$$

With the psevdoviscosity in view, we replace everywhere in formulas (36) the pressure $p$ by the approved rule $g = p + \omega$, leaving us with

$$v_t = -g^{(\sigma_1)}_{\bar{s}}\,, \quad \eta_t = v^{(0.5)}_s\,, \quad \varepsilon_t = -g^{(\sigma_1)}\,v^{(0.5)}_s\,, \quad g = p + \omega\,. \tag{43}$$

For the ideal gas and the linear viscosity, this amounts to

$$p\,\eta = (\gamma - 1)\,\varepsilon\,, \qquad \omega = -\frac{\nu}{\eta}\,v_s\,. \tag{44}$$

Putting these together with the supplementary boundary conditions for $i = 0$ and $i = N$, say with the values $p_0^j$, $p_N^j$ of pressure, we must write down the equation of the motion for $v_i^j$ not only at the inner nodes, but also on the boundary for $i = 0$ and $i = N$:

$$\frac{v_0^{j+1} - v_0^j}{\tau} = -\left(\frac{p_{1/2} - p_0}{0.5\,h}\right)^{(\sigma_1)}, \qquad \frac{v_N^{j+1} - v_N^j}{\tau} = -\left(\frac{p_N - p_{N-1/2}}{0.5\,h}\right)^{(\sigma_1)},$$

giving $v_0^{j+1}$ and $v_N^{j+1}$. The remaining quantities $\eta$, $\varepsilon$, $p$ are sought only at the inner half-integer points $s_{1/2}, s_{3/2}, \ldots, s_{N-1/2}$.

It seems worthwhile giving one possible example of nonconservative schemes. A "cross" scheme was very popular and much applicable in recent decades. Any scheme of this structure can be written on a "chess" grid by regarding $\varepsilon$, $p$, $\eta$ to half-integer nodes $(s_{i+1/2}, t_{j+1/2})$ and $v$, $x$ to integer nodes $(s_i, t_j)$ of the grid at hand.

Within the notations $p_{i+1/2}^{j+1/2} = \bar{p}_i^j = \bar{p}$ and $\eta_{i+1/2}^{j+1/2} = \bar{\eta}_i^{j+1} = \hat{\bar{\eta}}$, etc., the "cross" scheme

$$\frac{v_i^{j+1} - v_i^j}{\tau} = -\frac{p_{i+1/2}^{j+1/2} - p_{i-1/2}^{j+1/2}}{h},$$

$$\frac{\eta_{i+1/2}^{j+3/2} - \eta_{i+1/2}^{j-1/2}}{\tau} = \frac{v_{i+1}^{j+1} - v_i^{j+1}}{h},$$

$$\frac{\varepsilon_{i+1/2}^{j+3/2} - \varepsilon_{i+1/2}^{j+1/2}}{\tau} = -p_{i+1/2}^{j+3/2}\,\frac{v_{i+1}^{j+1} - v_i^{j+1}}{h}$$

admits an alternative form

$$v_t = -\bar{p}_{\bar{s}}\,, \qquad \bar{\eta}_t = \hat{v}_s\,, \qquad \bar{\varepsilon}_t = -\hat{\bar{p}}\,\hat{v}_s\,, \tag{45}$$

showing the new members to be sensible ones. In this line the values on every new layer is found by the explicit formulas. No wishing to load the book down with full details on this point, we cite only a final result after multiplying $v_t = -\bar{p}_{\bar{s}}$ by $v^{(0.5)}$ and repeating the preceding manipulations in such a setting:

$$\left(\bar{\varepsilon} + 0.5\,v^2\right)_t = -\left(\bar{p}_{(-1)}\,v^{(0.5)}\right)_s - \delta E\,,$$

where $\delta E = \tau \bar{p}_t v_s^{(0.5)} + 0.5 \tau p v_{st} + 0.5 \tau^2 \bar{p}_t v_{st}$, that is, scheme (45) is not conservative, thus causing some limitations in practical implementations.

**5. Numerical solution of difference equations by Newton's method.** As can readily be observed, the system of nonlinear equations capable of specifying the values $v^{j+1}$, $g^{j+1}$ and $\eta^{j+1}$ on every new layer will be solved by making several iterations of Newton's method. This can be done by reducing equations (43)–(44) to the following ones:

$$\hat{v} + \sigma_1 \tau \hat{g}_{\bar{s}} = v - (1 - \sigma_1)\, \tau\, g_{\bar{s}} \,,$$

$$\hat{\eta} - 0.5\, \tau\, \hat{v}_s = \eta + 0.5\, \tau\, v_s \,,$$

$$\hat{\varepsilon} + \sigma_1 \hat{g}\, (\hat{\eta} - \eta) + (1 - \sigma_1)\, g\, \hat{\eta} = \varepsilon + (1 - \sigma_1)\, g\, \eta \,,$$

$$\hat{g}\, \hat{\eta} - \hat{\varepsilon}\, (\gamma - 1) + \nu\, \hat{v}_s = 0 \,.$$

After that, applying Newton's method yields

$$\text{(46)} \qquad \begin{aligned} &\Delta \overset{k+1}{v} + \sigma_1 \tau\, \Delta \overset{k+1}{g}_{\bar{s}} = \overset{k}{f}_1 \,, \\ &\Delta \overset{k+1}{\eta} - 0.5\, \tau\, \Delta \overset{k+1}{v} = \overset{k}{f}_2 \,, \end{aligned}$$

$$\text{(47)} \qquad \Delta \overset{k+1}{\varepsilon} + \overset{k}{g}{}^{(\sigma_1)} \Delta \overset{k+1}{\eta} + \sigma_1 (\overset{k}{\eta} - \eta)\, \Delta \overset{k+1}{g} = \overset{k}{f}_3 \,,$$

$$\text{(48)} \qquad -\Delta \overset{k+1}{\varepsilon} + a\, \overset{k}{g}\, \Delta \overset{k+1}{\eta} + a\, \overset{k}{\eta}\, \Delta \overset{k+1}{g} + a\, \nu\, \Delta \overset{k+1}{v}_s = \overset{k}{f}_4 \,,$$

$$k = 0, 1, 2, \ldots \,,$$

where

$$a = 1/(\gamma - 1) \,,$$

$$\overset{k}{f}_1 = \overset{k}{f}_2 = 0 \quad \text{for} \quad k > 0 \,,$$

$$\overset{0}{f}_1 = v - \overset{0}{v} - (1 - \sigma_1)\, \tau\, g_{\bar{s}} - \sigma_1 \tau\, \overset{0}{g}_{\bar{s}} \,,$$

$$\overset{0}{f}_2 = \eta - \overset{0}{\eta} + 0.5\, \tau\, (v_s + \overset{0}{v}_s) \,,$$

$$\overset{k}{f}_3 = -\overset{k}{\varepsilon} + \varepsilon - \overset{k}{g}{}^{(\sigma_1)} (\overset{k}{\eta} - \eta) \,,$$

$$\overset{k}{g}{}^{(\sigma_1)} = \sigma_1 \overset{k}{g} + (1 - \sigma_1)\, g \,,$$

$$\overset{k}{f_4} = \overset{k}{\varepsilon} - a\,\overset{k}{g}\,\overset{k}{\eta} - a\,\nu\,\overset{k}{v_s}\,,$$

$$\Delta \overset{k+1}{v} = \overset{k+1}{v} - \overset{k}{v}\,,$$

$$\Delta \overset{k+1}{\eta} = \overset{k+1}{\eta} - \overset{k}{\eta}\,,\text{etc.}$$

During the course of the elimination of $\Delta \overset{k+1}{\varepsilon}$, $\Delta \overset{k+1}{\eta}$ and $\Delta \overset{k+1}{v}$ from the foregoing we obtain the three-point difference equation for determination of $\overset{k+1}{y} = \Delta \overset{k+1}{g}$

$$\left(\sigma_1(\overset{k}{\eta} - \eta) + a\,\overset{k}{\eta}\right)\overset{k+1}{y} - \sigma_1\tau\left((\overset{k}{g}{}^{(\sigma_1)} + a\,\overset{k}{g})\,0.5\,\tau + a\,\nu\right)\overset{k+1}{y}_{\bar{s}s} = \overset{k}{F}\,,$$

where $\overset{k}{F}$ is expressed in terms of $\overset{k}{f_1}$, $\overset{k}{f_2}$, $\overset{k}{f_3}$ and $\overset{k}{f_4}$. The elimination method can be employed for the last equation, permitting one to find, first, $\Delta \overset{k+1}{v}$, $\Delta \overset{k+1}{\eta}$, $\Delta \overset{k+1}{\varepsilon}$ with knowledge of $\Delta \overset{k+1}{g} = \overset{k+1}{y}$ and, second, $\overset{k+1}{g} = \Delta \overset{k+1}{g} + \overset{k}{g}$, $\overset{k+1}{v} = \Delta \overset{k+1}{v} + \overset{k}{v}$, etc.

**6. Convergence of Newton's method.** We are now in a position to find out the conditions under which Newton's method converges. With this aim, the differences

$$\delta \overset{k+1}{g} = \overset{k+1}{g} - \hat{g}\,, \quad \delta \overset{k+1}{\eta} = \overset{k+1}{\eta} - \hat{\eta}\,, \quad \delta \overset{k+1}{v} = \overset{k+1}{v} - \hat{v}\,,$$

where $\hat{g}$, $\hat{\eta}$ and $\hat{v}$ are the exact solutions to equations (43), will be given special investigation. For this, we write down a typical equation related to such a difference. By the linearity of equations (46) the homogeneous equations

$$\text{(49)} \qquad \delta \overset{k+1}{v} = -\sigma_1\tau\,\delta \overset{k+1}{g}_{\bar{s}}\,, \qquad \delta \overset{k+1}{\eta} = 0.5\,\tau\,\delta \overset{k+1}{v}_s\,, \quad k = 0, 1, 2, \dots,$$

follow immediately from the foregoing. Putting equations (47)–(48) together with the newly formed differences

$$\Delta \overset{k+1}{\varepsilon} = \delta \overset{k+1}{\varepsilon} - \delta \overset{k}{\varepsilon}\,, \quad \Delta \overset{k+1}{\eta} = \delta \overset{k+1}{\eta} - \delta \overset{k}{\eta}\,,$$

$$\Delta \overset{k+1}{v} = \delta \overset{k+1}{v} - \delta \overset{k}{v}\,, \quad \Delta \overset{k+1}{g} = \delta \overset{k+1}{g} - \delta \overset{k}{g}\,,$$

one succeeds in showing that equation (47) becomes

$$\delta \overset{k+1}{\varepsilon} + \overset{k}{g}{}^{(\sigma_1)} \delta \overset{k+1}{\eta} + \sigma_1 (\overset{k}{\eta} - \eta)\, \delta \overset{k+1}{g} = \overset{k}{F}_3 ,$$

$$\begin{aligned} \overset{k}{F}_3 &= \overset{k}{g}{}^{(\sigma_1)} \delta \overset{k}{\eta} + \delta \overset{k}{\varepsilon} + \sigma_1 (\overset{k}{\eta} - \eta)\, \delta \overset{k}{g} \\ &\quad - \left[ \delta \overset{k}{\varepsilon} + (\hat{\varepsilon} - \varepsilon) + (\sigma_1 \delta \overset{k}{g} + g^{(\sigma_1)}) (\delta \overset{k}{\eta} + \hat{\eta} - \eta) \right] \\ &= \sigma_1 (\overset{k}{\eta} - \eta)\, \delta \overset{k}{g} - \sigma_1 (\hat{\eta} - \eta)\, \delta \overset{k}{g} - \left[ (\hat{\varepsilon} - \varepsilon) + g^{(\sigma_1)} (\hat{\eta} - \eta) \right] \\ &= \sigma_1 \delta \overset{k}{\eta}\, \delta \overset{k}{g} , \end{aligned}$$

since $\hat{\varepsilon} - \varepsilon + g^{(\sigma_1)} (\hat{\eta} - \eta) = 0$ on account of (43). Because of this, we thus have

$$\delta \overset{k+1}{\varepsilon} + \overset{k}{g}{}^{(\sigma_1)} \delta \overset{k+1}{\eta} + \sigma_1 (\overset{k}{\eta} - \eta)\, \delta \overset{k+1}{g} = \sigma_1 \delta \overset{k}{\eta}\, \delta \overset{k}{g} . \tag{50}$$

An alternative form of equation (48) is

$$-\delta \overset{k+1}{\varepsilon} + a \overset{k}{g}\, \delta \overset{k+1}{\eta} + a \overset{k}{\eta}\, \delta \overset{k+1}{g} + a \nu \delta \overset{k+1}{v_s} = a\, \delta \overset{k}{\eta}\, \delta \overset{k}{g} . \tag{51}$$

Indeed, from (48) it follows that

$$\begin{aligned} 0 &= -(\delta \overset{k+1}{\varepsilon} - \delta \overset{k}{\varepsilon}) + a \overset{k}{g} (\delta \overset{k+1}{\eta} - \delta \overset{k}{\eta}) + a \overset{k}{\eta} (\delta \overset{k+1}{g} - \delta \overset{k}{g}) \\ &\quad + a \nu (\delta \overset{k+1}{v_s} - \delta \overset{k}{v_s}) - (\delta \overset{k}{\varepsilon} + \hat{\varepsilon} - a \overset{k}{g} \overset{k}{\eta} - a \nu \overset{k}{v_s}) \\ &= \left[ -\delta \overset{k+1}{\varepsilon} + a \overset{k}{g}\, \delta \overset{k+1}{\eta} + a \overset{k}{\eta}\, \delta \overset{k+1}{g} + a \nu \delta \overset{k+1}{v_s} \right] - \overset{k}{F}_4 , \end{aligned}$$

where

$$\overset{k}{F}_4 = a \overset{k}{g}\, \delta \overset{k}{\eta} + a \overset{k}{\eta}\, \delta \overset{k}{g} + a \nu \delta \overset{k}{v_s} + \hat{\varepsilon} - a \overset{k}{\eta} \overset{k}{g} - a \nu \overset{k}{v_s} .$$

Substituting here $\hat{\varepsilon} = a \hat{g} \hat{\eta} + a \nu \hat{v}_s$ yields $\overset{k}{F}_4 = a\, \delta \overset{k}{\eta}\, \delta \overset{k}{g}$.

Having completed the elimination of $\delta \overset{k+1}{\varepsilon}$ from (50) and (51) both, we arrive at

$$\begin{aligned} (a \overset{k}{g} + \overset{k}{g}{}^{(\sigma_1)})\, \delta \overset{k+1}{\eta} + \left( (a + \sigma_1) \overset{k}{\eta} - \sigma_1 \eta \right) \delta \overset{k+1}{g} + a \nu \delta \overset{k+1}{v_s} \\ = (a + \sigma_1)\, \delta \overset{k}{\eta}\, \delta \overset{k}{g} . \end{aligned} \tag{52}$$

By inserting now $\delta \overset{k+1}{v_s} = -\sigma_1 \tau \delta \overset{k+1}{g_{\bar{s}s}}$ and $\delta \overset{k+1}{\eta} = 0.5\tau\delta \overset{k+1}{v_s}$ we deduce with regard to

$$\alpha_k = \sigma_1 \tau \left[a\,\nu + 0.5\,\tau\,(a\overset{k}{g} + \overset{k}{g}{}^{(\sigma_1)})\right] / \left[(a+\sigma_1)\overset{k}{\eta} - \sigma_1 \eta\right] \geq 0$$

that

$$\delta \overset{k+1}{g} - \alpha_k \delta \overset{k+1}{g_{\bar{s}s}} = q_k \delta \overset{k}{g}\,, \quad q_k = \frac{(a+\sigma_1)\,\delta\,\overset{k}{\eta}}{(a+\sigma_1)\,\overset{k}{\eta} - \sigma_1 \eta}\,, \tag{53}$$

provided the condition $(a+\sigma_1)\overset{k}{\eta} - \sigma_1\eta > 0$ holds. The meaning of this is that

$$\overset{k}{\eta} > \frac{\sigma_1}{a+\sigma_1}\,\eta \quad \text{for all} \quad k = 0, 1, 2, \ldots \tag{54}$$

was supposed before proceeding to further derivations.

When the pressure is prescribed for $i = 0$ and $i = N$, the boundary conditions for $\delta \overset{k+1}{g}$ are certainly homogeneous:

$$\delta \overset{k+1}{g_0} = 0\,, \qquad \delta \overset{k+1}{g_N} = 0\,. \tag{55}$$

After scrutinising the canonical form of equation (53) with respect to $\delta \overset{k+1}{g}(s_i)$

$$A(P)\,y(P) = \sum_{Q \in III'(P)} B(P,Q)\,y(Q) + F(P)$$

we can be pretty sure that

$$A(P) > 0\,, \quad B(P,Q) > 0\,, \quad D(P) = A(P) - \sum_{Q \in III'(P)} B(P,Q) = 1\,.$$

This serves to motivate the validity of the maximum principle with regard to equation (58) supplied by the homogeneous boundary conditions (55). By utilizing this fact it is plain to show that the estimate holds:

$$\|\delta \overset{k+1}{g}\|_C \leq |q_k|\, \|\delta \overset{k}{g}\|_C\,, \tag{56}$$

whence it follows that the iterations converge under either of the following conditions:

$$\begin{cases} |q_k| \leq q < 1 \quad \text{for all} \quad k = 0, 1, 2, \ldots\,, \\[2ex] \dfrac{|\overset{k}{\eta} - \hat{\eta}|}{\overset{k}{\eta} - b\,\eta} \leq q\,, \qquad b = \dfrac{\sigma_1}{a+\sigma_1}\,. \end{cases} \tag{57}$$

The preceding is equivalent to the inequalities

$$\frac{\hat{\eta} + b\,q\,\eta}{1+q} \le \overset{k}{\eta} \le \frac{\hat{\eta} - b\,q\,\eta}{1-q}\,, \qquad \hat{\eta} > b\,q\,\eta\,, \tag{58}$$

thus causing, in fact, some restrictions on the step in time in connection with the dependence upon the variations of volume $\eta$ (or density $\rho = 1/\eta$). By setting $k = 0$ and choosing $\overset{0}{\eta} = \eta$ we obtain

$$\frac{1}{\eta}\,|\hat{\eta} - \eta| \le q\,(1-b)$$

or, what amounts to the same,

$$\big(1 - q\,(1-b)\big)\,\eta \le \hat{\eta} \le \big(1 + q\,(1-b)\big)\,\eta\,. \tag{59}$$

When the isotermic flow of the ideal gas (17) is considered, scheme (43) can be written in simplified form, since the energy equation was disappeared because $T =$ const. A proper iteration process is governed by the same rule as in the adiabatic case, the convergence of which can be established in a similar way without difficulties. In the isotermic case with the assigned values $\gamma = 1$, $a = \infty$, $b = 0$ we deduce instead of (58) that

$$\frac{\hat{\eta}}{1+q} \le \overset{k}{\eta} \le \frac{\hat{\eta}}{1-q} \qquad \text{and} \qquad \frac{|\hat{\eta} - \eta|}{\eta} \le q\,.$$

Under the first or second condition (57) the following relations occur:

$$\|\delta \overset{k+1}{g}\|_C \le q^{k+1}\,\|\delta \overset{0}{g}\|_C\,, \qquad \overset{k+1}{g} \longrightarrow 0 \quad \text{as} \quad k \to \infty\,,$$

thereby justifying the convergence rate as a geometric progression for the iterations just established.

Numerical calculations for $\gamma = 5/3$ ($a = 1.5$) showed that the iterations within the framework of Newton's method converge even if the steps $\tau$ are so large that the shock wave runs over two-three intervals of the grid $\omega_h$ in one step $\tau$. Of course, such a large step is impossible from a computational point of view in connection with accuracy losses. Thus, the restrictions imposed on the step $\tau$ are stipulated by the desired accuracy rather than by convergence of iterations.

**7. Equations of gas dynamics with heat conductivity.** We are now interested in a complex problem in which the gas flow is moving under the heat conduction condition. In conformity with (1)–(7), the system of differential equations for the ideal gas in Lagrangian variables acquires the form

$$\frac{\partial v}{\partial t} = -\frac{\partial g}{\partial s}, \quad \frac{\partial \eta}{\partial t} = \frac{\partial v}{\partial s}, \quad \frac{\partial \varepsilon}{\partial t} = -g\,\frac{\partial v}{\partial s} - \frac{\partial w}{\partial s},$$

$$p\eta = RT, \quad \varepsilon = c_v T, \quad \omega = \omega\Big(\eta, \frac{\partial v}{\partial s}, \nu\Big),$$

where $w = -\varkappa(\rho, T)\dfrac{\partial T}{\partial s}$ is the heat flow and $g = p + \omega$.

It is plain to create for this system of equations a fully conservative scheme such as

$$\begin{aligned} &v_t = -g_{\bar{s}}^{(\alpha)}, \quad \eta_t = v_s^{(0.5)}, \\ &\varepsilon_t = -g^{(\alpha)}\, v_s^{(0.5)} - w_s^{(\beta)}, \\ &w = -k\, T_{\bar{s}}, \quad g = p + \omega, \quad \omega = \omega(\eta, v_s, \nu), \\ &p = RT/\eta, \quad \varepsilon = c_v T, \end{aligned} \tag{60}$$

where $k = k_i = \varkappa(0.5\,(\eta_{i-0.5} + \eta_{i+0.5}), 0.5\,(T_{i-0.5} + T_{i+0.5}))$. Here $\alpha > 0$ and $\beta > 0$ are arbitrary numbers.

Generally speaking, Newton's method may be employed for nonlinear difference equations on every new layer, but the algorithm of the matrix elimination for a system of two three-point equations (see Chapter 10, Section 1) suits us perfectly for this exceptional case. We will say more about this later.

The main idea behind this approach is to accelerate and simplify the algorithms by means of the method of separate or successive eliminations. To that end, the difference equations (60) are divided into the following groups:

I: **"dynamical group"**

$$v_t = -g_{\bar{s}}^{(\alpha)}, \qquad \eta_t = v_s^{(0.5)}, \qquad g = p + \omega,$$

$$\omega = \omega(\eta, v_s, \nu), \qquad p = p(\eta, T).$$

II: "heat group"

$$\varepsilon_t = -g^{(\alpha)}\, v_{\bar{s}}^{(0.5)} - w_{s}^{(\beta)}\,, \qquad w = -k\, T_{\bar{s}}\,,$$

$$\varepsilon = \varepsilon(\eta, T)\,, \qquad k = \varkappa\left(0.5\,(\rho + \rho_{(-1)}), 0.5\,(T + T_{(-1)})\right).$$

After that, Newton's method of iterations applies equally well to either of these groups independently. By analogy with the isotermic case the first group of equations is to be solved with a prescribed temperature, while the second one needs the assigned values of $\eta$ and $v$. The essence of the matter in the last case is that the origin of the heat conduction equation is stipulated by the available sources of a dynamical nature.

The iterations in the first and second groups can be found successively by the elimination method. Having completed the $k$th iteration, for which the condition of termination $\|\Delta \overset{k}{v}\|_C = \|\overset{k}{v} - \overset{k-1}{v}\|_C \le \varepsilon_0 \|\overset{k}{v}\|_C$ is fulfilled, where $\varepsilon_0 > 0$ is a prescribed accuracy, there is no doubt that the values $\overset{k}{v}$ and $\overset{k}{\eta}$ are known.

With knowledge of $\overset{k}{v}$ and $\overset{k}{\eta}$ the $m$th iteration $\overset{m}{T}$ is recovered from the equations of the second group by Newton's method.

The process of the exterior iterations continues to develop prior to the occurrence of the convergence conditions.

Another way of going further is connected with re-ordering of these groups in reverse direction or inserting $k = 1$ and $m = 1$ in all of the iterations. The separate elimination method may be of assistance in minimizing the total volume of the available information in the storage of high-performance computers.

# 9

# Economical Difference Schemes for Multidimensional Problems in Mathematical Physics

One of the serious developments in computational mathematics owes a debt to economical difference methods available for solving partial differential equations of several spatial variables. Recent years have seen the publications of numerous papers on this subject for multiple equations of parabolic, hyperbolic and elliptic types as well as the constructions of various economical schemes. The general stability theory lies in the foundations of the possible theory of economical methods which will be given special investigation throughout the entire chapter. Two classes of admissible economical schemes are of great importance: schemes with a factorized operator on the upper layer and additive schemes generating a summarized approximation in a certain up-agreed sense. These can depend on the range of situations to be considered.

## 9.1 THE ALTERNATING DIRECTION METHOD (THE LONGITUDINAL-TRANSVERSE SCHEME) FOR THE HEAT CONDUCTION EQUATION

**1. Some preliminary information on economical schemes.** One of the most important issues in numerical methods is the well-founded choice of economical computational algorithms, the realization of which requires a minimal execution time in giving an approximate solution with a prescribed accuracy $\varepsilon > 0$. The total number of the necessary arithmetic operations $Q(\varepsilon)$ for doing so is the main characteristic of the algorithms in question, since other characteristics such as the quality of the related software and the availability of advanced-architecture computers are beyond our control.

In view of this, the economy requirement becomes rather urgent and extremely important in numerical solution of multidimensional problems arising time and again in mathematical physics.

To understand the nature of this a little better, we focus our attention on the simplest examples serving to motivate what is done with economical difference schemes and regarding to some preliminaries. The object of investigation is the heat conduction equation in the space $\mathbf{R}^p$:

$$\frac{\partial u}{\partial t} = L\,u\,, \quad L\,u = \sum_{\alpha=1}^{p} L_\alpha\,u\,, \quad L_\alpha\,u = \frac{\partial^2 u}{\partial x_\alpha^2}\,, \tag{1}$$

$$x \in G\,, \quad t \in (0, t_0]\,, \quad u|_\Gamma = 0\,, \quad u(x,0) = u_0(x)\,.$$

Let $G = G_{0p} = \{0 \le x_\alpha \le 1,\ \alpha = 1, 2, \dots, p\}$ be a cube of the dimension $p$; $\bar{\omega}_h = \{(i_1 h, \dots, i_p h) \in G\}$ be a cubic grid with step $h$ in all directions $x_\alpha$, $\alpha = 1, 2, \dots, p$, and $\bar{\omega}_\tau$ be the grid with step $\tau = t_0/n_0$ on the segment $0 \le t \le t_0$. At the next stage the operator

$$L_\alpha u = \frac{\partial^2 u}{\partial x_\alpha^2}$$

is approximated by the difference operator $\Lambda_\alpha y = y_{\bar{x}_\alpha x_\alpha}$, so that $\Lambda = \sum_{\alpha=1}^{p} \Lambda_\alpha$, leaving us with the weighted two-layer scheme

$$(2) \qquad \begin{aligned} & y_t = \Lambda \left( \sigma \hat{y} + (1 - \sigma)\, y \right), \quad x \in \omega_h\,, \quad 0 \le t = n\tau < t_0\,, \\ & y|_{\gamma_h} = 0\,, \quad y(x,0) = y_0(x)\,. \end{aligned}$$

We recall from Chapter 5, Section 3 that scheme (2) is stable with respect to the initial data under the constraint

$$\sigma \ge \frac{1}{2} - \frac{h^2}{4\,p\,\tau} = \sigma_0\,.$$

By merely setting $\sigma = 0$ we have at our disposal the explicit scheme

$$y_t = \Lambda\, y \quad \text{or} \quad \hat{y} = y + \tau\,\Lambda\,y\,, \tag{3}$$

which is stable for $\tau \leq 0.5\,h^2/p$. If equation (1) contains the variable coefficients, that is, it acquires the form

$$L_\alpha u = \frac{\partial}{\partial x_\alpha}\left(k_\alpha(x,t)\,\frac{\partial u}{\partial x_\alpha}\right), \quad 0 < k_\alpha \leq c_1\,,$$

then

$$\Lambda_\alpha u = \left(a_\alpha\, y_{\bar{x}_\alpha}\right)_{x_\alpha}, \quad \Lambda = \sum_{\alpha=1}^{p} \Lambda_\alpha\,, \quad 0 < a_\alpha \leq c_2\,,$$

and the explicit scheme (3) is stable for $\tau \leq 0.5\,h^2/(p\,c_2)$.

From here it seems clear that the admissible step in the explicit scheme is yet to be refined along with increasing the maximum value of the coefficient of heat conductivity. As a matter of fact, the last requirement is unreal for the problems with fastly and widely varying coefficients. Just for this reason explicit schemes are of little use not only for multidimensional problems, but also for one-dimensional ones ($p = 1$). On the other hand, the explicit schemes offer real advantages that the value $\hat{y} = y_{n+1}$ on every new layer $t_{n+1} = t_n + \tau$ is found by the explicit formulas (3) with a finite number of operations at every node of the grid $\omega_h$, so that the amount of arithmetic operations required in passing from one layer to another is proportional to the total number of the grid nodes and so it is a quantity of $O(1/h^p)$.

Being concerned with the implicit scheme for $\sigma = 1$, we may set up the problem for determination of $y^{n+1}$:

$$y^{n+1} - \tau\,\Lambda\,y^{n+1} = y^n\,, \quad y^{n+1}\big|_{\gamma_h} = 0\,, \quad y(x,0) = u_0(x)\,.$$

Numerical solution of this system containing $1/h^p$ equations requires, for example, during the course of Gaussian elimination $O(1/h^{3p-2})$ operations in connection with a special structure of the matrix $E - \tau\Lambda$.

Some consensus of opinion is desirable in this matter, since a smaller number of operations is performed in the explicit scheme, but it is stable only for sufficiently small values of $\tau$. In turn, the implicit scheme being absolutely stable requires much more arithmetic operations.

What schemes are preferable for later use? Is it possible to bring together the best qualities of both schemes in line with established priorities? In other words, the best scheme would be absolutely stable as the implicit schemes and schould require in passing from one layer to another exactly $Q$ arithmetic operations. As in the case of the explicit schemes, $Q$ would be proportional to the total number of the grid nodes so that $Q = O(1/h^p)$.

Because of these facts, the number of the necessary operations at every node of the grid is independent of the total number of the grid nodes. All the schemes with the indicated property are said to be **economical.**

In what follows one possible example demonstrates for a system of ordinary differential equations that there is an implicit scheme which is rather economical than the explicit ones requiring the additional operations.

**Example** With this aim, it seems worthwhile giving the following system of differential equations:

$$\frac{du}{dt} + A\,u = 0\,, \qquad t > 0\,, \qquad u(0) = u_0\,,$$

where $u = \big(u^{(1)}(t), \ldots, u^{(m)}(t)\big)$ is a vector of order $m$ and $A = (a_{ij})$ is a symmetric positive definite matrix. In passing from one layer to another the explicit scheme $y_{n+1} = y_n - \tau A y_n$ requires $2m^2 + 2m$ arithmetic operations.

Furthermore, let $A^- = (a_{ij}^-)$ and $A^+ = (a_{ij}^+)$ be an upper and a lower tridiagonal matrices with coinciding main diagonals $a_{ii}^- = a_{ii}^+ = 0.5\,a_{ii}$. Both matrices (operators) are positive definite in the sense of the usual inner product in the space $\mathbf{R}^m$, since $A^- = (A^+)^*$ and

$$(Ax, x) = (A^+x, x) + ((A^+)^*x, x) = 2(A^+x, x) = 2(A^-x, x)\,.$$

Before going further, we initiate the construction of the scheme

$$\text{(4)} \qquad \frac{y_{2n+1} - y_{2n}}{\tau} + A^- y_{2n+1} + A^+ y_{2n} = 0\,,$$

$$\text{(5)} \qquad \frac{y_{2n+2} - y_{2n+1}}{\tau} + A^- y_{2n+1} + A^+ y_{2n+2} = 0\,, \quad n = 0, 1, \ldots,$$

in which it is necessary to perform the inversion of both tridiagonal matrices $(E + \tau A^-)$ and $(E + \tau A^+)$ in determining $y_{2n+1}$ and $y_{2n+2}$.

It is plain to show that the scheme in view is absolutely stable for any $\tau > 0$, permitting one to eliminate $y_{2n+1}$ from the difference equations (4) and (5). By subtracting equation (5) from equation (4) we find that

$$2\,y_{2n+1} = y_{2n} + y_{2n+2} + \tau\,A^+(y_{2n+2} - y_{2n})\,.$$

Upon substituting the resulting expression into (5) we obtain through such an analysis the scheme

$$\text{(6)} \qquad B\,\frac{y_{2n+2} - y_{2n}}{2\,\tau} + A\,y_{2n} = 0\,, \qquad A = A^- + A^+\,,$$

where the operator $B = (E+\tau A^-)(E+\tau A^+)$ is a product of two self-adjoint "triangle" operators, since $A^+ = (A^-)^*$.

Observe here that the factorized operator $B$ also is self-adjoint. By means of these operators a sufficient condition of stability becomes

$$B - 0.5\,(2\,\tau\, A) = E + \tau A + \tau^2 A^- A^+ - \tau\, A > E$$

in light of the relations $(A^- A^+ x, x) = \|\, A^+ x\,\|^2 > 0$ and $A^- A^+ > 0$. This supports the view that scheme (6) is absolutely stable and it is of second-order accuracy.

Let $\widetilde{A}^{\pm}$ be new tridiagonal matrices differing from $A^{\pm}$ solely by the zero elements on the main diagonals. While solving equation (4) and equation (5) we should save in the storage the vectors $\widetilde{A}^- y_{2n+1}$ and $\widetilde{A}^+ y_{2n+2}$, respectively. All this enables us to evaluate the number $Q_1$ relating to the necessary operations in passing from the layer $t_{2n}$ to the layer $t_{2n+2}$. For scheme (4)–(5) we have $Q_1 = 2m^2 + 12m$, while for the explicit scheme – $Q_0 = 4m^2 + 4m$, that is, $Q_1 \le Q_0$ for $m \ge 4$.

**2. An alternating direction scheme.** Further developments are concerned with the heat conduction equation of two independent variables that can serve as test vehicles for the difference schemes to be presented:

$$\frac{\partial u}{\partial t} = L\,u + f(x,t)\,, \quad x \in G_{02}\,, \quad t \in (0, t_0]\,, \tag{7}$$

$$u|_{\Gamma} = \mu(x,t)\,, \quad u(x,0) = u_0(x)\,,$$

$$L\,u = \Delta\,u = (L_1 + L_2)\,u\,, \quad L_\alpha\, u = \frac{\partial^2 u}{\partial x_\alpha^2}\,, \quad \alpha = 1, 2\,.$$

Here $\bar{G}_{02} = \bar{G}_0 = \{0 \le x_\alpha \le l_\alpha,\ \alpha = 1, 2\}$ is the rectangle of sides $l_1$ and $l_2$ with the boundary $\Gamma$.

As a first step towards the solution of this problem, it will be sensible to introduce an equidistant grid $\bar{\omega}_h$ in the direction $x_\alpha$ with steps $h_1 = l_1/N_1$ and $h_2 = l_2/N_2$ and the boundary $\gamma_h$ containing all the nodes lying on the sides of the rectangle except its vertices, $\bar{\omega}_h = \omega_h + \gamma_h$. We will use them for later approximation of the operator $L_\alpha$ by the difference operator

$$\Lambda_\alpha y = y_{\bar{x}_\alpha x_\alpha}\,, \qquad \Lambda = \Lambda_1 + \Lambda_2\,.$$

Recall that in the case of the one-dimensional heat conduction equation a similar implicit scheme is associated on every layer with the difference boundary-value problem of the form

$$(8)\qquad A_i y_{i-1} - C_i y_i + B_i y_{i+1} = -F_i\,, \quad i = 1, 2, \ldots, N-1\,,$$

$$y_0 = \mu_1\,, \quad y_N = \mu_2\,, \quad A_i > 0\,, \quad B_i > 0\,, \quad C_i \geq A_i + B_i\,.$$

This problem can be solved by the standard elimination method requiring $O(1/h) = O(N)$ operations, the amount of which is proportional to the total number $N$ of nodal points of the grid $\bar{\omega}_h = \{x_i = ih,\ 0 \leq i \leq N\}$.

With regard to problem (7) posed in the rectangle, it is worth noting several things. The grid $\bar{\omega}_h$ at hand may be treated either as a collection of the nodes along the rows $i_2 = 0, 1, \ldots, N_2$ or as a collection of the nodes along the columns $i_1 = 0, 1, \ldots, N_1$, thereby providing for subsequent compositions the availability of $N_1 + 1$ nodal points along every row as well as $N_2 + 1$ nodal points along every column.

In trying to solve a typical problem like (8) by the elimination method for fixed $i_2$ (or $i_1$), exactly $O(N_1 N_2)$ arithmetic operations are needed in giving a solution at all the nodes of the grid. Their amount is proportional to the total number of the grid nodes in the plane. The main idea behind economical methods lies in successive solution of one-dimensional problems of the type (8) along rows and along columns in passing from one layer to another.

The **scheme** ascribed to **Peaceman and Rachford** provides some realization of this idea and refers to implicit alternating direction schemes. The present values $y = y^n$ and $\hat{y} = y^{n+1}$ of this difference scheme are put together with the intermediate value $\bar{y} = y^{n+1/2}$, a formal treatment of which is the value of $y$ at moment $t = t_{n+1/2} = t_n + \tau/2$. The passage from the $n$th layer to the $(n+1)$th layer can be done in two steps with the appropriate spacings $0.5\,\tau$:

$$(9)\qquad \frac{y^{n+1/2} - y^n}{0.5\,\tau} = \Lambda_1 y^{n+1/2} + \Lambda_2 y^n + \varphi^n\,,$$

$$(10)\qquad \frac{y^{n+1} - y^{n+1/2}}{0.5\,\tau} = \Lambda_1 y^{n+1/2} + \Lambda_2 y^{n+1} + \varphi^n\,.$$

These equations are written at all inner nodes $x = x_i$ of the grid $\omega_h$ and for all $t = t_n > 0$. Let us stress here that the first scheme is implicit along the direction $x_1$ and it is explicit along the direction $x_2$, while the second one is explicit in the direction $x_1$ and it is implicit in the direction $x_2$. Equations (9)–(10) are supplemented with the initial conditions

$$(11)\qquad y(x, 0) = u_0(x)\,, \qquad x \in \bar{\omega}_h\,,$$

and the difference boundary conditions, for example, of the special type

$$y^{n+1} = \mu^{n+1} \quad \text{for } i_2 = 0 \text{ and } i_2 = N_2 , \tag{12}$$

$$y^{n+1/2} = \bar{\mu} \quad \text{for } i_1 = 0 \text{ and } i_1 = N_1 , \tag{13}$$

where

$$\bar{\mu} = \frac{1}{2}\,(\mu^{n+1} + \mu^n) - \frac{\tau}{4}\,\Lambda_2(\mu^{n+1} - \mu^n) . \tag{14}$$

The meaning of the boundary condition (12) is known to us. On the other hand, condition (13), which assigns the boundary value $\bar{y}$, needs certain clarification. In this way, the difference boundary-value problem (9)–(14) can be put in correspondence with problem (7). The method for solving this difference problem is mostly based on alternative forms of equations (9)–(10):

$$\begin{aligned} &\frac{2}{\tau}\,\bar{y} - \Lambda_1\,\bar{y} = F , \qquad F = \frac{2}{\tau}\,y + \Lambda_2\,y + \varphi , \\ &\frac{2}{\tau}\,\hat{y} - \Lambda_2\,\hat{y} = \bar{F} , \qquad \bar{F} = \frac{2}{\tau}\,\bar{y} + \Lambda_1\,\bar{y} + \varphi . \end{aligned} \tag{15}$$

To make our exposition more transparent, it is more convenient to introduce the new members

$$x_i = (i_1 h_1, i_2 h_2) , \qquad F = F_{i_1 i_2} , \qquad y = y_{i_1 i_2}$$

and approve the following rule: when one of the subscripts is kept fixed, we omit it for a while in relevant expressions. This should cause no confusion and guides a proper choice of alternative forms of equation (15) for later use:

$$\begin{aligned} &\frac{1}{h_1^2}\,\bar{y}_{i_1-1} - 2\left(\frac{1}{h_1^2} + \frac{1}{\tau}\right)\bar{y}_{i_1} + \frac{1}{h_1^2}\,\bar{y}_{i_1+1} = -F_{i_1} , \\ &i_1 = 1, 2, \dots , N_1 - 1 , \quad \bar{y} = \bar{\mu} \quad \text{for} \quad i_1 = 0, N_1 , \end{aligned} \tag{16}$$

$$\begin{aligned} &\frac{1}{h_2^2}\,\hat{y}_{i_2-1} - 2\left(\frac{1}{h_2^2} + \frac{1}{\tau}\right)\hat{y}_{i_2} + \frac{1}{h_2^2}\,\bar{y}_{i_2+1} = -\bar{F}_{i_2} , \\ &i_2 = 1, 2, \dots , N_2 - 1 , \quad y = \hat{\mu} \quad \text{for} \quad i_2 = 0, N_2 . \end{aligned} \tag{17}$$

Let a value $y = y^n$ be given. Starting from $F$, we move further along the rows $i_2 = 1, 2, \ldots, N_2 - 1$ to solve problem (16) by the standard elimination method, whose use permits us to determine the values $\bar{y}$ at all the nodes of the grid $\omega_h$. After that, we calculate $\bar{F}$ and move along the columns $i_1 = 1, 2, \ldots, N_1 - 1$ in an attempt to solve problem (17) and find the values $\hat{y} = y^{n+1}$. In passing from the $(n+1)$th layer to the $(n+2)$th layer the same procedure is workable, thus causing the alternating directions.

Since the elimination method requires several operations at one node, the total number of which is independent of the grid step, the algorithm just established will be economical if one succeeds in showing that scheme (9)–(14) is absolutely stable. The following sections place a special emphasis on stability and convergence of the scheme concerned.

**3. Stability.** Subsequent considerations of stability of scheme (9)–(14) are conducted with a priori elimination of the intermediate value $\bar{y}$. This can be done by subtracting equation (10) from equation (9) and re-ordering of the relevant one as

$$2\,\bar{y} = \hat{y} + y - 0.5\,\tau\,\Lambda_2(\hat{y} - y)\,, \qquad x \in \omega_h\,, \tag{18}$$

Substituting (18) into (9) yields

$$\frac{\hat{y} - y}{\tau} - \frac{1}{2}\,\Lambda_2(\hat{y} - y) = \frac{1}{2}\,\Lambda_1(\hat{y} + y) - \frac{\tau}{4}\,\Lambda_1\Lambda_2(\hat{y} - y) + \Lambda_2 y + \varphi\,. \tag{19}$$

With the relation $\hat{y} = y + \tau y_t$ in view, the intention is to use (19) in the canonical form

$$(E - 0.5\,\tau\,\Lambda_1)\,(E - 0.5\,\tau\,\Lambda_2)\,y_t = \Lambda\,y + \varphi\,. \tag{20}$$

Under such an approach formula (18) should also be valid for $x_1 = 0$ and $x_1 = l_1$, since otherwise $(\Lambda_1\bar{y})_{i_1}$ refers to the undetermined values for $i_1 = 1$ and $i_1 = N_1 - 1$. Knowing $y = \mu$ and $\hat{y} = \hat{\mu}$ for $x_1 = 0$ and $x_1 = l_1$, we deduce from (18) that

$$\bar{y} = \frac{1}{2}\,(\hat{\mu} + \mu) - \frac{\tau^2}{4}\,\Lambda_2\mu_t = \bar{\mu} \quad \text{for} \quad x_1 = 0 \quad \text{and} \quad x_1 = l_1\,,$$

by observing that these coincide with the boundary conditions (13)–(14). Thus, we have proved that a solution of problem (9)–(14) satisfies equation (20) subject to the supplementary conditions

$$y|_{\gamma_h} = \mu\,, \quad \hat{y}|_{\gamma_h} = \hat{\mu}\,, \quad y(x, 0) = u_0(x)\,. \tag{21}$$

On the other hand, a solution of problem (20)–(21) applies equally well to problem (9)–(14). Indeed, by specifying $\bar{y}$ by formula (18) we deduce from (18) that

$$(E - 0.5\,\tau\,\Lambda_2)\,\hat{y} = 2\,\bar{y} - (E + 0.5\,\tau\,\Lambda_2)\,y$$

and insert then the resulting expression in (20). By minor changes we are led to equation (9), which in combination with (18) gives (10). This provides enough reason to establish the equivalence between problem (9)–(14) and problem (20)–(21) with compliance of the boundary values $\bar{y}$ assigned by formulas (13)–(14). Careful analysis of scheme (9)–(14) is accompanied by more a detailed exploration of scheme (20)–(21) in "integer steps".

The general theory of two-layer schemes is quite applicable in such a setting. By regarding the boundary conditions to be homogeneous we turn to the problem

$$(22) \qquad (E - 0.5\,\tau\,\Lambda_1)\,(E - 0.5\,\tau\,\Lambda_2)\,y_t = \Lambda\,y + \varphi\,, \quad t \geq 0\,,$$

$$y(x,0) = u_0(x)\,, \qquad y\big|_{\gamma_h} = 0\,,$$

with further reference to the space $H$ of all grid functions given on the set $\bar{\omega}_h$ and vanishing on the boundary $\gamma_h$ under the inner product structure

$$(y,v) = \sum_{x\in\omega_h} y(x)\,v(x)\,h_1\,h_2 = \sum_{i_1=1}^{N_1-1}\sum_{i_2=1}^{N_2-1} y(i_1h_1, i_2h_2)\,v(i_1h_1, i_2h_2)\,h_1\,h_2\,.$$

The associated norm is taken, as usual, to be $\|\,y\,\| = \sqrt{(y,y)}$. We refer to the operator $A = -\Lambda = -(\Lambda_1 + \Lambda_2)$, which, by definition, is self-adjoint and positive in the space $H$. The norm on the energetic space $H_A$ is defined either by

$$\|\,y\,\|_A^2 = \sum_{i_1=1}^{N_1}\sum_{i_2=1}^{N_2-1} \big(y_{\bar{x}_1}(i_1h_1, i_2h_2)\big)^2 h_1\,h_2 + \sum_{i_1=1}^{N_1-1}\sum_{i_2=1}^{N_2} \big(y_{\bar{x}_2}(i_1h_1, i_2h_2)\big)^2 h_1\,h_2$$

or by

$$(23) \qquad \|\,y\,\|_A^2 = \|y_{\bar{x}_1}\|_1^2 + \|y_{\bar{x}_2}\|_2^2\,.$$

When treating $y = y(t)$ as an abstract function of the argument $t \in \omega_\tau$ with the values in the space $H$, scheme (22) admits an alternative form

$$(24) \qquad B\,y_t + A\,y = \varphi(t)\,, \quad 0 \leq t = n\tau < t_0\,, \quad y(0) = u_0\,,$$

where $B = (E + 0.5\,\tau A_1)(E - 0.5\,\tau A_2)$, $A_\alpha = -\Lambda_\alpha$, $A = A_1 + A_2$.

Because of the rectangular form of the initial domain, the operators $A_1$ and $A_2$ are self-adjoint, positive and commutative. It is straightforward to verify the relations $A_1 A_2 = A_2 A_1$ and $\Lambda_1 \Lambda_2 y = \Lambda_2 \Lambda_1 y = y_{\bar{x}_1 x_1 \bar{x}_2 x_2}$ at all the inner nodes of the grid. In view of this, the strict inequality $A_1 A_2 > 0$ is simple to follow. From the form (24) it seems clear that

$$B > E + 0.5\,\tau\,, \tag{25}$$

thereby clarifying that scheme (24) is stable in the space $H_A$. Indeed,

$$B - 0.5\,\tau A = \Big(E + \frac{\tau}{2}\,A + \frac{\tau^2}{4}\,A_1\,A_2\Big) - \frac{\tau}{2}\,A$$

$$= E + \frac{\tau^2}{4}\,A_1\,A_2 > E\,.$$

From condition (25) it follows that for scheme (24) Theorem 7 from Chapter 6, Section 2 is still valid with $\varepsilon = 1$, due to which a solution of problem (22) satisfies the inequality

$$\|y(t+\tau)\|_A \le \|y(0)\|_A + \frac{1}{\sqrt{2}} \left( \sum_{t'=0}^{t} \tau \,\|\, \varphi(t') \,\|^2 \right)^{1/2} . \tag{26}$$

Also, the a priori estimate holds true:

$$\|\, y(t+\tau)\,\| \le \|\, y(0)\,\| + \frac{1}{\sqrt{2}} \left( \sum_{t'=0}^{t} \tau \,\|\varphi(t')\|_{A^{-1}} \right)^{1/2} . \tag{27}$$

To make sure of it, we apply the operator $A^{-1} > 0$ to both parts of equation (24). The outcome of this

$$\tilde{B}\, y_t + \tilde{A}\, y = \tilde{\varphi}\,, \quad \tilde{A} = E\,, \quad \tilde{\varphi} = A^{-1} \varphi\,,$$

$$\tilde{B} = A^{-1} + \frac{\tau}{2}\,E + \frac{\tau^2}{4}\,A^{-1} A_1\,A_2\,. \tag{28}$$

Since the operators $A_1$, $A_2$ and $A^{-1} > 0$ are commutative and self-adjoint, the relations $A^{-1} A_1 A_2 > 0$ and $\tilde{B} > A^{-1} + 0.5\,\tau E$ occur. Applying Theorem 10 from Chapter 6, Section 2 yields estimate (27).

**Theorem 1** *Scheme (22) is stable with respect to the initial data and the right-hand side. A solution of problem (22) satisfies a priori estimates (26)–(27).*

**4. Stability and accuracy.** By utilizing the fact that scheme (9)–(14) is equivalent to scheme (20)–(21) subsequent considerations of convergence and accuracy of the first scheme will appear for the second one. Let $u = u(x,t)$ be a solution of problem (7) and $y = y(x_i, t_n)$ be a solution of problem (9)–(14) and scheme (20)–(21). Upon substituting $y = z + u$ into (20) we may set up the problem for the error of this scheme:

$$(29) \qquad B\, z_t = \Lambda\, z + \psi\,, \quad x \in \omega_h\,, \quad 0 \le t = n\tau < t_0\,,$$

$$z\big|_{\gamma_h} = 0\,, \quad z(x,0) = 0\,,$$

where $B = (E - 0.5\,\tau\,\Lambda_1)(E - 0.5\,\tau\,\Lambda_2)$ and $\psi$ is the error of approximation equal to

$$(30) \qquad \psi = \varphi + \Lambda\, u - B\, u_t = 0.5\,\Lambda\,(\hat{u} + u) - u_t - \tau^2 \Lambda_1\, \Lambda_2\, u_t + \varphi\,.$$

From such reasoning it seems clear that

$$\psi = O(|h|^2 + \tau^2)\,, \qquad |h|^2 = h_1^2 + h_2^2\,,$$

under the condition that the solution $u = u(x,t)$ possesses in the region $\bar{Q}_T = \bar{G}_0 \times [0, t_0]$ the derivatives

$$(31) \quad \left|\frac{\partial^3 u}{\partial t^3}\right| \le M\,, \quad \left|\frac{\partial^5 u}{\partial x_1^2\, \partial x_2^2\, \partial t}\right| \le M\,, \quad \left|\frac{\partial^4 u}{\partial x_1^4}\right| \le M\,, \quad \left|\frac{\partial^4 x}{\partial x_2^4}\right| \le M\,.$$

The final result is an immediate implication of the asymptotic relations $0.5\,(\hat{u} + u) = \bar{u} + O(\tau^2)$ and $u_t = \bar{\dot{u}} + O(\tau^2)$, where $\bar{u} = u(x, t_n + 0.5\,\tau)$ and the quantity $\Lambda_1 \Lambda_2 u$ is bounded. Since for problem (29) estimate (26) is valid for the assigned value $z(0) = z_0 = 0$, we might formulate the following assertion.

**Theorem 2** *Under conditions (31) scheme (20)–(21) converges in the grid norm (23) with the rate $O(\tau^2 + |h|^2)$.*

**5. A scheme for the governing equation with variable coefficients.** The intention is to use the alternating direction scheme for the heat conduction equation with variable coefficients

$$(32) \qquad \frac{\partial u}{\partial t} = L\,u + f\,, \qquad (x,t) \in Q_T\,,$$

$$u|_{\Gamma} = \mu(x,t)\,, \qquad u(x,0) = u_0(x)\,,$$

$$L\,u = L_1\,u + L_2\,u\,, \qquad L_\alpha\,u = \frac{\partial}{\partial x_\alpha}\left(k_\alpha(x,t)\,\frac{\partial u}{\partial x_\alpha}\right),$$

$$k_\alpha(x,t) > 0\,.$$

In such setting, for any $t$, the operator $L_\alpha$ is approximated by the difference operator

$$\Lambda_\alpha y = \Lambda_\alpha(t) y = \big(a_\alpha(x,t)\,y_{\bar{x}_\alpha}\big)_{x_\alpha}\,, \qquad \alpha = 1, 2\,,$$

where $a_\alpha$ can be founded either by the formula $a_\alpha = k_\alpha^{(-0.5_\alpha)}$ or by the formula $a_\alpha = 0.5\,(k_\alpha + k_\alpha^{(-1_\alpha)})$, $\alpha = 1, 2$, which guarantee a second-order approximation provided by the operator $\Lambda_\alpha$:

$$\Lambda_\alpha u - L_\alpha u = O(h_\alpha^2)\,.$$

Instead of (9)–(10) we are trying to adopt another scheme

$$(33) \qquad \frac{\bar{y} - y^n}{0.5\,\tau} = \Lambda_1(\bar{t})\,\bar{y} + \Lambda_2(t_n)\,y^n + \varphi^n\,,$$

$$\frac{y^{n+1} - \bar{y}}{0.5\,\tau} = \Lambda_1(\bar{t})\,\bar{y} + \Lambda_2(t_{n+1})\,y^{n+1} + \varphi^n\,,$$

$$\bar{t} = t_{n+1/2}\,, \qquad y(x,0) = u_0(x)\,,$$

with the boundary conditions

$$(34) \qquad \begin{aligned} y^{n+1} &= \mu^{n+1} && \text{for} \quad i_2 = 0\,,\ i_2 = N_2\,, \\ \bar{y} &= \bar{\mu} && \text{for} \quad i_1 = 0\,,\ i_1 = N_1\,, \end{aligned}$$

where

$$\bar{\mu} = \frac{\mu^{n+1} + \mu^n}{2} - \frac{\tau^2}{4} (\Lambda_2 \mu)_{t,n}$$

and

$$(\Lambda_2 \mu)_{t,n} = \left(\Lambda_2(t_n)\, \mu(t_n)\right)_t = \frac{\Lambda_2(t_{n+1})\, \mu(t_{n+1}) - \Lambda_2(t_n)\, \mu(t_n)}{\tau} .$$

The value $\bar{\mu}$ is put in correspondence with the expression

$$\bar{y} = \frac{y^{n+1} + y^n}{2} - \frac{\tau^2}{4} (\Lambda_2 y)_{t,n} , \tag{35}$$

arising from the statement of the difference problem (33) during the course of the elimination of $\Lambda_1 \bar{y}$.

In this regard, it should be taken into account that scheme (33)–(34) is equivalent to the factorized scheme (20)–(21) with the member $\Lambda_\alpha y = \left(a_\alpha(x)\, y_{\bar{x}_\alpha}\right)_{x_\alpha}$ in special cases: either in scheme (33) the operator $\Lambda_2(\bar{t})$ is involved at one and the same moment of time $\bar{t}$ in place of $\Lambda_2(t_n)$ and $\Lambda_2(t_{n+1})$ or $k_\alpha(x)$ and, hence, $\Lambda_\alpha$ are independent of $t$. Recall that scheme (20) provides on a solution $u(x,t)$ an approximation of $O(|h|^2) + \tau^2$ if the usual requirements of smoothness of $k_\alpha(x)$ in the variables $x_1$ and $x_2$ are imposed in addition to conditions (31). The principal difference from the case of constant coefficients is discovered in the stability analysis of scheme (20) by observing that the operators $A_1 = -\Lambda_1$ and $A_2 = -\Lambda_2$ are positive and self-adjoint. But, unfortunately, they are non-commutative:

$$\Lambda_1 \Lambda_2\, y = \left(a_1 (a_2\, y_{\bar{x}_2})_{x_2 \bar{x}_1}\right)_{x_1} \neq \Lambda_2 \Lambda_1\, y = \left(a_2 (a_1\, y_{\bar{x}_1})_{x_1 \bar{x}_2}\right)_{x_2} ,$$

where $a_\alpha = a_\alpha(x,t)$. Just for this reason the product $\Lambda_1 \Lambda_2$ is not obliged to be positive, thus causing some difficulties. In view of this, it is possible to establish the stability only for sufficiently small values of $\tau \leq \tau_0(c_1)$, where $c_1$ depends on the maximum of the derivatives of $k_\alpha$ with respect to $x_1$ and $x_2$. Let us stress here that $\tau \leq \tau_0(c_1)$ is a severe restriction and, as a matter of experience, it is connected with the available methods of special investigations of stability. In what follows we will show that scheme (33) is absolutely stable in another norm.

With this aim, we first write down the equation for the error by introducing the new variables

$$z^n = y^n - u^n , \quad z^{n+1} = y^{n+1} - u^{n+1} , \quad \bar{z} = \bar{y} - \bar{u} ,$$

where $u^n = u(x, t_n)$, $u^{n+1} = u(x, t_{n+1})$ and $\bar{u}$ is calculated by the formula

$$\bar{u} = \frac{u^n + u^{n+1}}{2} - \frac{\tau^2}{4} (\Lambda_2 u)_{t,n} \,. \tag{36}$$

This is consistent with (35) as stated before. Under such a choice of $\bar{u}$ we obtain the homogeneous boundary conditions for specifying $\bar{z}$.

By substituting into equation (33) $y^n = z^n + u^n$, $\bar{y} = \bar{z} + \bar{u}$ and $y^{n+1} = z^{n+1} + u^{n+1}$ we are led to the problem statement

$$\begin{aligned} \frac{\bar{z} - z^n}{0.5\,\tau} &= \Lambda_1(t_{n+1/2})\,\bar{z} + \Lambda_2(t_n)\,z^n + \psi_1^n \,, \\ \frac{z^{n+1} - \bar{z}}{0.5\,\tau} &= \Lambda_1(t_{n+1/2})\,\bar{z} + \Lambda_2(t_{n+1})\,z^{n+1} + \psi_2^n \,, \end{aligned} \tag{37}$$

$$z\big|_{\gamma_h} = 0 \,, \qquad z(x, 0) = 0 \,, \tag{38}$$

where $\psi_1^n$ and $\psi_2^n$ are the appropriate errors of approximation:

$$\psi_1^n = \Lambda_1(t_{n+1/2})\,\bar{u} + \Lambda_2(t_n)\,u^n + \varphi^n - \frac{\bar{u} - u^n}{0.5\,\tau} \,,$$

$$\psi_2^n = \Lambda_1(t_{n+1/2})\,\bar{u} + \Lambda_2(t_{n+1})\,u^{n+1} + \varphi^n - \frac{u^{n+1} - \bar{u}}{0.5\,\tau} \,.$$

Before giving further motivations, it is worth noting here that $\psi_1^n = \psi_2^n$. This fact can readily be verified by substituting expression (36) into the formula

$$\psi_2^n - \psi_1^n = \Lambda_2(t_{n+1})\,u^{n+1} - \Lambda_2(t_n)\,u^n - \frac{u^{n+1} - 2\,\bar{u} + u^n}{0.5\,\tau} = 0 \,.$$

At the next stage expression (36) is needed in the formula for the residual $\psi_1^n$

$$\begin{aligned} \psi_1^n = \Lambda_1(t_{n+1/2}) \Big( \frac{u^n + u^{n+1}}{2} - \frac{\tau^2}{4} (\Lambda_2 u)_{t,n} \Big) + \Lambda_2(t_n)\,u^n \\ + \varphi^n - \frac{u^{n+1} - u^n}{\tau} + \frac{\tau}{2} (\Lambda_2 u)_{t,n} \,. \end{aligned}$$

with further reference to the following asymptotic relations:

$$\frac{u^n + u^{n+1}}{2} = u(x, t_{n+1/2}) + O(\tau^2)\,,$$

$$\frac{u^{n+1} - u^n}{\tau} = \frac{\partial u}{\partial t}\bigg|_{t=t_{n+1/2}} + O(\tau^2)\,,$$

$$\Lambda_2(t_{n+1/2})\,\bar{u} = \Lambda_2(t_n)\,u^n + \frac{\tau}{2}\,(\Lambda_2 u)_{t,n} + O(\tau^2)\,,$$

$$\Lambda_1\,u + \Lambda_2\,u = L_1\,u + L_2\,u + O(|h|^2)\,,$$

$$\varphi^n = f(x, t_{n+1/2}) + O(\tau^2 + |h|^2)\,.$$

As a final result we obtain

$$\psi_1^n = (\Lambda_1\,u + \Lambda_2\,u)|_{t=t_{n+1/2}} - \frac{u^{n+1} - u^n}{\tau} + \varphi^n + O(\tau^2)$$

$$= O(\tau^2 + |h|^2)\,,$$

thereby justifying that scheme (33)–(34) generates a second-order approximation:

$$\psi_1^n = \psi_2^n = O(\tau^2 + |h|^2)\,.$$

Further development of some a priori estimate for a solution of the problem concerned is mostly based on an operator-difference analog of problem (37)–(38) such as

$$\frac{\bar{z} - z^n}{0.5\,\tau} + A_1(t_{n+1/2})\,\bar{z} + A_2(t_n)\,z^n = \psi_1^n\,,$$

(39)

$$\frac{z^{n+1} - \bar{z}}{0.5\,\tau} + A_1(t_{n+1/2})\,\bar{z} + A_2(t_{n+1})\,z^{n+1} = \psi_2^n\,,$$

$$n = 0, 1, 2, \ldots\,, \qquad z(0) = 0\,,$$

where $A_1(t)$ and $A_2(t)$ are linear operators in a Hilbert space $H$:

$$A_1 : H \mapsto H\,, \qquad A_2 : H \mapsto H\,.$$

In dealing with non-negative and, generally speaking, non-self-adjoint operators $A_1 \geq 0$ and $A_2 \geq 0$, it will be sensible to omit for a while any subscripts and superscripts

$$z = z^n , \qquad \hat{z} = z^{n+1} ,$$

$$\hat{A}_2 = A_2(t_{n+1}) , \quad A_2 = A_2(t_n) , \quad \bar{A}_1 = A_1(t_{n+1/2}) ,$$

allowing a simpler writing of the ensuing formulas:

$$(E + 0.5\,\tau \bar{A}_1)\,\bar{z} = (E - 0.5\,\tau A_2)\,z + 0.5\,\tau\,\psi_1 ,$$

$$(E + 0.5\,\tau \hat{A}_2)\,\hat{z} = (E - 0.5\,\tau \bar{A}_1)\,\bar{z} + 0.5\,\tau\,\psi_2 .$$

The triangle inequality gives

$$\|\,(E + 0.5\,\tau \bar{A}_1)\,\bar{z}\,\| \leq \|\,(E - 0.5\,\tau A_2)\,z\,\| + 0.5\,\tau\,\|\,\psi_1\,\| , \tag{40}$$

$$\|\,(E + 0.5\,\tau \hat{A}_2)\,\hat{z}\,\| \leq \|\,(E - 0.5\,\tau \bar{A}_1)\,\bar{z}\,\| + 0.5\,\tau\,\|\,\psi_2\,\| . \tag{41}$$

An auxiliary lemma may be useful in the sequel.

**Lemma 1** *If $A \geq 0$ is a linear operator in a Hilbert space $H$, then*

$$\|\,(E - (1-\sigma)\,\tau A)\,y\,\| \leq \|\,(E + \sigma\,\tau A)\,y\,\| \quad \text{for} \quad \sigma \geq 0.5 . \tag{42}$$

This assertion is an immediate implication of the chain of the relations

$$\|\,(E + \sigma\,\tau A)\,y\,\|^2 - \|\,(E - (1-\sigma)\,\tau A)\,y\,\|^2$$
$$= 2\,\tau\,(Ay, y) + 2\,(\sigma - 0.5)\,\tau^2\,\|\,Ay\,\|^2 \geq 0 ,$$

which are valid for $\sigma \geq 0.5$ and the operator $A \geq 0$.

Collecting inequalities (40) and (41) and taking into account estimate (42) with the value $\sigma = 0.5$ incorporated, we proceed to the elimination of $\bar{z}$. Following established practice, we arrive at

$$\|\,(E + 0.5\,\tau \hat{A}_2)\,\hat{z}\,\| \leq \|\,(E + 0.5\,\tau A_2)\,z\,\| + \frac{\tau}{2}\,(\|\,\psi_1\,\| + \|\,\psi_2\,\|) ,$$

yielding with the aid of inequality (42) either

$$\|\,(E + 0.5\,\tau \hat{A}_2)\,\hat{z}\,\| \leq \|\,(E + 0.5\,\tau A_2)\,z\,\| + \frac{\tau}{2}\,(\|\,\psi_1\,\| + \|\,\psi_2\,\|)$$

or

$$\| (E+0.5\,\tau\, A_2(t_{k+1}))\, z^{k+1} \| \le \| (E+0.5\,\tau\, A_2(t_k))\, z^k \| + \frac{\tau}{2} (\| \psi_1^k \| + \| \psi_2^k \|)\,.$$

Summing up over $k = 0, 1, \ldots, n$ and inserting $z^0 = 0$, we find that

$$\|z^{n+1}\|_{(1)} = \| (E + 0.5\,\tau\, A_2(t_{n+1}))\, z^{n+1} \| \le \frac{1}{2} \sum_{k=0}^{n} \tau\, (\| \psi_1^k \| + \| \psi_2^k \|)\,, \tag{43}$$

where

$$\|z\|_{(1)}^2 = \| (E + 0.5\,\tau\, A_2)\, z \|^2 = \| z \|^2 + \tau\, (A_2 z, z) + \frac{\tau^2}{4} \| A_2 z \|^2\,. \tag{44}$$

Observe that estimate (43) remains unchanged if the norm $\|z\|_{(1)}$ is replaced either by the norm $\| z \|$ or by the norm

$$\|z\|_{(2)} = \Big( \| z \|^2 + \frac{\tau^2}{4} \| A_2 z \|^2 \Big)^{1/2}$$

by virtue of the relations

$$\| z \|^2 \le \|z\|_{(1)}^2 \quad \text{and} \quad \|z\|_{(2)}^2 \le \|z\|_{(1)}^2 .$$

This type of situation is covered by the following assertion.

**Theorem 3** *Scheme* (33)–(34) *is absolutely stable (for any* $h_1$, $h_2$ *and* $\tau$*) and converges with the rate* $O(|h|^2) + \tau^2$ *in the norm* $\| \cdot \|_{(1)}$ *given by formula* (44) *under the conditions which guarantee a second-order approximation on a solution* $u = u(x, t)$ *of problem* (32).

Recall that a second-order accuracy of scheme (33)–(34) is ensured by making a special choice of the boundary conditions for the intermediate value $y = \bar{y}$ such as

$$\bar{y} = \bar{\mu} \quad \text{for} \quad i_1 = 0, N_1\,, \quad \bar{\mu} = \frac{\mu^n + \mu^{n+1}}{2} - \frac{\tau^2}{4} (\Lambda_2 \mu)_{t,n}\,.$$

It is possible to demonstrate that the accuracy $O(|h|^2) + \tau^2$ of this scheme remains unchanged if we might agree to consider

$$\bar{y} = \frac{\mu^n + \mu^{n+1}}{2} \quad \text{for} \quad i_1 = 0, N_1\,. \tag{45}$$

This means that the second term $O(\tau^2)$ in the available exppression for $\bar{\mu}$ will be excluded from further consideration. The proof of this statement is omitted here. As a matter of fact, the stability of the scheme at hand with respect to the boundary conditions is revealed through such a stability analysis.

It is worth mentioning here several things for later use. Scheme (33) with the boundary conditions (45) is in common usage for "step-shaped" regions $\bar{G}$, whose sides are parallel to the coordinate axes. In the case of an arbitrary domain this scheme is of accuracy $O(|h|^2 + \tau^2\sqrt{h})$. Scheme (9)–(10) cannot be formally generalized for the three-dimensional case, since the instability is revealed in the resulting scheme.

**6. A higher-accuracy scheme.** By minor changes we are led to a higher-accuracy scheme such as

$$\frac{y^{n+1/2} - y^n}{\tau} = \sigma_1 \Lambda_1 y^{n+1/2} + (1-\sigma_2)\Lambda_2 y^n + \sigma_1 \varphi^n ,$$

(46)

$$\frac{y^{n+1} - y^{n+1/2}}{\tau} = (1-\sigma_1)\Lambda_1 y^{n+1/2} + \sigma_2 \Lambda_2 y^{n+1} + (1-\sigma_1)\varphi^n ,$$

$$x \in \omega_h , \quad n = 0, 1, 2, \ldots , \quad y(x, 0) = u_0(x) , \quad x \in \bar{\omega}_h ,$$

(47)
$$y^{n+1} = \mu^{n+1} \quad \text{for} \quad i_2 = 0 , \; i_2 = N_2 ,$$

$$y^{n+1/2} = \bar{\mu} \quad \text{for} \quad i_1 = 0 , \; i_1 = N_1 ,$$

where

$$\bar{\mu} = \sigma_1 \mu^{n+1} + (1-\sigma_1)\mu^n - \tau \Lambda_2 \Big( \sigma_1 \sigma_2 \mu^{n+1} - (1-\sigma_1)(1-\sigma_2)\mu^n \Big) ,$$

$$\sigma_\alpha = \frac{1}{2} - \frac{h_\alpha^2}{12\,\tau} , \quad \varphi^n = \left( f + \frac{h_1^2}{12} \Lambda_1 f + \frac{h_2^2}{12} \Lambda_2 f \right) \Bigg|_{t=t_{n+1/2}} .$$

It is plain to show that the scheme in view is stable and generates an approximation of $O(\tau^2 + |h|^4)$. With this aim, it seems worthwhile to design a factorized scheme for the same reason as before. The starting point is to eliminate $y_{n+1/2}$ from equations (46) in the process of transformations

(48) $B_1 y^{n+1/2} = C_2 y^n + \sigma_1 \tau \varphi^n , \quad C_1 y^{n+1/2} = B_2 y^{n+1} - (1-\sigma_1)\tau \varphi^n ,$

where $B_\alpha = E - \sigma_\alpha \tau \Lambda_\alpha$ and $C_\alpha = E + (1-\sigma_\alpha) \tau \Lambda_\alpha$, $\alpha = 1, 2$.

Having multiplied the first equation (48) by $1-\sigma_1$ and the second one by $\sigma_1$ and having added the final result, we find with the aid of the relation $(1-\sigma_1)B_1 + \sigma_1 C_1 = E$ the intermediate value

$$y^{n+1/2} = \sigma_1 B_2 y^{n+1} + (1-\sigma_1) C_2 y^n . \tag{49}$$

Substituting this expression into the first equation (48) yields

$$\sigma_1 B_1 B_2 y^{n+1} = \big(C_2 - (1-\sigma_1) B_1 C_2\big) y^n + \sigma_1 \tau \varphi^n$$

or

$$B_1 B_2 \frac{y^{n+1} - y^n}{\tau} = \big(\Lambda_1 + \Lambda_2 + (1-\sigma_1-\sigma_2) \tau \Lambda_1 \Lambda_2\big) y^n + \varphi^n . \tag{50}$$

By observing that

$$(1-\sigma_1-\sigma_2)\tau = \frac{h_1^2 + h_2^2}{12}$$

this provides enough reason to rewrite the preciding scheme as

$$B_1 B_2 \frac{y^{n+1} - y^n}{\tau} = \Lambda' y^n + \varphi^n , \quad x \in \omega_h , \tag{51}$$

$$y(x,0) = u_0(x) \quad \text{for} \quad x \in \bar{\omega}_h , \qquad y^n = \mu^n \quad \text{for} \quad x \in \gamma_h , \tag{52}$$

where

$$\Lambda' y = (\Lambda_1 + \Lambda_2)\, y + \frac{h_1^2 + h_2^2}{12} \Lambda_1 \Lambda_2\, y .$$

Simple algebra gives

$$\psi = \Lambda' u^n + \varphi^n - B_1 B_2 \frac{u^{n+1} - u^n}{\tau} = O(\tau^2 + |h|^4) , \tag{53}$$

thereby justifying that scheme (51)–(52) generates an approximation of

$O(\tau^2 + |h|^4)$. This is due to the asymptotic expansions

$$\Lambda' u^n + \varphi^n = \left( Lu + \frac{h_1^2}{12} L_1^2 u + \frac{h_2^2}{12} L_2^2 u + \frac{h_1^2 + h_2^2}{12} L_1 L_2 u \right)\Bigg|_{t=t_n} + \varphi^n + O(|h|^4)$$

$$= \left[ \left( \frac{h_1^2}{12} L_1 + \frac{h_2^2}{12} L_2 \right) \frac{\partial u}{\partial t} + Lu - \frac{h_1^2}{12} L_1 f - \frac{h_2^2}{12} L_2 f \right]_{t=t_n} + \varphi^n + O(|h|^4) ,$$

$$B_1 B_2 \frac{u^{n+1} - u^n}{\tau} = \left( u_t - \frac{\tau}{2} (\Lambda_1 + \Lambda_2)\, u_t + \frac{h_1^2}{12} \Lambda_1 u_t + \frac{h_2^2}{12} \Lambda_2 u_t \right)\Bigg|_{t=t_n} + O(\tau^2 + |h|^4)$$

$$= \left( \frac{\partial u}{\partial t} + \frac{\tau}{2} \frac{\partial^2 u}{\partial t^2} - \frac{\tau}{2} L \frac{\partial u}{\partial t} + \frac{h_1^2}{12} L_1 \frac{\partial u}{\partial t} + \frac{h_2^2}{12} L_2 \frac{\partial u}{\partial t} \right)\Bigg|_{t=t_n} + O(\tau^2 + |h|^4) ,$$

arising from the well-established relations

$$L_\alpha^2 u = -L_1 L_2 u + L_\alpha \frac{\partial u}{\partial t} - L_\alpha f , \quad \alpha = 1, 2 .$$

Further stability analysis is connected with the homogeneous boundary conditions. In preparation for this, we introduce the space $H = \overset{\circ}{\Omega}_h$ of grid functions and refer to the operators $A_\alpha y = -y_{\bar{x}_\alpha x_\alpha}$, $y \in \overset{\circ}{\Omega}_h$, as suggested before. All this enables us to concentrate on an operator-difference scheme

$$B y_t + A' y = \varphi(t) , \quad 0 \le t = n\tau , \quad y(0) = u_0 , \tag{54}$$

with the members

$$B = (E + \sigma_1 \tau A_1)(E + \sigma_2 \tau A_2) , \quad A' = A_1 + A_2 - (\varkappa_1 + \varkappa_2) A_1 A_2 ,$$

$$\varkappa_\alpha = h_\alpha^2 / 12 , \quad \sigma_\alpha = 0.5 - \varkappa_\alpha / \tau , \quad \alpha = 1, 2 .$$

Here the operators $A_1$ and $A_2$ are supposed to be self-adjoint, positive and commuting:

$$A_1^* = A_1 > 0\,, \quad A_2^* = A_2 > 0\,, \quad A_1\,A_2 = A_2\,A_1\,,$$

implying that $(A_1\,A_2)^* = A_1\,A_2 > 0$.

The simple observations that $\|\,A_\alpha\,\| < 4/h_\alpha^2$ and $\varkappa_\alpha\,\|\,A_\alpha\,\| < 1/3$ may be of help in verifying the stability condition in the space $H^{A'}$: $B \geq 0.5\,\tau\,A'$. Furthermore, with the aid of the inequality $A_\alpha \leq \|\,A_\alpha\,\|\,E$ we obtain

$$\begin{aligned} B - \frac{\tau}{2}\,A' &= E - \varkappa_1 A_1 - \varkappa_2 A_2 + \frac{\tau}{2}\,(\varkappa_1 + \varkappa_2)\,A_1 A_2 \\ &\quad + \left(\frac{\tau}{2} - \varkappa_1\right)\left(\frac{\tau}{2} - \varkappa_2\right) A_1 A_2 \\ &= E - \varkappa_1 A_1 - \varkappa_2 A_2 + \left(\frac{\tau^2}{4} + \varkappa_1 \varkappa_2\right) A_1 A_2 \\ &> E - \varkappa_1\,\|\,A_1\,\|\,E - \varkappa_2\,\|\,A_2\,\|\,E < \frac{1}{3}\,E\,, \end{aligned}$$

meaning $B > 0.5\,\tau\,A' + E/3$ and confirming the stability of the scheme in view in the space $H_{A'}$.

In particular, Theorem 7 in Chapter 6, Section 2 states that scheme (54) satisfies the a priori estimate

$$\|y^{j+1}\|_{A'} \leq \|y^0\|_{A'} + \sqrt{\frac{3}{2}}\left(\sum_{j'=0}^{j} \tau\,\|\,\varphi^{j'}\,\|^2\right)^{1/2}. \tag{55}$$

In trying to evaluate the accuracy of scheme (53) we set up the problem for the error $z^{n+1} = y^{n+1} - u^{n+1}$:

$$B\,z_t + A'z = \psi\,, \qquad z(0) = 0\,,$$

whose solution obeys estimate (55)

$$\|z^{j+1}\|_A \leq \frac{3}{2}\left(\sum_{j'=0}^{j} \tau\,\|\,\psi^{j'}\,\|^2\right)^{1/2}. \tag{56}$$

In passing from (55) to (56) we have taken into account that $z^0 = y^0 - u^0 = 0$ and that the operator $A'$ must satisfy the inequalities

$$A' > \frac{2}{3}\,A \qquad \text{or} \qquad \|z\|_{A'}^2 > \frac{2}{3}\,\|z\|_A^2\,,$$

which are immediate implications of the chain of the relations

$$A' = A - (\varkappa_1 + \varkappa_2)\, A_1 A_2 \geq A - \varkappa_1 \,\|\, A_1 \,\|\, A_2 - \varkappa_2 \,\|\, A_2 \,\|\, A_1 > \frac{2}{3}\, A\,.$$

Estimate (56) serves to motivate that scheme (51)–(52) and one of its equivalent schemes, namely scheme (46)–(47) converge in the space $H_A$ with the rate $O(\tau^2 + |h|^4)$.

**Remark** By analogy with the preceding section it is possible to evaluate the errors of approximation for either of the equations (48) such as

$$\psi_1 = \frac{C_2\, u^n - B_1\, \bar{u}}{\tau} + \sigma_1\, \varphi^n, \quad \psi_2 = \frac{C_1\, \bar{u} - B_2\, u^{n+1}}{\tau} + (1 - \sigma_1)\, \varphi^n .$$

This can be done using formula (49) for the intermediate value $\bar{u}$ of artificial character

$$\bar{u} = \sigma_1 B_2 u^{n+1} + (1 - \sigma_1)\, C_2 u^n .$$

Upon substituting this fictitious value into the above formula for $\psi_1$ we find that

$$\frac{\psi_1}{\sigma_1} = \Lambda' u^n - B\, \frac{u^{n+1} - u^n}{\tau} + \varphi^n = \psi^n\,,$$

thereby clarifying that the quantity $\psi_1 \sigma_1$ coincides with the error of approximation for the factorized scheme (51) that is known to us from formula (53).

By virtue of the relation

$$\frac{\psi_1}{\sigma_1} = \frac{\psi_2}{1 - \sigma_2}$$

we can be pretty sure that every of the equations (48) generates an approximation of $O(|h|^4 + \tau^2)$.

## 9.2 ECONOMICAL FACTORIZED SCHEMES

**1. Schemes with factorized operator.** We now consider the two-layer difference scheme

$$(1) \qquad B\, y_t + A\, y = \varphi\,, \quad 0 \leq t = j\tau < t_0\,, \quad j = 0, 1, \ldots\,, \quad y(0) = y_0\,.$$

Knowing the value $y = y^j$ on the $j$th layer, it is required to find the value $y^{j+1}$. In preparation for this, we derive the equation related to $y^{j+1}$ with a known right-hand side $F^j$:

$$(2) \qquad B\,y^{j+1} = F^j, \quad F^j = (B - \tau\,A)/y^j + \tau\,\varphi^j, \quad j = 0, 1, \dots, .$$

As can readily be observed, $O(N)$ operations are needed in giving $F^j$ and their amount is proportional to the total number of the grid nodes. This is certainly so with any difference scheme, whose pattern is independent of the grid. From equation (2) it is easily seen that the stable scheme (2) will be "economical" once the users perform $O(N)$ operations while solving equation (2).

Let "economical" operators $B_\alpha$, $\alpha = 1, 2, \dots, p$, be such that $O(N)$ operations are necessary in connection with solving the equation

$$(3) \qquad B_\alpha\,v = F\,.$$

Then scheme (1) with a factorized operator $B$ of the structure

$$(4) \qquad B = B_1\,B_2\,\cdots\,B_p$$

will also be "economical" in line with established priorities: the numerical solution to equation (2) with operator (4) requires $O(N)$ operations. Indeed, a solution to the equation

$$(5) \qquad B_1\,B_2\,\cdots\,B_p\,y^{j+1} = F^j$$

can be found by successive solution of $p$ equations of the form (3):

$$(6) \qquad B_1\,y_{(1)} = F^j, \quad B_\alpha\,y_{(\alpha)} = y_{(\alpha-1)}, \quad \alpha = 2, 3, \dots, p,$$

so that $y^{j+1} = y_{(p)}$. Here $y_{(1)} = y^{j+1/p}$, ..., $y_{(\alpha)} = y^{j+\alpha/p}$, ..., $y_{(p-1)} = y^{j+(p-1)/p}$ stand for intermediate values arising in the process of calculations. It follows from the foregoing that the stable scheme (1) with the factorized operator $B$ being a product of a finite number of "economical" operators $B_1, \dots, B_p$ becomes "economical", so there is some reason to be concerned about this. All the schemes with a factorized operator $B$ are called **factorized schemes.** It was shown in Section 1 that the economical alternating direction scheme is equivalent to the factorized scheme with the operator

$$(7) \qquad B = B_1\,B_2\,, \quad B_\alpha = E - 0.5\,\tau\,\Lambda_\alpha\,, \quad \Lambda_\alpha\,y = y_{\bar{x}_\alpha\,x_\alpha}\,, \quad \alpha = 1, 2\,.$$

Especial attention is being paid to the factorized schemes with the members

$$(8) \qquad B = B_1\, B_2\,, \quad B_\alpha = E + \tau\, R_\alpha\,,$$

$$R_1\, y = \sum_{\alpha=1}^{p} \frac{y_{\bar{x}_\alpha}}{h_\alpha}\,, \qquad R_2\, y = -\sum_{\alpha=1}^{p} \frac{y_{x_\alpha}}{h_\alpha}\,,$$

where $R_1$ and $R_2$ are "triangle" operators with associated triangle matrices. These fall within the category of explicit alternating direction schemes. Let us stress here that the operators $R_1$ and $R_2$ may be, generally speaking, non-self-adjoint, but they are always mutually adjoint to each other. In that case a solution to equation (3) can be found by "through execution" formulas.

Undoubtedly, the reader comes across difference operators $B_\alpha$ of the structure $B_\alpha = E - \sigma\tau\Lambda_\alpha$, where $\Lambda_\alpha$ approximates the differential operator $L_\alpha$ with partial derivatives of one argument $x_\alpha$. For example, if $L_\alpha u = \partial^2 u/\partial x_\alpha^2$, then $\Lambda_\alpha y = y_{\bar{x}_\alpha x_\alpha}$ is a three-point operator, whose use permits us to solve equation (3) by the elimination method. It is worth mentioning here that any difference scheme can be reduced to a sequence of simpler schemes in a number of different ways. This is certainly so with scheme (1), implying that

$$y^{j+1} = y^j + \tau\, w^j\,,$$

where $w^j$ is a solution to the equation

$$(9) \qquad B_1\, B_2\, \dots\, B_p\, w = \Phi^j\,, \qquad \Phi^j = \varphi^j - A\, y^j\,.$$

The value $w^j$ can be recovered from the governing system of $p$ equations

$$(10) \qquad B_1\, w_{(1)} = \Phi^j, \quad B_\alpha\, w_{(\alpha)} = w_{(\alpha-1)}\,, \quad \alpha = 2, 3, \dots, p\,,$$

with further reference to

$$(11) \qquad w^j = w_{(p)}\,.$$

It is worth recalling here that the first economical schemes were intended for the elimination of intermediate values with no problems. The main idea behind this approach is to involve factorized schemes "in integer steps", a key role of which is to relate the values $y^j$ and $y^{j+1}$ in some or other convenient ways.

**2. The boundary conditions.** As one might expect, stability and approximation take place for the factorized scheme (1). In this view, it seems reasonable to adopt equations (6) or (10)–(11) as a perfect computational algorithm in designing the factorized scheme (1). But this equivalence can be established only with consistent boundary conditions and needs certain clarification.

In the forthcoming example the first boundary-value problem is posed for the heat conduction equation in the rectangle $\bar{G}_0 = \{0 \le x_1 \le l_1,\ 0 \le x_2 \le l_2\}$ with the boundary $\Gamma$:

$$\frac{\partial u}{\partial t} = (L_1 + L_2)\,u + f(x,t)\,, \quad u|_{\Gamma} = \mu(x,t)\,, \quad u(x,0) = u_0(x)\,, \tag{12}$$

where $L_\alpha u = \partial^2 u/\partial x_\alpha^2$, $\alpha = 1, 2$.

We begin by placing the factorized scheme (1) with $\Lambda_\alpha y = y_{\bar{x}_\alpha x_\alpha}$ on a rectangular grid $\bar{\omega}_h = \{(i_1 h_1, i_2 h_2)\}$ with steps $h_1$ and $h_2$ in the specified domain $\bar{G}_0$:

$$B_1\, B_2\, y_t = \Lambda\, y + \varphi\,, \quad y^j\,|_{\gamma_h} = \mu^j\,, \quad y^0 = u_0(x)\,. \tag{13}$$

Here and in all that follows $B_\alpha = E - \sigma\tau\Lambda_\alpha$, $\Lambda = \Lambda_1 + \Lambda_2$ and $\gamma_h$ is the boundary of the grid $\omega_h$. In passing from one layer to another algorithm (6) is performed for problem (13):

$$B_1\, y_{(1)} = F^j\,, \quad F^j = (B_1\, B_2 + \tau\,\Lambda)\, y^j + \tau\,\varphi^j\,, \quad B_2\, y^{j+1} = y_{(1)}\,, \tag{14}$$

which is put together with the boundary conditions $y^{j+1}|_{\gamma_h} = \mu^{j+1}$.

As far as the operator $B_1\, B_2$ on $\bar{\omega}_h$ is concerned (including the boundary $x_1 = 0$, $x_1 = l_1$), the equation $B_2\, y^{j+1} = y_{(1)}$ should be valid not only for $0 < x_1 < l_1$, but also on the boundary for $x_1 = 0$ and $x_1 = l_1$. Since the values $y^{j+1}|_{\gamma_h} = \mu^{j+1}$ are already known, it follows from the foregoing that

$$y_{(1)} = (E - \sigma\,\tau\,\Lambda_2)\,\mu^{j+1} = \mu^{j+1} - \sigma\,\tau\,\Lambda_2\,\mu^{j+1} \quad \text{for} \quad x_1 = 0, l_1\,. \tag{15}$$

When specifying $y_{(1)}$ for $x_1 = 0, l_1$ by the preceding formula, problems (13) and (14)–(15) become equivalent. This fact can easily be verified during the course of the elimination of $y_{(1)}$ from (14). In the framework of the second algorithm we accept

$$\begin{aligned} &B_1\, \omega_{(1)} = \Phi^j\,, \quad \Phi^j = \Lambda\, y^j + \varphi^j\,, \\ &B_2\, \omega_{(2)} = \omega_{(1)}\,, \quad y^{j+1} = y^j + \tau\,\omega_{(2)} \end{aligned} \tag{16}$$

and the boundary conditions are imposed to be

$$(17)\qquad \begin{aligned} \omega_{(1)} &= (E - \sigma\,\tau\,\Lambda_2)\,\frac{\mu^{j+1} - \mu^j}{\tau} \quad &&\text{for}\quad x_1 = 0, l_1\,, \\ \omega_{(2)} &= \frac{\mu^{j+1} - \mu^j}{\tau} &&\text{for}\quad x_2 = 0, l_2\,, \end{aligned}$$

that is, $\omega_{(1)} = 0$ and $\omega_{(2)} = 0$ on the boundary $\gamma_h$ if $\mu$ is independent of $t$.

Observe that in giving scheme (1) in matrix (operator) form it is possible to take into consideration the homogeneous boundary conditions by rearranging the right-hand side $\varphi$ at the near-boundary nodes. The design of the factorized scheme also involves the homogeneous boundary conditions ($y_{(1)} = y^j = 0$, $\omega_{(1)} = \omega_{(2)} = 0$ for $x \in \gamma_h$), but the retention of the approximation order necessitates imposing the extra member $-\sigma^2\tau^2 h_1^{-2}\Lambda_2\mu_t$ on the right-hand side of this scheme at the near-boundary nodes for $i_1 = 1$ and $i_1 = N_1 - 1$.

**3. Constructions of economical factorized schemes.** Using the regularization method behind, we try to develop the general method for constructing stable economical difference schemes on the basis of the primary stable scheme

$$(18)\qquad B\,\frac{y^{n+1} - y^n}{\tau} + A\,y^n = \varphi^n$$

with an operator of the structure

$$(19)\qquad B = E + \tau\,R\,.$$

The relation $B \geq 0.5\,\tau A$ is ensured by the stability property of this scheme.

In such a setting it is preassumed that $R$ is a sum of a finite number of "economical" operators $R_\alpha$, $\alpha = 1, 2, \ldots, p$:

$$(20)\qquad R = R_1 + \cdots + R_p\,.$$

The operator $B$ can be factorized by replacing $B = E + \tau\,(R_1 + \cdots + R_p)$ by the factorized operator

$$(21)\qquad \widetilde{B} = B_1 \cdots B_p\,, \qquad B_\alpha = E + \tau\,R_\alpha\,,$$

making it possible to ignore the primary scheme (1) in favour of the factorized scheme

$$(22)\qquad B_1 \cdots B_p\,y_t + A\,y = \breve{\varphi}\,.$$

Sometimes such a passage permits one to justify the approximation being $\tilde{\varphi}$ instead of $\varphi$ near the boundary of the grid domain.

> *If the primary scheme* (18) *is stable, then so is the factorized scheme* (22) *in the case, where the operators* $R_1$, $R_2, \ldots, R_p$ *are self-adjoint* ($R_\alpha = R_\alpha^*$), *non-negative* ($R_\alpha \geq 0$) *and pairwise commutative* ($R_\alpha R_\beta = R_\beta R_\alpha$, $\alpha, \beta = 1, 2, \ldots, p$).

By virtue of the indicated properties of the operators $R_\alpha$ their products $R_\alpha R_\beta$, $R_\alpha R_\beta R_\gamma$, etc., will be self-adjoint non-negative linear operators. This provides enough reason to conclude that

$$\widetilde{B} = B_1 B_2 = E + \tau (R_1 + R_2) + \tau^2 R_1 R_2$$

$$= B + \tau^2 R_1 R_2 \geq B \quad \text{for} \quad p = 2\,,$$

$$\widetilde{B} = B_1 B_2 \cdots B_p = E + \tau (R_1 + R_2 + \cdots + R_p) + \tau^2 Q_p$$

$$= B + \tau^2 Q_p \geq B\,,$$

where $Q_p^* = Q_p \geq 0$.

Thus, $\widetilde{B} \geq B \geq 0.5\, \tau A$, meaning the stability of the factorized scheme (22). The operators $R_\alpha$ are so chosen as to satisfy the condition of approximation, too. The forthcoming example helps clarify what is done.

**Example 1** We are looking for a solution of the first boundary-value problem for the heat conduction equation with variable coefficients

$$\frac{\partial u}{\partial t} = L_1 u + L_2 u + f(x,t)\,, \quad x \in G\,, \quad t > 0\,, \tag{23}$$

$$u|_\Gamma = \mu(x,t)\,, \quad u(x,0) = u_0(x)\,,$$

$$L_\alpha u = \frac{\partial}{\partial x_\alpha} \left( k_\alpha(x,t) \frac{\partial u}{\partial x_\alpha} \right), \quad 0 < c_1 \leq k_\alpha \leq c_2\,, \quad \alpha = 1, 2\,,$$

$$G = \{ 0 < x_\alpha < l_\alpha\,,\ \alpha = 1, 2 \}\,.$$

For this, we have occasion to use the two-layer factorized scheme

$$(24)\qquad \begin{gathered}(E+\tau R_1)(E+\tau R_2)\,y_t + A\,y = \varphi\,,\\ x\in\omega_h\,,\quad t=n\tau\,,\quad n=0,1,\dots\,,\\ y|_{\gamma_h} = \mu(x,t)\,,\quad t=n\tau\ge 0\,,\\ y(x,0)=u_0(x)\,,\quad x\in\bar\omega_h\,,\end{gathered}$$

where $\bar\omega_h = \{x_i=(i_1 h_1, i_2 h_2),\ i_\alpha = 0,1,\dots,N_\alpha,\ h_\alpha N_\alpha = l_\alpha,\ \alpha=1,2\}$ is the grid in the rectangle $\bar G$ with the boundary $\gamma_h$.

The operator $L_\alpha$ is approximated to second order by the difference operator

$$\Lambda_\alpha y = \left(a_\alpha(x,t)\,y_{\bar x_\alpha}\right)_{x_\alpha}\,,\quad 0<c_1\le a_\alpha(x,t)\le c_2\,,\quad \alpha=1,2\,.$$

Let

$$A=-(\Lambda_1+\Lambda_2)\,,\quad R_\alpha = -\sigma\,c_2\,\overset{\circ}{\Lambda}_\alpha\,,\quad \alpha=1,2\,,$$

where $\overset{\circ}{\Lambda}_\alpha y = y_{\bar x_\alpha x_\alpha}$.

Stability analysis is mostly based on the assumptions that the boundary conditions are homogeneous and $H=\overset{\circ}{\Omega}_h$ is the space of all grid functions given on the grid $\bar\omega_h$ and vanishing on the boundary $\gamma_h$. The inner product structure is the same as suggested before for problem (22). A brief survey of the properties of the operators $A$ and $R_\alpha$ as operators in that space $H$ is presented below:

$$(25)\qquad \begin{gathered}A=A^*>0\,,\quad A_\alpha\le c_2\,\overset{\circ}{A}_\alpha\,,\quad \overset{\circ}{A}_\alpha y = -\overset{\circ}{\Lambda}_\alpha y\ \ \text{for any}\ \ y\in\overset{\circ}{\Omega}_h\,,\\ R_\alpha=\sigma\,c_2\,\overset{\circ}{A}_\alpha\,,\quad R_\alpha=R_\alpha^*>0\,,\quad \alpha=1,2\,,\quad R_1R_2=R_2R_1\,.\end{gathered}$$

As can readily be observed, the stability condition

$$B=(E+\tau R_1)(E+\tau R_2)\ge 0.5\,\tau A$$

will be ensured if $\sigma\ge 0.5$ or even if $\sigma\ge 0.5-1/(\tau\|A\|)$. It seems clear that the three-point difference operators $B_\alpha = E+\tau R_\alpha$ with constant coefficients are "economical", since the equations

$$B_\alpha\,\omega = (E+\tau R_\alpha)\,\omega = F_\alpha\,,\quad \alpha=1,2\,,$$

can be solved using Gaussian elimination along the rows for $\alpha = 1$ as well as along the columns for $\alpha = 2$.

Under such an approach the factorized scheme is of no less than first-order accuracy in $\tau$. In a similar way an economical factorized scheme can be designed in the $p$-dimensional case when

$$\frac{\partial u}{\partial t} = \sum_{\alpha=1}^{p} L_\alpha u + f \tag{23'}$$

and the operator $L_\alpha$ is specified by formula (23). If so, we might agree to consider

$$B = \prod_{\alpha=1}^{p} (E + \tau R_\alpha) .$$

Here the operators $R_\alpha$ are of the same structure as was chosen for $p = 2$.

**Example 2** The statement of the first boundary-value problem for the parabolic equation with mixed derivatives in the parallelepiped $\bar{G}_0 = \{0 \le x_\alpha \le l_\alpha,\ \alpha = 1, 2, \ldots, p\}$ is

$$\frac{\partial u}{\partial t} = L u + f(x,t) , \quad u|_\Gamma = \mu(x,t) , \quad u(x,0) = u_0(x) , \tag{26}$$

$$L u = \sum_{\alpha,\beta=1}^{p} \frac{\partial}{\partial x_\alpha} \Big( k_{\alpha\beta}(x,t) \frac{\partial u}{\partial x_\beta} \Big) ,$$

$$0 < c_1 \sum_{\alpha=1}^{p} \xi_\alpha^2 \le \sum_{\alpha,\beta=1}^{p} k_{\alpha\beta}(x,t)\, \xi_\alpha \xi_\beta \le c_2 \sum_{\alpha=1}^{p} \xi_\alpha^2 .$$

As the operators $R_\alpha$ involved, we take once again the operators specified by formulas (25) that are self-adjoint, positive (for $\sigma > 0$) and pairwise commutative in the space $\overset{\circ}{\Omega}_h$, since $\bar{G}_0$ is a parallelepiped. The scheme with these members is stable for $\sigma \ge 0.5$. As far as $B_\alpha = E + \tau R_\alpha$, $\alpha = 1, 2, \ldots, p$, are three-point difference operators with constant coefficients, the possible follow up is the algorithm suggested in Example 1 for determination of $y^{j+1}$ on the basis of $y^j$. We will not pursue analysis of this: the ideas needed to do so have been covered.

**Example 3** As a matter of experience, the intention is to use a higher-accuracy scheme similar to (12) being used for the heat conduction equation. As we have stated in Section 1, the factorized scheme

$$(27) \qquad (E - \sigma_1 \tau \Lambda_1)(E - \sigma_2 \tau \Lambda_2) y_t = \Lambda' y + \varphi ,$$

$$x \in \omega_h , \quad 0 \le t = n\tau ,$$

$$y|_{\gamma_h} = \mu(t) , \quad t = n\tau , \quad y(x, 0) = u_0(x) , \quad x \in \bar{\omega}_h ,$$

where

$$(28) \qquad \Lambda' = \Lambda_1 + \Lambda_2 + \frac{h_1^2 + h_2^2}{12} \Lambda_1 \Lambda_2 ,$$

$$\Lambda_\alpha y = y_{\bar{x}_\alpha x_\alpha} , \quad \sigma_\alpha = \frac{1}{2} - \frac{h_\alpha^2}{12} , \quad \alpha = 1, 2 ,$$

$$\varphi = \left( f + \frac{h_1^2}{12} \Lambda_1 f + \frac{h_2^2}{12} \Lambda_2 f \right)^{n+1/2} ,$$

generates an approximation of $O(\tau^2 + |h|^4)$.

Also, the scheme so constructed is equivalent to the alternating direction scheme (46)–(47) from Section 1. Among other schemes, the users prefer two alternating direction schemes, either of which is equivalent to scheme (27).

*The first scheme:*

$$(E + \sigma_1 \tau \Lambda_1) \bar{y} = ((E - \sigma_1 \tau \Lambda_1)(E - \sigma_2 \tau \Lambda_2) + \tau \Lambda') y + \tau \varphi ,$$

$$(E - \sigma_2 \tau \Lambda_2) \hat{y} = \bar{y} .$$

Every equation refers to two-layer schemes with a common canonical form

$$(29) \qquad (E - \sigma_1 \tau \Lambda_1) \frac{\bar{y} - y}{\tau} = \Lambda_1 y + (1 - \sigma_2) \Lambda_2 y$$

$$+ (\varkappa_1 + \varkappa_2 + \tau \sigma_1 \sigma_2) \Lambda_1 \Lambda_2 y + \varphi ,$$

$$(E - \sigma_2 \tau \Lambda_2) \frac{\hat{y} - \bar{y}}{\tau} = \sigma_2 \Lambda_2 \bar{y} , \quad \varkappa_\alpha = \frac{h_\alpha^2}{12} , \quad \alpha = 1, 2 .$$

In turn, the supplementary boundary conditions become

$$(30)\qquad \begin{aligned} y^n &= \mu^n && \text{for } i_2 = 0, N_2, \\ \bar{y} &= (E - \sigma_2\,\tau\,\Lambda_2)\,\mu^{n+1} && \text{for } i_1 = 0, N_1. \end{aligned}$$

*The second scheme:*

$$(31)\qquad \begin{aligned} (E - \sigma_1\,\tau\,\Lambda_1)\,\frac{\bar{y} - y}{\tau} &= \Lambda'\,y + \varphi\,, \\ (E - \sigma_2\,\tau\,\Lambda_2)\,\frac{\hat{y} - \bar{y}}{\tau} &= \sigma_2\,\Lambda\,(\bar{y} - y) \end{aligned}$$

with the supplementary boundary conditions for $\bar{y}$:

$$(32)\qquad \bar{y} = \mu^n + \tau\,(E - \tau\,\sigma_2\,\Lambda_2)\,\frac{\mu^{n+1} - \mu^n}{\tau} \quad \text{for} \quad i_1 = 0, N_1\,.$$

Then problems (29)–(30) and (31)–(32) can be solved by the alternating direction method in just the same way as was done in Section 1 for problem (46)–(47) using Gaussian elimination along the rows as well as along the columns of the grid $\bar{\omega}_h$.

Thus, three different examples of interest bring out the indisputable merit of the alternating direction method and unveil its potential. From what has been said above it follows that the factorized schemes find some range of applications solely in rectangles and parallelepipeds and no more. The only exception is the case $B = B_1 B_2$, $B_\alpha = E + \tau R_\alpha$, where $R_1$ and $R_2$ are "triangle" operators, by means of which it is possible to generate a lower-order approximation only under the condition $\tau = O(h^2)$.

**4. Three-layer factorized schemes.** Being concerned with economical three-layer schemes, we confine ourselves here to

$$B y_{\overset{\circ}{t}} + \tau^2 R\, y_{\bar{t}t} + A\, y = \varphi\,, \quad 0 < t = j\tau < t_0\,, \quad y(0) = y_0\,, \quad y(\tau) = y_1\,.$$

A first step towards the solution of this difference problem is to solve it with respect to $y^{j+1}$, leading to

$$(B + 2\,\tau R)\,y^{j+1} = 2\,\tau\,(2R - A)\,y^j + (B - 2\,\tau R)\,y^{j-1} + 2\,\tau\,\varphi^j\,.$$

Because of this form, the operator $B + 2\,\tau R$ on the upper layer is yet to be factorized for economical reasons.

In what follows the weighted scheme

$$(33) \qquad y_{\overset{\circ}{t}} + A\left(\sigma_1 \hat{y} + (1-\sigma_1-\sigma_2)\, y + \sigma_2 \check{y}\right) = \varphi\,,$$

$$y(0) = y_0\,, \qquad y(\tau) = y_1\,,$$

is treated as a primary one and it will be written in canonical form for later use:

$$(34) \qquad (E + \tau(\sigma_1 - \sigma_2)\, A)\, y_{\overset{\circ}{t}} + 0.5\,(\sigma_1+\sigma_2)\,\tau^2 A\, y_{\bar{t}t} + A\, y = \varphi\,,$$

where the unknown operator $B + 2\,\tau R = E + 2\,\sigma_1 \tau A$ is sought. Let now $A = A_1 + A_2$. The factorized operator

$$\widetilde{B} + 2\,\tau \widetilde{R} = (E + 2\,\sigma_1 \tau A_1)(E + 2\,\sigma_1 \tau A_2) = E + 2\,\sigma_1 \tau A + 4\,\sigma_1^2 \tau^2 A_1 A_2$$

will appear in place of the operator $B + 2\tau R$ as one possible way of connecting two operators $\widetilde{B}$ and $\widetilde{R}$ by a unique condition and it may be of assistance in designing many factorized schemes. Later we will expound certain exploratory devices for obtaining them. For example, this can be done using (34) in the form

$$(B + 2\,\tau R)\, y_t + (B - 2\,\tau R)\, y_{\bar{t}} + 2\, A\, y = 2\,\varphi$$

and replacing the operator $B + 2\,\tau R$ by the factorized operator $\widetilde{B} + 2\,\tau \widetilde{R}$, making it possible to write the scheme at hand in canonical form with the aid of the well-known relations

$$y_t = y_{\overset{\circ}{t}} + 0.5\,\tau\, y_{\bar{t}t}\,, \qquad y_{\bar{t}} = y_{\overset{\circ}{t}} - 0.5\,\tau\, y_{\bar{t}t}\,.$$

The outcome of this is

$$(35) \qquad \left(E + \tau(\sigma_1-\sigma_2)\, A + 2\,\sigma_1^2\, \tau^2 A_1\, A_2\right) y_{\overset{\circ}{t}}$$

$$+ \tau^2 \left(\frac{\sigma_1+\sigma_2}{2}\, A + \sigma_1^2\, \tau A_1\, A_2\right) y_{\bar{t}t} + A\, y = \varphi\,,$$

$$y(0) = y_0\,, \qquad y(\tau) = y_1\,,$$

so that

$$\widetilde{B} = B + 2\,\sigma_1^2\, \tau^2 A_1\, A_2\,, \qquad \widetilde{R} = R + \sigma_1^2\, \tau A_1\, A_2\,.$$

Knowing the values $y^{j-1}$ and $y^j$ we do follow the same algorithm for determination of the value $y^{j+1}$, whose use permits us to find that

$$B_1 \omega_{(1)} = F^j, \qquad F^j = (2\tau R - B)\, y^j_{\bar{t}} - 2A y^j + 2\varphi^j,$$

$$B_2 \omega_{(2)} = \omega_{(1)}, \qquad y^{j+1} = y^j + \tau \omega_{(2)},$$

$$B_1 = E + 2\sigma_1 \tau A_1, \quad B_2 = E + 2\sigma_1 \tau A_2.$$

A similar procedure works for the boundary conditions imposed for $\omega_{(1)}$ as was done for a two-layer factorized scheme and for this reason it is omitted here.

Stability of the factorized scheme (35) can be established on account of the general theorems from Chapter 6, Section 3, due to which it follows from the foregoing that the conditions

$$\sigma_1 \ge \sigma_2, \quad \sigma_1 + \sigma_2 > 0.5, \quad A_\alpha = A^*_\alpha > 0, \quad A_1 A_2 = A_2 A_1$$

are sufficient for the stability of the scheme concerned. If $\sigma_1 \ge \sigma_2$, $\sigma_1 + \sigma_2 > 0.5$, $A_\alpha = A^*_\alpha > 0$, then the primary scheme is stable, since $B \ge E$ and $4R > A$. As far as the operators $A_1$ and $A_2$ are commuting, we deduce that $A_1 A_2 > 0$, meaning $\widetilde{B} > B$ and $\widetilde{R} > R$. Due to this fact the stability of the primary scheme implies that of the factorized scheme (35).

A particular case where $R = \sigma A$, $\sigma = 0.5\,(\sigma_1 + \sigma_2)$, is showing the gateway to the future research, whose aims and scope are connected with the general method for constructing three-layer economical factorized schemes by means of the **regularization principle** of difference schemes. A simple example

$$(36)\quad y_{\overset{\circ}{t}} + \tau^2 R\, y_{\bar{t}t} + A y = \varphi, \quad 0 < t = j\tau < t_0, \quad y(0) = u_0, \quad y(\tau) = \bar{u}_0,$$

can add interest and help in understanding. Later we will elaborate on this for rather complicated cases. Here the value $y(\tau) = \bar{u}_0$ for $t = \tau$ is so taken as to provide a second-order approximation in $\tau$. Also, the stability property guides a proper choice of the operator $R$.

That is why a reasonable form of the primary scheme is

$$(37)\qquad (E + 2\tau R)\, y_t = -F, \quad F = (2\tau R - E)\, y_{\bar{t}} - 2Ay + 2\varphi.$$

Let $R$ be a sum of "economical" operators such that $R = R_1 + R_2 + \cdots + R_p$. By replacing in (37) the operator

$$E + 2\tau R = E + 2\tau \sum_{\alpha=1}^{p} R_\alpha$$

by the factorized operator

$$E + 2\tau\widetilde{R} = \prod_{\alpha=1}^{p} (E + 2\tau R_\alpha) = E + 2\tau R + 4\tau^2 Q_p, \quad \widetilde{R} = R + 2\tau Q_p,$$

where $Q_p = R_1 R_2$ for $p = 2$ and $Q_p = R_1 R_2 + R_1 R_3 + R_2 R_3 + 2\tau R_1 R_2 R_3$ for $p = 3$, etc., the structure of the factorized scheme is

$$(38) \qquad B_1 \cdots B_p\, y_t = -F\,, \qquad B_\alpha = E + 2\tau R_\alpha\,.$$

The canonical form is given by

$$(39) \qquad B\, y_{\overset{\circ}{t}} + \tau^2 \bar{R}\, y_{\bar{t}t} + A\, y = \varphi\,,$$

where $B = E + 2\tau^2 Q_p$ and $\bar{R} = R + \tau Q_p$.

Let now $\psi_0(u)$ be the error of approximation in the class of solutions $u = u(x,t)$ of the continuous problem for the primary scheme (36) and $\psi_1(u)$ be the same quantity for the factorized scheme (39). In this regard, it should be noted that

$$\psi_1(u) = \psi_0(u) + \psi^*, \qquad \psi^* = 2\tau^2 Q_p u_t\,.$$

When $\|Q_p\, u_t\|_{(2^*)} = O(1)$ is accepted in some suitable grid norm $\|\cdot\|_{(2^*)}$ built into stability theorems, we might achieve $\|\psi^*\|_{(2^*)} = O(\tau^2)$ and in passing from the primary scheme to the economical factorized scheme (38) the error of approximation changes within a quantity of $O(\tau^2)$. Following these procedures, we obtain economical factorized schemes of second-order accuracy in $\tau$ as stated before due to the extra smoothness of the solution $u$. Such a stability analysis of schemes (36) and (39) is mostly based on the further treatment of the operators $R$ and $A$ as linear operators acting from the space $H = \overset{\circ}{\Omega}_h$ into the space $H$. In particular, this means that the boundary conditions on $\gamma_h$ are homogeneous for a scheme approximating (26).

Under the natural premises

$$A = A^* > 0\,, \quad R_\alpha = R_\alpha^* > 0\,, \quad \alpha = 1, 2, \ldots, p\,,$$

another conclusion can be drawn that the primary scheme (36) is stable, provided the condition

$$(40) \qquad R \geq \frac{1+\varepsilon}{4} A\,, \qquad \varepsilon > 0\,,$$

holds.

In the case of a variable operator $A = A(t)$ it is required, in addition, that $A(t)$ is Lipshitz continuous in $t$, allowing the operator $R$ to be constant.

In the situation when the operators $R_\alpha$ are pairwise commutative, the stability of the primary scheme implies that of the factorized scheme (39), since $Q_p > 0$, $B = E + 2\,\tau^2 Q_p > E$ and $\bar{R} = R + \tau\, Q_p$, giving

$$\bar{R} > \frac{1+\varepsilon}{4} A\,.$$

Let us stress that the well-founded choice of the regularizer $R$ is in complete agreement with an approved principle governing what can happen and depends on the range of situations to be considered. This is certainly so with two-layer and three-layer schemes and, therefore, *one and the same regularizer $R$ could be useful and perceived to be useful for different operators $A$.*

**Example 4** By having recourse to problem (23) associated with the heat conduction equation with variable coefficients for the same choice of the operators $A$, $R_1$ and $R_2$ as in Example 1 for the two-layer economical scheme (24) we concentrate on the primary scheme (36)

$$y_{\overset{\circ}{t}} + \tau^2 R\, y_{\bar{t}t} + A\, y = \varphi\,, \qquad R = R_1 + R_2\,,$$

which generates an approximation of order 2 and possesses the residual $\psi = O(\tau^2 + |h|^2)$. This scheme is stable for $\sigma \geq (1+\varepsilon)/4$. Observe that the factorized scheme of the type (39) with the members

$$B = E + 2\,\tau^2 R_1\, R_2, \qquad \bar{R} = R + \tau\, R_1\, R_2$$

turns out to be absolutely stable for $\sigma \geq (1+\varepsilon)/4$, $\varepsilon > 0$. This is due to the fact that the operators $R_1$ and $R_2$ are commuting.

Also, the operator $A(t)$ will be Lipshitz continuous if we agree to consider

$$|(a_\alpha)_t| \leq c_3 a_\alpha, \quad c_3 = \text{const} > 0\,, \quad \alpha = 1, 2\,.$$

Since every operator $B_\alpha = E + 2\tau R_\alpha$ is "economical", the same property holds true for scheme (39).

In an attempt to solve the system of difference equations

$$B_1 B_2\, y_t = -F\,, \qquad B_\alpha = E + 2\,\tau R_\alpha, \qquad y|_{\gamma_h} = \mu\,,$$

the following algorithm is recommended for the possible applications:

$$B_1 \, \omega_{(1)} = -F \,, \quad x \in \omega_h \,, \quad \omega_{(1)} = B_2 \, \mu_t \quad \text{for} \quad x_1 = 0, l_1 \,,$$

$$B_2 \, \omega_{(2)} = \omega_{(1)} \,, \quad x \in \omega_h \,, \qquad \omega_{(2)} = \mu_t \quad \text{for} \quad x_2 = 0, l_2 \,,$$

$$y^{j+1} = y^j + \tau \, \omega_{(2)} \,.$$

In this context, we should focus the reader's attention on the process of specifications of boundary conditions by doing Gaussian elimination along the rows for $\omega_{(1)}$ and along the columns for $\omega_{(2)}$. When $\mu$ happens to be independent of $t$, that is, $\mu = \mu(x)$, the quantities $\omega_{(1)}$ and $\omega_{(2)}$ satisfy the homogeneous boundary conditions.

**Example 5** In tackling problem (26) the operator $L$ is approximated by the difference operator

$$\Lambda \, y = 0.5 \sum_{\alpha,\beta=1}^{p} \Big[ \big( k_{\alpha\beta}(x,t) \, y_{\bar{x}_\beta} \big)_{x_\alpha} + \big( k_{\alpha\beta}(x,t) \, y_{x_\beta} \big)_{\bar{x}_\alpha} \Big]$$

with the coefficients $k_{\alpha\beta}$ still subject to the following conditions with constants $c_2 \geq c_1 > 0$:

$$c_1 \sum_{\alpha=1}^{p} \xi_\alpha^2 \leq \sum_{\alpha,\beta=1}^{p} k_{\alpha\beta}(x,t) \, \xi_\alpha \, \xi_\beta \leq c_2 \sum_{\alpha=1}^{p} \xi_\alpha^2 \quad \text{for all } x \in \bar{G}_0, \quad t \geq 0 \,.$$

Having stipulated the condition $|(k_{\alpha\beta})_t| \leq c_3$, where $c_3 = \text{const}$, the operator $A(t)y = -\Lambda(t)y$ is Lipshitz continuous in the space $H = \overset{\circ}{\Omega}_h$. Also, it will be sensible to approve the same selection rule for the regularizer as in the preceding example:

$$R = \sum_{\alpha=1}^{p} R_\alpha \,, \qquad R_\alpha \, y = -\sigma \, c_2 \, \overset{\circ}{\Lambda}_\alpha \, y \,, \qquad \overset{\circ}{\Lambda}_\alpha \, y = y_{\bar{x}_\alpha x_\alpha} \,.$$

In that case the factorized scheme (39) is stable under condition (40) or, what amounts to the same, for $\sigma \geq (1+\varepsilon)/4$, $\varepsilon > 0$, and it is of second-order accuracy with respect to all the variables.

In this direction, the intention is to construct in a similar manner the factorized scheme related to another primary scheme

$$(41) \qquad (E+\tau^2 R)\, y_{\bar{t}t} + A\, y = \varphi\,, \qquad R = \sum_{\alpha=1}^{p} R_\alpha\,,$$

where the operators $R_\alpha$ are pairwise commutative, self-adjoint and positive. This scheme is certainly associated with the hyperbolic type equation $\partial^2 u/\partial t^2 = Lu + f$. There are several mechanisms for passing from scheme (41) to the factorized scheme with the accompanying factorized operator at the members $\hat{y}$, $y_{\bar{t}t}$ or $y_t$. The usual way of covering this is connected with further replacement of the operator $E+\tau^2 R$ by the factorized operator

$$E+\tau^2 \widetilde{R} = \prod_{\alpha=1}^{p} (E+\tau^2 R_\alpha)\,, \quad \widetilde{R} = R + \tau^2 Q_p\,, \quad Q_p = Q_p^* > 0\,,$$

leaving us with the scheme

$$(E+\tau^2 \widetilde{R})\, y_{\bar{t}t} + A\, y = \varphi\,,$$

which is stable only if the primary scheme is stable, since $\widetilde{R} = \widetilde{R}^* > R$. The resulting scheme differs from the primary one within a quantity $O(\tau^2)$.

**Example 6** To avoid generality for which we have no real need, the object of investigation is the equation of hyperbolic type in the rectangle $\bar{G}_0 = \{0 \le x_\alpha \le l_\alpha,\ \alpha = 1, 2\}$. The boundary conditions of the first kind are specified on its boundary $\Gamma$. The complete posing of the problem is described by

$$(42) \qquad \frac{\partial^2 u}{\partial t^2} = (L_1 + L_2)\, u + f(x,t)\,, \quad L_\alpha\, u = \frac{\partial^2 u}{\partial x_\alpha^2}\,, \quad \alpha = 1, 2\,,$$

$$x \in G_0\,, \quad t \in (0, T]\,,$$

$$u|_\Gamma = \mu(x,t)\,, \quad t \ge 0\,, \quad u(x,0) = u_0(x)\,,$$

$$\frac{\partial u}{\partial x}(x,0) = \bar{u}_0(x)\,, \quad x \in \bar{G}_0\,.$$

We contrived to do the necessary factorization in a number of different ways. First, we initiated the construction of a primary weighted scheme on an equidistant rectangular grid

$$\bar{\omega}_h = \{x_i = (i_1 h_1, i_2 h_2),\ i_\alpha = 0, 1, \ldots, N_\alpha,\ h_\alpha N_\alpha = l_\alpha,\ \alpha = 1, 2\}\,,$$

on which the statement of the difference problem is

$$y_{\bar{t}t} = \Lambda\left(\sigma\,\hat{y} + (1-2\,\sigma)\,y + \sigma\,\check{y}\right) + \varphi(x,t)\,, \tag{43}$$

$$t = j\tau\,,\quad j = 1,2,\dots\,,\quad x \in \omega_h\,,$$

$$y|_{\gamma_h} = \mu(x,t)\,,\quad t = i\tau\,,\quad j \geq 0\,,$$

$$y(x,0) = u_0(x)\,,\quad y_t(x,0) = \tilde{u}_0(x)\,,\quad x \in \bar{\omega}_h\,,$$

where $\tilde{u}_0(x) = \bar{u}_0(x) + 0.5\,\tau\big(Lu_0 + f(x,0)\big)$, $\Lambda = \Lambda_1 + \Lambda_2$ and $\Lambda_\alpha y = y_{\bar{x}_\alpha x_\alpha}$.

Assuming the primary scheme (43) to be stable, that is, granting $4\sigma \geq 1 + \varepsilon$, $\varepsilon > 0$, we rewrite scheme (43) in the canonical form

$$(E - \sigma\,\tau^2\Lambda)\,y_{\bar{t}t} = \Lambda\,y + \varphi\,,$$

whose use permits us to turn to the economical factorized scheme

$$(E - \sigma\,\tau^2\Lambda_1)(E - \sigma\,\tau^2\Lambda_2)\,y_{\bar{t}t} = \Lambda\,y + \varphi\,. \tag{44}$$

Another way of proceeding is to reduce scheme (43) to

$$(E - \sigma\,\tau^2\Lambda)\,y_t = F\,,\quad F = (E - \sigma\,\tau^2\Lambda)\,y_{\bar{t}} + \tau\,(\Lambda\,y + \varphi)\,.$$

Having completed the factorization of the operator $E - \sigma\,\tau^2\Lambda$ at the member $y_t$ so that

$$(E - \sigma\,\tau^2\Lambda_1)(E - \sigma\,\tau^2\Lambda_2)\,y_t = F\,,$$

we deduce that

$$\sigma^2\tau^3\Lambda_1\,\Lambda_2\,y_{\overset{\circ}{t}} + (E - \sigma\,\tau^2\Lambda + 0.5\,\sigma^2\tau^4\Lambda_1\,\Lambda_2)\,y_{\bar{t}t} = \Lambda\,y + \varphi\,. \tag{45}$$

Both factorized schemes (44) and (45) generate second-order approximations in $\tau$ for any $\sigma$ and they are stable under the condition $4\sigma \geq 1 + \varepsilon$, $\varepsilon > 0$, since the operators $R_\alpha = -\Lambda_\alpha$ are self-adjoint, positive and commutative in the space $H = \overset{\circ}{\Omega}_h$ of all grid functions given on the grid $\bar{\omega}_h$ and vanishing on the boundary $\gamma_h$ of the grid.

In an attempt to recover $\hat{y} = y^{j+1}$ from the difference equations just established, we shall need the boundary conditions for an intermediate value.

With this aim, the equation $B_1\, B_2\, y_{\bar{t}t} = \Lambda y$ can be solved in such a way that

$$B_1\, \omega_{(1)} = \Lambda\, y + \varphi\,, \qquad \omega_{(1)}\big|_{\gamma_h} = B_2\, \mu_{\bar{t}t}\,, \quad x_1 = 0, l_1,$$

$$B_2\, \omega_{(2)} = \omega_{(1)}, \qquad \omega_{(2)}\big|_{\gamma_h} = \mu_{\bar{t}t}\,, \quad x_2 = 0, l_2,$$

$$y^{j+1} = 2\, y^j - y^{j-1} + \tau^2 \omega_{(2)}\,.$$

In so doing three sequences of the values $y^{j-1}$, $y^j$ and $\omega_{(2)}$ are yet to be saved in the storage in the process of calculations of $y^{j+1}$. One trick we have encountered is connected with a prior reduction of the preceding scheme to

$$B_1\, B_2\, y_t = \Phi\,, \qquad \Phi = \tau\,(\Lambda\, y + \varphi) + B_1\, B_2\, y_{\bar{t}}\,,$$

thereby increasing the total volume of computations, but saving only two sequences of the values $y^j$ and $\omega_{(2)}$.

The algorithm of solving equation (45) was demonstrated before and so it remains only to construct economical factorized schemes associated with problem (42) by means of the operator $L_\alpha$ acting in accordance with the rule

$$L_\alpha u = \frac{\partial}{\partial x_\alpha}\Big(k_\alpha(x,t)\,\frac{\partial u}{\partial x_\alpha}\Big), \qquad 0 < c_1 \le k_\alpha \le c_2\,,$$

and adopt as a primary one in that case

$$(E + \tau^2 R)\, y_{\bar{t}t} = \Lambda\, y + \varphi\,,$$

where

$$\Lambda\, y = \sum_{\alpha=1}^{2} \big(a_\alpha\, y_{\bar{x}_\alpha}\big)_{x_\alpha}, \qquad R\, y = -\sigma\, \mathring{\Lambda}\, y\,, \quad \mathring{\Lambda}\, y = y_{\bar{x}_1 x_1} + y_{\bar{x}_2 x_2}\,.$$

The parameter $\sigma$ is so chosen as to satisfy the stability condition

$$(R y, y) \ge \frac{1+\varepsilon}{4}\,(-\Lambda y, y)\,, \qquad \varepsilon > 0\,,$$

for any $y \in H = \mathring{\Omega}$, where $\mathring{\Omega}$ is the set of all functions given on the grid $\bar{\omega}_h$ and vanishing on the boundary $\gamma_h$ of the grid. True, it is to be shown that the choice $\sigma = (1+\varepsilon)c_2/4$ is sufficient for doing so.

Having replaced the operator $E + \tau^2(R_1 + R_2)$ by the factorized operator $(E + \tau^2 R_1)(E + \tau^2 R_2)$, where $R_\alpha y = -\sigma\, y_{\bar{x}_\alpha x_\alpha}$, $\alpha = 1, 2$, we obtain under such an approach the economical difference scheme

$$(E + \tau^2 R_1)(E + \tau^2 R_2)\, y_{\bar{t}t} = \Lambda\, y + \varphi\,,$$

$$y|_\gamma = \mu\,, \quad y(x, 0) = u_0(x)\,, \quad y_t(x, 0) = \tilde{u}_0(x)\,,$$

with $\tilde{u}_0 = \bar{u}_0 + 0.5\,\tau(Lu + f)|_{t=0}$ incorporated. This scheme is of second-order accuracy in $\tau$ and $|h|$. What is more, it is absolutely stable and appears preferable in practical implementations.

**5. Economical schemes for a system of equations of parabolic and hyperbolic types.** Let $\bar{G} = \{0 \le x_\alpha \le l_\alpha,\ \alpha = 1, 2, \ldots, p\}$ be a parallelepiped in the space $\mathbf{R}^p$,

$$\bar{Q}_T = \bar{G} \times [0 \le t \le T]\,, \qquad Q_T = G \times (0 \le t \le T]$$

and $k = (k_{\alpha\beta}) = (k_{\alpha\beta}^{sm})$, $s, m = 1, 2, \ldots, n$, be a matrix of size $p \times p$ with square blocks of size $n \times n$ satisfying the condition of symmetry

$$k_{\alpha\beta}^{sm}(x, t) = k_{\beta\alpha}^{ms}(x, t) \quad \text{for all } (x, t) \in \bar{Q}_T \tag{46}$$

as well as the condition of positive definiteness

$$c_1 \sum_{s=1}^{n} \sum_{\alpha=1}^{p} (\xi_\alpha^s)^2 \le \sum_{s,m=1}^{n} \sum_{\alpha,\beta=1}^{p} k_{\alpha\beta}^{sm}(x, t)\, \xi_\beta^m\, \xi_\alpha^s \le c_2 \sum_{s=1}^{n} \sum_{\alpha=1}^{p} (\xi_\alpha^s)^2\,, \tag{47}$$

where $c_1$ and $c_2$ are positive constants and $\xi_\alpha = (\xi_\alpha^1, \ldots, \xi_\alpha^s, \ldots, \xi_\alpha^n)$ is an arbitrary real vector. The positive definiteness of the matrix $k$ is equivalent to being strongly elliptic of the operator $L$ with the values

$$L\mathbf{u} = \sum_{\alpha,\beta=1}^{p} L_{\alpha\beta}\, \mathbf{u}\,, \quad L_{\alpha\beta}\, \mathbf{u} = \frac{\partial}{\partial x_\alpha} \Big( k_{\alpha\beta}(x, t) \frac{\partial \mathbf{u}}{\partial x_\beta} \Big)\,, \tag{48}$$

where $\mathbf{u} = (u^1, \ldots, u^s, \ldots, u^n)$ is a vector of order $n$. The meaning of this property is that we should have

$$c_1\, (-L^{(0)}\mathbf{u}, \mathbf{u}) \le (-L\mathbf{u}, \mathbf{u})\,, \tag{49}$$

where

$$(\mathbf{u},\mathbf{v}) = \sum_{s=1}^{n} \int_G u^s(x)\, v^s(x)\, dx\,, \quad dx = dx_1 \ldots dx_p,$$

$$L^{(0)}\mathbf{u} - \Delta\mathbf{u} - \sum_{\alpha=1}^{p} \frac{\partial^2 \mathbf{u}}{\partial x_\alpha^2}\,,$$

and $\mathbf{u}$ is an arbitrary sufficiently smooth vector-function vanishing on the boundary $\Gamma$.

The problem here consists of finding a continuous in $\bar{Q}_T$ solution of the system of parabolic equations

$$\frac{\partial \mathbf{u}}{\partial t} = L\,\mathbf{u} + \mathbf{f}(x,t)\,, \quad (x,t) \in Q_T\,, \tag{50}$$

$$\mathbf{u} = \mu(x,t) \text{ for } x \in \Gamma\,, \quad t \in [0,T]\,,$$

$$\mathbf{u}(x,0) = \mathbf{u}_0(x)\,, \quad x \in \bar{G}\,.$$

Before going further, it will be sensible to introduce in $\bar{G}$ a grid

$$\bar{\omega}_h = \{x_i = (i_1 h_1, \ldots, i_p h_p)\ 0 \le i_\alpha \le N_\alpha\,,\ h_\alpha = l_\alpha / N_\alpha\,,\ \alpha = 1,2,\ldots,p\}$$

and on the segment $0 \le t \le T$ a grid $\bar{\omega}_\tau = \{t_j = j\tau,\ j = 0,1,\ldots\}$. The operator $L_{\alpha\beta}$ is approximated by the difference operator $\Lambda_{\alpha\beta}$ acting in accordance with the rule

$$\Lambda_{\alpha\beta}\,\mathbf{u} = 0.5\,\left[\left(k_{\alpha\beta}\,\mathbf{u}_{\bar{x}_\beta}\right)_{x_\alpha} + \left(k_{\alpha\beta}\,\mathbf{u}_{x_\beta}\right)_{\bar{x}_\alpha}\right] \tag{51}$$

with the well-established notation

$$\Lambda\,\mathbf{u} = \sum_{\alpha,\beta=1}^{p} \Lambda_{\alpha\beta}\,\mathbf{u}\,. \tag{52}$$

Along these lines, we obtain for $\beta = \alpha$

$$\Lambda_{\alpha\alpha}\,\mathbf{u} = \left(a_\alpha\,\mathbf{u}_{\bar{x}_\alpha}\right)_{x_\alpha}, \qquad a_\alpha = 0.5\left(k_{\alpha\alpha} + k_{\alpha\alpha}^{(-1_\alpha)}\right).$$

The inner product in the space $\overset{\circ}{\Omega}_h$ of all grid vector-functions given on the grid $\bar{\omega}_h$ and vanishing on the boundary $\gamma_h$ is defined by

$$(\mathbf{y},\mathbf{v}) = \sum_{s=1}^{n} (y^s, v^s)\,, \quad (y^s, v^s) = \sum_{x \in \omega_h} y^s(x)\, v^s(x)\, h_1 \cdots h_p\,.$$

On the strength of (46) the operator $\Lambda$ is self-adjoint, meaning

$$(\Lambda \mathbf{y}, \mathbf{v}) = (\mathbf{y}, \Lambda \mathbf{v}) .$$

From (47) we derive the inequalities

$$(53) \qquad c_1 (-\Lambda^{(0)}\mathbf{y}, \mathbf{y}) \le (-\Lambda \mathbf{y}, \mathbf{y}) \le c_2 (-\Lambda^{(0)}\mathbf{y}, \mathbf{y}) , \quad \mathbf{y} \in \overset{\circ}{\Omega}_h ,$$

where

$$\Lambda^{(0)}\mathbf{y} = \sum_{\alpha=1}^{p} \mathbf{y}_{\bar{x}_\alpha x_\alpha} , \quad (-\Lambda^{(0)}\mathbf{y}, \mathbf{y}) = \sum_{\alpha=1}^{p} \sum_{s=1}^{n} (1, (y^s_{\bar{x}_\alpha})^2] .$$

As further developments occur, the operator

$$(54) \qquad R = \sum_{\alpha=1}^{p} R_\alpha , \quad R_\alpha \mathbf{y} = -\sigma \mathbf{y}_{\bar{x}_\alpha x_\alpha} , \quad \alpha = 1, 2, \dots, p ,$$

will be declared to be the regularizer $R$. Here $\sigma$ is a numerical parameter, the choice of which is stipulated by the stability criteria in the sequel.

In view of this, a reasonable form of the primary two-layer economical scheme is

$$(E + \tau R) \mathbf{y}_t = \Lambda \mathbf{y} + \varphi ,$$

where

$$\varphi = \mathbf{f}^j + O(|h|^2 + \tau^2) .$$

Upon replacing $E + \tau R = E + \tau \sum_{\alpha=1}^{p} R_\alpha$ by the newly formed factorized operator

$$\prod_{\alpha=1}^{p} (E + \tau R_\alpha) = E + \tau \widetilde{R} ,$$

$$\widetilde{R} = R + \tau Q_p , \qquad Q_p = \sum_{\alpha<\beta} R_\alpha R_\beta + \cdots ,$$

we are led by exactly the same reasoning as before to the economical factorized scheme

$$(55) \qquad \prod_{\alpha=1}^{p} (E + \tau R_\alpha) \mathbf{y}_t = \Lambda \mathbf{y} + \varphi , \quad x \in \omega_h , \quad t \in \omega_\tau ,$$

$$\mathbf{y}(x, t) = \boldsymbol{\mu}(x, t) , \quad x \in \gamma_h , \quad t \in \bar{\omega}_\tau ,$$

$$\mathbf{y}(x, 0) = \mathbf{u}_0(x) , \quad x \in \bar{\omega}_h .$$

In what follows the computational algorithm capable of describing

$$(E + \tau R_1)\,\boldsymbol{\omega}_{(1)} = \Lambda\,\mathbf{y} + \boldsymbol{\varphi}\,,$$

$$\boldsymbol{\omega}_{(1)} = \prod_{\beta=2}^{p} (E + \tau R_\beta)\,\boldsymbol{\mu}_t \quad \text{for} \quad x_1 = 0, l_1\,,$$

$$(E + \tau R_\alpha)\,\boldsymbol{\omega}_{(\alpha)} = \boldsymbol{\omega}_{(\alpha-1)}\,, \qquad \alpha = 2, 3, \dots, p\,,$$

$$\boldsymbol{\omega}_{(\alpha)} = \prod_{\beta=\alpha+1}^{p} (E + \tau R_\beta)\,\boldsymbol{\mu}_t \quad \text{for} \quad x_\alpha = 0, l_\alpha\,, \quad \alpha = 2, 3, \dots, p-1\,,$$

will be performed for determination of a vector-function $\mathbf{y}^{j+1} = \hat{\mathbf{y}}$. Scheme (55) is absolutely stable for $\sigma \le 0.5\,c_2$ and converges with the rate $O(\tau + |h|^2)$.

The three-layer scheme

(56) $$\mathbf{y}_{\overset{\circ}{t}} + \tau^2 R\,\mathbf{y}_{\bar{t}t} = \Lambda\,\mathbf{y} + \boldsymbol{\varphi}$$

is of second-order accuracy in $\tau$, so there is some reason to be concerned about this. Rewrite it in the form

$$(E + 2\tau R)\,\mathbf{y}_t = \mathbf{F}\,, \quad \mathbf{F} = 2\,(\Lambda\,\mathbf{y} - \boldsymbol{\varphi}) - (E - 2\tau R)\,\mathbf{y}_t$$

with further replacement of the operator

$$E + 2\tau R = E + 2\tau \sum_{\alpha=1}^{p} R_\alpha$$

by the newly formed factorized operator

$$\prod_{\alpha=1}^{p} (E + 2\tau R_\alpha) = E + 2\tau R + 4\tau^2 Q_p\,.$$

By these changes we are led to the economical factorized scheme

(57) $$\prod_{\alpha=1}^{p} (E + 2\tau R_\alpha)\,\mathbf{y}_t = F\,,$$

whose canonical form is

$$(58) \qquad (E + 2\,\tau^2\, Q_p)\, \mathbf{y}_{\overset{\circ}{t}} + \tau^2 (R + \tau\, Q_p)\, \mathbf{y}_{\bar{t}t} = \Lambda\, \mathbf{y} + \varphi\,,$$

$$\mathbf{y} = \boldsymbol{\mu} \quad \text{for } x \in \gamma_h\,, \quad t \in \bar{\omega}_\tau\,,$$

$$\mathbf{y}(x,0) = \mathbf{u}_0(x)\,, \quad \mathbf{y}_t(x,0) = \tilde{\mathbf{u}}_0(x)\,, \quad x \in \bar{\omega}_h\,,$$

where $\tilde{\mathbf{u}}_0(x) = L\mathbf{u}_0 + \mathbf{f}(x,0)$.

The following algorithm is performed for the recovery of $\hat{\mathbf{y}} = \mathbf{y}^{j+1}$ from scheme (58):

$$(E + \tau\, R_1)\, \omega^s_{(1)} = F^s\,, \quad s = 1, 2, \ldots, n\,,$$

$$\omega^s_{(1)} = \prod_{\beta=2}^{p} (E + \tau\, R_\beta)\, \mu^s_t \quad \text{for} \quad x_1 = 0, l_1\,,$$

$$(E + \tau\, R_\alpha)\, \omega^s_{(\alpha)} = \omega^s_{(\alpha-1)}\,, \quad \alpha = 2, 3, \ldots, p\,, \quad s = 1, 2, \ldots, n\,,$$

$$\omega^s_{(\alpha)} = \prod_{\beta=\alpha+1}^{p} (E + \tau\, R_\beta)\, \mu^s_t \quad \text{for} \quad x_\alpha = 0, l_\alpha\,, \quad \alpha = 2, 3, \ldots, p-1\,,$$

$$\mathbf{y}^{j+1} = \mathbf{y}^j + \tau\, \omega_{(p)}\,.$$

As can readily be observed, the components $\omega^s_{(\alpha)}$ can be found independently. The primary scheme (56) becomes stable by merely setting

$$\sigma = \frac{c_2\,(1+\varepsilon)}{4}\,, \qquad \varepsilon = \text{const} > 0\,.$$

The operators $R_\alpha$ are positive and pairwise commutative, assuring us of the validity of the inequalities $Q_p > 0$ and $\widetilde{R} > R$, where $\widetilde{R} = R + \tau Q_p$ is the regularizer of scheme (58). This supports the view that scheme (58) is absolutely stable in the space $\overset{\circ}{\Omega}_h$.

Let $\mathbf{y}$ be a solution of problem (58) and $\mathbf{u}$ be a solution of the original problem (50). Upon substituting $\mathbf{y} = \mathbf{z} + \mathbf{u}$ into (58) we establish for the error the following conditions:

(59) $$(E + 2\tau^2 Q_p)\,\mathbf{z}_{\overset{\circ}{t}} + \tau^2(R + \tau Q_p)\,\mathbf{z}_{\bar{t}t} = \Lambda\,\mathbf{z} + \psi\,,$$

$$\mathbf{z} = 0 \quad \text{for } x \in \gamma_h\,, \quad t \in \bar{\omega}_\tau\,, \quad \mathbf{z}(x,0) = 0 \quad \text{for } x \in \bar{\omega}_h\,,$$

$$\mathbf{z}_t(x,0) = \nu(x) \quad \text{for } x \in \bar{\omega}_h\,,$$

where

(60) $$\psi = \Lambda\,\mathbf{u} + \varphi - \mathbf{u}_{\overset{\circ}{t}} - \tau^2 R\,\mathbf{u}_{\bar{t}t} - 2\,\tau^2\,Q_p\,\mathbf{u}_t = \psi_0 - 2\,\tau^2\,Q_p\,\mathbf{u}_t\,,$$

$\psi_0$ is the error of approximation of the primary scheme (56) and

$$\mathbf{v} = \bar{\mathbf{u}} = \mathbf{u}_t = O(\tau)\,.$$

Since $B = E + 2\tau^2 Q_p > E$, Theorems 6 and 9 in Chapter 6, Section 3 are still valid for scheme (59), on account of which the error of approximation $\nu$ to the second initial condition can be most readily evaluated in the norm $\|\nu\|_D$, where

$$\|\nu\|_D^2 = (D\,\nu, \nu) = \tau^2\,(R\,\nu, \nu) + \tau^3\,(Q_p\,\nu, \nu) = O(\tau^4)\,,$$

$$\|\nu\|_D = O(\tau^2)\,, \quad \text{since} \quad \nu = O(\tau)\,.$$

From (60) it is easily seen that $\psi = O(\tau^2 + |h|^2)$. The smoothness properties due to which we would have $\psi = O(\tau^2 + |h|^2)$ and $\|\nu\|_D = O(\tau^2)$ will become more stronger in connection with increasing the number of observations $p$. These restrictions can be relaxed with the aid of the chain of the inequalities

$$2\,\tau\,(\psi, \mathbf{z}_{\overset{\circ}{t}}) = 2\,\tau^3\,(Q_p\mathbf{v}, \mathbf{z}_{\overset{\circ}{t}}) = 2\,\tau^3\,(Q_p^{(2)}\mathbf{v}, \mathbf{z}_{\overset{\circ}{t}}) + 2\,\tau^3 \sum_{s=3}^{p} \tau^{s-2}(Q_p^{(s)}\mathbf{v}, \mathbf{z}_{\overset{\circ}{t}})$$

$$\leq \tau\,\|\,\mathbf{z}_{\overset{\circ}{t}}\,\|^2 + \tau^5\,\|\,Q_p^{(2)}\mathbf{v}\,\|^2 + \sum_{s=3}^{p} \tau^s(Q_p^{(s)}\mathbf{z}, \mathbf{z}_{\overset{\circ}{t}}) + \sum_{s=3}^{p} \tau^{s+2}(Q_p^{(s)}\mathbf{v}, \mathbf{v})$$

$$\leq \tau\,(B\mathbf{z}_t, \mathbf{z}_{\overset{\circ}{t}}) + \tau^5\,\|\,Q_p^{(2)}\mathbf{v}\,\|^2 + \sum_{s=3}^{p} \tau^{s+2}(Q_p^{(s)}\mathbf{v}, \mathbf{v})\,,$$

which are put together with a priori estimates for equation (59) with the right-hand side

$$\psi = \tau^2 \, Q_p \mathbf{v} = \tau^2 \sum_{s=2}^{p} \tau^{s-2} Q_p^{(s)} \mathbf{v} \,, \qquad \mathbf{v} = \mathbf{u}_{\bar{t}} \,.$$

Two last summands in the preceding inequality are some quantities of $O(\tau^5)$, so there is some reason to make their contributions to the accurate account of the accuracy $\mathbf{z}$. Thus, scheme (58) converges in the grid space $\overset{\circ}{W}{}_2^1$ with the rate $O(\tau^2 + |h|^2)$.

The object of investigation is the system of hyperbolic equations

$$\frac{\partial^2 \mathbf{u}}{\partial t^2} = L\mathbf{u} + \mathbf{f}(x,t) \,, \quad (x,t) \in Q_T \,, \quad L = \sum_{\alpha,\beta=1}^{p} L_{\alpha\beta} \,, \tag{61}$$

with the supplementary conditions

$$\begin{gathered} \mathbf{u} = \boldsymbol{\mu}(x,t) \quad \text{for} \quad x \in \Gamma \,, \quad t \in [0,T] \,, \\ \mathbf{u}(x,0) = \mathbf{u}_0(x) \,, \quad \frac{\partial \mathbf{u}(x,0)}{\partial t} = \bar{\mathbf{u}}_0(x) \quad \text{for} \quad x \in \bar{G} \,, \end{gathered} \tag{62}$$

under which it is required to find a continuous in the cylinder $\bar{Q}_T$ solution. Here the operator $L = \sum_{\alpha,\beta=1}^{p} L_{\alpha\beta}$ is specified by formula (48). Observe that a system arising from elasticity theory such as

$$\frac{\partial^2 \mathbf{u}}{\partial t^2} = \mu \, \Delta \, \mathbf{u} + (\lambda + \mu) \text{ grad div } \mathbf{u} + \mathbf{f} \,, \quad \Delta \mathbf{u} = \sum_{\alpha=1}^{p} \frac{\partial^2 \mathbf{u}}{\partial x_\alpha^2} \,,$$

where $\lambda = \text{const} > 0$ and $\mu = \text{const}$ are Lame's coefficients, $\mathbf{u} = (u^1, u^2, \ldots, u^p)$ is a vector-function of order $p$, can be viewed as one particular case of the system (61) for $n = p$ and

$$k_{\alpha\beta}^{sm} = \mu \, \delta_{\alpha\beta} \, \delta_{sm} + (\lambda + \mu) \left[ \theta \, \delta_{\alpha s} \delta_{\beta m} + (1 - \theta) \, \delta_{\alpha m} \delta_{\beta s} \right] ,$$

$$\delta_{ij} = \begin{cases} 1 \,, & i = j \,, \\ 0 \,, & i \neq j \,, \end{cases}$$

Here $\theta$ is an arbitrary constant. In this context, several things are worth noting. Condition (46) is automatically fulfilled. Condition (47) continues to hold with constants $c_1 = \mu$ and $c_2 = \lambda + 2\mu$, when providing current manipulations:

$$J = \sum_{s,m=1}^{p} \sum_{\alpha,\beta=1}^{p} k_{\alpha\beta}^{sm} \xi_\alpha^s \xi_\beta^m$$

$$= \mu \sum_{\alpha,s=1}^{p} (\xi_\alpha^s)^2 + (\lambda+\mu) \left[ \theta \sum_{\alpha,s=1}^{p} \xi_\alpha^\alpha \xi_s^s + (1-\theta) \sum_{\alpha,s=1}^{p} \xi_\alpha^s \xi_s^\alpha \right].$$

Indeed, accepting here $\theta = 1$, we find that $c_1 = \mu$, since

$$J = \sum_{\alpha,\beta=1}^{p} (k_{\alpha\beta} \xi_\alpha, \xi_\beta)$$

$$= \mu \sum_{\alpha=1}^{p} |\xi_\alpha|^2 + (\lambda+\mu) \left( \sum_{\alpha=1}^{p} \xi_\alpha^\alpha \right)^2$$

$$\geq \mu \sum_{\alpha=1}^{p} |\xi_\alpha|^2 .$$

For $\theta = 0$ we are led to $c_2 = \lambda + 2\,\mu$ by virtue of the relations

$$J = \mu \sum_{\alpha=1}^{p} |\xi_\alpha|^2 + (\lambda+\mu) \sum_{\alpha,s=1}^{p} \xi_\alpha^s \xi_s^\alpha$$

$$\leq \mu \sum_{\alpha=1}^{p} |\xi_\alpha|^2 + \frac{\lambda+\mu}{2} \left[ \sum_{\alpha,s=1}^{p} (\xi_\alpha^s)^2 + \sum_{\alpha,s=1}^{p} (\xi_s^\alpha)^2 \right]$$

$$= \mu \sum_{\alpha=1}^{p} |\xi_\alpha|^2 + (\lambda+\mu) \sum_{\alpha=1}^{p} \left( \sum_{s=1}^{p} (\xi_\alpha^s)^2 \right)$$

$$= (\lambda + 2\,\mu) \sum_{\alpha=1}^{p} |\xi_\alpha|^2 .$$

Thus, $c_1 = \mu$ and $c_2 = \lambda + 2\mu$. The same operator $R$ will be adopted as a regularizer in the further development:

$$R = \sum_{\alpha=1}^{p} R_\alpha , \qquad R_\alpha\, y = -\sigma\, y_{\bar{x}_\alpha x_\alpha} .$$

Observe that the primary scheme $\mathbf{y}_{\bar{t}t} + \tau^2 R \mathbf{y}_{\bar{t}t} = \Lambda \mathbf{y} + \varphi$ is stable under the restriction

$$\sigma = \frac{c_2 (1+\varepsilon)}{4}, \qquad \varepsilon = \text{const} > 0 .$$

By replacing

$$E + \tau^2 R = E + \tau^2 \sum_{\alpha=1}^{p} R_\alpha$$

by the factorized operator

$$D = \prod_{\alpha=1}^{p} (E + \tau^2 R_\alpha)$$

we obtain the economical factorized scheme

$$\prod_{\alpha=1}^{p} (E + \tau^2 R_\alpha) \mathbf{y}_{\bar{t}t} = \Lambda \mathbf{y} + \varphi \quad \text{for} \quad x \in \omega_h, \quad t \in \omega_\tau ,$$

$$\mathbf{y} = \boldsymbol{\mu} \quad \text{for} \quad x \in \gamma_h, \quad t \in \bar{\omega}_\tau ,$$

$$\mathbf{y}(x,0) = \mathbf{u}_0(x), \quad \mathbf{y}_t(x,0) = \tilde{\mathbf{u}}_0(x) \quad \text{for} \quad x \in \bar{\omega}_h ,$$

where $\tilde{\mathbf{u}}_0(x) = \bar{\mathbf{u}}_0 + 0.5\,\tau\,(L\mathbf{u}_0 + \mathbf{f}(x,0))$.

It is plain to show that the scheme concerned is absolutely stable and it generates an approximation of order 2: $\boldsymbol{\psi} = O(\tau^2 + |h|^2)$, $\boldsymbol{\nu} = O(\tau^2)$. Whence the convergence with the rate $O(\tau^2 + |h|^2)$ immediately follows.

The search for $\mathbf{y}^{j+1}$ amounts to successive solution of three-point equations of the form $(E + \tau^2 R_\alpha)\mathbf{w} = \mathbf{F}_\alpha$ by the elimination method for every component of the vector $\mathbf{w}$ with the index account from $\alpha$ to $\alpha + 1$. One possible way of covering this is connected with the performance of the following algorithm:

$$(E + \tau^2 R_1)\mathbf{w}_{(1)} = \mathbf{F}, \quad \mathbf{F} = \prod_{\alpha=1}^{p} (E + \tau^2 R_\alpha)\, \mathbf{y}_{\bar{t}} + \tau\, (\Lambda\, \mathbf{y} + \varphi),$$

$$(E + \tau^2 R_\alpha)\mathbf{w}_{(\alpha)} = \mathbf{w}_{(\alpha-1)}, \quad \alpha = 2, \dots, p, \quad \mathbf{y}^{j+1} = \mathbf{y}^j + \tau\, \mathbf{w}_{(p)},$$

which includes as part the supplementary boundary conditions for the vector-functions $\mathbf{w}_{(\alpha)}$, $\alpha = 1, 2, \ldots, p-1$, in the case $x_\alpha = 0, l_\alpha$:

$$\mathbf{w}_{(1)} = (E + \tau^2 R_2) \cdots (E + \tau^2 R_p)\, \boldsymbol{\mu}_t\,, \quad x_1 = 0, l_1\,,$$

$$\mathbf{w}_{(\alpha)} = \prod_{\beta=\alpha+1}^{p} (E + \tau^2 R_\beta)\, \boldsymbol{\mu}_t\,, \quad x_\alpha = 0, l_\alpha\,.$$

From here the components of the vector $\mathbf{w}_{(\alpha)}$, $\alpha = 1, 2, \ldots, p$, can be recovered independently, since the operators $D_\alpha = E + \tau^2 R_\alpha$ possess diagonal matrices of coefficients with diagonal blocks.

## 9.3 THE SUMMARIZED* APPROXIMATION METHOD

**1. The problem statement.** First of all, it should be noted that it is impossible to generalize directly the alternating direction method for three and more measurements as well as for parabolic equations of general form. Second, economical factorized schemes which have been under consideration in Section 2 of the present chapter are quite applicable under the assumption that the argument $x = (x_1, x_2, \ldots, x_p)$ varies within a parallelepiped.

Because of this, there is a real need for designing the general method, by means of which economical schemes can be created for equations with variable and even discontinuous coefficients as well as for quasilinear non-stationary equations in complex domains of arbitrary shape and dimension. As a matter of experience, the universal tool in such obstacles is the method of summarized approximation, the framework of which will be explained a little later on the basis of the heat conduction equation in an arbitrary domain $G$ of the dimension $p$ with the boundary $\Gamma$

$$(1) \qquad \frac{\partial u}{\partial t} = L\,u + f(x,t)\,, \quad x = (x_1, x_2, \ldots, x_p) \in G\,, \quad t > 0\,,$$

$$L\,u = \sum_{\alpha=1}^{p} L_\alpha\, u\,, \quad L_\alpha u = \frac{\partial}{\partial x_\alpha}\Big(k_\alpha(x,t)\,\frac{\partial u}{\partial x_\alpha}\Big)\,, \quad k_\alpha \geq c_1 > 0\,,$$

provided that the conditions hold:

$$(2) \qquad u|_\Gamma = \mu(x,t)\,, \quad t \geq 0\,, \quad u(x,0) = u_0(x)\,, \quad x \in \bar{G}\,.$$

---

******Editor's note: Summarized = summed.*

The quasilinear heat conduction equation reproduces the case when $k_\alpha = k_\alpha(x, t, u)$ and $f = f(x, t, u)$.

Of course, the words "arbitrary domain" cannot be understood in a literal sense. Before giving further motivations, it is preassumed that the boundary $\Gamma$ is smooth enough to ensure the existence of a smooth solution $u = u(x, t)$ of the original problem (1)–(2). In the accurate account of the approximation error and accuracy we always take for granted that the solution of the original problem associated with the governing differential equation exists and possesses all necessary derivatives which do arise in the further development.

A common algorithmic idea behind available economical methods is connected with further reduction of numerical solution of a multidimensional problem to the process of solving a few simpler problems. In order to understand the nature of this a little better, we focus the reader's attention on second-order equations of hyperbolic and parabolic types for which the "basic algebraic problem" is related to a three-point difference problem (a second-order difference equation). The three-point difference problem obtained through such an approximation can be solved by the elimination method and it can be treated, as a rule, as a difference approximation to the one-dimensional (in $x_\alpha$) differential equation. Some consensus of opinion is to create on this basis a chain of simpler algorithms which constitute what is called an economical algorithm for complex problems. This idea lies in the origin and terminology of many economical methods available for solving multidimensional problems. Among them, the alternating direction method permits us to solve at every stage a one-dimensional problem along a fixed direction $x_\alpha$, the method of "fractional steps" necessitates placing in the storage intermediate (non-integer) values at every stage of a complex computational procedure, the method of separation of variables in a common setting of the problem reduces to a number of particular simpler tasks, etc. All these terms reflect one of the real advantages and the essence of economical methods.

However, throughout this book, the classification of difference methods is mostly based on the origin of difference schemes rather than on a possible way of constructing them and a perfect tool for solving this or that difference scheme (equation).

**2. The notion of summarized approximation.** In the preceding sections and chapters the basic fundamental property of difference schemes is to generate approximations on a solution to the governing differential equation. In what follows we get rid of the classical notion by introducing a more weaker condition of summarized approximation, expanding our possibilities and leading to **additive schemes**. The describing schemes of this type will be studied in more detail in Section 10, but a great deal of work still needs to be done in adopting those ideas. It is worth mentioning here their main peculiarities:

- the passage from the $j$th layer to the $(j+1)$th layer can be performed through the use of a sequence of the usual (two-point, three-point, etc.) schemes;
- the error of approximation provided by an additive scheme is taken to be a sum of the residuals of all auxiliary schemes, that is, any such scheme generates a **summarized approximation**.

In this regard, we should take into account that auxiliary schemes are not obliged to approximate the original problem; the approximation here is ensured by summarizing all the residuals obtained.

In Chapter 2 we came across the necessity of generalizing the notion of approximation in the real situations when a difference scheme cannot provide on the grid $\omega_h$ local approximations with a desired order in the norm of the space $C$, but it does the same in one of the negative norms, that is, in a certain sense of summarizing.

Likewise, it may happen that a scheme on the grid $\omega_\tau$ cannot provide local approximations in $t$, but at the final stage the approximation will be achieved once we bring together the residuals over several time layers. The notion of summarized approximation needs certain clarification. It seems worthwhile giving simple examples.

**Example 1** The Cauchy problem, being the most familiar one, comes first:

$$\frac{du}{dt} + a\,u = 0\,, \qquad t > 0\,, \quad u(0) = u_0\,.$$

Common practice involves for solving it the difference scheme

$$(3) \qquad \frac{y^{j+1/2} - y^j}{\tau} + a_1 y^j = 0\,, \qquad \frac{y^{j+1} - y^{j+1/2}}{\tau} + a_2 y^{j+1/2} = 0\,,$$

$$j = 0, 1, 2, \ldots\,, \qquad y^0 = u_0\,, \quad a_1 + a_2 = a\,,$$

which consists of two explicit schemes with residuals $\psi_1$ and $\psi_2$, respectively. Within more compact notations

$$y^j = z^j + u^j\,, \qquad y^{j+1/2} = z^{j+1/2} - \bar{u}\,, \qquad \bar{u} = 0.5\,(u^j + u^{j+1})\,,$$

we find that

$$\frac{z^{j+1/2} - z^j}{\tau} + a_1 z^j = -\psi_1^j\,, \qquad \frac{z^{j+1} - z^{j+1/2}}{\tau} + a_2 z^{j+1/2} = -\psi_2^j\,,$$

where

$$\psi_1 = \frac{u^{j+1} - u^j}{2\,\tau} + a_1 u^j\,, \qquad \psi_2 = \frac{u^{j+1} - u^j}{2\,\tau} + a_2\,\frac{u^j + u^{j+1}}{2}\,.$$

By substituting here the expressions

$$u^{j+1} = u^{j+1/2} + \frac{\tau}{2}\,\dot{u}^{j+1/2} + \frac{\tau^2}{8}\,\ddot{u}^{j+1/2} + O(\tau^3)\,,$$

$$u^j = u^{j+1/2} - \frac{\tau}{2}\,\dot{u}^{j+1/2} + \frac{\tau^2}{8}\,\ddot{u}^{j+1/2} + O(\tau^3)\,,$$

$$u^{j+1/2} = u(t_j + 0.5\,\tau)\,, \quad \dot{u} = \frac{du}{dt}\,, \quad \ddot{u} = \frac{d^2u}{dt^2}$$

we are led to

$$\psi_1 = \left(\tfrac{1}{2}\,\dot{u} + a_1 u\right)^{j+1/2} - \tfrac{1}{2}\,a_1 \tau\,\dot{u}^{j+1/2} + O(\tau^2)\,,$$

$$\psi_2 = \left(\tfrac{1}{2}\,\dot{u} + a_2 u\right)^{j+1/2} + O(\tau^2)\,.$$

Whence it follows that $\psi_1 = O(1)$, $\psi_2 = O(1)$ and $\psi_1 + \psi_2 = O(\tau)$, meaning that none of the auxiliary schemes concerned provides an approximation, but the triplex composition generates a summarized approximation of $O(\tau)$.

**Example 2** Of special interest is the heat conduction equation

$$\frac{\partial u}{\partial t} = L\,u\,, \qquad L\,u = \frac{\partial^2 u}{\partial x^2}\,,$$

$$0 < x < 1\,, \quad u(0,t) = u_1(t)\,, \quad u(1,t) = u_2(t)\,.$$

In dealing with the grid $\bar{\omega}_h = \{x_i = ih,\ i = 0, 1, \dots, N,\ Nh = 1\}$ and the operator $\Lambda u = u_{\bar{x}x} \sim Lu$, it seems reasonable to employ the explicit scheme on the odd layers and the implicit one on the even layers. The outcome of this is

$$
(4)\qquad
\begin{aligned}
&\frac{y^{2j+1} - y^{2j}}{\tau} = 2\,(1-\sigma)\,\Lambda\, y^{2j}\,,\\
&\frac{y^{2j+2} - y^{2j+1}}{\tau} = 2\,\sigma\,\Lambda\, y^{2j+2}\,, \quad j = 0, 1, 2, \ldots\,,
\end{aligned}
$$

where $\sigma \geq 0$ is an arbitrary parameter. In order to calculate the residuals

$$
\psi_1 = \frac{u^{2j+1} - u^{2j}}{\tau} - 2\,(1-\sigma)\,\Lambda\, u^{2j}\,, \quad \psi_2 = \frac{u^{2j+2} - u^{2j+1}}{\tau} - 2\,\sigma\,\Lambda\, u^{2j+2},
$$

we insert in them the expressions

$$
u^{2j+2} = u^{2j+1} + \tau\,\dot{u}^{2j+1} + 0.5\,\tau^2\,\ddot{u}^{2j+1} + O(\tau^3)\,,
$$

$$
u^{2j} = u^{2j+1} - \tau\,\dot{u}^{2j+1} + 0.5\,\tau^2\,\ddot{u}^{2j+1} + O(\tau^3)\,,
$$

$$
\Lambda u = L\,u + O(h^2)\,, \qquad L\,u = \dot{u}\,,
$$

implying that

$$
\psi_1 = \overset{\circ}{\psi}_1 + \psi_1^*\,, \quad \psi_2 = \overset{\circ}{\psi}_2 + \psi_2^*\,, \quad \psi_\alpha^* = O(\tau^2 + h^2)\,, \quad \alpha = 1, 2\,,
$$

$$
\overset{\circ}{\psi}_1 = \big((2\,\sigma - 1)\,\dot{u} + (1.5 - 2\,\sigma)\,\tau\,\ddot{u}\big)^{2j+1}\,,
$$

$$
\overset{\circ}{\psi}_2 = \big(-(2\,\sigma - 1)\,\dot{u} + (0.5 - 2\,\sigma)\,\tau\,\ddot{u}\big)^{2j+1}\,.
$$

From such reasoning it seems clear that $\psi_1 = O(1)$ and $\psi_2 = O(1)$ for $\sigma \neq 0.5$, but for any $\sigma$ there is no doubt that

$$
\psi = \psi_1 + \psi_2 = O\big((\sigma - 0.5)\,\tau + h^2 + \tau^2\big)\,.
$$

Collection of equations (4) permits us to eliminate the value $y^{j+1}$, leaving us with the weighted scheme with step $2\tau$:

$$
\frac{y^{2j+2} - y^{2j}}{2\,\tau} = \Lambda\,\big(\sigma\, y^{2j+2} + (1-\sigma)\, y^{2j}\big)\,, \quad j = 0, 1, \ldots
$$

An alternative form of writing includes the intermediate value $y^{j+1/2}$ and a smaller step $\tau$ in two times:

$$\frac{y^{j+1/2}-y^j}{\tau}=(1-\sigma)\,\Lambda\,y^j\,,\qquad \frac{y^{j+1}-y^{j+1/2}}{\tau}=\sigma\,\Lambda\,y^{j+1}.$$

Here the passage from the $j$th layer to the $(j+1)$th layer is carried out in the following two steps: the first one involves the explicit scheme and the second one – the implicit scheme as suggested above.

Of course, this example is not of global character and can serve mainly as an illustration of such theory. It should be noted that in the general case the elimination of intermediate values with further reduction to a scheme including the values of $y$ only at integer steps may be impossible and even meaningless in the theoretical research.

**3. Reduction of a multidimensional problem to a chain of one-dimensional problems.** The multiple equation we must solve is

$$(5)\qquad \frac{\partial u}{\partial t}=L\,u+f(x,t)\,,\quad 0\le t\le t_0\,,$$

$$u(x,0)=u_0(x)\,,\quad x=(x_1,x_2,\dots,x_p)\,,$$

where $L$ is a linear differential operator acting on $u(x,t)$ as a function of $x$, $x=(x_1,x_2,\dots,x_p)$ is a point in the $p$-dimensional domain $G$ with the boundary $\Gamma$, on which proper boundary conditions are imposed in one or another convenient way. That does not matter for us in subsequent discussions. An effective tool in designing economical methods is the accepted decomposition

$$L=L_1+L_2+\cdots+L_p$$

with simplified operators. For example, if $Lu=\Delta u$ and $L_\alpha u=\partial^2u/\partial x^2$, then $L_\alpha$ is the operator of the second derivative with respect to the argument $x_\alpha$ (operator of one variable).

Furthermore, to problem (5) there corresponds the first chain of "one-dimensional" equations by reducing either (5) or

$$\mathcal{P}\,u=\frac{\partial u}{\partial t}-L\,u-f(x,t)=0$$

to

$$\sum_{\alpha=1}^{p}\mathcal{P}_\alpha\,u=0\,,\qquad \mathcal{P}_\alpha\,u=\frac{1}{p}\,\frac{\partial u}{\partial t}-L_\alpha\,u-f_\alpha\,,$$

where $f_\alpha(x,t)$, $\alpha = 1,2,\dots,p$, are arbitrary functions with the same smoothness properties as the function $f(x,t)$, satisfying, in addition, the normalization condition

$$f_1 + f_2 + \cdots + f_p = f .$$

In working on the segment $0 \le t \le t_0$ with an equidistant grid

$$\bar{\omega}_\tau = \{t_j = j\tau\,,\ j = 0,1,\dots,j_0\}$$

with step $\tau$ it seems reasonable to divide every interval into $p$ parts by recording the points $t_{j+\alpha/p} = t_j + \alpha\tau/p$, $\alpha = 1,2,\dots,p-1$. Half-intervals $t_{j+(\alpha-1)p} < t \le t_{j+\alpha/p}$ denoted by $\Delta_\alpha$ are made up, as usual, by those points. By successive solution of the equations starting from $\alpha = 1,2,\dots$

$$\mathcal{P}_\alpha\, v_{(\alpha)} = 0\,, \quad x \in G\,, \quad t \in \Delta_\alpha\,, \quad \alpha = 1,2,\dots,p\,, \tag{6}$$

under the additional assumptions

$$v_{(1)}(x,0) = u_0(x)\,, \quad v_{(\alpha)}(x,t_{j+(\alpha-1)/p}) = v_{(\alpha-1)}(x,t_{j+(\alpha-1)/p})\,, \tag{7}$$

$$\alpha = 1,2,\dots,p\,,$$

we find the values $v(x,t_j) = v_{(p)}(x,t_j)$, $j = 0,1,2,\dots,j_0$, with further treatment as a solution of the problem concerned. For the sake of simplicity, it is supposed that we have on the boundary the homogeneous boundary condition of the first kind and this should confine no generality of further motivations.

Every equation $\mathcal{P}_\alpha v_\alpha = 0$ or

$$\frac{1}{p}\,\frac{\partial v_{(\alpha)}}{\partial t} - L_\alpha\, v_{(\alpha)} + f_\alpha\,, \qquad \alpha = 1,2,\dots,p\,, \tag{6'}$$

is replaced by the newly formed difference scheme in which $\partial u/\partial t$ and $L_\alpha$ are approximated by the appropriate difference expressions of the general form on a grid $\omega_h$ with steps $h_1, h_2, \dots, h_p$

$$\Pi_\alpha\, y_{(\alpha)} = 0\,, \qquad \alpha = 1,2,\dots,p\,. \tag{8}$$

In the simplest case the resulting two-layer scheme is aimed at connecting the values

$$y_{(\alpha)} = y^{j+\alpha/p} \quad \text{and} \quad y_{(\alpha-1)} = y^{j+(\alpha-1)/p}\,.$$

For example, the weighted scheme

$$\frac{y^{j+\alpha/p} - y^{j+(\alpha-1)/p}}{\tau} = \Lambda_\alpha \left(\sigma_\alpha y^{j+\alpha/p} + (1-\sigma_\alpha)\, y^{j+(\alpha-1)/p}\right) + \varphi^{j+\alpha/p}$$

suits us for doing so. Here $\Lambda_\alpha \sim L_\alpha$ and $\sigma_\alpha$ is an arbitrary parameter.

The governing equation $\mathcal{P}_\alpha v_{(\alpha)} = 0$ is approximated by scheme (8) in the usual sense so that

$$\Psi_\alpha = \Pi_\alpha u^{j+\alpha/p} - (\mathcal{P}_\alpha u)^{j+\alpha/p} \tag{9}$$

tends to zero in some suitable norm as $\tau \to 0$ and $h_\alpha \to 0$.

The difference equations (8) constitute what is called an **additive scheme**. Indeed, let $\psi_\alpha = \Pi_\alpha u^{j+\alpha/p}$ be the residuals of the same scheme (8) with the number $\alpha$ attached.

Arranging $\psi_\alpha$ as a sum

$$\psi_\alpha = (\mathcal{P}_\alpha u)^{j+\alpha/p} + \Psi_\alpha \tag{10}$$

and taking into account that

$$(\mathcal{P}_\alpha u)^{j+\alpha/p} = (\mathcal{P}_\alpha u)^{j+1/2} + O(\tau)\,,$$

we deduce that

$$\psi_\alpha = \overset{\circ}{\psi}_\alpha + \psi^*_\alpha\,, \qquad \overset{\circ}{\psi}_\alpha = (\mathcal{P}_\alpha u)^{j+1/2}$$

and $\| \psi^*_\alpha \| \to 0$ as $\tau \to 0$, $h \to 0$, where $\| \cdot \|$ is some suitable norm on the space of all grid functions given on the grid $\omega_h$. It follows from the foregoing that

$$\sum_{\alpha=1}^{p} \overset{\circ}{\psi}_\alpha = 0\,, \quad \| \psi \| = \left\| \sum_{\alpha=1}^{p} \psi_\alpha \right\| \longrightarrow 0 \quad \text{as} \quad \tau \to 0, \quad |h| \to 0\,,$$

that is,

> *scheme (8) generates a summarized approximation if either of the schemes (8) with the number $\alpha$ approximates the corresponding equation (6) in the usual sense.*

As a matter of experience, the estimation of the nearness of a solution of the difference problem amounts to the proximity between a solution of the original problem (5) and a solution of the chain of problems (6)–(7). The main idea behind this approach is connected with the obvious relation

$$\| y^j - u^j \| \le \| y^j - v^j_{(\alpha)} \| + \| v^j_{(\alpha)} - u^j \| .$$

In addition to (6)–(7), the second chain of the equations

$$\frac{\partial v_{(\alpha)}}{\partial t} = L_\alpha \, v_{(\alpha)} + f_\alpha \,, \quad x \in G \,, \quad t_j < t \le t_{j+1} \,, \tag{11}$$

under the additional conditions of conjugation complements subsequent studies

$$\begin{gathered} v_{(\alpha)}(x, t_j) = v_{(\alpha-1)}(x, t_{j+1}) \,, \quad \alpha = 2, 3, \dots, p \,, \\ v_{(1)}(x, t_j) = v(x, t_j) \,, \quad j = 1, 2, \dots \,, \quad v_{(1)}(x, 0) = u_0(x) \,. \end{gathered} \tag{12}$$

The solution of this problem is the function $v(x,t) = v_{(p)}(x,t)$.

In contrast to (6)–(7), every equation with the number $\alpha$ is solved here on the whole interval $t_j < t \le t_{j+1}$. It is interesting to note that in some particular cases solutions of problem (6)–(7) and problem (11)–(12) will coincide. This is certainly true in the situation when both $f_\alpha \equiv 0$ and $L_\alpha$ are independent of $t$.

Along these lines, both chains generate approximations on the solution $u = u(x,t)$ of the original problem (5). Indeed, it is straightforward to verify for problem (6)–(7) with the aid of the relation $\mathcal{P}_\alpha \, u = (\mathcal{P}_\alpha \, u)^{j+1/2} + O(\tau)$ that

$$\psi_\alpha = \overset{\circ}{\psi}_\alpha + \psi^*_\alpha \,, \quad \text{where} \quad \overset{\circ}{\psi}_\alpha = (\mathcal{P}_\alpha \, u)^{j+1/2} \,, \quad \psi^*_\alpha = O(\tau) \,,$$

where $\psi_\alpha = \mathcal{P}_\alpha u(x,t)$ is the residual for equation (6) with the number $\alpha$.

In view of this, it follows from the foregoing that

$$\sum_{\alpha=1}^{p} \overset{\circ}{\psi}_\alpha = 0 \,, \quad \psi = \sum_{\alpha=1}^{p} \psi_\alpha = \sum_{\alpha=1}^{p} \psi^*_\alpha = O(\tau) \,,$$

it being understood that the system (6)–(7) approximates equation (5) in a summarized sense.

Thus, the attainable summarized approximation of the additive scheme (8) owes a debt to the simultaneous usual approximations and a summarized approximation. In accordance with what has been said above, equations (5) are approximated by the chain of the difference equations (6)–(7) in a summarized sense and every scheme (8) with the number $\alpha$ approximates the corresponding equation involved in collection (6) in the usual sense.

All this enables us to obtain through such procedures a perfect approximation. Let us stress here that the summarized approximation is ensured by the following conditions:

- the operator $L$ is representable by a sum $L = L_1 + L_2 + \cdots + L_p$;
- the right-hand side $f$ contains exactly $p$ functions such that $f = f_1 + f_2 + \cdots + f_p$.

These conditions will be relaxed once we consider

$$L\,u - \sum_{\alpha=1}^{p} L_\alpha\, u = O(\tau)\,, \qquad f - \sum_{\alpha=1}^{p} f_\alpha = O(\tau)\,.$$

If the operator $L_\alpha$ includes the derivatives with respect to only one variable $x_\alpha$, we call it a one-dimensional operator and the equations $\mathcal{P}_\alpha v_{(\alpha)} = 0$ refer,correspondingly, to equations of one variable. The additive scheme (8) is termed a **locally one-dimensional scheme** (LOS).

Section 5 of the present chapter will be devoted to such schemes relating to the heat conduction equation.

**4. Examples of reduction of multidimensional problems to chains of one-dimensional ones.** It is apparent from that discussion that some class of problems for which a solution of problem (6) or problem (11) coincides on the grid $\omega_h$ with the exact solution of the multidimensional problem (5) plays an important role.

**Example 1** The Cauchy problem

$$\frac{du}{dt} + a\,u(t) = 0\,, \quad t \geq 0\,, \quad u(0) = u_0\,,$$

where $a > 0$, is good enough for the purposes of the present section. With the representation $a = a_1 + a_2$ in view, we may set up the problem

$$\frac{dv_{(1)}}{dt} + a_1\, v_{(1)}(t) = 0\,, \quad v_{(1)}(0) = u_0\,, \quad 0 \leq t \leq t^*\,,$$

$$\frac{dv_{(2)}}{dt} + a_2\, v_{(2)}(t) = 0\,, \quad v_{(2)}(0) = v_{(1)}(t^*)\,, \quad 0 \leq t \leq t^*\,,$$

where $t^* > 0$ is an arbitrary number. By resolving these equations we obtain

$$v_{(1)}(t) = u_0 \exp(-a_1 t) \qquad \text{and}$$

$$v_{(2)}(t) = v_{(1)}(t^*) \exp(-a_2 t) = u_0 \exp(-a_2 t - a_1 t^*),$$

yielding $v_2(t^*) = u(t^*)$.

**Example 2** Recall now the statement of the Cauchy problem for the transfer equation

$$\frac{\partial u}{\partial t} + L_1 u + L_2 u = 0, \quad L_\alpha u = \frac{\partial u}{\partial x_\alpha}, \quad \alpha = 1, 2,$$

$$-\infty < x_\alpha < \infty, \quad t \geq 0, \quad u(x, 0) = \mu(x),$$

whose solution $u(x,t) = \mu(x_1 - t, x_2 - t)$ is a "travelling wave" if $\mu(x)$ is a twice differentiable function. Since the operators $L_1$ and $L_2$ are commuting, it is plain to show that

$$u(x, t^*) = v_{(2)}(x, t^*),$$

where $v_{(2)}(x, t^*)$ is a solution of the system of equations

$$\frac{\partial v_{(1)}}{\partial t} + \frac{\partial v_{(1)}}{\partial x_1} = 0, \quad 0 < t \leq t^*, \quad v_{(1)}(x, 0) = \mu(x),$$

$$\frac{\partial v_{(2)}}{\partial t} + \frac{\partial v_{(2)}}{\partial x_2} = 0, \quad 0 < t \leq t^*, \quad v_{(2)}(x, 0) = v_{(1)}(x, t^*).$$

Indeed, a solution to the first equation acquires the form

$$v_{(1)}(x, t) = \mu(x_1 - t, x_2).$$

On the other hand, we find from the second equation that

$$v_{(2)}(x, t) = \mu(x_1 - t^*, x_2 - t),$$

implying that $v_{(2)}(x, t^*) = u(x, t^*)$.

**Example 3** We learn from Ladyzhenskaya (1968) that a solution of the Cauchy problem for the heat conduction equation

$$(13)\qquad \frac{\partial u}{\partial t} = (L_1 + L_2 + L_3)\, u = 0\,, \qquad L_\alpha\, u = \frac{\partial^2 u}{\partial x_\alpha^2}\,, \qquad \alpha = 1, 2, 3\,,$$

$$-\infty < x_\alpha < \infty\,, \quad t \geq 0\,, \quad u(x, 0) = u_0(x)\,,$$

is given by the formula

$$(14)\quad u(x, t) = u(x_1, x_2, x_3, t)$$

$$= \int_{-\infty}^{\infty} \int_{-\infty}^{\infty} \int_{-\infty}^{\infty} G(x_1, x_2, x_3; \xi_1, \xi_2, \xi_3, t)\, u_0(\xi_1, \xi_2, \xi_3)\, d\xi_1\, d\xi_2\, d\xi_3\,,$$

where $G(x_1, x_2, x_3; \xi_1, \xi_2, \xi_3, t)$ is a source function such that

$$G(x_1, x_2, x_3; \xi_1, \xi_2, \xi_3, t) = G_0(x_1, \xi_1, t)\, G_0(x_2, \xi_2, t)\, G_0(x_3, \xi_3, t)\,,$$

$$G_0(x_\alpha, \xi_\alpha, t) = \exp\big(-(x_\alpha - \xi_\alpha)^2/(4\,t)\big)/(2\,\sqrt{\pi\, t})\,, \qquad \alpha = 1, 2, 3\,.$$

Here $G_0(x_\alpha, \xi_\alpha, t)$ is a function of the heat source of the Cauchy problem associated with the one-dimensional heat conduction equation

$$\frac{\partial v_{(\alpha)}}{\partial t} = L_\alpha\, v_{(\alpha)}\,, \quad t > 0\,, \quad v_{(\alpha)}(x, 0) \quad \text{given}\,.$$

The general methodology provides proper guidelines for the selection rules in studying one-dimensional heat conduction equations

$$\frac{\partial v_{(1)}}{\partial t} = L_1\, v_{(1)}\,, \qquad \frac{\partial v_{(2)}}{\partial t} = L_2\, v_{(2)}\,, \qquad \frac{\partial v_{(3)}}{\partial t} = L_3\, v_{(3)}\,,$$

$$0 < t \leq t^*\,,$$

$$v_{(1)}(x, 0) = u_0(x)\,, \quad v_{(2)}(x, 0) = v_{(1)}(x, t^*)\,, \quad v_{(3)}(x, 0) = v_{(2)}(x, t^*)\,.$$

It is plain to derive for this chain oof the equations the following expressions:

$$v_{(1)}(x,t) = \int_{-\infty}^{\infty} G_0(x_1, \xi_1, t)\, v_{(1)}(\xi_1, x_2, x_3, 0)\; d\xi_1 ,$$

$$v_{(2)}(x,t) = \int_{-\infty}^{\infty} G_0(x_2, \xi_2, t)\, v_{(2)}(x_1, \xi_2, x_3, 0)\; d\xi_2 ,$$

$$v_{(3)}(x,t) = \int_{-\infty}^{\infty} G_0(x_3, \xi_3, t)\, v_{(3)}(x_1, x_2, \xi_3, 0)\; d\xi_3 .$$

Upon substituting here

$$v_{(1)}(\xi_1, x_2, x_3, 0) = u_0(\xi_1, x_2, x_3) ,$$

$$v_{(2)}(x_1, \xi_2, x_3, 0) = v_{(1)}(x_1, \xi_2, x_3, t^*) ,$$

$$v_{(3)}(x_1, x_2, \xi_3, 0) = v_{(2)}(x_1, x_2, \xi_3, t^*)$$

we obtain formula (14) for $v_{(3)}(x, t^*)$ at moment $t = t^*$, meaning

$$v_{(3)}(x, t^*) = u(x, t^*) \quad \text{for any} \quad t^* > 0 .$$

The property of this sort is an immediate implication of the representation $G(x, \xi, t) = G(x_1, x_2, x_3; \xi_1, \xi_2, \xi_3, t)$ as a product of the functions of one variable of the special type $G_0(x_\alpha, \xi_\alpha, t)$. The well-established representation of the source function is still valid for equation (13) with zero boundary conditions of the first kind $u = 0$ for $x_\alpha = 0, l_\alpha$, $\alpha = 1, 2, 3$, relating to the boundary-value problems in the parallelepiped $\{0 \le x_\alpha \le l_\alpha,\ \alpha = 1, 2, 3\}$. For this reason identity (14) is certainly true in that case.

What is more, it seems clear from the preceding examples that $v_{(p)}$ coincides with $u(x, t)$ at all nodal points.

Among other things, Examples 2 and 3 show the ways of expanding some spatial process into a sequence of processes being one-dimensional ones and continuing along the coordinate axes. Following established practice, the three-dimensional heat conduction problem in a space or a parallelepiped with the zero temperature on the lateral surface reduces on the same grounds to the model problem concerned. If the initial distribution of temperature is known at moment $t = t_0$, then the heat conduction will be possible at the same moment only along one direction $x_1$ without concern of others. At moment $t = t_0 + \Delta t$ the same procedure works along one direction $x_2$ and at moment $t = t_0 + 2\,\Delta t$ – along another direction $x_3$. As a final result we obtain at $t = t_0 + 3\,\Delta t$ the same temperature distribution as in the three-dimensional case at moment $t = t_0 + \Delta t$. As a matter of fact, this is a way of reducing the three-dimensional process to a sequence of one-dimensional ones of prolongations as well as of a real physical process in three times.

Generally speaking, proper guidelines in such matters are not so obvious as the users might expect. For example, in isotropic media the operator $L_\alpha$ involved in (6′) is taken to be

$$L_\alpha u = \frac{\partial}{\partial x_\alpha}\left(k_\alpha(x)\,\frac{\partial u}{\partial x_\alpha}\right).$$

In spite of the fact that $v_{(p)}(x, t^*)$ does not coincide with $u(x, t^*)$, the asymptotics reveals itself as

$$v_{(p)}(x, t^*) - u(x, t^*) = O(t^*)\,.$$

**5. A locally one-dimensional scheme (LOS) for the heat conduction equation in an arbitrary domain.** The method of summarized approximation can find a wide range of application in designing economical additive schemes for parabolic equations in the domains of rather complicated configurations and shapes. More a detailed exploration is devoted to a locally one-dimensional problem for the heat conduction equation in a complex domain $\bar{G} = G + \Gamma$ of the dimension $p$. Let $x = (x_1, x_2, \ldots, x_p)$ be a point in the Euclidean space $\mathbf{R}^p$.

The problem statement for the heat conduction equation in the cylinder $\bar{Q}_{t_0} = \bar{G} \times [0 \le t \le t_0]$ is

$$(15)\qquad \frac{\partial u}{\partial t} = L\,u + f(x,t)\,,\quad L = \sum_{\alpha=1}^{p} L_\alpha\,,\quad (x,t) \in Q_{t_0}\,,$$

$$u|_\Gamma = \mu(x,t)\,,\quad t \ge 0\,,\quad u(x,0) = u_0(x)\,,\quad x \in \bar{G}\,.$$

Here $\Gamma$ is the boundary of the domain $G$ and $L$ is a second-order elliptic operator. For the sake of simplicity we agree to consider $L = \Delta$, that is, $L_\alpha u = \partial^2 u/\partial x_\alpha^2$, $\alpha = 1, 2, \ldots, p$ (Laplace operator). Also, we take for granted that

- problem (15) has a unique sufficiently small solution;
- the intersection of the domain $G$ with any straight lines $C_\alpha$ parallel to the axis $Ox_\alpha$ may consist only of a finite number of intervals;
- it is possible to compose in the domain $\bar{G}$ a connected grid $\bar{\omega}_h$ with steps $h_\alpha$, $\alpha = 1, 2, \dots, p$ (for more detail see Chapter 6, Section 1).

To make our exposition more transparent, it is supposed that the intersection of the domain $G$ with the line $C_\alpha$ is a unique interval.

Thus, the set $\omega_h$ of all inner nodes contains the points $x = (x_1, x_2, \dots, x_p) \in G$ of the intersection of the hyperplanes $x_\alpha = i_\alpha h_\alpha$, $i_\alpha = 0, \pm 1, \dots$, $\alpha = 1, 2, \dots, p$, while the set $\gamma_h$ of all boundary nodes consists of the points of the intersection of the lines $C_\alpha$, $\alpha = 1, 2, \dots, p$, passing through all inner nodes $x \in \omega_h$, with the boundary $\Gamma$.

Here we retain the same notations as was done in Chapter 6, Section 1: $\gamma_{h,\alpha}$ is the set of all boundary nodes in the direction $x_\alpha$; $\gamma_h$ is the set of all boundary nodes $x \in \Gamma$; $\omega^*_{h,\alpha}$ is the set of all near-boundary nodes in the direction $x_\alpha$; $\omega^*_h$ is the set of all near-boundary nodes; $\omega^{**}_{h,\alpha}$ is the set of all irregular nodes in the direction $x_\alpha$; $\omega^{**}_h$ is the set of all irregular nodes and $\omega_h$ is the set of all regular nodes.

The difference approximation of the operator $L_\alpha$ at a node $x$ is constructed on the pattern consisting of the three nodal points $x^{(-1_\alpha)}$, $x$, $x^{(+1_\alpha)}$. In view of this, the difference operator $\Lambda_\alpha \sim L_\alpha$ is taken to be

$$\Lambda_\alpha y = y_{\bar{x}_\alpha x_\alpha} = \frac{1}{h_\alpha^2} \left( y^{(+1_\alpha)} - 2y + y^{(-1_\alpha)} \right) \tag{16}$$

at the regular nodes and

$$\Lambda_\alpha y = y_{\bar{x}_\alpha \hat{x}_\alpha} = \begin{cases} \dfrac{1}{h_\alpha} \left( \dfrac{y^{(+1_\alpha)} - y}{h_\alpha} - \dfrac{y - y^{(-1_\alpha)}}{h^*_\alpha} \right), & x^{(-1_\alpha)} \in \gamma_{h,\alpha}, \\[2ex] \dfrac{1}{h_\alpha} \left( \dfrac{y^{(+1_\alpha)} - y}{h^*_\alpha} - \dfrac{y - y^{(-1_\alpha)}}{h_\alpha} \right), & x^{(+1_\alpha)} \in \gamma_{h,\alpha}, \end{cases} \tag{17}$$

at the irregular nodes, where $h^*_\alpha$ is the distance from an irregular node $x$ to one of the boundary nodes: $x^{(+1_\alpha)}$ or $x^{(-1_\alpha)}$. But it may happen that both nodes $x^{(+1_\alpha)}$ and $x^{(-1_\alpha)}$, which are neighboring to $x \in \omega^*_{h,\alpha}$, belong to the boundary, that is, $x^{(\pm 1_\alpha)} \in \gamma_{h,\alpha}$. In that case

$$\Lambda_\alpha y = \frac{1}{h_\alpha} \left( \frac{y^{(+1_\alpha)} - y}{h^*_{\alpha+}} - \frac{y - y^{(-1_\alpha)}}{h^*_{\alpha-}} \right), \tag{18}$$

where $h^*_{\alpha\pm}$ is the distance between $x$ and $x^{(\pm 1_\alpha)}$ ($h^*_{\alpha\pm} \le h_\alpha$). The preceding expression for $\Lambda_\alpha y$ is in common usage. If $x$ is one of the regular nodes, then $h^*_{\alpha\pm} = h^*_{\alpha-} = h_\alpha$, yielding formula (16).

Observe that $\Lambda_\alpha$ provides a second-order approximation at the regular nodes $\Lambda_\alpha u - L_\alpha u = O(h_\alpha^2)$, while $\Lambda_\alpha u - L_\alpha u = O(1)$ at the irregular ones.

The intuition suggests some things in an attempt to write down the locally one-dimensional (LOS) scheme in conformity with available constructions of Section 3. In working on the segment $0 \le t \le t_0$ we introduce a grid $\bar{\omega}_\tau = \{t_j = j\tau\,,\ j = 0, 1, \dots, j_0\}$ with step $\tau = t_0/j_0$ and involve arbitrary functions $f_\alpha$ subject to the normalization condition

$$\sum_{\alpha=1}^{p} f_\alpha = f\,.$$

In line with established practice we replace the governing multidimensional equation by the chain of the one-dimensional heat conduction equations

$$\text{(19)}\qquad \frac{1}{p}\,\frac{\partial v_{(\alpha)}}{\partial t} = L_\alpha\, v_{(\alpha)} + f_\alpha \quad \text{for} \quad t_{j+(\alpha-1)/p} < t \le t_{j+\alpha/p}\,,$$

$$\alpha = 1, 2, \dots, p\,, \qquad x \in G\,,$$

with the supplementary conditions

$$\text{(20)}\quad v_{(1)}(x, 0) = u_0(x)\,, \quad v_{(\alpha)}(x, t_{j+(\alpha-1)/p}) = v_{(\alpha-1)}(x, t_{j+(\alpha-1)/p})\,,$$

$$\alpha = 1, 2, \dots, p\,, \qquad v_{(\alpha)} = \mu(x,t) \quad \text{for} \quad x \in \Gamma_\alpha\,,$$

where $t_{j+\alpha/p} = (j + \alpha/p)\,\tau$.

Here the boundary conditions for $v_{(\alpha)}$ may be imposed only on some part $\Gamma_\alpha$ of the entire boundary $\Gamma$ consisting of the points of the intersection of $\Gamma$ with possible lines $C_\alpha$ parallel to the axis $Ox_\alpha$ and passing through any inner point $x \in G$. The nodal points $x \in \gamma_{h,\alpha}$ belong to that part $\Gamma_\alpha$.

If, for example, $\bar{G} = \{0 \le x \le l_\alpha\}$ is a parallelepiped, then $\Gamma_\alpha$ comprises the planes $x_\alpha = 0$ and $x_\alpha = l_\alpha$.

Through the approximation of every heat conduction equation with number $\alpha$ on the half-interval $t_{j+(\alpha-1)/p} < t \le t_{j+\alpha/p}$ by the standard two-layer weighted scheme we arrive at the chain of $p$ one-dimensional schemes with arbitrary numbers $\sigma_\alpha$:

$$\frac{y^{j+\alpha/p} - y^{j+(\alpha-1)/p}}{\tau} = \Lambda_\alpha \left(\sigma_\alpha y^{j+\alpha/p} + (1-\sigma_\alpha)\, y^{j+(\alpha-1)/p}\right) + \varphi_\alpha^{j+\alpha/p},$$

$$\alpha = 1, 2, \ldots, p, \qquad x \in \omega_h .$$

The describing scheme is termed LOS. In what follows we confine ourselves to the applications of purely implicit LOS's with $\sigma_\alpha \equiv 1$:

$$\frac{y^{j+\alpha/p} - y^{j+(\alpha-1)/p}}{\tau} = \Lambda_\alpha y^{j+\alpha/p} + +\varphi_\alpha^{j+\alpha/p}, \tag{21}$$

$$\alpha = 1, 2, \ldots, p, \qquad x \in \omega_h .$$

In preparation for this, the preceeding is put together with the boundary condition

$$y^{j+\alpha/p} = \mu^{j+\alpha/p} \quad \text{for} \quad x \in \gamma_{h,\alpha}, \quad j = 0, 1, \ldots, j_0, \tag{22}$$

$$\alpha = 1, 2, \ldots, p,$$

and the initial condition

$$y(x, 0) = u_0(x) . \tag{23}$$

As we will see a little later, the right-hand side $\varphi_\alpha^{j+\alpha/p}$ and the boundary value $y^{j+\alpha/p}|_{\gamma_{h,\alpha}}$ can be expressed through the functions $f_\alpha(x,t)$ and $\mu(x,t)$ taken at arbitrary moments $t_\alpha^*$ and $t_\alpha^{**}$ from the segment $[t_j, t_{j+1}]$, so that $\varphi_\alpha^{j+\alpha/p} = f_\alpha(x, t_\alpha^*)$ and $\mu^{j+\alpha/p} = \mu(x, t_\alpha^{**})$, thereby retaining the accuracy order. For the sake of definiteness, we accept

$$\varphi_\alpha^{j+\alpha/p} = f_\alpha(x, t_{j+0.5}), \quad \mu^{j+\alpha/p} = \mu(x, t_{j+\alpha/p}), \quad \alpha = 1, 2, \ldots, p,$$

by regarding $y^j$ to the known values. The value $j^{j+1}$ can be found on every new layer from (21)–(22) by successive solution of $p$ equations of the form (21) with the boundary conditions (22) in a step-by-step fashion for

$\alpha = 1, 2, \ldots, p$. In determining $y^{j+\alpha/p}$ we must solve the boundary-value problem

$$(24)\qquad \begin{aligned} & A_{i_\alpha} y_{i_\alpha - 1}^{j+\alpha/p} - C_{i_\alpha} y^{j+\alpha/p} + A_{i_\alpha+1} y_{i_\alpha+1}^{j+\alpha/p} = -F_\alpha y^{j+\alpha/p} \quad \text{for} \quad x \in \omega_h, \\ & y^{j+\alpha/p} = \mu^{j+\alpha/p} \quad \text{for} \quad x \in \gamma_{h,\alpha}, \quad \alpha = 1, 2, \ldots, p, \end{aligned}$$

with varying subscripts only. The difference equation is written on the segment $\Delta_\alpha \in C_\alpha$ with the endpoints belonging to the boundary $\gamma_{h,\alpha}$ and it can be solved by the elimination method along all the segments $\Delta_\alpha$ for fixed $\alpha$. For doing so our expenses are not considerable, since the number of arithmetic operations required at every node of the grid $\omega_h$ are $O(1)$ solely in connection with successive determination of the values $y^{j+1/p}$, $y^{j+2/p}, \ldots, y^{j+\alpha/p}, \ldots, y^{j+1}$ by setting $\alpha = 1, 2, \ldots, p$ and changing the directions of the eliminations. Thus, the locally one-dimensional scheme (21)–(23) falls within the category of economical schemes.

**6. The error of approximation of a locally one-dimensional scheme.** Upon raising the question regarding the error of approximation provided by one or another LOS it is straightforward to verify that every separate equation (21) with the number $\alpha$ does not approximate equation (15) in spite of the fact that the sum of the residuals $\psi = \psi_1 + \psi_2 + \cdots + \psi_p$ involved tends to 0 as $\tau \to 0$ and $|h| \to 0$.

Let $u = u(x, t)$ be a solution of problem (15) with the operator $L_\alpha u = \partial^2 u / \partial x_\alpha^2$ and $y^{j+\alpha/p}$, $\alpha = 1, 2, \ldots, p$ be a solution of problem (21)–(23). The accuracy of LOS is characterized, as usual, by the difference $y^{j+1} - u^{j+1} = z^{j+1}$.

The intermediate values $y^{j+\alpha/p}$ need to be compared with $u^{j+\alpha/p} = u(x, t_{j+\alpha/p})$ by making the choice $z^{j+\alpha/p} = y^{j+\alpha/p} - u^{j+\alpha/p}$. Upon substituting $y^{j+\alpha/p} = z^{j+\alpha/p} + u^{j+\alpha/p}$ into equation (21) we may set up the problem for the error $z^{j+1}$:

$$(25)\qquad \frac{z^{j+\alpha/p} - z^{j+(\alpha-1)/p}}{\tau} = \Lambda_\alpha\, z^{j+\alpha/p} + \psi_\alpha^{j+\alpha/p},$$

$$j = 0, 1, \ldots, j_0, \qquad \alpha = 1, 2, \ldots, p,$$

$$z^{j+\alpha/p} = 0 \quad \text{for} \quad x \in \gamma_{h,\alpha}, \qquad z(x, 0) = 0,$$

where

$$(26)\qquad \psi^{j+\alpha/p} = \Lambda_\alpha\, u^{j+\alpha/p} + \varphi_\alpha^{j+\alpha/p} - \frac{u^{j+\alpha/p} - u^{j+(\alpha-1)/p}}{\tau}.$$

Further analysis relies on the quantity with omitted superscripts:

$$\overset{\circ}{\psi}_{\alpha} = \left( L_{\alpha}\, u + f_{\alpha} - \frac{1}{p}\, \frac{\partial u}{\partial t} \right)^{j+1/2} . \tag{27}$$

A simple observation that

$$\sum_{\alpha=1}^{p} \overset{\circ}{\psi}_{\alpha} = 0$$

in the case where

$$\sum_{\alpha=1}^{p} f_{\alpha} = f$$

serves as a basis for the representation of $\psi_{\alpha} = \psi_{\alpha}^{j+\alpha/p}$ as a sum

$$\psi_{\alpha} = \overset{\circ}{\psi}_{\alpha} + \psi_{\alpha}^{*} ,$$

where

$$\psi_{\alpha}^{*} = \left( \Lambda_{\alpha}\, u^{j+\alpha/p} - L_{\alpha}\, u^{j+1/2} \right) + \left( \varphi_{\alpha}^{j+\alpha/p} - f_{\alpha}^{j+1/2} \right)$$

$$- \left( \frac{u^{j+\alpha/p} - u^{j+(\alpha-1)/p}}{\tau} - \frac{1}{p} \left( \frac{\partial u}{\partial t} \right)^{j+1/2} \right) .$$

From the very definition of $\Lambda_{\alpha}$ and $\varphi_{\alpha}$ it follows that

$$\psi_{\alpha}^{*} = O(h_{\alpha}^{2} + \tau) \quad \text{at the regular nodes,}$$

$$\psi_{\alpha}^{*} = O(1) \qquad \text{at the irregular nodes.}$$

Thus, we might have

$$\psi = \sum_{\alpha=1}^{p} \psi_{\alpha} = \sum_{\alpha=1}^{p} \psi_{\alpha}^{*} = O(\tau + |h|^{2})$$

at the regular nodes, that is, LOS in question generates a summarized approximation of $O(\tau + |h|^2)$ at the regular nodes of the grid $\omega_h$, while $\psi = O(1)$ at the irregular nodes.

**7. Stability of LOS.** The main goal of stability consideration is to establish that the uniform convergence with the rate $O(\tau + |h|^2)$ follows from a summarized approximation obtained. This can be done using the maximum principle and a priori estimates in the grid norm of the space $C$ for a solution of problem (21)–(23) expressing the stability of the scheme concerned with respect to the initial data, the right-hand side and boundary conditions.

Recall that in Chapter 4, Section 2 we have proved the maximum principle and derived a priori estimates for a solution to the grid equation of the general form

$$A(P)\,y(P) = \sum_{Q \in III'(P)} B(P,Q)\,y(Q) + F(P) \quad \text{for} \quad P \in \Omega\,, \tag{28}$$

$$y(P) = \mu(P) \quad \text{for} \quad P \in S\,,$$

where $P$ and $Q$ are some nodes of a connected grid $\Omega + S$ and $III'(P)$ is a neighborhood of the node $P$ except the point $P$ itself. The coefficients $A(P)$ and $B(P,Q)$ must satisfy the conditions

$$A(P) > 0\,, \qquad B(P,Q) > 0 \tag{29}$$

$$D(P) = A(P) - \sum_{Q \in III'(P)} B(P,Q) \geq 0\,.$$

Applying theorems of Section 2, Chapter 4 to problem (21)–(23) yields the following result.

**Theorem 1** *The locally one-dimensional scheme (21)–(23) is uniformly stable in the metric of the space $C$ with respect to the initial data, the right-hand side and boundary conditions and a solution of problem (21)–(23) admits for any $\tau$ and $h$ the estimate*

$$\|y^j\|_C \leq \|u_0\|_C + \max_{0<t'\leq j\tau} \|\mu(x,t')\|_{C_\gamma} \tag{30}$$

$$+ \max_{0\leq t'\leq j\tau} h^2\, \|\varphi(x,t')\|_{C^*} + \sum_{j'=0}^{j-1} \tau \sum_{\alpha=1}^{p} \|\varphi_\alpha^{j'+\alpha/p}\|_{\overset{\circ}{C}}\,,$$

*where*

$$t' = t_{j'+\alpha/p}\,, \qquad \varphi_{x,t'} = \varphi_\alpha^{j'+\alpha/p}\,,$$

$$h = \max_{1\le\alpha\le p} h_\alpha\,, \quad \|y\|_C = \max_{x\in\omega_h} |y|\,, \quad \|y\|_{C'} = \max_{x\in\gamma_h} |y|\,,$$

$$\|\varphi\|_{C^*} = \max_{x\in\omega_h^*} |\varphi|\,, \quad \|\varphi\|_{\overset{\circ}{C}} = \max_{x\in\overset{\circ}{\omega}_h} |\varphi|\,.$$

To prove this assertion, we represent a solution of the problem concerned as a sum

$$y = \bar{y} + v + w\,,$$

where $\bar{y}$ is a solution to homogeneous equations (21) with boundary and initial conditions (22)–(23) and $v$ and $w$ are solutions to nonhomogeneous equations (21) with the homogeneous boundary and initial conditions:

$$\text{(31)} \qquad \frac{v^{j+\alpha/p} - v^{j+(\alpha-1)/p}}{\tau} = \Lambda_\alpha\, v^{j+\alpha/p} + \overset{\circ}{\varphi}{}_\alpha^{j+\alpha/p}\,, \quad x\in\omega_h\,,$$

$$\alpha = 1,2,\dots,p\,, \quad v(x,0) = 0\,, \quad v^{j+\alpha/p} = 0 \quad \text{for} \quad x\in\gamma_{h,\alpha}\,,$$

$$\text{(32)} \qquad \frac{w^{j+\alpha/p} - w^{j+(\alpha-1)/p}}{\tau} = \Lambda_\alpha\, w^{j+\alpha/p} + \varphi_\alpha^{*\,j+\alpha/p}\,, \quad x\in\omega_h\,,$$

$$\alpha = 1,2,\dots,p\,, \quad w(x,0) = 0\,, \quad w^{j+\alpha/p} = 0 \quad \text{for} \quad x\in\gamma_{h,\alpha}\,.$$

Here $\overset{\circ}{\varphi}_\alpha$ and $\varphi_\alpha^*$ are specified by the formulas

$$\overset{\circ}{\varphi}_\alpha = \begin{cases} \varphi_\alpha & \text{for} \quad x\in\overset{\circ}{\omega}_h\,, \\ 0 & \text{for} \quad x\in\omega_h^*\,, \end{cases} \qquad \varphi_\alpha^* = \begin{cases} 0 & \text{for} \quad x\in\overset{\circ}{\omega}_h\,, \\ \varphi_\alpha & \text{for} \quad x\in\omega_h^*\,, \end{cases}$$

so that

$$\overset{\circ}{\varphi}_\alpha + \varphi_\alpha^* = \varphi_\alpha \quad \text{for} \quad x\in\omega_h\,,$$

thereby clarifying that $\varphi_\alpha^*$ differs from zero only at the near-boundary nodes.

For convenience in analysis, the grid $\bar{\omega}'_\tau$ is made up by

$$\bar{\omega}'_\tau = \{0\,,\, t_{j+\alpha/p} = (j+\alpha/p)\,\tau\,,\, j = 0,1,2,\dots,j_0-1\,,\, \alpha = 1,2,\dots,p\}\,,$$

containing not only the nodes $t_j = j\tau$ of the grid $\bar{\omega}_\tau$, but also fictitious nodes $t_{j+\alpha/p}$, $\alpha = 1, 2, \dots, p-1$. Let $\omega'_\tau$ be a set of the nodes of the grid $\bar{\omega}'_\tau$, for which $t > 0$. Finally, let $P(x, t')$, where $x \in \omega_h$, $t' \in \omega'_\tau$, be a node of the $(p+1)$-dimensional grid $\Omega = \omega_h \times \omega'_\tau$; $S$ be the boundary of the grid $\Omega$ containing the nodes $P(x, 0)$ for $x \in \bar{\omega}_h$ and the nodes $P(x, t_{j+\alpha/p})$ for $t_{j+\alpha/p} \in \bar{\omega}'_\tau$ and $x \in \gamma_{h,\alpha}$ for all $\alpha = 1, 2, \dots, p$, $j = 0, 1, \dots, j_0$; $\Omega^*_\alpha$ be a set of the nodes $P(x, t_{j+\alpha/p})$, where $x \in \omega^*_{h,\alpha}$ is a near-boundary node in the direction $x_\alpha$ of the grid $\bar{\omega}_h$.

With these, we proceed to the complete posing of problems related to $\bar{y}$ and $w$. The intention is to use the equation for $y$ in the canonical form (25) in combination with expression (18) for the difference operator $\Lambda_\alpha$ not only at the regular nodes, but also at the irregular nodes:

$$\left[\frac{1}{\tau} + \frac{1}{h_\alpha}\left(\frac{1}{h^*_{\alpha+}} + \frac{1}{h^*_{\alpha-}}\right)\right]\bar{y}^{j+\alpha/p} = \frac{1}{h^*_{\alpha+}\,h_\alpha}\,\bar{y}^{j+\alpha/p}_{i_\alpha+1} + \frac{1}{h^*_{\alpha-}\,h_\alpha}\,\bar{y}^{j+\alpha/p}_{i_\alpha-1} + \frac{1}{\tau}\,\bar{y}^{j+(\alpha-1)/p}, \tag{33}$$

where $y^{j+\alpha/p}_{i_\alpha\pm1} = y(x^{(\pm1_\alpha)}, t_{j+\alpha/p})$. From here it seems clear that conditions (29) are satisfied and $D(P) = 0$. Because of this, Theorem 2 from Chapter 4, Section 2 asserts that for a solution to equation (33) the estimate is valid:

$$\max_{P\in\Omega+S} |\bar{y}(P)| \le \max_{P\in S} |\bar{y}(P)|\,.$$

Taking into account that

$$\max_{P\in\Omega+S} |\bar{y}(P)| = \max_{t'\in\omega'_\tau} \|\bar{y}(x, t')\|_C\,,$$

$$\text{where}\quad \|y(x)\|_C = \max_{x\in\omega_h} |y(x)|\,,$$

$$\max_{P\in S} |\bar{y}(P)| = \max\left(\max_{t'\in\omega'_\tau} \|\mu(x, t')\|_{C_\gamma}\,, \|u_0\|_C\right),$$

$$\text{where}\quad \|\mu(x)\|_{C_\gamma} = \max_{x\in\gamma_h} |\mu(x)|,$$

we eventually get

$$\|\bar{y}^{j_0}\|_C \le \|u_0\|_C + \max_{0<t'\le j_0\tau} \|\mu(x, t')\|_{C_\gamma}\,. \tag{34}$$

With regard to problem (32) for $w$, it will be sensible to rewrite statement (32) in the canonical form (28)

$$\left[\frac{1}{\tau}+\frac{1}{h_\alpha}\left(\frac{1}{h^*_{\alpha+}}+\frac{1}{h^*_{\alpha-}}\right)\right] w^{j+\alpha/p} = \frac{1}{h^*_{\alpha+}\,h_\alpha}\, w^{j+\alpha/p}_{i_\alpha+1}$$

$$+\frac{1}{h^*_{\alpha-}\,h_\alpha}\, w^{j+\alpha/p}_{i_\alpha-1} + \frac{1}{\tau}\, w^{j+(\alpha-1)/p} + \varphi^{*\,j+\alpha/p}_{\alpha}, \qquad \alpha = 1, 2, \dots, p,$$

$$w^{j+\alpha/p} = 0 \quad \text{for} \quad x \in \gamma_{h,\alpha}, \quad w(x,0) = 0,$$

meaning $w = 0$ on the boundary $S$ of the grid $\Omega$:

$$w(P) \equiv 0 \qquad \text{for} \quad P \in S.$$

The right-hand side $\varphi^*$ differs from zero only at such nodes $(x, t')$, where $x \in \omega^*_h$. The trace of the homogeneous boundary condition $w = 0$ should be clearly seen in

$$D(P) \geq \min_\alpha \frac{1}{h^*_{\alpha\pm}\,h_\alpha} = \frac{1}{h^2}, \qquad \text{where} \quad h = \max_\alpha h_\alpha.$$

Applying Theorem 4 from Chapter 4, Section2 yields

$$\max_{\Omega+S} |y(P)| \leq \max_{t'\in\omega'_\tau} \left\| \frac{\varphi^*(x,t')}{D} \right\|_{C^*} \leq \max_{t'\in\omega'_\tau} h^2 \|\varphi^*\|_{C^*}. \tag{35}$$

In the estimation of the function $v$ we write down equation (31) in the canonical form (28) by regarging $P = x$ to nodal points of the $p$-dimensional grid $\bar\omega_h$:

$$\left[\frac{1}{\tau}+\frac{1}{h_\alpha}\left(\frac{1}{h^*_{\alpha+}}+\frac{1}{h^*_{\alpha-}}\right)\right] v^{j+\alpha/p} = \frac{1}{h^*_{\alpha+}\,h_\alpha}\, v^{j+\alpha/p}_{i_\alpha+1}$$

$$+\frac{1}{h^*_{\alpha-}\,h_\alpha}\, v^{j+\alpha/p}_{i_\alpha-1} + F^{j+\alpha/p}_\alpha,$$

$$F^{j+\alpha/p}_\alpha = \frac{v^{j+(\alpha-1)/p}}{\tau} + \overset{\circ}{\varphi}{}^{j+\alpha/p}_\alpha,$$

$$v^{j+\alpha/p} = 0 \quad \text{for} \quad x \in \gamma_{h,\alpha} \quad v(x,0) = 0.$$

In that case $D(P) = 1/\tau$ and Theorem 3 in Chapter 4, Section 2 states that

$$\text{(36)} \qquad \|v^{j+\alpha/p}\|_C \leq \left\| \frac{F_\alpha^{j+\alpha/p}}{D} \right\|_C = \tau \, \|F_\alpha^{j+\alpha/p}\|_C$$

$$\leq \|v^{j+(\alpha-1)/p}\|_C + \tau \, \| \overset{\circ}{\varphi}{}_\alpha^{j+\alpha/p} \|_C \, .$$

The first summation of (36) over $\alpha = 1, 2, \ldots, p$ gives rise to the relation

$$\|v^{j+1}\|_C \leq \|v^j\|_C + \tau \sum_{\alpha=1}^{p} \| \overset{\circ}{\varphi}{}_\alpha^{j+\alpha/p} \|_C \, ,$$

which is followed upon summing up over $j = 0, 1, 2, , \ldots, j_0 - 1$ by the estimate

$$\text{(37)} \qquad \|v^{j_0}\|_C \leq \sum_{j'=0}^{j_0 - 1} \tau \sum_{\alpha=1}^{p} \| \overset{\circ}{\varphi}{}_\alpha^{j+\alpha/p} \|_C \, .$$

Estimate (30) for a solution of problem (21)–(23) is an immediate implication of the collection of relation (34), (35), (37) if we might invoke the arbitrariness in the choice of the number $j_0$.

**8. Uniform convergence of LOS.** This type of situation is covered by the following assertion.

**Theorem 2** *Let problem* (15) *possess a unique solution* $u = u(x, t)$ *continuous in* $\bar{Q}_T$ *and there exist continuous in* $Q_{t_0}$ *derivatives*

$$\frac{\partial^2 u}{\partial t^2} \, , \qquad \frac{\partial^4 u}{\partial x_\alpha^2 \, \partial x_\beta^2} \, , \qquad \frac{\partial^3 u}{\partial t \, \partial x_\alpha^2} \, , \qquad \frac{\partial^2 f}{\partial x_\alpha^2} \, , \qquad 1 \leq \alpha, \, \beta \leq p \, .$$

*Then scheme* (21)–(23) *converges uniformly with the rate* $O(h^2 + \tau)$ *(it is of first-order accuracy in* $\tau$ *and of second-order accuracy in* $h$*), so that*

$$\|y^j - u^j\|_C \leq M \, (h^2 + \tau) \, , \qquad j = 1, 2, \ldots \, ,$$

*where* $h = \max\limits_{1 \leq \alpha \leq p} h_\alpha$, $M = \text{const} > 0$ *is independent of* $\tau$ *and* $h_\alpha$.

**Proof** Let $z_{(\alpha)} = z^{j+\alpha/p}$. As usual, we may attempt a solution $z_{(\alpha)} = y_{(\alpha)} - u^{j+\alpha/p}$ of problem (25) as a sum $z_{(\alpha)} = v_{(\alpha)} + \eta_{(\alpha)}$ with $\eta_{(\alpha)}$ still subject to the conditions

$$(38) \qquad \frac{\eta_{(\alpha)} - \eta_{(\alpha-1)}}{\tau} = \overset{\circ}{\psi}_{\alpha} \quad \text{for} \quad x \in \omega_h + \gamma_{h,\alpha}, \quad \alpha = 1, 2, \dots, p,$$

$$\eta(x, 0) = 0 .$$

From here we deduce for $j = 0, 1, \dots, j_0$ that

$$\eta^{j+1} = \eta_{(p)} = \eta^j + \tau\,(\overset{\circ}{\psi}_1 + \overset{\circ}{\psi}_2 + \dots + \overset{\circ}{\psi}_p) = \eta^j = 0 ,$$

since $\eta^0 = 0$. With regard to $\eta_{(\alpha)}$ it is plain to show that

$$\eta_{(\alpha)} = \tau\,(\overset{\circ}{\psi}_1 + \overset{\circ}{\psi}_2 + \dots + \overset{\circ}{\psi}_\alpha) = -\tau\,(\overset{\circ}{\psi}_{\alpha+1} + \dots + \overset{\circ}{\psi}_p) .$$

The function $v_{(\alpha)}$ is yet to be recovered from the conditions

$$(39) \qquad \frac{v_{(\alpha)} - v_{(\alpha-1)}}{\tau} = \Lambda_\alpha\, v_{(\alpha)} + \tilde{\psi}_\alpha , \quad x \in \omega_h , \quad \alpha = 1, 2, \dots, p ,$$

$$v_{(\alpha)} = -\eta_{(\alpha)} \quad \text{for} \quad x \in \gamma_{h,\alpha} , \quad v(x, 0) = 0 ,$$

where $\tilde{\psi}_\alpha = \psi^*_\alpha + \Lambda_\alpha \eta_{(\alpha)}$.

A solution of problem (39) can readily be evaluated on account of Theorem 1 in Section 7 with $v = 0$ incorporated for $t = 0$, stating that

$$(40) \qquad \|v^j\|_C \le \max_{0<j'+\alpha/p<j} \left(h^2\, \|\tilde{\psi}^{j'+\alpha/p}\|_{C^*} + \|\eta^{j'+\alpha/p}\|_{C_\gamma}\right)$$

$$+ \sum_{j'=0}^{j-1} \tau \sum_{\alpha=1}^{p} \|\tilde{\psi}^{j+\alpha/p}\|_{\overset{\circ}{C}} .$$

If there exist for $\alpha \ne \beta$ the derivatives $\partial^4 u / \partial x_\alpha^2\, \partial x_\beta^2$ continuous in the closed domain $\bar{Q}_T$, then at all the nodes $x \in \omega_h$

$$\Lambda_\alpha \eta_{(\alpha)} = -\tau\, \Lambda_\alpha\, (\overset{\circ}{\psi}_{\alpha+1} + \dots + \overset{\circ}{\psi}_p) = O(\tau) ,$$

since the quantity $\eta_{(\alpha)}$ is recovered from equation (38) everywhere on $\omega_h + \gamma_{h,\alpha}$. On the other hand, $\psi^*_\alpha = O(h^2_\alpha + \tau)$ at the regular nodes of the grid $\omega_h$ and $\psi^*_{\tilde{\alpha}} = O(h^2_\alpha + \tau)$ at the irregular nodes. Therefore, $h^2 \|\tilde{\psi}\|_{C^*} = O(h^2 + \tau)$ and $\|\tilde{\psi}\|_{\overset{\circ}{C}} = O(h^2 + \tau)$, so that estimate (40) in such a setting gives

$$\|z^j\|_C = \|v^j\|_C \leq M\,(h^2 + \tau)\,,$$

since $\eta^j = 0$ for all $j = 0, 1, \ldots, j_0$.

Observe that the stability with respect to the right-hand side and boundary conditions implies that the moments $t^*_\alpha$ and $t^{**}_\alpha$ can be arbitrarily taken from the interval $(t_j, t_{j+1})$.

**9. LOS for equations with variable coefficients.** One way of covering equations with variable coefficients is connected with possible constructions of locally one-dimensional schemes and the main ideas adopted for problem (15). It sufficies to point out only the necessary changes in the formulas for the operators $L_\alpha$ and $\Lambda_\alpha$, which will be used in the sequel, and then bear in mind that any locally one-dimensional scheme can always be written in the form (21)–(23). Several examples add interest and help in understanding.

1) *A linear equation of parabolic type.*

Let in the statement of problem (15) involves

$$L_\alpha\, u = \frac{\partial}{\partial x_\alpha} \left( k_\alpha(x, t)\, \frac{\partial u}{\partial x_\alpha} \right), \quad 0 < c_1 \leq k_\alpha \leq c_2\,.$$

Minor changes in the complete posing of problem (21)–(23) are based on the formula for the difference operator $\Lambda_\alpha$ acting in accordance with the rule

$$\Lambda_\alpha\, y_{(\alpha)} = \left( a_\alpha(x, \bar{t})\, y_{\bar{x}_\alpha} \right)_{x_\alpha}, \quad 0 < c_1 \leq a_\alpha \leq c_2\,, \quad \bar{t} = t_{j+1/2}\,.$$

A second-order approximation provided by the difference operator $\Lambda_\alpha$ on a regular pattern

$$\Lambda_\alpha\, u - L_\alpha\, u = O(h^2_\alpha)$$

is ensured by a proper choice of the coefficient $a_\alpha$. This can be done using, for example,

$$a_\alpha = k_\alpha \left( x^{(-0.5\,\alpha)}, \bar{t} \right),$$

$$x^{(-0.5\,\alpha)} = \left( x_1, \ldots, x_{\alpha-1}, x_\alpha - 0.5\, h_\alpha, x_{\alpha+1}, \ldots, x_p \right),$$

allowing Theorems 1–2 to hold.

2) *A quasilinear equation of parabolic type.*

The complete posing of problem (15) includes

$$L_\alpha u = \frac{\partial}{\partial x_\alpha}\Big(k_\alpha(x,t,u)\,\frac{\partial u}{\partial x_\alpha}\Big), \quad 0 < c_1 \le k_\alpha, \quad \frac{\partial k_\alpha}{\partial u} \ge c_0 > 0 .$$

Two possible ways of approximating the operator $L_\alpha$ are:

a) $$\Lambda_\alpha y_{(\alpha)} = \Big(a_\alpha\big(x,\bar t, 0.5\,(y_{(\alpha)} + y_{(\alpha)}^{(-1_\alpha)})\big)\, y_{\bar x_\alpha}\Big)_{x_\alpha}, \quad \bar t = t_{j+1/2},$$

b) $$\Lambda_\alpha y_{(\alpha)} = \Big(a_\alpha\big(x,\bar t, 0.5\,(y_{(\alpha-1)} + y_{(\alpha-1)}^{(-1_\alpha)})\big)\, y_{\bar x_\alpha}\Big)_{x_\alpha}, \quad \bar t = t_{j+1/2}.$$

In the first case a nonlinear equation is aimed at determining $y_{(\alpha)}$ that can be solved by one or another iterative method. Every iteration can be found during the course of the elimination.

In the second case we obtain a linear equation related to $y_{(\alpha)}$ and then use the elimination method for solving it. The uniform convergence with the rate $O(\tau + h^2)$ takes place under the extra restrictions concerning the boundedness of the derivatives $\partial^2 k_\alpha/\partial u^2$, $\partial^2 k_\alpha/\partial x_\alpha\,\partial u$, $\partial^2 k_\alpha/\partial x_\alpha^2$.

Locally one-dimensional schemes find a wide range of applications in solving the third boundary-value problem. If, for example, $G$ is a rectangle of sides $l_1$ and $l_2$ or a "step-shaped" domain, then equations (21) should be written not only at the inner nodes of the grid, but also on the appropriate boundaries. When the boundary condition $\partial u/\partial x_1 = \sigma_1^- u + \nu_1^{(-)}$ is imposed on the side $x_1 = 0$ of the rectangle $\{0 \le x_\alpha \le l_\alpha,\ \alpha = 1,2\}$, the main idea behind this approach is to write for $\alpha = 1$ equations (21) at the node $x_1 = 0$ as well. This can be done by setting

$$\Lambda_1 y_{(1)} = \frac{y_{(1)x_1} - \sigma_1^{(-)} y_{(1)}}{0.5\,h_1}, \qquad \varphi_1 = \frac{\nu_1^{(-)}}{0.5\,h_1},$$

assuring the uniform convergence of the describing locally one-dimensional scheme with the rate $O(\tau + |h|^2)$.

**10. Additive schemes. The general formulations and statements.** Considerable effort is devoted to a discussion of additive schemes after introducing the notion of summarized approximation. With this aim, we recall the notion of the $n$-layer difference scheme as a difference equation with respect to $t$ of order $n-1$ with operator coefficients:

$$\sum_{\beta=0}^{n-1} C_\beta(t_j)\, y(t_{j+1} - \beta\,\tau) = f(t_j), \qquad (n-1)\,\tau \le t_j \le t_0,$$

where $C_\beta$ are linear operators acting in a vector normed space $H_h$. We shall need yet $n-1$ initial vectors $y(0)=y_0$, $y(\tau)=y_1$, ..., $y((n-2)\tau)=y_{n-2}$ for its numerical solution.

By the **$n$-layer composite scheme of period** $m$ (of order $m$) we generally mean a system of differential equatioins with operator coefficients

$$(41) \qquad \sum_{\beta=1}^{m} C_{\alpha\beta}(t_j)\, y(t_j+\beta\,\tau) = \sum_{\beta=0}^{n-2} D_{\alpha\beta}(t_j)\, y(t_j-\beta\,\tau) + f_\alpha(t_j)\,,$$

$$\alpha = 1,2,\ldots,m\,, \qquad (n-1)\,\tau \le t_j \le t_0\,,$$

with known initial values $y(k\,\tau)$, $k=0,1,\ldots,n-2$. Here $t_j$ takes on the values

$$t_j = (n-1)\,\tau + k\,m\,\tau\,, \qquad k=0,1,\ldots,$$

and the total number of layers is equal to the amount of the initial conditions.

With knowledge of the values $y_{j-\beta}$, $\beta=0,1,\ldots,n-2$, where $t_j = (m+n-1)\,\tau$, it is possible to find $y(t_j+m\,\tau)=y_{j+m}$ in the process of solving a system of equations with the operator matrix $C=(C_{\alpha\beta})$ of size $m\times m$.

Several particular cases will be given special investigation. For $m=1$ the composite scheme (41) falls within the category of standard $n$-layer schemes. For $n=2$ the describing scheme is termed a **two-layer composite scheme of period** $m$

$$(42) \qquad \sum_{\beta=1}^{m} C_{\alpha\beta}(t_j)\, y(t_j+\beta\,\tau) = D_{\alpha 0}\, y(t_j) + f_\alpha(t_j)\,,$$

$$1 \le \alpha \le m\,, \qquad y(0)=y_0\,.$$

If for the composite scheme (41) the error of approximation $\psi$ is adopted as a sum of the residuals $\psi_\alpha$ of separate equations, that is, $\psi=\psi_1+\cdots+\psi_m$, the composite scheme (41) is called an **additive scheme**.

By replacing $\tau$ by $\tau/m$ scheme (42) admits an alternative form

$$(42') \qquad \sum_{\beta=1}^{m} C_{\alpha\beta}(t_j)\, y(t_j+\beta\,\tau/m) = D_{\alpha 0}\, y(t_j) + f_\alpha(t_j)\,,$$

$$\alpha = 1,2,\ldots,m\,.$$

A two-layer additive scheme can always be written in the canonical form

$$B\,\frac{y^{j+\alpha/m} - y^{j+(\alpha-1)/m}}{\tau} + \sum_{\beta=0}^{m} A_{\alpha\beta}\, y^{j+\beta/m} = \varphi_\alpha^j\,, \tag{43}$$

$$\alpha = 1, 2, \ldots, m\,,$$

where $B$ and $A_{\alpha\beta}$ are some linear operators. It is straightforward to verify that all of the available economical methods with canonical form (43) can generate a summarized approximation.

Furthermore, let $u(t) \in H_0$ be an abstract function of the argument $t \in [0, t_0]$ with the values in a normed space $H_0$ and $u_h = \mathcal{P}_h\, u \in H_h$ be the projection of $u$ onto $H_h$,

$$\psi_\alpha(u_h^j) = \varphi_\alpha^j - B\,\frac{u_h^{j+\alpha/m} - u_h^{j+(\alpha-1)/m}}{\tau} - \sum_{\beta=0}^{m} A_{\alpha\beta}\, u_h^{j+\beta/m}$$

is the residual for equation (43) with the number $\alpha$ and $u^{j+\alpha/m} = u(t_j + \tau\alpha/m)$. Assuming this to be the case, the sum

$$\psi(u_h^j) = \sum_{\alpha=1}^{m} \psi_\alpha(u_h^j)$$

is of our initial concern.

By definition, the additive scheme (42) provides a summarized approximation on a function $u(t) \in H_0$ if

$$\max_{0 \le j \le j_0} \left\| \psi(u_h^j) \right\|_{(2_h)} \longrightarrow 0 \quad \text{as} \quad \tau \to 0\,,\ h \to 0\,,$$

where $\|\cdot\|_{(2_h)}$ is some suitable norm on the space $H_h$. In conformity with Section 2 the additive scheme (42) is said to be **economical** if the operator (matrix) $C$ is economical. That is to say, the work and storage necessary in the numerical solution of the system of operator equations

$$\sum_{\beta=1}^{m} C_{\alpha\beta}\, y^{j+\beta/m} = \Phi_\alpha^j \tag{44}$$

require a minimal number (in some up-agreed sense) of arithmetical operations. For example, it may be proportional to the dimension $N$ of the space $H_h$ (it is equal to the total number of the grid nodes $\omega_h$).

If $(C_{\alpha\beta}) = C^-$ is a lower triangle matrix and all the operators $C_{\alpha\alpha}$ are invertible, then the procedure of solving equations (42′) can be reduced to successive solution of the equations

$$C_{11}\, y^{j+1/m} = D_{10}\, y^j + f_1^j \,,$$

$$C_{\alpha\alpha}\, y^{j+\alpha/m} = D_{\alpha 0}\, y^j - \sum_{\beta=1}^{\alpha-1} C_{\alpha\beta}\, y^{j+\beta/m} + f_\alpha^j \,, \quad \alpha = 2, 3, \dots, m \,.$$

Such a triangle additive scheme will be economical once we involve economical diagonal operators $C_{\alpha\alpha}$, $\alpha = 1, 2, \dots, m$. Economical schemes arising in practical implementations of multidimensional mathematical-physics problems turn out to be triangle additive schemes (usually lower, but sometimes upper), whose matrices are of a special structure. As a rule, nonzero elements of the matrix $(C_{\alpha\beta})$ stand only on one or two diagonals adjacent to the main diagonal. With this in mind, the scheme

$$C_{\alpha\alpha}\, y^{j+\alpha/m} + C_{\alpha\alpha-1}\, y^{j+(\alpha-1)/m} = D_{\alpha 0}\, y^j + f_\alpha^j \,, \quad \alpha = 1, 2, \dots, m \,,$$

may be of assistance in achieving the final aims. In particular, when $D_{\alpha 0} = 0$ for all the values $\alpha = 1, 2, \dots, m$, the preceding reduces to

$$C_{\alpha\alpha}\, y^{j+\alpha/m} + C_{\alpha\alpha-1}\, y^{j+(\alpha-1)/m} = f_\alpha^j \,,$$

one special case of which is the weighted scheme

$$\frac{y^{j+\alpha/m} - y^{j+(\alpha-1)/m}}{\tau} + A_\alpha \left( \sigma_\alpha\, y^{j+\alpha/m} + (1-\sigma_\alpha)\, y^{j+(\alpha-1)/m} \right) = \varphi_\alpha^j \,.$$

Such locally one-dimensional schemes were investigated before, all the triks and turns remain here unchanged. The following issues are yet to be answered in the possible theory:

1) the estimation of stability and accuracy of an additive scheme;
2) the design of an economical additive scheme for a multidimensional problem in mathematical physics.

**11. Methods for the convergence rates of additive schemes.** So far we have established many times that approximation and stability of a difference scheme provide its convergence. For additive schemes we shall need stability with respect to the right-hand side so that it follows from the condition of summarized approximation

$$\left\| \sum_{\alpha=1}^{p} \psi_\alpha \right\| \longrightarrow 0$$

that a solution of the relevant difference problem supplied by the zero initial condition approaches zero. Such a priori estimates hinging on the summarized approximation properties hold true for additive schemes associated with systems of parabolic and hyperbolic equations.

We outline in what follows the general theory for an additive typical scheme in a Hilbert space $H_h$ such as

$$B\,\frac{z^{j+\alpha/p}-z^{j+(\alpha-1)/p}}{\tau}+\sum_{\beta=1}^{p}A_{\alpha\beta}\,z^{j+\beta/p}=\psi_\alpha^j\,, \tag{45}$$

$$\alpha=1,2,\ldots,p\,,\quad j=0,1,\ldots\,,\quad z^0=0\,.$$

**Theorem 3** *If $B=B^*$ is a positive definite constant operator and the matrix-operator $A$ is non-negative $A=(A_{\alpha\beta})\geq 0$, that is, for any vectors $\xi_\alpha, \xi_\beta\in H$*

$$\sum_{\alpha,\beta=1}^{p}(A_{\alpha\beta}\,\xi_\alpha,\xi_\beta)\geq 0\,, \tag{46}$$

*then a solution of problem* (45) *satisfies the a priori estimate*

$$\|z^j\|_B\leq\max_{0\leq k\leq j}\left\{\sqrt{e\,t_j}\left\|\sum_{\alpha=1}^{p}\psi_\alpha^k\right\|_{B^{-1}}+p\sqrt{\tau\,t_j}\sum_{\alpha=1}^{p}\left\|\psi_\alpha^k\right\|_{B^{-1}}\right\}. \tag{47}$$

The proof of this formula is omitted here. It should be noted that from such reasoning it seems clear that due to the summarized approximation in the space $H_{B^{-1}}$ the convergence occurs in the space $H_B$. That is to say, the conditions

$$\left\|\sum_{\alpha=1}^{p}\psi_\alpha\right\|_{B^{-1}}=\|\psi\|_{B^{-1}}\to 0\,,\quad \|\psi_\alpha\|_{B^{-1}}=O(1)\,,\quad \tau\to 0\,,\ |h|\to 0 \tag{48}$$

guarantee the convergence $\|z^j\|_B\to 0$ for all $j=1,2,\ldots$ by observing that estimate (47) is valid under rather mild conditions: $B$ is a positive definite self-adjoint operator and $A$ is a non-negative matrix-operator. But in a Banach space $H_h$ another method of further derivation of a priori estimates is employed for scheme (45).

In what follows scheme (45) is supposed to be stable so that

$$\|z^j\|_{(1)}\leq M\max_{0\leq k<j}\sum_{\alpha=1}^{p}\|\psi_\alpha^k\|_{(2)}\,, \tag{49}$$

where $\|\cdot\|_{(1)}$ and $\|\cdot\|_{(2)}$ are some suitable norms on the space $H_h$. The usual trick we have encountered is to represent the residual $\psi_\alpha$ by

$$\psi_\alpha = \overset{\circ}{\psi}_\alpha + \psi^*_\alpha, \quad \text{so that} \quad \sum_{\alpha=1}^{p} \overset{\circ}{\psi}_\alpha = 0. \tag{50}$$

By setting $z^{j+\alpha/p} = \eta^{j+\alpha/p} + v^{j+\alpha/p}$, where $\eta^{j+\alpha/p}$ is determined from the conditions

$$B\,\frac{\eta^{j+\alpha/p} - \eta^{j+(\alpha-1)/p}}{\tau} = \overset{\circ}{\psi}{}^j_\alpha, \quad \alpha = 1, 2, \dots, p, \quad \eta^0 = 0,$$

it is plain to show that

$$B\,\eta^{j+\alpha/p} = B\,\eta^j + \tau \sum_{\beta=1}^{\alpha} \overset{\circ}{\psi}{}^j_\beta, \quad B\,\eta^{j+1} = B\,\eta^j = \dots = B\,\eta^0 = 0,$$

giving $\eta^j = 0$ and $z^j = v^j$ for all $j = 1, 2, \dots$ and

$$\eta^{j+\alpha/p} = \tau \sum_{\beta=1}^{\alpha} B^{-1} \overset{\circ}{\psi}{}_\beta{}^j = -\tau \sum_{\beta=\alpha+1}^{p} B^{-1} \overset{\circ}{\psi}{}^j_\beta, \quad \alpha = 1, 2, \dots, p-1.$$

In turn, $v^{j+\alpha/p}$ satisfies equation (45) with the right-hand side

$$\tilde{\psi}^j_\alpha = \psi^{*j}_\alpha + \tau \sum_{\beta=1}^{\alpha} A_{\alpha\beta} \sum_{\beta'=\beta+1}^{p} B^{-1} \overset{\circ}{\psi}{}^j_\beta$$

and the initial condition $v^0 = 0$. Having stipulated condition (49), the following estimate is valid:

$$\|z^j\|_{(1)} = \|v^j\|_{(1)} \le M \max_{0 \le j' < j} \sum_{\alpha=1}^{p} \|\tilde{\psi}^{j'}_\alpha\|_{(2)}.$$

The reader is invited to prove this assertion on his/her own. The summarized approximation condition means that

1) the residual $\psi_\alpha$ admits representation (50),

2) $\|\tilde{\psi}^j_\alpha\|_{(2)} \to 0$ as $\tau \to 0$ and $|h| \to 0$.

The second condition is ensured if

$$\|\psi^*_\alpha\|_{(2)} \longrightarrow 0\,, \qquad \|A_{\alpha\,\beta}\, B^{-1}\, \overset{\circ}{\psi}{}^{j}_{s}\| = O(1) \quad \text{as} \quad \tau \to 0\,,\ |h| \to 0\,.$$

The second method of special investigations with concern of additive schemes was demonstrated in Section 8 in which convergence in the space $C$ of a locally one-dimensional scheme associated with the heat conduction equation was established by means of this method. Let us stress that in such an analysis we assume, as usual, the existence, uniqueness and a sufficient smoothness of a solution of the original multidimensional problem under consideration.

Let, for example, $u$ be a solution of problem (5) and $y = y_h$ be a solution provided by this or that additive scheme, $y_h \in H_h$, where $H_h$ is the set of grid functions. Following established practice, the difference $z^j_h = y^j_h - u^j_h$, where $u_h = \mathcal{P}_h\, u$ and $\mathcal{P}_h$ is a linear operator from the space $H_0$ into the space $H_h$ ($u \in H_0$, $u_h \in H_h$), needs investigation. To be more specific, we are interested in the possible estimates of the quantity $\|y^j_h - u^j_h\|_{(1_h)}$ in some suitable norm $\|\cdot\|_{(1_h)}$ on the space $H_h$. The traditional ways of covering this are to set up the relevant problem for the error $z_h$, calculate the residual $\psi_\alpha = \overset{\circ}{\psi}_\alpha + \psi^*_\alpha$ and then adapt one of the well-developed methods for determination of $z_h$.

**12. An approximation of the "multidimensional" abstract Cauchy problem by a chain of the "one-dimensional" Cauchy problems.** It is to be hoped that the forthcoming reduction helps clarify what is done. The problem statement involves problem (5) with the homogeneous boundary conditions on the boundary $\Gamma$ under the agreement that the function $u(x,t)$ as a function of the argument $x$ can be treated in a common setting as an element of some vector normed space $H_0$. Then $L$ refers to a linear operator in that space and $u = u(t)$ may be viewed as an abstract function of the argument $t$ with the values $H_0$, it being understood that $u(t) \in H_0$ for all $t \in [0, t_0]$. In this view, it is possible to write in problem (5) the usual derivative in $t$ in place of the partial derivative, making our exposition more transparent. As a final result we get the abstract Cauchy problem

$$(51) \qquad \frac{du}{dt} + \mathcal{A}\, u = f(t)\,, \qquad 0 \le t \le t_0\,, \qquad u(0) = u_0 \in H_0\,,$$

where $\mathcal{A}$ is a linear operator in the Banach space $H_0$. The domain $\mathcal{D}(\mathcal{A}) \subset H_0$ of the operator $\mathcal{A}$ is everywhere dense in the space $H_0$ and comprises all the functions satisfying the homogeneous boundary conditions on the boundary $\Gamma$ and its range $\Delta(\mathcal{A})$ belongs to the space $H_0$.

Also, the operator $\mathcal{A}$ will be taken to be

$$\mathcal{A} = \sum_{\alpha=1}^{p} \mathcal{A}_\alpha , \tag{52}$$

where linear operators $\mathcal{A}_\alpha$ are so chosen as to provide the relation

$$\bigcap_{\alpha=1}^{p} \mathcal{D}(\mathcal{A}_\alpha) = \mathcal{D}(\mathcal{A}) ,$$

making it possible to reduce the solution of the Cauchy problem (51) to solving successively the Cauchy problems of the same type, but with operators $\mathcal{A}_\alpha$ standing in place of the operator $\mathcal{A}$. We confine ourselves here to two possible ways of such a reduction.

For later use, it will be sensible to introduce on the segment $0 \le t \le t_0$ a grid $\bar{\omega}_\tau = \{t_j = j\tau,\ j = 0, 1, \dots, j_0\}$ with step $\tau$ and to attempt the function $f$ in the form

$$f = \sum_{\alpha}^{p} f_\alpha .$$

*The first reduction* (for more detail see Section 3). The object of investigation is a chain of the equations

$$\frac{1}{p}\, \frac{dv_{(\alpha)}}{dt} + \mathcal{A}_\alpha\, v_{(\alpha)} = f_\alpha , \qquad \alpha = 1, 2, \dots, p, \tag{53}$$

$$t_{j+(\alpha-1)/p} < t \le t_{j+\alpha/p} ,$$

with the supplementary initial conditions

$$v_{(1)}(0) = u_0 , \qquad v_{(1)}(t_j) = v_{(p)}(t_j) , \qquad j = 1, 2, \dots , \tag{54}$$

$$v_{(\alpha)}(t_{j+(\alpha-1)/p}) = v_{(\alpha-1)}(t_{j+(\alpha-1)/p}) ,$$

$$j = 0, 1, \dots , \qquad \alpha = 2, 3, \dots, p .$$

The function $v(t_{j+1}) = v_{(p)}(t_{j+1})$ refers to a solution of this problem for $t = t_{j+1}$. In the general case we might have

$$\| v(t_j) - u(t_j) \| = O(\tau) \qquad \text{for all} \quad j = 1, 2, \dots$$

*The second reduction* of the Cauchy problem. On the whole segment $t_j \le t \le t_{j+1}$ we must solve sequentially $p$ Cauchy problems

$$\text{(55)} \qquad \begin{aligned} &\frac{dv_{(1)}}{dt} + \mathcal{A}_1(t)\, v_{(1)}(t) = f_1(t)\,, \qquad t_j \le t \le t_{j+1}\,,\\ &\qquad \dots\\ &\frac{dv_{(\alpha)}}{dt} + \mathcal{A}_\alpha(t)\, v_{(\alpha)}(t) = f_\alpha(t)\,, \qquad t_j \le t \le t_{j+1}\,,\\ &\qquad \dots\\ &\frac{dv_{(p)}}{dt} + \mathcal{A}_p(t)\, v_{(p)}(t) = f_p(t)\,, \qquad t_j \le t \le t_{j+1}\,, \end{aligned}$$

with the initial data

$$\text{(56)} \qquad \begin{aligned} &v_{(1)}(0) = u_0\,, \qquad v_{(1)}(t_j) = v_{(p)}(t_j)\,, \qquad j = 1, 2, \dots\,,\\ &v_{(\alpha)}(t_j) = v_{(\alpha-1)}(t_{j+1})\,, \quad j = 0, 1, \dots,\,, \quad \alpha = 2, 3, \dots, p\,. \end{aligned}$$

By definition, an element

$$v(t_{j+1}) = v_{(p)}(t_{j+1})\,, \qquad j = 0, 1, 2, \dots\,,$$

gives a solution of problem (55) for $t = t_{j+1}$. For $t = 0$ we agree to consider

$$\text{(57)} \qquad v_{(1)}(0) = u(0) = u_0\,.$$

Knowing $v(t_j)$, it is possible to determine $v_{(1)}(t_{j+1})$ from the first equation entering the above collection with $v_{(1)}(t_j) = v(t_j)$ incorporated. At the next stage $v_{(1)}(t_{j+1})$ is taken as the initial value of $v_{(2)}(t)$ for $t = t_j$, allowing to solve the second equation for $\alpha = 2$, etc. The outcome of solving all the $p$ problems is $v_{(p)}(t_{j+1}) = v(t_{j+1})$, giving a solution of the system of equations (55)–(57) for $t = t_{j+1}$.

When the operators $\mathcal{A}_\alpha$ happen to be independent of $t$ and $f = 0$, problems (53)–(54) and (55)–(57) become equivalent. With this in mind, we are going to show that problem (51) is approximated by problem (55)–(56) in a summarized sense. To that end, the differences

$$\begin{aligned} &z_{(\alpha)}(t) = v_{(\alpha)}(t) - u(t_{j+1}) \quad \text{for} \quad \alpha = 2, 3, \dots, p\,, \quad t \in [t_j, t_{j+1}]\,,\\ &z_{(1)}(t) = v_{(1)}(t) - u(t) \qquad \text{for} \quad t \in [t_j, t_{j+1}] \end{aligned}$$

will be given special investigation, where $u(t)$ is a solution of the Cauchy problem (51) and $v_{(\alpha)}(t)$, $\alpha = 1, 2, \dots, p$, is a solution of problem (55)-(56). By substituting here $v_{(\alpha)}(t) = z_{(\alpha)}(t) + u^{j+1}$, $u^{j+1} = u(t_{j+1})$, $\alpha = 2, 3, \dots, p$ and $v_{(1)} = z_1(t) + u(t)$ into (55)-(56) we are led to

$$\frac{dz_{(\alpha)}}{dt} + \mathcal{A}_\alpha(t)\, v_{(\alpha)}(t) = \psi_\alpha(t)\,, \quad t_j \le t \le t_{j+1}\,, \quad 1 \le \alpha \le p\,,$$

$$z_{(1)}(t_j) = z_{(p)}(t_j)\,, \qquad j = 1, 2, \dots\,, \qquad z_1(0) = 0\,,$$

$$z_{(\alpha)}(t_j) = z_{(\alpha-1)}(t_{j+1})\,, \qquad j = 0, 1, \dots\,, \qquad \alpha = 2, 3, \dots, p\,,$$

$$z(t_{j+1}) = z_{(p)}(t_{j+1})\,,$$

where

$$\psi_\alpha(t) = -\mathcal{A}_\alpha(t)\, u^{j+1} + f_\alpha(t)\,, \qquad \alpha = 2, 3, \dots, p\,,$$

$$\psi_1(t) = -\mathcal{A}_1(t)\, u(t) - \frac{du}{dt} + f_1(t)\,, \qquad t \in [t_j, t_{j+1}]\,.$$

From such reasoning it seems clear that

$$\psi = \psi_1(t) + \dots + \psi_p(t)$$

$$= f(t) - \frac{du}{dt} - \mathcal{A}_1(t)\, u(t) - \sum_{\alpha=2}^{p} \mathcal{A}_\alpha(t)\, u^{j+1}.$$

With the aid of the relations $u^{j+1} = u(t) + O(\tau)$, valid for any $\alpha = 2, 3, \dots, p$ on the whole segment $t \in [t_j, t_{j+1}]$, we finally get

$$\psi_\alpha = \overset{\circ}{\psi}_\alpha + \psi_\alpha^*\,, \quad \psi_\alpha^* = O(\tau)\,, \quad \overset{\circ}{\psi}_\alpha = f_\alpha(t) - \mathcal{A}_\alpha(t)\, u(t) - \delta_{\alpha,1}\, \frac{du}{dt}\,,$$

where $\delta_{\alpha,1}$ is, as usual, Kronecker's delta. By the same token,

$$\sum_{\alpha=1}^{p} \overset{\circ}{\psi}_\alpha = \sum_{\alpha=1}^{p} f_\alpha(t) - \sum_{\alpha=1}^{p} \mathcal{A}_\alpha(t)\, u(t) - \frac{du}{dt} = 0\,,$$

yielding

$$\psi = \sum_{\alpha=1}^{p} \psi_\alpha^* = O(\tau)\,.$$

This means that the system of differential equations (55)–(56) generates an approximation of order 1 in a summarized sense to the Cauchy problem (51) under the extra restrictions on the existence and boundedness of the derivative $\mathcal{A}(t)d^2u/dt^2$ in some suitable norm.

Further comparison of the solution $v(t_j)$ of problem (55)–(57) with the solution $u(t_j)$ of the original problem allows to cite without proofs several interesting remarks.

a) Let $f = 0$ and all the functions $f_\alpha = 0$. If constant operators $\mathcal{A}_\alpha$ are pairwise commutative: $\mathcal{A}_\alpha \mathcal{A}_\beta = \mathcal{A}_\beta \mathcal{A}_\alpha$, $\alpha, \beta = 1, 2, \ldots, p$, then for any $\tau$ the equality holds:

$$v(t_j) = u(t_j) \quad \text{for all} \quad j = 0, 1, \ldots, j_0\,, \tag{58}$$

where $v$ is a solution of problem (55)–(57) and $u$ is a solution of problem (51).

When the operators $\mathcal{A}_\alpha = \mathcal{A}_\alpha(t)$ happen to be dependent on the time $t$, equality (58) is still valid for commutative operators $\mathcal{A}_\alpha(t')$ and $\mathcal{A}_\beta(t'')$, $\alpha \neq \beta$, taken at different moments $t' \neq t''$, so that

$$\mathcal{A}_\alpha(t')\,\mathcal{A}_\beta(t'') = \mathcal{A}_\beta(t'')\,\mathcal{A}_\alpha(t')\,, \quad \alpha\,, \beta = 1, 2, \ldots, p\,,$$

for any $t', t'' \in [0, t_0]$.

In this regard, we refer the readers to a few examples of Section 4 in which equality (58) holds true for commutative operators $\mathcal{A}_\alpha\,\mathcal{A}_\beta = \mathcal{A}_\beta\,\mathcal{A}_\alpha$.

b) If the operators $\mathcal{A}_\alpha(t)$ and $\mathcal{A}_\beta(t)$ are non-commutative, then estimate

$$\|\, v(t_j) - u(t_j) \,\| = O(\tau)\,, \qquad j = 1, 2, \ldots, \tag{59}$$

will be valid under the additional condition of smoothness:

$$\|\, \mathcal{A}_\alpha\,\mathcal{A}_\beta\, u \,\| \leq M\,, \qquad \alpha\,, \beta = 1, 2, \ldots, p\,.$$

Is it possible or not to improve the accuracy in $\tau$ without essential modifications of the composite Cauchy problem? In an attempt to give a definite answer to this question, the composite Cauchy problem (55) is designated by the symbolism

$$\mathcal{A}_1 \longrightarrow \mathcal{A}_2 \longrightarrow \cdots \longrightarrow \mathcal{A}_p\,.$$

The symmetrized problem consists of $2p$ Cauchy problems such as

$$0.5\,\mathcal{A}_1 \longrightarrow 0.5\,\mathcal{A}_2 \longrightarrow \cdots \longrightarrow 0.5\,\mathcal{A}_p$$

$$\longrightarrow 0.5\,\mathcal{A}_p \longrightarrow 0.5\,\mathcal{A}_{p-1} \longrightarrow \cdots \longrightarrow 0.5\,\mathcal{A}_1\,.$$

This chain can be associated with the representation of the operator $\mathcal{A}$ as a sum

$$\mathcal{A} = \sum_{\alpha=1}^{2p} \mathcal{A}'_\alpha , \quad \text{where} \quad \mathcal{A}'_\alpha = \begin{cases} 0.5\,\mathcal{A}_\alpha & \text{for} \quad 1 \le \alpha \le p , \\ 0.5\,\mathcal{A}_{2p-\alpha+1} & \text{for} \quad p < \alpha \le 2p . \end{cases}$$

The problem we have posed above is of second-order accuracy in $\tau$:

$$\| v^j - u^j \| = O(\tau^2) ,$$

provided that some smoothness property of the initial vector $u_0$ of the type

$$\| \mathcal{A}_\alpha^2 \, \mathcal{A}_\beta \, u_0 \| \le M , \quad \alpha , \beta = 1, 2, \dots , p ,$$

holds together with the smoothness property of the operators $\mathcal{A}_\alpha(t)$ in $t$.

In such a way, the procedure of solving problem (51) reduces to a sequence of simpler problems (55)–(57), whose solution can be obtained by means of exact or approximate methods. In particular, the finite difference method suits us perfectly for doing so. If the operators $\mathcal{A}_\alpha$ are pairwise commutative, the accuracy of an approximate method available for solving problem (51) depends on how well we are able to solve every auxiliary problem with the number $\alpha$ from sequence (55). The above exploration is still valid for the case of the homogeneous boundary conditions. In dealing with the nonhomogeneous boundary conditions the accuracy of the composite Cauchy problem (55)–(57) depends significantly on the possible ways of specifying the boundary conditions for $v_{(\alpha)}$. The same remark applies equally well to difference analogs of problem (55)–(57).

The difference approximation of every auxiliary problem from collection (55) through the use fo the simplest two-layer scheme with weights leads to an additive scheme. If either of the auxiliary schemes with the number $\alpha$ is economical, then so is the resulting difference scheme.

**Remark** The accurate account of error $z_h = y_h - u_h$ can be done in a number of different ways. In concluding Section 11 the usual way of proper evaluation of the error $z_h$ was recommended for an additive scheme. Another way of proceeding is connected with the triangle inequality

$$\| z_h \| = \| y_h - u_h \| \le \| y_h - v_h \| + \| v_h - u_h \| ,$$

where $v$ is a solution of the locally one-dimensional problem (53)–(54) or (55)–(57). From such reasoning ir seems clear that the further estimation of the error $z_h$ amounts to evaluating the proximity between $y_h$ and $v_h$, $v_h$ and $u_h$. Some progress in such matters can be achieved by the subsidiary information about the smoothness of the functions $u$ and $v$, thus causing some cumbersome exposition in connection with more a detailed exploration of the properties of the solution $v$ of the composite Cauchy problem (53)–(54) or (55)–(57).

**13. Alternating direction methods as additive schemes.** The starting point for special investigations is the alternating direction method developed in Section 1:

$$(60)\qquad \begin{aligned} \frac{y^{j+1/2}-y^j}{\tau} &= 0.5\,(\Lambda_1\, y^{j+1/2} + \Lambda_2\, y^j\,) + 0.5\,\varphi^j\,, \\ \frac{y^{j+1}-y^{j+1/2}}{\tau} &= 0.5\,(\Lambda_1\, y^{j+1/2} + \Lambda_2\, y^{j+1}) + 0.5\,\varphi^j\,. \end{aligned}$$

As can readily be observed, this system of equations is equivalent to

$$(61)\qquad \begin{aligned} \frac{y^{j+1/4}-y^j}{\tau} &= 0.5\,\Lambda_2\, y^j + 0.5\,\varphi^j\,, \\ \frac{y^{j+1/2}-y^{j+1/4}}{\tau} &= 0.5\,\Lambda_1\, y^{j+1/2}\,, \\ \frac{y^{j+3/4}-y^{j+1/2}}{\tau} &= 0.5\,\Lambda_1\, y^{j+1/2}\,, \\ \frac{y^{j+1}-y^{j+3/4}}{\tau} &= 0.5\,\Lambda_2\, y^{j+1} + 0.5\,\varphi^j\,, \end{aligned}$$

relating to locally one-dimensional schemes. Both intermediate values $y^{j+1/4}$ and $y^{j+3/4}$ can be eliminated in the usual way without any difficulties. It is straightforward to verify that scheme (61) generates a summarized approximation such that

$$\psi = \psi_1 + \psi_2 + \psi_3 + \psi_4 = O(\tau^2 + |h|^2)\,.$$

The alternating direction **scheme** ascribed to **Duglas** and **Rachford** and associated with the difference operator $\Lambda_\alpha\, y = y_{\bar{x}_\alpha x_\alpha}$ comes second:

$$(62)\qquad \begin{aligned} \frac{y^{j+1/2}-y^j}{\tau} &= \Lambda_1\, y^{j+1/2} + \Lambda_2\, y^j\,, \\ \frac{y^{j+1}-y^{j+1/2}}{\tau} &= \Lambda_2\,(y^{j+1}-y^j\,)\,. \end{aligned}$$

Simple algebra gives the residuals

$$\psi_1 = \Lambda_1\,\frac{\hat{u}+u}{2} + \Lambda_2\, u - \frac{\hat{u}-u}{2\,\tau}\,, \qquad \psi_2 = \Lambda_2\,(\hat{u}-u) - \frac{\hat{u}-u}{2\,\tau}\,,$$

where $u = u^j$ and $\hat{u} = u^{j+1}$. From here another conclusion can be drawn that $\psi_1 = O(1)$ and $\psi_2 = O(1)$, but

$$\psi = \psi_1 + \psi_2 = O(|h|^2 + \tau^2),$$

thereby clarifying a summarized approximation provided by scheme (62). Upon eliminating the intermediate value $y^{j+1/2}$ we have at our disposal the factorized scheme containing the values $y^j$ and $y^{j+1}$ and approximating the heat conduction equation to $O(\tau + |h|^2)$ in the usual sense. But this scheme is stable only for commutative operators $\Lambda_1$ and $\Lambda_2$. No restrictions of this sort is made for the additive scheme concerned.

**14. LOS for a multidimensional hyperbolic equation of second order.** The method of summarized approximation offers more advantages in designing absolutely stable and convergent locally one-dimensional schemes for equations of hyperbolic type. The object of investigation is the equation

$$\frac{\partial^2 u}{\partial t^2} = \sum_{\alpha=1}^{p} L_\alpha u + f(x,t), \quad L_\alpha u = \frac{\partial}{\partial x_\alpha}\left(k_\alpha(x,t)\frac{\partial u}{\partial x_\alpha}\right), \tag{63}$$

$$k_\alpha(x,t) \geq c_1 > 0, \quad c_1 = \text{const},$$

where $x = (x_1, \ldots, x_p) \in \mathbf{R}^p$ and $G$ is an arbitrary domain in the space $\mathbf{R}^p$ with the boundary $\Gamma$, $\bar{G} = G + \Gamma$ and

$$\bar{Q}_T = \bar{G} \times [0 \leq t \leq T], \qquad Q_T = G \times (0 < t \leq T].$$

The problem statement here consists of finding a continuous in the cylinder $\bar{Q}_T$ solution to equation (63) satisfying the boundary condition

$$u = \mu(x,t) \quad \text{for} \quad x \in \Gamma, \quad 0 \leq t \leq T, \tag{64}$$

and the initial conditions

$$u(x,0) = u_0(x), \qquad \frac{\partial u(x,0)}{\partial t} = \bar{u}_0(x) \quad \text{for} \quad x \in \bar{G}. \tag{65}$$

As usual, it is preassumed in a common setting that the problem concerned is uniquely solvable and its solution $u = u(x,t)$ possesses all necessary derivatives which do arise in all that follows. The domain of interest $G$ is still subject to the same conditions as we imposed in Section 5 for parabolic equations. Also, let $\omega_\tau = \{t_j = j\tau, \ j = 0, 1, \ldots\}$ be a uniform grid in $\tau$ on the segment $0 \le t \le T$ and a grid $\omega_h$ in $\bar{G}$ remains unchanged (for more detail see Chapter 5).

If $G$ is a parallelepiped in the space $\mathbf{R}^p$, problem (63)–(65) can be solved through the use of an economical factorized scheme with accuracy $O(\tau^2 + |h|^2)$. The design of such a scheme was made in Section 2 and it was investigated there in full details. Applying the same procedure serves ti motivate that, first, the operators

$$\mathcal{P}_\alpha u = \frac{1}{p} \frac{\partial^2 u}{\partial t^2} - (L_\alpha u + f_\alpha), \quad \alpha = 1, 2, \dots, p, \tag{66}$$

with $f_\alpha$ still subject the normalization condition

$$\sum_{\alpha=1}^{p} f_\alpha = f$$

are approximated successively with step $\tau/p$.

Second, the difference expresions

$$u_{\bar{t}_\alpha \bar{t}_\alpha} = \frac{u_{(\alpha)} - 2\, u_{(\alpha-1)} + \check{u}_{(\alpha)}}{\tau^2} \sim \frac{1}{4} \left( \frac{\partial^2 u}{\partial t^2} \right)^{j+(\alpha-1)/p} \tag{67}$$

are aimed at approximating to the derivative $\partial^2 u/\partial t^2$ with step $\tau/p$, where

$$u_{(\alpha)} = u^{j+\alpha/2}, \quad \check{u}_{(\alpha)} = u^{j-1+\alpha/2}, \quad u_{(0)} = \check{u} = u^{j-1}, \quad u_{(2)} = u^j,$$

$$u_{\bar{t}_\alpha \bar{t}_\alpha} = \frac{u_{(\alpha)} - u_{(\alpha-1)} - u_{(\alpha-2)} + \check{u}_{(\alpha)}}{\tau^2} \sim \frac{2}{9} \frac{\partial^2 u}{\partial t^2}, \tag{68}$$

$$\alpha = 1, 2, 3 \quad \text{for} \quad p = 3,$$

with the members $u_{(-1)} = \check{u}_{(2)} = u^{(j-1)+2/3}$ and $u_{(-2)} = \check{u}_{(1)} = u^{j-2/3}$.

Third, the operator $L_\alpha u + f_\alpha$ is approximated to second order by the homogeneous difference operator $\Lambda_\alpha y + \varphi_\alpha$ on the grid $\omega_h$ in the space $\mathbf{R}^p$. The coefficients at the member $\Lambda_\alpha$ and the righ-hand side $\varphi_\alpha$ are taken at the moment

$$t'_\alpha = 0.5\,(t_{j+\alpha/p} + t_{j-1+\alpha/p}) = t_{j+\alpha/p-0.5} = t_j + (\alpha/p - 0.5)\,\tau,$$

so that $\Lambda_\alpha = \Lambda_\alpha(t'_\alpha)$ and $\varphi_\alpha = \varphi_\alpha(x, t'_\alpha)$.

"Locally one-dimensional" schemes for hyperbolic equations acquires the form

$$(69)\qquad y_{\bar{t}_\alpha \bar{t}_\alpha} = \sigma_p\, \Lambda_\alpha \left(y_{(\alpha)} + \check{y}_{(\alpha)}\right) + 2\,\sigma_p\, \varphi_\alpha\,, \qquad \alpha = 1, \ldots, p\,, \qquad p = 2, 3\,,$$

where

$$\sigma_p = \begin{cases} 1/4 & \text{for} \quad p = 2\,, \\ 1/3 & \text{for} \quad p = 3\,, \end{cases}$$

and the left-hand side $y_{\bar{t}_\alpha \bar{t}_\alpha}$ is given by formula (67) for $p = 2$ and by formula (68) for $p = 3$. With the detailed forms in mind, the final schemes refer to three-point additive ones ($p = 2$) and four-point ones ($p = 3$). The principal difference from the case of parabolic equations lies in the dependence on the number of measurements $p$.

Equation (69) can be rewritten as

$$(70)\qquad (E - \sigma_p \tau^2 \Lambda_\alpha)\,(y_{(\alpha)} + \check{y}_{(\alpha)})$$

$$= \begin{cases} 2\, y_{(\alpha-1)} + 2\,\sigma_2 \tau^2 \varphi_\alpha & \text{for} \quad p = 2\,, \\ y_{(\alpha-1)} + y_{(\alpha-2)}) + 2\,\sigma_3 \tau^2 \varphi_\alpha & \text{for} \quad p = 3\,. \end{cases}$$

The solution $y_{(\alpha)}$ is sought from the three-point equation

$$(E - \sigma_p \tau^2 \Lambda_\alpha)\, y_{(\alpha)} = F_\alpha$$

along segments parallel to the axis $Ox_\alpha$ with the boundary condition

$$(71)\qquad y_{(\alpha)} = \mu(x, t_{j+\alpha/p}) \qquad \text{for} \quad x \in \gamma_h^\alpha$$

with further reference to the elimination method.

First of the initial conditions $u(x, 0) = u_0(x)$ is approximated exactly:

$$(72)\qquad y(x, 0) = u_0(x)\,.$$

The intermediate values $y^{1/2} = y(x, \tau/2)$ for $p = 2$ and $y^{1/3} = y(x, \tau/3)$, $y^{2/3} = y(x, 2\tau/3)$ for $p = 3$ are found, respectively, from the equations

$$(73)\qquad \left(E - \frac{\tau^2}{4}\,\Lambda_1\right) y^{1/2} = F_1\,,$$

$$F_1 = u_0 + \frac{\tau}{2}\,\bar{u}_0 + \frac{\tau^2}{4}\,\Lambda_1\, u_0 + \tau^2 \left(f_1 - \frac{1}{8}\,(\Lambda\, u + f)\right)_{t=0}$$

if $p = 2$ or

$$(74)\qquad \Big(E - \frac{\tau^2}{4}\Lambda_1\Big)\, y^{1/3} = F_1\,,$$

$$\Big(E - \frac{\tau^2}{4}\Lambda_2\Big)\,(y^{2/3} + u_0) = 2\, y^{1/3} + F_2\,,$$

$$F_1 = u_0 + \frac{\tau}{3}\,\bar{u}_0 + \tau^2\,\Lambda_1\,u_0 + \tau^2\Big(\frac{2}{3}\,f_1 - \frac{1}{6}\,(\Lambda\,u + f)\Big)_{t=0}\,,$$

$$F_2 = \tau^2\Big(\frac{2}{3}\,f_2 - \frac{1}{9}\,(\Lambda\,u + f)\Big)_{t=0} \quad \text{if} \quad p = 3\,.$$

We are now interested in the more detailed designs of LOS for $p = 2$:

$$(75)\qquad u(x,0) = u_0(x), \quad \Big(E - \frac{\tau^2}{4}\Lambda_1\Big)\, y^{1/2} = F_1 \quad \text{for} \quad t = 0.5\,\tau\,,$$

$$(76)\qquad \begin{aligned} \frac{y^{j+1/2} - 2\,y^j + y^{j-1/2}}{\tau^2} &= \frac{1}{4}\,\Lambda_1\,(y^j + y^{j-1/2}) + \frac{1}{2}\,\varphi_1^j\,,\\ \frac{y^{j+1} - 2\,y^{j+1/2} + y^j}{\tau^2} &= \frac{1}{4}\,\Lambda_2\,(y^{j+1} + y^j) + \frac{1}{2}\,\varphi_2^j\,, \quad j = 1, 2, \ldots\,. \end{aligned}$$

In that case the boundary conditions become

$$(77)\qquad \begin{aligned} y^{j+1/2} &= \mu(x, t_{j+1/2}) \quad \text{for} \quad x \in \gamma_h^1\,,\\ y^{j+1} &= \mu(x, t_{j+1}) \qquad \text{for} \quad x \in \gamma_h^2\,, \end{aligned}$$

and the function $y^{j+1/2}$ can be recovered from the equation

$$y^{j+1/2} - \frac{\tau^2}{4}\,\Lambda_1\,y^{j+1/2} = \Phi_1^j$$

with the right-hand side $\Phi_1^j$ and the known boundary conditions (77). In turn, the function $y^{j+1}$ is found from the equation

$$y^{j+1} - \frac{\tau^2}{4}\,\Lambda_2\,y^{j+1} = \Phi_2^{j+1/2}$$

with the right-hand side $\Phi_2^{j+1/2}$ and the known boundary conditions (77). Either of these equations can be solved by the standard elimination method.

With the basic tools in hand, we proceed to carry out the accurate account of the approximation error $z_{(\alpha)} = z^{j+\alpha/2} = y^{j+\alpha/2} - u(x, t_{j+\alpha/2})$ for scheme (75)–(77), where $u$ is a solution of problem (63)–(65) and $y$ is a solution of problem (75)–(77), by inserting the value $y_{(\alpha)} = z_{(\alpha)} + u^{j+\alpha/2}$ in equation (75). As a final result we obtain

$$z_{\bar{t}_\alpha \bar{t}_\alpha} = \frac{1}{4}\,\Lambda_\alpha (z_{(\alpha)} + \check{z}_{(\alpha)}) + \psi_\alpha \quad \text{for} \quad t \geq \tau\,, \tag{78}$$

$$\left(E - \frac{\tau^2}{4}\,\Lambda_1\right) \frac{z_{(1)}}{\tau^2} = \psi_1 \quad \text{for} \quad t = 0.5\,\tau\,,$$

$$z(x, 0) = 0, \quad x \in \bar{\omega}_h, \quad z_{(\alpha)} = z^{j+\alpha/2} = 0 \quad \text{for} \quad x \in \gamma_h^\alpha, \quad \alpha = 1, 2\,,$$

with the member

$$\psi_\alpha = \frac{1}{4}\,\Lambda_\alpha\,(u_\alpha + \check{u}_\alpha) - u_{\bar{t}_\alpha \bar{t}_\alpha} + 0.5\,\varphi_\alpha\,, \tag{79}$$

serving as one possible representation for the residual of equation (66) with the number $\alpha = 1, 2$. The error of approximation for LOS of the form (75 )–(77) is viewed as a sum

$$\psi = \psi_1 + \psi_2\,. \tag{80}$$

Further progress in this area will be achieved by utilizing the fact that scheme (76) approximates problem (63)–(65) in a summarized sense and $\psi = O(\tau + |h|^2)$. Indeed, taking into account that

$$0.5\,\Lambda_\alpha\,(u_\alpha + \check{u}_\alpha) = (L_\alpha\, u)^{j+(\alpha-1)/2} + O(h_\alpha^2) \quad \text{for} \quad x \in \overset{\circ}{\omega}_{h,\alpha}\,,$$

$$0.5\,\Lambda_\alpha\,(u_\alpha + \check{u}_\alpha) = (L_\alpha\, u)^{j+(\alpha-1)/2} + O(h_\alpha) \quad \text{for} \quad x \in \omega_{h,\alpha}^*\,,$$

$$u_{\bar{t}_\alpha \bar{t}_\alpha} = \frac{1}{4}\left(\frac{\partial^2 u}{\partial t^2}\right)^{j+(\alpha-1)/2} + O(\tau^2)\,,$$

$$\varphi_\alpha = f_\alpha^{j+(\alpha-1)/2} + O(\tau^2)\,,$$

we find that $\psi_\alpha = \overset{\circ}{\psi}_\alpha + \psi_\alpha^*$, where

$$\overset{\circ}{\psi}_\alpha = \frac{1}{2}\left(L_\alpha\, u - \frac{1}{2}\,\frac{\partial^2 u}{\partial t^2} + f_\alpha\right)^{j+(\alpha-1)/2}, \qquad \alpha = 1, 2\,,$$

and

$$\psi^*_\alpha = \begin{cases} O(h^2_\alpha + \tau^2) & \text{for} \quad x \in \overset{\circ}{\omega}_{h,\alpha}\,, \\ O(h_\alpha + \tau^2) & \text{for} \quad x \in \overset{\circ}{\omega}{}^*_{h,\alpha}\,. \end{cases}$$

Whence it follows that

$$\overset{\circ}{\psi}{}^j_1 + \overset{\circ}{\psi}{}^{j+1/2}_2 = 0.5\,(L_1\,u - 0.5\,\ddot{u} + f_1)^j$$

$$+ 0.5\,(L_2\,u - 0.5\,\ddot{u} + f_2)^{j+1/2}$$

$$= 0.5\,\big((L_1 + L_2)\,u - \ddot{u} + f_1 + f_2\big)^j + 0.5\,\tau\,\overset{\circ}{\psi}_{2\,\bar{t}_1}\,.$$

With equation (63) in view, the first summand becomes zero:

$$(L_1 + L_2)\,u - \ddot{u} + f = 0\,, \qquad f = f_1 + f_2\,,$$

yielding

$$\overset{\circ}{\psi}{}^j_1 + \overset{\circ}{\psi}{}^j_2 = 0\,. \tag{81}$$

By definition, this means that scheme (75)–(77) generates a summarized approximation. The accepted view is to involve the sum

$$\overset{\circ}{\psi}_2 + 2\overset{\circ}{\psi}_1 + \breve{\overset{\circ}{\psi}}_2 = 0.5\,\big[(L_2\,u - \ddot{u} + f_2)^{j+1/2} + (L_2\,u - \ddot{u} + f_2)^{j-1/2}\big]$$

$$+ (L_1\,u - \ddot{u} + f_1)^j$$

$$= \big((L_1 + L_2)\,u - \ddot{u} + f_1 + f_2\big)^j + \big(\overset{\circ}{\psi}{}^{j+1/2}_2 - 2\overset{\circ}{\psi}{}^j_2 + \overset{\circ}{\psi}{}^{j-1/2}_2\big)$$

$$= \tau^2\,\big(\overset{\circ}{\psi}{}_2\big)_{\bar{t}_1\bar{t}_1}\,,$$

making it possible to deduce that

$$\overset{\circ}{\psi}_2 + 2\,\overset{\circ}{\psi}_1 + \breve{\overset{\circ}{\psi}}_2 = \tau^2\,\big(\overset{\circ}{\psi}_2\big)_{\bar{t}_1\bar{t}_1} = O(\tau^2)\,. \tag{82}$$

No wishing to load the book down with full details on this point, we cite here only final results: LOS of the form (75)–(77) converges in the grid norm of the space $W^1_2$ with the rate $O(\tau + |h|^2)$ if the solution $u = u(x,t)$ has in $\bar{Q}_T$ continuous derivatives of the first four orders, the derivatives $\partial^4 u/\partial x^4_\alpha$ satisfy the Lipshitz condition in $t$ and the right-hand side $f$ is twice differentiable in $t$.

**15. Additive schemes for a system of equations.** Later in this section we will survey some devices that can be used in trying to produce additive schemes for systems of parabolic equations. With this aim, problem (50) we have completely posed in Section 2 will serve as a basis for the up-to-date presentation of tools and techniques, their theory and applications. In this connection we may attempt the operator $L$ in the form $L = L^- + L^+$ with "triangle" operators $L^-$, $L^+$, the associated matrices $k_{\alpha\alpha}$ of which arrange themselves as sums $k_{\alpha\alpha} = k^-_{\alpha\alpha} + k^+_{\alpha\alpha}$, where $k^-_{\alpha\alpha} = (k^{-sm}_{\alpha\alpha})$ and $k^+_{\alpha\alpha} = (k^{+sm}_{\alpha\alpha})$ are triangle matrices with entries

$$k^{-sm}_{\alpha\alpha} = k^{sm}_{\alpha\alpha}\,, \quad k^{+sm}_{\alpha\alpha} = k^{sm}_{\alpha\alpha}\,, \quad k^{-ss}_{\alpha\alpha} = k^{+ss}_{\alpha\alpha} = 0.5\, k^{ss}_{\alpha\alpha}\,,$$

$$k^{-sm}_{\alpha\alpha} = 0 \quad \text{for} \quad m > s\,, \quad k^{+sm}_{\alpha\alpha} = 0 \quad \text{for} \quad m < s\,.$$

Observe that the matrices $k^-_{\alpha\alpha}$ and $k^+_{\alpha\alpha}$ are symmetric each to other, since $k^{-sm}_{\alpha\alpha} = k^{+ms}_{\alpha\alpha}$, whence it follows that

$$L_{\alpha\alpha} = L^-_{\alpha\alpha} + L^+_{\alpha\alpha}\,, \qquad L^\pm_{\alpha\alpha} = \frac{\partial}{\partial x_\alpha} \left( k^\pm_{\alpha\alpha} \frac{\partial u}{\partial x_\alpha} \right].$$

By introducing a few auxiliary operators with the properties

$$L^-_\alpha u = L^-_{\alpha\alpha} u + \sum_{\beta=1}^{\alpha-1} L_{\alpha\beta} u = \sum_{\beta=1}^{\alpha} L^-_{\alpha\beta} u\,, \quad L^-_{\alpha\beta} u = L_{\alpha\beta} u \quad \text{for} \quad \beta < \alpha\,,$$

$$L^+_\alpha u = L^+_{\alpha\alpha} u + \sum_{\beta=\alpha+1}^{p} L_{\alpha\beta} u = \sum_{\beta=\alpha}^{p} L^+_{\alpha\beta} u\,, \quad L^+_{\alpha\beta} u = L_{\alpha\beta} u \quad \text{for} \quad \beta > \alpha\,,$$

and representing the operator $L$ by a sum

$$L = L^- + L^+\,, \quad L^- u = \sum_{\alpha=1}^{p} L^-_\alpha u\,, \quad L^+ u = \sum_{\alpha=1}^{p} L^+_\alpha u\,, \tag{83}$$

the solution of the system of equations (50) or

$$\sum_{\alpha=1}^{p} \left[ \frac{1}{2p} \frac{\partial \mathbf{u}}{\partial t} - (L^-_\alpha \mathbf{u} + L^+_\alpha \mathbf{u}) - (\mathbf{f}^-_\alpha + \mathbf{f}^+_\alpha) \right] = 0\,, \tag{84}$$

where

$$\sum_{\alpha=1}^{p} (\mathbf{f}^-_\alpha + \mathbf{f}^+_\alpha) = \mathbf{f}\,,$$

reduces to successive solution of the system of $2p$ equations

$$\text{(85)}\qquad \frac{1}{2p}\frac{\partial \mathbf{v}}{\partial t} = L_\alpha^- \mathbf{v} + \mathbf{f}_\alpha^- , \quad t_{j+(\alpha-1)/(2p)} \le t \le t_{j+\alpha/(2p)} ,$$

$$\frac{1}{2p}\frac{\partial \mathbf{v}}{\partial t} = L_\alpha^+ \mathbf{v} + \mathbf{f}_\alpha^+ , \quad t_{j+1-\alpha/(2p)} \le t \le t_{j+1-(\alpha-1)/(2p)} ,$$

$$\alpha = 1, 2, \dots, p .$$

The usual starting procedure is connected with approximating the operators $L_\alpha^\pm$ by the difference operators $\Lambda_\alpha^\pm$ acting in accordance with the rules

$$\Lambda_\alpha^- = \sum_{\beta=1}^{\alpha} \Lambda_{\alpha\beta}^- , \qquad \Lambda_\alpha^+ = \sum_{\beta=\alpha}^{p} \Lambda_{\alpha\beta}^+ ,$$

$$\Lambda_{\alpha\beta}^\pm u = 0.5 \left[ \left(k_{\alpha\beta}^\pm u_{\bar{x}_\beta}\right)_{x_\alpha} + \left(k_{\alpha\beta}^\pm u_{x_\beta}\right)_{\bar{x}_\alpha} \right] .$$

In this regard, it seems clear that the operator $L_\alpha^\pm$ is approximated to second order by the difference operator $\Lambda_\alpha^\pm$, the structure of which involves the coefficients $k_{\alpha\beta}^\pm$ taken for all $\alpha$ and $\beta$ either at one and the same moment $t = t_{j+1/2}$ or at any another moment $t^* \in [t_j, t_{j+1}]$. With these members, the additive scheme in question acquires the form

$$\text{(86)}\qquad \frac{\mathbf{y}^{j+\alpha/(2p)} - \mathbf{y}^{j+(\alpha-1)/(2p)}}{\tau} = \sum_{\beta=1}^{\alpha} \Lambda_{\alpha\beta}^- \mathbf{y}^{j+\beta/(2p)} + (\varphi_\alpha^-)^{j+\alpha/(2p)} ,$$

$$\frac{\mathbf{y}^{j+\alpha_1/(2p)} - \mathbf{y}^{j+(\alpha_1-1)/(2p)}}{\tau} = \sum_{\beta=\alpha}^{p} \Lambda_{\alpha\beta}^+ \mathbf{y}^{j+\beta_1/(2p)} + (\varphi_\alpha^+)^{j+\alpha_1/(2p)} ,$$

$$\alpha = 1, 2, \dots, p ,$$

where $\alpha_1 = 2p + 1 - \alpha$ and $\beta_1 = 2p + 1 - \beta$. Also, we should indicate the direction of index account: $\alpha_1$ is being increased from $p + 1$ to $2p$ along with decreasing $\alpha$ in reverse order from $p$ to 1.

The usual boundary conditions are imposed on the boundary $x \in \gamma_h^\alpha$:

$$\text{(87)}\qquad \begin{aligned} \mathbf{y}^{j+\alpha/(2p)} &= \mu^{j+\alpha/(2p)} \quad \text{for} \quad x \in \gamma_h^\alpha , \quad \alpha = 1, 2, \dots, p , \\ \mathbf{y}^{j+\alpha_1/(2p)} &= \mu^{j+\alpha_1/(2p)} \quad \text{for} \quad x \in \gamma_h^{\alpha_1} , \quad \alpha_1 = p+1, \dots, 2p . \end{aligned}$$

The initial condition is satisfied exactly:

$$\mathbf{y}(x,0)=\mathbf{u}_0(x)\,. \tag{88}$$

The following system of equations emerges in determining $\mathbf{y}^{j+\alpha/(2p)}=\mathbf{y}_{(\alpha)}$ and $\mathbf{y}^{j+\alpha_1/(2p)}=\mathbf{y}_{(\alpha_1)}$ during the course of the elimination:

$$(E-\tau\,\Lambda^-_{\alpha\alpha})\,\mathbf{y}_{(\alpha)}=\mathbf{F}^-_\alpha\,,\qquad (E-\tau\,\Lambda^+_{\alpha\alpha})\,\mathbf{y}_{(\alpha_1)}=\mathbf{F}^+_{\alpha_1}\,,$$

$$\mathbf{F}^-_\alpha=\tau\sum_{\beta=1}^{\alpha-1}\Lambda^-_{\alpha\beta}\,\mathbf{y}_{(\beta)}+\tau\,\varphi^-_\alpha+\mathbf{y}_{(\alpha-1)}\,,$$

$$\mathbf{F}^+_{\alpha_1}=\tau\sum_{\beta=\alpha+1}^{p}\Lambda^+_{\alpha\beta}\,\mathbf{y}_{(\beta_1)}+\tau\,\varphi^+_\alpha+\mathbf{y}_{(\alpha_1-1)}$$

Having completed the elimination, we must solve the system of equations

$$(E-\tau\,\Lambda^-_{\alpha\alpha})\,\mathbf{y}_{(\alpha)}=\mathbf{F}^-_\alpha\,.$$

In light of the special structure of the diagonal matrices $k^-_{\alpha\alpha}$, whose blocks are lower triangle, the components $y^s_{(\alpha)}$, $s=1,2,\dots,n$, of the unknown vector $\mathbf{y}_{(\alpha)}$ are to be determined successively by the elimination method in passing from $\alpha$ to $\alpha+1$ and from $s$ to $s+1$. By the elimination formulas for a three-point equation we constitute in a term-by-term fashion the vectors $\mathbf{y}_{(\alpha)}$, $\alpha=1,2,\dots,p$. Moving in reverse order from $\alpha+1$ to $\alpha$ and from $s+1$ to $s$ the vectors $\mathbf{y}_{(p+1)},\dots,\mathbf{y}_{(2p)}$ are recovered from the system

$$(E-\tau\,\Lambda^+_{\alpha\alpha})\,\mathbf{y}_{(\alpha_1)}=\mathbf{F}^+_{\alpha_1}\,.$$

Upon terminating this process the resultant vector $\mathbf{y}_{(2p)}$ is just the solution $\mathbf{y}^{j+1}=\mathbf{y}_{(2p)}$ on the layer $t=t_{j+1}$.

Since the system of differential equations (56) approximates equation (50) in a summarized sense in compliance with approximating equation (85) by equation (86) with the number $\alpha$, the additive scheme (86)–(88) generates an approximation of $O(\tau+|h|^2)$:

$$\psi=\sum_{\alpha=1}^{p}(\psi^-_\alpha+\psi^+_\alpha)=O(\tau+|h|^2)\,.$$

In concluding this chapter it will be sensible to introduce the space $\overset{\circ}{\Omega}$ of all grid vector-functions given on the grid $\bar{\omega}_h$ and vanishing on the boundary $\gamma_h$ of the grid under the inner product structure

$$(\mathbf{y}, \mathbf{v}) = \sum_{s=1}^{n} (y^s, v^s), \quad (y^s, v^s) = \sum_{x \in \omega_h} y^s(x)\, v^s(x)\, h_1 h_2 \cdots h_p .$$

Having involved the operators

$$A^- = \sum_{\alpha=1}^{p} A_\alpha^-, \qquad A^+ = \sum_{\alpha=1}^{p} A_\alpha^+,$$

$$A_\alpha^- \mathbf{y} = -\sum_{\beta=1}^{\alpha} \Lambda_{\alpha\beta}^- \mathbf{y}, \quad A_\alpha^+ \mathbf{y} = -\sum_{\beta=\alpha}^{p} \Lambda_{\alpha\beta}^+ \mathbf{y}, \quad \mathbf{y} \in \overset{\circ}{\Omega},$$

we are going to show that these operators $A^-$ and $A^+$ are mutually adjoint each to other:

$$(A^- \mathbf{y}, \mathbf{v}) = (\mathbf{y}, A^+ \mathbf{v}) \quad \text{for all} \quad \mathbf{y}, \mathbf{v} \in \overset{\circ}{\Omega},$$

if the matrix $k = (k_{\alpha\beta}^{sm})$ is symmetric, that is, under the condition of symmetry $k_{\alpha\beta}^{sm} = k_{\beta\alpha}^{ms}$. Indeed, $k_{\beta\alpha}^- = k_{\alpha\beta}^+$ and, because of this, we might have

$$\begin{aligned}
(A^- \mathbf{y}, \mathbf{v}) &= 0.5 \sum_{\alpha=1}^{p} \sum_{\beta=1}^{\alpha} \left[ (k_{\alpha\beta}^- \mathbf{y}_{\bar{x}_\beta}, \mathbf{v}_{\bar{x}_\alpha}) + (k_{\alpha\beta}^- \mathbf{y}_{x_\beta}, \mathbf{v}_{x_\alpha}) \right] \\
&= 0.5 \sum_{\beta=1}^{p} \sum_{\alpha=\beta}^{p} \left[ (k_{\alpha\beta}^- \mathbf{y}_{\bar{x}_\beta}, \mathbf{v}_{\bar{x}_\alpha}) + (k_{\alpha\beta}^- \mathbf{y}_{x_\beta}, \mathbf{v}_{x_\alpha}) \right] \\
&= 0.5 \sum_{\alpha=1}^{p} \sum_{\beta=\alpha}^{p} \left[ (k_{\beta\alpha}^- \mathbf{y}_{\bar{x}_\alpha}, \mathbf{v}_{\bar{x}_\beta}) + (k_{\beta\alpha}^- \mathbf{y}_{x_\alpha}, \mathbf{v}_{x_\beta}) \right] \\
&= 0.5 \sum_{\alpha=1}^{p} \sum_{\beta=\alpha}^{p} \left[ (k_{\alpha\beta}^+ \mathbf{v}_{\bar{x}_\beta}, \mathbf{y}_{\bar{x}_\alpha}) + (k_{\alpha\beta}^+ \mathbf{v}_{x_\beta}, \mathbf{y}_{x_\alpha}) \right] \\
&= (\mathbf{y}, A^+ \mathbf{v}) .
\end{aligned}$$

It follows from the foregoing that

$$(A^- \mathbf{y}, \mathbf{y}) = (A^+ \mathbf{y}, \mathbf{y}) = 0.5\, (A\, \mathbf{y}, \mathbf{y}) \geq 0.5\, c_1\, (\overset{\circ}{A}\, \mathbf{y}, \mathbf{y}),$$

where

$$\overset{\circ}{A} y = -\sum_{\alpha=1}^{p} \mathbf{y}_{\bar{x}_\alpha x_\alpha},$$

yielding

$$(\overset{\circ}{A}\, \mathbf{y}, \mathbf{y}) = \sum_{\alpha=1}^{p} (1, \mathbf{y}_{\bar{x}_\alpha}^2]_\alpha \geq 8 \sum_{\alpha=1}^{p} \frac{1}{l_\alpha^2} \|\, \mathbf{y} \,\|^2 .$$

Thus, the operators $A^-$ and $A^+$ so defined are positive definite:

$$A^- \geq \delta\, E\,, \quad A^+ \geq \delta\, E\,, \quad \delta = 4\, c_1 \sum_{\alpha=1}^{p} l_\alpha^{-2} .$$

Stability of scheme (86)–(87) with zero boundary conditions is asserted by Theorem 3 in Section 11, due to which a solution of the auxiliary problem

$$\mathbf{z}_{\bar{t}_\alpha} = \sum_{\beta=1}^{\alpha} \Lambda_{\alpha\beta}^- \, \mathbf{z}_{(\beta)} + \psi_\alpha^-\,, \quad \mathbf{z}_{\bar{t}_{\alpha_1}} = \sum_{\beta=\alpha}^{p} \Lambda_{\alpha\beta}^+ \, \mathbf{z}_{(\beta_1)} + \psi_\alpha^+\,,$$

$$z_{(\alpha)} = 0\,, \quad z_{(\alpha_1)} = 0\,, \quad \mathbf{z}(x, 0) = 0 \quad \text{for} \quad x \in \gamma_h^\alpha\,,$$

where

$$\mathbf{z}_{\bar{t}_\alpha} = \frac{\mathbf{z}_{(\alpha)} - \mathbf{z}_{(\alpha-1)}}{\tau}\,, \qquad \mathbf{z}_{\bar{t}_{\alpha_1}} = \frac{\mathbf{z}_{(\alpha_1)} - \mathbf{z}_{(\alpha_1-1)}}{\tau}\,,$$

satisfies the a priori estimate

$$\|\, \mathbf{z}^{j+1} \,\| \leq M_1 \max_{0 \leq j' \leq j} \left\| \sum_{\alpha=1}^{p} \left( (\psi_\alpha^-)^{j'+\alpha/(2p)} + (\psi_\alpha^+)^{j'+1-(\alpha-1)/(2p)} \right) \right\|$$

$$+ M_2 \sqrt{\tau} \max_{0 \leq j' \leq j} \sum_{\alpha=1}^{p} \left( \|(\psi_\alpha^-)^{j'+\alpha/(2p)}\| + \|(\psi_\alpha^+)^{j'+1-(\alpha-1)/(2p)}\| \right).$$

This supports the view that the additive scheme (86)–(88) converges in the grid norm of the space $L_2$ with the rate $O(\sqrt{\tau} + |h|^2)$.

We will pursue the discussion of additive schemes further with regard to problem (61)–(62) capable of describing the system of hyperbolic equations

$$\sum_{\alpha=1}^{p}\left[\frac{1}{p}\frac{\partial^2\mathbf{u}}{\partial t^2}+(L_\alpha^-+L_\alpha^+)\,\mathbf{u}+\mathbf{f}_\alpha\right]=0\,,\qquad \sum_{\alpha=1}^{p}\mathbf{f}_\alpha=\mathbf{f}\,. \tag{89}$$

What has been done is to reduce this system to successive solution of simpler equations moving from $\alpha$ to $\alpha+1$:

$$\frac{1}{p}\frac{\partial^2\mathbf{u}}{\partial t^2}=L_\alpha^-\,\mathbf{u}+L_\alpha^+\,\mathbf{u}+\mathbf{f}_\alpha\,,\qquad \alpha=1,2,\dots,p\,. \tag{90}$$

One possible additive scheme

$$\frac{1}{\sigma_p}\,\mathbf{y}_{\bar t_\alpha \bar t_\alpha}=\sum_{\beta=1}^{\alpha}\Lambda_{\alpha\beta}^-\,\mathbf{y}_{(\beta)}+\sum_{\beta=\alpha}^{p}\Lambda_{\alpha\beta}^+\,\check{\mathbf{y}}_{(\beta)}+\varphi_\alpha\,, \tag{91}$$

$$\alpha=1,2,\dots,p\,,\quad (x,t)\in\omega_h\times\omega_\tau\,,$$

$$\mathbf{y}_{(\alpha)}=\mu(x,t'_\alpha)\,,\quad x_\alpha=0,l_\alpha\,,\quad \alpha=1,2,\dots,p\,,$$

$$\mathbf{y}(x,0)=\mathbf{u}_0(x)\,,$$

can be obtained through the usual approximations of $p$ equations, where $\varphi_\alpha=\mathbf{f}_\alpha(x,t'_\alpha)$ and $t'_\alpha=t_{j+(\alpha/p-0.5)\tau}$, the coefficients $k_{\alpha\beta}$ are taken at moment $t'_\alpha$, $y_{\bar t_\alpha \bar t_\alpha}$ is determined by formula (67) or formula (68), $\sigma_p=0.5$ for $p=2$ and $\sigma_p=1.5$ for $p=3$.

The second initial condition is approximated by setting

$$\mathbf{y}^{\alpha/p}=\mathbf{u}_0(x)+\frac{\alpha\,\tau}{p}\,\mathbf{u}_0(x)+\frac{\alpha^2\,\tau^2}{2\,p^2}\,(L\,\mathbf{u}_0+\mathbf{f}(x,0))\,,$$

$$\alpha=1,2,\dots,p-1\,.$$

Because of these facts, the describing scheme generates a summarized approximation

$$\psi=\sum_{\alpha=1}^{p}\psi_\alpha=O(\tau+|h|^2)\,.$$

With regard to $\mathbf{y}^{j+1} = \mathbf{y}_{(p)}$ we obtain the system of equations

$$(E - \sigma_p \tau^2 \Lambda_{\alpha\alpha}^-)\, \mathbf{y}_{(\alpha)} = \mathbf{F}_\alpha \,,$$

where the right-hand side $\mathbf{F}_\alpha$ is expressed in terms of the vectors $\mathbf{y}_{(\beta)}$, $\beta < \alpha$. Such a system can be solved successively moving from $\alpha$ to $\alpha + 1$ and from $s$ to $s + 1$ during the course of the elimination. By interchanging the positions of $\Lambda_\alpha^-$ and $\Lambda_\alpha^+$ we arrive at the second scheme

$$\frac{1}{\sigma_p}\, \mathbf{y}_{\bar{t}_\alpha \bar{t}_\alpha} = \sum_{\beta=1}^{\alpha} \Lambda_{\alpha\beta}^- \, \check{\mathbf{y}}_{(\beta)} + \sum_{\beta=\alpha}^{p} \Lambda_{\alpha\beta}^+ \, \mathbf{y}_{(\beta)} + \varphi_\alpha \,. \tag{92}$$

In this case the system can be solved successively moving from $\alpha + 1$ to $\alpha$ and from $s+1$ to $s$. Alternation of schemes (91) and (92) leads to the third scheme, which interests us.

By means of the energy method the reader can derive on his/her own an a priori estimate for the error $z = y - u$ by exactly the same reasoning as before with further reference to the property of summarized approximation. On this basis the convergence of additive schemes can be obtained through such an analysis.

# 10

# Methods for Solving Grid Equations

In this chapter economical direct and iterative methods are designed for numerical solution of difference elliptic equations.

In Section 1 we confine ourselves to direct economical methods available for solving boundary-value problems associated with Poisson's equation in a rectangle such as the decomposition method and the mathod of separation of variables.

The general theory of iterative methods is presented in the next sections with regard to an operator equation of the first kind $Au = f$, where $A$ is a self-adjoint operator in a finite-dimensional Euclidean space. The applications of such theory to elliptic grid equations began to spread to more and more branches as they took on an important place in "real-life" situations.

## 10.1 DIRECT METHODS

**1. Direct and iterative methods.** Recall that the final results of the difference approximation of boundary-value problems associated with elliptic equations from Chapter 4 were various systems of linear algebraic equations (difference or grid equations). The sizes of the appropriate matrices are extra large and equal the total number $N$ of the grid nodes. For example, for any grid with steps $h$ along all the directions $x_1, x_2, \dots, x_p$ $(h_1 = h_2 = \dots = h_p = h)$ the amount of nodal points is $N = O(h^{-p})$, where $p$ is the total number of observations. In the two-dimensional and three-dimensional cases $N \approx 10^4 - 10^6$, for example, for $h = 1/100$. More specifically, such sparse matrices are of the special band structure and the characteristic rations between their greatest and the smallest eigenvalues are high-order quantities ($\sim 10^3 - 10^4$ of order $O(h^{-2})$).

Numerical solution of elliptic grid equations necessitates, in view of their peculiarities, creating special economical algorithms, because direct economical methods are applicable only in some narrow classes of grid equations. We will elaborate on this later.

When solving difference boundary-value problems for Poisson's equation in rectangular, angular, cylindrical and spherical systems of coordinates direct economical methods are widely used that are known to us as the decomposition method and the method of separation of variables. The calculations in both methods for two-dimensional problems require $Q$ arithmetic operations, $Q = O(N^2 \log_2 N)$, where $N$ is the number of the grid nodes along one of the directions.

As a matter of fact, the first method is some modification of Gaussian elimination relating to the odd-even elimination with the accompanying factorization, the second one is mostly based on the algorithm of the fast Fourier transform. The third method of the matrix elimination seems to offer more advantages in the domains of rather complicated configurations, but the work during the course of the matrix elimination is done with $Q = O(N^4)$ arithmetic operations and the extra storage in connection with emerging intermediate values.

Iterative methods of successive approximation are in common usage for rather complicated cases of arbitrary domains, variable coefficients, etc. Throughout the entire section, the Dirichlet problem for Poisson's equation is adopted as a model one in the rectangle $\bar{G} = \{0 \le x_\alpha \le l_\alpha,\ \alpha = 1, 2\}$ with the boundary $\Gamma$:

$$(1) \qquad \Delta u = \frac{\partial^2 u}{\partial x_1^2} + \frac{\partial^2 u}{\partial x_2^2} = -f(x)\,,$$

$$x = (x_1, x_2) \in G\,, \qquad u|_\Gamma = \mu(x)\,,$$

and a rectangular grid in $\bar{G}$ with steps $h_1$ and $h_2$ is taken to be

$$\omega_h = \left\{ x_{i_1 i_2} = (i_1 h_1, i_2 h_2) \in \bar{G},\ i_\alpha = 0, 1, \dots, N_\alpha,\ h_\alpha N_\alpha = l_\alpha,\ \alpha = 1, 2 \right\}.$$

The statement of the difference Dirichlet problem associated with problem (1) is

$$(2) \qquad \Lambda y = -f(x), \qquad x \in \omega_h, \qquad y|_{\gamma_h} = \mu(x),$$

$$\Lambda = \Lambda_1 + \Lambda_2, \quad \Lambda_\alpha y = y_{\bar{x}_\alpha x_\alpha}, \quad \alpha = 1, 2, \quad y_{i_1 i_2} = y(i_1 h_1, i_2 h_2).$$

Problem (2) was the object of special investigations in Chapter 4.

**2. The decomposition method.** Common practice involves the reduction of problem (2) to the system of vector equations

$$(3) \qquad -\mathbf{Y}_{j-1} + C\,\mathbf{Y}_j - \mathbf{Y}_{j+1} = \mathbf{F}_j, \qquad j = 1, 2, \dots N_2 - 1,$$

$$\mathbf{Y}_0 = \mathbf{F}_0, \qquad \mathbf{Y}_{N_2} = \mathbf{F}_{N_2},$$

where $\mathbf{Y}_j$ and $\mathbf{F}_j$ are vectors, whose components are the values of the solution $y_{ij} = y(ih_1, jh_2)$ and the right-hand side $f_{ij} = f(ih_1, jh_2)$ on the $j$th column of the grid $\omega_h$ and the difference operator $C$ will be specified in the sequel. By re-ordering of the right-hand sides of equations (2) at the near-boundary nodes we might agree to consider $y_{ij} = 0$ at the boundary nodes for $i = 0$ and $i = N_1$.

An alternative form of writing equation (2) may be useful in the further development:

$$(4) \qquad -y_{i,j-1} + (2y - h_2^2\, y_{\bar{x}_1 x_1})_{ij} - y_{i,j+1} = h_2^2\, \varphi_{ij},$$

$$1 < i < N_1 - 1, \qquad 1 < j < N_2 - 1,$$

$$y_{0j} = y_{N_1 j} = 0, \qquad 0 < j < N_2,$$

$$y_{i0} = \mu_{i0} = 0, \quad y_{iN_2} = \mu_{iN_2}, \quad 0 < i < N_1,$$

where

$$\varphi_{ij} = f_{ij} \quad \text{for } 1 < i < N_1 - 1, \quad 1 \le j \le N_2 - 1,$$

$$\varphi_{1j} = f_{1j} + \frac{1}{h_1^2}\,\mu_{0j}, \quad \varphi_{N_1-1,j} = f_{N_1-1,j} + \frac{1}{h_1^2}\,\mu_{N_1,j}.$$

Also, it will be sensible to deal with the newly formed vectors

$$(5)\quad \mathbf{Y}_j = (y_{1j}, y_{2j}, \ldots, y_{N_1-1,j}), \quad j = 0, 1, \ldots, N_2,$$

$$\mathbf{F}_j = \left( h_2^2 f_{1j} + \frac{h_2^2}{h_1^2}\mu_{0j}, h_2^2 f_{2j}, \ldots, h_2^2 f_{N_1-2,j}, h_2^2 f_{N_1-1,j} + \frac{h_2^2}{h_1^2}\mu_{N_1 j}, \right),$$

$$j = 1, 2, \ldots, N_2 - 1,$$

$$\mathbf{F}_j = (\mu_{1j}, \mu_{2j}, \ldots, \mu_{N_1-1,j}), \quad j = 0, N_2,$$

and a difference operator $C$ acting in accordance with the rule

$$\left(C\mathbf{Y}_j\right)_i = \left(2y - h_2^2 y_{\bar{x}_1 x_1}\right)_{ij}, \quad 0 < i < N_1, \quad y_{0j} = y_{N_1 j} = 0.$$

From such reasoning it seems clear that problem (2) in view is tantamount to the system of vector equations (3).

Let $N_2 = 2^n$ for the clarity only. The main idea behind the decomposition method is the further successive elimination from the governing equations of the vectors $\mathbf{Y}_j$ with odd numbers and, after this, with even numbers divisible 2, 4, 8 etc. Other ideas are connected with setting the following equations for $j = 2, 4, 6, \ldots, N_2 - 2$, where $N_2 = 2^n$:

$$-\mathbf{Y}_{j-2} + C\mathbf{Y}_{j-1} - \mathbf{Y}_j = \mathbf{F}_{j-1},$$

$$-\mathbf{Y}_{j-1} + C\mathbf{Y}_j - \mathbf{Y}_{j+1} = \mathbf{F}_j,$$

$$-\mathbf{Y}_j + C\mathbf{Y}_{j+1} - \mathbf{Y}_{j+2} = \mathbf{F}_{j+1}.$$

Applying the operator $C$ to the second equation and summing up three resultant equations yield a revised "short" system

$$(6)\qquad -\mathbf{Y}_{j-2} + C^{(1)}\mathbf{Y}_j - \mathbf{Y}_{j+2} = \mathbf{F}_j^{(1)}, \quad j = 2, 4, 6, \ldots, N_2 - 2,$$

$$\mathbf{Y}_0 = \mathbf{F}_0, \qquad \mathbf{Y}_{N_2} = \mathbf{F}_{N_2},$$

containing only the unknowns with even numbers and involving the members

$$C^{(1)} = \left[C^{(0)}\right]^2 - 2E, \quad \mathbf{F}_j^{(1)} = \mathbf{F}_{j-1}^{(0)} + C^{(0)}\mathbf{F}_j^{(0)} + \mathbf{F}_{j+1}^{(0)},$$

$$C^{(0)} = C, \qquad \mathbf{F}_j^{(0)} = \mathbf{F}_j.$$

Having recovered from the system (6) the vectors $\mathbf{Y}_j$ with even numbers, we succeed in finding these with odd numbers from the following equations:

$$C^{(0)}\,\mathbf{Y}_j = \mathbf{F}_j^{(0)} + \mathbf{Y}_{j+1} + \mathbf{Y}_{j-1}, \quad j = 1, 3, 5, \dots, N_2 - 1\,.$$

The same procedure works albeit with obvious modifications for the vectors $\mathbf{Y}_j$ subject to the system (6), whose subscripts $j$ are oddly (but unevently) even, etc. As a final result we obtain a proper system of equations with regard to all the unknowns:

$$(7) \qquad C^{(k-1)}\,\mathbf{Y}_j = \mathbf{F}_j^{(k-1)} + \mathbf{Y}_{j-2^{k-1}} + \mathbf{Y}_{j+2^{k-1}},$$

$$j = 2^{k-1}, 3\cdot 2^{k-1}, 5\cdot 2^{k-1}, \dots, N_2 - 2^{k-1},$$

$$k = n, n-1, \dots, 2, 1\,,$$

$$\mathbf{Y}_0 = \mathbf{F}_0, \qquad \mathbf{Y}_{N_2} = \mathbf{F}_{N_2},$$

where $C^{(k)}$ and $\mathbf{F}_j^{(k)}$ must satisfy the recurrence formulas

$$(8) \qquad C^{(k)} = \left[C^{(k-1)}\right]^2 - 2\,E\,, \quad k = 1, 2, \dots, n-1\,, \quad C^{(0)} = C\,,$$

$$\mathbf{F}_j^{(k)} = \mathbf{F}_{j-2^{k-1}}^{(k-1)} + C^{(k-1)}\mathbf{F}_j^{(k-1)} + \mathbf{F}_{j+2^{k-1}}^{(k-1)},$$

$$j = 2^k, 2\cdot 2^k, 3\cdot 2^k, \dots, N_2 - 2^k, \quad k = 1, 2, \dots, n-1\,.$$

The decomposition algorithm necessitates performing the factorization of the operator $C^{(k)}$ of the special structure

$$(9) \qquad C^{(k)} = \prod_{l=1}^{2^k} (C - \mu_l\,E)\,, \qquad \mu_l = 2\,\cos\frac{(2l-1)\,\pi}{2^{k+1}}\,,$$

making it possible to reduce the further inversion of the operator $C^{(k)}$ to successive inversion of difference operators by the elimination method. In what follows one simple equation

$$C^{(k)} v = \varphi$$

serves to motivate what is done. With representation (9) in view, the sought function $v = v^{(2^k)}$ will appear as the outcome of successive solution of three-point difference equations

$$(C - \mu_1\,E)\,v^{(1)} = \varphi\,, \quad (C - \mu_l E)\,v^{(l)} = v^{(l-1)}, \quad l = 2, 3, \dots, 2^k\,,$$

or, what amounts to the same,

$$2\,v^{(l)} - h_2^2\,v^{(l)}_{\bar{x}_1 x_1} - \mu_l\,v^{(l)} = v^{(l-1)}, \quad h_1 \le x_1 \le l_1 - h_1,$$

$$v^{(l)}(0) = v^{(l)}(l_1) = 0\,.$$

The preceding typical equation can be solved by the elimination method with regard to three-point equations. In giving $\mathbf{F}_j^{(k)}$ the intention is to use the formula

$$\mathbf{F}_j^{(k)} = C^{(k)}\,\mathbf{p}_j^{(k)} + \mathbf{q}_j^{(k)} \tag{10}$$

with new vectors $\mathbf{p}_j^{(k)}$ and $\mathbf{q}_j^{(k)}$ still subject to (8):

$$\begin{aligned} C^{(k)}\mathbf{p}_j^{(k)} + \mathbf{q}_j^{(k)} &= C^{(k-1)}\big[\mathbf{p}^{(k-1)}_{j-2^{k-1}} + \mathbf{p}^{(k-1)}_{j+2^{k-1}} + \mathbf{q}_j^{(k-1)}\big] \\ &\quad + \big(C^{(k-1)}\big)^2\,\mathbf{p}_j^{(k-1)} + \mathbf{q}^{(k-1)}_{j-2^{k-1}} + \mathbf{q}^{(k-1)}_{j+2^{k-1}}\,. \end{aligned} \tag{11}$$

Granting the decomposition

$$\mathbf{q}_j^{(k)} = 2\,\mathbf{p}_j^{(k)} + \mathbf{q}^{(k-1)}_{j-2^{k-1}} + \mathbf{q}^{(k-1)}_{j+2^{k-1}} \tag{12}$$

and taking into acount that

$$C^{(k)} = \big(C^{(k-1)}\big)^2 - 2\,E\,,$$

we recast equation (11) as

$$\big(C^{(k-1)}\big)^2\,\big(\mathbf{p}_j^{(k)} - \mathbf{p}_j^{(k-1)}\big) = C^{(k-1)}\,\big(\mathbf{p}^{(k-1)}_{j-2^{k-1}} + \mathbf{p}^{(k-1)}_{j+2^{k-1}} + \mathbf{q}_j^{(k-1)}\big)$$

with further elimination of $C^{(k-1)}$. The outcome of this is

$$C^{(k-1)}\,\mathbf{S}_j^{(k-1)} = \mathbf{p}^{(k-1)}_{j-2^{k-1}} + \mathbf{p}^{(k-1)}_{j+2^{k-1}} + \mathbf{q}_j^{(k-1)}, \quad \mathbf{p}_j^{(k)} = \mathbf{p}_j^{(k-1)} + \mathbf{S}_j^{(k-1)}$$

with $\mathbf{q}_j^{(k)}$ still subject to (12). Substitution of formula (10) into (7) yields

$$C^{(k-1)}\,\big[\mathbf{Y}_j - \mathbf{p}_j^{(k-1)}\big] = \mathbf{q}_j^{(k-1)} + \mathbf{Y}_{j-2^{k-1}} + \mathbf{Y}_{j+2^{k-1}}\,,$$

showing the new vectors to be sensible ones. Summarizing, the numerical solution of problem (2) can be done using only two operation: inversion of the operators $C^{(k-1)}$ and summation of relevant vectors in the process of calculation of the right-hand sides of these equations. Thus, the computational procedures include the following steps:

- specifications of the initial values $\mathbf{p}_j^{(0)}$ and $\mathbf{q}_j^{(0)}$ so that

$$\mathbf{q}_j^{(0)} = \mathbf{F}_j\,, \qquad \mathbf{p}_j^{(0)} = 0\,, \qquad j = 1, 2, \dots, N_2 - 1\,;$$

- for all $k = 1, 2, \dots, n-1$ solution of the equations

$$C^{(k-1)}\,\mathbf{S}_j^{(k-1)} = \mathbf{p}_{j-2^{k-1}}^{(k-1)} + \mathbf{p}_{j+2^{k-1}}^{(k-1)} + \mathbf{q}_j^{(k-1)}$$

  and calculations of the vectors $\mathbf{p}_j^{(k)}$ and $\mathbf{q}_j^{(k)}$ by the recurrence formulas

$$\mathbf{p}_j^{(k)} = \mathbf{p}_j^{(k-1)} + \mathbf{S}_j^{(k-1)}, \qquad \mathbf{q}_j^{(k)} = 2\,\mathbf{p}_j^{(k)} + \mathbf{q}_{j-2^{k-1}}^{(k-1)} + \mathbf{q}_{j+2^{k-1}}^{(k-1)}$$

  for all $j = 2^k, 2\cdot 2^k, 3\cdot 2^k, \dots, N_2 - 2^k$;
- the solution of the equations

$$C^{(k-1)}\,\mathbf{S}_j^{(k-1)} = \mathbf{q}_j^{(k-1)} + \mathbf{Y}_{j-2^{k-1}} + \mathbf{Y}_{j+2^{k-1}}\,,$$

$$\mathbf{Y}_0 = \mathbf{F}_0\,, \quad \mathbf{Y}_{N_2} = \mathbf{F}_{N_2}\,,$$

  for determination of the unknown vectors by the formula

$$\mathbf{Y}_j = \mathbf{p}_j^{(k-1)} + \mathbf{S}_j^{(k-1)}$$

  for all $j = 2^{k-1}, 3\cdot 2^{k-1}, 5\cdot 2^{k-1}, \dots, N_2 - 2^{k-1}$, $k = n, n-1, \dots, 1$.

During the course of the decomposition method the users will perform $Q = O(N_1\,N_2\,\log_2 N_2)$ arithmetic operations with the extra storage about $1.5N$, where $N$ is the total number of the unknowns. Some modification of the preceding algorithm with insignificant prolongations in time may be of assistance in mastering the last difficulty involved.

**3. The method of separation of variables.** The problem we must solve is problem (2) with the homogeneous boundary conditions

$$\Lambda y = -\varphi , \qquad y|_{\gamma_h} = 0 , \tag{13}$$

where the function $\varphi$ differs from the right-hand side $f$ of problem (2) only at the near-boundary nodes in the following way: within the quantity $\frac{1}{h_1^2}$ for $i_1 = 1$, $i_1 = N_1 - 1$ and within the quantity $\frac{1}{h_2^2}$ for $i_2 = 1$, $i_2 = N_2 - 1$.

Before giving further motivations, it will be sensible to introduce the eigenfunction $\mu_k(jh_2)$ and the eigenvalue $\lambda_k$ with the number $k$ of the problem

$$\Lambda_2 \mu_k + \lambda_k \mu_k = 0 , \quad h_2 \le x_2 \le l_2 - h_2 , \quad \mu_k(0) = \mu_k(l_2) = 0 . \tag{14}$$

We learn from Chapter 2 that

$$\mu_k(jh_2) = \sqrt{\frac{2}{l_2}} \sin \frac{k \pi j}{N_2} , \qquad \lambda_k = \frac{4}{h_2^2} \sin^2 \frac{k \pi h_2}{2 l_2} , \qquad k = 1, 2, \ldots , N_2 - 1 ,$$

and may attempt a solution of problem (13) in the form

$$y_{ij} = \sum_{k=1}^{N_2 - 1} c_k(ih_1)\, \mu_k(jh_2) , \tag{15}$$

$$i = 1, 2, \ldots , N_1 - 1 , \qquad j = 1, 2, \ldots , N_2 - 1 ,$$

where the Fourier coefficient $c_k$ depends on $x_1 = ih_1$.

Upon substituting representation (15) into equation (13) we obtain

$$\begin{aligned} \Lambda y &= \Lambda_1 y + \Lambda_2 y \\ &= \sum_{k=1}^{N_2 - 1} \left[ \mu_k(jh_2)\, \Lambda_1 c_k(ih_1) + \mu_k(jh_2)\, \Lambda_2 c_k(ih_1) \right] \\ &= - \sum_{k=1}^{N_2 - 1} \varphi_k(ih_1)\, \mu_k(jh_2) , \end{aligned} \tag{16}$$

where $\varphi_k(ih_1)$ is the Fourier coefficient of the function $\varphi(x)$:

$$\varphi_k(ih_1) = \sum_{j=1}^{N_2-1} \varphi(ih_1, jh_2)\, \mu_k(jh_2)\, h_2 .$$

Due to the problem statement (14) and the orthogonality of the functions $\mu_k$ we derive from (16) the problem statement for determination of the numbers $c_k$ for all $k = 1, 2, \dots, N_2 - 1$:

$$(17) \qquad \Lambda_1 c_k - \lambda_k c_k = -\varphi_k, \quad h_1 \le x_1 \le l_1 - h_1, \quad c_k(0) = c_k(l_1) = 0 .$$

Because of this form, the applications of the elimination method for $N_2 - 1$ times to $c_k(ih_1)$ as a function of the argument $x_1 = ih_1$ for fixed $k$ permit us to find a solution of problem (13) by means of formula (15). As can readily be observed, the calculations of the Fourier coefficients $\varphi_k$ and solutions $y_{ij}$ can be carried out by the same formulas related to common sums of the special type

$$v_j = \sum_{k=1}^{N-1} z_k \sin \frac{k\pi j}{N}, \qquad j = 1, 2, \dots, N-1 .$$

Omitting more details on this point, we refer the readers to the well-developed algorithm of the fast Fourier transform, in the framework of which $Q$ arithmetic operations, $Q \approx 2N \log_2 N$, $N = 2^n$, are necessary in connection with computations of these sums (instead of $O(N^2)$ in the case of the usual summation), thus causing $O(n_1 N_2 \log_2 N_2)$ arithmetic operations performed in the numerical solution of the Dirichlet problem (2) in a rectangle.

In such matters some progress can be achieved by combinations of the decomposition method and the method of separation of variables. For example, this can be done using the method of separation of variables for the "reduced" system (6) upon eliminating the unknown vectors with odd subscripts $j$. This trick allows one to solve problem (2): here the expenditures of time are $Q \approx 2 n_1 n_2 \log_2 N_2$ arithmetic operation, half as much than required before in the method of separation of variables.

**4. The method of matrix elimination.** The system of equations (3) is one particular case of the following problem:

$$(18) \quad -A_j \mathbf{Y}_{j-1} + C_j \mathbf{Y}_j - B_j \mathbf{Y}_{j+1} = \mathbf{F}_j, \qquad j = 1, 2, \dots, N-1,$$

$$C_0 \mathbf{Y}_0 - B_0 \mathbf{Y}_1 = \mathbf{F}_0, \qquad -A_N \mathbf{Y}_{N-1} + C_N \mathbf{Y}_N = \mathbf{F}_N,$$

where $\mathbf{Y}_j$ and $\mathbf{F}_j$ are vectors of the same order $M_j$, $C_j$ is a square matrix of size $M_j \times M_j$, $A_j$ is a rectangular matrix of size $M_j \times M_{j-1}$ and $B_j$ is a rectangular matrix of size $M_j \times M_{j+1}$. As usual, we may attempt a solution of this problem in the form

$$(19) \qquad \mathbf{Y}_j = \alpha_{j+1}\,\mathbf{Y}_{j+1} + \beta_{j+1}\,, \quad j = N-1, N-2, \dots, 1, 0\,,$$

where $\alpha_j$ is a rectangular $M_{j-1} \times M_j$-matrix and $\beta_j$ is an $M_{j-1}$-dimensional vector. Following established practice with Gaussian elimination, we derive from (18)–(19) the recurrence relations for finding $\alpha_j$ and $\beta_j$ both:

$$\alpha_{j+1} = (C_j - A_j\,\alpha_j)^{-1} B_j, \quad j = 1, 2, \dots, N-1\,,$$

$$\alpha_1 = C_0^{-1} B_0,$$

$$\beta_{j+1} = (C_j - A_j\,\alpha_j)^{-1} (\mathbf{F}_j + A_j\,\beta_j)\,, \quad j = 1, 2, \dots, N-1, N\,,$$

$$\beta_1 = C_0^{-1}\mathbf{F}_0,$$

$$\mathbf{Y}_N = (C_N - A_N\,\alpha_N)^{-1}(\mathbf{F}_N + A_N\,\beta_N) = \beta_{N+1}\,,$$

$$\mathbf{Y}_j = \alpha_{j+1}\,\mathbf{Y}_{j+1} + \beta_{j+1}, \quad j = N-1, N-2, \dots, 1, 0\,.$$

For a complete and rigorous treatment, it is required that

$$\|\, C_0^{-1} B_0 \,\| \le 1\,, \qquad \|\, C_N^{-1} A_N \,\| \le 1\,,$$

$$\|\, C_j^{-1} A_j \,\| + \|\, C_j^{-1} B_j \,\| \le 1\,, \qquad 1 \le j \le N-1\,,$$

and, moreover, at least one of these inequalities should be strict. This provides the sufficient background for the stability of the matrix elimination method with respect to random errors, meaning $\|\,\alpha_j\,\| \le 1$, $j = 1, 2, \dots, N$. In the case (3), which interests us, the members become $A_j = B_j = E$, $C_j = C$ for $1 \le j \le N-1$ and $B_0 \doteq A_N = 0$, $C_0 = C_N = E$, by means of which the ensuing formulas can be written in simplified form:

$$\alpha_{j+1} = (C - \alpha_j)^{-1}, \qquad \alpha_1 = 0\,, \qquad j = 1, 2, \dots, N_2 - 1\,,$$

$$\beta_{j+1} = \alpha_{j+1}\,(\mathbf{F}_j + \beta_j)\,, \quad \beta_1 = \mathbf{F}_0, \qquad j = 1, 2, \dots, N_2 - 1\,,$$

$$\mathbf{Y}_j = \alpha_{j+1}\,\mathbf{Y}_{j+1} + \beta_{j+1}, \quad \mathbf{Y}_{N_2} = \mathbf{F}_{N_2}, \quad j = N_2 - 1, \dots, 2, 1\,.$$

Here $\mathbf{Y}_j$ and $\boldsymbol{\beta}_j$ are vectors of order $N_1 - 1$, $\alpha_j$ and $C$ are square matrices of the same size $(N_1 - 1) \times (N_1 - 1)$. Then the unique sufficient condition of stability is

$$\| C^{-1} \| \leq 0.5 .$$

This is certainly true, since $C \geq 2E$.

Proper evaluation of the necessary actions in solving problem (5) by the matrix elimination method is stipulated, as usual, by the special structures of the matrices involved. Because all the matrices $\alpha_j$ are complete in spite of the fact that $C$ is a tridiagonal matrix, $O(N_1^3)$ arithmetic operations are required for determination of one matrix $\alpha_{j+1}$ on the basis of $\alpha_j$ all of which are known to us in advance. Thus, it is necessary to perform $O(N_1^3 N_2)$ operations in practical implementations with all the matrices $\alpha_j$, $j = 1, 2, \dots, N_2$. Further, $O(N_1^2)$ arithmetic operations are required for determination of one vector $\beta_{j+1}$ with knowledge of $\beta_j$ and $O(N_1^2 N_2)$ operations for determination of all vectors $\beta_j$.

With regard to all the $\mathbf{Y}_j$'s, the same number of operations suffice and so the total volume is still $Q = O(N_1^3 N_2)$.

## 10.2 TWO-LAYER ITERATION SCHEMES

**1. Two-layer iteration schemes. The problem statement.** In what follows it is required to solve a first kind equation of the form

$$(1) \qquad A u = f ,$$

where $A : H \mapsto H$ is a linear operator in a finite-dimensional real space $H$ of the dimension $N$ with an inner product $( \, , \, )$ and associated norm $\| y \| = \sqrt{(y, y)}$. In a common setting it is supposed that $A = A^* > 0$, where $f \in H$ is an arbitrtary vector. From the viewpoint of possible applications, any iterative method provides proper guidelines for successive determination of approximate solutions $y_1, y_2, \dots, y_k, y_{k+1}, \dots$ to equation (1) with the starting point (the initial approximation) $y_0 \in H$. Any such approximations are known as **iterations** with relevant iteration numbers $k = 1, 2, \dots$. The essence of these methods is that the value $y_{k+1}$ can be obtained through the preceding iterations $y_{k-1}, y_k, \dots$. An iterative method is called a **one-step method** or a **two-step method** if only one or two preceding iterations are needed in finding every value $y_{k+1}$. As we will see later, these methods fall within the category of two-layer and three-layer methods, respectively, on the same footing as in Chapter 6. What is more, the one-step iterative method coincides in form with a two-layer scheme designed in Chapter 6.

Before going further, we recall that any linear one-step iterative method can be written as

$$B_k\, y_{k+1} = C_k\, y_k + F_k\,, \qquad k = 0, 1, 2, \dots, \tag{2}$$

where $B_k$ and $C_k$ are linear operators from the space $H$ into the space $H$ depending, generally speaking, on the iteration number $k$, $F_k \in H$ is a known function of $k$ and $y_k$ is the $k$th iteration, under the agreement that all the inverses $B_k^{-1}$ exist. A natural requirement in the further development is that the exact solution $u$ to equation (1), not depending on $k$, should identically satisfy equation (2):

$$(B_k - C_k)\, u = F_k .$$

But it is possible only if $(B_k - C_k)\, A^{-1} f = F_k$, implying that

- the inverse operator $(B_k - C_k)^{-1}$ exists;
- $f = A\,(B_k - C_k)^{-1} F_k$.

This is acceptable if we agree to consider

$$\tau_{k+1}^{-1}(B_k - C_k) = A\,, \qquad F_k = f\,\tau_{k+1}\,, \qquad k = 0, 1, 2, \dots,$$

where $\tau_{k+1} > 0$ is a numerical parameter. Under such an approach the canonical form of two-layer iteration schemes is

$$B_k\, \frac{y_{k+1} - y_k}{\tau_{k+1}} + A\, y_k = f\,, \qquad k = 0, 1, 2, \dots, \tag{3}$$

where the initial approximation $y_0 \in H$ is free to be chosen in any convenient way. Since the inverse $B_k^{-1}$ exists, it follows from the foregoing that

$$y_{k+1} = y_k - \tau_{k+1}\, B_k^{-1}\,(A\, y_k - f) \tag{4}$$

or, what amounts to the same,

$$y_{k+1} = y_k - \tau_{k+1}\, B_k^{-1}\, r_k = y_k - \tau_{k+1}\, w_k\,,$$

where $r_k = A\, y_k - f$ is the residual and $w_k = B_k^{-1} r_k$ is the correction. With knowledge of $y_k$ the value of $y_{k+1}$ can be recovered from equation (4). Knowing $y_0$, it is plain to determine successively $y_1, y_2 \dots$. Of course, it is meaningful only for convergent iterative methods:

$$\|\, y_k - u\,\| \longrightarrow 0 \quad \text{as} \quad t \to \infty\,. \tag{5}$$

The usual practice involves the numerical solution of problem (5) with a prescribed accuracy $\varepsilon > 0$ (a relative accuracy $\| y_k - u \| / \| y_0 - u \|$), it being understood that the calculations should be terminated if

(6) $$\| y_k - u \| \leq \varepsilon \| y_0 - u \| .$$

In connection with inconvenience caused by the unknown vector $u$, it seems reasonable to replace this condition by the inequality for the residual

(7) $$\| A y_k - f \| \leq \varepsilon \| A y_0 - f \| .$$

In the general case the accepted view is the termination condition of the type

(8) $$\|y_k - u\|_D \leq \varepsilon \; \|y_0 - u\|_D \, ,$$

where $D = D^* > 0$ is some operator. By merely setting $D = A^2$ we deduce from (8) inequality (7).

We are now interested in the governing equation related to the residual $z_k = y_k - u$. Since $Au = f$, we might have

(9) $$B_k \, \frac{z_{k+1} - z_k}{\tau_{k+1}} + A \, z_k = 0 \, , \qquad k = 0, 1, 2, \ldots ,$$

where $z_0 \in H$ is known. As far as $B_k = B$ is independent of the subscript $k$, the correction $w_k = B^{-1} r_k$ satisfies the homogeneous equation

$$B \, \frac{w_{k+1} - w_k}{\tau_{k+1}} + A \, w_k = 0 \, .$$

Indeed, (4) implies that

$$y_{k+1} - y_k = -\tau_{k+1} \, B^{-1} \, r_k = -\tau_{k+1} \, w_k \, .$$

Applying the operator $A$ twice to both sides of the preceding equality and taking into account that

$$A \, y_{k+1} - A \, y_k = (A \, y_{k+1} - f) - (A \, y_k - f) = r_{k+1} - r_k \, ,$$

$$r_{k+1} - r_k = B \, (B^{-1} r_{k+1} - B^{-1} r_k) = B \, (w_{k+1} - w_k) \, ,$$

we establish the homogeneous equation for the correction $w_k$.

Also, it seems clear from (9) that

$$z_{k+1} = S_{k+1} z_k , \qquad S_{k+1} = E - \tau_{k+1} B_k^{-1} A ,$$

where $S_{k+1}$ is the transition operator from the layer $k$ to the layer $k+1$. Having completed the elimination of $z_k, z_{k-1}, \dots, z_1$, we find for $k = n-1$ that

$$z_n = T_n z_0 , \qquad T_n = S_n S_{n-1} \cdots S_2 S_1 ,$$

where $T_n$ is the resolving operator of scheme (9). This serves to motivate the estimates

$$\|z_n\|_D = \|T_n z_0\|_D \le \|T_n\|_D \cdot \|z_0\|_D \qquad \text{or}$$

$$\|z_n\|_D \le q_n \|z_0\|_D , \qquad q_n = \|T_n\|_D .$$

From such reasoning it seems clear that the condition of termination is ensured if $q_n \le \varepsilon$, thereby reducing the question of convergence of the iterations to the norm estimation of the resolving operator $T_n$.

Scheme (3) generates an exact approximation on a solution $u$ of the equation $Au = f$ for any operators $\{B_n\}$ and any choice of the parameters $\{\tau_{k+1}\}$, but the quantity $q_n$ depends on $\{B_n\}$ and $\{\tau_{k+1}\}$ both. Some consensus of opinion here is that $\{B_n\}$ and $\{\tau_{k+1}\}$ should be so chosen as to minimize the norm $\|T_n\|_D = q_n$ of the resolving operator $T_n$ of scheme (3) and to minimize the total number of arithmetic operations which will be needed for recovering the value $y_{k+1}$ from the equation

$$B_k y_{k+1} = F_k , \qquad F_k = B_k y_k - \tau_{k+1} (A y_k - f) ,$$

with a known value $y_{k+1}$.

In accordance with what has been said above, any iteration scheme (3) can be treated as a two-layer scheme being used for solving the nonstationary problem

$$B \frac{du}{dt} + A u = f ,$$

where the parameter $\tau_{k+1}$ regards to one possible step in a nonreal time $t_{k+1} = \sum_{m=1}^{k+1} \tau_m$. The main differences between iteration schemes and available schemes for nonstationary problems are:

- the iteration scheme (3) approximates exactly equation (1), since a solution $u$ to equation (1) satisfies equation (3) for any $B_k$ and $\tau_{k+1}$;

- proper choices of the parameter $\tau_{k+1}$ and the operators $B_k$ are caused only by the necessity of convergence of the iterations and the economy requirements in trying to solve the original problem with a prescribed accuracy, while the restrictions on the steps for non-stationary problems are connected with the approximations which do arise in such matters.

Let now $Q(\varepsilon)$ be the total number of arithmetic operations necessary for obtaining a solution to equation (1) with a prescribed accuracy $\varepsilon > 0$ regardless of the initial approximation in the iteration scheme (3). Its ingredients $B_k$ and $\tau_k$ should be so chosen as to minimize the quantity $Q(\varepsilon)$. If the desirable accuracy can be attained in a minimal number of the iterations $n = n(\varepsilon)$, then

$$Q(\varepsilon) = \sum_{k=1}^{n(\varepsilon)} Q_k = \bar{Q}_n n ,$$

where $Q_k$ is the number of the necessary actions during the course of $k$th iteration. Thus, the minimum problem for $Q(\varepsilon)$ reduces to the minimum problems for $n(\varepsilon)$ and the number $Q_k$, which depends on $B_k$.

In this context, if $B_k = E$ is the identity operator, then scheme (3) refers to **explicit iteration schemes** of the structure

$$(10) \qquad \frac{y_{k+1} - y_k}{\tau_{k+1}} + A\, y_k = f\,, \qquad k = 0, 1, 2, \ldots\,, \qquad \text{for any } y_0 \in H\,.$$

If $B_k \neq E$, then scheme (3) is termed an **implicit** iteration scheme.

**2. A stationary scheme. The main theorem on the convergence of iterations.** Quite often, the iteration schemes such as

$$(3') \qquad B\, \frac{y_{k+1} - y_k}{\tau} + A\, y_k - f\,, \qquad k - 0, 1, \ldots\,,$$

with a constant operator $B$ and constant $\tau$ are called **stationary methods of iterations**. In particular, the upper relaxation method and Seidel method fall within the category of such methods. In that case equation (9) related to the error of approximation $z_k = y_k - u$ takes the form

$$(9') \qquad B\, \frac{z_{k+1} - z_k}{\tau} + A\, z_k = 0\,, \quad k = 0, 1, \ldots\,, \quad z_0 = y_0 - u\,,$$

and it remains valid for the correction $w_k = B^{-1}(A\, y_k - f)$. The operator $B$ is, generally speaking, non-self-adjoint and possesses the own inverse $B^{-1}$. This type of situation is covered by the following assertion.

**Theorem 1** *If $A$ is a self-adjoint operator ($A = A^* > 0$), then*

$$B > \frac{1}{2}\tau A \quad \text{or} \quad (Bx, x) > \frac{1}{2}\tau (Ax, x) \quad \text{for all} \quad x \in H \tag{11}$$

*is a sufficient condition for the convergence of the method of iterations (3′) in the space $H_A$ with the rate of convergence of a geometric progression*

$$\| z_{k+1} \|_A \le \rho \| z_k \|_A, \qquad k = 0, 1, \dots, \qquad \rho < 1, \tag{12}$$

*where $\rho = (1 - 2\tau\delta_*\delta/\|B\|^2)^{1/2}$ is its denominator and $\delta = \min_k \lambda_k(A)$, $\delta_* = \min_k \lambda_k(B_0 - \tau A/2)$, $B_0 = (B + B^*)/2$ is the symmetric part of the operator $B$.*

**Proof** Knowing from (3′) $z_{k+1} = S z_k$ with the operator $S = E - \tau B^{-1} A$, we find that

$$\begin{aligned} \| z_{k+1} \|_A^2 &= (A z_{k+1}, z_{k+1}) = (A S z_k, S z_k) \\ &= \big(A(E - \tau B^{-1} A) z_k, (E - \tau B^{-1} A) z_k)\big) \\ &= \| z_k \|_A^2 - \tau \left[ (A B^{-1} A z_k, z_k) + (B^{-1} A z_k, A z_k) \right] \\ &\quad + \tau^2 (A B^{-1} A z_k, B^{-1} A z_k). \end{aligned}$$

With the relation $A = A^*$ in view, we deduce upon substituting here $A z_k = -B v_k$ and $v_k = -B^{-1} A z_k$, where $v_k = \frac{1}{\tau}(z_{k+1} - z_k)$, that

$$\| z_{k+1} \|_A^2 = \| z_k \|_A^2 - 2\tau \big( (B - \tau A/2) v_k, v_k \big). \tag{13}$$

Because of (11), by utilizing the fact that the operator $P = B - \tau A/2$ is positive we establish its positive definiteness in a finite-dimensional space $H$ (for more detail see Chapter 2, Section 1):

$$B - \frac{1}{2}\tau A \ge \delta_* E, \qquad \delta_* > 0, \tag{11′}$$

where $\delta_*$ is the smallest eigenvalue of the operator $P_0 = B_0 - \frac{1}{2}\tau A$, so that

$$2\tau \big( (B - \frac{1}{2}\tau A) v_k, v_k \big) \ge 2\tau \delta_* \| v_k \|^2. \tag{13′}$$

On the other hand, it follows from the foregoing that

$$
\begin{aligned}
\| z_k \|_A^2 &= (A z_k, z_k) = (B v_k, A^{-1} B v_k) \\
&\le \| A^{-1} \| \cdot \| B v_k \|^2 \\
&\le \| A^{-1} \| \cdot \| B \|^2 \cdot \| v_k \|^2 \\
&= \frac{\| B \|^2 \cdot \| v_k \|^2}{\delta} ,
\end{aligned}
$$

yielding

$$
\| v_k \|^2 \ge \frac{\delta}{\| B \|^2} \, \| z_k \|_A^2 . \tag{14}
$$

By inserting (13′) and (14) in (13), it is possible to show that the bilaterial estimate

$$
\| z_{k+1} \|_A^2 = \| S z_k \|_A^2 \le \rho^2 \, \| z_k \|_A^2
$$

becomes valid with $\rho^2 = 1 - 2\tau\delta\delta_* / \| B \|^2 < 1$, assuring estimate (12) and the inequality $\| z_n \|_A \le \rho^n \, \| z_0 \|_A$ and justifying the convergence of the iterations, since $\rho^n \to 0$ as $n \to \infty$. The same estimate is certainly true with the correction $w_n = B^{-1}(A y_n - f)$.

**Remark** Condition (11) for fixed $B$ may be viewed as a selection rule for those values of $\tau$ for which the iterations converge. For example, for the explicit scheme with the identity operator $B = E$ condition (11′) is ensured if all the eigenvalues are subject to the relation

$$
\lambda_k \left( E - \frac{1}{2} \tau A \right) = 1 - \frac{1}{2} \tau \lambda_k(A) > 0
$$

or, what amounts to the same,

$$
1 - \frac{1}{2} \tau \| A \| > 0 .
$$

Thus, the iterations converge for any $\tau < 2/\| A \|$. Let us stress here that the estimate obtained for $\rho$ is too rough for determination of the total number $n(\varepsilon, N)$ of the necessary iterations and indicates mainly the true order in $n$ as $N \to \infty$.

**3. The explicit scheme with optimal set of Chebyshev's parameters.** In what follows the intention is to use the explicit scheme (10) without concern for how the parameters $\tau_1, \tau_2, \dots, \tau_n$ will be chosen in trying to minimize the total number of iterations $n = n(\varepsilon)$. Also, under the agreement that the operator $A$ is self-adjoint and positive we operate with its smallest $\gamma_1 > 0$ and greatest $\gamma_2$ eigenvalues:

$$(15) \qquad A = A^* > 0\,, \qquad \gamma_1\, E \le A \le \gamma_2\, E\,, \qquad \gamma_1 > 0\,.$$

The meaning of this is that we should have

$$\gamma_1 \,\|\, x \,\|^2 \le (Ax, x) \le \gamma_2 \,\|\, x \,\|^2 \quad \text{for any } x \in H\,.$$

If the parameter $\tau = \text{const}$ is independent of the subscript $k$, that is, $\tau_1 = \tau_2 = \cdots = \tau_n = \tau$, scheme (10) is called the simple iteration scheme:

$$(16) \qquad y_{k+1} = y_k - \tau\,(A\, y_k - f)\,.$$

In Section 1 of the present chapter we have established that the residual $r_k = A\, y_k - f$ satisfies the homogeneous equation

$$(17) \qquad \frac{r_{k+1} - r_k}{\tau_{k+1}} + A\, r_k = 0\,, \quad k = 0, 1, 2, \dots\,, \quad r_0 = A\, y_0 - f \in H\,,$$

or $r_{k+1} = S_{k+1} r_k$, $S_{k+1} = E - \tau_{k+1} A$, showing a way of relating $r_n$ and $r_0$:

$$(18) \qquad r_n = T_n\, r_0\,, \qquad t_n = S_1\, S_2\, \cdots\, S_n\,.$$

Here $T_n$ is the resolving operator being a polynomial of degree $n$ with respect to the operator $A$:

$$(19) \qquad T_n = \mathcal{P}_n(A) = (E - \tau_1 A)\,(E - \tau_2 A)\, \cdots\, (E - \tau_n A)\,,$$

so that $r_n = \mathcal{P}_n(A) r_0$. On this basis the residual $r_n$ obeys the estimate

$$(20) \qquad \|\, r_n \,\| \le \|\, \mathcal{P}_n(A) \,\| \cdot \|\, r_0 \,\| = q_n \,\|\, r_0 \,\|\,.$$

The next step is to evaluate the quantity $\|\, \mathcal{P}_n(A) \,\|$ of interest in terms of $\gamma_1$ and $\gamma_2$, making it possible to extract those parameters $\tau_1, \tau_2, \dots, \tau_n$, for which the minimal value of $q_n = \|\, \mathcal{P}_n(A) \,\|$ is attained. The preceding polynomial

$$\mathcal{P}_n(A) = \prod_{m=1}^{n} (E - \tau_m A) = \sum_{k=0}^{n} c_k\, A^k, \quad c_0 = 1\,, \quad \mathcal{P}_n(0) = 1\,,$$

refers to self-adjoint operators, since any degree of the operator $A$ also is a self-adjoint operator: $A^m = (A^m)^*$.

Let $\{\lambda_s, \xi_s\}$ be eigenvalues and orthonormal eigenfunctions of the operator $A$:

$$A\xi_s = \lambda_s \xi_s, \quad s = 1, 2, \dots, N, \quad 0 < \lambda_1 \le \lambda_2 \le \dots \le \lambda_N,$$

where $N$ is the dimension of the space $H$ and $\lambda_1 = \min_s \lambda_s = \gamma_1$, $\lambda_N = \max_s \lambda_s = \gamma_2$. By definition,

$$A^k \xi_s = \lambda_s A^{k-1} \xi_s = \lambda_s^k \xi_s,$$

meaning that $\lambda_s^k$ is one of the eigenvalues of the operator $A^k$. This serves as a basis for the representations

$$\mathcal{P}_n(A)\xi_s = \sum_{k=0}^{n} c_k A^k \xi_s = \left(\sum_{k=0}^{n} c_k \lambda_s^k\right) \xi_s = \mathcal{P}_n(\lambda_s)\xi_s$$

and, on the same grounds, $\lambda(\mathcal{P}_n(A)) = \mathcal{P}_n(\lambda(A))$.

Thus, the eigenvalues of the polynomial $\mathcal{P}_n(A)$ are equal to the polynomial $\mathcal{P}_n(\lambda)$ of the eigenvalues $\lambda = \lambda(A)$ of the operator $A$. With the relation $(\mathcal{P}_n(A))^* = \mathcal{P}_n(A)$ established, we find that

$$\| \mathcal{P}_n(A) \| \le \max_{\gamma_1 \le x \le \gamma_2} |\mathcal{P}_n(x)|. \tag{21}$$

Because of this, the problem of searching for $\min_{\tau_1, \tau_2, \dots, \tau_n} \| \mathcal{P}_n(A) \|$ can be reduced to the well-known minimax problem for the polynomial $\mathcal{P}_n(x)$ in question. By interchanging the variables by the rule

$$x = 0.5\left[(\gamma_2 - \gamma_1)t + \gamma_2 + \gamma_1\right] \tag{22}$$

the segment $[\gamma_1, \gamma_2]$ carries into the segment $[-1, 1]$, so that $\mathcal{P}_n(x) = \widetilde{\mathcal{P}}_n(t)$, $t \in [-1, 1]$, and $\mathcal{P}_n(0) = 1$.

With the detailed forms in mind, a revised statement of the problem consists of finding a polynomial with minimal deviations from zero on the segment $[-1, 1]$ such that $\max_{-1 \le t \le 1} |\widetilde{\mathcal{P}}_n(t)|$ is minimal under the additional condition of normalization $\widetilde{\mathcal{P}}_n(t_0) = 1$, where the point $t_0$ corresponds to the point $x = 0$. From formula (22) it follows for $x = 0$ that

$$t_0 = -\frac{\gamma_2 + \gamma_1}{\gamma_2 - \gamma_1}. \tag{23}$$

Thus, the well-known Chebyshev polynomial defined by

$$\widetilde{\mathcal{P}}_n(t) = \frac{T_n(t)}{T_n(t_0)},$$

where

$$T_n(t) = \cos(n \arccos t) \quad \text{for } |t| \le 1, \tag{24}$$

is just the solution of the original problem concerned. For $|t| \ge 1$ the polynomial of interest is specified by the formula

$$T_n(t) = 0.5\left[(t+\sqrt{t^2-1})^n + (t-\sqrt{t^2-1})^n\right], \quad |t| > 1. \tag{25}$$

Since $\max\limits_{|t|\le 1} |T_n(t)| = 1$, the relations occur:

$$\min_{\{\tau_k\}} \max_{\gamma_1 \le x \le \gamma_2} |\mathcal{P}_n(x)| = \min_{\{\tau_k\}} \max_{-1\le t\le 1} |\widetilde{\mathcal{P}}_n(t)| = \frac{1}{|T_n(t_0)|} = q_n. \tag{26}$$

In an attempt to find the unknown parameters $\tau_1, \tau_2, \dots, \tau_n$ by the approved rule saying that the zeroes of the sought polynomial $\widetilde{\mathcal{P}}_n(t)$ should coincide with known zeroes of Chebyshev's polynomial such as

$$t_k = \cos\frac{2k-1}{2n}\pi, \qquad k = 1, 2, \dots, n; \tag{27}$$

we recall from calculus that the polynomial

$$\mathcal{P}_n(x) = (1-\tau_1 x)(1-\tau_2 x)\cdots(1-\tau_n x)$$

has zeroes at the points $x_k = 1/\tau_k$, $k = 1, 2, \dots, n$. By formula (22), relating $x$ and $t$, we deduce that

$$2 = \left[(\gamma_1+\gamma_2) + (\gamma_2-\gamma_1)t_k\right]\tau_k,$$

giving

$$\tau_k = \frac{2}{(\gamma_1+\gamma_2+(\gamma_2-\gamma_1)t_k)}, \qquad k = 1, 2, \dots, n.$$

Also, it will be sensible to introduce more compact notations

$$\xi = \frac{\gamma_1}{\gamma_2}, \quad \rho_0 = \frac{1-\xi}{1+\xi}, \quad \rho_1 = \frac{1-\sqrt{\xi}}{1+\sqrt{\xi}}, \quad \tau_0 = \frac{2}{\gamma_1+\gamma_2}. \tag{28}$$

All this enables us to write down

$$\tau_k = \frac{\tau_0}{1+\rho_0\, t_k}, \qquad k=1,2,\ldots,n, \tag{29}$$

thereby completing the task of motivating the choice of the parameters $\tau_1, \tau_2, \ldots, \tau_n$. The expression for $q_n = 1/|T_n(t_0)|$, $t_0 = -1/\rho_0$, is needed as further developments occur. Since $|t|_0 > 1$, applying formula (25) to $T_n(t_0)$ yields

$$|T_n(t_0)| = \frac{1}{2}\left[\left(\frac{1}{\rho_0} + \sqrt{\frac{1}{\rho_0^2} - 1}\right)^n + \left(\frac{1}{\rho_0} - \sqrt{\frac{1}{\rho_0^2} - 1}\right)^n\right].$$

By minor changes the expressions in various brackets are modified into

$$\frac{1}{\rho_0} + \sqrt{\frac{1}{\rho_0^2} - 1} = \frac{1+\sqrt{1-\rho_0^2}}{\rho_0} = \frac{1}{\rho_0}\left(1 + \sqrt{\frac{4\,\xi}{(1+\xi)^2}}\right)$$

$$= \frac{1+\xi+2\sqrt{\xi}}{1-\xi} = \frac{1+\sqrt{\xi}}{1-\sqrt{\xi}} = \frac{1}{\rho_1},$$

$$\frac{1}{\rho_0} - \sqrt{\frac{1}{\rho_0^2} - 1} = \frac{1+\xi-2\sqrt{\xi}}{1-\xi} = \frac{1-\sqrt{\xi}}{1+\sqrt{\xi}} = \rho_1$$

and, hence,

$$q_n = \frac{2\,\rho_1^n}{1+\rho_1^{2n}}. \tag{30}$$

Thus, for scheme (14) with optimal set of the parameters $\tau_1, \tau_2, \ldots, \tau_n$ given by formula (29) the relation

$$\|\,A\,y_n - f\,\| \le q_n\,\|\,A\,y_0 - f\,\| \tag{31}$$

is valid with $q_n$ still subject to (30).

The number $n = n(\varepsilon)$ is so chosen as to satisfy the condition

$$q_n = \frac{2\,\rho_1^n}{1+\rho_1^{2n}} \le \varepsilon.$$

For this, it suffices to require in the further derivation that $\rho_1^n \le \varepsilon/2$ or, what amounts to the same, that

$$n \ge \frac{\ln(2\,\varepsilon)}{\ln(1/\rho_1)}. \tag{32}$$

The well-established expansion of the function

$$\ln \frac{1+x}{1-x} = 2\,x + \frac{1}{3}\left(\frac{1}{(1+\bar{x})^3} + \frac{1}{(1-\bar{x})^3}\right)x^3, \quad 0 < x < 1,$$

where $0 < \bar{x} < x$, is aimed at establishing the relations

$$\ln \frac{1+x}{1-x} > 2\,x, \qquad \ln \frac{1}{\rho_1} = \ln \frac{1+\sqrt{\xi}}{1-\sqrt{\xi}} > 2\,\sqrt{\xi}$$

and, hence, inequality (32) holds true for

$$n \geq n_0(\varepsilon), \quad n_0(\varepsilon) = \frac{\ln(2/\varepsilon)}{2\,\sqrt{\xi}} = \frac{1}{2}\sqrt{\frac{\gamma_2}{\gamma_1}}\,\ln\frac{2}{\varepsilon}. \tag{33}$$

This estimate is more convenient in practical implementations than estimate (32).

**4. The simple iteration scheme.** By formally setting $n = 1$ in formula (29) the preceding is referred to as the **simple iteration method**

$$\frac{y_{k+1} - y_k}{\tau_0} + A\,y_k = f \tag{34}$$

with the parameter $\tau_0$ incorporated:

$$\tau_0 = \frac{2}{\gamma_1 + \gamma_2}. \tag{34'}$$

Here $t_1 = \cos\frac{\pi}{2} = 0$, $\tau_1 = \tau_0$ and

$$q_1 = \frac{2\,\rho_1}{1+\rho_1^2} = \rho_0\,. \tag{35}$$

The equation for the residual $r_k = A\,y_k - f$ reduces to $y_{k+1} = Sy_k$, $S = E - \tau_0 A$. As far as the operator $T_1 = S$ is concerned, formulas (20) and (35) together imply the estimate for the norm of the transition operator

$$\|\,S\,\| = \rho_0 = \frac{1-\xi}{1+\xi}\,.$$

By making $n$ iterations of the simple iteration method we find that

$$r_n = S^n r_0, \qquad \|\,r_n\,\| \leq \rho_0^n\,\|\,r_0\,\|\,.$$

The condition $\rho_0^n \le \varepsilon$ is ensured if

$$n \ge \frac{\ln(1/\varepsilon)}{\ln(1/\rho_0)},$$

which is certainly true for

$$n \ge n_0(\varepsilon), \qquad n_0(\varepsilon) = \frac{\ln(1/\varepsilon)}{2\xi}. \tag{36}$$

**5. A model problem. Comparison of methods.** Further comparison of various iterative methods will be conducted by having recourse to the Dirichlet problem associated with Poisson's equation in the square $\{0 \le x_1 \le 1, 0 \le x_2 \le 1\}$ of the unit sides $l_1 = l_2 = 1$ and posed on a square grid $\omega_h$ with steps $h_1 = h_2 = h$. As a special case of problem (2) in Section 2, the problem of interest is characterized by the grid equations

$$\Lambda y = \Lambda_1 y + \Lambda_2 y = -f(x), \quad x \in \omega_h, \quad y|_{\gamma_h} = 0, \tag{37}$$

where

$$\Lambda_\alpha y = y_{\bar{x}_\alpha x_\alpha} = \frac{y^{(+1_\alpha)} - 2y + y^{(-1_\alpha)}}{h^2}, \qquad \alpha = 1, 2.$$

The system of equations (37) can be recast in the operator form as $Ay = f$, where $Ay = -\Lambda y$ in the space $H = \overset{\circ}{\Omega}$ of all grid functions given on the grid $\bar{\omega}_h$ and vanishing on its boundary $\gamma_h$. An inner product and associated norm in that space $H$ are defined by

$$(y, v) = \sum_{x \in \omega_h} y(x)\, v(x)\, h^2, \qquad \| y \| = \sqrt{(y, y)}.$$

We spoke about the operator $A$ a lot of time in Chapter 4: it is self-adjoint, positive and possesses the eigenvalues

$$\lambda_{k_1 k_2} = \frac{4}{h^2} \left( \sin^2 \frac{\pi k_1 h}{2} + \sin^2 \frac{\pi k_2 h}{2} \right),$$

$$k_\alpha = 1, 2, \dots, N-1, \qquad \alpha = 1, 2,$$

so that

$$\begin{aligned} \gamma_1 &= \min \lambda_{k_1 k_2} = \frac{8}{h^2} \sin^2 \frac{\pi h}{2}, \\ \gamma_2 &= \max \lambda_{k_1 k_2} = \frac{8}{h^2} \cos^2 \frac{\pi h}{2}. \end{aligned} \tag{38}$$

In what follows problem (37) will be treated as a **model** one in the further comparison of various methods in a step-by-step fashion in line with established priorities and answering real needs. We concentrate primarily on the total number of the iterations required in the simple iteration method (34)–(34′) and the method with optimal set of Chebyshev's parameters (14), (29).

Recall that it is fairly common to write the iteration number $k$ over the sought function $y$ within the frameworks of iterative methods available for difference equations. The same procedure works in the simple iteration scheme (SIS) which has been designed for problem (37):

$$\overset{k+1}{y} = \overset{k}{y} + \tau_0 \left( \Lambda \overset{k}{y} + f \right), \tag{39}$$

where

$$\tau_0 = \frac{2}{\gamma_1 + \gamma_2} = \frac{h^2}{4}.$$

Upon substituting the assigned value into (39) we derive the formula

$$\overset{k+1}{y}_{i_1 i_2} = \frac{\overset{k}{y}_{i_1-1,i_2} + \overset{k}{y}_{i_1+1,i_2} + \overset{k}{y}_{i_1,i_2-1} + \overset{k}{y}_{i_1,i_2+1}}{4} + \frac{h^2}{4} f_{i_1 i_2}, \tag{40}$$

by means of which the $(k+1)$th iteration is completed.

For the explicit scheme (14) with Chebyshev's parameters (SCP) the calculations are performed by the formula

$$\begin{aligned} \overset{k+1}{y}_{i_1 i_2} = \overset{k}{y}_{i_1 i_2} &+ \frac{\tau_{k+1}}{h^2} \\ &\times \left( \overset{k}{y}_{i_1-1,i_2} + \overset{k}{y}_{i_1+1,i_2} + \overset{k}{y}_{i_1,i_2-1} + \overset{k}{y}_{i_1,i_2+1} - 4 \overset{k}{y}_{i_1 i_2} \right) \\ &+ \tau_{k+1} f_{i_1 i_2}, \end{aligned}$$

where

$$\tau_{k+1} = \frac{\tau_0}{1 + \rho_0 t_k} = \frac{h^2}{4} \frac{1}{1 + \rho_0 t_k}.$$

Since the volumes of computations in determining $\overset{k}{y}$ in these methods differ slightly, the main criterion in such matters is the total number of the iterations. By formula (38) we find for $h \ll 1$ that

$$\xi = \frac{\gamma_1}{\gamma_2} = \operatorname{tg}^2 \frac{\pi h}{2} \approx \frac{\pi^2 h^2}{4}.$$

For example, the reasonable accuracy $\varepsilon = 2 e^{-10} \approx 10^{-4}$ is attained for $n_0(\varepsilon) \approx \frac{3.2}{h}$ in the case of SCP and for $n_0(\varepsilon) \approx \frac{2}{h^2}$ in the case of SIS, making it possible to fill in the following table regardless of the initial approximation, that is, for any element $y_0$ of the space $H$.

| $h$ | SIS | SCP |
|---|---|---|
| 1/10 | 200 | 32 |
| 1/50 | 5000 | 160 |
| 1/100 | 20000 | 320 |

From here it seems clear that much more iterations are needed in SIS than in SCP which is, generally speaking, preferable.

However, some progress may be achieved in reducing the total number of operations by making the well-founded choice of the initial approximation.

**6. On computational stability of iterative methods.** Until recent years the iterative method with optimal set of Chebyshev's parameters was of little use in numerical solution of grid equations. This can be explained by real facts that various sequences turn out to be nonequivalent in computational procedures.

At the initial stage such of such an analysis of algorithms it is usually supposed that a computing process is ideal, that is, computations are carried out with an infinite number of significant digits. But any computer makes calculations with a finite speed and a finite number of digits. Not all numbers are accessible to computers, there are computer null and computer infinity. For instance, abnormal termination occurs when computer infinity arises during the course of execution. A computing process may become unstable, thus causing difficulties. In such cases rounding errors may accumulate to a considerable extent so that the algorithm will be useless in practical applications

For example, for doing so with the set of Chebyshev's parameters $t_k$ in increasing order

$$t_k = \cos \frac{2k-1}{2n} \pi , \qquad k = 1, 2, \dots , n , \tag{41}$$

or in reverse order

$$t_k = - \cos \frac{2k-1}{2n} \pi , \qquad k = 1, 2, \dots , n , \tag{42}$$

abnormal terminations may occur for sufficiently small $\xi$ in connection with growing intermediate values $y_k$ for $k \le n$. Such a danger is caused by a nonmonotone character of the approximation $y_k$ to $u$, since the norm of the operator $S_k = E - \tau_k A$ associated with the transition from the $(k-1)$th iteration to the $k$th iteration may be greater than 1 for negative values of $t_k$.

Before proceeding to further discussions, let us give the details of a simple example.

**Example** The system of equations to be solved is

$$v(x_{i-1}) - 2\,v(x_i) + v(x_{i+1}) = 0\,, \quad x_i = i\,h\,, \quad 1 \le i \le N-1\,,$$

$$v(0) = 1\,, \quad v(1) = 0\,, \quad h = 1/N\,,$$

relating to the problem

$$u'' = 0\,, \quad 0 < x < 1\,, \quad u(0) = 1\,, \quad u(1) = 0\,,$$

for which the exact solutions are known:

$$u(x) = 1 - x\,, \qquad v(x_i) = 1 - x_i\,.$$

In that case the ingredients become

$$A\,y = -y_{\bar{x}x}\,, \quad \gamma_1 = \frac{4}{h^2}\sin^2\frac{\pi\,h}{2}\,, \quad \gamma_2 = \frac{4}{h^2}\cos^2\frac{\pi\,h}{2}\,, \quad \xi = \operatorname{tg}^2\frac{\pi\,h}{2}\,.$$

For convenience in analysis and clarity, we take into consideration 19 equations (N=20) and set $\varepsilon = 10^{-4}$. The analytical estimate (33) gives $n_0(\varepsilon) = 63.2$, so that $n = 64$ and the parameters $\tau_1, \tau_2, \dots, \tau_n$ for $n = 64$ are specified by formulas (29) and (41). The final results of computational procedures are presented below in Table 3.

**Table 3**

| $k$ | $\Delta_k$ |
|---|---|
| 53 | 0.12 |
| 54 | 1.5 |
| 55 | 27 |
| 56 | $6.3 \cdot 10^2$ |
| 57 | $1.9 \cdot 10^4$ |
| 58 | $7.2 \cdot 10^5$ |
| 59 | $3.7 \cdot 10^7$ |
| 60 | $2.6 \cdot 10^9$ |
| 61 | $2.5 \cdot 10^{11}$ |
| 62 | $3.3 \cdot 10^{13}$ |
| 63 | $5.0 \cdot 10^{15}$ |

**Table 4**

| $k$ | $\Delta_k$ |
|---|---|
| 1 | 39.6 |
| 2 | $2.6 \cdot 10^3$ |
| 3 | $1.6 \cdot 10^5$ |
| 4 | $8.2 \cdot 10^6$ |
| 5 | $3.7 \cdot 10^8$ |
| 6 | $1.2 \cdot 10^{10}$ |
| 7 | $3.3 \cdot 10^{11}$ |
| 8 | $7.0 \cdot 10^{12}$ |
| 9 | $1.2 \cdot 10^{14}$ |
| 10 | $1.7 \cdot 10^{15}$ |
| 11 | $1.9 \cdot 10^{16}$ |

The iteration number $k$ is recorded along the first column and the quantity of interest

$$\Delta_k = \|y_k - y_{k-1}\|_C = \max_{0 \le x_i \le 1} |y_k(x_i) - y_{k-1}(x_i)|$$

is placed along the second one. From here it is easily seen that the iteration process is divergent, thus causing abnormal termination for $k=64$.

By obvious rearranging of the parameters $\tau_k$ in reverse order in conformity with formula (42) the inherent instability of this process is more significant due to the fact that abnormal termination occurs very fastly for $k=12$ (see Table 4).

In this regard, rounding errors can be treated as possible perturbations of the right-hand side of equation (1) at every step. The iteration scheme (14) with parameters (29), (41) or (42) becomes unstable with respect to the right-hand side by exactly the same reasoning as before: the norm of the operator $S_k = E - \tau_k A$ for the transition from the $(k-1)$th iteration to the $k$th iteration may exceed 1 for negative values of $t_k$, since

$$\| S_k \| = \frac{\rho_0 (1 + |t_k|)}{1 - \rho_0 |t_k|} \quad \text{for} \quad t_k < 0 ,$$

$$\| S_k \| > 1 \qquad \text{for} \quad \rho_0 (1 + 2 |t_k|) > 1 .$$

By the same token,

$$\| S_k \| = \frac{\rho_0 (1 + t_k)}{1 + \rho_0 t_k} < 1 \quad \text{for} \quad t_k > 0 .$$

The preceding formulas need certain clarification. Since $S_k = S_k^*$, we obtain

$$\| S_k \| = \sup_{\| x \| = 1} |(S_k x, x)| .$$

By virtue of the relations $\gamma_1 E \le A \le \gamma_2 E$ we find that

$$(\tau_k \gamma_1 - 1) E \le \tau_k A - E \le (\tau_k \gamma_2 - 1) E ,$$

which are followed by

$$\tau_k \gamma_1 - 1 = -\frac{\rho_0 (1 + t_k)}{1 + \rho_0 t_k} , \qquad \tau_k \gamma_2 - 1 = \frac{\rho_0 (1 - t_k)}{1 + \rho_0 t_k}$$

upon substituting

$$\tau_k = \frac{\tau_0}{1+\rho_0\, t_k}\,, \qquad \tau_0\gamma_1 = 1-\rho_0 \quad \text{and} \quad \tau_0\gamma_2 = 1+\rho_0\,.$$

From such reasoning it seems clear that $\| S_k \| = \tau_k\gamma_2 - 1$ for $t_k < 0$ and $\| S_k \| = 1 - \tau_k\gamma_1$ for $t_k > 0$. If $t_k < 0$ and $k \geq k_0$, where $k_0$ is the minimal number for which $t_{k_0} < 0$ and $\rho_0(1+2|t_{k_0}|) > 1$, then

$$\| S_{k+1} \| > \| S_k \| > 1\,.$$

This provides enough reason to conclude that

$$\prod_{j=k_0}^{n} \| S_j \| > \| S_{k_0} \|^{n-k_0} > 1$$

and rounding errors in specifying $y_{k_0}$ will grow with increasing $k$ from $k_0$ to $n$.

Let $\xi = \dfrac{\gamma_1}{\gamma_2} \ll 1$, so that $\rho_0 = 1-2\xi+O(\xi^2)$ and $t_{k_1} = \cos\frac{2k_1-1}{2n}\pi < 0$. All this enables us to deduce that

$$\| S_{k_1} \| = \frac{\rho_0\,(1+|t_{k_1}|)}{1-\rho_0\,|t_{k_1}|} = \frac{1+|t_{k_1}|}{1-|t_{k_1}|}\left(1-2\,\xi\left(1+\frac{1}{1-|t_{k_1}|}\right)\right) + O(\xi^2)\,.$$

Under the assumptions $|t_{k_1}| > 0.5$ and $\xi < 0.01$, we might have

$$\| S_{k_1} \| = 3\,(1-6\,\xi) + O(\xi^2) > 3\,\cdot\,0.9 = 2.7\,,$$

$$\prod_{j=k_1}^{n} \| S_j \| > 2.7^{n-k_1}.$$

As a matter of fact, formulas (30)–(31) express the stability of scheme (14) with parameters (29) with respect to the initial data. Those ideas supported by the preceding calculations provide proper guidelines for the further stability analysis of computational procedures with respect to the right-hand side as well as with respect to the initial data in passing from $y_0$ to $y_k$ for any $k = 1, 2, \ldots, n$. From the general stability theory outlined in Chapter 6, Section 1 we recall that stability with respect to the right-hand side is a corollary to the uniform stability with respect to the initial data, meaning stability in the process of moving from any $y_j$ to any $y_k$ with $k > j > 0$.

**7. Re-ordering of iteration parameters.** With regard to scheme (14) one interesting problem arises in connection with re-ordering of the iteration parameters $\{\tau_k\}$ so as to minimize as much as possible the influence of rounding errors and to avoid large intermediate values dependent on $n$.

The main goal of subsequent considerations is to constitute a "stable collection" $M_n$ of parameters $\tau_1, \tau_2, \dots, \tau_n$ for which scheme (14) becomes stable and then show the way this result is used in practical implementations. We improve our chances of ordering the set

$$\mathcal{M}_n = \left\{ -\cos\beta_i, \quad \beta_i = \frac{\pi}{2n}\,\theta_n(i), \quad i = 1, 2, \dots, n \right\}$$

if a sequence of odd integers will be available such as

$$\theta_n = \{\theta_n(1), \theta_n(2), \dots, \theta_n(n)\}, \quad 1 \le \theta_n(i) \le 2n - 1,$$

$$\text{for } i = 1, 2, \dots, n.$$

With the aid of its members the parameters $\{\tau_k\}$ are calculated by the formula

$$(43) \qquad \tau_k = \frac{\tau_0}{1 + \rho_0 \sigma_k}, \quad \sigma_k = -\cos\left[\frac{\pi}{2n}\,\theta_n(k)\right], \quad k = 1, 2, \dots, n.$$

Thus, the further composition of the set consisting of $n$ numbers $\theta_n$ may be of help in achieving these aims.

The traditional tool for carrying out this work is step-by-step transitions from the sets $\theta_m$ to the sets $\theta_{2m}$ and from the sets $\theta_{2m}$ to the sets $\theta_{2m+1}$. The intention in this direction is to use the formulas

$$(44) \qquad \theta_{2m}(2i-1) = \theta_m(i), \qquad \theta_{2m}(2i) = 4m - \theta_{2m}(2i-1),$$

$$i = 1, 2, \dots, m,$$

or, what amounts to the same,

$$(45) \qquad \theta_{2m}(2i-1) = \theta_m(i), \qquad \theta_{2m}(2i) = 4m + 2 - \theta_{2m}(2i-1),$$

$$i = 1, 2, \dots, m$$

for the first operation in passing from the sets $\theta_m$ to the sets $\theta_{2m}$ with placing the extra member $\theta_{2m+1}(2m+1)$ for the second one in passing from the sets $\theta_{2m}$ to the sets $\theta_{2m+1}$:

$$(46) \qquad \theta_{2m+1}(i) = \theta_{2m}(i), \qquad i = 1, 2, \dots, 2m,$$

$$\theta_{2m+1}(2m+1) = 2m + 1.$$

When $n = 2^p$, where $p > 0$ is an integer, $\theta_n$ is found by successively applying formula (44) for $m = 1, 2, \ldots, 2^{p-1}$ with the starting point $\theta_1 = \{1\}$.

Before taking up the general case, we give below the final result of this procedure for $n = 16 = 2^4$

$$\theta_2 = \{1, 3\},$$

$$\theta_4 = \{1, 7, 3, 5\},$$

$$\theta_8 = \{1, 15, 7, 9, 3, 13, 5, 11\},$$

$$\theta_{16} = \{1, 31, 15, 17, 7, 25, 9, 23, 3, 29, 13, 19, 5, 27, 11, 21\}.$$

When moving from $\theta_m$ to $\theta_{2m}$, the member $\theta_{2m}(2i) = 4m - \theta_m(i)$ stands on the right from the member $\theta_m(i)$ by the approved rule (44).

Among other things, we may attempt an arbitrary positive integer $n > 0$ in the form

$$n = 2^{k_1} + 2^{k_2} + \cdots + 2^{k_t},$$

where $t > 0$ is an integer, $k_i \geq$ are integers such that $k_i \geq k_{i+1} + 1$, $i = 1, 2, \ldots, t-1$, and $k_t \geq 0$, constitute a new sequence of odd integers

$$(47) \qquad n_j = \sum_{i=1}^{j} 2^{k_i - k_j}, \qquad j = 1, 2, \ldots, t, \qquad n_1 = 1,$$

and then set $n_{t+1} = 2n+1$, which formally corresponds to the value $k_{t+1} = -1$. It follows from the foregoing that

$$\frac{n_{j+1} - 1}{4} = 2^{k_j - k_{j+1} - 2}\, n_j .$$

In terms of how $n_{j+1}$ may be affected, some of the possibilities of interest are:

1) $r_j = k_j - k_{j+1} \geq 2$, that is, $n_{j+1} = 2^{r_j} n_j + 1$,

2) $r_j = k_j - k_{j+1} < 2$, that is, $r_j = 1$, $n_{j+1} = 2n_j + 1$.

The algorithm of obtaining the ordered set $\theta_n$ is mostly based on the approved rules for the transition from the sets $\theta_{n_j}$ to the sets $\theta_{n_{j+1}}$. Having formed the set $\theta_{n_j}$, the differences $r_j$ are sought in two ways:

(1) if $r_j \geq 2$, then the chain of the available sets

$$\theta_{2n_j} \to \theta_{2^2 n_j} \to \cdots \to \theta_{2^{r_j} n_j}$$

does follow (44) in conformity with the values $m = n_j, 2n_j, \ldots, 2^{r_j-1}n_j$. Since $n_{j+1} = 2^{r_j}n_j + 1$, the set $\theta_{n_{j+1}}$ is described by formula (46) under the condition $2m = 2^{r_j}n_j$;

(2) if $r_j = 1$, that is, $n_j = (n_{j+1} - 1)/2$, then the set $\theta_{2n_j}$ is given by formulas (45) and the set $\theta_{n_{j+1}} = \theta_{2n_{j+1}}$ is in line with formulas (46). After that, we are moving from $\theta_{n_{j+1}}$ to $\theta_{n_{j+2}}$, etc.

Starting from $j = 1$ with $n_1 = 1$ and $\theta_{n_j} = \theta_1 = \{1\}$, we proceed to calculations by a number of different formulas: if $k_t = 0$, then $n = n_t$ is an odd number and the computations must terminate for $j = t - 1$ upon receipt of formulas (46); if $k_t > 0$, then $n$ is an even number such that $n = 2^{k_t}n_t$ and $n_{t+1} = 2n + 1$. Since $k_{t+1} = -1$, we employ $r_t \geq 2$ and formulas (44) in passing from $\theta_{n_t}$ to $\theta_n$ for the values

$$m = n_t, 2\,n_t, \ldots, 2^{k_t-1}n_t = \tfrac{1}{2}\,n\,.$$

We will elaborate on this later for several particular cases. In conclusion it should be noted that formulas (44) must be used $2^{r_j-1}$ times as opposed to a single application of formulas (45) with the accompanying formula (46). Thus, the algorithm is completely described.

**Example 1** Let $n = 90$, $n = 2^6 + 2^4 + 2^3 + 2^1$, that is, $k_1 = 6$, $k_2 = 4$, $k_3 = 3$, $k_4 = 1$; $n_1 = 1$, $n_2 = 5$, $n_3 = 11$, $n_4 = 45$; $r_1 = 2$, $r_2 = 1$, $r_3 = 2$.

The chain of describing sets is designated by the symbols

$$\theta_1 \to \theta_2 \to \theta_4 \to \theta_5 \to \tilde{\theta}_{10} \to \theta_{11} \to \theta_{22} \to \theta_{44} \to \theta_{45} \to \theta_{90}\,,$$

leading to $\theta_n = \theta_{90}$. Here $\tilde{\theta}_{2m}$ for $m = 5$ refers to the set specified by formulas (45) in contrast to the others $\theta_{2m}$. The transition to the next member of that chain can be done using formulas (44), (45) or (46) as suggested before.

**Example 2** Let $n = 25$, $n = 2^4 + 2^3 + 2^0$, that is, $k_1 = 4$, $k_2 = 3$, $k_3 = 0$; $n_1 = 1$, $n_2 = 3$, $n_3 = 25$; $r_1 = 1$, $r_2 = 3$. The sets

$$\theta_1 \to \tilde{\theta}_2 \to \theta_3 \to \theta_6 \to \theta_{12} \to \theta_{24} \to \theta_{25}$$

constitute a perfect chain for the purposes of the present section.

Summarizing, the preceding algorithm is showing a way of obtaining the ordered set of $n$ zeroes of Chebyshev's polynomial $T_n(t)$ and a stable collection in a certain sense of the parameters $\tau_1, \tau_2, \ldots, \tau_n$ specified by formula (43). In what follows the set so formed is denoted by

$$\mathcal{M}_n^* = \left\{ -\cos\left(\frac{\pi}{2\,n}\,\theta_n^*(i)\right), \quad i = 1, 2, \ldots, n \right\}.$$

Scheme (14) with new parameters

$$\tau_k = \tau_k^* = \frac{\tau_0}{1 + \rho_0\,\sigma_k^*}, \quad \sigma_k^* = -\cos\left(\frac{\pi}{2\,n}\,\theta_n^*(k)\right), \quad k = 1, 2, \ldots, n,$$

possesses the property of computational stability. Assuming that the effect of rounding errors is equivalent to possible perturbations of input data: the initial approximation, the right-hand side and the operator $A$ involved in the iteration scheme (14). Under such an approach a numerical solution $\tilde{y}_k$ of problem (14) can be treated as the exact solution of the problem

$$\frac{\tilde{y}_{k+1} - \tilde{y}_k}{\tau_{k+1}} + \tilde{A}\,\tilde{y}_k = \tilde{f}_{k+1} + \frac{1}{\tau_{k+1}}\,\tilde{\omega}_{k+1}, \quad k = 0, 1, 2, \ldots, n.$$

In this context, it should be noted that the member $\tilde{\omega}_{k+1}$ covers the error of $\dfrac{1}{\tau_{k+1}} y_k$.

When possible perturbations of the operator $A$ are neglected for one reason or other, the intermediate solutions $\tilde{y}_m$ turn out to be bounded in norm:

$$\|\,\tilde{y}_m\,\| \le \frac{1}{\xi}\,\|\,\tilde{y}_0\,\| + \left(1 + \frac{c_m}{\xi^2}\right)\left(\frac{1}{\gamma_1}\max_{1\le i\le m}\,\|\,\tilde{f}_i\,\| + \frac{1}{\xi}\max_{1\le i\le m}\,\|\,\tilde{\omega}_i\,\|\right),$$

where $c_m = 1$ if $m \neq 2^p$; $c_m = 0$ if $m = 2^p$ and the error $z_n = \tilde{y}_n - u$ of $n$ iterations satisfies the estimate

$$\|\,\tilde{y}_n - u\,\| \le q_n\,\|\,\tilde{y}_0 - u\,\| + \frac{1 - q_n}{\gamma_1}\max_{1\le i\le n}\,\|\,\tilde{f}_i - f\,\| + \frac{4}{3\sqrt{\xi}}\max_{1\le i\le n}\,\|\,\tilde{\omega}_i\,\|.$$

The derivation of these estimates espressing computational stability of iterative methods with optimal sets of Chebyshev's parameters $\{\tau_k^*\}$ is omitted in the present book. In the sequel we involve only the collection $\{\tau_k^*\}$, allowing a simpler writing of the ensuing formulas without concern of symbols "*".

The results of calculations for the example from Section 5 through the use of the explicit scheme (14) with ordered sets of parameters $\{\tau_k^*\}$ are presented in the Table 5.

**Table 5**

| | | | | | | |
|---|---|---|---|---|---|---|
| $k$ | 1 | 4 | 5 | 6 | 8 | 9 |
| $\Delta_k$ | 39.6 | 4.7 | 7.4 | 3.2 | 1.1 | 6.7 |
| $k$ | 10 | 16 | 17 | 18 | 24 | 25 |
| $\Delta_k$ | 3.2 | 0.2 | 3.1 | 1.5 | 0.1 | 0.8 |
| $k$ | 26 | 32 | 33 | 34 | 48 | 49 |
| $\Delta_k$ | 0.1 | 0.04 | 0.3 | 0.14 | $1.5 \cdot 10^{-3}$ | $1.3 \cdot 10^{-2}$ |
| $k$ | 50 | 56 | 57 | 58 | 59 | 60 |
| $\Delta_k$ | $6.7 \cdot 10^{-3}$ | $2.2 \cdot 10^{-4}$ | $1.5 \cdot 10^{-3}$ | $7.2 \cdot 10^{-4}$ | $6.5 \cdot 10^{-4}$ | $2.1 \cdot 10^{-4}$ |
| $k$ | 61 | 62 | | | | |
| $\Delta_k$ | $1.1 \cdot 10^{-4}$ | $8.7 \cdot 10^{-5}$ | | | | |

In that case we are looking for a solution to the system of equations

$$v(x_{i-1}) - 2\,v(x_i) + v(x_{i+1}) = 0\,,$$

$$x_i = ih\,, \qquad i = 1, 2, \ldots, N-1\,,$$

$$h = 1/N\,, \qquad v(0) = 1\,, \qquad v(1) = 0\,,$$

under the following conditions: $N - 1 = 19$, $\varepsilon = 10^{-4}$, $n = 64 = 2^6$. The chain in question

$$\theta_1 \to \theta_2 \to \theta_4 \to \theta_8 \to \theta_{16} \to \theta_{32} \to \theta_{64}$$

is constituted in agreement with formula (44) and it reveals a nonmonotone character of convergence of the iterations. True, it is to be shown that during the transition from the $k$th iteration ($k = 4j$) to the $(k+1)$th iteration ($k + 1 = 4j + 1$) the error of approximation $\Delta_k = \|y_k - y_{k-1}\|_C$ is being increased and, after this, is held down in passing to $k = 4j + 2$, $4j + 3$, $4j + 4$.

## 10.3 THE ALTERNATIVE–TRIANGULAR METHOD

**1. Seidel method.** As we have mentioned above, implicit schemes are rather stable in comparison with explicit ones. Seidel method, being the simplest implicit iterative one, is considered first. The object of investigation here is the system of linear algebraic equations

$$Au = f \tag{1}$$

or

$$\sum_{j=1}^{N} a_{ij} u_j = f_i, \qquad i = 1, 2, \dots, N,$$

with nonzero diagonal elements $a_{ii} \neq 0$. When this is the case, the iterative **method** ascribed to **Seidel** amounts to

$$\sum_{j=1}^{i} a_{ij} \overset{k+1}{y_j} + \sum_{j=i+1}^{N} a_{ij} \overset{k}{y_j} = f_i, \qquad a_{ii} \neq 0, \tag{2}$$

where $\overset{k}{y_j}$ is the $k$th iteration. Furthermore, starting from $i = 1$, the $(k+1)$th iteration is to be formed in line with

$$a_{11} \overset{k+1}{y_1} + \sum_{j=2}^{N} a_{1j} \overset{k}{y_j} = f_1, \qquad a_{11} \neq 0,$$

giving $\overset{k+1}{y_1}$. At the next stage, knowing $\overset{k+1}{y_1}$ and setting $i = 2$, we find $\overset{k+1}{y_2}$ from

$$a_{21} \overset{k+1}{y_1} + a_{22} \overset{k+1}{y_2} + \sum_{j=3}^{N} a_{2j} \overset{k}{y_j} = f_2, \qquad a_{22} \neq 0.$$

The matrix $A$ rearranges itself as a sum

$$A = A^- + A^+ + D, \tag{3}$$

where $A^-$ is a lower triangle matrix with zero elements on the main diagonal: $A^- = (a_{ij}^-)$, $a_{ij}^- = a_{ij}$ for $j < i$, $a_{ij}^- = 0$ for $j \geq i$; $A^+$ is an upper triangle matrix with zero elements on the main diagonal: $A^+ = (a_{ij}^+)$, $a_{ij}^+ = a_{ij}$ for $j > i$, $a_{ij}^+ = 0$ for $j \leq i$, and, finally, $D$ is a diagonal matrix: $D = (a_{ii} \delta_{ij})$. Here $\delta_{ij}$ stands, as usual, for Kronecker's delta: $\delta_{ij} = 1$ for $j = i$ and $\delta_{ij} = 0$ for $j \neq i$.

Within these notations, Seidel method can be written as follows:

$$(4) \qquad (A^- + D)\overset{k+1}{y} + A^+\overset{k}{y} = f\,, \qquad y = (y_1, y_2, \dots, y_N)\,.$$

Alternative forms of such a two-layer scheme are

$$(A^- + D)\,(\overset{k+1}{y} - \overset{k}{y}) + (A^- + D + A^+)\,\overset{k}{y} = f$$

and

$$(5) \qquad (A^- + D)\,(\overset{k+1}{y} - \overset{k}{y}) + A\,\overset{k}{y} = f\,.$$

Further identification with the preceding canonical form reveals

$$(6) \qquad B\,\frac{\overset{k+1}{y} - \overset{k}{y}}{\tau_{k+1}} + A\,\overset{k}{y} = f\,, \quad k = 0, 1, \dots, n-1\,, \quad \overset{0}{y} \in H\,,$$

with $B = A^- + D$ and $\tau_k = 1$. This scheme is certainly implicit. The matrix $B$ concerned is triangular and, hence, it is not symmetric, because the operator is non-self-adjoint: $B \neq B^*$.

On the basis on the model problem

$$\Lambda\, y = -\varphi\,, \quad x \in \omega_h\,, \quad y|_{\gamma_h} = 0\,,$$

arising from Section 2, Seidel method acquires the form

$$\overset{k+1}{y}_{i_1-1} + \overset{k+1}{y}_{i_2-1} - 4\,\overset{k+1}{y} + \overset{k}{y}_{i_1+1} + \overset{k}{y}_{i_2+1} = -h^2\varphi\,,$$

so that

$$(7) \qquad \overset{k+1}{y} = \frac{\overset{k+1}{y}_{i_1-1} + \overset{k+1}{y}_{i_2-1} + \overset{k}{y}_{i_1+1} + \overset{k}{y}_{i_2+1} + h^2\varphi}{4}\,.$$

In such a setting it seems reasonable to begin operations at the node $i_1 = 1$, $i_2 = 1$ in terms of known values $\overset{k+1}{y}_{i_1-1}$ and $\overset{k+1}{y}_{i_2-1}$ and the right-hand side of (7) at the boundary nodes $(0,1)$ and $(1,0)$. With knowledge of $\overset{k+1}{y}$ at the node $i_1 = 1$, $i_2 = 1$ the values $\overset{k+1}{y}$ are yet to be determined along the lower row by setting $i_2 = 2, 3, \dots$ for $i_1 = 1$ with further transition to the rows with $i_1 = 2, 3, \dots$. During the course of Seidel method it is possible to get the values $\overset{k+1}{y}$ at all grid nodes.

On account of the basic theorem proved in Section 1 of the present chapter Seidel method converges if the operator $A$ is self-adjoint and positive. More specifically, the sufficient stability condition (11) for the convergence of iterations in scheme (3′) with a non-self-adjoint operator $B$ takes the form

$$B_0 - \frac{\tau}{2} A > 0 . \tag{8}$$

Within the framework of Seidel method we thus have $B = A^- + D$, $\tau = 1$ and

$$B_0 = \tfrac{1}{2} \left[ (A^- + D) + (A^- + D)^* \right] = \tfrac{1}{2} (A^- + A^+ + 2\, D) ,$$

since $A^+ = (A^-)^*$ and $D^* = D > 0$ in light of the properties of the operator $A$: $A = A^* > 0$. With this in mind, condition (8) becomes

$$B_0 - \frac{\tau}{2} A = B_0 - \frac{1}{2} A > 0 = \frac{1}{2} D > 0 ,$$

thereby justifying that Seidel method converges with the rate of a geometric progression.

**2. The upper relaxation method.** In order to accelerate the iteration process in view, we are forced to revise Seidel method by inserting in (5) the iteration parameter $\omega$ so that

$$\left( A^- + \frac{1}{\omega} D \right) \left( \overset{k+1}{y} - \overset{k}{y} \right) + A \overset{k}{y} = f . \tag{9}$$

This method falls within the category of relaxation methods and gives rise to Seidel method in one particular case where $\omega = 1$. In the modern literature the iteration process (9) with $\omega > 1$ is known as the **upper relaxation method**.

By comparing of (9) with (6) it is easily seen that

$$B = A^- + \frac{1}{\omega} D , \qquad \tau_k = 1$$

or, what amounts to the same,

$$B = \omega\, A^- + D , \qquad \tau_k = \omega .$$

As far as a non-self-adjoint operator $B$ is concerned, the workable procedure reduces to inversion of a lower triangle matrix.

As shown above, Seidel method is quite applicable for any operator $A = A^* > 0$. However, the extra restriction $0 < \omega < 2$ is necessary for the convergence of the upper relaxation method. This is certainly true under condition (8) with a known operator $B_0$. Along these lines, it is straightforward to verify that $B = \omega A^- + D$, $\tau_k = \omega$ and

$$B_0 = \tfrac{1}{2}\left((\omega A^- + D) + (\omega A^+ + D)\right) = \tfrac{1}{2}\left(\omega A + (2-\omega) D\right).$$

This supports the view that $B_0 > 0$ for $0 < \omega < 2$ and condition (8) is fulfilled for $\omega < 2$, since

$$B_0 - \frac{\tau}{2} A = B_0 - \frac{\omega}{2} A = \frac{(2-\omega)}{2} D > 0 \quad \text{for } \omega < 2 .$$

But the convergence rate depends on the parameter $\omega$. However, there are analytical estimates for $\omega$ and the convergence rate when the subsidiary information on the spectral bounds of the operator $D^{-1}(A^- + A^+)$ is available, but their determination is some problem in itself. Just for this reason the parameter $\omega$ is so chosen as to minimize the total number of iterations. In dealing with numerous similar problems this approach is more convenient and effective.

By appeal once again to the model problem of interest it is plain to show that the upper relaxation method is a perfect tool in such matters, since the work and storage require

$$n \geq n_0(\varepsilon), \qquad n_0(\varepsilon) = O\Big(\frac{1}{h} \ln \frac{1}{\varepsilon}\Big).$$

**3. Implicit iteration schemes.** Convergence of implicit iteration schemes was the subject of investigation in Section 2 for the special case

$$\frac{y_{k+1} - y_k}{\tau_{k+1}} + A\, y_k = f, \quad y_0 \in H \text{ given}, \tag{10}$$

when the iterations $y_{k+1}$ can be immediately found by the formula

$$y_{k+1} = y_k - \tau_{k+1}(A\, y_k - f).$$

As a matter of fact, the upper relaxation method and Seidel method are nothing more than the implicit scheme (6) with $B \neq E$ incorporated. Still using the usual framework of implicit iterative methods, the value $y_{k+1}$ is determined from the equation

$$B\, y_{k+1} = F_k, \qquad F_k = B\, y_k - \tau_{k+1}(A\, y_k - f), \tag{11}$$

with a known right-hand side $F_k$. In both cases the matrix $B$ falls within the category of triangle matrices, since $B = A^- + D$ and $B = A^- + \frac{1}{\omega} D$ do arise during the course of Seidel and upper relaxation methods, respectively. Due to this property every new iteration $y_{k+1}$ can be obtained by making a minimal number of iterations. For the model problem concerned, the number of the necessary iterations is equivalent to the total number $N$ of the grid nodes.

Proper guidelines for the well-founded choice of the operator $B$ are provided by the following requirements:

- a minimal number of iterations;
- the operator $B$ is economical, the work and storage require $O(N)$. The meaning of the latter property for difference second-order equations of the elliptic type is that the equation $By_{k+1} = F_k$ with a known right-hand side $F_k$ must be solved in a minimal number of operations.

The main result in Section 2 regarding the optimal set of parameters $\{\tau_k\}$ can be generalized directly for implicit schemes with $B \neq E$ as follows:

$$B\,\frac{y_{k+1} - y_k}{\tau_{k+1}} + A\,y_k = f\,, \quad k = 0, 1, \ldots, n-1\,, \quad y_0 \in H \text{ given,}$$

under the conditions

$$(12) \qquad B = B^* > 0\,, \quad A = A^* > 0\,, \quad \gamma_1 B \leq A \leq \gamma_2 B\,, \quad \gamma_1 > 0\,,$$

whose key role is to specify the primary family (6) of iterative methods at the very beginning. But this is not the case for the upper relaxation method and Seidel method both in connection with the property that the operator $B$ is non-self-adjoint: $B \neq B^*$.

In mastering the difficulties involved, the intention is to use the homogeneous equation related to the correction $w_k = B^{-1}(A\,y_k - f)$:

$$(13) \qquad B\,\frac{w_{k+1} - w_k}{\tau_{k+1}} + A\,w_k = 0\,, \quad k = 0, 1, \ldots, n-1\,,$$

$$w_0 = B^{-1}(A\,y_0 - f) \in H \text{ given.}$$

The preceding scheme is equivalent to the explicit scheme

$$(14) \qquad \frac{x_{k+1} - x_k}{\tau_{k+1}} + C\,x_k = 0\,, \quad k = 0, 1, \ldots, n-1\,, \quad x_0 \in H \text{ given,}$$

where $x_k = B^{1/2} w_k$ and $C = B^{-1/2} A B^{-1/2}$.

Indeed, because the operator $B$ is self-adjoint and positive: $B = B^* > 0$, the square root $B^{1/2}$ does exist with the property

$$(B^{1/2})^* = B^{1/2} > 0 .$$

By applying the operator $B^{-1/2}$ to equation (13) and inserting $w_k = B^{-1/2} x_k$ we arrive at scheme (14). Arguing in reverse order we might obtain the same result without any difficulties.

**Lemma 1** *Given operators*

$$A = A^* > 0 , \qquad B = B^* > 0 , \qquad C = B^{-1/2} A B^{-1/2} ,$$

*the operator inequalities*

$$\gamma_1 B \le A \le \gamma_2 B , \qquad \gamma_1 E \le C \le \gamma_2 E , \qquad \gamma_1 > 0 , \tag{15}$$

*are equivalent.*

**Proof** A reasonable form of the functional at hand is

$$\begin{aligned} I &= ((A - \gamma\, B)\, y, y) = (A\, y, y) - \gamma\, (B\, y, y) \\ &= (AB^{-1/2}(B^{1/2} y), B^{-1/2}(B^{1/2} y)) - \gamma\, (B^{1/2} y, B^{1/2} y) \\ &= (C\, x, x) - \gamma\, (x, x) = ((C - \gamma\, E)\, x, x) , \end{aligned}$$

where $x = B^{1/2} y$ is an arbitrary element of the space $H$ due to the arbitrariness in the choice of the element $y \in H$. Because of this, the equality

$$I = ((A - \gamma\, B)\, y, y) = ((C - \gamma\, E)\, x, x) \tag{16}$$

implies that the operators $A - \gamma\, B$ and $C - \gamma\, E$ are of the same sign. For the sake of simplicity we may assume $A - \gamma_1 B \ge 0$ and then insert $\gamma = \gamma_1$ in (16), leaving us with the inequality $I = ((C - \gamma_1 E)\, x, x) \ge 0$. This means that $C \ge \gamma_1 E$, etc. Thus, the assertion of the lemma is completely proved.

From what has been said above another conclusion can be drawn in this direction: the possible applications of the implicit scheme (6) in solving the original equation $Au = f$ are equivalent to the numerical solution of the auxiliary equation $Cv = \varphi$ through the use of the explicit scheme

$$\frac{x_{k+1} - x_k}{\tau_{k+1}} + C\, x_k = \varphi , \quad k = 0, 1, \dots, n , \quad x_0 \in H \text{ given,} \tag{17}$$

if we accept $C = B^{-1/2} A B^{-1/2}$ and $\varphi = B^{-1/2} f$. Under this agreement all the results obtained in Section 2 for a particular implicit scheme can be covered by the following statement concerning explicit schemes.

**Theorem 2** *Let conditions (12) hold. Then an optimal set of parameters* $\{\tau_k\}$ *specified by formulas (27)–(29) exists and may be of assistance for solving problem (6), providing the validity of the estimate*

$$\|A\,y_n - f\|_{B^{-1}} \le q_n\,\|A\,y_0 - f\|_{B^{-1}} \tag{18}$$

*with*

$$q_n = \frac{2\,\rho_1^n}{1+\rho_1^{2n}}\,, \qquad \rho_1 = \frac{1-\sqrt{\xi}}{1+\sqrt{\xi}}\,, \qquad \xi = \frac{\gamma_1}{\gamma_2}\,. \tag{19}$$

Recall that Theorem 2 has been proved in Section 2 placing a special emphasis on one particular case of explicit schemes with the identity operator $B = E$ involved. To make our exposition more transparent, the implicit scheme transforms into the explicit scheme (17). Having stipulated the conditions $\gamma_1 E \le C \le \gamma_2 E$, the estimate

$$\|\,C\,x_n - \varphi\,\| \le q_n\,\|\,C\,x_0 - \varphi\,\| \tag{20}$$

is an immediate implication of estimate (31). The forthcoming substitutions $\varphi = B^{-1/2} f$, $B^{-1/2} x_n = y_n$ and $C = B^{-1/2} A B^{-1/2}$ may be useful when providing current manipulations:

$$\begin{aligned}
\|\,C\,x_n - \varphi\,\|^2 &= (C\,x_n - \varphi, C\,x_n - \varphi)\\
&= \Big(B^{-1/2}(AB^{-1/2}\,x_n - f), B^{-1/2}(AB^{-1/2}\,x_n - f)\Big)\\
&= \Big(B^{-1}(A\,y_n - f), A\,y_n - f\Big)\\
&= \|A\,y_n - f\|_{B^{-1}}^2\,.
\end{aligned}$$

By inserting $\|A\,y_n - f\|_{B^{-1}}$ in (20) in place of the norm $\|\,Cx_n - \varphi\,\|$ we arrive at inequality (18), thereby completing the proof of the theorem.

In concluding this discussion it is worth noting that the type of the original equation $Au = f$ and the operator $B$ have no influence on a universal method of numbering the parameters $\tau_1, \ldots, \tau_n$ that can be obtained through the use of the ordered set $\mathcal{M}_n^*$ of zeroes of Chebyshev's polynomial of degree $n$, whose description and composition were made in Section 2 of the present chapter.

**4. The alternative triangle method.** Making a substantiated choice of the operator $B$ will be justified in more detail a little latter. Recall that any product $B$ of "economical" operators is also an "economical" operator. This is certainly true for "triangle" operators $B_1$ and $B_2$ for which the operator $B = B_1 B_2$ would be "economical".

Any such operator $R = R^* > 0$ is representable by

$$R_1 + R_2 = R\,, \qquad R = R^* > 0\,, \qquad R_1^* = R_2, \tag{21}$$

where $R_1$ and $R_2$ are "triangle" operators. The associated matrix $\mathcal{R} = (r_{ij})$ is symmetric, that is, $r_{ij} = r_{ji}$. It is obvious that the appropriate matrices $\mathcal{R}_1$ and $\mathcal{R}_2$ are given by:

$$\mathcal{R}_1 = (r_{ij}^-)\,, \quad r_{ij}^- = \begin{cases} r_{ij} & \text{for } j < i\,, \\ 0 & \text{for } j > i\,, \end{cases}$$

$$r_{ii}^- = r_{ii}^+ = 0.5\, r_{ii}\,.$$

$$\mathcal{R}_2 = (r_{ij}^+)\,, \quad r_{ij}^+ = \begin{cases} 0 & \text{for } j < i\,, \\ r_{ij} & \text{for } j > i\,, \end{cases}$$

From here and the symmetry condition $r_{ij} = r_{ji}$ it follows that $\mathcal{R}_1^* = \mathcal{R}_2$.

The operator $B$ built into scheme (6) arranges itself as a product of "triangle" operators:

$$B = (E + \omega\, R_1)\,(E + \omega\, R_2)\,, \tag{22}$$

where $\omega > 0$ is a numerical parameter. One succeeds in showing that $B$ is a self-adjoint positive operator: $B = B^* > 0$. Due to these properties scheme (6) with operator (22) will belong to the primary family of schemes (12).

Indeed, it turns out that the operators

$$B_1 = E + \omega\, R_1\,, \qquad B_2 = E + \omega\, R_2$$

are adjoint and positive:

$$B_1^* = (E + \omega\, R_1)^* = E + \omega\, R_2 = B_2, \quad B_1 > E, \quad B_2 > E \text{ for } \omega > 0\,,$$

since $R_1 > 0$ and $R_2 > 0$:

$$(R\, y, y) = (R_1\, y, y) + (R_2\, y, y) = 2\,(R_1\, y, y) = 2\,(R_2\, y, y) > 0\,.$$

This implies that

$$(By, v) = (B_1 B_2\, y, v) = (y, B_2^* B_1^* v) = (y, B_1 B_2 v)\,,$$

meaning $B = B^*$. By the same token,

$$(By, y) = (B_1 B_2\, y, y) = (B_2\, y, B_2, y) = \| B_2\, y \|^2 > 0.$$

It seems clear from the equation

$$(23)\quad (E + \omega\, R_1)(E + \omega\, R_2)\, \frac{y_{k+1} - y_k}{\tau_{k+1}} + A\, y_k = f\,, \qquad k = 0, 1, \ldots, n\,,$$

$$\text{given } y_0 \in H\,,$$

that the value $y_{k+1}$ remains as yet unknown and can be found from the equation

$$(E + \omega\, R_1)(E + \omega\, R_2)\, y_{k+1} = F_k$$

with the right-hand side $F_k = By_k - \tau_{k+1}(Ay_k - f)$.

As a matter of experience, by solving successively two equations

$$(24)\qquad (E + \omega\, R_1)\, \bar{y} = F_k\,, \qquad (E + \omega\, R_2)\, y_{k+1} = \bar{y}$$

with upper and lower triangle matrices we get a final result. So, the title and origin of the **alternative-triangle method (ATM)** owe a debt to scheme (23) with operator (22) that can add interest and help in understanding of how one could progress in this direction. Before going further, we need to know the values of parameters $\gamma_1$ and $\gamma_2$.

**Lemma 2** *Let operators $R$, $R_1$ and $R_2$ be in line with the conditions $R = R_1 + R_2$, $R_2^* = R_1$ and*

$$(25)\qquad R = R^*\,, \qquad R \geq \delta\, E\,, \qquad \delta > 0\,,$$

$$(26)\qquad R_1 R_2 \leq \frac{\Delta}{4}\, R\,, \qquad \Delta > 0\,.$$

*Then the estimates are valid*

$$(27)\qquad \mathring{\gamma}_1\, B \leq R \leq \mathring{\gamma}_2\, B$$

*with*

$$\overset{\circ}{\gamma}_1 = \frac{\delta}{1 + \omega\,\delta + \frac{1}{4}\omega^2\delta\,\Delta}\,, \qquad \overset{\circ}{\gamma}_2 = \frac{1}{2\,\omega}\,, \tag{28}$$

*where an operator $B$ is specified by formula* (22). *The ratio*

$$\overset{\circ}{\xi} = \frac{\overset{\circ}{\gamma}_1(\omega)}{\overset{\circ}{\gamma}_2(\omega)}$$

*attains its maximal value for*

$$\omega = \omega_0 = \frac{2}{\sqrt{\delta\,\Delta}}\,, \tag{29}$$

*giving in that case*

$$\overset{\circ}{\xi} = \frac{2\sqrt{\eta}}{1 + \sqrt{\eta}}\,, \quad \eta = \frac{\delta}{\Delta}\,, \quad \overset{\circ}{\gamma}_1 = \frac{\delta}{2\,(1 + \sqrt{\eta})}\,, \quad \overset{\circ}{\gamma}_2 = \frac{\delta}{4\sqrt{\eta}}\,. \tag{30}$$

**Proof** Inequalities (25) and (26) together imply that

$$(R\,y, y) \geq \delta\,(y, y)\,,$$

$$(R_1 R_2\, y, y) = (R_2\, y,\, R_1^* y) = \|\, R_2\, y\,\|^2 \leq \frac{\Delta}{4}\,(R\,y, y)\,,$$

whence it follows that $\delta \leq \Delta$ and $\eta \leq 1$.

Before giving further motivations, we may attempt operator (22) in the form

$$B = E + \omega\,(R_1 + R_2) + \omega^2\, R_1\, R_2 = E + \omega\, R + \omega^2\, R_1\, R_2\,. \tag{31}$$

With the relations $R \geq \delta E$ or $E \leq \frac{1}{\delta} R$ in view, we obtain for the operator $B$ the upper estimate

$$B \leq \frac{1}{\delta} R + \omega\, R + \frac{\omega^2\,\Delta}{4}\, R = \left(\frac{1}{\delta} + \omega + \frac{\omega^2\,\Delta}{4}\right) R = \frac{1}{\overset{\circ}{\gamma}_1}\, R\,,$$

yielding the relation $R \geq \overset{\circ}{\gamma}_1 B$. Formula (31) transforms into

$$B = E - \omega\,(R_1 + R_2) + \omega^2\, R_1\, R_2 + 2\,\omega\,(R_1 + R_2)$$

$$= (E - \omega\, R_1)\,(E - \omega\, R_2) + 2\,\omega\, R\,,$$

when performing current manipulations with the operators $B$ and $R$:

$$\begin{aligned}(By,y) &= \big((E-\omega R_1)(E-\omega R_2)\,y,y\big) + 2\omega\,(Ry,y)\\ &= \big((E-\omega R_2)\,y,(E-\omega R_2)\,y\big) + 2\omega\,(Ry,y)\\ &= \|\,(E-\omega R_2)\,y\,\|^2 + 2\omega\,(Ry,y)\\ &\ge 2\omega\,(Ry,y) = \frac{1}{\overset{\circ}{\gamma}_2}\,(Ry,y)\,.\end{aligned}$$

From such reasoning it seems clear that $(Ry,y) \le \overset{\circ}{\gamma}_2\,(By,y)$.

The next step is to find a maximal value of the ratio

$$\overset{\circ}{\xi}(\omega) = \frac{\overset{\circ}{\gamma}_1}{\overset{\circ}{\gamma}_2} = \frac{2\,\omega\,\delta}{1+\omega\delta+(\omega^2\delta\,\Delta)/4}$$

by observing that

$$\frac{d\overset{\circ}{\xi}}{d\omega} = 2\,\delta\,\frac{1-\omega^2\delta\,\Delta/4}{\Big(1+\omega\delta+(\omega^2\delta\,\Delta)/4\Big)^2}\,.$$

From calculus it is known that the maximal value of $\overset{\circ}{\xi}(\omega)$ is attained for $\omega = \omega_0 = \dfrac{2}{\sqrt{\delta\Delta}}$, since $\overset{\circ}{\xi}{}''(\omega) < 0$ for $\omega = \omega_0$. Upon substituting $\omega_0 = 2\sqrt{\eta}/\delta$ into (28) we derive formulas (30).

**Theorem 2** *Let under the conditions of Lemma 2 the inequalities*

$$c_1\,R \le A \le c_2\,R\,, \qquad c_1 > 0\,, \tag{32}$$

*hold simultaneously for operators $A = A^* > 0$ and $R = R^* > 0$. Then for ATM given by formula (23) with optimal set of Chebyshev's parameters*

$$\tau_k^* = \frac{\tau_0}{1+\rho_0\sigma_k^*}\,, \quad \tau_0 = \frac{2}{\gamma_1+\gamma_2}\,, \quad \rho_0 = \frac{1-\xi}{1+\xi}\,, \quad \xi = \frac{\gamma_1}{\gamma_2}\,, \tag{33}$$

*where*

$$\gamma_1 = c_1\,\overset{\circ}{\gamma}_1\,, \quad \gamma_2 = c_2\,\overset{\circ}{\gamma}_2\,, \quad \overset{\circ}{\gamma}_1 = \frac{\delta}{2\,(1+\sqrt{\eta})}\,, \quad \overset{\circ}{\gamma}_2 = \frac{\delta}{4\sqrt{\eta}}\,,$$

$$\eta = \frac{\delta}{\Delta}\,, \qquad \sigma_k^* \in \mathcal{M}_n^*\,,$$

*estimate* (18) *is true and it suffices to perform* $n$ *iterations for the validity of the inequality*

$$\|A\, y_n - f\|_{B^{-1}} \le \varepsilon\, \|A\, y_0 - f\|_{B^{-1}},$$

*where*

$$n \ge n_0(\varepsilon), \qquad n_0(\varepsilon) = \sqrt{\frac{c_2}{c_1}}\, \frac{\ln(2/\varepsilon)}{2\sqrt{2}\,\sqrt[4]{\eta}}. \tag{34}$$

**Proof** In complete agreement with the preceding theorem with known coefficients $\gamma_1$ and $\gamma_2$, arising from the operator inequalities $\gamma_1 B \le A \le \gamma_2 B$, we deduce from (27) and (32) that

$$A \ge c_1 R \ge c_1 \overset{\circ}{\gamma}_1 B = \gamma_1 B, \quad \text{meaning} \quad \gamma_1 = c_1 \overset{\circ}{\gamma}_1,$$

$$A \le c_2 R \le c_2 \overset{\circ}{\gamma}_2 B = \gamma_2 B, \quad \text{meaning} \quad \gamma_2 = c_2 \overset{\circ}{\gamma}_2.$$

The conditionality parameter for $\omega = \omega_0$ is equal to

$$\xi = \frac{\gamma_1}{\gamma_2} = \frac{c_1}{c_2}\,\overset{\circ}{\xi} = \frac{c_1}{c_2}\,\frac{2\sqrt{\eta}}{1+\sqrt{\eta}}.$$

Theorem 2 and the results of Section 2 apply equally well to such a setting. As stated in Section 2, the condition $q_n \le \varepsilon$ is ensured by $n \ge \ln \frac{2}{\varepsilon} \Big/ (2\sqrt{\xi})$. Substituting here $\xi = \gamma_1/\gamma_2$ and taking into account that $\xi < \frac{c_1}{c_2} 2\sqrt{\eta}$, we derive the sufficient condition (34), which can easily be verified in one particular case of interest:

$$A = R = R_1 + R_2, \qquad R_2^* = R_1, \qquad c_1 = c_2 = 1.$$

The total number of iterations required in this connection is no less than

$$n \ge n_0(\varepsilon), \qquad n_0(\varepsilon) = \frac{\ln(2/\varepsilon)}{2\sqrt{2}\,\sqrt[4]{\eta}}.$$

**5. ATM for the difference Dirichlet problem.** The trace of those ideas can clearly be seen in tackling the difference Dirichlet problem associated with Poisson's equation in the rectangle $\bar{G} = \{0 \le x_\alpha \le l_\alpha,\ \alpha = 1, 2\}$:

$$\Lambda\, y = y_{\bar{x}_1 x_1} + y_{\bar{x}_2 x_2} = -f(x), \quad x \in \omega_h, \quad y|_{\gamma_h} = \mu(x). \tag{35}$$

Here the composition of the grid is

$$\bar{\omega}_h = \{x_i = (i_1 h_1, i_2 h_2),\ i_\alpha = 0, 1, 2, \ldots, N_\alpha,\ N_\alpha h_\alpha = l_\alpha,\ \alpha = 1, 2\}$$
$$= \omega_h \cup \gamma_h ,$$

where $\gamma_h$ is its boundary. In that case

$$A\,y = -\Lambda y , \quad y \in H = \overset{\circ}{\Omega} ,$$

$$R = A , \quad R_1\, y = y_{\bar{x}_1}/h_1 + y_{\bar{x}_2}/h_2 , \quad R_2\, y = -y_{x_1}/h_1 - y_{x_2}/h_2 ,$$

and instead of (35) we obtain

$$(35') \qquad A\,y = \varphi , \qquad y \in \Omega ,$$

where the right-hand side $\varphi$ differs from the function $f$ only at the near-boundary nodes. Simple algebra gives

$$(36) \qquad R \geq \delta\, E , \qquad \delta = 4\left(\frac{1}{h_1^2}\sin^2\frac{\pi\, h_1}{2\, l_1} + \frac{1}{h_2^2}\sin^2\frac{\pi\, h_2}{2\, l_2}\right),$$

$$(R_1\, R_2\, y, y) = \|\, R_2\, y\,\|^2 = \left\|\frac{y_{x_1}}{h_1} + \frac{y_{x_2}}{h_2}\right\|^2 = \left(\left(\frac{y_{x_1}}{h_1} + \frac{y_{x_2}}{h_2}\right)^2, 1\right)$$
$$\leq \left(\left(\frac{1}{h_1^2} + \frac{1}{h_2^2}\right)(y_{x_1}^2 + y_{x_2}^2), 1\right)$$
$$\leq \left(\frac{1}{h_1^2} + \frac{1}{h_2^2}\right)(R\,y, y) ,$$

implying that

$$(37) \qquad \Delta = \frac{4}{h_1^2} + \frac{4}{h_2^2} .$$

Here we should take into account that

$$(a_1\, b_1 + a_2\, b_2)^2 \leq (a_1^2 + a_2^2)\ (b_1^2 + b_2^2)$$

and

$$(R\,y, y) = \left(y_{\bar{x}_1}, y_{\bar{x}_1}\right]_1 + \left(y_{\bar{x}_2}, y_{\bar{x}_2}\right]_2 \geq \|\, y_{x_1}\,\|^2 + \|\, y_{x_2}\,\|^2$$

under the inner product structure

$$(u,v]_1 = \sum_{i_1=1}^{N_1} \sum_{i_2=1}^{N_2-1} u_{i_1 i_2} \, v_{i_1 i_2} \, h_1 \, h_2 \, .$$

The inner product $(u,v]_2$ can be introduced in a similar way.

Knowing $\delta$ and $\Delta$, it is not difficult to determine the values of $\eta$, $\gamma_1$, $\gamma_2$ and $\xi$ and, after this, to estimate the number of the iterations required in such matters.

Available data processing is a special starting procedure in the further comparison of three methods in line with established priorities by having recourse to the model problem (35) that we have set up on a square grid $h_1 = h_2 = h$ in the square $G$ with the unit sides ($l_1 = l_2 = 1$). Plain calculations show that

$$\eta = \sin^2 \frac{\pi h}{2} \, , \qquad \xi = \frac{2\sqrt{\eta}}{1+\sqrt{\eta}} \approx 2\sqrt{\eta} = 2 \sin \frac{\pi h}{2} \, .$$

For small values $\frac{\pi h}{2} \ll 1$ we might have

$$\sqrt{2} \, \sqrt[4]{\eta} \approx \sqrt{\pi h} \approx 1.77 \sqrt{h} \, ,$$

$$n_0(\varepsilon) \approx \frac{0.29}{\sqrt{h}} \ln \frac{2}{\varepsilon} = \frac{2.9}{\sqrt{h}} \quad \text{for} \quad \varepsilon = 2 e^{-10} \approx 10^{-4} \, .$$

The first criterion among others is the number of the iterations in the following three schemes:

| $h$ | SIS | SCP | ATM |
|---|---|---|---|
| 1/10 | 200 | 32 | 9 |
| 1/50 | 5000 | 160 | 21 |
| 1/100 | 20000 | 320 | 29 |

We present below the algorithm of finding the $(k+1)$th iteration from the equation

$$B \overset{k+1}{y} = (E + \omega_0 R_1)(E + \omega_0 R_2) \overset{k+1}{y} = \overset{k}{F} \, , \tag{38}$$

$$\overset{k}{F} = B\overset{k}{y} - \tau_{k+1} \, (A \overset{k}{y} - \varphi) = B\overset{k}{y} + \tau_{k+1} \, (\Lambda \overset{k}{v} + f) \, ,$$

with the supplementary conditions

$$\overset{k+1}{y} = \overset{k}{y} = 0 \text{ on } \gamma_h, \quad \overset{k}{v} = \overset{k}{y} \text{ for } x \in \omega_h, \quad \overset{k}{v} = \mu \text{ for } x \in \gamma_h .$$

A solution $\overset{k+1}{v}$ of problem (38) on $\omega_h + \gamma_h$ can be obtained through the use of such a procedure: $\overset{k+1}{v} = \overset{k+1}{y}$ on $\omega_h$ and $\overset{k+1}{v} = \mu$ on $\gamma_h$. The recovery of $\overset{k+1}{y}$ requires successive solution of the problems

$$(39) \qquad (E + \omega_0 R_1)\, \bar{y} = \overset{k}{F}, \quad (E + \omega_0 R_2)\, \overset{k+1}{y} = y, \quad x \in \omega_h,$$

where

$$(40) \qquad R_1\, y = \frac{y - y_{i_1-1}}{h_1^2} + \frac{y - y_{i_2-1}}{h_2^2}$$

$$= \left(\frac{1}{h_1^2} + \frac{1}{h_2^2}\right) y - \left(\frac{1}{h_1^2}\, y_{i_1-1} + \frac{1}{h_2^2}\, y_{i_2-1}\right),$$

$$(41) \qquad R_2\, y = -\frac{y_{i_1+1} - y}{h_1^2} - \frac{y_{i_2+1} - y}{h_2^2}$$

$$= \left(\frac{1}{h_1^2} + \frac{1}{h_2^2}\right) y - \left(\frac{1}{h_1^2}\, y_{i_1+1} + \frac{1}{h_2^2}\, y_{i_2+1}\right).$$

In preparation for this, we have at our disposal

$$y = y(i_1 h_1, i_2 h_2), \quad y_{i_1 \pm 1} = y((i_1 \pm 1) h_1, i_2 h_2), \quad y_{i_2 \pm 1} = y(i_1 h_1, (i_2 \pm 1) h_2).$$

The operator $E + \omega_0 R_1$ is specified on a three-point pattern $(i_1 h_1, i_2 h_2)$, $((i_1 - 1)\, h_1, i_2\, h_2)$, $(i_1\, h_1, ((i_2 - 1)\, h_2)$, while the operator $E + \omega_0 R_2$ – on a three-point pattern $(i_1\, h_1, i_2\, h_2)$, $((i_1 + 1)\, h_1, i_2\, h_2)$, $(i_1\, h_1, ((i_2 + 1)\, h_2)$. Upon substituting the above expressions for $R_1\, y$ and $R_2\, y$ into (39) we establish the recurrence relations which will be needed in the sequel:

$$(42) \qquad y = \frac{\varkappa_1\, y_{i_1-1} + \varkappa_2\, y_{i_2-1} + \overset{k}{F}}{1 + \varkappa_1 + \varkappa_2}, \qquad \bar{y}|_{\gamma_h} = 0,$$

$$(43) \qquad \overset{k+1}{y} = \frac{\varkappa_1\, \overset{k+1}{y}_{i_1+1} + \varkappa_2\, \overset{k+1}{y}_{i_2+1} + y}{1 + \varkappa_1 + \varkappa_2}, \qquad \overset{k+1}{y}|_{\gamma_h} = 0,$$

$$\varkappa_1 = \frac{\omega_0}{h_1^2}, \qquad \varkappa_2 = \frac{\omega_0}{h_2^2},$$

giving the values $\bar{y}$ and $\overset{k+1}{y}$ at the centre $(i_1 h_1, i_2 h_2)$ of the suitably chosen pattern.

Further development with $\bar{y}$ on the grid $\omega_h$ is due to a special choice of the near-boundary node $i_1 = 1$, $i_2 = 1$ at the left lower corner of the domain of interest so that two other nodes $((i_1 - 1) h_1, i_2 h_2)$ and $(i_1 h_1, ((i_2 - 1) h_2)$ should belong to the boundary on which the values $\bar{y}_{i_1-1}$ and $\bar{y}_{i_2-1}$ are already known. Formula (42) gives the value $\bar{y}$ with further contingency either along rows or along columns.

The directions of row account are from the left to the right: for fixed $i_2 = 1$ we are moving from $i_1 = 2, 3, \ldots$ to $i_1 = N_1 - 1$ and, after this, for fixed $i_2 = 2$ – from $i_1 = 1, 2 \ldots$ to $i_1 = N_1 - 1$. The directions of column account are bottom up. The value $\overset{k+1}{y}$ depends on $\overset{k+1}{y}_{i_1+1}$ and $\overset{k+1}{y}_{i_2+1}$, therefore account starts at the right upper corner of the domain: $i_1 = N_1 - 1$ and $i_1 = N_1 - 1$, so that two adjacent nodes should belong to the boundary on which the values $\overset{k+1}{y}_{i_1+1}$ and $\overset{k+1}{y}_{i_2+1}$ are already known. Further calculations are performed either along rows (from right to left) or along columns (from top to bottom).

All of the calculations are best conducted by recurrence formulas, whose algorithm is certainly stable and is called **"through execution" algorithm**.

Within its framework it is necessary to perform 4 operations of addition and 6 operations of multiplication at every node of the grid for determination of $\overset{k+1}{y}$ with knowledge of $\overset{k}{F}$. In giving $\overset{k}{F}$ 10 operations of addition and 10 operations of multiplication should be performed. Summarazing, it is required to carry out 14 operations of addition and 16 operations of multiplication in passing from $\overset{k}{y}$ to $\overset{k+1}{y}$.

In trying to make these numbers small enough, it is highly recommended to save two numerical sequences instead of $\overset{k}{y}$. This can be done using the algorithm

$$(44) \qquad \begin{aligned} &(E + \omega_0 R_1)\bar{\omega} = \Phi_k, \qquad \bar{\omega}|_{\gamma_h} = 0, \\ &(E + \omega_0 R_2)\overset{k}{\omega} = \bar{\omega}, \qquad \overset{k}{\omega}|_{\gamma_h} = 0, \\ &\overset{k+1}{y} = \overset{k}{y} + \tau_{k+1} \overset{k+1}{\omega}, \end{aligned}$$

where $\Phi_k = \Lambda \overset{k}{v} + f$ and $\overset{k}{v}|_{\gamma_h} = \mu$, making it possible to perform 10 operations of addition and 10 operations of multiplication at one node for determination of $\overset{k+1}{y}$. In so doing the passage from the $k$th iteration to the $(k+1)$th iteration necessitates the storage not only for the value $\overset{k}{y}(i_1 h_1, i_2 h_2)$, but also for the value $\overset{k}{w}(i_1 h_1, i_2 h_2)$.

**6. A higher-accuracy scheme in a rectangle.** In Chapter 4, Section 5, the Dirichlet problem

$$\Delta u = -f(x), \quad x \in G, \quad u = \mu(x), \quad x \in \Gamma, \tag{45}$$

was completely posed, for which the difference scheme of fourth-order accuracy was composed in the following way:

$$\Lambda' y = -\varphi(x), \quad x \in \omega_h, \quad y = \mu(x), \quad x \in \gamma_h, \tag{46}$$

where

$$\Lambda' y = \Lambda y + \frac{h_1^2 + h_2^2}{12} \Lambda_1 \Lambda_2 y,$$

$$\Lambda = \Lambda_1 + \Lambda_2, \qquad \Lambda_\alpha y = y_{\bar{x}_\alpha x_\alpha}, \qquad \alpha = 1, 2,$$

$$\varphi = f + \frac{h_1^2}{12} \Lambda_1 f + \frac{h_2^2}{12} \Lambda_2 f.$$

Let now $H = \overset{\circ}{\Omega}$ be a space of all grid functions given on the grid $\bar{\omega}_h$ and vanishing on the boundary $\gamma_h$. Having involved the operators $Ay = -\Lambda' y$ and $Ry = -\Lambda y$ specified for any $y \in H$, we deduce instead of (46) that $Ay = \tilde{\varphi}$, where $\varphi \neq \tilde{\varphi}$ only at the near-boundary nodes. Applying the estimates obtained in Chapter 4 for the operator $A$ yields

$$\frac{2}{3} R \le A \le R,$$

and reveals the coefficients to be $c_1 = \frac{2}{3}$ and $c_2 = 1$. The operator $R$ arranges itself as a sum

$$R = R_1 + R_2, \quad R_1 y = \frac{y_{\bar{x}_1}}{h_1} + \frac{y_{\bar{x}_2}}{h_2}, \quad R_2 y = -\frac{y_{x_1}}{h_1} - \frac{y_{x_2}}{h_2}.$$

Being concerned with the same quantities $\delta$, $\Delta$ and $\omega_0$ as in Section 5, we might agree to consider $\gamma_1 = \frac{2}{3} \overset{\circ}{\gamma}_1$ and $\gamma_2 = \overset{\circ}{\gamma}_2$, where $\overset{\circ}{\gamma}_1, \overset{\circ}{\gamma}_2$ were specified in Section 4:

$$\overset{\circ}{\gamma}_1 = \frac{\delta}{2(1+\sqrt{\eta})}, \qquad \overset{\circ}{\gamma}_2 = \frac{\delta}{4\sqrt{\eta}}.$$

Knowing $\gamma_1$ and $\gamma_2$, it remains to find the iteration parameters $\{\tau_k^*\}$ with a simple observation that formulas (42)–(43) are still valid with another member $\overset{k}{F}$:

$$\overset{k}{F} = \overset{k}{F}_{(38)} + \overset{k}{\Phi}\,, \qquad \overset{k}{\Phi} = -\tau_{k+1}\,\frac{h_1^2 + h_2^2}{12}\,\Lambda_1\Lambda_2\,\overset{k}{y}\,, \tag{47}$$

where $\overset{k}{F}_{(38)}$ is given by formula (38). The total number of the iterations is equal to

$$n \geq n_0(\varepsilon)\,, \quad n_0(\varepsilon) = \sqrt{\frac{3}{2}}\; n_0^*(\varepsilon)\,,$$

where $n_0^*(\varepsilon)$ is the total number of the iterations for the Dirichlet problem associated with Poisson's equation (35). The total number of the iterations required in ATM for a higher-order scheme is being increased in $\sqrt{1.5} \approx 1.22$ times as compared with a scheme of second-order accuracy.

In the two-dimensional case the iterative alternating direction method or the direct decomposition method turns out to be more economical, but for the multidimensional Dirichlet problem ATM is the most economical one among other available methods. This advantage is stipulated by the special structure of the operator $\Lambda'$ (see Chapter 4, Section 5):

$$\Lambda' y = \sum_{\alpha=1}^{p} \Lambda_\alpha \prod_{\beta=1,\,\beta\neq\alpha}^{p} (E + \varkappa_\beta \Lambda_\beta)\,, \quad \varkappa_\beta = \frac{h_\beta^2}{12}\,, \quad \Lambda_\alpha\, y = y_{\bar{x}_\alpha x_\alpha}\,, \tag{48}$$

assuring the operator inequalities

$$\left(\frac{2}{3}\right)^{p-1} R \leq A \leq R\,, \tag{49}$$

where

$$A\,y = -\Lambda' y\,, \quad R\,y = -\Lambda y = -\sum_{\alpha=1}^{p} \Lambda_\alpha y\,, \quad y \in H\,,$$

$$n_0(\varepsilon) = \left(\frac{3}{2}\right)^{(p-1)/2} n_0^*(\varepsilon)\,,$$

and $n_0^*(\varepsilon)$ is the total number of the iterations required in connection with solving the equation $Ry = f$, not depending, in fact, on the number of observations $p$.

**7. Difference schemes for elliptic equations of general form.** As we have mentioned above, the applications of ATM in the numerical solution of an operator equation of the first kind consist of several steps:

- determination of the operator $R$;
- design of special "triangle" operators $R_1$ and $R_2$;
- calculations of the equivalence constants $c_1$ and $c_2$ and $\delta$, $\Delta$;
- selection of the number $n = n(\varepsilon)$ of the iterations and a set of parameters $\{\tau_k^*\}$.

We begin our exposition with a discussion of examples that make it possible to draw fairly accurate outlines of the possible theory regarding these questions and with a listing of the basic results together with the development desired for them. Common practice involves the Laplace operator as the operator $R$ in the case of difference elliptic operators $A$. The present section is devoted to rather complicated difference problems of the elliptic type. Here and below it is supposed that the domain of interest is a $p$-dimensional parallelepiped $\bar{G} = \{0 \le x_\alpha \le l_\alpha,\ \alpha = 1, 2, \dots, p\}$ with the boundary $\Gamma$ (a rectangle for the case $p = 2$), on which the boundary condition of the first kind is imposed:

$$u|_\Gamma = \mu(x)\,.$$

In what follows it is required to find a continuous in $\bar{G}$ solution of the Dirichlet problem

$$Lu = -f(x)\,, \quad x \in G\,, \quad u|_\Gamma = \mu(x)\,, \tag{50}$$

with a second-order elliptic operator $L$ involved.

With this aim, we introduce in the domain $\bar{G}$ the usual grid

$$\bar{\omega}_h = \{x_i = (i_1 h_1, i_2 h_2, \dots, i_p h_p) \in \bar{G},\ i_\alpha = 0, 1, 2, \dots, N_\alpha,$$
$$h_\alpha N_\alpha = l_\alpha,\ \alpha = 1, 2, \dots, p\}$$

with the boundary $\gamma_h$, so that $\bar{\omega}_h = \omega_h \cup \gamma_h$. As usual, an inner product in the space $H = \overset{\circ}{\Omega}$ of all grid functions defined on $\bar{\omega}_h$ and vanishing on the boundary $\gamma_h$ is taken to be

$$(y, v) = \sum_{\omega_h} y(x)\, v(x)\, h_1 h_2 \cdots h_p\,.$$

It seems worthwhile giving several particular cases.

a) *An equation of the elliptic type with mixed derivatives.* In that case we are agree to consider

$$(51) \qquad L\,u = \sum_{\alpha,\beta=1}^{p} L_{\alpha\beta}u\,, \quad L_{\alpha\beta}u = \frac{\partial}{\partial x_\alpha}\Big(k_{\alpha\beta}(x)\frac{\partial u}{\partial x_\beta}\Big)\,,$$

$$(52) \qquad c_1\sum_{\alpha=1}^{p}\xi_\alpha^2 \le \sum_{\alpha,\beta=1}^{p} k_{\alpha\beta}(x)\,\xi_\alpha\,\xi_\beta \le c_2\sum_{\alpha=1}^{p}\xi_\alpha^2\,,$$

where $\xi = (\xi_1, \xi_2, \dots, \xi_p)$ is an arbitrary vector and $c_1 > 0$ and $c_2 > 0$ are constants.

In dealing with difference operators on the grid $\omega_h$ at hand such as

$$(53) \qquad \Lambda y = \sum_{\alpha,\beta=1}^{p} \Lambda_{\alpha\beta}\,y\,, \quad \Lambda_{\alpha\beta}\,y = \frac{1}{2}\Big[\big(k_{\alpha\beta}\,y_{\bar{x}_\beta}\big)_{x_\alpha} + \big(k_{\alpha\beta}\,y_{\bar{x}_\alpha}\big)_{x_\beta}\Big]\,,$$

we may set up in conformity with problem (50) the difference Dirichlet problem

$$(54) \qquad \Lambda\,y = -\varphi(x)\,, \quad x\in\omega_h, \quad y = \mu(x)\,, \quad x\in\gamma_h\,,$$

and follow established practice: introduce in the space $H = \overset{\circ}{\Omega}$ the operators $A$ and $R$ acting in accordance with the rules

$$A\,y = -\Lambda\,y\,, \qquad R\,y = -\overset{\circ}{\Lambda}\,y = -\sum_{\alpha=1}^{p} y_{\bar{x}_\alpha x_\alpha}\,,$$

where $\overset{\circ}{\Lambda}$ is a difference $(2p+1)$-point Laplace operator, and accept instead of (54) the governing equation $A\,y = \tilde{\varphi}$, where $\tilde{\varphi} \ne \varphi$ only at the near-boundary nodes. Recall that

$$(55) \qquad c_1 R \le A \le c_2 R\,,$$

where $c_1$ and $c_2$ are constants arising from the ellipticity condition (52).

By analogy with Section 5 the operators $R_1$ and $R_2$ are specified by the formulas

$$R_1\,y = \sum_{\alpha=1}^{p}\frac{y_{\bar{x}_\alpha}}{h_\alpha}\,, \qquad R_2\,y = -\sum_{\alpha=1}^{p}\frac{y_{x_\alpha}}{h_\alpha}\,, \qquad R_1 + R_2 = R\,.$$

The constants $\delta$ and $\Delta$ remain as yet unknown. Since $\delta$ is the smallest eigenvalue, it is not difficult to show that

$$\delta = \sum_{\alpha=1}^{p} \frac{4}{h_\alpha^2} \sin^2 \frac{\pi h_\alpha}{2 l_\alpha} .$$

By virtue of the relations

$$\| R_2 y \|^2 = \left\| \sum_{\alpha=1}^{p} \frac{y_{x_\alpha}}{h_\alpha} \right\|^2 \le \sum_{\alpha=1}^{p} \frac{1}{h_\alpha^2} \sum_{\alpha=1}^{p} \| y_{x_\alpha} \|^2 \le \sum_{\alpha=1}^{p} \frac{1}{h_\alpha^2} (R y, y)$$

plain calculations permit one to conclude that the constant $\Delta$ is equal to $\sum_{\alpha=1}^{p} \frac{4}{h_\alpha^2}$. With knowledge of $\delta$, $\Delta$, $c_1$ and $c_2$ simple algebra gives $\omega$, $n(\varepsilon)$ and $\{\tau_k^*\}$, which constitute what is needed in the applications of ATM. The total number of the iterations

$$n(\varepsilon) \ge n_0(\varepsilon) = \sqrt{\frac{c_2}{c_1}} \frac{\ln(2/\varepsilon)}{2\sqrt{2}\sqrt[4]{\eta}} , \qquad \eta = \frac{\delta}{\Delta} , \tag{56}$$

is proportional to $\sqrt{\frac{c_2}{c_1}}$.

In the case of a cubic grid ($h_1 = h_2 = \cdots = h_p = h$) in a unit $p$-dimensional cube we might have

$$\delta = \frac{4p}{h^2} \sin^2 \frac{\pi h}{2} , \qquad \Delta = \frac{4p}{h^2} , \qquad \eta = \sin^2 \frac{\pi h}{2} ,$$

thereby justifying that the total number of iterations is independent of the number of observations.

The iteration $\overset{k+1}{y}$ is recovered from the equations

$$B_1 \bar{y} = (E + \omega R_1) \bar{y} = \overset{k}{F} , \quad B_2 \overset{k+1}{y} = (E + \omega R_2) \overset{k+1}{y} = \bar{y} , \tag{57}$$

$$\overset{k}{F} = B\overset{k}{y} - \tau_{k+1}(A\overset{k}{y} - \tilde{\varphi}) = B\overset{k}{y} + \tau_{k+1}(\Lambda\overset{k}{v} + \varphi) . \tag{58}$$

A solution of problem (57) is given by $\overset{k+1}{v} = \overset{k+1}{y}$ for $x \in \omega_h$ and by $\overset{k}{v} = \mu$ for $x \in \gamma_h$, so that $\overset{k+1}{v} - y = \overset{k+1}{y} - y$. The preceding difference operators $B_1$ and $B_2$ can be written as

$$B_1 y = y + \omega R_1 y = \left( 1 + \omega \sum_{\alpha=1}^{p} \frac{1}{h_\alpha^2} \right) y - \omega \sum_{\alpha=1}^{p} \frac{1}{h_\alpha^2} y^{(-1_\alpha)} , \tag{59}$$

$$B_2 y = y + \omega R_2 y = \left( 1 + \omega \sum_{\alpha=1}^{p} \frac{1}{h_\alpha^2} \right) y - \omega \sum_{\alpha=1}^{p} \frac{1}{h_\alpha^2} y^{(+1_\alpha)} . \tag{60}$$

The calculations start from the corner $i_1 = 1, i_2 = 1, \ldots, i_p = 1$, of the parallelepiped of interest for which all of the adjacent nodes $x^{(-1_\alpha)} \in \gamma_h$ fall into the boundary ones at which the values $y^{(-1_\alpha)} = \mu$ are already known. With knowledge of $\bar{y}$ at that node we fix $i_2 = 1, \ldots, i_p = 1$ and change $i_1 = 1, 2, \ldots, N_1 - 1$. After that, continuing this process for $i_1 = 1, 2, \ldots, N_1 - 1$ and for fixed $i_2 = 2$, etc. we find at all the nodes $x \in \omega_h$ that

$$\bar{y} = \left( \omega \sum_{\alpha=1}^{p} \frac{1}{h_\alpha^2} \, \bar{y}^{(-1_\alpha)} + \overset{k}{F} \right) \left( 1 + \frac{\omega \, \Delta}{4} \right)^{-1} .$$

Along these lines, the starting nodes $i_\alpha = N_\alpha - 1$, $\alpha = 1, 2, \ldots, p$, suit us perfectly for determination of $\overset{k+1}{y}$ on the grid $\omega_h$ by the rule

$$\overset{k+1}{y} = \left( \omega \sum_{\alpha=1}^{p} \frac{1}{h_\alpha^2} \, \overset{k+1}{y}{}^{(+1_\alpha)} + \bar{y} \right) \left( 1 + \frac{\omega \, \Delta}{4} \right)^{-1} .$$

b) *A system of elliptic equations.* Let $\mathbf{u} = (u^1, u^2, \ldots, u^{m_0})$ be a vector and let a block $p \times p$-matrix $k = (k_{\alpha\beta}^{sm})$ with blocks of size $m_0 \times m_0$ will be so chosen as to obtain some suitable matrix $k_{\alpha\beta} = (k_{\alpha\beta}^{sm})$ of size $m_0 \times m_0$ for later use. The Dirichlet problem for a system of equations is first considered in the parallelepiped $G$:

$$(61) \qquad L u^s = -f^s , \quad x \in G , \quad u^s = \mu^s , \quad x \in \Gamma , \quad s = 1, 2, \ldots, m_0 ,$$

where

$$(62) \qquad L u^s = \sum_{\alpha,\beta=1}^{p} \sum_{m=1}^{m_0} \frac{\partial}{\partial x_\alpha} \left( k_{\alpha\beta}^{sm} \, \frac{\partial u^m}{\partial x_\beta} \right) .$$

In such a setting the condition of ellipticity becomes

$$(63) \qquad c_1 \sum_{\alpha=1}^{p} \sum_{s=1}^{m_0} (\xi_\alpha^s)^2 \le \sum_{\alpha,\beta=1}^{p} \sum_{s,m=1}^{m_0} k_{\alpha\beta}^{sm}(x) \, \xi_\alpha^s \, \xi_\beta^m \le c_2 \sum_{\alpha=1}^{p} \sum_{s=1}^{m_0} (\xi_\alpha^s)^2 ,$$

where $\boldsymbol{\xi}_\alpha = (\xi_\alpha^1, \xi_\alpha^2, \ldots, \xi_\alpha^{m_0})$, $\alpha = 1, 2, \ldots, p$, are arbitrary vectors and $c_1 > 0$, $c_2 > 0$ are constants. A reasonable form of the difference operator is

$$(64) \qquad \Lambda y^s = \sum_{\alpha,\beta=1}^{p} \sum_{m=1}^{m_0} \Lambda_{\alpha\beta}^{sm} \, y^m$$

with

$$\Lambda_{\alpha\beta}^{sm}\, y^m = \frac{1}{2}\left[\left(k_{\alpha\beta}^{sm}\, y_{\bar{x}_\beta}^m\right)_{x_\alpha} + \left(k_{\alpha\beta}^{sm}\, y_{x_\beta}^m\right)_{\bar{x}_\alpha}\right], \tag{65}$$

making it possible to set up the difference Dirichlet problem

$$\Lambda\, y^s = -\varphi^s\,, \quad x \in \omega_h\,, \quad y^s = \mu^s\,, \quad x \in \gamma_h\,, \tag{66}$$

in the space $H = \overset{\circ}{\Omega}$ of all grid functions under the inner product structure

$$(\mathbf{y}, \mathbf{v}) = \sum_{s=1}^{m_0} (y^s, v^s)\,, \quad (y^s, v^s) = \sum_{x\in\omega_h} y^s(x)\, v^s(x)\, h_1 h_2 \cdots h_p\,,$$

$$\mathbf{y} = (y^1, y^2, \ldots, y^s, \ldots, y^{m_0})\,, \quad \mathbf{v} = (v^1, v^2, \ldots, v^s, \ldots, v^{m_0})\,,$$

$$y^s \in H\,, \quad v^s \in H\,, \quad s = 1, 2, \ldots, m_0\,.$$

Being concerned with the operator $Ay^s = -\Lambda y^s$ and the regularizer $Ry^s = -\overset{\circ}{\Lambda}\, y^s = -\sum_{\alpha=1}^{p} y_{\bar{x}_\alpha x_\alpha}^s$ in the space $H$, where $\overset{\circ}{\Lambda}$ is a $(2p+1)$-point difference Laplace operator, we rely in the further derivation on the Green formula and condition (63), whose combination gives the operator inequalities (32). Having involved the same operator $R$ as was done in problem (53)–(54), we obtain the constant $\omega$ and the operator $B$ in terms of known members $\delta$ and $\Delta$. Just for this reason the same algorithm as in a) is workable for determination of the $(k+1)$th iteration for either of the components $\overset{k+1}{y^s}$ and so it is omitted here.

c) *A system of equations in elasticity theory.* The system of Lame's equations arose from the stationary elasticity theory:

$$L\,\mathbf{u} = \mu\,\Delta\,\mathbf{u} + (\lambda + \mu)\operatorname{grad}\operatorname{div}\mathbf{u} = -\mathbf{f}(x) \tag{67}$$

with vectors $\mathbf{u} = (u^1, u^2, \ldots, u^p)$ and $\mathbf{f} = (f^1, f^2, \ldots, f^p)$ and Lame's constants $\lambda > 0$ and $\mu > 0$. We may attempt the preceding system in the form

$$\mu \sum_{\alpha=1}^{p} \frac{\partial^2 u^s}{\partial x_\alpha^2} + (\lambda+\mu) \sum_{\beta=1}^{p} \frac{\partial^2 u^\beta}{\partial x_\beta \partial x_s} = -f^s\,, \quad s = 1, 2, \ldots, p\,. \tag{68}$$

Further comparison of this with (62) allows us to deduce that

$$k_{\alpha\beta}^{sm} = \mu\,\delta_{\alpha\beta}\,\delta_{sm} + (\lambda+\mu)\,\delta_{\alpha s}\,\delta_{\beta m}\,, \tag{69}$$

where $\delta_{ij}$ is, as usual, Kronecker's delta.

We learn from Chapter 9, Section 2 that the constants $c_1$ and $c_2$ involved in inequalities (63) are equal to

$$c_1 = \mu\,, \qquad c_2 = \lambda + 2\,\mu\,.$$

With these, the difference Dirichlet problem associated with (68) is described by

$$\text{(70)} \qquad \Lambda^s\, y^s = -f^s\,, \quad x \in \omega_h\,, \quad y^s = \mu^s\,, \quad x \in \gamma_h\,, \quad s = 1, 2, \ldots, p\,,$$

where

$$\Lambda^s\, y^s = \mu \sum_{\alpha=1}^{p} \Lambda_\alpha y^s + (\lambda + \mu) \sum_{\beta=1}^{p} \Lambda_{\beta s} y^\beta\,,$$

$$\Lambda_\alpha\, y = y_{\bar{x}_\alpha x_\alpha}\,, \quad \Lambda_{\beta s}\, y = \tfrac{1}{2}\left( y_{\bar{x}_\beta x_s} + y_{x_\beta \bar{x}_s} \right).$$

The above framework necessitates specifying for any $y^s \in H$ the operator $Ay^s = -\Lambda^s y^s$ and the regularizer $Ry^s = -\sum_{\alpha=1}^{p} \Lambda_\alpha y^s$, thereby clarifying that the same operator $R$ is adopted for these purposes as before.

By Green's formula it is straightforward to verify the inequalities $c_1 R \le A \le c_2 R$ with constants $c_1 = \mu$ and $c_2 = \lambda + 2\mu$ incorporated. By exactly the same reasoning as in the preceding examples ATM requires $n_0(\varepsilon)$ iterations, where

$$\text{(71)} \qquad n_0(\varepsilon) = \sqrt{\frac{\lambda + 2\,\mu}{\mu}}\; n_0^*(\varepsilon)$$

and $n_0^*(\varepsilon)$ is the total number of the necessary iterations in solving the Laplace equation.

**8. ATM for solving grid elliptic equations in an arbitrary complex domain.** Two lines of research in subsequent considerations of such equations are evident in available publications in this area over recent years. No much is known in the case of an arbitrary complex domain. The first one is connected with nonequidistant grids, while the second one deals with some modifications of available methods. Even a constant step in each direction does not allow to overcome the difficulties in the near-boundary zones as a result of emerging non-uniform grids.

It turns out that the convergence of the aforementioned methods become worse on account of widely varying bounds of the spectra of difference operators.

Current explanations need certain clarification by having recourse to one simplest example devoted to the governing equation

$$u'' = -f(x)\,, \quad 0 < x < 1\,, \quad u(0) = u(1) = 0\,,$$

and associated three-point scheme

$$\Lambda\, y = -f(x)\,, \quad x \in \omega_h\,, \quad y(0) = 0\,, \quad y(1) = 0\,, \tag{72}$$

which is constructed on the grid

$$\omega_h = \{x_0 = 0,\ x_1 = h_1, x_2 = h_1 + h, x_3 = 1\}\,,$$

so that at two inner points

$$\Lambda\, y_1 = \frac{1}{h}\left(\frac{y_2 - y_1}{h} - \frac{y_1}{h_1}\right), \qquad \Lambda\, y_2 = \frac{-2\, y_2 + y_1}{h^2}\,,$$

since $y_0 = y_3 = 0$. For clarity only, we take $h_1 < h$ and $2h + h_1 = 1$.

The next step is to find the eigenvalues of the operator $\Lambda$. By definition, $\Lambda y + \lambda y = 0$ or, what amounts to the same,

$$\frac{1}{h}\left(\frac{y_2 - y_1}{h} - \frac{y_1}{h_1}\right) + \lambda\, y_1 = 0\,, \qquad \frac{-2\, y_2 + y_1}{h^2} + \lambda\, y_2 = 0\,.$$

Having completed the elimination of $y_2$, we derive the quadratic equation related to $\mu = \lambda h^2$:

$$t\, \mu^2 - (1 + 3t)\, \mu + 2 + t = 0\,, \qquad t = h_1/h\,, \tag{73}$$

with the roots

$$\mu_{(1,2)} = \frac{1 + 3t \pm \sqrt{(1 + 3t)^2 - 4t\,(2 + t)}}{2\,t}\,, \qquad \lambda_{(1,2)} = \frac{\mu_{(1,2)}}{h^2}$$

and the characteristic relation between $\lambda_{(1)} = \lambda_{\min} = \delta$ and $\lambda_{(2)} = \lambda_{\max} = \Delta$:

$$\eta = \frac{\delta}{\Delta} = \frac{2\,t\,(2 + t)}{(1 + 3t)^2 - 2t\,(2 + t) + (1 + 3t)\,\sqrt{(1 + 3t)^2 - 4t\,(2 + t)}}\,.$$

Whence it follows that for $t = 1$, that is, on an equidistant grid $\eta = 1/3$, while $\eta \approx 2t = 2h_1/h$ for small ratios $t = h_1/h \ll 1$; making worse the conditionality of the system (72) of equations along with decreasing the ratio $h_1/h$.

It seems clear that in solving the system (72) the number of the iterations within the framework of the explicit scheme with optimal set of Chebyshev's parameters or of the simple iteration scheme is proportional to $1/\sqrt{\eta}$ or $1/\eta$, thus causing an enormous growth as $h_1 \to 0$.

The generalized problem on eigenvalues of the same operator $\Lambda$ is of the form

$$\Lambda\, y + \lambda\, D\, y = 0\,,$$

where $D$ is a diagonal matrix such that $Dy = d(x)y$, $d(x) > 0$. In the case of interest

$$D = \begin{pmatrix} d_1 & 0 \\ 0 & d_2 \end{pmatrix}$$

and the problem of determining $\lambda$ amounts to

$$\frac{1}{h}\left(\frac{y_2 - y_1}{h} - \frac{y_1}{h_1}\right) + \lambda\, d_1 y_1 = 0\,, \qquad -\frac{2\, y_2 + y_1}{h^2} + \lambda\, d_2\, y_2 = 0\,.$$

Instead of (73) we must solve the quadratic equation

$$d_1\, d_2\, t\, \mu^2 - \left[(1+t)\, d_2 + 2t\, d_1\right] \mu + 2 + t = 0\,.$$

For the choice $d_1 = 1/t$ and $d_2 = 1$ we deduce from the foregoing that $\eta = 1/(2+t)$ and $\eta \approx 0.5$ when $t \ll 1$, that is, $\eta$ remains finite as $h_1 \to 0$. From such reasoning it seems clear that the scheme

$$D\, \frac{y_{k+1} - y_k}{\tau_{k+1}} = \Lambda\, y_k + f \tag{74}$$

offers more advantages in comparison with the explicit scheme.

A similar situation exists during the course of ATM for solving elliptic equations on nonequidistant grids or in arbitrary complex domains, giving rise to obvious modifications of ATM with the intervention of the operator $D = D^* > 0$ built into the structure of the operator $B$. Making a substantiated choice of the operator $D$ is stipulated by economy reasoning for every iteration as well as by a minimal number of iterations. Let $D = D^* > 0$ be an arbitrary operator and the operator $A = A^* > 0$ from the equation $Au = f$ be a sum of mutually adjoint operators $A_1$ and $A_2$:

$$A = A^* > 0\,, \quad A = A_1 + A_2\,, \quad A_1^* = A_2\,, \tag{75}$$

by means of which the factorized operator $B$ is selected by the rule

$$B = (D + \omega A_1)\, D^{-1}(D + \omega A_2)\,, \tag{76}$$

where $\omega > 0$ is the iteration parameter. The operator $B$ so defined is self-adjoint and positive. Once supplemented with the conditions

$$A \geq \delta D \quad \text{or} \quad (Ax, x) \geq \delta\,(Dx, x)\,, \quad \delta > 0\,, \tag{77}$$

$$A_1 D^{-1} A_2 \leq \frac{\Delta}{4} A \quad \text{or} \quad (D^{-1} A_2\, x, A_2\, x) \leq \frac{\Delta}{4}\,(Ax, x)\,, \quad \Delta > 0\,, \tag{78}$$

which are valid for all $x \in H$, Theorem 2 and formulas (27)–(30) continue to hold with the operator $A$ standing in place of the operator $R$, so there is no need to rewrite them once again in a common setting.

Of special interest is the well-founded choice of the operator $D$ under the agreement that the appropriate matrix $D$ is diagonal:

$$Dy = d(x)\, y\,, \qquad d(x) > 0\,. \tag{79}$$

Because of this form, the member $d(x)$ is so chosen as to maximize the ratio $\eta = \delta/\Delta$ in the modified alternative-triangular method (MATM) with reasonable efficiency.

Adopting those ideas, some progress has been achieved by means of MATM in tackling the Dirichlet problem in an arbitrary complex domain $G$ with the boundary $\Gamma$ for an elliptic equation with variable coefficients:

$$Lu = \frac{\partial}{\partial x_1}\left(k_1(x)\,\frac{\partial u}{\partial x_1}\right) + \frac{\partial}{\partial x_2}\left(k_2(x)\,\frac{\partial u}{\partial x_2}\right) = -f(x)\,, \tag{80}$$

$$x = (x_1, x_2) \in G\,, \qquad u(x) = \mu(x)\,, \qquad x \in \Gamma\,,$$

$$k_\alpha(x) \geq c_1 > 0\,, \qquad \alpha = 1, 2\,.$$

The usual assumptions are made here saying that the boundary $\Gamma$ is smooth enough and the intersection of the domain $G$ and a straight line passing through any point $x \in G$ and in parallel to the axes $Ox_\alpha$, $\alpha = 1, 2$, consists of a unique interval. The latter should not confine generality.

The design of a difference scheme for problem (80) is mostly based on a nonequidistant grid $\bar{\omega}_h$ (generally speaking, nonequidistant everywhere, not only near the boundary $\Gamma$). When drawing up the family of straight lines $x_\alpha = x_\alpha^{(i_\alpha)}$, $i_\alpha = 0, \pm 1, \pm 2, \dots$, $\alpha = 1, 2$, the points $x_i = \left(x_1^{(i_1)}, x_2^{(i_2)}\right)$ constitute the basic pattern $R_2^h$ in the plane $(x_1, x_2)$. Any such point $x_i$ belonging to $G$ is called an **inner node** of the grid, the total collection of which is denoted by $\omega_h$, that is, $\omega_h = \{x_i \in G \cap R_2^h\}$.

The intersection of the domain $G$ and any straight line passing through a point $x \in \omega_h$ and in parallel to the axis $Ox_\alpha$ consists of the interval $\Delta_\alpha(x_i)$. The endpoints of this interval are called **boundary nodes** in the $x_\alpha$-direction. A set of all boundary nodes in the $x_\alpha$-direction is designated by $\gamma_a$. The boundary $\gamma_h$ of the grid $\omega_h$ is just the union $\gamma_h = \gamma_1 \cup \gamma_2$, making it possible to write down $\bar{\omega}_h = \omega_h \cup \gamma_h$.

Furthermore, let $\omega_\alpha(x_\beta)$, $\beta = 3 - \alpha$, $\alpha = 1, 2$, be a set of nodal points from the interval $\Delta_\alpha$; let $\omega_\alpha^+(x_\beta)$ be a set containing $\omega_\alpha$ and the right endpoint of the interval $\Delta_\alpha$; let $\bar{\omega}_\alpha(x_\beta)$ be a set containing $\omega_\alpha(x_\beta)$ and both endpoints of the interval $\Delta_\alpha$. For an inner node $x \in \omega_\alpha(x_\beta)$, we denote by $x^{(+1_\alpha)}$ and $x^{(-1_\alpha)}$ adjacent nodes belonging to $\bar{\omega}_\alpha(x_\beta)$. If $x^{(+1_\alpha)} \in \gamma_\alpha$, it may happen that this node does not coincide with the node $x^{(i_\alpha+1)}$ of the grid at hand. The accepted view is connected with steps $h_\alpha^\pm(x)$ of the grid as possible spacings between the nodal points $x \in \omega_h$ and the grid nodes $x^{(\pm 1_\alpha)} \in \bar{\omega}_h$.

What is more, at all inner nodes of the grid $\omega_h$ the mean steps are taken to be

$$\hbar_\alpha(x_\alpha) = 0.5\left(x_\alpha^{(i_\alpha+1)} - (x_\alpha^{(i_\alpha-1)}\right).$$

When the pattern of interest happens to be nonuniform in either of the directions $x_\alpha$, that is,

$$x_\alpha^{(i_\alpha)} = i_\alpha h_\alpha\,, \qquad i_\alpha = 0, \pm 1, \pm 2, \dots,$$

the mean steps $\hbar_\alpha = h_\alpha$ become constant and $\hbar_\alpha^\pm$ at a near-boundary zone differ from $h_\alpha$:

$$\text{if} \quad h_\alpha^+ < h_\alpha\,, \quad \text{then} \quad x^{(+1_\alpha)} \in \gamma_h\,, \quad x_\alpha^{(+1_\alpha)} \neq x_\alpha^{(i_\alpha+1)} = (i_\alpha + 1)\, h_\alpha\,.$$

With these ingredients, a reasonable difference scheme is

$$\Lambda y = \varphi(x)\,, \quad x \in \omega_h\,, \quad y(x) = \mu(x)\,, \quad x \in \gamma_h\,, \tag{81}$$

where

$$\Lambda = \widetilde{\Lambda}_1 + \widetilde{\Lambda}_2 ,$$

$$\widetilde{\Lambda}_\alpha y = \left(a_\alpha\, y_{\bar{x}_\alpha}\right)_{\hat{x}_\alpha} = \frac{1}{\hbar_\alpha}\left(a_\alpha^+ \frac{y^{(+1_\alpha)} - y}{h_\alpha^+} - a_\alpha \frac{y - y^{(-1_\alpha)}}{h_\alpha^-}\right),$$

$$y^{(\pm 1_\alpha)} = y\left(x^{(\pm 1_\alpha)}\right), \qquad a_\alpha^+ = a_\alpha\left(x^{(+1_\alpha)}\right),$$

$$y_{\bar{x}_\alpha} = \frac{1}{h_\alpha^-}\left(y - y^{(-1_\alpha)}\right), \qquad y_{x_\alpha} = \frac{1}{h_\alpha^+}\left(y^{(+1_\alpha)} - y\right),$$

$$y_{\hat{x}_\alpha} = \frac{1}{\hbar_\alpha}\left(\hat{y}^{(+1_\alpha)} - y\right).$$

Here the coefficients $a_\alpha$ and $\varphi(x)$ are so chosen as to provide on a uniform grid a local approximation of order 2. By analogy with Chapter 4, Section 3 it is plain to justify the uniform convergence of scheme (81) with the rate $O(|h|^2)$.

We proceed to more a detailed exploration of MATM for solving the system (81) of difference equations just established. For this, the sum $\Lambda\, y = \Lambda_1\, y + \Lambda_2\, y$ involves the members

$$\Lambda_1 y = -\sum_{\alpha=1}^{2}\left[\frac{a_\alpha}{\hbar_\alpha}\, y_{\bar{x}_\alpha} + \frac{1}{2\,\hbar_\alpha}\left(\frac{a_\alpha^+}{h_\alpha^+} - \frac{a_\alpha}{h_\alpha^-}\right) y\right], \tag{82}$$

$$\Lambda_2 y = \sum_{\alpha=1}^{2}\left[\frac{a_\alpha^+}{\hbar_\alpha}\, y_{x_\alpha} + \frac{1}{2\,\hbar_\alpha}\left(\frac{a_\alpha^+}{h_\alpha^+} - \frac{a_\alpha}{h_\alpha^-}\right) y\right], \tag{83}$$

governing what can happen: $y_{\bar{x}_\alpha} = \dfrac{1}{h_\alpha} y$ if $x^{(-1_\alpha)} \in \gamma_h$ and $y_{x_\alpha} = \dfrac{1}{h_\alpha} y$ if $x^{(+1_\alpha)} \in \gamma_h$.

Other ideas are connected with operators $A_1$ and $A_2$ such that $A_\alpha y = -\Lambda_\alpha y$, $\alpha = 1, 2$, for any $y \in \overset{\circ}{\Omega} = H$, where $\overset{\circ}{\Omega}$ is the set of all grid functions vanishing on the boundary $\gamma_h$. By the same token, $A = A_1 + A_2 = -\Lambda$.

Under the inner product structure in the space $H = \overset{\circ}{\Omega}$

$$(y, v) = \sum_{x \in \omega_h} y(x)\, v(x)\, \hbar_1\, \hbar_2$$

the operators $A_1$ and $A_2$ are mutually adjoint to each other: $(A_1 u, v) = (y, A_2 v)$. Because of this fact, the operator $A = A_1 + A_2$ is self-adjoint.

Cumbersome calculations for a proper choice of $d(x)$ will be excluded from further consideration. It seems worthwhile giving only the final results for the later use

$$(84)\qquad d(x)=\sum_{\alpha=1}^{2}\left(\frac{a_\alpha^+}{\hbar_\alpha\,h_\alpha^+\,\sqrt{\varphi_\alpha}}+\frac{1}{2\,\hbar_\alpha}\left|\frac{a_\alpha^+}{h_\alpha^+}-\frac{a_\alpha}{h_\alpha^-}\right|\right)\frac{1}{\sqrt{\varphi_\alpha}+\psi_\alpha}\,,$$

$$(85)\qquad \varphi_\alpha(x_\beta)=\max_{x_\alpha\in\omega_\alpha} v_1^{(\alpha)}(x)\,,\qquad \psi_\alpha(x_\beta)=\max_{x_\alpha\in\omega_\alpha} v_2^{(\alpha)}(x)\,,$$

$$\beta=3-\alpha\,,\qquad \alpha=1,2\,.$$

The functions $v_1^{(\alpha)}(x)$ and $v_2^{(\alpha)}(x)$, $\alpha=1,2$, are declared to be solutions of the relevant problems

$$(86)\qquad \begin{aligned} &\left(a_\alpha\,v_{1\bar{x}_\alpha}^{(\alpha)}\right)_{\hat{x}_\alpha}=-\rho_1^{(\alpha)},\quad x_\alpha\in\omega_\alpha(x_\beta)\,,\\ &v_1^{(\alpha)}\big|_{\gamma_\alpha}=0\,,\quad \rho_1^{(\alpha)}=\frac{a_\alpha^+}{\hbar_\alpha\,h_\alpha^+}\,,\end{aligned}$$

$$(87)\qquad \begin{aligned} &\left(a_\alpha\,v_{2\bar{x}_\alpha}^{(\alpha)}\right)_{\hat{x}_\alpha}=-\rho_2^{(\alpha)},\qquad x_\alpha\in\omega_\alpha(x_\beta)\,,\\ &v_2^{(\alpha)}\big|_{\gamma_\alpha}=0\,,\qquad \rho_2^{(\alpha)}=\frac{1}{2\,\hbar_\alpha}\left|\frac{a_\alpha^+}{h_\alpha^+}-\frac{a_\alpha}{h_\alpha^-}\right|.\end{aligned}$$

These will be given special investigation by means of the elimination method with $O(1)$ operations required at every grid node. With the intervention of four new functions $v_1^{(\alpha)}(x)$, $v_2^{(\alpha)}(x)$ it is not difficult to develop the grid functions $\varphi_\alpha(x_\beta)$ and $\psi_\alpha(x_\beta)$ of one variable and then specify $d(x)$ by formula (84). Under such a choice we obtain

$$(88)\qquad \delta=1\,,\qquad \Delta=4\max_{\alpha=1,2}\left[\max_{x_\beta\in\omega_\beta}\left(\sqrt{\varphi_\alpha(x_\beta)}+\psi_\alpha(x_\beta)\right)^2\right],$$

leaving us with the iteration parameters $\omega_0$ and $\{\tau_k\}$ and the iteration scheme

$$(D+\omega\,A_1)\,D^{-1}(D+\omega\,A_2)\,\frac{\overset{k+1}{y}-\overset{k}{y}}{\tau_{k+1}}+A\overset{k}{y}=\varphi\,,\quad k=0,1,\ldots\,.$$

For determination of $\overset{k+1}{y}$ we must solve the equation

$$(D+\omega A_1)\, D^{-1}(D+\omega A_2)\overset{k+1}{y} = \overset{k}{F}$$

with the right part $\overset{k}{F} = B\overset{k}{y} + \tau_{k+1}(\Lambda \overset{k}{v} + \varphi)$ ($\overset{k}{v} = \overset{k}{y}$ on $\omega_h$ and $\overset{k}{v} = \mu$ on $\gamma_h$). As usual, this amounts to successive solution of the following equations:

$$(D+\omega A_1)\,\bar{y} = \overset{k}{F}\,, \qquad \bar{y}\big|_{\gamma_h} = 0\,,$$

$$(D+\omega A_2)\overset{k+1}{y} = D\,\bar{y} = d(x)\,\bar{y}\,, \qquad \overset{k+1}{y}\big|_{\gamma_h} = 0\,.$$

In this regard, the more detailed forms of the members may be useful in subsequent constructions:

(89)
$$\bar{y} = \frac{1}{K}\left[\sum_{\alpha=1}^{2} \frac{\omega\, a_\alpha}{\hbar_\alpha\, h_\alpha^-}\, \bar{y}^{(-1_\alpha)} + \overset{k}{F}\right],$$

$$\overset{k+1}{y} = \frac{1}{K}\left[\sum_{\alpha=1}^{2} \frac{\omega\, a_\alpha^+}{\hbar_\alpha\, h_\alpha^+}\, \overset{k+1}{y}{}^{(+1_\alpha)} + d\,\bar{y}\right],$$

where

(90)
$$K = d + \frac{1}{2}\sum_{\alpha=1}^{2} \frac{\omega}{\hbar_\alpha}\left(\frac{a_\alpha^+}{h_\alpha^+} + \frac{a_\alpha}{h_\alpha^-}\right).$$

The same procedure is workable here as was done in Section 5 of the present chapter.

**9. The Dirichlet problem for Poisson's equation in an arbitrary complex domain.** The algorithm of MATM is demonstrated by appeal to the Dirichlet problem associated with Poisson's equation

$$\Delta u = \frac{\partial^2 u}{\partial x_1^2} + \frac{\partial^2 u}{\partial x_2^2} = -f(x)\,, \quad x \in G\,, \quad u = \mu(x)\,, \quad x \in \Gamma\,.$$

In working on a square pattern

$$\omega_h = \{x_i = (i_1 h, i_2 h) \in G,\ i_\alpha = 0, \pm 1, \ldots,\ \alpha = 1, 2\}\,,$$

$$x_\alpha^{(i_\alpha+1)} = x_\alpha^{(i_\alpha)} + h,\ i_\alpha = 0, \pm 1, \ldots,\ \alpha = 1, 2\,,$$

where $\hbar_\alpha = h$ and $h^{\pm}_\alpha$ differ from $h$ only at the near-boundary nodes, the intention is to use scheme (81) with the members $a_\alpha(x) \equiv 1$ and $\hbar_\alpha \equiv h$, so that

$$\Lambda\, y = y_{\bar{x}_1 x_1} + y_{\bar{x}_2 x_2} = -\varphi\,, \qquad x \in \omega_h\,, \quad y\big|_{\gamma_h} = \mu\,.$$

In an attempt to adapt here MATM, the functions $v_1(x)$ and $v_2(x)$ are declared to be solutions of one and the same equation

$$\Lambda_\alpha\, v = v_{\bar{x}_\alpha x_\alpha} = -\rho(x)\,, \quad x_\alpha \in \omega_\alpha(x_\beta)\,, \quad v\big|_{\gamma_\alpha} = 0\,, \quad \beta = 3-\alpha\,, \quad \alpha = 1, 2\,,$$

but with different right-hand sides

$$\rho_1(x) = \frac{1}{h\,\hbar^{+}_\alpha}\,, \qquad\qquad \rho_2(x) = \frac{1}{2\,h}\left|\frac{1}{h^{+}_\alpha} - \frac{1}{h^{-}_\alpha}\right|.$$

Denote by $x_\alpha = l_\alpha(x_\beta)$ and $x_\alpha = L_\alpha(x_\beta)$ the endpoints of an interval $\Delta_\alpha$ and by $h^{-}_\alpha$ and $h^{+}_\alpha$ irregular steps at the left and the right ends of this interval; in so doing

$$\Lambda_\alpha\, v = \begin{cases} \dfrac{1}{h}\left(\dfrac{v^{(+1_\alpha)} - v}{h} - \dfrac{v - v^{(-1_\alpha)}}{h^{-}_\alpha}\right), & x_\alpha = l_\alpha + h^{-}_\alpha\,, \\[2ex] \dfrac{1}{h}\left(\dfrac{v^{(+1_\alpha)} - v}{h} - \dfrac{v - v^{(-1_\alpha)}}{h}\right), & l_\alpha + h^{-}_\alpha < x_\alpha < L_\alpha - h^{+}_\alpha\,, \\[2ex] \dfrac{1}{h}\left(\dfrac{v^{(+1_\alpha)} - v}{h^{+}_\alpha} - \dfrac{v - v^{(-1_\alpha)}}{h}\right), & x_\alpha = L_\alpha - h^{+}_\alpha\,. \end{cases}$$

The functions $v_1^{(\alpha)}(x)$ and $v_2^{(\alpha)}(x)$ can be found in explicit form, while the functions $\varphi_\alpha(x_\beta)$ and $\psi_\alpha(x_\beta)$ are expressed by

$$\varphi_\alpha(x_\beta) = \frac{1}{2\,h^2}\left(\frac{L_\alpha(x_\beta) - l_\alpha(x_\beta)}{2}\right)^2 + 1\,,$$

$$\psi_\alpha(x_\beta) = \frac{1}{2}\,, \quad \beta = 3 - \alpha\,, \quad \alpha = 1, 2\,,$$

giving $\Delta$ due to (88) and evaluating the number of the iterations:

$$\text{(91)} \qquad n \geq n_0(\varepsilon)\,, \qquad n_0(\varepsilon) = \frac{\ln(2/\varepsilon)}{3.4\,\sqrt{h/l_0}}\,,$$

where $l_0$ is the diameter of the domain $G$ of interest.

This provides enough reason to conclude that in the case of any complex domain the number of the iterations depends only on the main step $h$ of the grid $\omega_h$ regardless of near-boundary nodes.

Applying (91) to the model problem in a square of sides $l_0$ yields

$$n_0(\varepsilon) = \frac{\ln(2/\varepsilon)}{3.54\sqrt{h/l_0}},$$

whence it follows that the number of the iterations in an arbitrary domain $G$ is being increased in $3.54/3.4 = 1.04$ times, that is, in 4% in comparison with those performed in a square, whose sides are equal to its diameter, thus causing one-two iterations in practical implementations for $\varepsilon = 10^{-4}$ and $h = 1/100$.

Summarizing, the number of the iterations required during the course of MATM in an arbitrary complex domain is close to the number of the iterations performed for the same Dirichlet problem in a minimal rectangle containing the domain $G$ and numerical realizations confirm this statement.

**10. On solving difference equations for problems with variable coefficients.** In the preceding sections this trend of research was due to serious developments of the Russian and western scientists. Specifically, the method for solving difference equations approximating an elliptic equation with variable coefficients in complex domains $G$ of arbitrary shape and configuration is available in Section 8 with placing special emphasis on real advantages of MATM in the numerical solution of the difference Dirichlet problem for Poisson's equation in Section 9.

One of the most important issues is concerned with a smaller number of the iterations performed in the numerical solution of equations with variable coefficients. It was shown in Section 7 that the number of the iterations required during the course of ATM is proportional to $\sqrt{c_1/c_2}$, where $c_1$ and $c_2$ are the smallest and the greatest values of coefficients, respectively. The operator $R$ in question can be put in correspondence with the operator $A$ with variable coefficients such that

$$c_1 R \le A \le c_2 R .$$

One way of covering this is connected with further intervention of the difference Laplace operator $\overset{\circ}{\Lambda}$: $R = -\overset{\circ}{\Lambda}$. A key role of $R$ owes a debt to the structure of the factorized operator $B$:

$$B = (E + \omega R_1)(E + \omega R_2), \quad R_1 + R_2 = R, \quad R_1^* = R_2 .$$

But it may happen that the coefficients of the governing differential equation are varying very fastly, but locally in a small region so that the spectral bounds of the operator $A$ change insignificantly.

In the case where $G$ is a rectangle, it is possible to adopt $B = R$ with further reference to one of the direct methods available for determination of $\overset{k+1}{y}$, thus causing the ratio $\xi = c_1/c_2$ and the independence of the total number of the iterations upon $h$:

$$n(\varepsilon) \approx \frac{1}{1}2\sqrt{\frac{c_2}{c_1}}\ \ln\frac{2}{\varepsilon}\,.$$

Recall that in the modification of ATM (MATM, see for more detail Section 8) there is no need for involving the operator $R$ and so a proper choice of parameters $\{\tau_n\}$ will be independent of constants $c_1$ and $c_2$ both. Here the operator $A$ arranges itself as a sum $A = A_1 + A_2$.

By having recourse to problem (81) in a unit square we will illustrate the performance of MATM in the sequel. In preparation for this, we introduce a square grid

$$\bar\omega_h = \{x_i = (i_1 h, i_2 h)\,,\ i_\alpha = 0, 1, \ldots, N\,,\ hN = 1\,,\ \alpha = 1, 2\}$$

and write down on it the equations

$$\left(a_1\, y_{\bar x_1}\right)_{x_1} + \left(a_2\, y_{\bar x_2}\right)_{x_2} = -\varphi\,, \qquad x \in \omega_h\,, \qquad y\big|_{\gamma_h} = 0\,.$$

The coefficients $a_1(x)$ and $a_2(x)$ are given by the formulas

$$a_1(x) = 1 + K_0\left[(x_1 - 0.5)^2 + (x_2 - 0.5)^2\right],$$

$$a_2(x) = 1 + K_0\left[0.5 - (x_1 - 0.5)^2 - (x_2 - 0.5)^2\right],$$

$$K_0 = \text{const} > 0\,.$$

A proper choice of the right-hand side $\varphi(x)$ is stipulated by the fact that $y(x) = x_1\,(1 - x_1)\,x_2\,(1 - x_2)$ is the exact solution of the problem concerned. Also, it will be sensible to introduce

$$c_1 = \min_{x,\alpha=1,2} a_\alpha(x) = 1\,, \qquad c_2 = \max_{x,\alpha=1,2} a_\alpha(x) = 1 + 0.5\,K_0$$

and then take the operators $R_1$ and $R_2$ with the values

$$R_1\, y = \frac{y_{\bar x_1}}{h} + \frac{y_{\bar x_2}}{h}\,, \qquad R_2\, y = -\frac{y_{x_1}}{h} - \frac{y_{x_2}}{h}\,.$$

Adopting those ideas, the framework of ATM (see Section 7) involves the operator $B = (E + \omega R_1)(E + \omega R_2)$. By going through the matter chronologically we rely on $A = A_1 + A_2$ with the members

$$A_1 y = \sum_{\alpha=1}^{2} \left( \frac{a_\alpha}{h} y_{\bar{x}_\alpha} + \frac{y}{2h^2} \left( a_\alpha^+ - a_\alpha \right) \right),$$

$$A_2 y = - \sum_{\alpha=1}^{2} \left( \frac{a_\alpha^+}{h} y_{x_\alpha} + \frac{y}{2h^2} \left( a_\alpha^+ - a_\alpha \right) \right).$$

The operator $D$ arose in Section 8, by means of which it is plain to form $B = (D + \omega A_1) D^{-1} (D + \omega A_2)$.

No wishing to cover the computational procedures of MATM once again, we fill in the table on the basis of numerical experiments for $\varepsilon = 10^{-4}$.

**Table 6**

| $c_2/c_1$ | h=1/32 | | h=1/64 | | h=1/128 | |
|---|---|---|---|---|---|---|
| | MATM | ATM | MATM | ATM | MATM | ATM |
| 2 | 20 | 23 | 28 | 32 | 39 | 45 |
| 8 | 23 | 46 | 33 | 64 | 47 | 90 |
| 32 | 25 | 92 | 37 | 128 | 53 | 180 |
| 128 | 26 | 184 | 39 | 256 | 57 | 360 |
| 512 | 26 | 367 | 39 | 512 | 59 | 720 |

From here it is easily seen that MATM offers more advantages not only in an arbitrary domain, but also in the case of variable coefficients.

## 10.4 ITERATIVE ALTERNATING DIRECTION METHODS

**1. The alternating direction method for solving the difference Dirichlet problem in a rectangle.** So far we have considered the Dirichlet problem for Poisson's equation in a rectangle $\bar{G} = (0 \le x_\alpha \le l_\alpha,\ \alpha = 1, 2)$:

$$\text{(1)} \qquad \Lambda u = \Lambda_1 u + \Lambda_2 u = -f(x), \quad x \in \omega_h, \quad u\big|_{\gamma_h} = \mu(x),$$

$$\Lambda_\alpha y = y_{\bar{x}_\alpha x_\alpha}, \qquad \alpha = 1, 2,$$

$$\bar{\omega}_h = \omega_h + \gamma_h = \{x_i = (i_1 h_1, i_2 h_2) \in \bar{G}\}.$$

The numerical solution of problem (1) by means of iteration schemes can be done using the alternating direction scheme for the heat conduction equation $\partial u/\partial t = \Delta u + f(x)$, taking now for $j = 0, 1, \ldots$ the form

$$\text{(2)} \qquad \frac{y^{j+1/2} - y^j}{\tau_{j+1}^{(1)}} = \Lambda_1\, y^{j+1/2} + \Lambda_2\, y^j + f(x)\,,$$

$$x \in \omega_h\,, \quad y^{j+1/2}\big|_{\gamma_h} = \mu(x)\,,$$

$$\frac{y^{j+1} - y^{j+1/2}}{\tau_{j+1}^{(2)}} = \Lambda_1\, y^{j+1/2} + \Lambda_2\, y^{j+1} + f(x)\,,$$

$$x \in \omega_h\,, \quad y^{j+1}\big|_{\gamma_h} = \mu(x)\,,$$

with any initial data $y^0 = y(x,0)$. As usual, $j$ is the number of the iteration, $y^{j+1/2}$ is the intermediate iteration, $\tau_{j+1}^{(1)} > 0$ and $\tau_{j+1}^{(2)} > 0$ are the iteration parameters which will be so chosen as to provide a minimal number of iterations.

The transition from the $i$th iteration to the $(j+1)$th iteration can be done by the elimination method along the rows as well as along the columns for the following three-point equations:

$$y^{j+1/2} - \tau_{j+1}^{(1)}\, \Lambda_1\, y^{j+1/2} = F^j\,,$$

$$F^j = y^j + \tau_{j+1}^{(1)}\, \Lambda_2\, y^j + \tau_{j+1}^{(1)}\, f\,, \qquad \text{(along the rows)}$$

$$y^{j+1} - \tau_{j+1}^{(2)}\, \Lambda_2\, y^{j+1} = F^{j+1/2}\,,$$

$$F^{j+1/2} = y^{j+1/2} + \tau_{j+1}^{(2)}\, \Lambda_1\, y^{j+1/2} + \tau_{j+1}^{(2)}\, f\,. \qquad \text{(along the columns)}$$

Thus, the users must perform $O(1/(h_1\, h_2))$ arithmetic operations in calculating one iteration or $O(1)$ arithmetic operations at every node of the grid $\omega_h$.

By analogy with the nonstationary heat conduction equation the iterative process (2) is referred to as the **alternating direction method** (ADM), the convergence of which can be established on account of the homogeneous equations for the error $z^{j+1} = y^{j+1} - u$:

$$\frac{z^{j+1/2} - z^j}{\tau^{(1)}_{j+1}} = \Lambda_1 z^{j+1/2} + \Lambda_2 z^j , \quad x \in \omega_h , \quad z^{j+1/2}\big|_{\gamma_h} = 0 ,$$

$$\frac{z^{j+1} - z^{j+1/2}}{\tau^{(2)}_{j+1}} = \Lambda_1 z^{j+1/2} + \Lambda_2 z^{j+1} , \quad x \in \omega_h , \quad z^{j+1}\big|_{\gamma_h} = 0 ,$$

$$z^0 = y^0 - u .$$

Here we accept, on the same grounds as before, $z^{j+1/2} = y^{j+1/2} - u$.

In this view, it seems reasonable to turn to operator-difference schemes and then follow the usual practice: the operators $A_1 y = -\Lambda_1 y$ and $A_2 y = -\Lambda_2 y$ are introduced for any $y$ from the space $H = \overset{\circ}{\Omega}$ of all grid functions defined on the grid $\bar{\omega}_h$ and vanishing on the boundary $\gamma_h$ under the inner product structure

$$(y, v) = \sum_{x \in \omega_h} y(x)\, v(x)\, h_1 h_2$$

$$= \sum_{i_1=1}^{N_1-1} \sum_{i_2=1}^{N_2-1} y(i_1 h_1, i_2 h_2)\, v(i_1 h_1, i_2 h_2)\, h_1 h_2 .$$

It is well known that the operators so defined possess some remarkable properties:

$$A^*_\alpha = A_\alpha , \quad \delta_\alpha E \le A_\alpha \le \Delta_\alpha E , \quad \delta_\alpha > 0 , \quad \alpha = 1, 2 ,$$

(3)

$$\delta_\alpha = \frac{4}{h_\alpha^2} \sin^2 \frac{\pi h_\alpha}{2 l_\alpha} , \quad \Delta_\alpha = \frac{4}{h_\alpha^2} \cos^2 \frac{\pi h_\alpha}{2 l_\alpha} , \quad \alpha = 1, 2 .$$

What is more, the relation $A_1 A_2 = A_2 A_1$ takes places in various rectangles only.

**2. The general framework of ADM.** The object of subsequent discussions is an operator equation

(4) $$A u = f , \qquad A = A_1 + A_2 ,$$

where $A : H \mapsto H$, $H$ is a finite-dimensional Euclidean space with an inner product $(y, v)$ and associated norm $\| y \| = \sqrt{(y, y)}$.

Also, we take for granted that

(5) 1) $A = A_1 + A_2\,, \quad A_1^* = A_1\,, \quad A_2^* = A_2\,,$

(6) 2) $\delta_\alpha E \le A_\alpha \le \Delta_\alpha E\,, \quad \delta_\alpha > 0\,, \quad \alpha = 1, 2\,,$

3) the operators $A_1$ and $A_2$ are commuting: $A_1 A_2 = A_2 A_1\,,$

and develop the iterative alternating direction method with these members in just the same way as was done before for scheme (2):

$$\frac{y_{j+1/2} - y_j}{\tau_{j+1}^{(1)}} + A_1\, y_{j+1/2} + A_2\, y_j = f\,, \tag{7}$$

$$\frac{y_{j+1} - y_{j+1/2}}{\tau_{j+1}^{(2)}} + A_1\, y_{j+1/2} + A_2\, y_{j+1} = f\,,$$

$$\text{given} \quad y_0 \in H\,, \quad j = 0, 1, 2, \ldots$$

For the error $z_{j+1} = y_{j+1} - u$, we must solve the homogeneous equations

$$(E + \tau_{j+1}^{(1)}\, A_1)\, z_{j+1/2} = (E - \tau_{j+1}^{(1)}\, A_2)\, z_j\,,$$

$$(E + \tau_{j+1}^{(2)}\, A_2)\, z_{j+1} = (E - \tau_{j+1}^{(2)}\, A_1)\, z_{j+1/2}\,,$$

$$z_0 = y_0 - u \in H\,, \quad j = 0, 1, 2, \ldots\,.$$

Having completed the elimination of $z_{j+1/2}$, we find that

$$(E + \tau_{j+1}^{(1)}\, A_1)\,(E + \tau_{j+1}^{(2)}\, A_2)\, z_{j+1} = (E - \tau_{j+1}^{(2)}\, A_1)\,(E - \tau_{j+1}^{(1)}\, A_2)\, z_j$$

by observing that

$$z_{j+1} = S_{j+1}\, z_j\,, \quad S_{j+1} = S_{j+1}^{(1)}\, S_{j+1}^{(2)}\,,$$

$$S_j^{(1)} = (E + \tau_j^{(1)}\, A_1)^{-1}\,(E - \tau_j^{(2)}\, A_1)\,,$$

$$S_j^{(2)} = (E + \tau_j^{(2)}\, A_2)^{-1}\,(E - \tau_j^{(1)}\, A_2)\,,$$

since the operators $A_1$ and $A_2$ are commuting. As a final result we get

$$(8) \qquad z_n = T_n\, z_0\,, \quad T_n = \prod_{j=1}^{n} S_j\,,$$

where $T_n$ is a self-adjoint operator ($T_n^* = T_n$) as a product of commutative self-adjoint operators involved.

From (8) the estimate $\| z_n \| \le \| T_n \| \cdot \| z_0 \|$ immediately follows, in which the quantity $\| T_n \|$ depends on the parameters $\tau_j^{(1)}$ and $\tau_j^{(2)}$. Roughly speaking, a proper choice of such parameters is stipulated by the minimum condition for the norm $\| T_n \|$ in connection with a minimal number of the necessary iterations. To be more specific, when making a substantiated choice, we have at our disposal two collections of parameters $\tau_1^{(1)}$, $\tau_2^{(1)}$, $\dots$, $\tau_n^{(1)}$ and $\tau_1^{(2)}$, $\tau_2^{(2)}$, $\dots$, $\tau_n^{(2)}$ with a prescribed number $n = n(\varepsilon)$ that are known to us in advance:

$$\min_{\{\tau_j^{(1)}, \tau_j^{(2)}\}} \| T_n \| = q_n\,.$$

**3. A proper choice of iteration parameters by Jordan's rule.** First of all, observe that the spectra of the operators $A_1$ and $A_2$ are located, because of (6), on different segments $\delta_\alpha \le \lambda(A_\alpha) \le \Delta_\alpha$ with $\delta_1 \ne \delta_2$ and $\Delta_1 \ne \Delta_2$. One trick we have encountered is to replace $A_1$ and $A_2$ by the newly formed operators $A_1'$ and $A_2'$ with coinciding bounds:

$$\eta\, E \le A_\alpha' \le E\,, \qquad \alpha = 1, 2\,, \qquad \eta > 0\,.$$

This can be done using the decompositions

$$(9) \quad A_1 = (q\, E - r\, A_1')^{-1} (A_1' - p\, E)\,, \quad A_2 = (q\, E + r\, A_2')^{-1} (A_2' + p\, E)\,,$$

where the numbers $r$, $q$, $p$ are free to be chosen in any convenient way and the new parameters $\omega^{(1)}$, $\omega^{(1)}$ are taken to be

$$(10) \qquad \omega^{(1)} = \frac{\tau^{(1)} - r}{q - \tau^{(1)} p}\,, \qquad \omega^{(2)} = \frac{\tau^{(2)} + r}{q + \tau^{(2)} p}\,.$$

With these, we arrive at

$$S_j = \tilde{S}_j^{(1)}\, \tilde{S}_j^{(2)}\,,$$

$$\tilde{S}_j^{(1)} = (E + \omega_j^{(1)}\, A_1')^{-1} (E - \omega_j^{(2)}\, A_1')\,,$$

$$\tilde{S}_j^{(2)} = (E + \omega_j^{(2)}\, A_2')^{-1} (E - \omega_j^{(1)}\, A_2')\,.$$

The norm $\| T_n \|$ can be expressed through the eigenvalues of the operators $A_1'$ and $A_2'$ such as

$$\alpha_{k_1} = \lambda_{k_1}^{(1)}(A_1'), \quad \beta_{k_2} = \lambda_{k_2}^{(2)}(A_2'), \quad k_\alpha = 1, 2, \dots, N_\alpha, \quad \alpha = 1, 2.$$

As far as the operators $A_1'$ and $A_2'$ are commuting, other operators $A_1$, $A_2$, $A$ and $T_n$ possess common systems of eigenfunctions. By utilizing this fact we denote by $\lambda_k(T_n)$ the eigenvalues of the operator $T_n$ and take into account that

$$\lambda(\tilde{S}^{(1)}) = (1 - \omega^{(2)}\alpha)/(1 + \omega^{(1)}\alpha), \quad \lambda(\tilde{S}^{(2)}) = (1 - \omega^{(1)}\beta)/(1 + \omega^{(2)}\beta)$$

with missing subscripts $j$ and $k$. All this enables us to find that

$$\lambda(T_n) = \prod_{j=1}^{n} \frac{1 - \omega_j^{(2)}\alpha}{1 + \omega_j^{(1)}\alpha} \, \frac{1 - \omega_j^{(1)}\beta}{1 + \omega_j^{(2)}\beta} \tag{11}$$

with

$$0 < \eta \le \alpha_{k_1} \le 1, \quad 0 < \eta \le \beta_{k_2} \le 1, \quad k_\alpha = 1, 2, \dots, N_\alpha, \quad \alpha = 1, 2.$$

As known, the norm of the operator $T_n$ is equal to the greatest eigenvalue $\max_k \lambda_k(T_n)$. Further replacements of $\alpha_{k_1}$ and $\beta_{k_2}$ in (11) by continuous variables $\alpha$ and $\beta$ lead to increased maximum of the right-hand side of (11); meaning

$$\| T_n \| \le \max_{\alpha \in [\eta, 1]} \left| \prod_{j=1}^{n} \frac{1 - \omega_j^{(2)}\alpha}{1 + \omega_j^{(1)}\alpha} \, \frac{1 - \omega_j^{(1)}\beta}{1 + \omega_j^{(2)}\beta} \right| . \tag{12}$$

Also, we may accept $\omega^{(1)} = \omega^{(2)}$ and $\alpha = \beta$, since the arguments $\alpha$ and $\beta$ run over one and the same segment $[\eta, 1]$ and their positions in formula (12) are certainly symmetric. The problem statement here is to find the parameters $\omega_1, \omega_2, \dots, \omega_n$ for which

$$\min_{\{\omega_j\}} \max_{\alpha \in [\eta, 1]} \prod_{j=1}^{n} \left( \frac{1 - \omega_j \alpha}{1 + \omega_j \alpha} \right)^2 . \tag{13}$$

A solution of such a problem is already known and so it remains only to write the final expressions for optimal parameters $\tau_j^{(1)}$ and $\tau_j^{(1)}$ in question. Having stipulated the conditions $\alpha = \beta = \eta$ for $\lambda(A_1) = \delta_1$, $\lambda(A_2) = \delta_2$

and $\alpha = \beta = 1$ for $\lambda(A_1) = \Delta_1$, $\lambda(A_2) = \Delta_2$, the constants $p$, $q$, $r$, $\eta$ are found by the formulas

$$(14) \qquad \eta = \frac{1-t}{1+t}, \qquad t = \sqrt{\frac{(\Delta_1 - \delta_1)(\Delta_2 - \delta_2)}{(\Delta_1 + \delta_2)(\Delta_2 + \delta_1)}},$$

$$(15) \qquad \begin{aligned} p &= \frac{\varkappa - t}{\varkappa + t}, \qquad \varkappa = \frac{(\Delta_1 - \delta_1)\Delta_2}{(\Delta_2 + \delta_1)\Delta_1}, \\ r &= \frac{\Delta_1 - \Delta_2 + (\Delta_1 + \Delta_2)p}{2\Delta_1\Delta_2}, \qquad q = r + \frac{1-p}{\Delta_1} \end{aligned}$$

with $\varkappa > t$ and $p > 0$.

At the next stage, with reasonable accuracy $\varepsilon > 0$ and knowledge of the spectral bounds $\delta_\alpha$, $\Delta_\alpha$ of the operators $A_\alpha$ simple algebra gives $\eta$, $p$, $q$, $r$ by formulas (14)–(15). When providing a prescribed accuracy $\varepsilon > 0$, that is, $\| y_n - u \| \le \varepsilon \| y_0 - u \|$, it is necessary to perform $n = n(\varepsilon)$ iterations that can be most readily evaluated by the approximate formula

$$(16) \qquad n(\varepsilon) \approx \frac{1}{\pi^2} \ln\frac{4}{\varepsilon} \ln\frac{4}{\eta}.$$

In the new notations

$$\theta = \frac{1}{16}\eta^2\left(1 + \frac{1}{2}\eta^2\right), \qquad \sigma = \frac{2j-1}{2n}, \qquad j = 1, 2, \dots, n,$$

the quantities $\omega_j$ are specified by the formulas

$$\omega_j = \frac{(1+2\theta)(1+\theta^\sigma)}{2\theta^{\sigma/2}(1+\theta^{1-\sigma}+\theta^{1+\sigma})}, \qquad j = 1, 2, \dots, n.$$

In agreement with (10) the undetermined parameters become

$$\tau_j^{(1)} = \frac{q\omega_j + r}{1 + \omega_j p}, \qquad \tau_j^{(2)} = \frac{q\omega_j - r}{1 - \omega_j p},$$

making it possible to solve problem (7).

It should be noted that for particular values $\delta_1 = \delta_2 = \delta$ and $\Delta_1 = \Delta_2 = \Delta$ formulas (14)–(15) give $\varkappa = \xi$, $p = r = 0$, $q = 1/\Delta$ and $\xi = (1-\eta)/(1+\eta)$, $\eta = \delta/\Delta$. Via transforms (9)–(10), amounting for now to

$A_1 = \Delta A_1'$, $A_2 = \Delta A_2'$, $\omega^{(1)} = \Delta\tau^{(1)}$ and $\omega^{(2)} = \Delta\tau^{(2)}$, we might have $\tau^{(1)} = \tau^{(2)} = \tau$ under the condition $\omega^{(1)} = \omega^{(2)}$.

For the model problem (37) posed completely in Section 2 we obtain

$$\eta = \mathrm{tg}^2 \frac{\pi h}{2} \approx \frac{\pi^2 h^2}{4}, \qquad \frac{4}{\eta} = \frac{16}{\pi^2} \frac{1}{h^2} \approx \frac{1.6}{h^2},$$

giving in combination with (16)

$$n(\varepsilon) \approx 0.2 \ln \frac{1.27}{h} \ln \frac{4}{\varepsilon}.$$

Thus, for example, we find that $n(\varepsilon) \approx 6$ for $h = 1/10$, $n(\varepsilon) \approx 9$ for $h = 1/50$ and $n(\varepsilon) \approx 11$ for $h = 1/100$ with reasonable accuracy $\varepsilon = 2\,e^{-10} \approx 10^{-4}$.

**4. ADM for the case of noncommutative operators.** Of special interest is the equation of the form

$$A\,u = (A_1 + A_2)\,u = f$$

with noncommutative operators $A_1$ and $A_2$ still subject to conditions (5)–(6):

$$A_\alpha = A_\alpha^* > 0\,, \quad \delta_\alpha E \le A_\alpha \le \Delta_\alpha E\,, \quad \delta_\alpha > 0\,, \quad \alpha = 1, 2\,.$$

These properties have had a significant impact on modifications of iterative methods and provide the possibility of applications of the two-parameter iterative ADM:

$$(E + \omega_1 A_1)\, y_{k+1/2} = (E - \omega_1 A_2)\, y_k + \omega_1 f\,, \tag{17}$$

$$(E + \omega_2 A_2)\, y_{k+1} = (E - \omega_2 A_1)\, y_{k+1/2} + \omega_2 f\,,$$

$$\text{given} \quad y_0 \in H\,, \quad k = 0, 1, 2, \ldots\,,$$

where a proper choice of parameters $\omega_1$ and $\omega_2$ will be substantiated a little later for some reason or another.

Consistent with $z_{k+1} = y_{k+1} - u$ and $z_{k+1/2} = y_{k+1/2} - u$ is the idea of setting up the homogeneous equations for the error such as

$$(E + \omega_1 A_1)\, z_{k+1/2} = (E - \omega_1 A_2)\, z_k\,, \tag{18}$$

$$(E + \omega_2 A_2)\, z_{k+1} = (E - \omega_2 A_1)\, z_{k+1/2}\,,$$

$$\text{given} \quad z_0 \in H\,, \quad k = 0, 1, 2, \ldots\,.$$

Having involved $v_k = (E + \omega_2 A_2)\, z_k$ with furter elimination of $z_{k+1/2}$ from (18), we derive the equation

$$v_{k+1} = S_1\, S_2\, v_k\,, \quad v_0 \in H\,, \quad k = 0, 1, 2, \dots\,, \tag{19}$$

whose transition operator $S$ arranges itself as a product $S = S_1\, S_2$ with the multipliers

$$\begin{aligned} S_1 &= (E + \omega_1 A_1)^{-1}\,(E - \omega_2 A_1)\,, \\ S_2 &= (E + \omega_2 A_2)^{-1}\,(E - \omega_1 A_2)\,. \end{aligned} \tag{20}$$

This is showing the gateway to subsequent considerations: since

$$\|\, v_{k+1} \,\| \leq \|\, S_1 S_2 \,\| \cdot \|\, v_k \,\|\,,$$

it is necessary to evaluate the norm $\|\, S_1 S_2 \,\|$ and find $\min_{\omega_1, \omega_2} \|\, S_1 S_2 \,\|$ with the aid of the well-established relation.

As further developments occur, we need an auxiliary lemma.

**Lemma** *Let an operator $A : H \mapsto H$ be in line with the conditions*

$$A = A^* > 0\,, \quad \delta E \leq A \leq \Delta E\,, \quad \delta > 0\,. \tag{21}$$

*Then the norm $\|\, S(\omega) \,\|$ of the operator $S(\omega) = (E + \omega A)^{-1}(E - \omega A)$ attains for $\omega = \omega_0 = \dfrac{1}{\sqrt{\delta\, \Delta}}$ the minimal value*

$$\min_{\omega} \|\, S(\omega) \,\| = \|\, S(\omega_0) \,\| = \frac{1 - \sqrt{\eta}}{1 + \sqrt{\eta}}\,, \quad \text{where} \quad \eta = \frac{\delta}{\Delta}\,.$$

For the most part, the proof is connected with further treatment of $S(\omega)$ as the transition operator of the two-layer scheme

$$(E + \omega\, A)\, \frac{y_{k+1} - y_k}{2\, \omega} + A\, y_k = 0\,, \quad k = 0, 1, \dots\,, \quad y_0 \in H\,, \tag{22}$$

with the operator $B = E + \omega A$ involved. This supports the view that

$$y_{k+1} = S(\omega)\, y_k\,, \qquad \|\, y_{k+1} \,\| \leq \rho\, \|\, y_k \,\|\,,$$

where $\rho = \rho(\omega) = \|\, S(\omega) \,\|$ needs to be minimized.

Along these lines, it is worth noting that we are still in the framework of the general stability theory outlined in Chapter 6, Section 2 for difference schemes like (22) asserting that a necessary and sufficient condition for the estimate $\|y_{k+1}\|_D \le \rho \, \|y_k\|_D$ to be valid for any $0 < \rho < 1$ and any operator $D = D^* > 0$ commuting with $A$ is

$$\frac{1-\rho}{2\,\omega}\,(E + \omega\, A) \le A \le \frac{1+\rho}{2\,\omega}\,(E + \omega\, A)\,.$$

This is acceptable if we agree to consider $D = E$ or $D = A$. In this regard, it should be noted that the preceding is equivalent to the bilateral operator estimate

$$\frac{\xi}{\omega}\, E \le A \le \frac{1}{\xi\,\omega}\, E\,, \quad \text{where} \quad \xi = \frac{1-\rho}{1+\rho} \quad \text{and} \quad \rho = \frac{1-\xi}{1+\xi}\,.$$

Putting these together with (21) we conclude that

$$(23) \qquad \frac{\xi}{\omega} \le \delta\,, \quad \frac{1}{\xi\,\omega} \ge \Delta \quad \text{or} \quad \xi\,\omega \le \frac{1}{\Delta}\,, \quad \text{so that} \quad \xi^2 \le \eta\,,$$

it being understood that the minimum of $\rho$ will be attainable in the case of the maximal value of $\xi$ in (23). This is certainly so with

$$\frac{\xi}{\omega} = \delta\,, \qquad \frac{1}{\xi\,\omega} = \Delta\,, \qquad \xi = \sqrt{\eta}\,,$$

whose use permits us to establish the chain of the relations

$$\omega = \frac{\xi}{\delta} = \frac{\sqrt{\eta}}{\delta} = \frac{1}{\sqrt{\delta\,\Delta}} = \omega_0$$

and arrive at

$$\min_{\omega}\, \rho(\omega) = \rho(\omega_0) = \frac{1-\sqrt{\eta}}{1+\sqrt{\eta}}\,.$$

However, the same result can be obtained by a simple observation that

$$\rho = \|\, S(\omega)\,\| = \max_{\delta \le \lambda \le \Delta} \left| \frac{1-\omega\,\lambda}{1+\omega\,\lambda} \right|,$$

where $\lambda = \lambda(A)$ is an eigenvalue of the operator $A$. Plain calculations show that

$$\min_{\omega}\, \rho(\omega) = \min_{\omega}\, \max_{\delta \le \lambda \le \Delta} \left| \frac{1-\omega\,\lambda}{1+\omega\,\lambda} \right|$$

$$= \min_{\frac{1}{\Delta} \le \omega \le \frac{1}{\delta}}\, \max \left( \frac{1-\omega\,\delta}{1+\omega\,\delta}\,, \frac{\omega\,\Delta - 1}{\omega\,\Delta + 1} \right)$$

$$= \rho(\omega_0) = \frac{1-\sqrt{\eta}}{1+\sqrt{\eta}} \quad \text{for} \quad \omega = \omega_0\,,$$

thereby completing the task of motivating the desired minimum.

Via transforms (9)–(10) available for the operators $A_1$ and $A_2$ and the parameters $\omega_1$ and $\omega_2$ we accept

$$\frac{\omega_1 - r}{q - \omega_1 p} = \omega\,, \qquad \frac{\omega_2 + r}{q + \omega_2 p} = \omega\,,$$

making it possible to reduce the operators specified by (20) to

$$S = \tilde{S}_1 \tilde{S}_2\,,$$

$$\tilde{S}_1 = (E + \omega A_1')^{-1} (E - \omega A_1')\,,$$

$$\tilde{S}_2 = (E + \omega A_2')^{-1} (E - \omega A_2')\,.$$

Their members satisfy the relations

$$\eta E \le A_\alpha' \le E\,, \qquad \alpha = 1, 2\,, \quad \eta > 0\,.$$

In the preceding reasoning the member $\eta$ and the parameters $p$, $q$, $r$ are given by formulas (14)–(15). With the inequality $\| S \| \le \| \tilde{S}_1 \| \cdot \| \tilde{S}_2 \|$ in view, the lemma asserts that for $\omega = \omega_0 = \sqrt{\eta}/\delta$

$$\min_\omega \| \tilde{S}_1 \| = \min_\omega \| \tilde{S}_2 \| = \frac{1 - \sqrt{\eta}}{1 + \sqrt{\eta}}$$

and a solution of problem (18) obeys the a priori estimate

$$\text{(24)} \qquad \| (E + \omega_2 A_2)\, z_n \| \le \rho^n \, \| (E + \omega_2 A_2)\, z_0 \|\,, \quad \rho = \Big( \frac{1 - \sqrt{\eta}}{1 + \sqrt{\eta}} \Big)^2\,,$$

with $\omega_1$ and $\omega_2$ still subject to the relations

$$\text{(25)} \qquad \omega_1 = \frac{r + q\,\omega_0}{1 + p\,\omega_0}\,, \quad \omega_2 = \frac{q\,\omega_0 - r}{1 - p\,\omega_0}\,, \quad \omega_0 = \frac{\sqrt{\eta}}{\delta}\,.$$

In particular, $\eta = \delta / \Delta$ for $\delta_1 = \delta_2 = \delta$ and $\Delta_1 = \Delta_2 = \Delta$.

From (24) it follows that the condition $\rho^n < \varepsilon$ of the termination of iterations is ensured if

$$n \ge n^{(1)}(\varepsilon)\,, \quad n^{(1)}(\varepsilon) = \frac{\ln(1/\varepsilon)}{2\,\xi}\,, \quad \text{where} \quad \xi = \frac{2\sqrt{\eta}}{1 + \eta}\,.$$

In order to understand the nature of this a little better, we refer to a model problem posed in Section 2 all over again for which

$$\delta_1 = \delta_2 = \delta = 4\, \sin^2\Big(\frac{\pi\, h}{2}\Big)\Big/ h^2\,,$$

$$\Delta_1 = \Delta_2 = \Delta = 4\, \cos^2\Big(\frac{\pi\, h}{2}\Big)\Big/ h^2\,,$$

$$\sqrt{\eta} = \sqrt{\operatorname{tg}^2 \frac{\pi\, h}{2}} \approx \frac{\pi\, h}{2} \approx 1.57\, h\,.$$

The main goal of subsequent considerations is the comparison between ADM of the type (17) with parameters (25) and the explicit method with optimal set of Chebyshev's parameters

$$\frac{y_{k+1} - y_k}{\tau_{k+1}} + A\, y_k = f\,, \quad k = 0, 1, \dots\,,$$

arising in Section 2 and requiring no less than $n^{(0)}(\varepsilon)$ iterations:

$$n \geq n^{(0)}(\varepsilon)\,, \quad n^{(0)}(\varepsilon) = \frac{\ln(2/\varepsilon)}{2\sqrt{\xi}}\,, \quad \xi = \gamma_1/\gamma_2\,.$$

In that case $\gamma_1 = \delta_1 + \delta_2 = 2\,\delta$, $\gamma_2 = \Delta_1 + \Delta_2 = 2\,\Delta$ and $\xi = \delta/\Delta = \eta$. Thus, we might have

$$n^{(0)}(\varepsilon) \approx \frac{\ln(2/\varepsilon)}{2\sqrt{\eta}}\,, \qquad n^{(1)}(\varepsilon) \approx \frac{\ln(1/\varepsilon)}{4\sqrt{\eta}}\,,$$

thereby justifying that $n^{(0)}(\varepsilon) \approx 2\, n^{(1)}(\varepsilon)$.

When solving the model problem concerned, the transition from the $k$th iteration to the $(k+1)$th iteration is performed either in 9 steps or in 26 steps: 5 operations of addition and 4 operations of multiplication during the course of the explicit Chebyshev's method and 12 operations of addition and 14 operations of multiplication in the case of ADM in connection with the double elimination (first, along the rows and then along the columns). This provides reason enough to conclude that in the case of noncommutative operators the first method is rather economical than the second one. Both methods require $O(\frac{1}{h}\,\ln\frac{1}{\varepsilon})$ iterations in the process of solving the model problem under consideration.

**5. Factorized iteration schemes and ADM.** The main idea behind this approach is connected with the equivalence between ADM from the preceding sections and the two-layer iteration scheme

$$B\,\frac{y_{k+1}-y_k}{\tau}+A\,y_k=f \tag{26}$$

with the factorized operator

$$B=(E+\omega_1 A_1)(E+\omega_2 A_2) \tag{27}$$

and the iteration parameter

$$\tau=\omega_1+\omega_2\,. \tag{28}$$

To make sure of it, we rewrite (17) as

$$B_1\,y_{k+1/2}=C_2\,y_k+\omega_1\,f\,,\quad B_2\,y_{k+1}=C_1\,y_{k+1/2}+\omega_2\,f\,, \tag{29}$$

where $B_1=E+\omega_1 A_1$, $B_2=E+\omega_2 A_2$, $C_1=E-\omega_2 A_1$ and $C_2=E-\omega_1 A_2$. The elimination of the iteration $y_{k+1/2}$ can be done by successively applying the operator $C_1$ to the first equation and the operator $B_2$ to the second one. Combination of the resulting equations with further reference to the useful relations $B_1C_1=C_1B_1$, $\omega_1 C_1+\omega_2 B_1=\omega_1(E-\omega_2 A_1)+\omega_2(E+\omega_1 A_1)=(\omega_1+\omega_2)E$ gives

$$B_1B_2\,y_{k+1}=C_1C_2\,y_k+(\omega_1+\omega_2)\,f\,. \tag{30}$$

It seems worthwhile to reduce equation (30) to the canonical form (26) by observing that $B_1B_2-C_1C_2=(\omega_1+\omega_2)(A_1+A_2)$. The arguments in reverse order are obvious.

Let us stress here that the applications of the above framework to noncommutative operators $A_1$ and $A_2$ could result in wrong reasoning in light of the property that operator (27) is non-self-adjoint and scheme (26) does not fall within the category of two-layer iteration schemes lying in the fundamentals of the general theory.

Before going further in more detail on this point, it is worth mentioning that the factorized operator (27) is self-adjoint and positive: $B=B^*>0$. To decide for yourself whether the obtained results are acceptable for commutative operators $A_1$ and $A_2$ and conditions (5)–(6), the first step is to discover from (30) the structure of the transition operator of scheme (26) such as

$$S=S_1S_2\,,\qquad S_\alpha=B_\alpha^{-1}C_\alpha\,,\qquad \alpha=1,2\,.$$

We learn from Section 4 the values of parameters $\omega_1$ and $\omega_2$ for which the minimum $\rho$ of the norm $\| S(\omega) \|$ is attained:

$$\min_{\omega_1, \omega_2} \| S \| = \rho , \quad \rho = \left( \frac{1 - \sqrt{\eta}}{1 + \sqrt{\eta}} \right)^2 , \quad = \frac{1 - \xi}{1 + \xi} , \quad \xi = \frac{2 \sqrt{\eta}}{1 + \eta} ,$$

where $\eta$, $\omega_1$ and $\omega_2$ are given by formulas (14), (15) and (25). Knowing $\rho$, $\omega_1$ and $\omega_2$, it is simple to compute the constants of equivalence $\overset{\circ}{\gamma}_1$ and $\overset{\circ}{\gamma}_2$ for the operators $B$ and $A$ such that

$$\overset{\circ}{\gamma}_1 B \leq A \leq \overset{\circ}{\gamma}_2 , \tag{31}$$

which will be needed in subsequent discussions of the operator $B$ from a viewpoint of the possible general theory.

A necessary and sufficient condition for the $\rho$-stabity of scheme (26) is

$$\frac{1 - \rho}{\tau} B \leq A \leq \frac{1 + \rho}{\tau} B$$

with assigned values of $\rho$ and $\tau = \omega_1 + \omega_2$. The outcome of this is

$$\begin{aligned} \overset{\circ}{\gamma}_1 &= \frac{1 - \rho}{\omega_1 + \omega_2} = \frac{2 \xi}{(1 + \xi)(\omega_1 + \omega_2)} , \\ \overset{\circ}{\gamma}_2 &= \frac{1 + \rho}{\omega_1 + \omega_2} = \frac{2}{(1 + \xi)(\omega_1 + \omega_2)} . \end{aligned} \tag{32}$$

A keystone in the design of an operator of the type (27) is the possible structure of an auxiliary operator $R = R^* > 0$ being a sum of two operators $R_1$ and $R_2$ such that

$$R = R_1 + R_2 , \qquad R_1 R_2 = R_2 R_1 ,$$

$$R_\alpha = R^*_\alpha , \qquad \delta_\alpha E \leq R_\alpha \leq \Delta_\alpha E , \qquad \alpha = 1, 2 ,$$

by means of which the factorized operator in question reveals to be

$$B = (E + \omega R_1)(E + \omega R_2) . \tag{33}$$

This serves to motivate instead of (31) the relations $\overset{\circ}{\gamma}_1 B \leq R \leq \overset{\circ}{\gamma}_2 B$ with constants $\overset{\circ}{\gamma}_1$ and $\overset{\circ}{\gamma}_2$ arising from (32).

When the operator $R$ happens to be a regularizer for the operator $A$, that is, $c_1 R \le A \le c_2 R$, the constants of equivalence for the operators $A$ and $B$ such that $\gamma_1 B \le A \le \gamma_2 B$ become

$$\gamma_1 = c_1 \overset{\circ}{\gamma}_1 , \qquad \gamma_2 = c_2 \overset{\circ}{\gamma}_2 .$$

The intention is to use such a factorized operator in the explicit method with optimal set of Chebyshev's parameters:

$$B \frac{y_{k+1} - y_k}{\tau_{k+1}} + A y_k = f , \quad k = 1, 2, \ldots, n , \quad y_0 \in f ,$$

which requires no less than $n_0(\varepsilon)$ iterations:

$$n \ge n_0(\varepsilon) , \quad n_0(\varepsilon) = \frac{\ln (2/\varepsilon)}{2 \sqrt{\xi}} , \quad \xi = \frac{2 \sqrt{\eta}}{1 + \eta} \frac{c_1}{c_2} .$$

In complete agreement with (22), $\eta$ is expressed through the parameters $\delta_1$, $\delta_2$, $\Delta_1$ and $\Delta_2$ of the operators $R_1$ and $R_2$ involved.

It is worth noting here that the same estimate for $n_0(\varepsilon)$ was established before for ATM with optimal set of Chebyshev's parameters, but other formulas were used to specify $\eta$ in terms of $\delta_\alpha$ and $\Delta_\alpha$. If $R = -\Lambda$, where $\Lambda$ is the difference Laplace operator, and the Dirichlet problem is posed on a square grid in a unit square, then

$$\eta = \delta / \Delta = \operatorname{tg}^2 \frac{\pi h}{2}$$

for both cases: ATM and the factorized scheme (26)–(28) relating to ADM. Just for this reason these methods require the same number of iterations. As a matter of experience, ATM is more economical and preferable, since one iteration necessitates performing a smaller number of arithmetic operations. Some consensus of opinion here is to accept $B = R$ and then find $y_{k+1}$ from the equation

$$R y_{k+1} = F_k , \qquad F_k = R y_k - \tau_{k+1} (A y_k - f) ,$$

by one of the available direct methods, say by the decomposition method or Fourier fast transform method.

Summarizing, we are somewhat uncertain in which situations scheme (26)–(28) would be more better than, for example, ATM.

**Remark 1** From such reasoning it seems clear that ADM with variable parameters is equivalent to the two-layer scheme (26) with iteration parameter $\tau = \tau_k = \tau_k^{(1)} + \tau_k^{(2)}$ and factorized operator

$$B_k = (E + \tau_k^{(1)} A_1)\,(E + \tau_k^{(2)} A_2)\,.$$

We will not pursue analysis of this: the ideas needed to do so have been covered.

**Remark 2** If $A$ is a sum of $p > 2$ pairwise commutative operators such that

$$A = \sum_{\alpha=1}^{p} A_\alpha\,, \quad A_\alpha^* = A_\alpha > 0\,, \quad \delta_\alpha E \le A_\alpha \le \Delta_\alpha E\,, \quad \delta_{\dot{\alpha}} > 0\,,$$

$$\alpha = 1, 2, \ldots, p\,, \quad A_\alpha A_\beta = A_\beta A_\alpha\,, \quad \alpha, \beta = 1, 2, \ldots, p\,,$$

direct applications of ADM described by (17) is impossible in principle, so there is a real need for constructing scheme (26) with the factorized operator

$$B_k = \prod_{\alpha=1}^{p} (E + \tau_k^{(\alpha)} A_\alpha)\,, \qquad \tau_l^{(\alpha)} > 0\,.$$

In this regard, we are unaware of any exact solution to minimax problem and the so-called cyclic set of the ensuing parameters may be of help in achieving the final aims.

The main idea behind this approach is connected with the equation related to a new iteration $y_{k+1}$:

$$\prod_{\alpha=1}^{p} (E + \tau_k^{(\alpha)} A_\alpha)\, y_{k+1} = F_k\,,$$

which reduces to successive solution of $p$ equations

$$(E + \tau_k^{(1)} A_1)\, y^{(1)} = F_k\,, \quad (E + \tau_k^{(\alpha)} A_\alpha)\, y^{(\alpha)} = y^{(\alpha-1)}\,, \quad \alpha = 2, \ldots, p\,,$$

with $y_{k+1} = y^{(p)}$ incorporated.

The difference Dirichlet problem for Poisson's equation in a $p$-dimensional unit cube such as

$$A_\alpha y = -\Lambda_\alpha y\,, \qquad \Lambda_\alpha y = y_{\bar{x}_\alpha x_\alpha}\,,$$

helps clarify what is done. As a matter of fact, searching for $y_{k+1}$ amounts to successive elimination along the directions $x_1, x_2, \ldots, x_p$ relating to algorithms of the ADM-type. The availability of the cyclic set of parameters gives us hopes to achieve a prescribed accuracy $\varepsilon$ by making $n_0(\varepsilon) = O(\ln(1/h)\ \ln(1/\varepsilon))$ iterations, where $h = h_1 = h_2 = \cdots = h_p$ is the grid step. On the other hand, it is no difficult to achieve what we suggest by means of ATM with the operators

$$R_1\, y = \sum_{\alpha=1}^{p} \frac{1}{h_\alpha}\, y_{\bar{x}_\alpha}, \qquad R_2\, y = -\sum_{\alpha=1}^{p} \frac{1}{h_\alpha}\, y_{x_\alpha}.$$

This applies equally well to the problem posed above and requires no less than $n_0(\varepsilon)$ iterations, where

$$n_0(\varepsilon) \approx \frac{\ln(2/\varepsilon)}{3.54\sqrt{h}} = O\Big(\frac{1}{\sqrt{h}}\ \ln\frac{2}{\varepsilon}\Big),$$

meaning that the asymptotic behavior of ATM becomes worse in comparison with ADM. However, when $p = 3$ and $h \geq 1/60$, that is, the total number of the grid nodes $\leq 2.16 \cdot 10^5$, a smaller number of iterations is performed during the course of ATM in contrast to ADM with the cyclic set of parameters. With regard to the work and storage required, ATM being rather economical (in 2–2.5 times) offers more advantages than ADM on any admissible grid no matter how it is chosen.

**6. ADM for non-self-adjoint operators.** The equation we must solve is of the form

$$A\,u = (A_1 + A_2)\,u = f,$$

where $A_1$ and $A_2$ are non-self-adjoint positive definite operators subject to the conditions

$$A_\alpha \geq \delta_\alpha E, \quad A_\alpha^{-1} \geq \frac{1}{\Delta_\alpha} E, \quad \delta_\alpha > 0, \quad \Delta_\alpha > 0, \quad \alpha = 1, 2. \tag{34}$$

As can readily be observed, the second condition is equivalent to the inequality

$$\|\, A_\alpha y\,\|^2 \leq \Delta_\alpha\, (A_\alpha y, y). \tag{35}$$

Indeed, by merely setting $x = A_\alpha y$, we are led to

$$0 \leq (A_\alpha^{-1} x, x) - \frac{1}{\Delta_\alpha}(x, x) = (y, A_\alpha y) - \frac{1}{\Delta_\alpha}(A_\alpha y, A_\alpha y).$$

In trying to solve the equation $Au = f$ the iterative ADM with iteration parameter $\omega$ is good enough for our purposes:

$$(36)\qquad \begin{aligned} (E + \omega A_1)\, y_{k+1/2} &= (E - \omega A_2)\, y_k + \omega\, f\,, \\ (E + \omega A_2)\, y_{k+1} &= (E - \omega A_1)\, y_{k+1/2} + \omega\, f\,. \end{aligned}$$

By exactly the same reasoning as in Section 4 with regard to the error $z_{k+1} = y_{k+1} - u$ we deduce for $v_{k+1} = (E + \omega A_2)\, z_{k+1}$ that

$$v_{k+1} = S_1\, S_2\, v_k\,, \quad S_\alpha = (E + \omega A_\alpha)^{-1}(E - \omega A_\alpha)\,, \qquad \alpha = 1, 2\,,$$

$$\| v_{k+1} \| \le \| S_1\, S_2 \| \cdot \| v_k \| \le \| S_1 \| \cdot \| S_2 \| \cdot \| v_k \|\,.$$

We are going to show that under conditions (34) the following estimates hold true:

$$(37)\qquad \| S_\alpha \|^2 \le \frac{1 - \varkappa_\alpha}{1 + \varkappa_\alpha}\,, \quad \varkappa_\alpha = \frac{2\,\delta_\alpha\,\omega}{1 + \omega^2\,\delta_\alpha\,\Delta_\alpha}\,, \quad \alpha = 1, 2\,.$$

Indeed, from the obvious identities

$$\| (E - \omega A_\alpha)\, x \|^2 = \| (E + \omega A_\alpha)\, x \|^2 - 4\,\omega\, (A_\alpha x, x)\,,$$

$$\| (E + \omega A_\alpha)\, x \|^2 = \| x \|^2 + \omega^2 \| A_\alpha x \|^2 + 2\,\omega\, (A_\alpha x, x)\,,$$

which are put together with inequalities (34)–(35), it follows that

$$\| (E + \omega A_\alpha)\, x \|^2 \le (1/\delta_\alpha + \omega^2 \Delta_\alpha + 2\,\omega)\, (A_\alpha x, x)\,,$$

$$(A_\alpha x, x) \ge \frac{\delta_\alpha}{1 + \omega^2 \delta_\alpha\, \Delta_\alpha + 2\,\omega\,\delta_\alpha} \| (E + \omega A_\alpha)\, x \|^2\,,$$

$$\| (E - \omega A_\alpha)\, x \|^2 \le \frac{1 - \varkappa_\alpha}{1 + \varkappa_\alpha} \| (E + \omega A_\alpha)\, x \|^2\,.$$

By inserting here $x = (E + \omega A_\alpha)^{-1}$ we find that

$$\| S_\alpha y \|^2 \le \frac{1 - \varkappa_\alpha}{1 + \varkappa_\alpha} \| y \|^2$$

as required. By virtue of (37) it is plain to establish the relations

$$\| S_1 S_2 \|^2 \le \| S_1 \| \cdot \| S_2 \|^2 \le \frac{1-\varkappa_1}{1+\varkappa_1} \frac{1-\varkappa_2}{1+\varkappa_2} = F(\omega) .$$

Here the parameter $\omega$ is so chosen as to satisfy the minimum condition for the function $F(\omega)$ by discovering that the functions $F_1(\omega) = (1-\varkappa_1)(1+\varkappa_1)$ and $F_2(\omega) = (1-\varkappa_2)(1+\varkappa_2)$ attain the minimal values, respectively, for $\omega = 1/\sqrt{\delta_1 \Delta_1}$ and $\omega = 1/\sqrt{\delta_2 \Delta_2}$. These points are are identical when $\delta_1 \Delta_1 = \delta_2 \Delta_2$. In that case for $\omega = 1/\sqrt{\delta_1 \Delta_1} = 1/\sqrt{\delta_2 \Delta_2}$ the inequality

$$\| S_1 S_2 \|^2 \le \rho^2, \qquad \rho^2 = \frac{1-\sqrt{\eta_1}}{1+\sqrt{\eta_1}} \cdot = \frac{1-\sqrt{\eta_2}}{1+\sqrt{\eta_2}}, \tag{38}$$

$$\eta_\alpha = \frac{\delta_\alpha}{\Delta_\alpha}, \qquad \alpha = 1, 2,$$

holds and the error $z_n = y_n - u$ admits the estimate

$$\| (E + \omega A_2) z_n \| \le \rho^n \| (E + \omega A_2) z_0 \| . \tag{39}$$

If $\delta_1 \Delta_1 \ne \delta_2 \Delta_2$, then $\omega = 1/\sqrt{\delta \Delta}$, where $\delta = \min(\delta_1, \delta_2)$ and $\Delta = \max(\Delta_1, \Delta_2)$.

In comparison with the case of self-adjoint operators the number of the iterations is being increased in two times. This is clearly seen, for example, from (38) by setting $\delta_1 = \delta_2 = \delta$ and $\Delta_1 = \Delta_2 = \Delta$:

$$\rho = \frac{1-\sqrt{\eta}}{1+\sqrt{\eta}} .$$

It should be compared with

$$\rho = \left( \frac{1-\sqrt{\eta}}{1+\sqrt{\eta}} \right)^2 ,$$

emerging from estimate (39) with regard to self-adjoint operators $A_1$ and $A_2$. One needs to exercise good judgement in deciding which to consider.

## 10.5 OTHER ITERATIVE METHODS

**1. Three-layer iteration schemes.** So far we have considered two-layer iteration schemes available for solving operator equations of the form $Au = f$ with a self-adjoint operator $A$ under the assumption that the spectral bounds $\gamma_1$ and $\gamma_2$ for the operator $A$ are known in advance either in a space $H$ or in a space $H_B$, where $B = B^* > 0$ is some stabilizator. Other iterative methods find a wide range of applications in some or other aspects.

An excellent start in this direction is to describe three-layer (two-step) iteration schemes in the general setting due to which it is required to solve the equation

$$(1) \qquad A u = f , \qquad A : H \mapsto H ,$$

with a self-adjoint positive definite operator $A$, whose spectral bounds are already known:

$$(2) \qquad A = A^*, \qquad \gamma_1 E \le A \le \gamma_2 E , \qquad \gamma_1 > 0 .$$

The links between three iterations $y_{k-1}$, $y_k$ and $y_{k+1}$ are provided by three-layer iteration schemes, by means of which $y_{k+1}$ can be expressed through the values $y_{k-1}$ and $y_k$. The standard form of explicit schemes is

$$(3) \qquad y_{k+1} = (1+\alpha)\, S\, y_k - \alpha\, y_{k-1} + (1+\alpha)\, \tau_0\, f , \qquad k = 1, 2, \dots ,$$

$$y_1 = S\, y_0 + \tau_0\, f , \quad \text{given } y_0 \in H ,$$

where $S = E - \tau_0 A$ is the transition operator for the two-layer simple iteration scheme with the optimal parameter $\tau_0$

$$(4) \qquad \tau_0 = \frac{2}{\gamma_1 + \gamma_2}, \qquad \alpha = \rho_1^2, \quad \rho_1 = \frac{1 - \sqrt{\xi}}{1 + \sqrt{\xi}}, \quad \xi = \frac{\gamma_1}{\gamma_2} .$$

The two-layer simple iteration scheme permits us to find the first iteration $y_1$.

In an attempt to create scheme (3), equation (1) should be represented in the so-called "preliminary" form

$$u = u - \tau A u + \tau f = S(\tau)\, u + \tau f , \quad S(\tau) = E - \tau A ,$$

and the parameter $\tau$ should be so chosen as to minimize the norm $||S||$. We now know from Section 2 that this aim can be achieved by setting $\tau = \tau_0$, so that the preceding becomes

$$u = S(\tau_0)\, u + \tau_0\, f\,, \tag{5}$$

allowing alternative forms of writing:

$$(1+\alpha)\, u = (1+\alpha)\, S\, u + (1+\alpha)\, \tau_0 f\,,$$

$$u = (1+\alpha)\, S\, u - \alpha\, u + (1+\alpha)\, \tau_0 f\,.$$

Having replaced $(1+\alpha)Su$ by $(1+\alpha)Sy_k$ and $\alpha u$ by $\alpha y_{k-1}$, we try to adapt the explicit scheme, the parameter $\alpha$ of which needs to be selected by the approved rule in a minimal number of iterations. Unfortunately, more a detailed exploration on this point and the convergence of scheme (3) are not available in the present book. A final result can be obtained through such an analysis by utilizing the fact that the residual $r_k = A\, y_k - f$ satisfies the homogeneous equations

$$r_{k+1} = (1+\alpha)\, S\, r_k - \alpha r_{k-1}, \quad k = 1, 2, \dots, \tag{6}$$

$$r_1 = S\, r_0, \quad r_0 = A\, y_0 - f \in H\,,$$

no matter how the initial value $r_0$ is chosen.

In a revised statement of the problem for $\alpha = \rho_1^2$ the estimate

$$||\, A\, y_n - f\, || \le q_n\, ||\, A\, y_0 - f\, || \tag{7}$$

is valid with

$$q_n = \rho_1^n \left(1 + \frac{1-\rho_1^2}{1+\rho_1^2}\, n\right), \tag{8}$$

assuring

$$n \ge \frac{\ln \dfrac{1}{\varepsilon} + \ln \left(1 + \dfrac{1-\rho_1^2}{1+\rho_1^2}\, n\right)}{\ln \dfrac{1}{\rho_1}}$$

This is certainly so with

$$n \ge \frac{\ln \dfrac{1}{\varepsilon} + \ln \left(1 + \dfrac{2\sqrt{\xi}\, n}{1+\xi}\right)}{2\sqrt{\xi}}\,. \tag{9}$$

Comparison of (8) with the resulting expression for $q_n = \dfrac{2\,\rho_1^n}{1+\rho_1^{2n}}$, valid for Chebyshev's scheme, shows that both schemes are of the same asymptotic order as $\xi \to 0$:

$$n = n(\varepsilon) = O\Big(\frac{1}{\sqrt{\xi}}\,\ln\,\frac{1}{\varepsilon}\Big).$$

Here $n$ is, as usual, the number of iterations. But in practical implementations Chebyshev's scheme with known values $\gamma_1$ and $\gamma_2$ is preferable, because the extra iterations and storage are necessary for later use of the three-layer scheme concerned. What is more, the second scheme depends more significantly on the errors in specifying $\gamma_1$ and $\gamma_2$ than the first one.

**Remark** The passage from explicit three-layer schemes to implicit ones can be accomplished by the replacements of $A$ by $B^{-1}A$ and $f$ by $B^{-1}f$, so that

$$y_{k+1} = (1+\alpha)\,(E - \tau_0\,B^{-1}A)\,y_k - \alpha y_{k-1} + (1+\alpha)\,\tau_0\,B^{-1}f$$

or

(10) $$By_{k+1} = (1+\alpha)\,(B - \tau_0 A)\,y_k - \alpha B y_{k-1} + (1+\alpha)\,\tau_0\,f\,,$$

$$By_1 = B\,y_0 - \tau_0\,A\,y_0 + \tau_0\,f\,, k = 1, 2, \ldots\,, \quad \text{given } y_0 \in H\,.$$

Observe that equation (10) emerged from the identity

$$Bu = (1+\alpha)\,(Bu - \tau_0 A\,u) - \alpha Bu + (1+\alpha)\,\tau_0\,f$$

with further indication of iteration number in the appropriate positions, if any. formulas (4) for $\tau_0$ and $\alpha$ are still valid with known spectral bounds $\gamma_1$ and $\gamma_2$

(11) $$\gamma_1 B \le A \le \gamma_2 B\,, \quad \gamma_1 > 0\,, \quad B = B^* > 0\,,$$

for the operator $A$ acting in a certain space $H_B$ (not in the entire space $H$), making it possible to establish for a solution of problem (10) instead of (7) the estimate

(12) $$\|A\,y_n - f\|_{B^{-1}} \le q_n\,\|A\,y_0 - f\|_{B^{-1}}$$

with $q_n$ still subject to (8). This is acceptable if operator (76) is involved in the framework of ATM in Section 3.

**2. The minimal residual method.** Until now the bounds $\gamma_1$ and $\gamma_2$ of the operator $A$ were known before giving special investigations. But it may happen that these constants are either too rough or, generally speaking, indeterminate in advance. In mastering the difficulties involved, variational iterative methods such as the method of conjugate gradient, the minimal residual method and the method of steepest descent can be employed in the further elaboration on this subject.

We confine ourselves here to the minimal residual method and the method of steepest descent relating to two-layer schemes. As usual, the explicit scheme is considered first:

$$\frac{y_{k+1} - y_k}{\tau_{k+1}} + A\, y_k = f\,, \quad k = 0, 1, 2, \dots, \quad \text{given } y_0 \in H\,, \tag{13}$$

or

$$y_{k+1} = y_k - \tau_{k+1}\, r_k\,, \qquad r_k = A\, y_k - f\,, \tag{13'}$$

where $r_k$ is the residual.

The only difference between the methods we have mentioned above lies in the selection rules for the parameter $\tau_{k+1}$. In the minimal residual method the choice

$$\tau_{k+1} = \frac{(A\, r_k, r_k)}{\|\, A\, r_k \,\|^2}\,, \qquad \text{where} \quad r_k = A\, y_k - f\,, \tag{14}$$

is stipulated by the minimum condition for the norm $\|\, r_{k+1} \,\|$ of the residual. In this context, several questions are yet to be answered. The equation for the residual

$$\frac{r_{k+1} - r_k}{\tau_{k+1}} + A\, r_k = 0\,, \qquad k = 0, 1, 2, \dots, \tag{15}$$

implies that

$$\|\, r_{k+1} \,\|^2 = \|\, r_k \,\|^2 - 2\, \tau_{k+1} (A\, r_k, r_k) + \tau_{k+1}^2 \,\|\, A\, r_k \,\|^2\,. \tag{16}$$

The right-hand side of relation (16) is the polynomial $P_2(\tau_{k+1})$ of degree 2 with respect to the parameter $\tau_{k+1}$. Equating the derivatives $P_2'(\tau_{k+1})$ to zero reveals $\tau_{k+1}$ in complete agreement with (14). Because the second derivative for that value of the parameter $\tau_{k+1}$ is positive, the quantity $\|\, r_{k+1} \,\|$ of interest turns out to be minimal.

Adopting those ideas to the case of a non-self-adjoint operator $A$, we derive from the foregoing the a priori estimate

$$(17) \qquad \| r_{k+1} \| \le \rho_0 \| r_k \| , \quad \text{meaning} \quad \| A y_n - f \| \le \rho_0^n \| A y_0 - f \| ,$$

where $\rho_0 = (1-\xi)/(1+\xi)$, $\xi = \gamma_1/\gamma_2$ and $\gamma_1$, $\gamma_2$ are the accurate bounds of the operator $A = A^* > 0$.

Indeed, for the value $\tau_{k+1}$ assigned by (14) the right-hand side of (16) is minimal for fixed $r_k \in H$. Due to this property it is being increased for all other values and, in particular, for $\tau = \tau_0$. The meaning of this is that we should have

$$\| r_{k+1} \|^2 \le \| r_k \|^2 - 2\tau_0 (A r_k, r_k) + \tau_0^2 \| A r_k \|^2$$

$$\le \| r_k - \tau_0 A r_k \|^2 \le \| E - \tau_0 A \|^2 \cdot \| r_k \|^2 ,$$

$$\| r_{k+1} \| \le \| E - \tau_0 A \| \cdot \| r_k \| .$$

On the other hand, we learn from Section 2 that $\| E - \tau_0 A \| = \rho_0$ for $\tau_0 = 2/(\gamma_1 + \gamma_2)$, thereby justifying estimate (17) and the convergence of the minimal residual method with the same rate as occurred before for the simple iteration method with the exact values $\gamma_1$ and $\gamma_2$.

During the course of MRM the same procedures (13′) and (14) are workable with increased volume of calculations in connection with formula (14) for $\tau_{k+1}$ as compared with the simple iteration method.

The **implicit minimal residual method** can be designed in line with established practice:

$$(18) \qquad y_{k+1} = y_k - \tau_{k+1} w_k , \qquad w_k = B^{-1} (A y_k - f) ,$$

which is referred to as the **minimal correction method**. In that case instead of the governing equation $Au = f$ we are dealing with

$$(19) \qquad C v = \varphi , \quad v = B^{1/2} u , \quad C = B^{-1/2} A B^{-1/2} , \quad \varphi = B^{-1/2} f .$$

Applying here the explicit minimal residual method yields

$$x_{k+1} = x_k - \tau_{k+1} (C x_k - \varphi) ,$$

where

$$\tau_{k+1} = \frac{(C \tilde{r}_k, \tilde{r}_k)}{(C \tilde{r}_k, C \tilde{r}_k)} , \qquad \tilde{r}_k = C x_k - \varphi .$$

The forthcoming substitutions $x_k = B^{1/2} y_k$, $C = B^{-1/2} A B^{-1/2}$ and $\varphi = B^{-1/2} f$ allow us to modify these into

$$\tilde{r}_k = B^{-1/2}(A\, y_k - f) = B^{-1/2} r_k \,,$$

$$(C\, \tilde{r}_k, \tilde{r}_k) = (B^{-1/2} A B^{-1} r_k, B^{-1/2} r_k) = (A\, w_k, w_k)\,,$$

$$(C\, \tilde{r}_k, C\, \tilde{r}_k) = (B^{-1/2} A\, w_k, B^{-1/2} A\, w_k) = (B^{-1} A\, w_k, A w_k)\,,$$

leading to equation (18) with the correcting term $w_k$ and parameter $\tau_{k+1}$ such that

$$(20) \qquad w_k = B^{-1} r_k \,, \qquad \tau_{k+1} = \frac{(A\, w_k, w_k)}{(B^{-1} A\, w_k, A\, w_k)} \,.$$

Instead of (17) we eventually get one more useful estimate

$$(21) \qquad \|A\, y_n - f\|_{B^{-1}} \le \rho_0^n \, \|A\, y_0 - f\|_{B^{-1}} \,.$$

**3. The method of steepest descent.** The explicit method of steepest descent is given by the formulas

$$y_{k+1} = y_k - \tau_{k+1} (A\, y_k - f)\,, \quad k = 0, 1, 2, \dots\,, \quad \text{given} \quad y_0 \in H\,,$$

where

$$(22) \qquad \tau_{k+1} = \frac{(r_k, r_k)}{(A\, r_k, r_k)}\,, \qquad r_k = A\, y_k - f\,, \qquad k = 0, 1, 2, \dots\,,$$

arising from the minimum condition for the norm of the error $z_k = y_k - u$ in the space $H_A$, meaning

$$\min_{\{\tau_{k+1}\}} \| z_{k+1} \|_A \,, \qquad \| z \|_A = \sqrt{(Az, z)}\,.$$

The error $z_k = y_k - u$ satisfies the equation

$$z_{k+1} = z_k - \tau_{k+1}\, A z_k \,.$$

By interchanging the variables $v_k = A^{1/2} z_k$ we are led to

$$(23) \qquad v_{k+1} = v_k - \tau_{k+1}\, A\, v_k \,.$$

Having squared their norms, we find that

$$\| v_{k+1} \|^2 = \| v_k \|^2 - 2\,\tau_{k+1}(v_k, A\,v_k) + \tau_{k+1}^2 \| A\,v_k \|^2, \tag{24}$$

giving

$$\tau_{k+1} = \frac{(A\,v_k, v_k)}{\| A\,v_k \|^2} \tag{25}$$

under the above condition $\min\limits_{\{\tau_{k+1}\}} \| v_{k+1} \|^2$. The arguments of the preceding section serve to motivate the estimate

$$\| v_n \| \le \rho_0^n \| v_0 \| . \tag{26}$$

Thus, it remains to return to $z_k = A^{-1/2} v_k$ by observing furthermore that

$$Az_k = A(y_k - u) = Ay_k - f = r_k ,$$

$$(Av_k, v_k) = \| Az_k \|^2 = \| r_k \|^2 ,$$

$$\| Av_k \|^2 = (Ar_k, r_k) .$$

All this enables us to transform formula (25) into formula (22) for later use. Moreover, from inequality (26) it follows that

$$\| y_n - u \|_A \le \rho_0^n \| y_0 - u \|_A , \tag{27}$$

since

$$\| v_n \|^2 = (v_n, v_n) = (Az_n, Az_n) = \| z_n \|_A^2 .$$

No doubt, several conclusions can be drawn from such reasoning. First, the method being employed above converges in the space $H_A$ with the same rate as the simple iteration method although it occurs in one of the subordinate norms. Second, the minimal residual method converges in the space $H_{A^2}$, that is, in a more stronger norm.

By analogy with Section 2 the implicit method of steepest descent is described by

$$B\,\frac{y_{k+1} - y_k}{\tau_{k+1}} + A\,y_k = f\,, \quad k = 0, 1, 2, \dots , \quad \text{given} \quad y_0 \in H\,, \tag{28}$$

with parameters

$$(29)\qquad \tau_{k+1} = \frac{(r_k, w_k)}{(A w_k, w_k)}, \qquad w_k = B^{-1} r_k, \quad r_k = A y_k - f.$$

Estimate (29) is still valid in that case under the following conditions:

$$\gamma_1 B \le A \le \gamma_2 B, \qquad \gamma_1 > 0, \qquad B = B^* > 0.$$

**4. On solving equations with non-self-adjoint operators.** In dealing with an equation

$$A u = f, \qquad A: H \mapsto H,$$

where $A$ is a positive definite non-self-adjoint linear operator, the intention is to use one of the stationary iterative methods associated with the two-layer scheme, the parameter of which is constant:

$$(30)\qquad \frac{y_{k+1} - y_k}{\tau} + A y_k = f, \quad k = 0, 1, 2, \dots, \quad \text{given} \quad y_0 \in H,$$

The homogeneous equation for the error $z_k = y_k - u$ amounts to

$$z_{k+1} = S z_k, \quad S = E - \tau A, \quad k = 0, 1, 2, \dots, \quad z_0 \in H,$$

permitting to establish the convergence rate of iterations. At first glance, $\| z_{k+1} \| \le \| S \| \cdot \| z_k \|$. As before, a proper choice of the parameter $\tau$ is stipulated by the condition $\min_\tau \| S(\tau) \|$.

In such a setting the assumptions are made on the lower bounds of the operators $A$ and $A^{-1}$:

$$(31)\qquad \begin{aligned} &A \ge \gamma_1 E \quad \text{or} \quad (Ay, y) \ge \gamma_1 \| y \|^2, \quad \gamma_1 > 0, \\ &A^{-1} \ge \frac{1}{\gamma_2} E \quad \text{or} \quad \| Ay \|^2 \le \gamma_2 (Ay, y), \quad \gamma_2 > 0. \end{aligned}$$

A simple observation that the second condition for the case where $A = A^*$ is equivalent to the condition $A \le \gamma_2 E$ may be of help in further elaboration on this subject. Provided the condition $2 - \tau \gamma_2 \ge 0$ holds, we are able to arrive at the chain of the relations

$$\begin{aligned} \| S y \|^2 &= \| y - \tau A y \|^2 = \| y \|^2 - 2 \tau (A y, y) + \tau^2 \| A y \|^2 \\ &\le \| y \|^2 - 2 \tau (A y, y) + \tau^2 \gamma_2 (A y, y) = \| y \|^2 - \tau (2 - \tau \gamma_2) (A y) \\ &\le \| y \|^2 - \tau (2 - \tau \gamma_2) \gamma_1 \| y \|^2 = (1 - 2 \tau \gamma_1 + \tau^2 \gamma_1 \gamma_2) \| y \|^2, \end{aligned}$$

yielding

$$\| S \|^2 \le 1 - 2\tau\gamma_1 + \tau^2\gamma_1\gamma_2 .$$

Having stipulated the minimum condition for the preceding trinomial, the parameter $\tau$ is found to be $\tau = 1/\gamma_2$, so that $\| S \|^2 \le (1 - \gamma_1/\gamma_2)$ or, what amounts to the same,

$$\| S \| \le \sqrt{1-\xi} \quad \text{for} \quad \tau = 1/\gamma_2 , \quad \xi = \gamma_1/\gamma_2 . \tag{32}$$

In an attempt to generalize the results obtained to the case of three parameters $\gamma_1$, $\gamma_2$, $\gamma_3$, the operator $A$ arranges itself as a sum

$$A = A_0 + A_1 , \quad A_0 = \tfrac{1}{2}(A + A^*) , \quad A_1 = \tfrac{1}{2}(A - A^*) . \tag{33}$$

Here $A_0$ is a symmetric operator and $A_1$ is a sketch-symmetric operator, so that

$$A_0^* = A_0 , \quad A_1^* = -A_1 , \quad (A_1 x, x) = -(x, A_1 x) = 0 ,$$

meaning $(Ax, x) = (A_0 x, x)$. Let the the members of the operator $A$ be in line with the conditions

$$\gamma_1 E \le A_0 \le \gamma_2 E , \qquad \| A_1 \| \le \gamma_3 , \tag{34}$$

where $\gamma_2 > \gamma_1 > 0$ and $\gamma_3 \ge 0$ are known numbers.

An alternative form of the equation $z_{k+1} = (E - \tau A) z_k$ for the error $z_{k+1} = y_{k+1} - u$ is

$$z_{k+1} = (E - \tau A_0 - \tau A_1) z_k = (\theta E - \tau A_0) z_k + [(1-\theta) E - \tau A_1) z_k , \tag{35}$$

where the number $0 < \theta < 1$ is free to be chosen in any convenient way. The main goal of further development is to minimize the norm

$$\| S \| = \| E - \tau (A_0 + A_1) \|$$

by observing that on account of the triangle inequality

$$\| z_{k+1} \| \le \theta \left\| E - \frac{\tau}{\theta} A_0 \right\| \cdot \| z_k \| + \| (1-\theta) z_k - \tau A_1 z_k \| . \tag{36}$$

To this end, the numbers $\tau$ and $\theta$ are so chosen as to satisfy the minimum condition that we have mentioned above. As far as $\gamma_1 E \le A_0 \le \gamma_2 E$ for a self-adjoint operator $A_0$, we might have

$$\min_{\tau/\theta} \left\| E - \frac{\tau}{\theta} A_0 \right\| = \rho_0 \quad \text{for} \quad \frac{\tau}{\theta} = \tau_0 = \frac{2}{\gamma_1 + \gamma_2} , \tag{37}$$

where

$$\rho_0 = \frac{1-\xi}{1+\xi}, \qquad \xi = \frac{\gamma_1}{\gamma_2},$$

so that $\tau = \tau_0 \theta$. The second summand on the right-hand side of (36) will be the subject of special investigations. The outcome of this is

$$\begin{aligned} \| (1-\theta)\, y - \tau A_1 y \|^2 &= (1-\theta)^2 \| y \|^2 - 2\tau (1-\theta)(A_1 y, y) + \tau^2 \| A_1 y \|^2 \\ &= (1-\theta)^2 \| y \|^2 + \tau^2 \| A_1 y \|^2 \\ &\le \left[ (1-\theta)^2 + \tau^2 \gamma_3^2 \right] \cdot \| y \|^2 \\ &= \left[ (1-\theta)^2 + \tau_0^2 \theta^2 \gamma_3^2 \right] \cdot \| y \|^2 , \end{aligned}$$

which assures us of the validity of the following inequalities for $\tau = \tau_0 \theta$:

$$\| z_{k+1} \| \le \| S \| \cdot \| z_k \| , \qquad \| S \| \le f(\theta),$$

(38)

$$f(\theta) = \theta \rho_0 + \sqrt{(1-\theta)^2 + \theta^2 a^2}\,, \qquad a^2 = \tau_0^2 \gamma_3^2 .$$

In order to find the minimum of the function $f(\theta)$, we calculate and analyze its derivative

$$f'(\theta) = \rho_0 - \frac{1-\theta - a^2\theta}{\sqrt{(1-\theta)^2 + a^2\theta^2}} = \rho_0 - \frac{\alpha - a^2}{\sqrt{\alpha^2 + a^2}}, \qquad \alpha = \frac{1-\theta}{\theta}.$$

The equation $f'(0) = 0$ gives $\rho_0 \sqrt{\alpha^2 + a^2} = \alpha - a^2$. It may be viewed as a quadratic equation with respect to $\alpha$:

$$(1-\rho_0^2)\,\alpha^2 - 2a^2\alpha + a^2 (a^2 - \rho_0^2) = 0$$

with the first root

$$\alpha = a \, \frac{a + \rho_0 \sqrt{1 - \rho_0^2 + a^2}}{1 - \rho_0^2} .$$

We note in passing that the second root is unacceptable in connection with the possible negativity for some value of the parameters $a$ and $\rho_0$. Also, it will be sensible to introduce the notation $\varkappa = \gamma_3 / \sqrt{\gamma_1 \gamma_2 + \gamma_3^2}$, so that

$$\gamma_3^2 = \frac{\varkappa^2}{1-\varkappa^2}\, \gamma_1 \gamma_2 , \quad a^2 = \tau_0^2 \gamma_3^2 = \frac{4\gamma_1\gamma_2}{(\gamma_1+\gamma_2)^2} \, \frac{\varkappa^2}{1-\varkappa^2} = \frac{(1-\rho_0^2)\,\varkappa^2}{1-\varkappa^2},$$

$$\frac{a^2}{\varkappa^2} = \frac{1-\rho_0^2}{1-\varkappa^2} .$$

When providing current manipulations with

$$1-\rho_0^2+a^2=1-\rho_0^2+\frac{\varkappa^2\,(1-\rho_0^2)}{1-\varkappa^2}=\frac{1-\rho_0^2}{1-\varkappa^2}=\frac{a^2}{\varkappa^2}\,,$$

we find that

$$\alpha=\frac{a^2\,(\varkappa+\rho_0)}{\varkappa\,(1-\rho_0^2)}=\frac{\varkappa\,(\varkappa+\rho_0)}{1-\varkappa^2}\,,$$

$$1+\alpha=\frac{1+\varkappa\,\rho_0}{1-\varkappa^2}\,,$$

$$\theta=\frac{1}{1+\alpha}=\frac{1-\varkappa^2}{1+\varkappa\,\rho_0}$$

and, consequently, the expression for the function

$$f(\theta)=\frac{1}{1+\alpha}\,\rho_0+\frac{1}{1+\alpha}\,\sqrt{\alpha^2+a^2}=\frac{\rho_0^2+\alpha-a^2}{(1+\alpha)\,\rho_0}\,.$$

Taking into account that

$$\rho_0^2+\alpha-a^2=(1+\alpha)-(1-\rho_0^2+a^2)=1+\alpha-\frac{a^2}{\varkappa^2}$$

$$=\frac{1+\varkappa\,\rho_0}{1-\varkappa^2}-\frac{1-\rho_0^2}{1-\varkappa^2}=\rho_0\,\frac{\rho_0+\varkappa}{1-\varkappa^2}\,,$$

it is plain to show that

$$\|\,S\,\|\le\frac{\varkappa+\rho_0}{1+\varkappa\,\rho_0}\quad\text{for}\quad\tau=\tau_0\,\frac{1-\varkappa^2}{1+\varkappa\,\rho_0}\,.$$

The meaning of this is that a solution of problem (30) satisfies the estimate

$$\|\,y_n-u\,\|\le\rho^n\,\|\,y_0-u\,\|$$

with

$$(39)\qquad \rho=\frac{\varkappa+\rho_0}{1+\varkappa\,\rho_0}\,,\qquad \varkappa=\frac{\gamma_3}{\sqrt{\gamma_1\,\gamma_2+\gamma_3^2}}\,,\qquad \tau=\bar{\tau}=\frac{\tau_0\,(1-\varkappa^2)}{1+\varkappa\,\rho_0}\,,$$

provided that conditions (34) hold. The number $n$ of iterations for doing so is no less than $n_0(\varepsilon)$:

$$n \geq n_0(\varepsilon) \geq \ln \frac{1}{\varepsilon} \Big/ \ln \frac{1}{\rho} .$$

What can happen upon replacing the describing explicit scheme by an implicit one? Being concerned with one of the implicit schemes

$$(40) \qquad B \, \frac{y_{k+1} - y_k}{\tau_{k+1}} + A \, y_k = f \,, \quad k = 0, 1, 2, \ldots \,, \qquad y_0 \in H \,,$$

with a self-adjoint operator $B = B^* > 0$, the intention is to employ the explicit scheme

$$x_{k+1} = x_k - \tau \, (C x_k - \varphi) \,, \qquad x_k = B^{1/2} y_k \,,$$

with the members

$$C = B^{-1/2} A B^{-1/2} \,, \qquad \varphi = B^{-1/2} f$$

and rearrange conditions (34) with regard to the operator $C$ as revised conditions for the operators $A$ and $B$:

$$(41) \qquad \gamma_1 B \leq A_0 \leq \gamma_2 B \,, \qquad (B^{-1} A_1 y, A_1 y) \leq \gamma_3^2 (By, y) \,,$$

assuring

$$\| A_1 y \|_{B^{-1}} \leq \gamma_3 \| y \|_B \qquad \text{and} \qquad \| y_n - u \|_B \leq \rho^n \, \| y_0 - u \|_B$$

in contrast to the preceding estimate just established.

The minimal residual method can be employed for the equation $Au = f$ with a non-self-adjoint operator $A$, the convergence rate of which coincides with that of scheme (30) for $\tau = \bar{\tau}$.

For the explicit scheme (13) we have

$$\frac{y_{k+1} - y_k}{\tau_{k+1}} + A \, y_k = f \,, \quad k = 0, 1, 2, \ldots \,, \qquad y_0 \in H \quad \text{given},$$

where the parameter $\tau_{k+1}$ is chosen by the approved rule

$$\tau_{k+1} = \frac{(A \, r_k, r_k)}{\| A \, r_k \|^2} \,, \qquad r_k = A \, y_k - f \,,$$

(for more detail see Section 2, formula (14)) governing the minimum condition for $\| r_{k+1} \|^2$. No property of the self-adjointness of the operator $A$ applies here and below. In what follows minor changes will appear in connection with the necessity of proving its convergence. In particular, $\bar{\tau}$ will stand in place of $\tau_{k+1}$ on the right-hand side of identity (16) and will result in the relations

$$\| r_{k+1} \|^2 \leq \| r_k \|^2 - 2\bar{\tau}(A r_k, r_k) + \bar{\tau}^2 \| A r_k \|^2$$
$$= \| r_k - \tau A r_k \|^2 \leq \| E - \bar{\tau} A \|^2 \cdot \| r_k \|^2$$
$$\leq \rho^2 \| r_k \|^2 ,$$

thereby assuring $\| r_{k+1} \| \leq \rho \| r_k \|$ and justifying the estimate

$$\| A y_n - f \| \leq \rho^n \| A y_0 - f \| .$$

This is certainly true for the minimal residual method (13)–(14) under conditions (34). Here $y_n$ is a solution of problem (13) and $\rho$ is specified by formula (39). The fastest move method is useless in that case because the operator $A$ is non-self-adjoint in such a setting.

Observe that under conditions (34) the estimate

$$\|A y_n - f\|_{B^{-1}} \leq \rho^n \|A y_0 - f\|_{B^{-1}}$$

is valid for the implicit minimal correction method with $B = B^* > 0$ incorporated.

**5. Gibrid (combined) methods.** In mastering the difficulties involved in solving difference elliptic equations, some consensus of opinion is to bring together direct and iterative methods in some or other aspects as well as to combine iterative methods of various types (two-step methods). All the tricks and turns will be clarified for the iteration scheme

$$B_k \frac{y_{k+1} - y_k}{\tau_{k+1}} + A y_k = f , \qquad y_0 \in H \quad \text{given}, \tag{42}$$

yielding

$$y_{k+1} = y_k - \tau_{k+1} w_k , \tag{43}$$

where $w_k$ is the correction satisfying the equation $B_k w_k = r_k$ and $r_k = A y_k - f$ is the residual.

Let now $R = R^* > 0$ be a regularizer such that $c_1 R \le A \le c_2 R$, $c_1 > 0$. Knowing $w_k$, it is plain to find the $(k+1)$th iteration $y_{k+1}$ in line with (43), so there is some reason to be concerned about this.

The operator $B_k$ can be specified in a number of different ways and, in particular, in explicit form. One way of covering this is

$$B_k = R\,, \tag{44}$$

$$B_k = (E + \omega_k^{(1)} R_1)\,(E + \omega_k^{(2)} R_2)\,, \tag{45}$$

where $R_1$, $R_2$ are economical operators and $\omega_k^{(1)}$, $\omega_k^{(2)}$ are iteration parameters. We dealt with the factorized operator (45) during the course of ATM with $R_1 = R_2^*$, $\omega_k^{(1)} = \omega_k^{(2)} = \omega$ as well as of ADM with $R_\alpha^* = R_\alpha > 0$, $\alpha = 1, 2$, $R_1 R_2 = R_2 R_1$.

a) *A direct method for determination of the correction.* Let $B_k = R$ and one of the available direct methods will be employed for solving the system of algebraic equations $Rw = r_k$. We refer the users to Section 1 (items 2–3) in which such methods have been designed for elliptic grid equations. Among them the decomposition method with $\gamma_1 = c_1$ and $\gamma_2 = c_2$ is highly recommended for practical implementations. By the selection rule for Chebyshev's parameters $\tau_1, \tau_2, \dots, \tau_n$ we derive the estimate for the number of the necessary iterations

$$n \ge n_0(\varepsilon)\,, \qquad n_0(\varepsilon) = \frac{1}{2}\sqrt{\frac{c_2}{c_1}}\,\ln\frac{2}{\varepsilon}\,.$$

In this regard, it should be noted that for the difference elliptic problems posed in Section 3, item 7, $n_0(\varepsilon)$ is independent of the grid step. What is more, when $G$ is a rectangle and $h_1 = h_2 = h$, the work in doing this is

$$O\Big(\sqrt{\frac{c_2}{c_1}}\,\frac{1}{h^2}\,\ln\frac{1}{h}\,\ln\frac{2}{\varepsilon}\Big)\,.$$

b) *An iterative method for determination of the correction.* The starting point for subsequent considerations is the equation $Rw_k = r_k$, which can be solved by means of some (internal) iterative method, whose use permits us to find $w^{(m)} = w_k$, where $m$ is the number of the internal iteration.

Further reduction of a two-layer iteration scheme with the accompanying self-adjoint operators to an explicit one by replacing the variable $w$ by $R^{1/2} w$ leads to

$$R^{1/2}(w^{(m)} - w) = T_m\, R^{1/2}(w^{(0)} - w)\,,$$

where $w$ is the exact solution to the equation $Rw = r_k$ and $T_m$ is the resolving operator subject to the relations

$$(46) \qquad T_m^* = T_m , \qquad \| T_m \| \leq q < 1 .$$

Here the dependence $q$ upon $m$ is omitted. This should cause no confusion.

With the initial value $w^{(0)} = 0$ in view, we find that

$$w = R^{-1/2}(E - T_m)^{-1} R^{1/2} w^{(m)} .$$

Upon substituting this expression into the equation $Rw = r_k$ we obtain the equation $Bw_k = r_k$ with the members

$$B = R^{1/2}(E - T_m)^{-1} R^{1/2} , \qquad w_k = w^{(m)} ,$$

thereby completing the task of searching for the correction. By the same token, $y_{k+1} = y_k - \tau_{k+1} w_k$.

Still using the framework of the general theory, it is straightforward to verify that the operator $B$ is self-adjoint ($B = B^*$) with the aid of the relations $R = R^*$ and $T_m^* = T_m$. We shall need yet the constants $\gamma_1$ and $\gamma_2$ of the energetic equivalence of the operators $B$ and $A$. Because of (46), we thus have

$$(1 - q) E \leq E - T_m \leq (1 + q) E ,$$

$$(1 + q)^{-1} E \leq (E - T_m)^{-1} \leq (1 - q)^{-1} E ,$$

$$(Bx, x) = (R^{1/2}(E - T_m)^{-1} R^{1/2} x, x) = ((E - T_m)^{-1} y, y) ,$$

where $y = R^{1/2} x$. From such reasoning it seems clear that

$$(1 - q) B \leq R \leq (1 + q) E ,$$

giving $\overset{\circ}{\gamma}_1 = 1 - q$ and $\overset{\circ}{\gamma}_2 = 1 + q$ and implying that $\gamma_1 = c_1(1 - q)$ and $\gamma_2 = c_1(1 + q)$.

For Chebyshev's scheme with such an operator we obtain

$$n_0(\varepsilon) = \frac{1}{2} \sqrt{\frac{c_2\,(1 + q)}{c_1\,(1 - q)}}\, \ln \frac{2}{\varepsilon} ,$$

since

$$\xi = \frac{c_1}{c_2}\, \frac{(1 - q)}{(1 + q)} .$$

If $R = R_1 + R_2$, $R_\alpha^* = R_\alpha > 0$ and $R_1 R_2 = R_2 R_1$ ADM from Section 4, items 2–3 may be employed, where the minimum condition for the work in connection with solving the equation $Au = f$ guides a proper choice of the quantity $q$.

When the spectrum of $R$ happens to be unknown in advance, it seems worthwhile adapting iterative methods of variational type for determination of the correction. Unfortunately, in this book we have no possibility to say more about other combinations arising time and again in such matters.

# Symbols

$\omega_h = \{x_i = ih,\ h > 0,\ i = 1, 2, \dots, N-1,\ hN = l\}$ – an equidistant (uniform) grid on the interval $(0, l)$;

$\bar{\omega}_h = \{x_i = ih,\ h > 0,\ i = 0, 1, \dots, N,\ hN = l\}$ – an equidistant (uniform) grid on the segment $[0, l]$;

$h$ – step of the grid $\omega_h$;

$x = x_i$ – a node of the grid $\omega_h$;

$y = y_i = y(x_i)$ – a function defined on the grid $\omega_h$;

$y_x = (y_{i+1} - y_i)/h$ – the right difference derivative at a point $x_i$;

$y_{\bar{x}} = (y_i - y_{i-1})/h$ – the left difference derivative at a point $x_i$;

$y_{\bar{x}x} = (y_{i+1} - 2y_i + y_{i-1})/h^2$ – the second difference derivative at a point $x_i$;

$\hat{\omega}_h = \{x_i \in (0, l),\ i = 1, 2, \dots, N-1\}$ – a non-equidistant (nonuniform) grid on the interval $(0, l)$;

$\bar{\hat{\omega}}_h = \{x_i \in [0, l],\ i = 0, 1, \dots, N,\ x_0 = 0,\ x_N = l\}$ – a non-equidistant (nonuniform) grid on the segment $[0, l]$;

$h_i = x_i - x_{i-1}$ – step of the grid $\hat{\omega}_h$;

$\hbar_i = 0.5\,(h_i + h_{i+1})$;

$y_x = (y_{i+1} - y_i)/h_{i+1}$,

$y_{\bar{x}} = (y_i - y_{i-1})/h_i$;

$y_{\hat{x}} = (y_{i+1} - y_i)/\hbar_i$;

$$y_{\bar{x}\hat{x}} = \frac{1}{\hbar_i}\left(\frac{y_{i+1} - y_i}{h_{i+1}} - \frac{y_i - y_{i-1}}{h_i}\right);$$

A list of inner products and associated norms on a grid:

$$(y,\, v) = \sum_{i=1}^{N-1} y_i\, v_i\, h\,, \qquad \| y \| = \sqrt{(y,\, y)};$$

$$(\, y,\, v\,] = \sum_{i=1}^{N} y_i\, v_i\, h\,, \qquad \| y ]\| = \sqrt{(\, y,\, y\,]};$$

$$\| y \|_C = \max_{x_i \in \bar{\omega}_h} | y(x_i) |;$$

$$\| \varphi \|_{(-1)} = \left(\sum_{i=1}^{N-1} h \left(\sum_{k=1}^{i} h\varphi_k\right)^2\right)^{1/2};$$

$\omega_\tau = \{t_j = j\tau,\ \tau > 0,\ j = 0, 1, \dots\}$ – a grid in time;

$\tau$ – step of the grid $\omega_\tau$;

$y = y^j = y(t_j)$ – a function defined on the grid $\omega_\tau$;

$\hat{y} = y^{j+1} = y(t_{j+1}), \qquad \check{y} = y^{j-1} = y(t_{j-1})$;

$\bar{y}_t = (\hat{y} - y)/\tau, \qquad \bar{y}_{\bar{t}} = (y - \check{y})/\tau, \qquad y_{\overset{\circ}{t}} = (\hat{y} - \check{y})/(2\tau)$;

$y_{\bar{t}t} = (\hat{y} - 2y + \check{y})/\tau^2$;

$x = x_i = \left(x_1^{(i_1)}, \dots, x_\alpha^{(i_\alpha)}, \dots, x_p^{(i_p)}\right)$ – a node of the $p$-dimensional rectangular grid $\omega_h$;

$x_\alpha^{(i_\alpha)} = h_\alpha\, i_\alpha$;

$h_\alpha$ – step of the grid $\omega_h$ along the direction $\alpha$;

$\mathbf{x}_i^{(\pm 1_\alpha)} = \left(x_1^{(i_1)}, \dots, x_\alpha^{(i_\alpha)} \pm h_\alpha, \dots, x_p^{(i_p)}\right)$;

$y = y(x_i), \quad y^{(\pm 1_\alpha)} = y\left(\mathbf{x}_i^{(\pm 1_\alpha)}\right)$;

$y_{x_\alpha} = \left(y^{(+1_\alpha)} - y\right)/h_\alpha$;

$y_{\bar{x}_\alpha} = \left(y - y^{(-1_\alpha)}\right)/h_\alpha$;

$y_{\bar{x}_\alpha x_\alpha} = \left(y^{(+1_\alpha)} - 2y + y^{(-1_\alpha)}\right)/h_\alpha^2$;

$\overset{\circ}{\Omega}$ – set of all grid functions given on the grid $\omega_h$ and vanishing on its boundary;

$H$ – Hilbert space;

$(y,\, v)$ – inner product of elements $y,\, v \in H$, and associated norm $\| y \| = \sqrt{(y,\, y)}$;

$\mathcal{D}(A)$ – domain of definition of an operator $A$;

$\mathcal{R}(A)$ – range of values of an operator $A$;

$E$ – identity operator;

$A : H \mapsto H$ – operator $A$ with the domain $\mathcal{D}(A) = H$ and range $\mathcal{R}(A) \subseteq H$;

$A^*$ – adjoint operator;
$A^{-1}$ – inverse operator;
$S$ – transition operator from one layer to another;
$T$ – resolving operator;
$A > 0$ – positive operator;
$A \geq 0$ – nonnegative operator;
$A \geq \delta E$, $\delta > 0$, – positive definite operator;
$\| y \|_A = \sqrt{(Ay, y)}$, $y \in H$;
$\| \varphi \|_{A^{-1}} = \sqrt{(A^{-1}\varphi, \varphi)}$, $\varphi \in H$.

# Concluding Remarks

In concluding this book we give a brief commentary regarding the main results forth in the foregoing. In order to make the book accessible not only to specialists, but also to students and engineers, we give in Chapter 1 a complete account of definitions and notations and present a number of relevant topics from other branches of mathematics. The detailed outline of mathematical models leading to partial differential equations is available in many textbooks and monographs on equations of mathematical physics. In particular, we refer the reader to Courant and Hilbert (1953, 1962), Godunov (1971), Morse and Feshbach (1953), Tikhonov and Samarskiĭ (1963). Current exposition follows the best legacies of the past: my first book with my dear teacher – the late Academ. Tikhonov "Equations of Mathematical Physics", throughout which the reader can find thorough, advanced-undergraduate to graduate level treatments of problems leading to partial differential equations: hyperbolic, parabolic, elliptic equations; wave propagations in space, heat conduction in space, special functions, etc. with emphasis on the mathematical formulation of problems, rigorous solutions, physical interpretation of the results obtained.

The books by Gelfand (1967), Samarskiĭ and Nikolaev (1989) cover in full details the general theory of linear difference equations. Sometimes the elimination method available for solving various systems of algebraic equations is referred to, in the foreign literature, as Thomas' algorithm and this should cause no confusion. This method was independently proposed by many authors in the early fifties. The bibliographical sources of Godunov and Ryabenkiĭ (1964) and Richtmyer (1957) help clarify its appearance a little bit. Faddeev and Faddeeva (1963) have treated the elimination method as the well-established Gaussian elimination for systems of equations with a tridiagonal matrix.

Most of the propositions in Chapters 2–4 are of independent value although, for the present book they are used only as part of the auxiliary mathematical apparatus. Some of them were known earlier and the rest were discovered and proven in recent years in connection with the rapid development of the theory of difference schemes.

Relevant results from functional analysis, operator theory and the well-developed numerical methods are given at the very beginning of Chapter 2. All proofs af the main statements as well as a detailed exposition of the foundations of functional analysis, operator theory and the theory of difference schemes are outlined in the textbooks and monographs by Ames (1977), Mitchell and Griffits (1980), Morton and Mayers (1994), Ortega and Poole (1981), Samarskiĭ (1987), Samarskiĭ and Gulin (1989). The basic concepts, notions and mathematical apparatus of the contemporary theory of difference methods for solving mathematical physics problems completed in late fiftieth owe a debt to Forsythe and Wasow (1960), Richtmyer (1957), Richtmyer and Morton (1967), Ryabenkiĭ and Filippov (1956). The theorem on connection between approximation and stability of a difference scheme and its convergence to a solution of the original problem appeared in the works of Filippov (1955), Lax and Richtmyer (1956).

The theory of homogeneous difference schemes for ordinary differential equations of second order with variable coefficients and, in particular, with discontinuous ones, was developed by A.N. Tikhonov and A.A. Samarskiĭ in the middle of $50^{\text{th}}$ – in the early $60^{\text{th}}$. The first results in this area were obtained by Tikhonov and Samarskiĭ (1956). Later a complete and systematic study was carried out by Tikhonov and Samarskiĭ (1961) along these lines. Special investigations of convergence of difference schemes on non-equidistant grids have been done by Tikhonov and Samarskiĭ (1961ab, 1962). The monograph of Samarskiĭ, Lazarov and Makarov (1987) includes more a detailed exploration of the theory of explicit difference schemes and schemes of any accuracy order for ordinary differential equations, while the original works of Alekseevskiĭ (1984), Bagmut (1969), Chao Show (1963) and Prikazchikov (1965) concentrate on special cases in tackling singular-perturbed problems, equations with singularity, some problems associated with a four-order equation and the Sturm–Liouville problem, respectively.

The notion of the conservatism of a difference scheme played a crucial role in the further recognition and establishment of convergence of homogeneous difference schemes with discontinuous coefficients (for more detail see Tikhonov and Samarskiĭ (1961ab)). It was shown therein that conservatism is a necessary property for convergence of a homogeneous scheme in the class of discontinuous coefficients. Further development of the conservatism principle provided by Samarskiĭ and Popov (1969, 1992) gave rise to the notion of full conservatism.

Being concerned with homogeneous conservative difference schemes, A.N. Tikhonov and A.A. Samarskiĭ recommended the design of the balance method being an integro-interpolation one and providing the necessary framework for constructing difference schemes on arbitrary irregular grids. Samarskiĭ, Koldoba, Poveshchenko, Tishkin and Favorskiĭ (1966) followed these procedures in solving the global problems of mathematical physics under such an approach. We note in passing that the methods based on Marchuk's identity and well-established by Babuska et al. (1966) and Marchuk (1975) are conceptually close to the balance method mentioned above. A greater gain in accuracy of difference schemes on a sequence of grids were the main idea behind this approach justified in full details by Marchuk and Shajdurov (1979).

Some modification of the describing monotone difference scheme for "divergent" second-order equations was made by Golant (1978) and Karetkina (1980). In Andreev and Savin (1995) this scheme applies equally well to some singular-perturbed problems. Various classes of monotone difference schemes for elliptic equations of second order were composed by Samarskiĭ and Vabishchevich (1995), Vabishchevich (1994) by means of the regularization principle with concern of difference schemes.

# References

Akhiezer, N. and Glazman, I. (1966) The Theory of Linear Operators in Hilbert Space. Nauka: Moscow (in Russian).

Alekseevskiĭ, M. (1984) Difference schemes of higher-order accuracy for some singular-perturbed boundary-value problems. Differential Equations, 17, 1177–1183 (in Russian).

Ames, W. (1977) Numerical Methods for Partial Differential Equations. Academic Press: New York–London.

Andreev, V. and Savin, I. (1995) On the uniform convergence in a small parameter of Samarskiĭ' scheme and its modification. Zh. Vychisl. Mat. i Mat. Fiz., 35, 739–752 (in Russian); English transl. in USSR Comput. Mathem. and Mathem. Physics.

Babuska, I., Pager, M. and Vitasek, E. (1966) Numerical Processes in Differential Equations. Wiley: Chichester.

Bagmut, G. (1969) Difference schemes of higher-order accuracy for an ordinary differential equations with singularity. Zh. Vychisl. Mat. i Mat. Fiz., 9, 221–226 (in Russian); English transl. in USSR Comput. Mathem. and Mathem. Physics.

Balakrishnan, A. (1976) Applied Functional Analysis. Springer: New-York–Berlin–Heidelberg.

Berezanskij, Yu. (1968) Expansions in Eigenfunctions of Self-Adjoint Operator. AMS: Providence, RI.

Bitsadze, A. (1976) Equations of Mathematical Physics. Nauka: Moscow (in Russian).

Cannon, J. (1984) The one-dimensional heat equation. In: Encyclopedia of Mathematics and Its Applications. v. 23, Addison-Wesley: Reading, MA.

Chao Show (1963) Homogeneous difference schemes for a four order equation with discontinuous coefficients. Zh. Vychisl. Mat. i Mat. Fiz., 5, 123–132 (in Russian); English transl. in USSR Comput. Mathem. and Mathem. Physics.

Courant, R. and Hilbert, D. (1953) Methods of Mathematical Physics. Volume 1. Wiley: New York.

Courant, R. and Hilbert, D. (1962) Methods of Mathematical Physics. Volume 2. Wiley: New York.

Dezin, A. (1980) General Questions of the Theory of Boundary-Value Problems. Nauka: Moscow (in Russian).

Dunford, N. and Schwartz, J. (1971a) Linear Operators (second edition). Part I: General Theory. Wiley: New York.

Dunford, N. and Schwartz, J. (1971b) Linear Operators (second edition). Part II: Spectral Theory. Self-Adjoint Operators in Hilbert Space. Wiley: New York.

Faddeev, D. and Faddeeva, V. (1963) Computational Methods of Linear Algebra. Freeman&Co.: San Francisco.

Filippov, A. (1955) On stability of difference equations. Dokl. Akad. Nauk SSSR, 100, 1045–1048 (in Russian); English transl. in Soviet Mathem. Dokl.

Forsythe, G. and Wasow, W. (1960) Finite Difference Methods for Differential Equations. Wiley: New York.

Gelfand, A. (1967) Calculus of Finite Differences. Nauka: Moscow (in Russian).

Godunov, S. (1971) Equations of Mathematical Physics. Nauka: Moscow (in Russian).

Godunov, S. and Ryabenkiĭ, V. (1964) Introduction to the Theory of Difference Schemes. Interscience: New York.

Golant, E. (1978) On conjugate families of difference schemes for equations of parabolic type with lower members. Zh. Vychisl. Mat. i Mat. Fiz., 18, 1162–1169 (in Russian); English transl. in USSR Comput. Mathem. and Mathem. Physics.

Kantorovich, L. and Akilov, G. (1977) Functional Analysis in Normed Spaces. Nauka: Moscow (in Russian).

Karetkina, N. (1980) Nonconditionally stable difference schemes for parabolic equations with the first derivatives. Zh. Vychisl. Mat. i Mat. Fiz., 20, 236–240 (in Russian); English transl. in USSR Comput. Mathem. and Mathem. Physics.

Ladyzhenskaya, O. (1973) Boundary Value Problems of Mathematical Physics. Nauka: Moscow (in Russian).

Lax, P. and Richtmyer, R. (1956) A servey of stability of linear finite difference equations. Comm. Pure Appl. Mathem., 9, 267–293.

Marchuk, G. (1975) Methods of Computational Mathematics. Springer: New York.

Marchuk, G and Shajdurov, V. (1979) Gain in Accuracy of Difference Schemes. Nauka: Moscow (in Russian).

Mitchell, A. and Griffits, D. (1980) The Finite Difference Methods in Partial Differential Equations. Wiley: New York.

Morozov, V. (1984) Regularization Methods for Solving Improperly Posed Problems. Springer: New-York–Berlin–Heidelberg.

Morse, P. and Feshbach H. (1953) Methods of Theoretical Physics. Volumes 1,2. McGraw Hill: New York.

Morton, K. and Mayers, D. (1994) Numerical Solution of Partial Differential Equations. An Introduction. Cambridge University Press: Cambridge.

Ortega, Y. and Poole, W. (1981) An Introduction to Numerical Methods for Differential Equations. Pitman: London.

Prikazchikov, V. (1965) A difference problem on eigenvalues of an elliptic operator. Zh. Vychisl. Mat. i Mat. Fiz., 5, 648–657 (in Russian); English transl. in USSR Comput. Mathem. and Mathem. Physics.

Richtmyer, R. (1957) Difference Methods for Initial-Value Problems. Interscience: New York.

Richtmyer, R. (1978) Principles of Advanced Mathematical Physics. Volume 1. Springer: New York–Berlin–Heidelberg.

Richtmyer, R. (1981) Principles of Advanced Mathematical Physics. Volume 2. Springer: New York–Berlin–Heidelberg.

Richtmyer, R. and Morton, K. (1957) Difference Methods for Initial-Value Problems. Interscience: New York.

Ryabenkiĭ, V. and Filippov, A. (1956) On Stability of Difference Equations. Gostekhizdat: Moscow (in Russian).

Samarskiĭ, A. (1987) Introduction to Numerical Methods. Nauka: Moscow (in Russian).

Samarskiĭ, A. and Gulin, A. (1989) Numerical Methods. Nauka: Moscow (in Russian).

Samarskiĭ, A. and Popov, Yu. (1969) Totally conservative difference schemes. Zh. Vychisl. Mat. i Mat. Fiz., 9, 953–954 (in Russian); English transl. in USSR Comput. Mathem. and Mathem. Physics.

Samarskiĭ, A. and Popov, Yu. (1992) Difference Methods for Solving Problems in Gas Dynamics. Nauka: Moscow (in Russian).

Samarskiĭ, A. and Vabishchevich, P. (1995) Computational Heat Transfer (Volumes 1,2). Wiley: Chichester.

Samarskiĭ, A., Koldoba, A., Poveshchenko, Yu., Tishkin V. and Favorskiĭ, A. (1966) Difference Schemes on Irregular Grids. Nauka: Minsk (in Russian).

Samarskiĭ, A., Lazarov, R. and Makarov, V. (1987) Difference Schemes for Differential Equations with Generalized Solutions. Visha Schkola: Moscow (in Russian).

Tikhonov, A. and Samarskiĭ, A. (1956) On difference schemes for equations with discontinuous coefficients. Dokl. Akad. Nauk SSSR, 109, 393–396 (in Russian); English transl. in Soviet. Mathem. Dokl.

Tikhonov, A. and Samarskii, A. (1963) Equations of Mathematical Physics. Pergamon Press: New York–London.

Tikhonov, A. and Samarskiĭ, A. (1961a) On homogeneous difference schemes. Zh. Vychisl. Mat. i Mat. Fiz., 1, 5–63 (in Russian); English transl. in USSR Comput. Mathem. and Mathem. Physics.

Tikhonov, A. and Samarskiĭ, A. (1961b) Homogeneous difference schemes of higher-order accuracy on non-equidistant grids. Zh. Vychisl. Mat. i Mat. Fiz., 1, 425–440 (in Russian); English transl. in USSR Comput. Mathem. and Mathem. Physics.

Tikhonov, A. and Samarskiĭ, A. (1962) Homogeneous difference schemes on non-equidistant grids. Zh. Vychisl. Mat. i Mat. Fiz., 2, 812–832 (in Russian); English transl. in USSR Comput. Mathem. and Mathem. Physics.

Trenogin, V. (1980) Functional Analysis. Nauka: Moscow (in Russian).

Vabishchevich, P. (1994) Monotone difference schemes for convection-diffusion problems. Differential Equations, 30, 503–531 (in Russian).

Vladimirov, V. (1971) Equations of Mathematical Physics. Nauka: Moscow (in Russian).

Vladimirov, V. (1976) Distributions in Mathematical Physics. Nauka: Moscow (in Russian).

Yosida, K. (1965) Functional Analysis. Springer: New York–Berlin–Heidelberg.

# Index